THE CONTINENTAL DRIFT CONTROVERSY

Volume I: Wegener and the Early Debate

Resolution of the sixty year debate over continental drift, culminating in the triumph of plate tectonics, changed the very fabric of Earth Science. Plate tectonics can be considered alongside the theories of evolution in the life sciences and of quantum mechanics in physics in terms of its fundamental importance to our scientific understanding of the world. This four-volume treatise on *The Continental Drift Controversy* is the first complete history of the origin, debate and gradual acceptance of this revolutionary explanation of the structure and motion of the Earth's outer surface. Based on extensive interviews, archival papers, and original works, Frankel weaves together the lives and work of the scientists involved, producing an accessible narrative for scientists and non-scientists alike.

Wegener's theory of continental drift captured the attention of Earth Scientists worldwide. In the early 1900s he noticed that the Earth's major landmasses could be fitted together like a jigsaw and went on to propose that the continents had once been joined together in a single landmass, which became known as Pangaea, and that they had later drifted apart. This first volume describes the reception of Wegener's theory as it splintered into sub-controversies over the geometrical fit of continental margins and disjuncts between biotic and geologic provinces. Without a convincing resolution of any of the sub-controversies or physical measurement of continental drift, scientific opinion remained divided between the "fixists" and "mobilists."

Other volumes in *The Continental Drift Controversy*:

Volume II – Paleomagnetism and Confirmation of Drift

Volume III – Introduction of Seafloor Spreading

Volume IV – Evolution into Plate Tectonics

HENRY R. FRANKEL was awarded a Ph.D from Ohio State University in 1974 and then took a position at the University of Missouri–Kansas City where he became Professor of Philosophy and Chair of the Philosophy Department (1999–2004). His interest in the continental drift controversy and the plate tectonics revolution began while teaching a course on conceptual issues in science during the late 1970s. The controversy provided him with an example of a recent and major scientific revolution to test philosophical accounts of scientific growth and change. Over the next thirty years, and with the support of the United States National Science Foundation, National Endowment for the Humanities, the American Philosophical Society, and his home institution, Professor Frankel's research went on to yield new and fascinating insights into the evolution of the most important theory in the Earth Sciences.

"A well constructed and gripping narrative, which preserves the complex scientific detail, but invites one in to this fascinating world and helps the reader patiently to find a way through its labyrinth. Frankel is a wonderful guide and worthy of your trust."

MOTT GREENE, University of Puget Sound and University of Washington

"What is so impressive about this monumental work is its completeness. Frankel has gone back to the original sources and papers, to ensure complete understanding of the scientific issues involved. I recommend these volumes to anyone interested in the subject."

DAN MCKENZIE, University of Cambridge

"This is the definitive history of the way science really worked during the prolonged great geoscience debate of the twentieth century. ...Superb either for sampling, eased by excellent organization, or for a long, rewarding read."

WARREN HAMILTON, Colorado School of Mines

THE CONTINENTAL DRIFT CONTROVERSY

Volume I: Wegener and the Early Debate

HENRY R. FRANKEL

University of Missouri–Kansas City

To Paula

CAMBRIDGE
UNIVERSITY PRESS

CAMBRIDGE
UNIVERSITY PRESS

University Printing House, Cambridge CB2 8BS, United Kingdom

One Liberty Plaza, 20th Floor, New York, NY 10006, USA

477 Williamstown Road, Port Melbourne, VIC 3207, Australia

4843/24, 2nd Floor, Ansari Road, Daryaganj, Delhi - 110002, India

79 Anson Road, #06-04/06, Singapore 079906

Cambridge University Press is part of the University of Cambridge.

It furthers the University's mission by disseminating knowledge in the pursuit of education, learning and research at the highest international levels of excellence.

www.cambridge.org
Information on this title: www.cambridge.org/9781316616048

First published 2012
First paperback edition 2017

A catalogue record for this publication is available from the British Library

Library of Congress Cataloging in Publication data
Frankel, Henry R, 1944–
The continental drift controversy / Henry Frankel.
p. cm.
Includes bibliographical references and index
Contents: Machine generated contents note: 1. How the mobilism debate was structured;   2. Wegener and Taylor develop their theories of continental drift; 3. Sub-controversies in the drift debate, 1920s–1950s;   4. The mechanism sub-controversy: 1921–1951;   5. Arthur Holmes and his Theory of Substratum Convection, 1915–1955;   6. Regionalism and the reception of mobilism: South Africa, India and South America: 1920s through the early 1950s; 7. Regional reception of mobilism in North America: 1920s through the 1950s; 8. Reception and development of mobilism in Europe: 1920s through the 1950s; 9. Fixism's popularity in Australia: 1920s to middle 1960s;   Index.
ISBN 978-0-521-87504-2 Hardback
ISBN 978-1-316-61604-8 Paperback
1. Continental drift–Research–History–20th century.   2. Academic disputations–History–20th century.   1. Title
QE511.5.F73   2011
551.1'36–dc22   2011001412   CIP

ISBN 978-0-521-87504-2 Hardback
ISBN 978-1-316-61604-8 Paperback

Contents

Foreword

I have been asked by Prof. Frankel's publisher to provide a brief foreword to this first volume of his definitive history and philosophical study: *The Continental Drift Controversy*. I am well aware, speaking as a biographer of Alfred Wegener, of the immense difficulties that faced Prof. Frankel in undertaking to detail this complex and fascinating story.

The debate over continental drift has the same role and stature in the history of the earth sciences as the debate over Darwinian evolution in the history of the life sciences, and the debates over relativity and quantum theory in the history of physics. In the largest sense, the history of earth science, the history of biology, and the history of physics in the 20th century are all histories of the consolidation of opinion and the formation of broad consensus -- that these theories were the best way to organize and advance these sciences.

When we look at the ways the history of these three scientific realms has been written, we are immediately aware of a striking asymmetry. While the history of evolutionary biology, and the history of relativistic and quantum physics, are today conducted on an industrial scale, with (literally) thousands of active researchers, the history of earth sciences can boast (at best) a few score scholars at work at any one time. While the study of the lives of Darwin and of Einstein are burgeoning industries in and of themselves, with hundreds of contributors actively involved at any one time in sifting the most minute details of their thought and careers, most major figures in the earth sciences have never been considered biographically at all.

It is therefore the more remarkable that Henry Frankel has accomplished, in this and the three succeeding volumes of his *The Continental Drift Controversy*, an historical task that many would judge, on its face, to be impossible. Working as a single investigator for more than 35 years, he has produced *on his own*, a comprehensive multivolume history of a debate every bit as complex and intricate as those that characterize the emergence of modern evolutionary biology and modern relativity and quantum physics. The work before you is, however, not a preliminary study, not a tentative sketch, not an essay, but an abundantly documented and definitive history of the debate over continental drift from its beginnings to its final resolution.

There is more. Frankel's work here captures not only this fundamental transformation in the theoretical content of earth science in the 20th century, he also chronicles and captures an equally fundamental shift in the way the earth sciences, and indeed all sciences, are conducted. If the careers of Frank Taylor, and Alfred Wegener (the early exponents of the theory of continental motion in the 20th century) might have raised hopes that we could write the history of the earth sciences in the same way we write the history of biology and the history of physics, concentrating on Darwin, and concentrating on Einstein, these hopes gave way quickly. By the 1930s, and certainly in the postwar era, the debate over continental drift was no longer associated with the name of Wegener, or with his particular theoretical ideas. Working from this historical truth, Frankel demonstrates the amplitude and multi-focal character of the emergent debates in middle decades of the 20th century. There were many important players, but coordinated research efforts were rare. Problems and confusions were abundant; satisfactory solutions were elusive.

When Alfred Wegener wrote, in the final edition of his work: *Origin of Continents and Oceans* that: "the Newton of drift theory has not yet appeared," he expressed a widespread historical conviction concerning the outcome of significant debates in the history of science. According to this paradigm, eventually, in every science, all major problems will be resolved by the emergence of a single figure, a "Newton." Wegener was convinced that such a figure must appear. Yet this figure never emerged: there is no Newton of continental drift, no Newton of plate tectonics. There is no single theorist to whose name we may attach the solution of all the major difficulties, the resolution of all the significant anomalies, the pointing of the one way forward. Frankel has clearly seen this and not tried to invent a fictional Newton for continental drift.

Frankel's active grasp of the new way that major theoretical shifts happen – without a guiding genius – is the most remarkable aspect of this book. In modern and contemporary science, governed by multi-author papers, multiyear research programs, intercontinental consortia, coordination of disparate subfields, and science by committee, final agreement is achieved through allegiance of a vast community of investigators to a series of techniques and findings, not to a name or an individual. To tell the story this way is to tell the story of how science is *now* done, and not to wish for a fairytale history in the present, that would mimic heroic science in the past.

Faced with a field of scientific endeavor lacking a single dominant theorist, and therefore without a single individual whose papers one might study, whose work one might trace, whose correspondence one might follow, whose ideas one might highlight, Frankel undertook the necessary labor: he actually pursued the daunting task of reading almost all the theoretical literature pertinent to this question across a span of 60 years. Having oriented himself to the literature, he contacted every principal player in the world who was still living, and interviewed as many as would speak with him. Some were initially reluctant, but as the years went by it became more and more evident to everyone in the earth science community that Frankel's history would be the definitive history of that debate, and not to speak with him was to volunteer to be left out. The project grew in size, scope, and complexity as the years went on, but Frankel has resolutely pursued a consistent and measured strategy.

To deal with the manifold conceptual complexities of this continental drift debate Frankel has developed a typological approach to problems, difficulties, and solutions. Here the training and instinct of the philosopher have amply supported the work of the historian. Geologists are accustomed to thinking in terms of problems and solutions, and because there is no "lower bound" to their curiosity about the earth, they are exquisitely talented at generating challenges to any explanatory hypothesis on any scale -- right down to the molecular, and right up to the cosmic. The result is debates that are long, intricate, fractious, and difficult to follow. Frankel's typological approach renders these matters comprehensible where they might otherwise be bewildering.

The philosophical and historical development of a major scientific controversy can, of course, be told schematically and compactly, but to do so sacrifices nuance and complexity. Since this nuance and complexity is precisely what makes the debates interesting and allows us to see how sciences actually work, Frankel determined to produce a *histoire raisonée*, told as much as possible in the words of the principal thinkers and controversialists and preserving their unique diction and approach and the manifold variations in their particular concerns, while ordering their disputes in a way that is readily comprehensible. It is difficult for me to express how complicated this task must have been and how brilliantly Frankel has achieved it.

Readers will, I think be grateful to Frankel for the calm and measured manner in which he has written this work. Most previous writers, faced with the theoretical complexity of this debate, have exploded into adjectives and adverbs accompanied by much arm waving and antic expostulation. This curse has beset almost every popular book ever written on this topic. Here instead the reader will find a well constructed and gripping narrative, which preserves the complex scientific detail, but invites one in to this fascinating world and helps the reader patiently to find a way through its labyrinth. Frankel is a wonderful guide and worthy of your trust.

Mott Greene
John Magee Professor of Science and Values, University of Puget Sound
Affiliate Professor of Earth and Space Sciences, University of Washington

Acknowledgments

I could not have completed this book and undertaken this overall project without enormous help from many. I owe much to conversations over many years with historians of science Stephen Brush, Mott Greene, and Rachel Laudan. I also owe much to Edward Irving for critically reading earlier versions of this manuscript, and for providing updates of how several problems have been resolved. A. E. M. Celâl Şengör, Warren Hamilton, and Robert Fisher commented on several chapters. It is a pleasure to acknowledge their considerable help. I have benefited from Ursula Marvin's and Anthony Hallam's work on the drift controversy, which inspired me to learn more about it. I thank Cecil Schneer, Michelle Aldrich, and Alan Levitan for welcoming me into the community of historians of Earth science. I thank Bill Ashworth and Bruce Bradley of Linda Hall Library, Kansas City, Missouri, and Bruce Bubacz, Weihang Chen, George Gale, Clancy Martin, and Dana Tulodziecki, colleagues and former colleagues in the philosophy department at the University of Missouri-Kansas City, for suffering through many conversations about the drift controversy. I thank Ray Coveney in the geosciences department at UMKC for putting up with many questions, especially early on. I thank the late Robert Turnbull, Peter Machamer, Ron Giere, Tom Nickles, and Michael Ruse for early encouragement. I thank Deborah Dysart-Gale for translating into English from German key passages of several papers. I thank former Dean Karen Vorst of the College of Arts and Sciences at UMKC for her support of this project.

I am much indebted to the many Earth scientists who have answered questions about their work or that of others. Ken Caster, Brian Harland, Lester King, Edna Plumstead, Curt Teichert, and Eugene Wegmann, all deceased, were very helpful. I also thank Robert Dott Jr., William Chaloner, George Doumani, Warren Hamilton, William Long, Brian McGowran, Arthur Mirsky, Martin Rudwick, Rudolf Trümpy, and Albert Wolfson for discussing their work and that of others.

I thank Nancy V. Green and her digital imaging staff at Linda Hall Library for providing the vast majority of the images; Richard Franklin for the color image of Wegener's Pangea; and the Missouri Botanical Garden Library for several other figures. I should also like to thank the reference librarians at Linda Hall Library, and the interlibrary staff at the Miller Nicholas Library, UMKC.

I would like to thank former students Bob Arnold, Jim Blanton, Fang Chen, Jane Connolly, Julie Dunlap, Kathleen Higgins, Erin Lawrence, Andy Miller, Gary Moore, Tom Pickert, Megan Rickel, and Matt Seacord for comments and encouragement.

I owe much to Nanette Biersmith for serving as my longtime editor and proofreader.

I am indebted to the United States National Science Foundation, the National Endowment of the Humanities, and the American Philosophical Society for financial support. I also thank the University of Missouri Research Board and my own institution for timely grants to continue this project.

Finally, I wish to thank Susan Francis and her staff at Cambridge University Press for believing in this project and for their great assistance throughout its production.

Abbreviations

AAPG	American Association of Petroleum Geologists
AGU	American Geophysical Union
ANZAAS	Australian and New Zealand Association for the Advancement of Science
APW	Apparent polar wander
BAAS	British Association for the Advancement of Science
CSIRO	Commonwealth Scientific and Industrial Research Organisation
ETH Zurich	Eidgenössische Technische Hochschule Zürich
FRS	Fellow of the Royal Society (London)
GSA	Geological Society of America
IAU	International Astronomical Union
IGY	International Geophysical Year
IPS	Institute of Polar Studies
IUGG	International Union of Geodesy and Geophysics
Lamont	Lamont Geological Observatory
NAS	National Academy of Sciences (USA)
NSF	National Science Foundation (USA)
RAS	Royal Astronomical Society (UK)
RS1	Research Strategy 1
RS2	Research Strategy 2
RS3	Research Strategy 3
S_2A_3	South African Association for the Advancement of Science
SEPM	Society of Economic Paleontalogists and Mineralogists
Scripps	Scripps Institution of Oceanography
USGS	United States Geological Survey

Introduction

It was in the mid-1970s when I originally became interested in the controversy surrounding the continental drift theory of Alfred Wegener, thinking of it as a possible subject for testing philosophical accounts of scientific growth and change. There had just been a scientific revolution in the Earth sciences that no philosopher of science had even begun to examine, and in the late 1970s and 1980s I wrote several papers on the drift controversy testing some of these accounts.[1] I also became interested in the controversy for its own sake, and wrote papers concerned with key aspects: the very different reception of Wegener's ideas among specialists; debates over: the origin of the vast Permo-Carboniferous glaciations that intermittently clothed much of the southern continents from 300 to 250 million years ago; the broken (or disjunctive) distribution of past and present-day life forms; the early paleomagnetic work of the 1950s that re-energized the controversy; the development in the early 1960s of the notion of seafloor spreading and of its corollary the Vine–Matthews hypothesis.[2] I then thought of working them up into a book but realized that I still had only a minimal understanding of what later transpired, no feeling for the way the controversy was resolved. Like many others at the time, I underestimated paleomagnetism's support of continental drift, and I did not understand some subtle and some not so subtle features of plate tectonics.

Now, twenty-five years later, after studying the continental drift controversy to its conclusion, Cambridge University Press has brought out my account in four volumes. It is an attempt to tell the story from end to end in some detail, from its initiation in the early twentieth century to its conclusion in the late 1960s as a general theory describing the mobile nature of the Earth's surface features – plate tectonics. The story is of new discoveries and ideas, and it is also a social history in which the operative workers and institutions are identified as they appear and their stories told.

The continental drift or mobilism versus fixism controversy, as it is sometimes called, involved almost all branches of Earth science and no single person is competent in them all. So is it sensible that a philosopher, yes with a degree in biology but with no direct experience of research in Earth science, should attempt such a task? I do have a long-standing interest in scientific reasoning and in theory choice, the initial reasons why I was attracted to the controversy. Within the mobilism controversy there were many sub-controversies and I was struck by the similarities in the manner in which participants, whatever their interests, attacked their opponents' solutions and defended their own. I was also struck by how their arguments centered about the identification of difficulties faced by their opponents' arguments, and by proactively imagining the difficulties that might be raised against their own. My analysis is described in Volume I, Chapter 1.

I could not have written this account without the assistance of many of the participants. My approach was to read through their various papers, and then send them questions about how they came to undertake their investigations and why they made, or did not make, certain claims. But, they often did much more: they gave me tutorials, enabling me to understand better their various fields of enquiry. The narrative begins with an account of geological theorizing in the early twentieth century and is followed by an account of the drama of Alfred Wegener's life (Volume I, §2.2 to §2.5) and his theory of continental drift, which I summarize here.

Over a period of seventeen years in approximately a dozen publications between 1912 and 1929 Wegener described his revolutionary theory of continental drift. He imagined continents floating on a denser substrate through which they plowed, impelled by Earth's tidal and rotational forces. He placed his mechanism at the center of his theory. Working backwards in time he closed the Atlantic, Antarctic, and Indian oceans and assembled the continents like pieces of a jigsaw into a single landmass, which he called Pangea (Volume I, §3.2). According to Wegener, Pangea lasted from the Late Carboniferous to the Cretaceous, from about 300 million to 100 million years ago. He reconstructed the breakup of Pangea into the present continents and mapped their drift to present positions. At this stage his frame of reference, his grid of latitudes and longitudes, was arbitrary, and he adopted the convention of keeping Africa fixed and moved other continents relative to the grid (Volume I, Figures 3.1 and 3.2). This was a grid of convenience; it was not an authentic geographical grid appropriate to the times in question. Later, with his father-in-law Vladimir Köppen, Wegener used the distribution of climate-sensitive deposits to determine geographic latitudes for his maps (Volume I, §3.15; Figures 3.6, 3.7, and 3.8). The manner in which Wegener fitted continents together and justified that fit by appealing to evidence of ancient climates is of great interest historically. He fitted them geometrically by their shapes and matched them by their geological features, as workers do today. He noted that the matches in his reconstructed Pangea were generally excellent. He then placed them in appropriate latitude zones, relative, that is, to the geographical pole at that time. Jigsaw and latitude complemented one another. This agreement (or consilience) between data from diverse sources that were independent of one another was perhaps Wegener's strongest argument (Volume I, §3.2); in different contexts, consiliences such as this were a recurrent theme throughout the mobilism debate.

During the 1920s, the theory of continental drift was widely discussed. For a brief period, many saw virtue in Wegener's ideas. As things settled down, it became clear that its reception by individuals correlated strongly with their specialization and region of study – for example, those who worked on the Permo-Carboniferous glaciations of the continents of the Southern Hemisphere favored drift, whereas those who worked within what they saw as geologically self-contained regions, especially in North America and the Soviet Union, rejected drift. I describe these important relationships in Volume I, Chapters 6 through 9.

By the 1930s, Wegener's progressive ideas had lost ground to the doctrines of fixed continents and oceans. Especially hurtful was the demonstration that Wegener's mechanism would not work. As a result, over the next forty years few adopted continental drift theory or used it as a basis for research or teaching; it was widely ignored or reduced to a footnote in many geology courses and texts. Especially was this true not only in North America but elsewhere too. Volume I covers this "classical stage" in the drift debate.

It was in the early 1950s, when acceptance of continental drift was at a low point, and discussion of it going nowhere, when out of the blue, paleomagnetists breathed new life into an essentially moribund controversy. Paleomagnetists had taught themselves how to determine the history of the geomagnetic field as recorded by the magnetism of rocks, and how to construct motions of the migration of Earth's rotational axis pole relative to continental blocks. These motions were expressed as paths of apparent polar wander (APW) relative to each fixed continent (Volume II, Chapter 3): They can also be expressed as motions of continents relative to a fixed pole. They learned how to combine their work with the findings of paleoclimatologists, establishing generally excellent agreement between their inferred latitudes and the evidence of past climates in the same continent (Volume II, §3.12; Volume II, §5.10–§5.14; Volume III, §1.7; Volume III, §1.18). They found huge disagreements between polar wander paths from different continents that made no sense unless continents had moved relative to one another in much the same way as Wegener's theory required. It was a rough blow to fixism.

This work led to a revival of interest in continental drift in Britain, the Soviet Union, South Africa, and Australia and prompted R. A. Fisher to remark at the time, "I think a lot of geologists must be timidly peering out of their holes on hearing the strange news that geophysicists are talking about continental drift." This work led, by the end of the decade, to the first physical confirmation of continental drift (Volume III, §2.17).

Critics may say that I have given too much space, certainly, proportionately, more than others have, to the 1950s paleomagnetic revival of the drift debate. Anticipating work that I shall describe in a moment, there are four reasons for doing so. (1) Apart from a short early review, the history of this phase of the controversy has never been properly described. (2) The narrative would be truly incomplete were I not to record how those who did not favor continental drift (they were the majority) took little or no notice of its new paleomagnetic support; variously, they did not understand it, failed to read the key supportive papers, or rejected it on the basis of hearsay. (3) Reversals of the geomagnetic field were discovered by paleomagnetists working on land, and this was essential to what transpired later. Had paleomagnetists not discovered reversals by direct observation, first in stratigraphic sequence on land and then in deep ocean cores, showing them to be a general property of Earth's magnetic field, how would they have been recognized at sea and the kinematics of plate tectonics thereby quantified? It is hard to imagine plate tectonics without reversals of the geomagnetic field. Certainly progress would have been very slow. (4) In the 1920s Wegener and Köppen established the geographical frame of reference for continental drift. Likewise in the 1950s, paleomagnetists laid the groundwork for establishing quantitatively the geographic framework for plate tectonics by determining the motions of continents relative to the spin axis. Their work is described in Volume II and the first two chapters of Volume III.

While interest in continental drift was being revived by work on land, the tectonics of oceans floors was little known. In the later 1950s and 1960s there was a flood of data and ideas about the ocean floors.[3] Especially critical were studies of what turned out to be largest mountain ranges on Earth, albeit underwater – the mid-ocean ridges. Several suggestions were proposed to explain their origin (Volume III, Chapters 3 through 6). Seafloor spreading (1960) proposed that these ridges marked the places where hot material welled up, cooled and became rock, just as lavas on land do. To balance this new crust being created at the ridges, oceanic crust was thought to be descending in subduction zones beneath the numerous deep ocean trenches and being resorbed into Earth's mantle (Volume III, §3.20). This idea was confirmed in 1967 (Volume IV, §7.4). Soon the ridges were found to have quite remarkable tell-tale magnetic anomalies. So what was more natural (although revolutionary and not immediately accepted) than to propose, as it was in 1963 and confirmed in 1966, that these anomalies corresponded to reversals in the geomagnetic field recorded as these hot upwelling lavas cooled and became magnetized just as paleomagnetists on land had shown they did (Volume IV, Chapters 2 through 6). Furthermore, when these reversals recorded by seafloor spreading (Volume IV, §6.6, §6.8) were calibrated by studies, first on land, of radiometrically dated reversals (Volume II, §8.15; Volume IV, §6.4), and then with astonishing consistency of reversals in deep-sea sediment cores (Volume IV, §6.5), they were used to map the motions of ocean floor (Volume IV, §6.6, §6.8, §7.6). Once rates of seafloor spreading were determined and estimates were made to determine, for example, when the Atlantic opened, they were found to agree with estimates based on land-based paleomagnetic findings (Volume IV, §6.6, §6.8, §7.6). These consiliences enhanced the strengthened support for seafloor spreading and showed the success of the paleomagnetic case for mobilism developed during the 1950s.

Enigmatically the mid-ocean ridges were offset by huge fracture zones sometimes thousands of kilometers in length and the motion across them appeared to be in the wrong sense. In 1965 these fracture zones were recognized as a completely new sort of structure, called transform faults, and, in 1967, their existence was confirmed (Volume IV, Chapters 4 through 6). These allowed major structures to be kinematically linked and many fundamental crustal boundaries to be recognized.

The great fracture zones of the northeast Pacific Ocean were originally thought to be great circles, circles that have their centers at the center of the Earth (Volume IV, Chapter 7). On closer inspection, they were found to be not great, but small circles, circles which, like lines of latitude, had their centers along an axis of rotation. This was also found to be so for the fracture zones associated with active mid-ocean ridges. For any particular ridge, the small circles that define the fracture zones along it were found to be concentric about a point on the Earth's surface, called the Euler pole or pivot point (Volume IV, Chapter 7). Tellingly, fault plane solutions to the earthquakes occurring along these ridge-associated fracture zones (they are transform faults) gave slip vectors that were essentially horizontal and indicated that current relative motions across fracture zones were in a strike-slip sense and occurred about the same pivot point. Most tellingly, these present-day relative motions were found to be consistent with motions over the past tens of millions of years determined from the analysis of marine magnetic anomalies.

These discoveries led directly to the theory of plate tectonics (Volume IV, Chapter 7). It is a kinematic theory which says that the Earth's lithospheric shell is divided into a number of large plates that are moving relative to one another along three sorts of boundaries: extensional at the active mid-ocean ridges, compressional at the great subduction zones, and strike-slip along the great transform boundaries. Plates are composed mainly of oceanic lithosphere, although most of them contain a large segment of continental lithosphere, the great landmasses.

As just mentioned, land-based paleomagnetic techniques determine the position of land-masses relative to the geographical pole, most importantly their latitude (Volume II, Chapter 3). The techniques of plate tectonics determine the motions of plates relative to one another; the method is generally blind to past latitude and provides no record of position relative to Earth's axis of rotation, except in those situations when marine magnetic lineations or profiles can be exploited to provide paleolatitudes. Hence the two methods are complementary, land-based paleomagnetic results providing the geographic frame of reference for plate tectonics reconstructions.

Plate tectonics offers no explanation for the forces that drive plates, a point that was made abundantly clear by the discoverers of plate tectonics. The great irony of the mobilism controversy is that for over forty years the lack of an acceptable mechanism was generally regarded as a strong reason to reject continental drift. Ironically, it is remarkable that plate tectonics was accepted almost immediately even though it is a kinematic not a dynamic theory. Once accepted, the lack of mechanism was no longer a difficulty but an advantage, freeing the discussion of the relentless and unnecessary burden it had carried for so long. Many very different and independent measurements and analyses had showed that large-scale horizontal tectonics were a reality; they were no longer in doubt and objections to their existence because there was no acceptable mechanism became groundless. Indeed, in retrospect, the perceived lack of mechanism never was a good reason to reject drifting continents, even if there were good reason to reject certain proposed mechanisms.

Notes

1 Frankel (1978, 1979a, 1979b).
2 Frankel (1976, 1978, 1980b,1981, 1982, 1984a, 1984b).
3 This flood of new information about the ocean floors was made possible by massive governmental support of marine geology prompted by defense concerns. See Schlee (1973), Bullard (1975b), Menard (1986), Strommel (1994), Rainger (2000), Hamblin (2005), and Doel *et al.* (2006).

1

How the mobilism debate was structured

1.1 The three phases of the continental drift controversy

The *continental drift*, or more generally the *mobilism* controversy lasted sixty years. It was the longest and most important controversy of the last century in Earth science, and one of the more important in all of science in that period.

Within it were many sub-controversies and their important feature was that some were long-lived too, even surviving the controversy itself. For example, the apparent conflict noticed in the 1920s between the "Permian" glacial Squantum Tillite of the Boston region and the equatorial situation assigned to that part of North America by drift theory was not resolved until the beginning of the twenty-first century, when radiometric dating assigned it to the Late Precambrian glaciation (Snowball or Panglacial Earth) (Thompson and Bowring, 2000). Also outliving the controversy and of much more general significance was the debate about the mechanism of continental drift, which began in the early 1920s and, although much progress has been made, is still not entirely resolved ninety years later. Throughout much of the controversy a solution to the mechanism question was regarded by most workers as essential, but at the end it was jettisoned and left in the wake of plate tectonics. Not only was it left unresolved by plate tectonics, it had to be first set aside in order for progress to be made – set aside as a problem for the future. In the end, the avalanche of evidence for the geometrification of tectonics and the convincing kinematic picture it gave overwhelmed concerns about dynamics – the mechanism difficulty, as I shall call it.

The durability of the sub-controversies lends a certain repetitiveness to the history of the controversy. Repeatedly problems were thought to have been laid to rest and then revived and discussed all over again and so on, and this is why, in the narrative, they and the scientists involved appear and reappear time and time again. These are not inadvertent repetitions (although I am sure there are some of those too), they are a characteristic feature of the story – how it unfolds. In these three volumes, scientists and their work are introduced in rough chronological order beginning in Chapter 2.

In this first chapter I comment on what I believe is a remarkable similarity in the arguments used throughout the controversy. This common thread provides a degree of coherence to the entire debate, and, I think, is a means by which it can be understood. Some readers may find the formalities of this chapter a little tedious. Perhaps they should skip to Chapter 2, and if, after some time, they feel the need of help finding this common thread then they could return here.

It was Émile Argand, one of the first converts to continental drift, who first introduced the terms "mobilism" and "fixism," and I view the controversy in terms of these two competing traditions. Fixists maintained that, except perhaps very early in Earth's history, continents remained fixed in the same place relative to one another. Fixists did not however always agree among themselves. Some claimed that the axis of rotation moved relative to the Earth as a whole (polar wander) so that although continents remained fixed relative to each other, they have changed their positions relative to the geographical poles. Nor were fixists agreed as to how continents were formed – whether they increased in size, how mountain belts formed, and how intercontinental biotic and geologic disjunctions (disjuncts) arose. Disjuncts are occurrences of similar phenomena now separated by wide expanses of land or, more commonly, ocean where such occurrences are wholly absent. Some fixists, the landbridgers, explained biotic disjuncts by supposing that species of life migrated across former transoceanic landbridges. Others, who for other reasons were called permanentists and who were especially prevalent in North America, abandoned landbridges in favor of long isthmian connections and island hopping as a means by which land organisms migrated across oceans. In contrast, most mobilists declared that continents have changed their position relative to one another and relative to the geographic poles, that continents have undergone horizontal displacement, changing their latitude and longitude over time. A very few mobilists, who mustered under the flag of mobilism, favored Earth expansion, claiming that even though the latitude and longitude of continents have not changed, their distances from one another have increased as Earth has expanded.[1]

The debate between fixism and mobilism evolved through three phases. Fixism was almost universally assumed until Frank Taylor in 1910 and Alfred Wegener in 1912 introduced their mobilist theories of continental drift (Taylor, 1910; Wegener, 1912a, 1912b) and inaugurated the first or classical phase. Throughout this phase, fixism remained ascendant, although a number of old-time drifters and a few new converts carried the flag for mobilism. During this phase, which is the subject of Volume I, participants argued over who had the better solution to a nest of problems, among them: explaining the congruency of opposing continental margins especially across the Atlantic basin, explaining biotic and geologic disjuncts, and explaining the origin of Tertiary mountain belts and the vast far-flung Permo-Carboniferous glaciation. Although mobilists were very greatly outnumbered, neither tradition gained a decisive overall advantage during the classical phase. Mobilists offered better solutions than fixists to some problems, and fixists offered better solutions to others, but every solution offered was plagued with difficulties. The standard operating procedures of both mobilists and fixists was to propose new solutions to problems in terms of the basic tenets of their own tradition, to attack their opponents' solutions by raising difficulties against them, to defend their solutions against attacks by attempting to remove difficulties, and to argue that their own solutions were preferable to those of their opponents. These procedures were so prevalent that I shall call them research strategies (RS), and later in this chapter describe some of their more prominent features. Introducing the idea of a difficulty-free solution, I shall argue that none were produced during the first phase. As a result, neither mobilists nor fixists had to admit defeat; they were never obliged to acknowledge that their position was no longer tenable because their opponents had succeeded, whereas they had failed to produce a difficulty-free solution.

The second phase of the controversy, the subject of Volume II and the first two chapters of Volume III, is marked by the rise in the early 1950s of paleomagnetism, initially in the United Kingdom, as a result of information obtained at the Department of Geodesy and Geophysics at the University of Cambridge and the Physics Department of Imperial College, London. Paleomagnetists quickly developed and articulated their new procedures. They acted in accordance with these standard research strategies to garner support for mobilism, to anticipate difficulties that might be raised against their work, and to remove difficulties as they were raised by others. By about 1959, British, Australian, and South African paleomagnetists had developed an explanation of the accumulating paleomagnetic data which showed that the continents had changed their positions relative to each other more or less as Wegener had proposed; they gave fulsome support to mobilism. Their work signaled that all might not be well with the doctrine of fixism. It definitely rekindled interest in mobilism, but not many outside paleomagnetism were convinced that mobilism merited acceptance. I shall argue that the paleomagnetists' explanation of their data, despite its often uncomprehending and hostile reception by all but a few, actually warranted the acceptance of mobilism. I shall argue that these paleomagnetists had developed an essentially difficulty-free mobilistic solution, while fixists still had not provided any such justification for their continued support of the traditional view.

The third and final phase of the controversy, the subject of most of Volume III and Volume IV, began in the mid-1950s when there was a massive influx of new information about the seafloor obtained through the use of new geophysical techniques and instruments made possible by extensive funding for defense purposes.[2] This phase began in earnest about a decade after World War II and intensified during the early stages of the Cold War. This new information was gathered primarily at Columbia University's Lamont Geological Observatory, Palisades, New York; Scripps Institution of Oceanography, La Jolla, California; Woods Hole Oceanographic Institute, Woods Hole, Massachusetts; and the Department of Geodesy and Geophysics at the University of Cambridge. Various fixist and mobilist theories were developed to explain this new information. One mobilist theory proposed that new seafloor is created at the center of mid-ocean ridges, where it separates, moving away sideways creating new seafloor. This theory, seafloor spreading, spawned two key hypotheses. One sought to explain the remarkable striped patterns of marine magnetic anomalies as records of reversals of the geomagnetic field resulting from seafloor spreading. The other explained the movement of seafloor between ridge-offsets by what became known as ridge-ridge transform faults. Confirmation of the former in 1966 and the latter the following year led to the overnight acceptance of mobilism by most scientists working within the controversy. I shall argue that both hypotheses became difficulty-free as a result of these confirmations. With this resolution, seafloor spreading and continental drift morphed into plate tectonics, which commanded swift approval throughout the Earth science community. It became the reigning theory in Earth science.

The ideas or concepts of difficulty-free solutions and research strategies employed by participants to defend their views and attack those of their opponents are embedded in my account. I now need to elaborate them.

1.2 Solutions, theories, hypotheses, and ideas or concepts

These words refer to explanations or proposals that were designed to solve problems in various ways. Solutions were proposed to solve one problem; theories too, like solutions, solved problems. Theories were designed to solve just one or, more commonly, to solve several problems; they generally provided a common solution to several problems, and within them each solution had a common element. Wegener's theory of continental drift sought to solve many problems and contained many solutions. Wegener's theory related together many problems by providing a common framework: Earth's continental crust was once united in Pangea, which fractured and the fragments drifted apart to form the present continents. Sometimes scientists expanded the range of a theory by using it to solve problems that it was not originally designed to solve; hypotheses may refer to solutions that were not established when proposed or to theories before they were established. Theories were hypothetical when proposed; solutions may have been. Solutions and theories may have been dismissed, or they may have become established. Sometimes, however, they still are habitually referred to as hypotheses even though they now are firmly established. In the Earth sciences, the Vine–Matthews hypothesis is perhaps the most famous example. Other proposals are labeled as concepts or ideas. Such proposals often hypothesized the existence of a new process or entity, and the most famous example from Earth science is the concept of transform faults. J. Tuzo Wilson, recognizing that there was a new kind of fault, wisely coined a new term – transform fault. When others discussed Wilson's idea, they often referred to it as a concept. Wilson's idea (like the Vine–Matthews hypothesis) was a corollary of seafloor spreading. Wilson explained the different types of transform faults, offered about a dozen examples, and argued that the existence of transform faults provided strong support for mobilism.

Readers will understand from the above that I shall not be so foolhardy as to attempt to strictly define the words in the title of this section. To attempt to do so would be to limit their usefulness and divorce them from the very varied literature that has grown up over the past half century around the fixism versus mobilism debate. Instead I shall try my best to make myself clear from the context.

1.3 Problems and difficulties[3]

I begin with a taxonomy of *problems* and *difficulties*. Scientists addressed problems by proposing solutions and theories. As already noted, solutions were designed to solve one or perhaps two closely related problems; theories were designed to solve numerous problems, and the number of problems they solved increased as workers found new ways to apply them. Other scientists questioned these proposals by raising difficulties. I then introduce the notion of a *difficulty-free solution*, and delineate a set of *research strategies* that participants in the controversy employed to defend their own solutions and to attack those of their opponents.

Difficulties were objections that were raised against these proposed solutions and theories, obstacles that were in their way all along or placed there later by opponents. Stated in this way, there can be no difficulties without problems and their proffered solutions. *Problems* arose when scientists became puzzled by phenomena they could not explain. Sometimes more data were gathered to establish the legitimacy of a problem and to clarify it. A solution was then offered. Difficulties were usually raised by scientists with opposing views. They were also raised by supporters in the same camp and even by scientists themselves against their own solutions. Difficulties were removed either by amending the flawed solution or theory, or by showing that the raised difficulty was itself unfounded, a phantom difficulty. The mobilism controversy was replete with proffered solutions to problems, and with real and phantom difficulties.

1.4 First and second stage problems

There are first and second stage problems. First stage problems arose through the discovery of a puzzling phenomenon. A scientist offered an explanation by postulating a hypothetical process. For instance, scientists noticed and were intrigued by the similarity in shape of the opposing coastlines of South America and Africa and proceeded to explain it by suggesting that they were once united into a single landmass, which split into two parts that drifted apart. Scientists then wondered, how did the single landmass split apart? This new problem was a consequence of solving the first stage problem. Secondary problems have as their subject matter entities or processes that are invoked to solve primary problems. Secondary problems cannot arise until a solution has been offered to a first stage problem.

1.5 Four examples of first stage problems

(i) *Permo-Carboniferous glaciation.* By 1910 many Earth scientists believed that large areas of the Southern Hemisphere continents and peninsular India had been glaciated at times during the later Carboniferous (~320–300 million years ago) and Permian Periods (~300–251 million years ago) because they had found thick deposits likely of glacial origin of this age in Australia, in southern Africa and South America, and in India. Evidence of Late Paleozoic glaciations had not, at the time, been described from Antarctica. The extent and location of these deposits was startling, and explanations were sought for them. Workers were particularly puzzled by the presence of glacial deposits in India within the present tropics. To make matters worse there was good reason to believe that the climate in the Northern Hemisphere continents, some of them approximately antipodal to glacial occurrences in the Southern Hemisphere, had been mild or even tropical throughout much of the Permo-Carboniferous because of the occurrence there of thick limestones, coral reefs, vast coal deposits of tropical aspect, and evaporites. Wegener solved this problem by invoking continental drift. He argued that the continents had been united during the Permo-Carboniferous with the northern continents in low northern latitudes and the South Pole just south and east of Africa. This arrangement placed the South Pole at the center of the glacial deposits and kept them within mid to high southern latitudes. Thus Wegener could claim that the glaciation had not extended to regions formerly located on or near the equator; such argument later allowed him and Wladimir Köppen, a leading climatologist, to argue that Earth's climate during the Permo-Carboniferous was divisible into latitudinal zones very much as it is today.

(ii) *Past and present biotic disjuncts.* Before Wegener began to speculate about continental drift, paleontologists and biogeographers had found that many similar ancient terrestrial life forms had inhabited regions now widely separated from one another. To explain these disjuncts, they proposed the existence of landbridges that served as migratory routes across oceans. The bridges later sank, becoming part of the seafloor. Wegener realized that continental drift offered an alternative solution, and argued that disjuncts arose because regions formerly close together had since moved apart.

(iii) *The scattered paleopole problem.* In the early 1950s the rapidly emerging field of paleomagnetism attracted attention. Paleomagnetists studied the natural remanent magnetism of rocks to obtain a record of the orientation of Earth's magnetic field at the time of formation. A pattern quickly emerged. Poles corresponding to the directions of remanent magnetism of samples from Upper Tertiary and Quaternary rocks (the past 25 million years) were, within error, coincident with the present geographic poles: the time-averaged geomagnetic field could be represented by a dipole at the Earth's center directed along the axis of rotation. Poles or paleopoles are the points at which the axis of this dipole intersects Earth's surface. Paleopoles calculated from the directions of Early Tertiary (Paleogene) and older rocks differed greatly from the present geographic pole, and it was soon found that poles from the same continent fell sequentially on curved paths, later called apparent polar wander (APW) paths; samples of the same age from the same landmass (strictly each continental craton) gave similar paleopoles, but samples of differing age did not. Paleogene and older samples from different landmasses gave different poles and APW paths. Paleomagnetists realized that these diverging APW paths from different landmasses could not be reconciled without invoking continental drift, and they very quickly recognized that Wegener's theory conceived forty years earlier could explain the main features of their observations.

(iv) *The reversal problem.* Another paleomagnetic problem that gained considerable attention throughout the 1950s arose from the discovery, dating back to the turn of the century, that many rocks have a remanent magnetism of polarity opposite (antiparallel) to that of the present Earth's magnetic field. This is called reversed remanent magnetism; rocks that have it are said to have *reversed polarity*. Two competing solutions were offered: reversed polarity could have been caused by some mechanism whereby rocks become magnetized spontaneously in the opposite sense from the ambient field (*self-reversal*), or by reversals in polarity of the Earth's magnetic field. By the late 1950s a very good case could be made that field reversal was the correct solution, but this was not generally accepted until the mid-1960s when it promptly became a cornerstone of the theory of plate tectonics.

1.6 Four examples of second stage problems

(i) *The mechanism problem of continental drift.* Wegener, having postulated extensive horizontal displacements of continents, sought to identify the forces that caused them. To solve this secondary problem, he postulated two mechanisms: flight from the poles, which sought to explain equator-ward drift of continents, and tidal force to explain their westward drift.

(ii) *The mechanism problem of plate tectonics.* What are the processes responsible for plate motions? Some workers proposed mantle convection that drags along lithospheric plates; others tied convection directly to lithosphere plates, invoking gravitational forces that directly pull lithospheric plates down subduction zones or forces that push them away from spreading ridges.

(iii) *The mechanism problem of landbridges.* Paleontologists and biogeographers who proposed landbridges to account for the primary problem of biotic disjuncts thought about what might have caused landbridges to sink after they were no longer needed as migration routes.

(iv) *What caused reversals of magnetization?* As noted above, reversals of magnetization could be caused by spontaneous self-reversal or by reversal of the geomagnetic field. Each spawned secondary problems. Workers who opted for spontaneous self-reversal had their various mechanisms. Workers who preferred field reversals offered theories of reversal of the geomagnetic field, although it is only very recently that realistic field reversal theories have become possible.

1.7 Difficulties

Scientists regularly identified and addressed problems during the controversy, and it was no small accomplishment to construct solutions and theories to explain them. The hardest task of all, however, was to construct solutions that were free from difficulties. Indeed, I claim that in the mobilism controversy the identification and removal of difficulties played a very central role. To think, as some have maintained, that the raising of difficulties was silly polemics, and their removal was a mopping-up operation left to the ungifted but hard-working scientists is, I believe, to misunderstand completely what actually happened during the mobilism debate; participants engaged in the controversy expended considerable skill and imagination raising difficulties against solutions proffered by their opponents and removing those raised against their own. Time and time again throughout the mobilist controversy, the identification and removal of difficulties were the keys to progress, not just routine filling in the gaps.

Difficulties raised fell into two main categories, *data* and *theoretical difficulties*. *Data difficulties* arose when the data used to evaluate a solution or theories were found to be suspect. *Theoretical difficulties* arose when a proposed solution or theory was plagued with inconsistencies or ambiguities. There were three general sorts of data difficulties, which I call *unreliability*, *anomaly*, and *missing-data difficulties*, and two sorts of theoretical difficulties – *external* and *internal*. I deal with each in turn.

1.8 Unreliability difficulties

Unreliability difficulties comprised a variety of real or imagined irregularities in the collection of data, its analysis, or interpretation. They include the questionable use of previously tested procedures, or of untested new ones, and the use of outdated procedures, procedures that had been superseded by new ones in the hope of raising standards. They may have involved alleged theoretical or sampling biases of the investigator. Unreliability difficulties were also directed against scientists who, while supporting their own theory, misused data of others; for example, scientists sometimes overstated the reliability of data (their own or that of others) that were particularly favorable to their solution or theory: not infrequently they ignored data not supportive of it. Here are four examples.

(i)   Wegener was accused of having created an unreliability difficulty when marshaling support for his explanation of the congruency of continental margins facing the Atlantic. Alex du Toit, although a vigorous drifter, claimed that Wegener had mistaken the actual shape of the continents, he had ignored the fact that they extend beyond their coastlines to the shelf edge; continental shapes could only be determined from bathymetric data. Actually, Wegener had used bathymetric data, and it was du Toit's criticism that was incorrect.

(ii) Paleomagnetists working during the late 1940s and 1950s who sought to design trustworthy techniques for collecting, measuring and analyzing the remanent magnetization of rock samples encountered an unusually large number of complex unreliability difficulties. Theirs was an uphill battle, and they went to great lengths to imagine and anticipate unreliability difficulties that might later be raised against them.

There were three reasons why paleomagnetism was plagued in this way, and I shall spend a moment explaining them. The first was that within a few short years in the early and mid-1950s paleomagnetists had launched physically based, global surveys and challenged on a broad front the monolithic fixism of the classical phase. Consequently there was general incredulity that results obtained so swiftly and so readily, with such apparent ease, could be reliable and yet conflict so dramatically with the widely believed doctrine of fixism, based, as it was then thought to be, on evidence from much of Earth science.

Second, it was a young discipline without preexisting standards. Early workers designed and built their own instruments, developed field criteria to recognize and weed out potentially magnetically unstable samples, and established their own procedures to obtain, to analyze, and to verify their data. For instance, they had to design procedures allowing for the magnetic heterogeneity of a sample to ensure that their measurements were representative of the sample as a whole, and they often thought it necessary to check the general reliability of a new magnetometer by comparing results with those obtained on previous instruments. Early workers also realized the need for sound statistical techniques to determine the accuracy of data and struggled to apply them.

The third general reason why paleomagnetists had to be sensitive to reliability issues pertains to the very nature of remanent magnetism and the difficulty of extracting information about the ancient field millions of years ago. Paleomagnetists had first to determine if the remanent magnetism was sufficiently strong and then if it was stable. They had to estimate also when it was acquired, no small task. In mid-century, because radiometric dating of rocks was still in its infancy, and igneous rocks at the time were often poorly dated, paleomagnetists turned to sedimentary rocks because in general they were better dated and promised to provide a more detailed record of field behavior. They first had to determine which of the many sorts of sedimentary rocks gave coherent results and were likely therefore to yield a record of the ancient field. They soon found that sedimentary rocks possess weak remanence, are commonly magnetically unstable, some even changing their magnetism after collection. However, British workers soon discovered that fine-grained red sandstones or siltstones regularly gave consistent results of great age, and they exploited their discovery focusing on that lithology. A further complication was that both igneous and sedimentary rocks can become partially or entirely re-magnetized, acquiring secondary magnetizations (overprints) of uncertain ages, and early paleomagnetists had to develop field and laboratory tests to identify them. Initially, data showing signs of overprinting were rejected. Eventually procedures were developed to correct for overprints and finally for removing them altogether.

Although these early paleomagnetists usually took great care to screen data, others often questioned the reliability of their opponents' methods, arguing from a perfectionist standpoint that insufficient care had been taken in doing this or that. For others, unfamiliar with these new techniques, it required a special effort to tell which studies, if any, were reliable. Consequently, Earth scientists unfamiliar with these new studies were generally wont to take little notice or even to dismiss them. Not only were they new and unfamiliar, and not only did the new results challenge long and generally accepted fixist norms, but Earth scientists generally were aware that within the quarrelsome paleomagnetic community there were differing opinions as to the reliability of results and what they signified.

(iii) To support his postulated westward drift of Greenland relative to Europe, Wegener invoked geodetic data that were later shown to be unreliable; he overestimated their reliability. This is an example of a theorist using, somewhat recklessly, someone else's data to support his solution. It was not uncommon for data to be assumed more reliable than they really were. Cautionary remarks by the original observers may go unheeded. Data may be unreliable for reasons unknown at the time. The theorist's understanding of the data gathering process may be limited.

(iv) Wegener and du Toit were criticized for emphasizing paleontological studies that supported mobilism, and muting or ignoring those that favored fixism. The American paleontologist, George Gaylord Simpson, claimed (*circa* 1940) that mobilists ignored studies which indicated that the number of biotic disjuncts, their stock-in-trade, had been overestimated.

1.9 Anomaly difficulties

Anomaly difficulties arose if there was a tension between a solution or theory and data on which it was based; that is, if the data indicated something highly unlikely given the solution or theory, or if an incompatibility arose between the data and a prediction of the solution or theory. During the debate, the critic's main purpose in raising anomaly difficulties was to show that the solution, although it *should be able* to explain the data, failed to do so. Consider the following critiques, the first aimed principally at Wegener.

(i) One of the more serious difficulties concerned his solution to the problem of the Permo-Carboniferous glaciation that invoked continental drift. Wegener united the glaciated regions into a single landmass. Assuming that the climate zones during the Permo-Carboniferous were similar to present ones, he located the South Pole near the center of the glaciated area which, on a united Pangea, positioned southern North America near the equator. Fixists contended that evidence of glaciation in the southern part of North America would be anomalous on this reconstruction, and pointed to Pennsylvanian (Late Carboniferous) conglomerates in Oklahoma and Kansas, and varved claystones near Boston (the supposed Permo-Carboniferous Squantum Tillite now known to be Precambrian), interpreting them as glacial deposits. As already noted, the Squantum Tillite remained bothersome to Wegener's theory almost to the very end of the controversy.

(ii) Fixists also raised anomaly difficulties against drift's solution to the problem of biotic disjuncts. The fern-like Permian *Glossopteris* land plants had been found in peninsular India and all southern continents, and Wegener made much of the fact that his positioning of the continents offered a wonderful explanation for their present widespread distribution. However, some fixists also made much of the identification (later disproved) of *Glossopteris* flora in northeastern Russia and Siberia, arguing that Wegener's reconstruction made it very difficult to account for these findings – more difficult than with the continents in their present positions.

1.10 Missing-data difficulties

Missing-data difficulties arose when a solution was based on insufficient or fragmentary data. Anomaly difficulties involve the discovery of data that are inconsistent or highly unexpected with a solution or its consequences; missing-data difficulties involve only the absence of, or failure to find, the needed data.

Both mobilist and fixist solutions to the problem of biotic disjuncts faced missing-data difficulties. Mobilists maintained that the ancestors of Australian marsupials had migrated from South America via Antarctica. Some fixists maintained that they migrated from Asia. But attempts to uncover marsupial fossils in either Antarctica or Asia were for many years unsuccessful, and hindered acceptance of mobilism as an explanation for marsupial distribution. Indeed, marsupial fossils were not found in Antarctica until after the mobilism debate was resolved (Woodburne *et al.*, 1982), and, much later somewhat surprisingly, were eventually found in Asia (Luo *et al.*, 2003). The failure to find evidence of Permo-Carboniferous glaciation in Antarctica until late in the debate is another clear-cut example of a missing-data difficulty for mobilism now removed.

1.11 Theoretical difficulties

Solutions or theories faced *theoretical difficulties*, difficulties other than those related to the data on which they are based. Theoretical difficulties may be external or internal.

External theoretical difficulties arose when a theory or solution stood in an uneasy or incompatible relationship with other solutions, theories, or methodological principles with which it was not in active competition. Some solutions conflicted with another theory or methodological principle that was well founded; the relationship between the theory or solution and other theory or principle was less compatible than desired. Some theories or solutions depended on other solutions, theories, or methodological principles that were not well founded. Here are five examples of theoretical difficulties: the first three illustrate clashes between solutions and relatively well founded theories; the others illustrate difficulties that were faced by solutions proposed on the basis of what critics deem were dubious theories.

(i) Wegener raised an external theoretical difficulty with landbridges that had been invoked to explain biotic disjuncts. Landbridges were widely accepted in the late nineteenth and early twentieth centuries prior to Wegener's introduction of continental drift. To explain the similarity of Paleozoic fossils in South America and Africa, landbridgers postulated the existence of a Paleozoic landbridge across the South Atlantic. Because these similarities later diminished, they supposed the landbridge to have sunk into the seafloor after the Paleozoic. Elsewhere, landbridgers erected them whenever and wherever they thought the record required them, and then sank them once the fossil record showed that migrations had ceased. Wegener (and others, too) raised an external theoretical difficulty with landbridges, arguing that they were inconsistent with the principle of isostasy that had come to enjoy strong support. According to isostasy (Archimedes' principle of buoyancy applied to a solid crust floating upon a liquid-like substrate) landmasses sink when they become denser either through internal densification or by being loaded with dense material. Landbridgers provided no evidence for such changes.

(ii) Critics of the Wegener/Köppen explanation of the problem of Permo-Carboniferous glaciation argued that it was in conflict with the theory of glaciation. Massive glaciation required massive snowfalls, whereas central Pangea, being far from moisture-producing oceans, would have been as dry as the Gobi Desert.

(iii) Fixists argued that Wegener's theory of continental drift – one vast episode of drift late in the Earth's history – was at odds with the principle of uniformitarianism. He was accused of reintroducing catastrophism into the Earth sciences.

(iv) Fixists raised a highly influential external theoretical difficulty with Wegener's solutions to the secondary problem of the mechanism of continental drift; they argued that the forces invoked by him were far too small.

(v) Arthur Holmes, a major proponent of mobilism, postulated (1928) that large-scale convection currents within Earth's mantle caused the drifting of the continents. At the time, fixists, on theoretical grounds, doubted the existence of mantle convection, and even if present, they considered his pattern of convection to be highly unlikely.

A theory or solution also encounters *internal theoretical difficulties* if it embodies an inconsistency. Here is an example. Fixists raised such a difficulty with Wegener's explanation of mountain building, arguing that it was not only inconsistent with itself but also with other solutions that he offered in his overall theory. Wegener claimed that continents plow through seafloor, sima (rock rich in silicon (Si) and magnesium (Mg) identified as the general constituent of ocean floors) being weaker than sial (rock rich in silicon and aluminum (Al) identified as the general constituent of continents). However, according to Wegener, mountains are formed at the leading edge of drifting continents as a result of crumbling and subsequent piling up of sial, so sial is weaker than sima because it crumbles when pushed into sima. Critics said that Wegener could not have it both ways; sial cannot be both weaker and stronger than sima.

1.12 Difficulty-free solutions

Difficulty-free solutions explain what they are supposed to explain without there being a reasonably conceived difficulty left standing against them. Difficulty-free solutions must be seen to be empirically sound, to be based on reliable data, to face no unanswered anomalies, and to lack no important supportive data. They must also be seen to be theoretically sound – not to clash with well-founded theories or methodological principles, not to utilize tenuous assumptions, and to be internally consistent.

During the mobilism controversy workers sought diligently for difficulty-free solutions. If after scrutiny a solution was acknowledged to be difficulty-free by its opponents, debate ceased and the solution was accepted. Although not called by that name, difficulty-free solutions were the Holy Grail of researchers during the mobilism debate. Difficulty-free solutions were very rare. Solutions were proposed, skeptical scientists scrutinized them. If the solution was to a new problem, it was not accepted without critical examination; critics searched relentlessly for difficulties that had been overlooked by the originators. If a new solution was proposed to a long-standing problem, scientists who had earlier proposed their own solutions used their knowledge of them to raise difficulties with the new one. If other solutions had not been proposed, those who had worked on the problem may have used what they learned to raise difficulties. Some difficulties were easily dismissed; others were not.

It may seem as if difficulty-free solutions are so rare as to be of no practical value; every solution will surely face difficulties sooner or later. I agree, and that is why

I have restricted difficulties to those that have actually been raised. It certainly is sometimes very challenging to tell for sure if a solution is difficulty-free. If a solution has no difficulties when proposed, or even when accepted, it may encounter them later; new discoveries may create difficulties for a solution that up to then had faced none. Difficulties may lurk in strange places. Whether a solution is difficulty-free is determined relative to what is known at that time. Potential difficulties, whatever their ontological status, do not count.

Even with this requirement, that difficulties must be raised to count, it may still seem that there can be no such real thing as a difficulty-free solution, that difficulty-free solutions are utopian. Suppose a formerly controversial solution is later considered difficulty-free. Surely some opponents somewhere will continue to raise what they perceive as difficulties. Opponents may mistakenly believe that they have found a new flaw; they may restate former difficulties because they do not know of them or understand the new work that has already removed them; such attacks do not alter the difficulty-free status of a solution. I think of them variously as illegitimate, phantom or pseudo-difficulties. They are not real difficulties, but may mislead unwary scientists into mistakenly believing that a solution is not difficulty-free even though it is. Workers who are not deeply familiar with the relevant field or are unsympathetic to the solution because it conflicts with their own views on this or related problems, are particularly prone to be led astray. I shall argue in Volume II that this is precisely what repeatedly happened with the paleomagnetists' mobilistic explanation of the vast differences they had observed among APW paths from different continents.

Is my idea of a difficulty-free solution too broad? Are not many youthful solutions difficulty-free when initially proposed because nobody has yet critically examined them sufficiently to determine if they face legitimate obstacles? However, it means little for a solution to be difficulty-free if opponents have not examined it. For example, when Wegener first introduced his theory, he argued strongly in favor of its solution to the problem of Permo-Carboniferous glaciation against which he himself raised no difficulties, and further he argued that it avoided the profound difficulties that any fixist solution faced. But fixists, some of whom were specialists in glaciology, proceeded to raise difficulties against Wegener's solution, some of which became widely accepted, rendering it far from difficulty-free. Thus I do not think that the above questions constitute a major stumbling block to the notion of difficulty-free. Difficulty-free solutions that have not been examined by opponents are given no special status. It is much harder to develop a solution that is difficulty-free after it has been critically scrutinized. Even though Wegener raised no difficulties with many of his solutions, his opponents certainly did. Participants in the debate typically claimed that competing solutions had difficulties, and argued that the difficulties faced by them were greater than those faced by their own solutions. Another reason I believe my insistence that difficulties must be identified in order to count as difficulties is not problematic is that participants during the mobilism controversy often realized that their own solutions were not difficulty-free, sometimes spelled out these difficulties, and suggested ways to remove them. For instance, the Cambridge University paleomagnetists Ken Creer, Edward Irving, and Keith Runcorn acknowledged throughout much of the mid-1950s that their mobilist explanation of scattered paleopoles faced as-yet unanswered difficulties. Most participants in the mobilism debate were realistic enough not to claim that their solutions were difficulty-free; sometimes they themselves raised difficulties with them, or sometimes simply proposed their solution and waited for others to comment.

There is, however, another way in which my definition of a difficulty-free solution may appear too broad. What is to stop defenders of a solution from removing difficulties by fiat, simply reclassifying them as unsolved problems? Nothing! However, once again, there is no guarantee that opponents will agree, thereby continuing the controversy. Saying a solution is difficulty-free does not make it so. But this leads to a more interesting question: why did reclassification of a difficulty as an unsolved problem sometimes work? Such attempts at reclassification were made during the continental drift controversy, notably during the development of plate tectonics. In the early and middle stages of the controversy several mobilists argued that the lack of a mechanism for continental displacement should not count as a difficulty against continents drifting, but should be considered an unsolved problem: electrical storms occur, but their cause is unknown. Fixists correctly dismissed the reclassification on the grounds that evidence favoring continental drift was substantially less than that favoring the existence of electrical storms. Late in the controversy, the developers of plate tectonics made a somewhat similar move. Jason Morgan, and Dan McKenzie, and Robert Parker made it clear that plate tectonics is a kinematic theory: it is the geometrification of tectonics that describes the kinematics of seafloor spreading and continental drift. By designating plate tectonics a kinematic not a dynamic theory, they made it clear that they were making no attempt whatsoever to explain the cause of plate movements. As far as they were concerned, the cause of plate movements was what I would call an unsolved second stage problem. What did unconverted fixists do? Hardly anyone voiced complaint. They accepted plate tectonics because they recognized that it contained several compatible difficulty-free solutions that were mutually supportive. Once these had been established there was no going back. This example shows that the reclassification of a former difficulty as an unsolved problem is permissible if there are difficulty-free solutions already associated with the theory in question. It also shows, as I shall later show, that in this instance reclassification was essential before progress could be made.

As already noted, I shall argue that only four difficulty-free solutions that remained difficulty-free after robust scrutiny were developed during the mobilism controversy. Paleomagnetists who explained scattered paleopoles in terms of continental drift developed the first. Its difficulty-free status was not at the time widely acknowledged. The key paleomagnetists who proposed and defended continental drift, Creer, Irving, Runcorn, P. M. S. Blackett, John Clegg, and Ernie Deutsch, who were all British, anticipated potential difficulties as they developed their pro-drift viewpoint. Their solution, initially proposed in 1956, became difficulty-free at the end of the decade. By 1959 almost every working paleomagnetist accepted the drift solution, and treated it as essentially difficulty-free. Three US paleomagnetists did not. They were John Graham, Richard Doell, and Alan Cox; who had very limited experience working directly on the drift problem. Harold Jeffreys, the great English theoretical geophysicist and long-time fixist, also dismissed paleomagnetism, and his views carried weight. Two other theoreticians, the Americans Walter Munk and Gordon Macdonald, who were admirers of Jeffreys, also were critical of the paleomagnetic method, even lampooning it and the case made from it for continental drift.

I shall argue, not just from hindsight but from a close reading of the paleomagnetic evidence published at the time, that these US dissenters were wrong; because of the prominent publication of their claims and the strong predisposition of fixists not even to consider continental drift, the difficulty-free status of the paleomagnetic case was not widely acknowledged. Fixists remained fixists, seeing no reason why they should change, and finding none in the prominent publications they were wont to read. Among those who actually read the relevant literature, few had the background to understand paleomagnetism, to question the legitimacy of the difficulties, and to appreciate its wider implications. Some did not grasp the idea of a time-averaged geomagnetic field; others underestimated the amount of care most paleomagnetists were taking to ensure the reliability of their data. Nor did they know that paleomagnetists in many scattered publications had already presented a stronger case than that described by the North Americans and dismissed by Jeffreys, Munk, and Macdonald.

The global attack on fixism launched by a handful of British paleomagnetists, I think it is fair to say, took fixists by surprise. To some fixists, paleomagnetists seemed like magicians with their bag full of tricks. They were even derisively called paleomagicians, full of hocus pocus. There were, however, a few insightful scientists, most notably Edward Bullard, Harry Hess, and Robert Dietz, who were neither paleomagnetists nor, at the time, mobilists, but who recognized the essential correctness of the paleomagnetists' pro-drift solution and began, as a result, to take mobilism seriously.

The second difficulty-free solution in the mobilism controversy was the Vine–Matthews hypothesis. It was independently proposed by Fred Vine and Drummond Matthews and by Lawrence Morley in 1963; Morley did not develop it further. It is a corollary of seafloor spreading and reversals of the geomagnetic field. Vine was then a graduate student working under Matthews on his Ph.D. at Cambridge University; Morley was employed by the Geological Survey of Canada. They proposed that if seafloor spreading and reversals of Earth's magnetic field occur, then the seafloor should possess alternating stripes of normally and reversed magnetized material. In short, when new seafloor is formed at ridge axes it becomes magnetized in the direction of the Earth's magnetic field. With repeated reversals of the field and continued spreading of the seafloor, blocks of alternately normal and reversely magnetized material become frozen in the seafloor. The seafloor acts like a tape recorder of the geomagnetic field. The hypothesis was designed to solve two problems: the documented presence of a large positive magnetic anomaly over ridge-axes, and the bizarre stripes of normally and reversed magnetic anomalies in the northeast Pacific basin. When first proposed, however, Vine and Matthews admitted that their hypothesis faced several substantial difficulties. Here are two: The telltale zebra pattern of magnetic anomalies in the northeastern Pacific basin did not appear to be associated with a mid-ocean ridge, and the central magnetic anomaly associated with the Mid-Atlantic Ridge did not seem to be flanked by the predicted pattern of magnetic anomalies. Other, more technical difficulties were raised by James Heirtzler, Xavier Le Pichon, and Manik Talwani. They were at Lamont Geological Observatory, and were probably the most vehement critics of the hypothesis. In 1966, however, Vine, Heirtzler, and Pitman, a research student at Lamont, found substantial support for the hypothesis, and Vine systematically removed essentially all other difficulties. With their work, the Vine–Matthews hypothesis became difficulty-free, and was quickly accepted by most marine geologists.

The third difficulty-free solution, I shall contend, was provided by the concept of ridge-ridge transform faults. Wilson and Alan Coode independently proposed the idea in 1965. Coode did not develop the idea any further; Wilson persisted. Wilson was impressed by the fact that movements of the Earth's crust appeared to be concentrated in three types of tectonic features, mid-ocean ridges, mountain ranges (including island arcs and trenches), and major faults with large horizontal displacements. These features seemed to end abruptly and be unconnected. Wilson proposed that they are connected by faults, which transfer motion from one feature to another. The upward motion of material from the mantle that formed new seafloor at a ridge is transformed into horizontal motion along a transform fault, and eventually is transformed into vertical motion beneath a trench. Wilson and Coode applied the idea of transform faults to the movement of seafloor between segments of mid-ocean ridges. Wilson called them transform faults. Mid-ocean ridges are not continuous features, but are made up of short segments, which are connected by faults. Before Wilson and Coode developed the idea of transform faults, these connecting faults were interpreted as transcurrent faults. Wilson predicted that motion across them should be opposite to that of transcurrent faults; this was confirmed in 1967 by Lynn Sykes, a seismologist at Lamont, after which no substantial difficulty was ever raised against the idea of transform faults.

The final difficulty-free solution was provided by the theory of plate tectonics. It was born difficulty-free. It extended to a sphere the geometric aspects of seafloor spreading, the Vine–Matthews hypothesis, and the concept of transform faults. Morgan, McKenzie, and Parker also proposed that the plates, which can be oceanic, continental or a combination, are essentially rigid. Once they moved to a sphere and assumed plate-rigidity, they were able to apply Euler's point theorem, and give the theory mathematical rigor. This allowed them to demonstrate that the seismically determined components of motion on faults bordering plates were tangential to small circles around the Euler pole of rotation; it allowed them to explain the variations in spreading rates along ocean ridges, which Hess's seafloor spreading and the Vine–Matthews hypothesis had not explained. As mentioned, there was the question of what caused the plates to move, which they deemed as an unsolved problem. Because their theory offered difficulty-free solutions, it encountered almost no resistance, and the way was open for studies of the dynamics of plate motions to proceed with no lingering doubts as to the correctness of mobilism. Their work was immediately further confirmed and developed by Lamont workers Le Pichon, Brian Isacks, Jack Oliver, and Sykes. When Morgan, McKenzie, and Parker first proposed plate tectonics, they took it to be a description of present-day motions, not past motions. Soon McKenzie and Morgan began to consider the evolution of plate boundaries, adding to plate tectonics its historical dimension.

There is a distinction to be made between difficulty-free solutions and theory choice. I am not proposing that in the mobilism debate a solution was unworthy of acceptance unless it had no difficulties. Being utterly difficulty-free was not a necessary condition to warrant acceptance. However, after intense discussion, a solution had been found that was free of all known difficulties: if it was a solution to a first stage problem, and if no competing solution was difficulty-free, then it merited acceptance. This is so, I suggest, even though, as in the paleomagnetism work of the 1950s, acceptance was delayed because only a few recognized its difficulty-free status. Thus, as intimated above, if scientists in the mobilism debate acknowledged that mobilism or fixism had a solution to a first stage problem that remained difficulty-free after careful scrutiny, but then refused to accept it in the absence of other competing difficulty-free solutions, they were being unreasonable.[4]

1.13 The three research strategies and how they gave structure to the debate

"Anything you can do I can do better. I can do anything better than you." "No you can't." "Yes I can."

(Irving Berlin, from Annie Get Your Gun)

During the mobilism controversy, participants devoted time and energy to showing that their solutions and theories were better than others. To do this, researchers acted in accordance with a number of *research strategies* that, I contend, pervaded the mobilism controversy, and provide a framework within which its evolution can, I believe, be satisfactorily described. During the controversy three general types of research strategies were employed, *improving* types used to *increase* the problem-solving effectiveness of solutions and theory, *attacking* types used to *decrease* the problem-solving effectiveness of competing solutions or theories, and *comparing* types by which the problem-solving effectiveness of the preferred solution or theory is argued to be *superior* to the competition.

Research Strategy 1 (hereafter, RS1) *was used to expand the problem-solving effectiveness of solutions and theories.* RS1 was used to improve, amend, and replace *existing* solutions, to develop *new* solutions to problems a theory had not previously addressed, and, as an ultimate goal, to develop what I describe as difficulty-free solutions. Du Toit, for example, amended Wegener's solution to Permo-Carboniferous glaciation to include inland seas within Pangea to avoid the difficulty fixists had raised about its likely arid climate. Holmes proposed an entirely new solution, mantle convection, to the second stage problem of mobilism's cause. Sometimes new solutions were developed by researchers who had not previously been involved in the debate. The paleomagnetists who developed the pro-drift solution to the problem of scattered paleopoles were not fully fledged mobilists when they began work or even when they realized that they could use paleomagnetism to test continental drift. However, once they realized that their work strongly indicated that the continents had moved more or less as Wegener had proposed, they became mobilists.

Participants sought to establish consilience between different sets of results, and their efforts were especially important in improving the effectiveness of solutions. Wegener was especially concerned to establish consilience between the marginal congruencies and geological similarities on either side of the Atlantic. Most paleomagnetists became mobilists once they had established agreement between their determination of past continental positions and Wegener's. Perhaps the most spectacular example, and surely gratifying to the workers involved, occurred in 1966 when consilience was established between the order of late Cenozoic geomagnetic reversals as determined from worldwide studies of radiometrically dated terrestrial lava flows, of deep-sea sediments sampled by cores, and of marine magnetic anomalies interpreted in terms of the Vine–Matthews hypothesis; three independent sources of information. This cemented the acceptance of the Vine–Matthews hypothesis and seafloor spreading, placing them and their supportive evidence beyond reasonable dispute. Around the same time, researchers also found excellent agreement between movements of the continents as determined by paleomagnetism and as indicated by seafloor spreading and the Vine–Matthews hypothesis.

Removal of difficulties was the usual method for improving problem-solving effectiveness. Proponents of solutions tried to show that contrary data were unreliable: mobilists showed that the fixists' claim that *Glossopteris* flora occurred in Russia was mistaken, being based on wrong fossil identifications, thus automatically enhancing mobilism's credibility. Proponents sidestepped anomalies by altering existing solutions: the fixist C. E. P. Brooks proposed that the Permo-Carboniferous glaciation of India could be explained by positing that India had been elevated and cold. Proponents often sought to remedy missing-data difficulties by attempting to obtain the missing data or by questioning how really necessary they were for their solution: fixists attempted to find ancestral marsupials in Asia (unsuccessfully until very recently), and mobilists attempted to find them in Antarctica (which they eventually did). Unreliability difficulties were often removed by showing that the procedures used to collect and process the data actually were reliable, by improving existing procedures to increase their reliability, or by developing new procedures: du Toit went to South America to secure more reliable comparisons between its geology and that of Africa, and paleomagnetists countered claims of unreliable data by continuing to develop new reliability criteria and policing their use.

Attempts were made to remove theoretical difficulties in a number of ways. If a solution had shaky assumptions, proponents attempted to bolster them with new arguments, to develop an entirely new solution based on other non-shaky assumptions, to argue that the difficulty had been incorrectly categorized by their opponents, or to eliminate the need for them altogether. Several mobilists argued that the difficulties posed by the impossibility of continents plowing through more rigid seafloor were toothless once they had made what they believed was the correct distinction between rigidity and strength (resistance to short- and long-term stress, respectively). Others responded to mobilism's lack of an acceptable mechanism by deeming it to be only an unsolved second stage problem, not, as critics argued, a fatal theoretical difficulty. Holmes tried a new approach: He proposed an entirely new cause, mantle convection, for continental drift, avoiding the theoretical difficulties faced by Wegener's solution. If a solution clashed with a well-founded theory, proponents of it often argued that the theory was, in fact, not well founded; alternatively, they sometimes developed a new solution that did not clash with the well-founded theory. If their opponents claimed that their solution contained an inconsistency, they attempted either to show that the claim was mistaken, to remove the inconsistency by altering their solution, or to develop an entirely new one. Proponents of a theory sought especially to remove theoretical difficulties that were considered to plague more than one of their solutions. Thus, in the classical stage of the debate, most mobilists were overly concerned that a plausible mechanism for continental drift be found.

Proponents also attempted to improve the explanatory power of a solution by having it explain more data than it previously explained. Du Toit's detailing of the geological similarities between Africa and South America generally enhanced the idea that they had once been together. Unexplained data are not anomalous with a solution; they just await a future explanation.

Researchers used RS1 to develop the prized difficulty-free solutions. It goes without saying that when a solution became difficulty-free, its problem-solving effectiveness increased greatly, and the problem-solving effectiveness of any theory that came to possess one or more of these difficulty-free solutions rose correspondingly. When proponents of a theory developed a difficulty-free solution, they employed it to support other solutions that were consilient with it. For example, Vine, although he seems to have never said as much, argued implicitly that the difficulty-free nature of the Vine–Matthews hypothesis greatly enhanced the case for seafloor spreading. This tactic increased the problem-solving effectiveness of seafloor spreading, itself, without Vine and Matthews, a non-difficulty-free solution. Proponents of seafloor spreading promptly argued that difficulties that had been raised against it were not nearly as serious as formerly thought. Another example is the welcome that some old-time mobilists extended to paleomagnetists' defense of continental drift, especially the consilience between the former latitudes of continents determined paleomagnetically and those inferred from paleoclimatic evidence.

Research Strategy 2 (hereafter, RS2), *was used to decrease the problem-solving effectiveness of competing solutions and theories*. It was an attacking strategy used to raise difficulties against opposing solutions, and to place all possible obstacles in their way. Raising legitimate difficulties ensured that the proffered solution was not difficulty-free, weakening its problem-solving effectiveness. Here are some by now familiar examples. In the 1920s, fixists raised anomaly difficulties against Wegener's explanation of the Permo-Carboniferous glaciation by pointing to what they considered glaciations during that time in North America; in the 1950s, fixists argued that paleomagnetic data were unreliable; in the 1940s and 1950s, fixists claimed that mobilists, to be believed, needed to find closely related ancestral marsupials in South America and Australia. Fixists also raised theoretical difficulties with mobilist solutions: Wegener's explanation of the Permo-Carboniferous glaciation clashed with the received theory of glaciation; his theory of mountain building clashed with his requirement that continents plowed their way through seafloor; his single very late episode of continental drift clashed with the principle of uniformitarianism; the forces he proposed to move continents were too small by several orders of magnitude.

Research Strategy 3 (hereafter, RS3) *was used to compare the problem-solving effectiveness of competing solutions and theories*. RS3 was used to *identify* and *emphasize* those aspects of a solution or theory that gave it a decided advantage over its competitors. When researchers had a solution that explained something which was anomalous or left unexplained by the competition, they emphasized their advantage. Holmes emphasized the fact that the mobilist solution to Permo-Carboniferous glaciation, unlike the fixist one, explained the glaciation in India. Vine emphasized the fact that competing solutions could not explain the magnetic anomalies on the seafloor without invoking unrealistic contrasts in the ability of adjacent blocks of rocks to acquire magnetizations.

Participants in the mobilism controversy generally did not always keep abreast of what was happening in the various sub-controversies. Those that did often participated directly in them. Not unnaturally, they paid particular attention to those sub-controversies in which their own solution or theory offered a substantial advantage. They made the most of it when they possessed a solution to a problem that the competition did not have, and they did not shrink from saying so when their theory possessed several consilient solutions. Correspondingly, they attempted to remedy situations when they were at a disadvantage. If their opponents had a solution to a problem that they had not solved, they attempted one. They attempted to raise difficulties against their opponents' solutions. They attacked if their opponents claimed to have consilient solutions; they began by raising difficulties against what their opponents regarded as consilient solutions, thereby weakening them, and then they argued that any advantage gained by possession of such wobbly consilient solutions was insecure, and they could no longer be regarded as being mutually supportive.

These standard research strategies were also misused, sometimes abused. RS1 was misused by offering a solution that was fraught with difficulties from the very beginning. This was especially detrimental if at the time there appeared to be no alternative solution to the problem, and its solution was central to the larger theory in question. This is essentially what Wegener did when he proposed that tidal force and flight from the poles caused continental drift. Once Jeffreys and other fixists had pointed out in the 1920s that both forces were several magnitudes too weak, the futility of the search for a plausible mechanism for continental drift became the mobilists' Achilles' heel. Then in the late 1960s, McKenzie and Morgan, standing advantageously on a mountain of evidence of large lateral displacements of continents and ocean floors, formulated the theory of plate tectonics. The unifying geometry of plate tectonics established quantitative consilience between several globally observed and independent kinematic indicators that had been obtained over the previous twenty years from seismology, marine geology, and paleomagnetism. Plate tectonics was and is not a dynamical theory; it was invulnerable to the mechanism difficulty. It paid to admit that a problem is unsolved and not court failure by attempting a solution.

If scientists attempt to solve a problem by offering a highly speculative solution, they may arm the opposition as Wegener did. He later admitted that the Newton of drift theory had not arrived, but the damage was already done. However, even if Wegener had simply invited physicists to speculate about the cause of continental drift, the inability to find its cause still may have become mobilism's Achilles heel. In contrast, Hess introduced seafloor spreading resulting from upwelling mantle convection as geopoetry, thereby emphasizing its speculative even metaphorical nature.

But Hess's mechanism had to be abandoned, while the process of seafloor spreading lived on, linked to reality by confirmation of the Vine–Matthews hypothesis and ridge-ridge transform faults.

RS2 may also be abused by proposing illegitimate or phantom difficulties. This is what happened in some of the attacks against the pro-drift solution to scattered paleopoles. Jeffreys, for example, attacked paleomagnetism by claiming that paleomagnetic data were unreliable based on his experiences fifty years earlier. He noted that careless handling of permanent magnets can cause change in their magnetization, and that rocks used for paleomagnetic studies are broken off with a geological hammer. He said he had not received an answer as to why the collecting and preparing of paleomagnetic samples does not alter their magnetization. I wonder who Jeffreys actually asked. Did he ask his neighboring Cambridge paleomagnetists, who had determined what sorts of rocks yield reliable data? Some rocks, when hit very hard with a hammer, do change their magnetization; others do not. Jeffreys knew Runcorn and Runcorn's students. Jeffreys also abused RS2 when repeatedly arguing that the fit of Africa to South America was poor. When asked by David Brown in 1959 if he had read a paper by S. Warren Carey, an avid mobilist at the University of Tasmania who had showed rigorously that the fit of Africa and South America was excellent, Jeffreys remarked that he had no intention of reading Carey's paper. Similarly, when invited by Jim Everett to examine his impressive computerized fit of the continents, he said that he did not have enough time. Marland Billings, a Harvard University structural geologist, abused RS2 during his Presidential Address before the Geological Society of America (hereafter, GSA) in 1959. He too attacked paleomagnetists' pro-drift support, claiming that their solution to the problem of scattered paleopoles rested on the mistaken assumption that the geomagnetic poles coincide with the geographic poles because this is certainly not so today. This was a silly mistake. He ought to have known that paleomagnetists had shown that when averaged over many thousand years the geomagnetic pole does coincide with the geographic pole, and it was part of their rubric to obtain such averages. Abuses of RS1 and RS2 were compounded when carried over to an RS3 evaluation of the relative effectiveness of competing solutions or theories as the narrative in these volumes makes clear.

1.14 Specialization and regionalism in the Earth sciences[5]

In his monograph, *A Geological Comparison of South America with South Africa*, du Toit wrote:

These pages have of necessity dealt in very summary fashion with this fascinating subject, more particularly as related to those territories beyond the respective opposed regions known personally to the writer. It is therefore to be hoped that scientists acquainted with those particular outside regions may be induced to set down their observations of those areas, supporting or else refuting the presumptions here put forward on behalf of this hypothesis, since a strict and impartial criticism is indeed required if this riddle in early history is to be deciphered. The coöperation of geologists, paleontologists, zoölogists, botanists, and physicists is urgently needed, the discussion, advanced here dealing admittedly with only a very limited, albeit extremely important, section of the globe, from which, under the eclectic hypothesis of crustal instability, corroborative or destructive evidence might reasonably be expected. By virtue of their enormous lengths of opposed coast-line and extraordinary geological parallelisms, the two most favored continents from which evidence is to be drawn would appear to be Africa and South America, and, should the details herein set forth appear to be worthy of serious consideration, it is to be hoped that more detailed investigations may shortly be instituted elsewhere on the two sides of the Atlantic for the purpose of clearing up some of the many crucial questions that must be regarded as *sub judice*.

(du Toit, 1927b: 119–120)

Du Toit envisaged studies of Earth in the broadest sense ranging from geophysics to biogeography. It is in this sense that I shall use the term "Earth sciences" and make these two points. The Earth sciences are a group of specialized, albeit somewhat overlapping, independent disciplines or individual sciences, and because the skills and knowledge needed to master each discipline vary greatly, mastery of one may be of limited help in understanding another. Second, most practicing Earth scientists concentrated on local or regional problems; some expanded their studies to a whole continent, few worked on global problems. Among those with a global outlook, many concentrated on problems about the Earth's interior, or on what happened worldwide during a particular geological interval. This distinction merits special attention because it has a direct bearing on what evidence was important vis-à-vis mobilism and fixism.

I begin with Earth science's many disciplines and, making no attempt to consider institutions elsewhere that played important roles in the mobilism debate, deal only with the United Kingdom and the United States. Upon receiving the Arthur L. Day Medal from the GSA, the British geophysicist Bullard said:

As far as I know, The Geological Society of America is the only body that has a proper appreciation of the real range of geology. The range of relevant subjects is, of course, the main difficulty in taking an intelligent interest in the Earth. Obviously one should know about stratigraphy, mineralogy, and palaeontology, and, of course, one should know a lot of physics and chemistry and a good deal about oceanography and astronomy, and be a pretty high-grade mathematician and statistician. It is also helpful to be a good draughtsman, a competent electronic designer and repair man, and to be able to code electronic computers. Obviously there are no well-qualified students of the Earth, and all of us, in different degrees, dig our own small specialised holes and sit in them. However, every month along comes the *Geological Society of America Bulletin* to remind us that we *are* sitting in a small hole, while all around us spread the subjects we should know about and don't. Belonging to this Society is the best cure for a belief that one knows a lot about the Earth.

(Bullard, 1960: 92)

Unraveling Earth's history and evolution requires the application of many disciplines. The Briton Arthur Holmes constructed a diagram (Figure 1.1) in his *Principles of Physical Geology* to display the relations among the major branches of the Earth sciences, and described the general science of geology and its four main branches as follows:

Modern geology has for its aim the deciphering of the whole evolution of the earth and its inhabitants from the time of the earliest records that can be recognized in the rocks right down to the present day. So ambitious a programme requires much subdivision of effort, and in practice it is convenient to divide the subject into a number of branches . . . The key worlds of the four main branches are the material of the earth's rock framework (Mineralogy and Petrology), and their dispositions, i.e., their forms, structures and inter-relationships (Structural Geology); the geological processes or machinery of the earth, by means of which changes of all kinds are brought about (Physical Geology); and finally the succession of these changes in time, or the history of the earth (Historical Geology).

(Holmes, 1944: 5; 1965: 9–10)

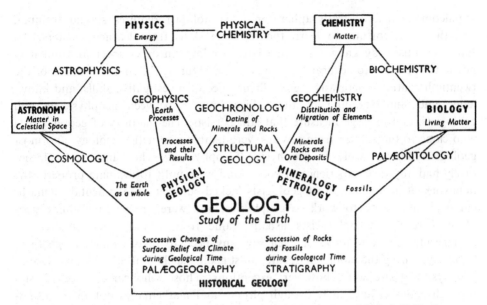

Figure 1.1 Holmes' Figure 1 from his *Principles of Physical Geology* (1945: 4; 1965: 9).

Within physical geology, the central subject of his book, he included geomorphology, oceanography, meteorology, and climatology. His divisions are just the beginning. On the other side of the Atlantic the GSA, when honoring Bullard, held multiple sessions at its 1959 annual meeting on areal geology, coal geology, crystallography, economic geology, engineering geology, geochemistry and geochronology, geochemistry and organic geochemistry, geochemistry and phase relations, geomorphology, geophysics, mineralogy, paleontology, petrology, Pleistocene and glacial geology, and sedimentation geology. Moreover, geophysics was itself subdivided into many disciplines. The American Geophysical Union (hereafter, AGU), which played a much more substantial role during the later two phases of the continental drift controversy than the GSA, had, in 1966, the year when the Vine–Matthews hypothesis gained acceptance among marine geologists, nine sections: geodesy; geomagnetism and aeronomy; hydrology; meteorology; oceanography; planetary sciences; seismology; tectonophysics; and volcanology, geochemistry, and petrology.[6] (Perhaps Bullard was being polite about membership in the GSA as "the best cure," for he was after all thanking its members for awarding him the Day Medal, and he certainly knew what was going on with the AGU.)

Although some Earth scientists become adept in more than one field, and some even know a fair amount about several, the diversity of Earth science is so great that few master more than a few of them. Becoming a good geochronologist, a paleomagnetist, a stratigrapher, or a seismologist requires special training, and the good fortune to learn the requisite skills from learned masters. An Earth scientist may know about the latest findings in other specialties but may not have the skills to accurately evaluate them. During the first two-thirds of the twentieth century, there was a great divergence of investigative skills, and knowledge and understanding of the subject matter between geophysicists and geologists, and in large measure that still persists. The majority of geophysicists who came to the subject from physics were unable to practice geology, and most geologists did not work in geophysics. Geophysicists who came from physics rarely had any field experience of rocks, and were unfit to evaluate properly the fieldwork of geologists. Most geologists lacked sufficient experimental or mathematical knowledge to work in geophysics, and were unable to evaluate geophysical results. Geophysicists usually came from physics, mathematics, or engineering, and few took, or even saw any need to take, courses in geology. Similarly, during the mobilism debate, most geologists had little or no exposure to geophysics. There were, however, exceptions. Those who came to geophysics through geology, and knew enough physics and mathematics not to be intimidated by their more mathematically inclined geophysical colleagues could understand developments in geophysics and evaluate their geological impact. In addition, some geophysicists realized they knew little geology and made a point of working with geologists.

I now turn to regionalism, so prevalent during the mobilist controversy. Although there were some Earth scientists whose professional interests extended well beyond the region in which they did fieldwork, they were the exception. Books on stratigraphy and historical geology typically were about specific regions or continents, although their titles sometimes indicated otherwise. Few books in either discipline were global in outlook.

Most Earth scientists worked on local and regional rather than global problems during the mobilism controversy, for which there are two major reasons. Most geological data are literally tied to Earth. Researchers, for the most part, must go to the field to examine and reexamine geological structures firsthand. Even though many fossils and rocks can be shipped across the world, rock formations, mountain ranges, sedimentary basins and other such geological features cannot. Of course, Earth scientists can bring rocks and fossils back to their laboratory, take photographs, draw maps, and write descriptions of what they observe. They can also send rocks and fossils to other scientists. Photographs and casts of fossils housed in a museum can be sent to interested workers. Detailed taxonomic work, however, is often difficult without examining and comparing actual specimens. Even when specimens are in museums, researchers often had to become world travelers in order to see them. The airplane eventually helped, but funding for extensive foreign fieldwork was difficult to obtain during the first half of the twentieth century. All these factors worked against the global approach.

Of course, there was an extensive body of literature that could be read pertaining to the geology of regions other than one's own; Earth scientists could subscribe to journals or walk over to the library and learn about distant lands. But geologists have a healthy mistrust of learning geology solely by reading reports about regions that they have not seen. This leads to the other main reason for the prevalence of regionalism. Fieldwork requires learning and honing a variety of observational skills that are best acquired through an apprenticeship with an experienced field geologist. There are competent and incompetent field geologists. How can readers unfamiliar with the author tell if a report is trustworthy? Making geological observations in the field is an interpretative activity. Inadvertently, guided by ideas they already have in mind, field geologists may to some degree see what they expect to see. As a result, scientists reading field reports by geologists associated with a particular point of view may often discount them as biased.

If all this sounds a little far-fetched, consider what Marland Billings wrote in the introduction to his textbook, *Structural Geology*, published first in 1942 and revised in 1954. A member of the National Academy of Sciences (NAS), he was a major figure in American geology, becoming President of the Geological Society of America in 1959, the year Bullard received the Arthur L. Day Medal. He wanted to make sure his readers understood the highly interpretative nature of fieldwork in geology.

Geologic mapping, when properly done, demands skill and judgment. Such mapping requires keen observation and knowledge of what data are significant. As the field work progresses and the larger geological picture begins to unfold, experience and judgment are essential if the geologist is to evaluate properly the vast number of facts gathered from thousands of outcrops. Above all, the field geologist must use the method of "working multiple hypotheses" to deduce the geological structure. While the field work progresses, he should conceive as many interpretations as are consistent with the known facts. He should then formulate tests for these interpretations, checking them by data already obtained, or checking them in the future by new data. Many of these interpretations will be abandoned, new ones will develop, and those finally accepted may bear little resemblance to hypotheses considered early during the field work.

Nothing is more naïve than to believe that a field geologist should gather only "facts," the interpretation of which is to be made at a later date. Because of his numerous tentative interpretations, the field geologist will know how to evaluate the facts; these hypotheses, moreover, will lead him to critical outcrops that might otherwise never have been visited. On the other hand, the field geologist should never let his temporary hypotheses become ruling theories, thus making him incapable of seeing contradictory facts.

(Billings, 1942: 3–4; 1954: 3)

Billings also let his readers know that the proper study of structural geology begins with the analysis of small structures, advances to the study of mountain ranges, and may eventually lead to a study of the structure and tectonic history of the entire Earth – a bottom-up process. Unraveling the tectonic history of the Earth required familiarity with the geological literature. Drawing together his comments about the interpretative nature of fieldwork, he claimed that only those who had undertaken detailed fieldwork of their own were able to judge "the reliability and importance" of publications describing the geology of regions they had not investigated themselves.

Before it is possible to analyze the structure of entire mountain ranges, it is essential to have precise information on the many small separate areas that comprise the range. These small areas may cover square miles, or they may be single mines or oil fields. This investigation of the structure of relatively local structure is the first and inevitable approach to the problem.

Equally important, and perhaps in some ways more fascinating, is the synthesis or weaving together of the many facts obtained from local areas into a unified picture of the structure and tectonic history of the outer shell of the whole earth. Such studies are necessarily based in large part upon an intimate knowledge of the literature of structural geology because it is manifestly impossible for one man to investigate many areas in detail. But in order that he may more judiciously evaluate the reliability and importance of the published information, such an investigator must have made detailed studies of his own.

(Billings, 1942: 4; 1954: 5–6)

Thus, it is no surprise that most Earth scientists during the mobilism debate worked on problems limited to those regions where they had actually done fieldwork.

1.15 Why regionalism and specialization affected theory preference during the mobilist debate

Now, in order to work hard on something, you have to get yourself believing that the answer's over *there*, so you'll dig hard there, right? So you temporarily prejudice or predispose yourself – but all the time, in the back of your mind, you're laughing. Forget what you hear about science without prejudice. Here in an interview . . . I have no prejudices – but when I'm working, I have lots of them.

But the thing that's unusual about good scientists is that while they're doing whatever they're doing, they're not so sure of themselves as others usually are. They can live with steady doubt, think "maybe it's so" and act on that, all the time knowing its only "maybe." Many people find that difficult; they think it means detachment or coldness. It's not coldness! It's a much deeper and warmer understanding, and it means you can be digging somewhere where you're temporarily convinced you'll find the answer, and somebody comes up and says, "Have you seen what they're coming up with over there?", and you look up and say *"Jeez! I'm in the wrong place!"* It happens all the time.

(Richard P. Feynman 1999: 199–200)

The two tendencies, specialization and regionalism, strongly affected theory preference during the mobilism debate. Specialists tended to prefer the overall theory (fixism or mobilism) that best accounted for the results of their specialty. Experts on the geology of certain regions, countries or even continents overwhelmingly tended to prefer the theory, whether fixist or mobilist, with the better regional solutions. These were consequences of the greater depth of knowledge and understanding that researchers had of their own specialty or region than of other specialties or regions.

If specialists preferred a certain theory they did so regardless of how well competitive theories or solutions solved problems in other specialties. It is safe to assume that specialists cared much more about what was going on in their own field than in others. They were also better equipped to judge whether solutions in their own field, as opposed to those in other fields, faced legitimate and, if legitimate, possibly serious difficulties. Scientists seem to have had little understanding about the relative merits of competing solutions in fields outside their own expertise. Not unnaturally, most of what they knew about other fields came from seminars, listening to colleagues, and general reading, not from doing systematic research. Specialists, who favored a particular theory because it possessed a better solution to a problem in their field of expertise, tended to minimize advantages amassed by competing theories that had achieved success in other fields. Sometimes they simply ignored what was happening outside their expertise. Sometimes they acknowledged what was out there, waited to see if others raised difficulties with the competing solution or devised an entirely new solution.

As noted already, workers in paleoclimatology or paleobotany who specialized in the Late-Paleozoic of the Southern Hemisphere tended to be mobilists. They were mainly concerned with the distribution of the Permo-Carboniferous glaciation and of the *Glossopteris* flora, two problems that mobilism seemed nicely to solve. However, once their counterparts in the Northern Hemisphere raised objections to the mobilist solutions, fixists outside of the specialties no longer felt the need to take them seriously. Paleontologists specializing in Cenozoic studies of the distribution of mammals tended to be fixists. Although mobilists countered that such life forms were not critical for mobilism because Cenozoic life forms only began to flourish after most continents had separated; during the classical phase most students of Cenozoic biota thought mobilism a waste of time.

The reaction to paleomagnetism of workers in other fields is a particularly telling example of the response by workers to studies outside their field of expertise during the drift controversy. Once Blackett, Clegg, Creer, Deutsch, Gough, K. W. T. Graham, Green, Hales, Irving, Nairn, and Runcorn and their colleagues and students had determined that magnetizations over several geological periods were inconsistent from continent to continent, they were quickly won over to mobilism; they also anticipated and removed difficulties that were soon raised, and by 1959 had, I shall argue, succeeded in elevating their solution to difficulty-free status. In diametrical contrast, the US paleomagnetists, Cox, Doell, and J. W. Graham raised objections that were irrelevant, had already been answered, or were actually illegitimate, and failed to explain the key advantages of the pro-drift solution as documented by British, South African, and Australian paleomagnetists. Jeffreys, the elderly Englishman, who was still regarded as one of the leading geophysicists in the world, summarily dismissed paleomagnetism.

As a result, only a handful of fixists outside of paleomagnetism became mobilists because of this successful confirmation of continental drift by paleomagnetism. Most fixists, buried in their beliefs and often taking their cue from the four objectors, dismissed the paleomagnetic case for mobilism. Non-specialists were in a particularly poor position to evaluate the legitimacy of the difficulties.

Paleomagnetism was a relatively new science, which to appreciate required a good understanding of geomagnetism. In 1950 the geomagnetic field was depicted as highly variable over centuries. First, Jan Hospers in Iceland and then others of the Cambridge group globally showed that by sampling stratigraphically and applying the new statistical techniques devised by R. A. Fisher, long-term information about the field could be obtained. This result was not easily achieved, nor was its general importance widely comprehended at the time. There was no royal road to paleomagnetism. Indeed, I gather from my reading that many who decided to ignore paleomagnetism's pro-drift results knew little about paleomagnetism and read little of the relevant literature. What about mobilists who worked outside paleomagnetism? They openly and enthusiastically welcomed its new pro-drift results and pointed to their consilience with their own findings; they rejoiced at mobilism's improved problem-solving effectiveness. I shall argue that the presentation of the paleomagnetic case for mobilism presented by the three North American paleomagnetists and their dismissal by Jeffreys were biased and unreasonable.

But what about those who simply dismissed this new paleomagnetic support and continued vehemently to argue against mobilism? Did they behave unreasonably? Perhaps they should have begun thinking about whether they could reinterpret their own work in terms of mobilism. Even if they were unable to evaluate whether the pro-drift solution to scattered paleopoles was difficulty-free, they should have realized that it was a possibility unwise to ignore. It might not be difficulty-free, but its remarkable consilience with paleoclimatic data and with other mobilist solutions could not, I shall argue, be reasonably denied. Moreover, there really were no competing fixist solutions for the paleomagnetic results that also explained this remarkable consilience. There were, however, some fixists who acted much more reasonably; they questioned their allegiance to fixism because of the new paleomagnetic data, and they began wondering if their own work could be reinterpreted mobilistically. Most fixists, however, did not, as Feynman suggested "happens all the time" to good scientists, "look up" at what paleomagnetists were digging up, and say "*Jeez! I'm in the wrong place!*" It was not easy for many scientists during the drift controversy to accept findings in a field other than their own that disagreed with what they had been taught at university and had, during their subsequent professional lives, found evidence to believe.

As indicated already, Jeffreys was severely critical of Wegener's drift mechanism but he was not alone; several North American and European geophysicists agreed with him. During the 1920s, the Irishman John Joly provided mobilism with another possible mechanism, and on cue Jeffreys soon raised objections to it. North American Earth scientists echoed Jeffreys' objections, and he carried the day. Holmes proposed mantle convection, and Jeffreys responded more cautiously, admitting that Holmes' solution was possible but highly unlikely. These attacks, especially by Jeffreys and echoed by others, definitely lessened support for mobilism, as evidenced by the very many times they were cited. Fixists enthusiastically welcomed Jeffreys' attacks, and mobilists were chastened by them, having to admit that indeed mobilism lacked a plausible mechanism. Why was this attack on mobilism's mechanism so effective? Why was mobilism, which had much evidence in its favor, so vulnerable on this issue? Jeffreys was certainly held in high esteem by most Earth scientists. And, so far as his arguments went, the consensus was that he was correct. Every mobilist solution had, at its core, relative displacement of continents, and therefore the lack of an acceptable mechanism created a general difficulty for all of its solutions. It could be brought up on any occasion when drift was discussed, and it was. It was a safe thing to say. Few geologists had enough physics or mathematics to seriously question Jeffreys' attack. Holmes was the only one to extract concessions.

Consider now the second tendency, regionalism, which is the overwhelming preference of either mobilism or fixism among participants who worked in a region that provided a decided advantage of one over the other. Most North American and Australian geologists prior to the middle 1960s were strongly opposed to mobilism, while most who worked in South Africa, South America, and India tended to support it. Participants from continental Europe and the British Isles were more varied in their allegiances. Participants who worked on problems connected with the geology of India, Southern Africa, or certain parts of South America supported mobilism because it explained better the Permo-Carboniferous glaciation, the strong similarities in the sedimentary history of the three regions, the congruency between facing margins of Africa and South America, and the origin of the Himalayas and Andes. Structural geologists studying orogenic belts in North America or Australia were much more likely to support fixism because the various solutions it offered worked particularly well when applied to regions characterized as they were by a Precambrian craton bordered by orogenic belts that decrease in age outwards. Confident in their own work, and familiar with the geology of their own region, they extrapolated outwards from it to other regions and continents. Theirs was a bottom-up approach.

Given that most Earth scientists who participated in the controversy worked on problems in the Northern Hemisphere, it is therefore no surprise that fixists greatly outnumbered mobilists. As already noted, geologists were often skeptical of field reports from regions other than their own that led to conclusions in disagreement with their own, and they raised difficulties accordingly (RS2). When considering regions they had not studied firsthand, they welcomed solutions that agreed with their own (RS1). Because most textbooks they read pertained primarily or exclusively to their own region, many fixists from elsewhere did not seem to appreciate how well mobilism solved major problems of African, South American, and Indian geology.

Moreover, because much of the early support of mobilism was based on far-flung intercontinental similarities, verification of these purported similarities meant studying seriously more than one region. Because of the differing training and skills of field geologists, the varying interpretative basis of fieldwork, and the absence of a uniform descriptive vocabulary, geologists tended to be skeptical of the fieldwork of others, especially if they were unknown to one another. Opportunities to undertake the fieldwork in regions outside their own were not common. This distrust of field reports of others tended to undermine claims of intercontinental geological similarities. Fixists generally had little interest in doing the needed comparative fieldwork, and the task was left to mobilists with their own vested interest in the results. Fixists then argued that the theoretical bias of mobilist fieldworkers led them to exaggerate the purported similarities. It was a vicious circle. However, some Northern Hemisphere geologists actually did work in the Southern Hemisphere. It is no accident that many of the British Earth scientists who favored mobilism, especially during the first phase of the controversy, spent time in India or southern Africa. Relying less on the field observations of others, they saw with their own eyes the southern evidence favoring mobilism.

Regionalism contaminated Earth scientists from regions where fixism enjoyed a decided advantage even if they themselves did not work on problems where fixism shone. This effect was to undermine or at least dampen the effect of specialization on theory preference. For example, it was no accident that the three major paleomagnetists who (*circa* 1960) did not think paleomagnetism's pro-drift solution was worthy of acceptance were from the United States. Most US geologists worked on regional problems and were, I believe as a consequence, unsympathetic toward mobilism. Their historical geology textbooks of the day hardly mentioned mobilism. They had little chance to see firsthand other continents where the more obvious consequences of mobilism are evident. They certainly were not encouraged by their teachers and colleagues to consider mobilism as a viable alternative. Graham at the Carnegie Institution had little and, at the end, no encouragement to continue working on the problem of scattered paleopoles, and Cox and Doell, at the United States Geological Survey (USGS), who worked very briefly on the problem, soon turned to working out the reversal timescale, which was more attractive because it swiftly became so brilliantly successful.

There were some Earth scientists who began their careers as specialists or regionalists but later graduated to generalists or globalists and, as a consequence, played substantial parts in the mobilism debate, and I want to say a few words about them. Generalists are interested in developments in several fields, and often are comfortable working in both geology and geophysics; globalists are interested in the geology of the Earth as a whole, in the evolution of its continents and oceans, and in its interior. Generalists and globalists were in a better position than specialists and regionalists to appreciate mobilism's merits. Generalists, knowledgeable as they were in several fields, provided a broader view than specialists; globalists, knowledgeable in the geology of several continents, understood far better than regionalists the force of inter-continental mobilistic arguments. Mobilists, who did work in regions or specialized in fields in which mobilism had a decided advantage, tended to be generalists and globalists. Versed in both geology and geophysics, they commanded, to a greater degree, several specialties and, although they concerned themselves with local or regional problems, they did not do so exclusively. In addition to Wegener himself, Earth scientists who were generalists and globalists and supported mobilism before, although sometimes not long before it was generally accepted, include du Toit, Holmes, Daly, Carey, Runcorn, Hess, Bullard, and Wilson. I add that, except for Runcorn and Bullard, they were well versed in geology, and did not enter geophysics solely from physics but solely or partly from geology. Runcorn, who was the least familiar with geology, had the good sense to recruit Irving, who was a geologist, and Creer, whose background was in geology and physics. They all stand in contrast to geophysicists such as Jeffreys and Maurice Ewing, who directed Lamont, and like Jeffreys was an adamant foe of mobilism. Jeffreys entered geophysics from mathematics, and learned little geology. Ewing seems to have learned what he knew from Bucher, a devout fixist, and from working with Heezen.

In concluding this chapter, I want to raise one other factor, moral courage, that I suspect negatively affected the reception of mobilism in communities in which it was ridiculed. R. A. Fisher raised such a possibility in 1956 when he reflected in a letter about the continental drift controversy and, in particular, the reception of mobilism's paleomagnetic support, which he himself helped to fortify. Fisher also compared the history of the drift controversy with that of evolution.

I think there is a parallelism in the nature of scientific controversy between continental drift in the last 80 years or so, and organic evolution about 100 years earlier. Each idea as it originated was necessarily speculative, and not accompanied certainly by sufficiently cogent evidence to carry final conviction. There were, however, many suggestive pointers. In consequence of this natural situation both questions have been argued with imperfect facts, incorrect theories, and often incompetent reasoning, over a long period during which many people have committed themselves to impossible positions, and many more, fearing to burn their fingers, have enclosed themselves in towers not of ivory, but of very solid wood. In the period 1800–1850, although geological specimens were being collected and described, experiments in plant hybridization carried out, the classification of animals and plants greatly improved, and men to run the risk of contemptuous ridicule, no one of consequence attempted to revive what had been left as speculative and almost poetical, ideas by Buffon, Erasmus Darwin and Lamarck. Darwin worked on the problem almost secretly from 1838, and only published in 1859 because he was forced to. After that time ice came down like Niagara, but of course the new idea was still ill-understood and ill-expounded for at least the next 50 years, during which a few subordinate causes of error had been removed by special research.

I think a lot of geologists must be timidly peering out of their holes on hearing the strange news that geophysicists are talking about continental drift, and I have often wondered how many scientific discoveries of importance have been left unmade for lack of the quality called moral courage.

(Fisher, 1956 letter to Irving; quoted in Box, 1978: 444–445)

Perhaps, many Earth scientists, who began in the 1950s to think in light of some new finding in paleomagnetism that the fixism they had been taught and believed might be wrong, and whose teachers, colleagues, and students continued to overwhelmingly reject mobilism, decided to hold their tongues, fearing that to do otherwise would hamper their career or subject them to ridicule. Scientists are human, and not immune to the prospect of being thought an oddball, frivolous, a lightweight. As Fisher said, not only would a lack of moral courage have "left unmade" scientific discoveries, but, I suspect, it would have "left unmade" declarations that an unpopular, purportedly ridiculous idea merited serious consideration. Group-think is not easy to combat either by those who rebel or by those who as part of the group are unaware of its blinding effect on the ability to evaluate opposing ideas impartially. Those who believe they have overcome or are immune to group-think are the worst afflicted.

By the same token, however, those who made the new pro-drift findings and had proposed correct pro-drift hypotheses, or spoken out early in their favor, especially if they were trained by fixists and worked almost entirely among them, were vindicated as the 1960s unfolded. Again I appeal to Fisher, whose early innovative ideas on statistical methods were first rejected. In 1947, reflecting on his own career, he had this to say.

A scientific career is peculiar in some ways. Its *raison d'entre* is the increase of natural knowledge. Occasionally, therefore, an increase of natural knowledge occurs. But this is tactless, and feelings are hurt. For in no small degree it is inevitable that views previously expounded are shown to be either obsolete or false. Most people, I think, can recognize this and take it in good part if what they have been teaching for ten years or so comes to need a little revision; but some undoubtedly take it hard, as a blow to their *amour propre*, or even as an invasion of the territory they had come to think of as exclusively their own, and they must react with the same ferocity as we can see in the robins and chaffinches these spring days when they resent an intrusion into their little territories. I don't think anything can be done about it. It is inherent in the nature of our possession; but a young scientist may be warned and advised that when he has a jewel to offer for the enrichment of mankind some certainly will wish to turn and rend him.

Another point to remember has to do with recognition, which the young clearly desire. A ballet dancer gets her ovation on the spot, while she is still warm from her efforts; the wit gets his laugh across the table, but a scientist must wait about five years for his laugh. Recognition in science, to the man who has something to give, is, I should guess, more just and more certain than in most occupations but it does take time.

(Box, 1978: 131)

Fisher underestimated; sometimes it takes more than a lifetime, as with Wegener and du Toit. Holmes, who died in 1965, did not live to see the avalanche of evidence for seafloor spreading, but he did see the piling up of paleomagnetic evidence for continental drift. All the paleomagnetists who developed this new support for mobilism saw their work vindicated. Hess, who died in 1969, also witnessed the triumph of seafloor spreading.

Notes

1 For variant uses of mobilism and fixism see Endnote 1, Chapter 8.
2 The flood of new information about the ocean floors simply would not have occurred without the massive governmental support of marine geology prompted by defense concerns. See Schlee (1973), Bullard (1975b), Menard (1986), Strommel (1994), Rainger (2000), Hamblin (2005), and Doel *et al.* (2006).
3 I first introduced this taxonomy in Frankel (1998), and it owes much to that introduced by L. Laudan in his 1977 *Progress and its Problems*. There are, however, significant differences between the two. In Laudan's taxonomy of scientific problems, the basic distinction is between empirical and conceptual problems. Empirical problems are first stage problems, substantive questions about the world that scientific theories are designed to solve. Laudan divides them into solved, unsolved, and anomalous problems. Conceptual problems are second-stage problems that have no existence independent of scientific theories. They are problems about the well-foundedness of solutions or theories. According to Laudan, there are internal and external conceptual problems. The former include internal inconsistencies and vague terms; the latter come about when a theory clashes with some other theory that is believed to be well founded. The basic distinction I make is between problems and difficulties; the former constitute the subject matter of solutions, and the latter are objections raised against solutions. Problems are divided into first and second stage problems. The former arise through the discovery of puzzling phenomena. The latter are questions about the origin or behavior of entities that are invoked to solve first stage problems. There are two basic kinds of difficulties. Data difficulties arise when some of the data used to support or defend a solution are claimed to be unreliable, anomalous, or missing. My theoretical difficulties are similar to Laudan's conceptual problems and I follow Laudan, dividing them into external and internal varieties. Laudan does not recognize what I describe as external theoretical difficulties that arise when a solution is closely tied to a questionable theory or principle. He also does not recognize reliability and missing-data difficulties.
4 To those readers who have noticed that I have smuggled in the condition that the difficulty-free solution must be to a first stage problem, I ask their indulgence, and will explain why in §5.14. In short, the stipulation that the solution be to a first stage problem is to avoid situations where one shows that a phenomenon may definitely happen but for which there is no evidence that it has happened.
5 The relevance of specialization and differential preference for mobilism or fixism was first suggested by Wegener (1929/1966: vii–viii). Du Toit (1927b: 119–120) agreed. Hallam (1973: 111), citing Wegener, agreed with him, and Frankel (1976) expanded the idea in a discussion of the reception of Wegener's mobilism. Le Grand (1988) further developed the idea. Frankel (1984a) put forth the thesis that regionalism, which he then called provincialism, goes a long way to explain the differential reception of mobilism by region that occurred during the classical period of the overall mobilist controversy. But this idea was not new with Frankel. Du Toit (1927b: 119–120) raised the same point about the controversy over mobilism. Le Grand (1986, 1988) agreed, but rejected Frankel's pejorative "provincialism" for the more neutral term "localism." Le Grand further included in his idea of localism, studying problems within a particular geological period. He also suggested that localism might be thought of in terms of a specialization, suggesting that there were regional specialties and discipline specialties. Thus, a geologist might be an expert on North America Pleistocene glaciation. Allègre (1988) also discussed the importance of specialization and regionalism, and declared that both were important factors in

understanding the controversy over mobilism. It should be obvious that more than one discipline may be used to attack a regional problem, that geologists in some disciplinary specialties, especially those falling within the field of geophysics, work on global problems, and that some geologists may concentrate on regional problems from different countries other than where they were trained. Oreskes (1999) disagrees, arguing that although Frankel and Le Grand have identified a real phenomenon, they have incorrectly described and understood it. Oreskes identifies the phenomenon as epistemological chauvinism. After pointing out that some of what I have claimed are obvious points, and adding that localism was also nationalistic, she bundles all into her notion of epistemological chauvinism or epistemological affinity.

Thus we might better call the phenomenon epistemological chauvinism, or perhaps less pejoratively, epistemological affinity. Geologists and geophysicists at the end of the nineteenth century responded to the difficulties in their discipline by relying on and emphasizing the information that was geographically, nationally, and/or disciplinarily closest to themselves. They did not necessarily ignore other information, but they did weigh the available evidence differentially, based on their affinities. And while these affinities expressed themselves epistemologically – in terms of differential weightings of evidence – their sources were largely social.

(Oreskes, 1999: 52–53)

I find Oreskes' epistemological chauvinism or affinity confusing. I do not want to conflate regionalism and specialization, and I do not want to conflate situations in which Earth scientists arrived at a conclusion based on the geology of their region, and those in which they arrived at a conclusion based on the prevailing opinion. However, she did point to phenomena that occurred during the continental drift controversy that I had ignored. During the latter two phases of the controversy, North American Earth scientists were inclined to support fixism even if they were working on problems in specialties in which regionalism does not apply. They were influenced by the powerful resistance to mobilism by their colleagues even if work in their area supported it, and they often were unfamiliar with much of the evidence in favor of mobilism based on Southern Hemisphere geology and paleoclimatology. I think that North American paleomagnetists and marine geophysicists at Lamont behaved in this manner. These cases should be distinguished from those in which Earth scientists worked on regional problems. There may, of course, be other factors, such as a strong personal feeling of nationalism or jingoism, that led particular scientists to embrace a theory or solution from their own country or dismiss a particular theory of solution from a different country.

6 Both the GSA and AGU have increased the number of their divisions, and AGU has now become the larger organization. In 2010, the GSA has approximately 22 000 members and divisions in archaeological geology; coal geology; engineering geology; geobiology and geomicrobiology; geoinfomatics; geology and health; geology and society; geophysics; geoscience education; history of geology; hydrogeology; limnogeology; organic geochemistry; mineralogy, geochemistry, petrology, and volcanology; planetary geology; Quaternary geology and geomorphology; sedimentary geology; structural geology and tectonics. The AGU has twice as many members as the GSA, with disciplinary sections in atmospheric sciences; biogeosciences; geodesy; geomagnetism and paleomagnetism; hydrology; ocean sciences; planetary sciences; seismology; space physics and aeronomy; tectonophysics; volcanology, geochemistry, and petrology. It also has twelve focus groups whose members engage in interdisciplinary work involving at least two disciplinary sections: atmospheric and space electricity; cryosphere sciences; Earth and planetary surface processes; Earth and space science informatics; mineral and rock physics; global and environmental change; natural hazards; near surface geophysics; nonlinear geophysics; paleoceanography and paleoclimatology; study of Earth's deep interior; and societal impacts and policy science. Other Earth science societies in the United States that serve as important organizations for particular disciplines include the American

Association of Petroleum Geologists, American Association of Stratigraphic Palynologists, Geochemical Society, Mineralogical Society of America, Paleontological Society, Society for Sedimentary Geology, Society of Economic Geologists, and the Society of Vertebrate Paleontology. This comparison reflects, I believe, the greater importance of geophysics over geology nowadays in understanding and unraveling processes responsible for Earth's evolution.

2

Wegener and Taylor develop their theories of continental drift

2.1 Introduction

I begin by briefly outlining the state of geological theorizing about the origin of Earth's surface features at the end of the nineteenth century. I then examine Alfred Lothar Wegener's theory of continental displacement as presented in 1912 and Frank Bursley Taylor's theory of continental creep presented in 1910. I shall then describe how they developed and defended their theories, the problems they sought to solve by introducing them, and the research strategies they adopted to accomplish their ends. Neither Taylor in 1910 nor Wegener in 1912 used the terms "continental drift" or "drift." Taylor in 1910 envisaged continents that were undeformed except at their edges.

Although Taylor's theory preceded Wegener's, I shall consider Wegener's first, being the more complete and influential. However, all elements of Taylor's theory are not dead. His idea of continental creep with the collision of Eurasia and formation of lobe-shaped structures such as the Malaysian Peninsula brings to mind the process of "extrusion" tectonics. I shall not cover what Newman (1995) eloquently describes as "embryonic expositions of a mobilist position" such as those of A. Snider (1858), or H. Baker (1912–1914, 1932) which have been reviewed by F. W. Hume (1948), Hallam (1973), Marvin (1973), and others.

Wegener developed his theory over a period of almost twenty years. In this chapter I am concerned only with his initial presentation of his theory of continental displacement in two 1912 papers. The first and shorter one was published in *Geologische Rundschau*; the second, more substantial one, in *Petermanns Geographische Mitteilungen*, and is now often referred to as the Petermanns paper. These were followed by four editions of *Die Entstehung der Kontinente und Ozeane* in 1915, 1920, 1922 (English translation, *The Origin of Continents and Oceans*, 1924), and 1929 (English translation, 1966). Wegener was also joint author with Köppen of a very important monograph on paleoclimates, *Die Klimate der Geologischen Vorzeit*, in 1924, and it was in this and in the 1922 edition of *Die Entstehung der Kontinente und Ozeane* and its English translation that the theory of continental drift appeared fully fledged, presenting a bold wholesale challenge to the prevailing fixist orthodoxy of the early twentieth century. It was this fully fledged challenge to the whole framework within much of geology and geophysics as then understood that launched the mobilism debate. This fully fledged challenge to the then current orthodoxy I consider in Chapters 3, 4, and 5. Following the initial 1912 papers, the editions of 1915 and 1920 describe the evolution of Wegener's thoughts leading up to the seminal 1922 edition, and I shall summarize this evolution at the end of this chapter (§2.15).

2.2 Geological theorizing at the turn of the twentieth century[1]

Mott Greene (1982) has shown that during the 1880s the prospect of finding an overall theory to explain a common nest of problems appeared excellent, and many Earth scientists believed that Eduard Suess (1831–1914), Europe's leading geologist, had provided them with the necessary basic framework for solving them. His theory, a version of contractionism, was a fixist theory. This general optimism was short-lived. Although many remained favorably inclined toward Suess' theory, others began to raise difficulties (RS2). Before Wegener and Taylor introduced their versions of continental drift, geologists had become uncertain what approach to take. Wegener and Taylor were well aware of the difficulties with Suess' contractionism, and they presented their mobilist theories as alternatives, theories they defended by stressing their problem-solving effectiveness. Other workers developed new versions of contractionism (RS1), which were designed to circumvent some of the difficulties with Suess' contractionism. However, neither the revised contractionism nor the new mobilism had anything approaching a recognized difficulty-free solution, and few, if any, Earth scientists were willing to accept either. It was against this background that the debate between mobilists and fixists began. However, the majority of Earth scientists did not engage in this debate at all but instead concentrated on their specialty or on local or regional problems.

2.3 The contractionism of Suess

What we are witnessing is the collapse of the world.

(Eduard Suess, 1885. Quoted by Wegener (1912a: 277; 1912b: 186))

Contractionism was the prevailing geotectonic doctrine during the nineteenth century. It was defended by such diverse thinkers as Elie de Beaumont (1798–1874) and James Dwight Dana (1813–95), prominent members at the time of the French and American geological communities. Under the influence of Charles Lyell (1797–1875) contractionism was less dominant in Great Britain, but the history of nineteenth-century geotectonics is, for the most part, that of ever more sophisticated and detailed versions of contractionism and a proliferation of the various controversies they provoked.

Suess' was the most comprehensive and authoritative version of contractionism. Many of his ideas first appeared in his *Die Enstehung der Alpen* (Suess, 1875), and they influenced the French geologist Marcel Bertrand (1847–1907) who became one of his major supporters. Suess began work on *Das Antlitz der Erde* in 1878, finishing it thirty-one years later. There were three volumes that appeared in full or in part in 1883, 1885, 1888, 1901, and 1909, over 2000 pages in all. In a review of the English translation, the British geologist J.W. Gregory (1864–1932), a later participant in the drift controversy, remarked:

English speaking geologists will be grateful to Dr. Hertha Sollas and the Clarendon Press for this excellent translation of the first volume of work which has probably had the deepest influence on geological thought since the publication of Lyell's "Principles."

(Gregory, 1905: 193–194)

Suess' influence was strongest in continental Europe. Albert Heim (1849–1937), who became Director of the Geological Survey of Switzerland, endorsed his work. His ideas were used by late nineteenth and early twentieth century Alpine geologists who developed the Nappe Theory of the Alps, and who extended his notion that the Alps had been formed by horizontal thrusting of the crust brought about by thermal contractions of the interior. The intervals between successive volumes allowed Suess to utilize ideas of his supporters, and he incorporated Nappe Theory into later volumes. His influence was also felt in North America and Great Britain, where he had many devotees who adopted his views and offered solutions to their own regional problems in accordance with his synthesis.[2] Suess even received telegrams signed by the entire memberships of the Geological Societies of America and London upon publication of the last volume of his monumental work (Greene, 1982: 274).

Suess' main analytical tool was trend-line analysis, his identification of the trends of large-scale geological structures, of whole mountain-systems. He began with the Alpine-system, turned next to continents and oceans, seeking similarities and differences among them, and so quite naturally developed a truly global theory.

Suess' contractionism had four major elements.[3] First there was his identification of two types of continental margins for which he offered differing explanations, both based on contraction theory. There are margins where geological features are abruptly truncated at the adjacent ocean basin. Suess proposed that such margins were caused by the collapse of part of a formerly larger continent (called a paleo-continent). The collapsed part becomes a new ocean basin, and the present margin is its marginal fracture. Geological features that had formerly extended across the entire paleocontinent now end abruptly at the new margin, which he called "Atlantic-type" margins. Continental margins surrounding the Atlantic Ocean are typically steep, and he claimed that the Atlantic basin had been formed by the collapse of continental material that once united the surrounding continents; mountain ranges on either side of the Atlantic had been severed by the foundering of the Atlantic basin. Bertrand later drew two important trans-Atlantic correlations: he matched the

Mid-Paleozoic Caledonian mountain system of Europe (Caledonides) with the Green Mountains in North America, and the more southerly Late-Paleozoic Hercynian system of Europe (Hercynides) with the Alleghenies of North America.

Suess' other type of continental margin had fold-mountain chains and island arcs running *parallel* to it, and because, typically, they bounded the Pacific basin, he called them "Pacific-type" margins. Because these mountain ranges are relatively young, Suess argued that they determined the shape and location of coastlines, and hence that the Pacific basin pre-dated Pacific-type margins. He considered the Pacific the oldest existing ocean – perhaps formed by the ripping away of the Moon early in Earth's history. The Indian Ocean had both Atlantic- and Pacific-type margins, indicating a complex history.

The second element in Suess' global theory was his adoption of the idea that Earth has always been cooling and contracting. He proposed that the Earth is composed of three concentric layers: outer or sal (Si-Al), a middle or sima (Si-Mg), and an inner or nife (Ni-Fe) layer, which I shall refer to as crust, substratum, and core, respectively. The crust is subject to tangential stress as the layers beneath cool and contract as Suess said, like the wrinkled skin of a drying apple. Horizontal thrusting and folding relieved tangential stress producing mountain-systems and island arcs. In this way he explained the extensive folding and arcuate shape of mountain ranges that his trend-line analysis had documented.

The third major element of Suess' system was his belief that arcuate mountain belts and island arcs generally owe their origin to horizontal unidirectional movement of continental crust. The Alps, for example, which are convex to the north, were formed by thrusting from the south. The mountains of Asia were formed by thrusting outward, southward toward the Indian Ocean, and, along with island arcs, eastward toward the Pacific Ocean.[4]

His hypothesis of northward thrusting of the Alps served as the basis for the Nappe Theory, which during the last decade of the nineteenth century became the chief achievement of Alpine structural geologists. They greatly expanded his idea of unidirectional thrusting, and by tracing the horizontal displacement of identifiable sedimentary nappes from the south to the north estimated that the Alps had been overthrust up to 100 km.

The last element of Suess' contractionism theory was that the present-day oceans were formed by sinking of parts of paleocontinents. Such massive collapses occurred as a consequence of the relief of radial tensions caused by Earth's contraction. According to Suess, past continents formerly stood where present-day oceans are now, but the present-day continents, which contain ancient shields or cratons, have remained continental throughout Earth's history; they have never foundered. He argued that the Atlantic Ocean was formed by the foundering of two paleocontinents – northern and southern Atlantic paleocontinents – separated by an ancient Mediterranean-type sea, the Tethys, extending from the present-day Gulf of Mexico through the present-day Mediterranean into the older part of the Indian Ocean.

Suess also recognized a huge southern continent, *Gondwana-Land*, comprising Africa, Madagascar, Arabia, peninsular India, the younger part of the Indian oceanic basin, New Guinea, and later, as his ideas developed, Australia, part of South America, and much of the south Atlantic.[5] There was also a northern paleocontinent, *Angara-Land*, spanning the present-day North Atlantic connecting North America with Europe via Greenland and including the whole of Asia north of the Tethys.

Suess supported the former existence of these now sunken paleocontinents with structural arguments; the correlation of European and North American Paleozoic mountain chains and their truncation at normal-faulted margins indicated a sunken paleocontinent and younger North Atlantic Ocean; the mountain ranges, originally single chains, were subsequently separated by a collapsed paleocontinent. Suess also used paleontological arguments, making much of the Permian *Glossopteris* flora, fern-like plants whose fossil remains had been found in India, South Africa, South America, Australia (and later in Antarctica).

Suess was not alone in postulating former paleocontinents extending across present-day oceans. Many European paleontologists had earlier done so in order to account for disjunctive biota: the presence, that is, of similar floras and faunas on landmasses now separated by wide oceans. (These tectonic and biotic disjuncts featured prominently throughout the mobilism debate.) Thus Suess not only used the idea of foundering paleocontinents to explain the origin of fractured coastlines and the geological similarities of continents on either side of the Atlantic, he also followed those paleontologists who had used the idea of landbridges to explain disjunct biota by having his paleocontinents serve as migratory conduits before they sank.

Thus Suess' theory explained the origin of mountain ranges, island arcs, the origin of "Atlantic-type" and "Pacific-type" margins, geological similarities between at least Europe and North America, and disjunctive biota. He also had an answer to the mechanism question: what is responsible for Earth's contraction?

2.4 The reception of Suess' contractionism and the difficulties it encountered

The contractional theory gives us a force having neither direction nor determinate mode of action, nor definite epoch of action. It gives us a force acting with a far greater intensity than we require, but with far less quantity. To provide a place for its action it must have recourse to an arbitrary postulate assuming for no independent reason the existence of areas of weakness in a supposed crust which would have no *raison d'etre* except that they are necessary for the salvation of the hypothesis.

(Clarence E. Dutton 1889: 62)

Forty-two years ago the contraction theory occurred to myself independently. I remember that in my youthful joy at what I thought a discovery, I forthwith vaulted over a gate! In 1868 I read my paper on "The Elevation of Mountains by Lateral Pressure," fully believing that I was elucidating the cause which had produced them in the contraction through secular cooling. In 1873, I began my paper on "The Inequalities of the Earth's Surface Viewed in Connection with the Secular Cooling," while still under the same impression. I first of all estimated the actual elevations, and, this done, I calculated the amount of those which would be formed upon Sir William Thomson's view of the mode of solidification. To my excessive surprise, the result showed the utter inadequacy of the contraction hypothesis. I thought I must have made some error in the calculations, but could find none. I still, however, adhered to the original idea of contraction, and suggested, towards the end of that paper, a fluid condition of the interior at some former period, thinking that sufficient contraction might be perhaps obtained by that means; for I had not yet dared to question Sir Wm. Thomson's dictum of the present complete solidity of the Earth. It was not until about a year ago, when I wrote the chapter in my book about the "Amount of Compression," that I perceived that even the condition of a liquid substratum would not give the necessary degree of contraction to produce the compression. I have thus been driven from the contraction hypothesis step by step, and have by no means been endeavouring to support a preconceived opinion against it.

(The Reverend Osmond Fisher (1883: 77))

Although contractionists understood none of the physical details, they possessed a general and basically geometrical solution to the problem of how continents founder and mountains rise: contraction of the Earth's interior, brought about by cooling, caused the build-up of radial and tangential tensions whose release collapsed continents and produced overthrust crustal folds. Nevertheless, in the 1880s their general account was plausible enough to satisfy most Earth scientists concerned with global questions. It was supported by Laplacian cosmogony. Equally important, little if anything was known about the physical nature of the Earth's interior and its relation to the crust, rendering any detailed mechanical account beyond the limits of prudent speculation.

Toward the end of the nineteenth century, contractionism became less favored; by the time Wegener and Taylor had introduced their mobilist theories in the early years of the twentieth century, geophysicists had already raised two very serious theoretical difficulties with Suess' theory (RS2). First, the discovery of isostasy made the sinking of paleocontinents or former wide landbridges unlikely; they could not have sunk into a denser seafloor; it was just as inconceivable that a block of ice would sink to the bottom of a lake; Suessian contractionism was inconsistent with the by now well-supported view that the continents are in approximate isostatic equilibrium with the substratum. Second, the discovery of heat-generating radioactive material within the crust meant Earth might not be cooling; radioactive decay might even be heating Earth causing it to expand.

Also, geologists identified two anomalies (RS2) in Suess' contractionist account of mountain building. Mountain belts are not evenly distributed over Earth, and critics argued that his model would not have restricted crustal folding to just a few places; wrinkles of a drying apple's skin are distributed generally throughout, not bunched in a few places. Others argued that the success of Alpine geologists in tracing the spectacular overthrusts in the Alps implied that the amount of crustal shortening was much greater than required by Suessian contractionism.

Although they diminished its overall problem-solving effectiveness, these attacks did not lead to a general rejection of Suess' theory. The overwhelming majority of continental European geologists continued to support contractionism, even the sinking of paleocontinents. They attempted to remove the difficulties raised (RS1) against Suess' theory, and in turn raised difficulties (RS2) with competing theories. For example, they questioned the reliability of the geodetic data on which isostasy was based. They argued that estimates of shortening needed to produce mountain belts had been exaggerated. They raised legitimate questions about the amount and location of Earth's radioactive material: perhaps there was not enough radioactive heat. They argued that supporters of isostasy were unable to construct an adequate theory of mountain building, and that their attempts were often not mutually compatible. Even though Suess' contractionism faced difficulties, his supporters argued that it was unwise to dismiss it in favor of more flawed theories.

Other Earth scientists, however, rather than rejecting Suess, developed (RS1) new versions of contractionism which, they believed, were consistent with isostasy and the presence of radioactivity. These new theories did not, however, include the notion of sinking paleocontinents. For example, Hans Stille (1876–1966) and Leopold Kober (1883–1970) from Germany developed contractionist views harking back to those of Elie de Beaumont (Şengör, 1979/1982a and 1982b). Harold Jeffreys (1891–1989) from the United Kingdom, and Bailey Willis (1857–1949) and Thomas Crowder Chamberlin (1843–1928) from the United States also developed contractionist theories that took account of both isostasy and radioactivity (Frankel, 1979; Greene, 1982; Brush, 1996a).

It is understandable why British and American geologists and geophysicists gave much greater weight to the difficulties raised against Suess' theory than did their continental European counterparts, because it was they who were primarily responsible for the establishment of isostasy and the application of radioactivity to geology. In 1885, the British geodesists George Airy (1801–92) and Archdeacon John Henry Pratt (1809–71) independently proposed the idea of isostasy to explain the mass deficiencies observed when determining the mass of mountains. The Reverend Osmond Fisher (1817–1914), a British geophysicist and mathematician, argued against the contractionist theory of mountain building and, in his 1881 *The Physics of the Earth's Crust*, attempted to draw out the geological implications of a buoyant crust floating on a denser interior. Clarence E. Dutton (1841–1912), an American geologist, coined the term "isostasy" in 1889, and agreed with Fisher's critique of contractionism. The initial application of radioactivity to geological problems originated primarily in Great Britain and Ireland, where Arthur Holmes and John Joly studied its implications for the evolution of Earth's geological features and raised difficulties against contractionism (§4.9, §5.3).

2.5 Wegener the man[6]

Out here, there is work worthy of a man; here life takes on meaning[7]
(Alfred Wegener, December 25, 1906, Greenland)

You know German and geology, and they influence a man very much.
(Oscar Wilde, The Importance of Being Earnest)

Wegener is the father of continental drift, or continental displacement as he called it, because he was the first to present an extended version of the theory providing solutions to many problems in geology and geophysics, yet he devoted only a small part of his professional life to its development and defense. I suspect that he viewed himself, as did his contemporaries, as a polar explorer and meteorologist/geophysicist rather than as the father of drift theory. Out of his approximately 150 publications, only about fifteen percent pertain to continental drift. Most concern meteorology, atmospheric physics, or his polar expeditions (see Benndorf, 1931, for a listing of Wegener's publications). Given Wegener's life, therefore, it is perhaps more fitting that he died in Greenland leading a scientific exhibition rather than in his study extending his drift theory and preparing further defenses of it. This is not to say that he did not take his work on drift theory seriously; he did, and he fully recognized its revolutionary character. One can only wonder what might have happened had he devoted himself exclusively to it.

Wegener was born in Berlin on November 1, 1880, to Richard and Anna (neé Schwarz) Wegener. He was one of five children, two of whom died during childhood. The others were his older brother, Kurt, and their younger sister, Tony. His father, a doctor of theology, was Director of the Schindler Orphanage from 1875 to 1904. After Alfred and Kurt had obtained their secondary education at the *Collnischen Gymnasium* in Berlin, they began to pursue scientific careers. Alfred studied at the Universities of Heidelberg and Innsbruck, and entered the University of Berlin in 1903, where, at the age of twenty-four, he earned his doctorate in astronomy in November of 1904. His dissertation topic, "The Alfonsine Tables for the Use of Modern Computers," involved the tedious conversion of planetary positional data, sexagesimally expressed in the thirteenth-century Alfonsine Tables, into the decimal system. But Wegener decided against a career in astronomy, later giving three reasons: "In astronomy everything has essentially been done. Only an unusual talent for mathematics together with specialized installations at observatories can lead to new discoveries; and besides, astronomy offers no opportunity for physical activity" (Schwarzbach, 1986: 16).

Unlike astronomy, a career combining meteorology and polar exploration allowed Wegener to work in a new field in which he could do original work without the use of higher mathematics, and to engage in more physical activity than even he eventually wanted.[8] One of his fellow students at the University of Berlin, W. Wundt, recalled that, in 1903, Wegener showed him the route he planned to take across the

Greenland ice-cap (see Georgi, 1962: 309). Although there is no independent corroboration of Wundt's recollection, Wegener certainly undertook the sort of physical training appropriate for anyone bent on becoming an Arctic explorer, and the dream of exploring regions around the North Pole was common enough at the time among adventurous, scientifically minded young men in northern Europe. At Innsbruck in the summer of 1901, he and his brother Kurt undertook a rigorous program of mountain climbing, and attended classes at the University of Innsbruck between their mountain climbs. In the winter of 1903–4 he became an adept skier while visiting a friend stationed at the observatory on Mt. Brocken, the highest area in the Harz mountains. Wundt also recalled how Wegener trained for polar exploration during the winter around Berlin by extensive skating and exhausting treks.

With his Ph.D. in hand, Wegener turned his attention to experimental meteorology. He joined his brother Kurt at the Aeronautic Observatory in Lindenberg, located on the outskirts of Berlin, where they were employed as technical aides. Here he learned to work with kites and captive balloons, the new experimental instruments of aerology. In the spring of 1906, the two brothers won the Gordon Bennett Contest for Free Balloons by staying aloft for fifty-two hours, breaking the previous record by thirty-five hours, a feat that attracted much attention in the popular press.

Wegener's work in experimental meteorology and his arduous physical training were rewarded in 1906 by an invitation to join the Danish *Danmark* Expedition to Greenland, led by Mylius-Erichsen. The expedition lasted two years, and Wegener spent part of two winters (1906–7 and 1907–8) at latitude higher than 77° N. He was the official meteorologist and undertook a number of field experiments, sending his kites and captive balloons to heights of 3000 meters. He photographed cloud formations and other meteorological phenomena.

Wegener wrote up the results of his kite and balloon experiments, the first of their kind performed in a polar region, which helped him obtain a teaching position in 1908 in meteorology at the University of Marburg in Germany (Schwarzbach, 1986: 19). His ten-year stay at Marburg was interrupted by a second expedition to Greenland (1912–13) and service in the armed forces (1914–18) during World War I. The period from his arrival at Marburg until his second expedition to Greenland was his most creative. He published over forty papers, most on his meteorological work on Greenland; wrote *Thermodynamics of the Atmosphere*, a standard textbook in meteorology (1911 and two later editions), and developed his theory of continental drift, publishing two papers on it in 1912. He taught courses in astronomy and meteorology, and was noted for his clear and unassuming lecture style. This period was equally important for Wegener's private life and future happiness. Ambitious as he was, in 1910, he took his manuscript of *Thermodynamics of the Atmosphere* to Professor Wladimir Köppen of Hamburg, a leading German meteorologist. Wegener had first contacted Köppen just prior to his leaving for Greenland on the *Danmark* Expedition, seeking advice on meteorological matters, and renewed his acquaintance after returning from Greenland.

Köppen praised the manuscript. More importantly, Wegener met Else, Köppen's eighteen-year-old daughter. They married in 1913.

In 1912–13, Wegener took part in his second expedition to Greenland, which was led by J. P. Koch, a Danish Arctic explorer, who Wegener had come to know during his first expedition. They left in the summer of 1912, spending two and a half weeks in Iceland. They were the first non-Greenlanders to overwinter on Greenland's ice-sheet, and in April 1913 he and Koch began a two-and-a-half month trek across its widest part, climbing to heights of 3000 meters. They almost starved. Inuits rescued them and they completed the longest recorded traverse of the Greenland ice-cap.

The expedition was a scientific success. Wegener made many meteorological observations, and became adept at using the Miethe Three-Color Method for taking photographs of clouds, ice, aurora, and mirages. His work was important; at the time there were few systematic observations of cloud formations. Wegener's work may be viewed as an attempt to capture and display photographically successive temporal slices of developing air masses.[9] Others, such as Vihelm Bjerknes (1862–1951), the Norwegian physicist and meteorologist who did much to modernize atmospheric science, preferred the medium of mathematics. (For a discussion of Bjerknes, see Friedman, 1982.)

Else pursued her own interests while her fiancé was in Greenland. She spent eleven months at Bjerknes' home in Oslo. Besides teaching Bjerknes' children German, she learned several Scandinavian languages and later translated two of Bjerknes' works into German. Alfred and Else returned to Marburg in 1913.

Wegener's promising career in meteorology and geophysics was interrupted in 1914 by war. Engaged in the German advance into Belgium, he was shot in the arm. Returning to combat fourteen days later, he was shot in the neck. This ended his combat duty, and he spent the remainder of his army service partly on the eastern (Russian) front as a meteorologist. While convalescing, he worked on his new theory of continental drift and the first edition of his *Die Entstehung der Kontinente und Ozeane* was published in 1915. Just slightly over ninety pages, it offered a more detailed presentation of the arguments he had already developed in his two earlier 1912 presentations.

Wegener succeeded Köppen (who retired at the age of seventy-two) as department head of theoretical meteorology at the German Marine Observatory in Hamburg, where his brother Kurt was employed. He transferred his right to deliver academic lectures from Marburg to the newly formed University of Hamburg. Because of the hard post-war times, the Wegeners and Köppens combined households. Köppen and Wegener also drew closer professionally and their co-authored work *Climates of the Geological Past* (*Die Klimate der Geologischen Vorzeit*) was published in 1924. Although Köppen had been worried about Wegener's spending too much time on his theory of continental drift at the expense of his future career in meteorology, Köppen soon thought his drift idea was worth serious consideration. Their book, unfortunately ignored by many and never translated into English, detailed the paleoclimatological case for continental drift, which constituted their attempt to pin down paleolatitudes via paleoclimatology (§3.14). During this Hamburg period, Wegener wrote the second (1920) and third (1922) editions of *Die Entstehung der Kontinente und Ozeane*, as well as a book (*Die Entstehung der Mondkrater*, 1921) on the origin of lunar craters, correctly arguing that they were of meteoric rather than volcanic origin.[10] He also found time to write papers on meteorology and to participate on an ocean voyage to Cuba and Mexico (1922) to measure upper air currents over the Atlantic.

Wegener finally secured a professorship in 1924 at the University of Graz in Austria as professor of meteorology and geophysics. With a light teaching load, Wegener had time to continue study of the vast literature relevant to his theory of continental drift, and it was there that he wrote the fourth edition (1929) of *Die Entstehung der Kontinente und Ozeane*.

Wegener, still not immune to the call of the wild, accepted an offer to lead a major expedition to Greenland.[11] The objectives of the expedition were: to determine the thickness of the ice-cap by explosion seismology; to determine whether Greenland is rising by the use of barometric, trigonometric, and gravimetric measurements; and to determine the general nature of Greenland's winter climate, paying special attention to the question of whether an area of high-pressure remained intact over the center of the Greenland ice-cap throughout the winter (Georgi, 1935: 12–15; E. Wegener, 1939: 2–3). In order to meet the meteorological objectives, Wegener decided to set up three stations along 71°N latitude, two on opposite coasts; the third, called "Mid-Ice," near the center of the ice-cap. Keeping Mid-Ice occupied throughout the winter made the plan very difficult. In 1928, he and three colleagues (Johannes Georgi, Fritz Loewe, and Ernst Sorge) made a seven-month reconnaissance trip to Greenland in preparation for the major effort in 1929.

Wegener never returned from the 1929 expedition. Keeping Mid-Ice open during the winter turned out to be his undoing. In a journey from there to the West Station he lost his life. Wegener, Loewe, and thirteen Inuks left the West Station on September 21, 1930, with a care-package of 4000 pounds. Because of heavy snow, half the cargo had to be abandoned and all but one of the Inuks, Rasmus Villumsen, were sent back to the coast. The three arrived at Mid-Ice on October 30. Wegener and Villumsen were in excellent condition, but Loewe arrived with frostbitten feet. Georgi, who was at Mid-Ice, summed up the situation in a letter written to his wife the day after their arrival.

At midday on Thursday a young Greenlander, Rasmus Villumsen, arrived, and soon afterwards Wegener and Loewe. They had begun the journey with fifteen sledges, and now arrived with only three – quite worn out by forty days' journey through awful weather, and without food or petroleum. A marvelous performance in a temperature of –61°! I am sure no one has

Figure 2.1 Last photograph of Wegener and Villumsen taken on November 1, 1930, before they set out from Mid-Ice on their unsuccessful attempt to reach West Station (Georgi, 1935: opposite p. 120 and reproduced in Schwarzbach, 1980/1986: 45). The temperature was approximately −50°C (Schwarzbach, 1980/1986: 45).

ever traveled in Greenland in such a temperature. All three of them are frost-bitten, Wegener and Rasmus only very slightly, on the fingers; but Loewe's feet are so bad that he cannot go back with the others, but must winter here with us.

(Georgi, 1935: 114)

Supplies were insufficient for the five to remain at Mid-Ice all winter, so Wegener and Villumsen decided to return to West Station. Georgi described his last meeting with Wegener in a letter to his wife, which he wrote on November 4, 1930.

So the two [Wegener and Villumsen] went off on Saturday, November 1. They were photographed and filmed at −40°. Both Wegener and we were much moved at saying good-bye, perhaps, for ever; and it was doubly painful for us, because we have to wait 'til May before we hear whether the two others got through, and can do nothing to help. I need not gloss over the fact that, after the two sledges had disappeared in the fog, I retired to the barometer-room to compose myself.

(Georgi, written November 4, 1930 at Mid-Ice; see Georgi, 1935: 121)

In 1962, Georgi later recalled the overall situation.

> I shall not forget the last meeting I had with him, when his successful journey [from West Station] ended on 30th October 1930 at "Mid-Ice" with the temperature −50 to −54°C. He greeted with beaming enthusiasm our dirty, low dug-out as a comfortable living room after the icy air in the tents during the preceding days, with joy shining in his face because Dr. Ernst Sorge and I, despite the lack of many provisions, were going to try to keep the station going until the following summer. On his 50th birthday, fully fit and with rested dogs, he [and his companion, Rasmus Villumsen] set off, certain of making a quick march back in comparison with the tedious journey out. Even with this difficult journey ahead of him he did not forget our state of mind, our downheartedness because, due to the absence of necessary scientific equipment, our work must remain patchy. "The fact that you have spent the winter here in the middle of Greenland, even without any particular results in research, doing only the simplest and most routine measuring, is something which is worth all that has gone into the expedition," he said; and who knows if it was not this encouraging word that helped us psychologically to get through that winter?

> *(Georgi, 1962: 322)*

Wegener died of a heart attack 120 miles from the western coast. His body, marked by his skis, was found on May 12, buried in the snow between two sleeping bag covers. Sorge and Karl Weiken, the two who discovered his body, left it as they had found it (see E. Wegener, 1939: 198–204, for Sorge and Weiken's account of their discovery of Wegener's body). Villumsen was never found (see E. Wegener, 1939: 205–208, for an account of the unsuccessful search for Villumsen's body). Apparently, he had buried Wegener, taken his diary and other personal effects, and attempted to make it back to the West Station. The German government wanted to bring the body of its hero home for a state funeral. Else Wegener refused. She felt it more fitting that her husband remain in Greenland and slowly drift out to sea entombed in ice.

2.6 Wegener's 1912 theory of partition and horizontal displacement of continents, from idea to working hypothesis [12]

Wegener's 1912 Petermanns paper makes it abundantly clear that he believed his notion of continental drift solved a very wide range of problems far more readily than other current theories. After conceiving the general idea in 1910, he began to familiarize himself with the relevant literature, and came to believe that a crisis had developed in the Earth sciences.[13] Although he acknowledged that Suess' theory of contractionism with its postulation of foundering paleocontinents explained the origin of disjunct biota, it clashed fatally with the principle of isostasy (RS2). Also, Wegener was dissatisfied with permanentism, whose proponents such as Willis argued that continents and oceans were permanent features. Willis, who renounced contractionism around 1907, became a major proponent of permanentism (Greene, 1982: 271–275) and, as will later become apparent, a vociferous critic of mobilism.

According to Wegener, permanentism could not explain the distribution of biotic or geologic disjuncts (RS2). While Suess' theory was geophysically untenable, permanentism, he thought, was no solution at all. He declared that his theory of continental drift offered a way out of this crisis because its solution to the problem of biota disjuncts was not in any way inconsistent with the principle of isostasy. Having established these two advantages, he began to treat his idea as a working hypothesis, expanding it to address and to solve other problems.

Wegner describes how he came up with the idea of continental drift in December 1910 in this way:

The first concept of continental drift came to me as far back as 1910, when considering the map of the world, under the direct impression produced by the congruence of the coastlines on either side of the Atlantic. At first I did not pay attention to the idea because I regarded it as improbable.

(Wegener, 1929: 1)

I believe that you consider my primordial continent to be a figment of my imagination, but it is only a question of interpretation of observations. I came to the idea on the grounds of the matching coastlines but the proof must come from geological observations. These compel us to infer, for example, a land connection between South America and Africa. This can be explained in two ways: the sinking of a connecting continent or separation. Previously, people have considered only the former and have ignored the latter possibility. But the modern teaching of isostasy, and more generally our current geophysical ideas oppose the sinking of a continent because it is lighter than the material on which it rests. Thus we are forced to consider the alternative interpretation. And if we now find many surprising simplifications and can begin at last to make real sense of an entire mass of geological data, why should we delay in throwing the old concept overboard? Is this revolutionary? I don't believe that the old ideas have more than a decade to live. At present the notion of isostasy is not yet thoroughly worked out; when it is, the contradictions involved in the old ideas will be fully exposed.

(Wegener, January 1911, letter to Köppen)[14]

In the fall of 1911, I came quite accidentally upon a synoptic report in which I learned for the first time of palaeontological evidence for a former land bridge between Brazil and Africa. As a result I undertook a cursory examination of relevant research in the fields of geology and palaeontology, and this provided immediately such weighty corroboration that a conviction of the fundamental soundness of the idea took root in my mind. On the 6th of January 1912 I put forward the idea for the first time ...

(Wegener, 1929: 1)

While studying a bathymetric world map, Wegener noticed the congruency of certain continental margins and shore lines.[15] He thought the similarity of the coasts of Africa and South America was sufficient to require an explanation, which he provided by placing them together in the Late Paleozoic. This congruency problem, he believed, could not be solved in any other way. The fit between Europe and North America was not nearly so good, but the similarity between Africa and South

America could hardly be accidental. He tells us, "At first I did not pay attention to the idea because I regarded it as improbable." However, he began to give it close attention. His letter to Köppen in January of 1911 shows that he already under-stood the need to postulate former land connections between South America and Africa, and the difficulties they faced from "the modern teaching of isostasy, and more generally our current geophysical idea." Then, in the fall of 1911, Wegener came across a paleontological report about the need for former landbridges between Africa and Brazil.[16] He already knew of the need to postulate some sort of trans-Atlantic land connections, but this report stimulated him to undertake "a cursory examination of the relevant research in the fields of geology and palaeontology." I suspect that he also read other reports to determine if his idea of what came in the 1920s to be known as continental drift could make sense of the geology and fossil distributions in regions other than the circum-Atlantic. Wegener was befriended by Hans Cloos at Marburg, and Cloos, a fixist, helped him with the geological literature (§8.3). Whatever the precise nature of his reading, it "provided immedi-ately such weighty corroboration that a conviction of the fundamental soundness of the idea took root in [his] mind" and he was prepared to defend it, in public, and did so on January 6, 1912. At the very least, continental drift was the best way to explain the congruencies between continental margins and some biotic disjuncts without violating isostasy.

2.7 Wegener presents and defends his drift theory in 1912: his six major arguments

In the following, a first tentative attempt will be made to give a genetic interpretation of the principal features of the earth's surface, i.e. continents and ocean basins, by a single universal principle of horizontal mobility of the continents. Whenever we used to have ancient land connections sink into the depth of the oceans, we shall now assume rifting and drifting of continental rafts.

(Wegener/Jacoby, 1912b/2001: 31)[17]

In the following section we shall discuss gravity and isostasy in more detail. A person, who does not close his eyes and is not prejudiced against the latter, can hardly hold on to the collapse hypothesis. Especially American researchers have frequently stressed this point. But since horizontal continental displacement was not taken into account, negation of collapse and the permanence of the continents leads to the dubious hypothesis of the permanence of the oceans, mainly connected with Dana (1985) and Wallace (1876), and recently expressed by Bailey Willis [1910] ... in very clear but also rigid form. The opposing school rejects the thesis with good reasons. We are forced to assume land bridges between distant continents separated by deep ocean to have existed in former geological periods. Many hundred paleontological discoveries are a continuously growing proof, that faunas and floras of these continents existed in totally uninhibited exchange directly across present deep sea. Does this not mean that both ideas are unacceptable: the permanence of the oceans and the collapse of the earth? The attempt to save the permanence of the oceans by the assumption of only island chains or

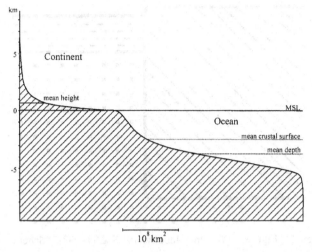

Figure 2.2 Wegener's Figure 1 (Wegener/Jacoby, 1912b/2001: 33). Hypsometric curve of the Earth's surface that "clearly demonstrates the existence of two preferred levels corresponding to the deep sea-floor and to the continental surface."

detours across presently still unexplored regions instead of genuine land bridges is so unsatisfactory that we need to deal with it further.

Thus these two ideas oppose each other without reconciliation. The two schools have good impeccable arguments, but they draw unacceptable conclusions. I shall attempt to show that the correct postulates of both are easily satisfied by the hypothesis of rifting and horizontal drift of the continents.

So stand these two views in opposition without mediation. Both parties have good, indisputable arguments, but both draw inadmissible conclusions from them. I wish to attempt to show that the justifiable claims of both can be more easily explained by the hypothesis of partition and horizontal displacement of continents.

(Wegener/Jacoby, 1912b/2001: 35–36)

Wegener began with the hypsometric curve, a plot of the elevations of the solid Earth (Figure 2.2). It "clearly demonstrates the existence of two preferred levels corresponding to the deep sea-floor [at 4300 m depth] and to the continental surface" at 700 m elevation extending out to the continental slope. Measurements of gravity "indicated that the oceans are not gravitationally deficient," continents and oceans being in approximate isostatic equilibrium (1912b: 33; my bracketed addition). He concluded (Figure 2.3) that the "lighter [sialic] continental rafts float, so to speak, on the heavier [simatic] mass," and "are positioned such that an equilibrium static pressure exists, similar to a floating iceberg in water" (Wegener/Jacoby, 1912b/ 2001: 37; my bracketed additions). Appealing to gravimetric and seismological studies, as known at the time, he estimated the thickness of the continents to be approximately 100 km. Oceanic crust, he claimed, is simply cooled and thus solidified sima. Wegener illustrated his model with an idealized figure of Earth's layers (Figure 2.4).

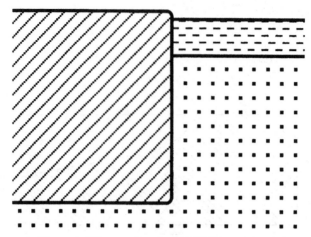

Figure 2.3 Wegener's Figure 3 (Wegener/Jacoby, 1912b/2001: 37). "Schematic cross-section through a continental margin." The diagonally hatched "lithosphere" or continental crust floats on the denser simatic layer (dotted). The dashed ocean is pictured atop the simatic layer. Not to scale.

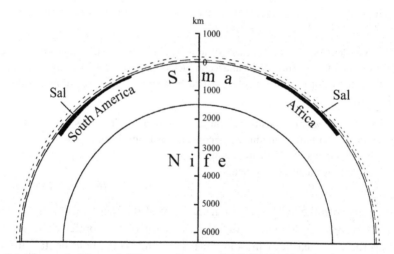

Figure 2.4 Wegener's Figure 5 (Wegener/Jacoby, 1912b/2001: 41). "Section along the great circle through South America and Africa in true proportions." The sima layer is atop the nife (nickel-iron) core. The dashed and dotted lines represent atmospheric layers.

Wegener then described the congruency of the opposed continental margins, especially, of South America and Africa, followed by an extensive literature review, which led him to argue that Suessian contractionism was in serious difficulty (RS2). He explained how his idea of lateral mobility of continents, what is now called continental drift, solved most of the problems Suess' theory did and some it did not, and argued that it avoided the difficulties faced by contractionism (RS1, RS2).

He came to believe that his theory solved more problems than either contractionism or permanentism (RS3), and merited further investigation. Wegener was not, however, ready to claim, at least in print, that his idea should be accepted as truly correct. Instead, he argued that it should be treated as a working hypothesis until confirmed through quantitative geodetic measurements by careful astronomical position-fixing of landmasses over time. In 1912, he thought correctly that the existing geodetic measurements were not reliable enough to confirm his idea, what I would call a difficulty-free solution (RS2). Wegener also addressed the problem (second stage) of the cause of continental drift. He suggested several approaches, but did not claim he had a fully satisfactory solution.

He went on to make six major points. (i) *He rejected Suessian theories of contractionism because they faced serious difficulties, difficulties that in Chapter 1 I categorize as anomaly and theoretical difficulties (RS2).* He referred to Fisher, Pratt, Airy, and Dutton, but relied most heavily on German geophysicists. He claimed, "Since Heim defended the contraction hypothesis, geophysics has piled up arguments against it" (Wegener/Jacoby, 1912b/2001: 34). He singled out the discovery of radioactive material in the Earth's crust indicating that Earth might not be cooling (Wegener/Jacoby, 1912b/2001: 34), the discovery that oceans and continents are in isostatic equilibrium (Wegener/Jacoby, 1912b/2001: 36–39), and the dearth of deep-sea sediments indicating that the continents had never been oceanic basins (Wegener/Jacoby, 1912b/2001: 35). For Wegener, the clash between isostasy and the proposed foundering of paleocontinents was the most serious; the recent geodetic surveys, which had been explicitly designed to test isostasy, demonstrated that continents and ocean basins were not the same material, the less dense sialic continents floated higher in isostatic equilibrium within the denser simatic substratum.[18]

We would have to reject the idea of the earth's collapses even without any of the above arguments. It contradicts the observations of gravity. If the ocean bottom were nothing but sunken continents, it would consist of the same material. Gravity, however, proves undeniably that the rocks below the oceans are denser than those of the continents. They are not just heavier, but exactly that much heavier that the elevation difference is compensated such as to achieve pressure equilibrium. In the following section we shall discuss gravity and isostasy in more detail. A person, who does not close his eyes and is not prejudiced against the latter, can hardly hold onto the collapse hypothesis.

(Wegener/Jacoby, 1912b/2001: 35)

(ii) *He rejected the permanency of ocean basins.* Although Wegener agreed with Willis and other American researchers that the foundering of paleocontinents or wide landbridges was inconsistent with isostasy, he thought little of their idea of replacing them with permanently positioned ocean basins because it did not address certain crucial problems. For example, it left unexplained both the geological similarities between the continents that face one another across the

Atlantic, and the biotic disjuncts. (iii) *He argued that continental drift both solved the problems and avoided the difficulties faced by Suessian contractionism (RS3).* Wegener argued that his theory solved the problems that were solved by Suess' theory, but unlike Suess' contractionism, it took account of the recent discovery of isostasy and was unaffected by the discovery of radioactivity. He thought that (i) and (ii) necessitated continental drift. Given (i), sinking landbridges were unacceptable. But, because of (ii), the substitution of permanent ocean basins for sinking landbridges or paleocontinents was also unacceptable. Intercontinental connections without landbridges were needed, and the partitioning of continental crust and subsequent horizontal dispersal of its fragments satisfied this need. Wegener sketched out the kinematic history, substituting partition and drift whenever and wherever Suessians had postulated former paleocontinents. He replaced Suess' northern and southern Atlantic paleocontinents with the formation of the Atlantic, and replaced Suess' Gondwana-Land with the breakup and drifting of India and the southern continents to their present positions (Wegener/Jacoby, 1912b/2001: 52). He stressed the close geological and paleontological comparisons across the Atlantic, citing Bertrand's matching of Paleozoic mountain ranges across the North Atlantic (Wegener/Jacoby, 1912b/2001: 49), and the need during the Mesozoic especially of a close connection between Africa and Brazil to account for their paleontological similarities (Wegener/Jacoby, 1912b/2001: 50). He emphasized the need for several land connections between Australia and other fragments of Gondwana-Land. However, he had not yet assembled the continents into a Late Paleozoic supercontinent. He did this later, and the supercontinent became known as Pangea, which swiftly became the trademark of his theory. (iv) *He extended drift theory's problem-solving effectiveness by providing solutions to other problems previously solved by Suessian contractionism (RS1).* Wegener linked his hypothesis to mountain belts. According to him, they formed on the leading edge of drifting continents as they plowed through the simatic ocean floor – the Andes for example (Wegener/Jacoby, 1912b/2001: 51). Wegener also suggested that his hypothesis could account for the distinctive Atlantic- and Pacific-type coasts as identified by Suess. The former were typically located on the trailing edges of drifting continents, and their fractured condition was caused not by the sinking of a paleocontinent but by splitting of the primordial supercontinent. The fold belts of Pacific-type coasts were located on the leading edges of drifting continents, as are the Andes, and were caused by the crumbling of continental sial (Wegener/Jacoby, 1912b/2001: 54–55). (v) *He extended continental drift's problem-solving effectiveness by solving problems that Suess' theory could not (RS1).* First, Wegener argued that continental drift explained the congruency of continental margins on either side of the Atlantic (RS1), whereas fixist theories did not (RS2). Second, there was the problem of the two preferred elevations, continental plateaus and ocean basins. Why had discussion of them caused so much confusion?

The opinions of today about the origin of the remarkable elevated plateaus of the earth's rind is an example of controversial confusion rare in science. Although I want to avoid controversy in this paper, I cannot help taking a short critical look at some of the ideas in order to show what we shall lose by replacing them with my hypothesis.

(Wegener/Jacoby, 1912b/2001: 33)

The two levels were, he argued, related to the origin of continents – recall that the title of both his 1912 papers was "The Origin of Continents" rather than, say, "The Horizontal Displacement of the Continents" – and directly reflected the hegemony of the notion of isostatic equilibrium between oceans and continents (RS1). Sialic continents remained higher than simatic ocean basins because they were less dense. Once isostasy was established, continents remained elevated above ocean basins. Although, so far as I can find, Wegener did not explicitly say so, these two dominant elevations presented a great difficulty for Suessian contractionism; assuming continuous contraction, as Suess did, there would be no reason for all continents (or ocean basins) to be at the same general level. The third problem that Wegener believed his theory solved (RS1), and no other came close to solving (RS2), was the origin of the extensive Late Paleozoic glaciation in the Southern Hemisphere.

The Permian Ice Age, therefore, presents an unsolvable problem for all models which do not dare to consider horizontal displacements of the continents.

(Wegener/Jacoby, 1912b/2001: 53)

He reviewed the evidence in favor of this widespread glaciation. Permian (and, as later became known, Late Carboniferous) glacial deposits had been described in South America, Africa, Australia, and, most interestingly, in peninsular India. But there was no indication of glaciation in antipodal regions of the Northern Hemisphere. Wegener cited the work of two German geologists, E. Koken and A. Penck, the latter a geomorphologist, both of whom claimed that no existing global theory could explain the existence of such a singularly vast ice-cap. Some researchers had postulated polar wandering. They placed the South Pole in the southern Indian Ocean central to the Gondwana ice sheets, but Wegener argued that this too faced what I would call an anomaly difficulty (RS2). The postulated ice-cap would have had to cover most of the Southern Hemisphere, seemingly inconsistent with paleoclimatological data which indicated that the Northern Hemisphere enjoyed a hot climate during the Permian. Additionally, placement of the South Pole in the Indian Ocean required a positioning of the North Pole in Mexico, which was at the time considered inconsistent with the paleoclimate of southern North America. The only adequate way to solve the problem, he argued, was to shift the South Pole to southern Africa, and then suppose that peninsular India and the continents in the Southern Hemisphere had been united around this region during the Permian. This required much less movement of poles (Wegener invoked some general polar wandering), greatly diminished the size of the former ice-cap, and placed the supposedly hot regions of the northern continents in tropical latitudes.

(vi) *On the basis of (i) through (v), Wegener deemed his hypothesis of continental drift worthy of serious consideration.* He believed that continental drift solved the following problems: the congruency of continental margins and geological similarities between the continents facing one another across the Atlantic Ocean; the origin of Atlantic and Pacific-type coasts; the origin of younger mountain ranges; the origin of island arcs; the distribution of disjunct biota past and present; the distribution of Permian glaciation; and explained the two dominant levels of the Earth's solid surface. Contractionism offered solutions to the geological similarities, disjunct biota, the two kinds of coastlines, and the origin of mountains and island arcs, but had no solution for the congruency of continental edges or for two preferred elevations of the solid Earth. Permanentism, he argued, offered no explanation for disjunct biota, intercontinental geological similarities, the two kinds of coastlines, and congruency of matching continental margins. Neither contractionism nor permanentism could, even if coupled with polar wandering, offer a solution to the origin of the Permian glaciation as effectively as Wegener's displacement hypothesis. Thus, Wegener argued that his theory was an effective problem solver and worthy of additional work.

2.8 Wegener's further arguments in 1912

Having summarized six major arguments on which Wegener based his case, I want now to comment on his convictions regarding his theory, on his recognition of the need for a definitive test, and on his discussion of possible mechanisms.

It is abundantly clear that Wegener was strongly inclined toward his theory; he was very enthusiastic about it, and he worked hard on it. But did he think it fully justified? Perhaps he thought it was true. In his January 1911 letter to Köppen (§2.6) he had firmly rejected contractionism. But it is unclear whether he accepted continental drift in his heart of hearts. I find it impossible to say.

Regardless of Wegener's private thoughts, he did not, in 1912, advocate, at least not in print, outright acceptance of his theory. Instead he argued that more work was needed and in particular a decisive test. The case was not foolproof; for example, the fit between continents across the North Atlantic was by no means perfect. He wanted, but did not yet have, a difficulty-free solution. But where was one to be found? Suppose he could obtain data that indicated the present displacement of one landmass relative to another in a direction and at a rate consilient with his above six arguments, and involving many independent lines of evidence? Fixism would then have a very serious, perhaps insurmountable obstacle. He might thereby be able to develop what I would call a difficulty-free solution, allowing him to argue for outright acceptance of continental drift. Fixing positions of landmasses by astronomical means might provide such decisive data.

In spite of its broad foundation, I call the new idea a working hypothesis and I wish it to be looked at as such, at least until it has been possible to prove by astronomical positioning with undoubtable accuracy that the horizontal displacements continue to this day.

(Wegener/Jacoby, 1912b/2001: 53)

No such reliable and sufficiently precise measurements existed in 1912, and they were not to do so for another seventy years, about twenty-five years after mobilism was accepted (Gordon and Stein, 1992; Kearey, Klepeis, and Vine, 2009). There were, however, three measurements from Europe and Greenland taken thirty-seven years apart, and Wegener claimed that they showed that the distance between Greenland and Europe had increased *"by about 950 m in 84 years or 11 m/a"* (Wegener/Jacoby, 1912b/2001: 59; Wegener's emphasis). However, he did temper his claim.

It is well known that longitude determinations with the aid of the moon are inaccurate. The probable error of each single observation series may be estimated to be several, perhaps many hundred meters. The difference of 950 m that has accumulated with time, however, appears to me to be a bit too large to be merely an error, caused by systematic summing in one direction. It appears much more probable that this is real displacement of the continents of the order indicated.

(Wegener/Jacoby, 1912b/2001: 59)

Regarding measurements between Greenwich (England) and Cambridge (Massachusetts), he again showed restraint: they "seem to indicate an increase in distance of about 0.01 seconds of time or about 4 m/a" (Wegener/Jacoby, 1912b/2001: 59). Again he admitted the possibility of errors in lunar observations. Such data did not provide a difficulty-free solution. But they were a start. There is no question that Wegener put a premium on this type of geodetic "proof" of continental drift. Later he would argue most confidently that observations acquired in the interim established Greenland's westward displacement relative to Europe (§3.15).

He had criticized the contractionist's mechanisms, and apparently felt it was incumbent on him to replace it with a more plausible one. So he tried, but he was not able to pinpoint the forces responsible for the displacement of continents. Perhaps, he thought, horizontal displacements are not physically impossible if there are weak but enduring forces, and it was to these that he appealed.

All these observations suggest that sima is plastic, not very highly fluid, and that the sialic rind possesses a considerably greater strength without lacking plasticity altogether. Thus from here we have no reason for objecting to the possibility of very slow, but great horizontal displacements of the continents if there are forces acting invariably in the same direction during geological time spans.

(Wegener/Jacoby, 1912b/2001: 44)

Wegener suggested that sial has greater strength than sima which, he argued, would yield to very slow movements of sial brought about by the action of weak but steady forces. He likened sima to pitch, which is resistant to rapidly acting forces (it shatters when hammered), but yields to weaker long-term forces; under gravity it flows slowly down inclines. But he frankly admitted that he had no fully developed mechanism. To be sure, there were a number of possible long-term driving forces, and Wegener referred to several of them: flight from the poles, tidal forces, meridional rifting, precessional forces, polar wandering, or some combination of them. He also mentioned sub-crust flow and upwelling sima at the Mid-Atlantic Ridge, albeit not as connected ideas.[19] However, he bracketed his brief discussion of possibilities by wisely pointing out that any serious attempt to answer this question would be premature until the "reality and nature of the displacements" have been observed, and established empirically.

> The question, which forces caused the proposed horizontal displacement of the continents, is so manifest that I cannot quite bypass it although I consider it premature. It will be necessary first to exactly determine the reality and the nature of the displacements before we can hope to discover their causes. Here we can only attempt, above all, to prevent wrong ideas, rather than suggest something that can already claim to be correct.
>
> *(Wegener/Jacoby, 1912b/2001: 47)*

He was arguing that his lack of an answer was not an obstacle. Mechanism was simply an unsolved problem, and it would be premature to argue that the absence of one was fatal for his theory, just as it was premature to offer a solution. A full account of Wegener's 1922 mechanism is given in Chapter 4.

However, Wegener's fixist opponents refused to consider the mechanism question as an unsolved problem (Chapters 4 and 5). They wanted it answered before drift discussions could proceed. This cast of mind survived until the mid-1960s and impeded progress in the mobilism debate for half a century; it dominated thinking especially among many geophysicists. I am left wondering if Wegener would have been wiser to have said, "I really have no idea what caused the partitioning of sial and the lateral displacement of the fragments, but the observational support for it is good, and studies of it should continue, meanwhile any serious attempt to determine its cause must wait until more is known about Earth's interior and the available forces." Whether such a statement would have altered the tone of the discussion will never be known, but it might have helped his case, rendering it less vulnerable. Probably fixists still would have attacked his theory for its lack of an acceptable mechanism, but Wegener could then have pointed out that the discovery of radioactivity in rocks meant that contractionism had no viable cause of its own. He would have avoided much ridicule had he resisted the temptation to put forward, however tentatively, his own fragile solution as he later did.

Wegener argued that his notion of the fragmentation of continental crust and the displacement of fragments laterally to form the present continents solved more problems than either Suess' contractionism or American permanentism. He explicitly discussed the seemingly fatal geophysical difficulties facing Suess' theory, and dismissed permanentism because it could not account for disjunct biota. He enumerated the problems that only his theory could solve.

Wegener's training placed him in a particularly advantageous position to make the most of his original insight, and does much to explain the direction his arguments took in his initial presentation. Because of his former work in positional astronomy and thermodynamics of the atmosphere, Wegener was well versed on the literature of geodesy and familiar with the geophysical literature. As a result, it took him less than a month to appreciate fully the crucial importance of the geophysical objections to Suess' view. At the same time, however, Wegener had little, if any, formal training in geotectonics before his eureka moment in 1910 and his 1912 presentations, and therefore he had no vested interest in defending any particular geotectonic theory. This freed him from any constraints he might have had as a devotee of Suess. These fortunate circumstances of his intellectual background ensured that he carried little baggage. It meant that he began his geological and paleontological review focused strongly on only one basic question: could his fragmentation and drift story explain anything else besides the congruency of the opposite margins of the Atlantic? Once he began his literature review and recognized the difficulties facing Suess' theory, it was immediately obvious to him that continental partition and subsequent drift were preferable to sinking paleocontinents. Finally, no doubt prompted by his experience in astronomy, he quickly identified the "critical" test of his theory as the astronomical position-fixing of the continents over many decades. He was not correct (§3.15), the critical test did not come first from tracking current continental motions geodetically. It came from geophysical (paleomagnetic) work on the continents during the 1950s, which most Earth scientists at the time failed to recognize as decisive, and geophysical study of the ocean floors during the 1960s, which most did see as decisive. The geodetic studies of current continental motion that Wegener dreamt about have, during the past two decades, finally begun to map continental drift occurring today. Because Wegener was not an outsider in the fields of geodesy and geophysics, he voiced authoritative objections to Suess' theory; because he was an outsider to geology and geotectonics in particular, he was not blinded by the earlier success of Suess' grand synthesis.

2.9 Taylor and his career

Taylor was born in Fort Wayne, Indiana, in 1860, twenty years before Wegener.[20] He was the only child of Fanny Wright and Robert Stewart Taylor. His father became a wealthy attorney, and an authority on patent law, and as a result of serving on the Mississippi River Commission for a third of a century, he became interested in geology of the mid-western United States. Taylor graduated from Fort Wayne High School in 1881. Ill health delayed his entrance to Harvard University for a year, and he did not complete a degree, again because of ill health. He took courses for two years, concentrating on astronomy and geology, which became his lifelong professional interests, and from which he developed his theory of continental drift creep or spreading. Leaving Harvard, he traveled to the upper Great Lakes region to improve his health and pursue geological interests. He began publishing papers as early as 1892, and eventually became expert on the glacial and post-glacial history of the region. Almost eighty percent of his approximately ninety publications are about the geology of the Great Lakes. He married Minetta Amelia Ketchum of Mackinac Island in 1899. She became his lifelong companion at home and on field trips, taking care of transportation, be it driving a team of horses or a car. Taylor's father covered his son's field and publishing expenses until 1900 when Taylor secured a job with the USGS. His USGS assignments took him to New England and back to the Great Lakes region. In 1915 he co-authored a monograph with Frank Leverett entitled *The Pleistocene of Indiana and Michigan and the History of the Great Lakes*; Leverett later became Director of the USGS. Taylor continued to work on regional questions concerned with Quaternary glaciation, and became a respected glaciologist. He was a fellow of the GSA, and published papers in major journals. He died in 1938, outliving Wegener by eight years, but not long enough to see continental drift accepted.

2.10 The emergence of Taylor's theory of creep and horizontal displacement

How Taylor's mobilistic theory of drift originated is obscured by the manner in which he presented and defended it. It grew out of his interest in astronomy, in particular, cosmogony. In an essay entitled "An endogenous planetary system," which was privately printed in 1898, he developed a hypothesis for the origin of the planets and their moons, and outlined in three pages his undeveloped ideas on the evolution of continents. These were extended ten years later at the 1908 Annual Meeting of the Geological Society of America when he also discussed some of his cosmogony. In 1925, he recalled:

At the 1908 meeting of the Geological Society of America referred to, the writer suggested what must at that time have seemed to be a very novel explanation. He expressed his belief that none of the current hypotheses of the origin of our great satellite, the Moon, is acceptable; but that another alternative remains open, namely: that the Moon may have been captured directly out of space . . . This suggestion caused only merriment in the meeting and apparently no serious thinking.

(Taylor, 1925: 16)

His first full paper (47 pages) on geotectonics was published in 1910. Submitted to the *Bulletin of the Geological Society of America* in October 1909, it contained a detailed account of his theory and its geological consequences, but without mentioning his cosmogony. In this extended essay, he defended and used Suess' method of trend-line analysis, and gave an explanation for Tertiary mountain belts, arguing that it was better than any contractionist version.

What are the reasons for these differences between his 1898 and 1910 versions, especially the absence of his cosmogony in the latter? This, I think, is easily explained. He wisely decided that if he wanted his paper to be accepted for publication or, at least, cause something other than merriment among its readers, he had to play it down and concentrate on geology. Moreover, his cosmogonical theory, even with its hypothesis of Earth's capture of the Moon, was distinct from his theory of continental drift; his theory of creep and drift could have been correct regardless of his explanation of creep and drift being a result of Earth's capture of the Moon.

How did Taylor come to develop his theory and his detailed explanation of the worldwide belt of Tertiary mountains? What was it between 1898 and the fall of 1909 when he submitted his "Bearing of the Tertiary Mountain Belt on the origin of the Earth's plan," that enabled Taylor to turn a simple notion into a theory of continental drift? It has to be his reading the English translation of Suess' *Das Antlitz der Erde.* [21] As Taylor put it in 1925:

Studies pursued for nearly twenty-five years previously gave the writer a keen interest in certain parts of Suess' *Face of the Earth*, which was read for the first time in 1908–9. Suess' analysis of the plan of the mountain ranges of Asia, and especially his explanation of the arcuate forms of the Tertic fold-mountains of the peripheral belt, seemed very attractive. His interpretation of the direction of crustal movements fitted exactly into the writer's scheme, but his explanation of the causes of those movements differed widely.

(Taylor, 1925: 15)

From reading Suess, especially his trend-line analysis, he identified the origin of the worldwide belt of Tertiary mountain ranges as a major geotectonic problem, which he thought his idea of drift could solve. I believe he correctly viewed his theory of continental drift as an extension of Suess' ideas of continental spreading.

2.11 Taylor's cosmogony and his notion of continental drift, 1898

This, then, is what I mean by an Endogenous Planetary System; one that grows by accessions at its center and which expands as it grows.

(Taylor, 1898: 12)

According to Frank Leverett (Leverett, 1939: 193), Taylor considered his 1910 paper as his first "important paper" on continental drift. He thought both his 1898 work and the 1903 privately printed volume, *The Planetary System*, incomplete. The latter was printed, most likely, at the insistence of Taylor's father (Leverett, 1939: 192). Taylor probably considered his cosmogony and his pre-1900 notion of continental drift to be no more than speculation.

His cosmogony was daring. It ran counter to Laplace's nebular hypothesis, which had been defended by Elie de Beaumont and Lord Kelvin, and to the fission theory of the origin of the Moon, suggested by G.H. Darwin and augmented by Osmond Fisher.[22] He thought his cosmogony preferable to theirs. Taylor's theory had, as its central element, the occasional transformation of comets into asteroids, planets, or planetary satellites. He believed that Neptune, the farthest planet then known from the Sun, was the oldest of the still-existing planets. Originally a comet, Neptune began its life as a planet much nearer the Sun in the orbit presently occupied by Mercury. The Sun subsequently captured other comets. Every new addition pushed out existing planets into larger orbits, and the newest planet then proceeded to occupy a position roughly like that occupied by Mercury. After the four outer planets, Neptune, Uranus, Saturn, and Jupiter, had formed, our solar system experienced what Taylor believed "to be by far the most wonderful event in the whole history of the solar system," namely, the invasion of "not one comet, but a horde, a cloud, a veritable storm of comets," which became asteroids (Taylor, 1898: 16). Like their predecessors, they took up planetary positions close to the Sun. Four further comets were captured, and became Mars, Earth, Venus, and Mercury. Taylor supposed that other planets may have existed before Mercury was captured by the Sun, but such planets would have escaped from the solar system once they had been pushed by incoming planets beyond the region where the Sun's gravitational attraction was able to hold them.

What about planetary satellites? "The assumption made above as to the capture of comets and their transformation into planets applies with appropriate modifications to their capture and transformation into satellites" (Taylor, 1898: 12). The Moon, according to Taylor, was a comet before being captured by Earth.

Turning to the geological consequences of Earth's capture of the Moon, he devoted no more than a page to his embryonic idea of what would now be called continental drift. With Earth's capture of the Moon, the gravitational forces between them caused Earth, at some unspecified later date, to increase its oblateness, and "produced a rearrangement of the relative positions of the land and sea" (Taylor, 1898: 29). Continental masses were pulled towards lower latitudes where they thickened, and folded mountain belts formed especially in middle latitudes. Taylor viewed his 1898 version of continental drift as little more than an interesting and possibly important geological consequence of his cosmogonical theory. By 1898, he had worked out few consequences of his notion of continental drift; it was not yet a theory with associated tests.

2.12 Taylor's 1910 presentation and defense of his creep and drift theory

The argument presented in this paper rests at last on the truth of Suess's interpretation of the mountain plan of Asia. The principles which he worked out there have been applied without important modification to the other continents, and the conclusions reached in this way appear to accord very closely with suggestions made by Suess himself in his later writings.

(Taylor, 1910: 226)

Deeply influenced by the first three volumes of Suess' *Das Antlitz der Erde*, Taylor turned his vague ideas of 1898 of "rearrangement of the relative positions of land and sea" into a theory of what would now be recognized as continental drift. Taylor did not use the term "drift" in his 1910 article but preferred the term "creep." However, he wrote of "the horizontal factor of displacement" in his work of 1898. He began with an appeal to Suess' trend-line analysis and his general description of the Tertiary belt of mountain ranges and island arcs along the southern and eastern borders of Asia. Trend-line analysis, he argued, agreeing with Suess, revealed the direction and scope of the forces required to form Asia's Tertiary belts; they had been formed through massive crustal folding toward the south and east produced by the southward and southeastward creep of Asia.

He then parted company with Suess and other contractionists, raising these difficulties (RS2):

All forms of the contraction hypothesis meet with two insurmountable difficulties with reference to the Tertiary period of mountain making. They fail to explain in a satisfactory way the distribution of Tertiary mountain ranges upon the Earth's surface, and they do not explain how so great a period of mountain-making could have occurred in so recent time. If due to contraction arising from cooling, it is necessary to suppose a very long period of accumulation and storage of mountain-making force before the beginning of the folding movement. The amount of crustal movement which occurred during the Tertiary period seems to be far in excess of the most that can be attributed to cooling and shrinking since the time of the Permo-Mesozoic (Appalachian) folding, even on the most liberal estimate. It is scarcely credible that any considerable mountain-making force derived from cooling before the time of the Permo-Mesozoic folding could have survived that event so as to be an important element in the Tertiary folding.

(Taylor, 1910: 225)

Suess and his followers, Taylor argued, had no explanation for the form and distribution of the Tertiary folded belts, and the length of time between them and previous mountain-building epochs was too short for contraction to build up sufficiently strong tangential stresses to cause the formation of folded belts as massive as those of the Tertiary. Something more powerful was required.

Taylor proposed very extensive continental creep leading to continental drift as an explanation of the origin of folded mountain Tertiary belts. Drawing on his field-work and understanding of glacial migration, he compared continental creep to the flow of glaciers. He declared that mountain belts and island arcs form through compression at the leading edge of a drifting continent. This contrasts with Wegener's explanation of island arcs as slivers broken off the trailing edge of drifting continents. Taylor regarded island arcs as partially submerged mountain ranges, which formed in the same way as fold mountain belts. He argued (1910: 181) that the degree of crustal deformation was primarily a function of the length and mass of the drifting continent.

Taylor used his knowledge of glaciology, comparing the movements of ice-sheets and crustal-sheets, remarking, "Suess himself suggests this comparison in one of his later works" (Taylor, 1910: 193). But Suess had underestimated the amount of crustal migration required and had been wrong to appeal to global contraction, which is far from sufficient. Nothing short of wholesale lateral motion of continents or what is now called continental drift would produce Tertiary fold belts.

Taylor (1910: 179–180) counted his explanation of the worldwide Tertiary fold belts as the centerpiece of his theory. He thought that the problem was of fundamental importance and that his explanation was the best. As explained below, he extended its problem-solving effectiveness by arguing that his theory provided explanations of the origin of the Atlantic Ocean and many other features (RS1). As Wegener was to do, but in far less detail, he cited pre-Tertiary geological disjuncts between Europe and North America, and argued that his theory explained the congruency of their margins. He paid special attention to the congruency between the margins of Greenland and North America.

Unlike Wegener, Taylor did not address the vast problems of inter-continental disjuncts, biotic, paleontologic and paleoclimatic, and his overall defense of continental drift was much less complete. Geologists concerned with pre-Tertiary Earth history found less of interest and value in Taylor's theory than in Wegener's. However, Taylor's account of Tertiary orogenesis was far more extensive than Wegener's. Earth scientists concerned with Tertiary events may have found Taylor's theory more promising than Wegener's. But this alone would not have encouraged geologists to work on Taylor's theory unless they agreed with his negative assessment of contractionism.

Foreshadowing Carey (1955, 1958) (II, §6.7; II, §6.14), Taylor used the trend-lines of the worldwide belt of Tertiary folds as his guide for determining when and which continents drifted. Eurasia and North America were originally united with Greenland.

Eurasia extended from Asia Minor to Canada. It included all of Europe with the Atlas ranges and Canary Islands of northwestern Africa; all of Asia excepting the peninsulas of Arabia and India, and in addition all that part of North America which lies west of the mountain angle of Alaska.

(Taylor, 1910: 200)

Eurasia broke away and drifted equatorward.

It seems certain that all this vast crustal sheet . . . in the Tertiary age . . . moved in a southerly direction substantially as a unit, and that the entire belt of Tertiary fold-mountains which forms its southern periphery was made at that time and by that movement.

(Taylor, 1910: 200).

He proposed that Eurasia eventually collided with Africa and India giving rise to the Himalayas, mountain ranges in northwest Africa and the northward bending of the

Alps. In those places where Eurasia met with no other continental masses, it continued to creep southward forming lobe-shaped structures such as the Malaysian Peninsula and Iran.[23] Eurasia also crept eastward, causing the vast chain of island arcs all along the Pacific coast of Asia, and eastward to the Aleutian arc.

When the Asiatic character of the Aleutian arc is recognized it becomes at once apparent that the crustal sheet of Eurasia is not limited on the east by a line through the Bering Strait dividing this arc and its backland, but includes the whole arc and the whole of its backland. Thus, Asiatic character is carried eastward to the heart of the Alaskan Mountains, where the curve of the Aleutian arc meets the Cordilleran ranges of North America in a sharp angle.

(Taylor, 1910: 199)

Taylor extended his theory to account for the origin of ocean trenches. After noting that they are often located on the ocean side of island arcs, he suggested that they

are of the nature of sunken or depressed forelands, and are apparently due to the stupendous weight and pressure of the adjacent ranges. These arcs are no doubt more or less overthrust upon the ocean floor, and the troughs are probably due in part to the elastic yielding and perhaps in part also to plastic flow.

(Taylor, 1910: 201)

Turning to North America and using the trendline of the Cordillera, the western ranges bordering the Pacific Ocean, he postulated southerly and westwardly creeping of North America to account for their formation. Next Taylor extended his theory to explain the formation of the shallow seas surrounding Greenland and the remarkable congruency between its western coast and the east coasts of the Canadian Arctic Islands and Labrador. He argued that the seas separating Eurasia and Greenland were larger than those between North America and Greenland because, relative to Greenland, Eurasia had drifted further than North America.

If, then, by reversing the process, North America be in imagination pressed back northeastward to a complete union with Greenland, it is evident that the Labrador Sea, Davis Strait, and Baffin Bay would be closed and obliterated entirely, and that Grant Land, Grinnell Land, and Ellesmere Island would be thrust northeastward past the north end of Greenland along the line of the rift – that is, along the line of Smith Sound and Kennedy and Robeson channels produced.

(Taylor, 1910: 208)

Taylor concluded his discussion of the "creeping" of Eurasia and North America by stressing the common equatorward movement of both land masses, and offered the diagrammatic summary shown in Figure 2.5.

Taylor also proposed Tertiary equatorward movements in the Southern Hemisphere. Australia and South America were originally united with Antarctica. Australia and South America broke away, moving to the northeast and northwest, respectively. Antarctica remained in its original position. Being less massive than Eurasia, Australia has smaller peripheral mountains and island arcs, which included New Guinea,

Figure 2.5 Taylor's Figure 6 (1910: 209) showing the movement of Eurasia and North America relative to Greenland. The light arrows display the direction of crustal creep. The heavy arrows radiating from the pole and Greenland show the relative distances that Eurasia and North America have moved toward lower latitudes. The broken line north of Eurasia marks the margin of its continental shelf.

New Zealand, and New Caledonia. Taylor attributed the complex nature of the East-Indian Archipelago to southeastward movement of Asia and northeastward movement of Australia. South America drifted to the northwest, creating the Andean Cordillera along its Pacific coast.

Shifting his attention to the Mid-Atlantic Ridge, Taylor suggested, "It is apparently a sort of horst ridge – a residual ridge along a line of parting or rifting – the Earth-crust having moved away from it on both sides" as Africa moved eastward during the pre-Mesozoic and South America drifted westward during the Tertiary (Taylor, 1910: 217). Taylor could not consistently postulate eastward shifting of Africa, for there is no Tertiary fold belt along its eastern coast. He needed Africa in its present position before the Tertiary because it served as a backstop for Eurasia, whose approach caused the Atlas Mountains along its northern coast.

As a partial answer to the mechanism problem, Taylor invoked the accumulation of tidal forces and Earth's change from a sphere to an oblate spheroid brought about by capture of the Moon. However, in his 1910 paper Taylor made no mention of the Moon's birth or his overall cosmogony. Instead, he gave a hint of things to come in the final sentence by stressing the predominance of equatorial movements of continents, relating them to the oblateness of the Earth, and suggesting that both arose from tidal forces.

For a change in the degree of oblateness . . . in deformations of the lithosphere, one is inclined to reject all internal causes and to look to some form of tidal force as the only possible agency.

(Taylor, 1910: 226)

This did not explain why drift did not occur until the Tertiary. Taylor left the mechanism question in the background rather than propose as he had in 1898 something as bizarre and likely unacceptable as capture of the Moon. He seems to have sensed that although capture of the Moon, if true, removed the difficulty that drift, in his scheme of things, did not occur until the Tertiary, to have included it would have meant repeating this cosmogony in its entirety – a general attack on contractionism and continental drift were novelty enough in a single paper.

Taylor also addressed, although in a minimal way, the issue of the location and nature of the surface over which continental sheets creep across Earth's solid substratum. He proposed a plane "situated beneath the Earth's crust just within the zone of rock flowage" over which crustal sheets of continental material are able to slide (Taylor, 1910: 224). He gave no estimate of the thickness of his crustal (continental) sheets. The seismological definition of "crust" was yet to come. Taylor, like Suess, but unlike Wegener, rejected the idea of isostatic equilibrium (Taylor, 1926). Nor did Taylor discuss the composition of his crustal sheets, although he presumably had in mind Suess' sial and sima. His concerns were primarily with tectonics and mountain-building.[24]

2.13 Wegener and Taylor: the independence of their inspiration

Taylor developed his theory of continental drift before Wegener had begun to think about it. Wegener never denied that. But Wegener continually maintained that he had the idea of continental drift independently of Taylor or anyone else. However, in December of 1931, about a year after Wegener's death, Taylor wrote a letter to the author of an article on continental drift, which appeared in the November 1931 issue of *Popular Science Monthly*. This November 1931 article made no mention of Taylor, was filled with historical inaccuracies, and made much of Wegener and his Arctic explorations. Understandably, Taylor was upset and, I imagine, already bitter about the lack of attention his work attracted compared to Wegener's. In his December 1931 response to this article, Taylor argued that Wegener's original inspiration for continental drift came from reading his own paper.

But let us go back to the beginning, or as near to it as need be just now, and note the order in which the works bearing on the subject appeared and Wegener's and my own relations to them. (1) My first paper "Bearing of the Tertiary Mountain Belt on the Origin of the Earth's Plan" was read before the Geological Society of America at Baltimore, December 29, 1908. . . There was an unusual delay in printing the *Bulletin* of that meeting, so my paper did not appear in print until June, 1910. The only person to take any notice of my paper was Alfred Wegener, a young (born, Dec. 1, 1880) professor of meteorology in the University of Hamburg, Germany. (2) In the spring of 1911, Wegener published a very brief note reviewing my paper, partly in terms of approval, but putting forth also some suggestions of his own. The page containing this note was sent to me and

I had it for several years, but it now appears to have been mislaid or lost; I have not been able to find it for three or four years. It was in fine type and was about twenty or twenty-five lines long. I am uncertain whether it was in *Petermanns Geogr. Mitteilungen*, early in 1911, or in some other journal. More than ten years ago I made a careful search in the German scientific journals in the library of the University of Michigan at Ann Arbor to see if Wegener had ever published anything before 1911 on this subject, and found nothing. From these facts, it seems practically certain to me that Wegener got his first inspiration of this subject from my first paper (1910). If you will consult his later papers and books up to seven or eight years ago you will see that he makes frank recognition of my earlier work with dates of publication. His comments about my views are not always fair or pleasing, but that might be expected. He did not follow my lead in some important respects and disagreed with me in many details, but that too is to be expected. (3) Wegener's first little notice (1911) of my paper was not worth anything as an expression of his own views. All of the authorities agree that his first publication of his theory was in 1912 following his first lectures on this subject in January of that year at Frankfurt-on-Main and at Hamburg.

(Taylor, 1931: 2–3)

Wegener never denied that he had read Taylor's 1910 paper. In fact, he referenced every one of his predecessors except Antonio Snider, an American who, in 1858, linked the separation of the new and old world with the Deluge.[25] However, Wegener maintained that he first thought of what is now called continental drift when he noticed the congruency of opposed continental margins while examining a bathymetric map of the Atlantic Ocean in the closing days of 1910 (1929/1966: 1). Moreover, Wegener, in 1912, mentioned Taylor's 1910 paper and explained its significance in the development of his own thinking.

. . . a recent contribution from F. B. Taylor can be considered as a forerunner of the present work, while they arose in complete independence from one another . . . It could not, as stated, give any stimulation for the present work as I came upon it too late.

(Wegener, 1912b: 185)

I think Taylor was incorrect; Wegener did not get the idea of continental drift by reading Taylor. In his December letter, Taylor claimed that Wegener had obtained his original inspiration from reading his 1910 paper, basing his claim on the existence of a note, Taylor contended, Wegener had written in the spring of 1911, in which he (Wegener) had spoken about Taylor's work. But Taylor could not find his copy of this note, and failed to locate it in a literature search at Ann Arbor.[26] Taylor felt strongly about the matter. He knew that Wegener had explicitly stated that his views "arose in complete independence" of Taylor's own, that he did not get "any stimulation" for his work from Taylor because he (Wegener) "came upon it too late," and that his inspiration came from noticing the congruencies of opposed continental margins. Taylor seems to have thought that Wegener either had lied or, more charitably, had forgotten how he had begun thinking about continental drift. I believe that Taylor's case against Wegener has about as much substance as his missing copy of Wegener's alleged note of 1911.

Although there is no extant copy, suppose Wegener actually had written a note regarding Taylor's work in spring 1911 and that it had been published. Wegener's eureka moment occurred around Christmas of 1910, and he already had sketched out an initial defense of his theory – a defense very different both in its origins and development from Taylor's – by January 1911. Taylor did not claim that Wegener said in his spring note that he got the idea of continental drift from reading his (Taylor's) work, and he did tell us that the note "was not worth anything as an expression of his [Wegener's] views." What could have been the purpose of the alleged note, if not for Wegener to lay out views? But why would Wegener submit it for publication, and why would anyone publish what had already been published? Perhaps Wegener got his idea about how mountain ranges form from Taylor, although Wegener did not accept Taylor's explanation of the formation of island arcs, which Taylor linked closely to mountain ranges. Perhaps Wegener benefited from Taylor's analysis of the similarities between Greenland's margins and those of surrounding lands. But this is not to say that he got his initial inspiration from reading Taylor. In addition, it would have been rather silly for Wegener to publish such a note in 1911 saying that his original inspiration for drift came from Taylor, and then deny it in his 1912 papers. It was at about this time, I suspect, that Wegener began reading journals such as the *Bulletin of the Geological Society of America* and he came across Taylor's paper, or perhaps one of Wegener's colleagues told him about it.[27]

Taylor had become despondent about the almost total lack of attention paid to his theory, was in poor health and at the end of his career. In his December 1931 letter to *Popular Science Monthly*, he was extremely upset:

I am much averse to blowing my own horn like this, but when you take account of all the references given above, I think you will agree that I am justified in objecting to untrue statements and in trying to hold onto what little I have gained by long years of work. If you wish to be on safe ground in writing on this theme please look up the references which I have given above. Your positive statements about Wegener quoted at the beginning of this letter are absolutely wrong and untrue. It seems to me that you owe me, but still more the public whom you have misinformed by what you have said, some sort of correction or reparation. I am getting old and am in feeble health. I cannot hope to accomplish much more that is worthwhile. I pray that in the future our younger men like yourself will have a more earnest regard for justice and truth even in their so-called popular writings. When I began this letter, I had no idea of running on to such great length, but while not exhaustive, I hope you may find this partial list of references on this interesting theme to be of some value to you in the future.

(Taylor, 1931: 5)

It is likely that Taylor, rightly very upset by the November 1931 *Popular Science Monthly* article, got carried away in the heat of the moment; the fault lay not with Wegener but with that article's author.

2.14 Wegener and Taylor compared

A comparison of Wegener's and Taylor's backgrounds, and theories, I think at least partly explains the widespread attention received by Wegener's theory and the almost total eclipse of Taylor's. I consider first their educational backgrounds, their fieldwork and their dependence on the fieldwork of others, and then describe the various stages in the development of their theories.

Taylor took courses for two years in astronomy and geology. Before he developed his 1898 version of continental drift, which was only a very rough sketch of his later, 1908–10 theory, his fieldwork in geology dealt primarily with the recent glacial history of the upper Great Lakes region. Thus, although Taylor had, unlike Wegener, experience in field geology before arriving at his theory, it was restricted to Quaternary deposits of a particular region. He was interested in glaciation and the geological effects of migrating glaciers. Taylor's interest in astronomy was cosmogonical and led to the development of his daring hypothesis of the endogenous origin of the solar system. I do not know whether he became interested in Tertiary mountain building while in college or after developing his cosmogonical theory. Tertiary mountain belts and Quaternary glaciation in the Great Lakes region were his only active geological interests. Taylor's cosmogonical interests were apparently central to his development of his drift theory, which apparently arose from his realization that Earth's capture of the Moon would increase tidal forces and have spectacular geological consequences.

Wegener's Ph.D. was in computational astronomy, and he became a meteorologist prior to the development of his theory. He did extensive meteorological fieldwork while in Greenland and elsewhere, work which included the study of glaciers. At the time he devised his theory, he was studying intensively the thermodynamics of the atmosphere, and was much familiar with current geophysical literature. Wegener, a geophysicist of the atmosphere, quickly informed himself about new developments in geophysics of the solid Earth, and used geophysical arguments against landbridges. Perhaps Wegener's background in meteorology was partially responsible for his quick appreciation of the central importance of Permo-Carboniferous glaciation, that its decidedly odd distribution required some radical global explanation, and was a central issue. Wegener's work in astronomy may partially explain his emphasis on the use of lunar fixing as a means for determining whether Greenland had drifted westward, and likely affected his defense of drift.

Both Wegener and Taylor engaged in fieldwork that had some bearing on their theories. Both compared the lateral motion of continents to ice movement. Wegener, in light of his geophysical orientation and awareness and acceptance of the principle of isostasy, drew an analogy between the floating of the sialic continents atop the denser sima of the upper substratum and seafloor, and ice floes floating in water; he stressed the idea that both systems were in equilibrium, brought about by vertical motions responding to density differences. Taylor, the tectonicist, focused on the horizontal movement of glaciers and how they react to impediments; just as migrating glaciers thicken and fold when meeting large obstacles, so sheets of continental sial form mountains when they collide.

Overwhelmingly both obtained the data in support of their theories from the work of others. Both gained much from reading Suess. Taylor's acknowledged debt to Suess was great. He used Suess' trend-line analysis, and adopted most of his geological descriptions. Taylor disagreed with Suess, arguing that contraction could not explain the formation of the Tertiary mountain belts, but nevertheless he seems to have regarded his theory as an outgrowth of Suess' work rather than new and radical.

Although beholden to Suess, Wegener, because of the importance he placed on geophysics and the new discoveries of isostasy and radioactivity, erected his theory in direct opposition to Suess' contractionism. He was proposing to reorganize solid Earth science over a broad front. Wegener had a much greater appreciation of geophysics than Taylor and Suess. Ironically, it was Wegener, the European, not Taylor, who appreciated the fundamental American work on isostasy. But it was Taylor, not Wegener, who became an Alpine geologist without even crossing the Atlantic. Wegener, unlike Taylor, recognized the novelty and revolutionary character of continental drift, and combined in a novel way the best of geophysics, biogeography, and geology that he could find in the literature.

I now look at the stages each went through in developing their theories. Wegener was swift to expand his early thoughts about continental drift, synthesizing much that was known about Earth into a full global hypothesis which he considered sufficiently plausible and detailed to put before the public. He conceived of the idea of continental drift in late December of 1910 and within a month he was prepared to argue that drift was geophysically preferable to landbridges to explain intercontinental disjuncts. Throughout the remainder of the year, he assembled from the literature abundant evidence in favor of his theory from paleontology, biogeography, paleoclimatology, geology, and tectonics, and began to speculate about the mechanism. Because of his background in geophysics and lack of geology, it is natural that he should have developed the geophysical arguments against contractionism before fleshing out the geological support for drift. It is also evident from his January 1911 letter to Köppen that, although he recognized that his idea was revolutionary, he believed it would soon be generally accepted as correct. This indicated a certain naiveté on Wegener's part; his underestimation of the resistance of others to new ideas and his sanguine belief in their willingness to accept criticism and their flexibility to adapt to it, perhaps, explain in part his eagerness to put before the public his ideas on such a broad topic within a year of their birth. He believed that he was on to something of profound importance; he was impatient to get it out into the public arena for discussion and no doubt eager to stake his claim.

In contrast, Taylor took almost ten years to develop the geological implications of his 1898 idea of continental drift before presenting them before the GSA. Why did it take him so long? He was occupied with fieldwork and plagued by ill health, but it is possible that he lacked inner confidence in his ideas and was also uncertain just what problems they could solve. In 1898, Taylor had a geological theory that offered a hint of a solution to the general problem of mountain building. The detailed solutions and field data came from reading Suess and, to a lesser extent, thinking of drifting continents as flowing glaciers. With Suess as his guide, he finally, after a decade, had enough to present his idea to geologists. Also, he conceived of continental drift through working on his theory of the endogenous development of the solar system, so, unlike Wegener, he had from the start, or so he thought, a solution to the mechanism problem. With Taylor there seems to have been no eureka moment, no sense of excitement or urgency to complete the task.

Taylor's theory received almost no attention from geologists, regardless of their specialty and nationality.[28] Wegener's theory was discussed, often feverishly, by geophysicists and geologists working in many fields and countries. As I later show, this was especially so after his remarkable maps reconstructing past continental positions were published in 1922. Wegener wrote books on the topic; Taylor, only a few articles. Wegener had students and a position in academe; Taylor worked for the USGS and had no students. No doubt these partially explain their different receptions, but they could be interpreted as effects rather than causes. A more likely cause lies in the explanatory scope and content of their theories. Taylor's public version of his theory primarily offered an explanation for the formation of Tertiary mountain ranges, especially in Europe and Asia. His answer was vintage Suess blended with his own drift theory. Under these circumstances, hardly anyone was interested in such an account by an American geologist who had never studied the Alps. If a European tectonicist had presented a theory of continental drift emphasizing the Alps, then perhaps others would have listened – indeed, the Swiss geologist Émile Argand (1879–1940) did just that in 1922 (Argand, 1924/1977), and his views were soon adopted by others working on the western Alps (§8.7). At the time, North American tectonicists were concerned mainly with the pre-Tertiary Appalachians, but Taylor hardly mentioned them. There was Taylor's cosmogony, but it was absent from his public version of drift theory. In addition, Taylor had little to say about geophysical matters: not one word, for example, about isostasy, and North American geophysicists consequently expressed little interest in his theory.

Wegener's theory as it came to be fully developed in the early 1920s eventually confronted much that was then known about Earth. It addressed fundamental geophysical and geological issues, and caught the attention of many Earth scientists; they may not have agreed with him but they could not readily entirely ignore it. His theory offered solutions to problems that were of intense interest to geophysicists, paleoclimatologists, paleontologists, geologists, geodesists, and those concerned with the origin of mountains. Whether they became interested was another matter. What he had to say was, in principle at least, relevant to their work; because he discussed the biotic and geologic disjuncts among landmasses in both hemispheres, especially the Permo-Carboniferous glaciation, his work by the mid-1920s, was relevant to the work of very many geologists worldwide. The different explanatory scope and content of their theories goes a long way in explaining why Taylor's was all but ignored and Wegener's caught the attention of many.

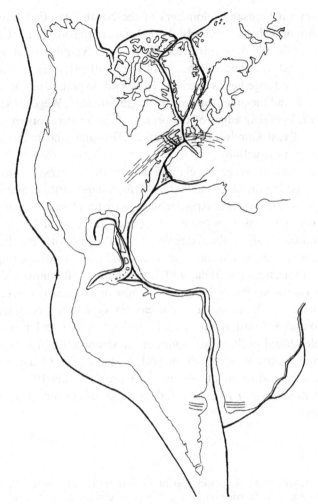

Figure 2.6 Wegener's Figure 17 (1915: 68). The continental margins are represented as heavy lines. The dashed lines represent the once connected Paleozoic orogenic belts of the southern Cape Province of South Africa and the eastern Buenos Aires Province of Argentina, and the more northern of eastern North America and western Europe.

2.15 Evolution of Wegener's theory, 1912–1922

The 1912 papers contained no tectonic or paleogeographic maps, maps such as the ones visible at www.cambridge.org/frankel1/. Wegener (1915, 1920, 1922) developed these maps progressively, some in collaboration with Köppen (1924), and they became central icons of Wegener's theory of continental drift.

The 1915 edition of his *Die Entstehung der Kontinente und Ozeane* gives a tectonic reconstruction of the peri-Atlantic continents into a single landmass (Figure 2.6). The Paleozoic orogenic belts of eastern North America and Western Europe, now separated by several thousand kilometers of the North Atlantic Ocean, he reconstructed as continuous belts. Similarly the Paleozoic orogenic belts of the southern Cape Province of South Africa and the Sierra de la Ventana of Argentina, now separated by several thousand kilometers of the South Atlantic, are on Wegener's reconstruction placed together as one belt. These two tie-points, one at the south end of the Atlantic rift and the other towards the north provided, Wegener argued, strong evidence that the opposing edges of the present Atlantic were formerly juxtaposed. The continents of east Gondwana (Antarctica, Australia, and India) were not yet included in this reconstruction.

In the 1920 edition Wegener elaborated on these trans-Atlantic structural comparisons, matching the Caledonides of Newfoundland, Britain, and Scandinavia, and immediately to the south the Appalachians of eastern United States and Canada with the contemporaneous Hercynides of western and central Europe. He added further comparative details in the Argentina–Cape Province orogenic belt and made extensive structural comparisons between Brazil and western Africa. Importantly he now included Antarctica, Australia, and India, uniting all continents in the Late Paleozoic into a single landmass, Pangea.[29] Because of the occurrences of glacial beds of Late Paleozoic age he clustered these eastern Gondwana continents together adjacent to southern South America and southern Africa and placed the south geographic pole central to them, just south of the southern tip of Africa. Given this pole he constructed latitude and longitude grids for the whole of Pangea. (See Marvin (1973: 74) for a reproduction of Wegener's figure from his 1920 edition of *Die Entstehung der Kontinente und Ozeane*, and for fuller discussion of his matching of geologic disjuncts.)

Notes

1 For an extensive treatment of geotectonics at the end of the nineteenth century see Greene (1982). My purpose is to provide only enough background for understanding how Wegener and Taylor came to develop their theories. Also see Şengör (1979/1982a, 1982b). Şengör discusses Suess' views and relates them to those of Wegener and Argand. He argues that Suess should be viewed as a predecessor to mobilism instead of fixism. I want to thank Celal Şengör for suggesting several improvements to this short synopsis of Suess.

2 To get a sense of the reverence with which Suess was held by continental European geologists, Şengör, who thinks that Suess is the greatest geologist of all time, presents the following anecdote:

Suess had such a spell-binding effect on his contemporaries that upon his return home from the 1903 International Geological Congress in Vienna, famous for its heated debates on the nappe theory, the well-known French geologist Charles Barrois' only response to those asking about his impressions of the congress was "j'ai vu Suess!" (I saw Suess!).

(Şengör, 2003: endnote 384, p. 328)

3 The following presentation is not intended to be a complete summary of his overall synthesis; it reveals little about the evolution of his thought, and says nothing about how he defended his position. Its only purpose is to provide enough background for understanding how the drift controversy began.

4 Şengör (1982b) notes that Suess claimed that there were a few exceptions such as the Caledonides (of Great Britain) and the Andes to the general rule that mountains formed asymmetrically.

5 There has been dispute about whether Gondwana-Land is redundant. *Gondwanaland* is supposedly redundant because *Gondwana* already means "land of the Gonds." Şengör (1983, 1991) argues that there is no redundancy. He (1991: 287–288) agrees that *Gondwána* means, in Dravidian, "the land of the Gonds," but argues that it "denotes a historical geographic/political unit that was occupied by the Gonds and formerly located in a particular region of what is now India." Suess, however, introduced the term Gondwana-Land to designate a hypothetical supercontinent once located in the Southern Hemisphere, where "Land" means, like the German "festland," mainland or continent that has no political dimension to its meaning. When discussing Suess, I shall, acknowledging Şengör's point, use Gondwana-Land when referring to the ancient supercontinent. However, I shall use Gondwana when discussing the work of other Earth scientists because it is short and simple and has become the accepted name of the supercontinent.

6 This brief portrait of Wegener contains no original research. I have drawn material from some of the biographical accounts on Wegener. See: Benndorf (1931), Georgi (1962), Hallam (1983), Greene (1984), and Schwarzbach (1986). For those interested in obtaining more details of Wegener's life, see, in addition to the above, E. Wegener (1960).

7 Quoted in Schwarzbach (1986: 55).

8 During his fourth and fatal expedition to Greenland he wrote in his journal, "when I dream, I actually dream of something completely different – the Adriatic, and vacation trips with no mountain climbing or other semi-polar adventures" (Schwarzbach, 1986: 40).

9 Wegener had an "artist's eye" for photography. Witness, for example, some of his photographs taken during his final expedition to Greenland reproduced in Else Wegener's (1939) *Greenland Journey*. The abstract quality of these photographs anticipates Edward Weston's work.

10 Greene (1998) has suggested that *Die Enstehung der Mondkrater* shows the connections within Wegener's work, and should not be looked on as superfluous to Wegener's other interests since it related to his work in meteorology and continental drift. The Moon also came into play in his attempt to use geodetic measurements to test drift theory. His suggestion that lunar craters are impact rather than volcanic craters sprung, at least in part, from his war experience of impact craters caused by heavy artillery and the bird's-eye view he got then while observing them from hot air balloons. I think this is an interesting idea, and also brings out the importance Wegener placed on "visual" and "geometrical" data. Greene also provides an excellent description of how Wegener, the *generalist*, mined data of others from various fields in developing his hypotheses of lunar meteor craters, continental drift, and past climates.

> Wegener's opportunity in this area [lunar craters] was strikingly similar to that presented to him in his earlier (and continuing) work on continental displacements, and his later work on the climates of the geological past. In all three areas he attacked the problem by coordinating recent but as yet uncombined results of specialized subfields, and drawing conclusions from the combinations of data not available to the workers in any one of the fields. . . . Wegener was a field scientist of distinction, but he was also a "data miner." In this he was very like Kepler, who worked with data gathered by others to develop his laws of planetary motion.
>
> *(Greene, 1988: 117; my bracketed addition)*

For an English translation of *Die Enstehung der Mondkrater* see Şengör (1975).

11 Two excellent accounts of Wegener's last two expeditions to Greenland are Johannes Georgi's *MID-ICE: The Story of the Wegener Expedition to Greenland*. Georgi's book, published in 1935, is a moving account of his winter at Mid-Ice with Ernst Sorge and Fritz Loewe. It includes entries from his diary as well as the one kept by Ernst Sorge, and it gives an account of Wegener's arrival at Mid-Ice on October 31, 1930, with Rasmus

Villumsen and Fritz Loewe; the making of the decision for Loewe, whose toes had become severely frostbitten, to remain at Mid-Ice over the winter with Sorge and Georgi; the removal of Loewe's toes by Georgi so as to save Loewe's lower limbs and, perhaps, even his life; the search for and discovery of Wegener's body and the failure to find Villumsen's body. The other account, *Greenland Journey*, edited by Else Wegener with help from Fritz Loewe and translated into English in 1939, contains articles by Else Wegener and many of the key members of the expedition along with extracts from some of Alfred's diaries.

12 As far as I know, there are no known notebooks in which Wegener details the construction of his theory of continental drift. What are known to exist are recollections of fellow scientists, letters between Wegener, Else, and Köppen that were written at the time he was constructing his hypothesis, and his own retrospective comments in the introduction to the fourth edition (1929) of *The Origin of Continents and Oceans*.

13 W. Wundt has reported that while both were at the University of Berlin in 1903, Wegener pointed out the congruency of the western and eastern Atlantic coasts. However, according to Wegener (1929), the idea of continental drift did not occur to him until 1910. Even if Wundt's report is accurate, it is clear that Wegener drew no conclusions about continental drift in 1903 that he thought worth pursuing (Georgi, 1962: 309).

14 Reprinted in Hallam (1983: 128).

15 Apparently, Wegener was so impressed with this bathymetric map that he described its beauty to Else in a letter of December, 1910. See Greene (1984).

16 In his 1912 papers in which he discusses the need for a paleontological connection between Africa and Brazil, Wegener cited the work of Theodor Arldt (1901), a German paleontologist who Wegener continued to rely on in his later works on continental drift. Apparently, it was this particular work of Arldt's that Wegener discovered in the fall of 1911 (Greene, 1984: 749). In addition, Greene has informed me (private communication) that Arldt's work contains a number of vivid paleontological reconstructions which colorfully display the migration of various fauna across landbridges connecting Africa and Brazil – just the sort of thing to incite Wegener to examine further the geological and paleontological literature.

17 This passage is from the *Petermanns* version, p. 185. Besides having the same title, both papers contain the same subsections: ones on the geophysical and geological arguments in support of his view, sandwiched by a historical introduction and a final section containing evidence derived from astronomical position-finding of the present-day horizontal displacements of continental crust. The *Geologische Rundschau* paper is a shortened version of the one that appeared in *Petermanns Geogr. Mitt.* I have used W. R. Jacoby's English translation of Wegener 1912b, which I shall refer to as Wegener/Jacoby, 1912b/2001. The page references are to Jacoby's translation.

18 Suess had also claimed that continents were sialic and oceans simatic.

19 Because of the later importance of subcrustal convection, it is worth examining what Wegener said. After mentioning lunar tides, which he suggested are probably "an essential cause," he added, "Probably it will be wise for now, to consider the continental displacements as the consequence of irregular currents inside the earth" (Wegener/Jacoby, 1912b/2001, part II: 47–48). Wegener did not provide a source for this idea. However, he probably got it from Otto Ampferer (1875–1947), an Austrian geologist who argued against contractionism and to whom Wegener appealed in his attack on contractionism (Wegener/Jacoby, 1912b/2001, part I: 35). In fact, Wegener referred to the paper in which Ampferer discussed subcrustal currents. Ampferer (1906) suggested that vertical movements of the Earth's crust might be the result of subcrustal currents. He first suggested that currents within the substratum could arise from regional thickening or subsidence of crust, from subcrustal density differences, and from temperature differences. He then suggested that convergence of subcrustal currents could cause formation of mountains, and their divergence could cause rift valleys (see Şengör, 2003: 237–238). Wegener later referred to Ampferer's subcrustal

currents in connection with mountain building (1924: 157).Wegener's mention of something that sounds like seafloor spreading appeared in his discussion of different depths of ocean floors. He suggested that the Atlantic and part of the Indian Oceans were less deep than the Pacific because their seafloor was less dense.

> Furthermore, we now seem to be able to explain the different ocean depths. Since for large areas we will have to assume isostatic compensation of the seafloor, the difference means that the seafloor we believe to be old is also denser than that believed to be young. Further, it seems undeniable that freshly exposed sima as in the Atlantic and western Indian Ocean will for a long time maintain not only a small rigidity but also higher temperatures . . . than old largely cooled seafloor . . . The depth variation appears also to suggest that the Mid Atlantic Ridge should be regarded as a zone in which the floor of the Atlantic, as it keeps spreading, is continuously tearing open and making space for fresh, relatively fluid and hot sima from depth.
>
> *(Wegener/Jacoby, 1912b/2001, part III: 55)*

In 1912, Wegener did not link this idea with subcrustal convection, and he never returned to it, as far as I can determine. He still suggested that upwelling sima fills cracks in the floor of the Atlantic but did not associate them with the Mid-Atlantic Ridge. Wegener (1929/1966: 210) later realized that ideas about the seafloor were in a state of flux. He was unsure what to do with the Mid-Atlantic Ridge; however, he (1929/1966: 209) admitted, "the wide mid-Atlantic ridge is a phenomenon which drift theory must cope with." He (1929/1966: 210) rejected the idea that the ridge is a sialic mass "which crumbled during the separation process" because it appeared to be too wide given the coastline congruencies of the circum-Atlantic continents. He proposed (1929/1966: 211) that the ocean floors are made of basalt overlaying an ultrabasic rock, probably dunite. So I don't think Wegener viewed the Mid-Atlantic Ridge as the locus of upwelling mantle convection with the creation of new seafloor. I shall, however, later suggest that he did believe that seafloor was newly exposed as continental blocks drifted apart. Furthermore, I shall suggest that Wegener did propose transform faults or something very much like them (III, §8.7).

20 This brief discussion of Taylor's life, education, and career is a synopsis of Aldrich (1976), Black (1979), and Leverett (1939). Taylor's idea of continental drift is discussed by Hallam (1973, 1983), Marvin (1973), Totten (1981), Greene (1982, 1984), and Laudan (1985).
I should like to thank Michelle Aldrich and Stanley Totten for sending me copies of some of Taylor's more obscure publications. My account of the development of Taylor's ideas owes much to that of Rachel Laudan (1985).

21 Taylor (1910: 224) tells us that he had read the first three volumes of the English translation of Suess' *Face of the Earth* by the time he submitted his paper for publication and that the fourth volume "reached the writer a few days before the proof of this paper." He added: "A hasty examination disclosed nothing that suggests any important change in the conclusions reached." Taylor relied much on Suess.

22 At the beginning of the twentieth century, the idea that the Moon had been formed through separation from the Earth was commonly accepted. The notion that the Moon had been captured by the Earth had been proposed by a few scientists before Taylor, but hardly anybody took it seriously (see Brush, 1986). Thus, Taylor's idea was both novel and unpopular.

23 This idea of Taylor's that the collision of Eurasia and India caused continued creep of Eurasia and formation of lobe-shaped structures such as the Malaysian Peninsula sounds somewhat like the process of "extrusion" tectonics. This idea, I believe, was introduced by Molnar and Tapponnier (1975). See also Tapponnier et al. (1982). They produced a model using plasticine and metal. India, a block of metal, collides with Eurasia, a block of plasticine. Given their view, intracontinental deformation occurred with the collision of the two landmasses. With the collision of and penetration of India into Eurasia, Asia rotated approximately 25 degrees clockwise, and Indochina was extruded to the southeast along a gigantic left-lateral strike-slip fault. Molnar (1988) provides an excellent review of the development of "extrusion" tectonics during the 1970s and 1980s.

24 Taylor preferred the expressions "crustal sheets" and "crustal flakes" when referring to moving continents or moving crust. In his 1910 article, he twice used the term "plates," but on both occasions he pointed out that the term was Suess' (see Taylor, 1910: 191, 198). I mention this to forestall any silly claim that Taylor invented plate tectonics in 1910 because of his use of the term "plate" when referring to moving crust.

25 Greene (1984) makes the point that Wegener referred to all of his known predecessors with the exception of Snider. See Marvin (1973: 42–43), for a discussion of Snider's version of continental drift.

26 The note is not in *Petermanns Geographische Meteilungen* or in *Geologische Rundschau*, the two journals that published Wegener's first two papers on continental drift. There is nothing by Wegener in 1910 or 1911 in the latter journal and there is only a short review of his book on the *Danmark* Expedition in *Petermanns Mitteilungen* (1910: 282), and nothing of any significance by or about him in the volume for 1912, except, of course, for Wegener (1912b).

27 I say this because Wegener, in one of his two initial papers on continental drift (1912b: 192), actually referred to a paper which appeared in the same issue of the *Bulletin* as Taylor's article. This paper by A. L. Day (1910) is a discussion of various laboratory experiments comparing the temperatures at which certain minerals melt or undergo changes in crystal structure.

28 Here I am interested principally in the amount of attention paid to their views rather than whether it was positive or negative. Of course, the two are not unrelated. If a view receives only strong criticism soon after it is proposed, then such criticism may be the only attention given to it. As I shall soon show, the general reception of Wegener's theory was mixed, although the majority opinion was contrary. Taylor's theory, however, received almost no attention whatsoever. Because his paper was published in the *Bulletin of the Geological Society of America*, a journal with a high profile, the response that hardly anyone saw his paper will not do.

29 Marvin (1973: 72–73) noted that Wegener "never used" the term "Pangea" "as a specific proper name or printed it on a map" but in "both the second and third editions" of the *Die Entstehung der Kontinente und Ozeane* "in an offhand manner, referred to such a continent as a 'Pangäa'." It was, she adds, his readers that quickly adopted "Pangea" as the name of the supercontinent.

3

Sub-controversies in the drift debate: 1920s–1950s

3.1 Introduction

The continental drift controversy became more widespread and intense after the publication in 1922 of the third edition of *Die Entstehung der Kontinente und Ozeane (The Origin of Continents and Oceans*; hereafter, *The Origin*) followed by its translation into at least English (1924), French (1924), Spanish (1924), and Russian (1925).[1] This marked the full flowering of the classical stage of the drift debate. The debate soon splintered into several sub-controversies, within each of which workers quickly coalesced around fixist or mobilist standpoints. The classical stage was terminated by the rise of paleomagnetism in the early 1950s, at which time all sub-controversies remained unresolved. During the classical stage, several Earth scientists developed new versions of mobilism, notably Argand, Daly, Joly, Holmes, and du Toit. Daly, Joly, and Holmes offered new solutions to the mechanism problem – what forces caused the horizontal displacement of continents relative to one another; Argand offered an overall account of the formation of mountains based on Wegener's theory, and du Toit improved several of Wegener's solutions.

In the previous chapter I described Wegener's theory as set out in his 1912 papers. Here I begin with a brief account of his theory as presented in the third edition a decade later. Although his basic tenets and defenses remained substantially unchanged, I review them because they triggered the main four sub-controversies.

The first sub-controversy concerned Wegener's explanation of biotic disjuncts, in which paleontologists and biogeographers engaged and which continued into the 1960s. The second was equally long-lived and concerned the distribution of Permo-Carboniferous glaciation, arguments regarding which were Wegener's strongest. The third, more short-lived, concerned an attempt to gather geodetic evidence in support of Greenland's westward drift relative to Europe. These three sub-controversies are now considered. The mechanism sub-controversy (the fourth) is described in Chapters 4 and 5, alongside later (post-Wegener) developments in drift theory before the rise of paleomagnetism in the early 1950s.

3.2 Wegener's theory as presented in 1922

The English version of Wegener's third edition of *The Origin* included a sympathetic introduction by J. W. Evans, President of the Geological Society of London (more on him in §8.3). Wegener also summarized the key tenets of his theory and its defense in a four-page paper, published in English in *DISCOVERY, a Monthly Popular Journal of Knowledge* (1922). Drawing on new supporting studies, he provided a more detailed and extensive account than previously.

It will be evident that this theory conflicts with the former fundamental views of several sciences, and especially those of geology. For a proper judgment upon it an enormous mass of facts must be collected together from such sciences as geophysics, geology, palaeontology, palaeoclimatology, animal and plant geography, and geodesy. In the decade since the first publication of the theory, much progress has been made towards a wide review of the facts. The theory offers solutions for so many apparent insoluble problems, and so simplifies our views, that the interest of many kindred sciences has been aroused, as is shown by the large and growing literature on the question.

(Wegener, 1922: 114–115)

Nevertheless, the statement and defense of his theory remained little changed from a decade earlier.

Wegener regarded the continents as rafts of sial floating on denser substratum or sima, which also formed the ocean floor, and the continents underwent horizontal displacement relative to one another by plowing through the sima. He argued that all of the continents once had been united, forming a supercontinent called "Pangea" which remained intact throughout the Late Paleozoic and most of the Mesozoic, and began to break up near the end of the Mesozoic.[2]

Wegener (1924: 6–7) illustrated the breakup of Pangea with tectonic maps, showing the position of the continents relative to each other during the Late Carboniferous, Eocene, and Early Quaternary periods. He proposed that the Atlantic began to open during the Cretaceous progressively from south to north and continued to widen, opening in the far north in the Quaternary (much later than is now accepted). Asia moved away from Antarctica and southern Africa, migrating northward and rotating counterclockwise relative to his frame of reference (Figures 3.1 and 3.2). By the beginning of the Quaternary (much later than is now accepted), Australia split off from Antarctica, and the Americas migrated westward relative to northern Europe. India slowly approached Asia, and the tract that formerly connected them was compressed and folded forming the Himalayas.

Echoing his 1912 presentation, Wegener argued that his theory solved six long-standing, first-stage problems.

(i)   *Why do the contours of the edges of eastern South America and western Africa fit together so well, and why are there so many similarities between the shapes of the*

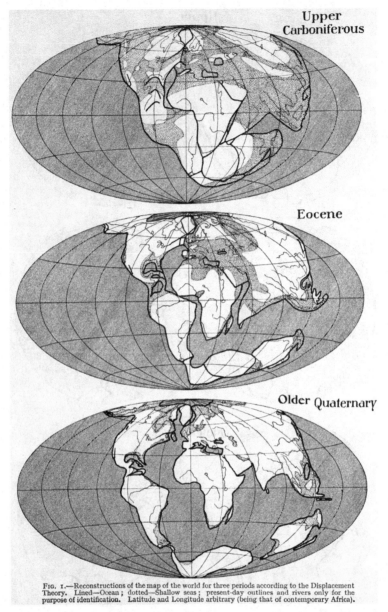

Fig. 1.—Reconstructions of the map of the world for three periods according to the Displacement Theory. Lined—Ocean ; dotted—Shallow seas ; present-day outlines and rivers only for the purpose of identification. Latitude and Longitude arbitrary (being that of contemporary Africa).

Figure 3.1 Wegener's Figure 1 (1924: 6). This and Figure 3.2 show his two "jigsaw" maps illustrating the breakup of Pangea and subsequent drifting of continents.

margins of North America and Europe? He argued that it was because these continents had originally been united and subsequently broken apart.

(ii) *Why are there so many geological similarities between Africa and South America, North America and Europe, and various other continents now separated by oceans?*

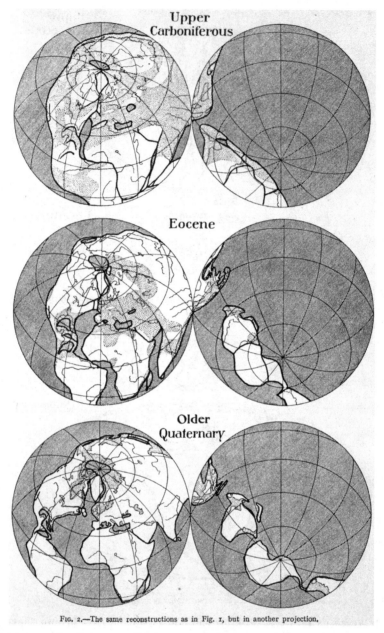

FIG. 2.—The same reconstructions as in Fig. 1, but in another projection.

Figure 3.2 Wegener's Figure 2 (1924: 7).

He elaborated on the similarities between the Cape Mountains of South Africa and the Sierras of Buenos Aires Province, between the gneiss plateaus of Brazil and Africa, and between the extensions of the Pleistocene terminal moraines of Europe and North America. He also emphasized the continuity of the gneiss mountains of Labrador with those of northwestern Scotland, of the Caledonian age fold belts of Newfoundland with those of Scotland and Ireland, and the continuity of the Appalachian fold belts of New England and eastern Canada with the Hercynian fold belts of Britain, southern Ireland, and western and central Europe. These likenesses arose because they were once contiguous. He cited similarities in the geological structure of India, Antarctica, Australia, New Zealand, and New Guinea, arguing that they too were once contiguous, uniting them along with the peri-Atlantic continents in a single supercontinent, Pangea.

(iii) *Why are many biologically similar occurrences now so widely separated geographically?* Because, Wegener argued, the ancestral life-forms lived contiguously in Pangea, now dismembered.

(iv) *Why are some mountain ranges located along the coastlines of continents, and why are orogenic zones long and narrow in shape?* Because, he proposed, mountain ranges formed along the leading edges of drifting continents which crumpled against the resisting ocean floor, the Cordilleran-Andean ranges being the most spectacular example. He also claimed that mountain ranges formed where continents collided. The Himalayas, for example, formed where India penetrated into Asia.[3]

(v) *Why does Earth's crust have two basic elevations, one corresponding to continental plateaus, the other to the ocean floor?* Following earlier workers, Wegener claimed that because the material of continents and ocean floors was fundamentally different, the lighter sialic material of continents and denser sima of ocean floors were in isostatic equilibrium. He declared that they had always been so because the major diastrophic motions were horizontal not vertical.

> The continental masses consist of comparatively light material (such as granite and gneiss) extending downward ... to a depth of 100 kilometers. The deep sea bottom is apparently composed of heavier material (such as basalt), in which the continents float like great ice-floes in water. The results of measurements of gravity, and of magnetic and seismic investigations, are in agreement with this conception, and the results of dredging do not contradict it.
>
> *(Wegener, 1922: 115)*

(vi) *How can we account for the Permo-Carboniferous glacial deposits of southern Africa, Argentina, southern Brazil, India, and Australia?* Wegener supposed that these respective continents had been united at the time with these places clustered around the South Pole near southern Africa.

Wegener's overall defense of his theory remained substantially the same as in 1912. He attempted to discredit various forms of fixism. He attacked contractionist accounts of mountain building by raising what eventually became a standard litany of difficulties (RS2). As fieldwork progressed, estimates of the amount of shortening in orogenic belts increased, and contractionists had to increase their estimate of the amount of global contraction, yet the discovery of radioactivity appeared to set limits to contraction. Also, contractionists could not explain the very uneven distribution of mountains over Earth's surface. And, they were unable to account for the continents and oceans themselves, because contractionism required the foundering of paleocontinents, which was inconsistent with "the more recent and increasingly well-proven doctrine of isostasy or the flotation of the crust of the earth on a plastic lower shell" (Wegener, 1924: 16). Wegener did however agree with the permanentists on one thing, continents could not have become ocean basins, and he repeated the standard objection that sinking landbridges were inconsistent with isostasy. Finally, Wegener noted the failure of contractionism to account for biotic disjuncts.

He then claimed (RS3) that the "theory of displacement of continents avoids all of these difficulties" (Wegener, 1924: 16), and concluded his general defense in this way:

Whilst we must totally reject the contraction theory, we need only reduce the doctrines of the land bridges and the permanence of oceans and continents to the conclusions that can be legitimately drawn from the arguments advanced for them, so as to reconcile both these apparently so opposite doctrines by means of the displacement theory. The latter says: Land connections there were, not through bridging continents which sank later, but by direct contact; permanence not of separate oceans and continents as such, but of the oceanic and of the continental areas as a whole.

(Wegener, 1924: 27)

Continental drift was, he thought, a wonderful compromise; it solved the problems addressed by contractionists without disregarding isostasy.

Wegener detailed his solutions to a number of key problems, devoting a chapter to each. Again he used arguments similar to those employed in 1912. In the next three sections (§3.3–§3.5), I consider three of these problems and sub-controversies that resulted from them. To end this section, I quote Wegener's overarching argument that embraced all sub-controversies, that mobilism was able to solve all these problems simultaneously thus enhancing its overall effectiveness. He especially emphasized how continental drift simultaneously explained the geological similarities and the similarity in the shapes of continental margins on either side of the Atlantic.

The correspondences of the Atlantic coasts already mentioned, namely, the folding of Cape Mountains and of the Sierras of Buenos Aires as well as the correspondence between the eruptive rocks, sediments, and strike-lines in the great gneissic plateaus of Brazil and Africa, the Armorican, Caledonian, and Algonkian systems of foldings, and the Pleistocene terminal moraines, in their sum-total, even if the conclusions may be still uncertain in particular questions, yield a proof, which is difficult to shake, of the validity of our supposition that the Atlantic must be considered as an expanded rift. We have also the circumstance of decisive importance that although the adjustment of the blocks must be made on the grounds of other phenomena, especially their outlines, yet by this adjustment the continuation of each structure on one side is brought into exact contact with the corresponding end on the other. It is just as if we put together the pieces of a torn newspaper by their ragged edges, and then ascertained if the lines of print ran evenly across. If they do, obviously there is no course but to conclude that the pieces were once actually attached in this way. If but a single line rendered a control possible, we should have already shown the great possibility of the correctness of our combination. But if we have n rows, then this probability is raised to the nth power. It is not a waste of time to make clear what this implies. We can assume, merely on the basis of our first "line," the folding of the Cape Mountains and the Sierras of Buenos Aires, that the chances are ten to one that the displacement theory is correct. Since there are at least six such independent controls, 10^6 or a million to one could be laid that our assumptions are correct. These figures may be exaggerated. But one must bear in mind, when passing judgment, what the increase in the number of independent controls really means.

(Wegener, 1924: 55–56)[4]

Wegener's critics agreed with one sentence only: "These figures may be exaggerated." They attacked every one of his solutions (RS2); to their minds, Wegener's torn newspaper page was a collection of badly fitting bits torn from different pages, and any consilience among Wegener's solutions carried little weight because each solution had its own, what they regarded as, fatal difficulties. Fixists also sought to improve their own explanations, or develop new ones (RS1). Mobilists countered by raising difficulties with both (RS2). Neither side developed explanations that were free of difficulties, that answered fully the objections raised.

3.3 Biotic disjuncts and Wegener's 1922 explanation of them

When Wegener was writing the third edition of *The Origin*, the need to account for disjunct biota had become urgent. As already mentioned, fixists offered two alternatives. Landbridgers, including Suess and other European geologists and paleontologists, proposed that former lands (Suess' broad paleocontinents) had extended across oceans and served as migratory routes. Permanentists, such as Willis and many other American geologists, proposed narrow volcanic isthmian links, like present-day Central America.

Wegener used the same general strategy that he had in 1912; he agreed with landbridgers that the paleontological and biological records required past intercontinental connections, but furnished them by placing continents together and then moving them apart. Unlike landbridgers, he imagined a situation where bridges were not needed. Moreover, he borrowed freely from their data compilations in order to set aside their arguments, and contend that his lateral displacement theory was superior to anything they had to offer.

Wegener (1924: 73) stated, "we can confine ourselves to a general synopsis and the selection of a few especially important facts" because the problems of biotic disjuncts "have already been dealt with in their relation to the geographical distribution of plants and animals, especially by the adherents of the land bridge hypothesis." Heeding his own advice, Wegener once again appealed to the work of T. Arldt,

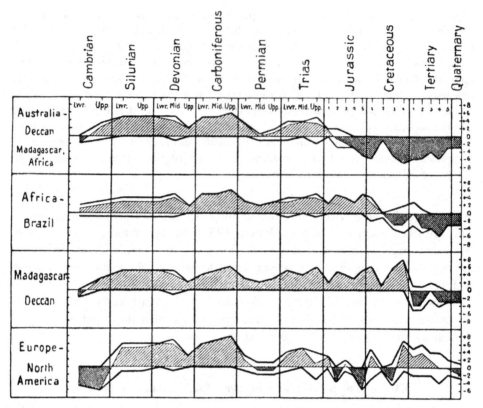

Figure 3.3 Wegener's Figure 15 (1924: 73). Votes on the question of the four post-Cambrian landbridges. The number of favorable votes is shown by the upper thick curve, the number of unfavorable votes by the lower thick curve. The difference is simply shaded if favorable; crosshatched if unfavourable.

who in 1917 provided a synopsis of the opinions of twenty paleontologists regarding the existence of eight hypothesized intercontinental landbridges (Figure 3.3).

Noting, but glossing over the "many differences of opinion," Wegener (1924: 75–76) claimed that the results were "very clear." According to landbridgers, the connection between Australia and Africa disappeared after the Early Jurassic; that between Africa and South America remained until the Lower to Middle Cretaceous; that connecting India and Madagascar lasted until the end of the Cretaceous; and that connecting North America and Europe did not entirely disappear until the Quaternary. (The latter is now recognized as wrong; it was based on the erroneous trans-Atlantic correlation of Quaternary glaciation limits.) The times and location of these breaking connections defined the timetable of Wegener's continental drift, his postulated fragmentation of Pangea.

Wegener then discussed proposed bridges connecting Antarctica with Patagonia and Australia; Arldt's summary revealed that some paleontologists did not believe in such bridges. However, Wegener believed all these regions were originally directly connected, and he dismissed the preponderance of negative votes in Arldt's opinion poll which were, for certain periods, as high as 6 to 1, on the shaky grounds of "slightness of our knowledge."

Of entirely different value are the two separations that we shall next discuss, which are concerned with the connections of the Antarctic with Patagonia on one side, and with Australia on the other. The strong majority of negative votes in this case evidently originates from the slightness of our knowledge concerning Antarctica, which has induced many authors to disregard a connection of this continent with the others, since no reason existed for such an hypothesis. Because of this, only the positive proportion of the votes will be considered.

(Wegener, 1924: 76)

In accordance with his proposed timetable and his alleged "positive proportion of votes," Wegener (1924: 76) suggested that the South America–Antarctica connection existed from Cretaceous to Pliocene, and Australian–Antarctica from Jurassic to Eocene. He also noted, "the very numerous faunal affinities of Australia to South America ... obviously used Antarctica as a bridge" (Wegener, 1924: 77). Fixists did not find his rather loose arguments difficult to dismiss.

Wegener turned to the remaining two landbridges discussed by Arldt, the present narrow land connection (Central America) between North and South America, and the recently operative much broader Bering landbridge between Siberia and Alaska (Beringia). Although "bridges of this sort naturally play no role in the displacement theory," Wegener (1924: 77) included them for completeness "to remove certain misunderstandings." In particular, he wanted to make clear that these bridges were not inconsistent with his theory of continental drift as C. Diener (1915) had said. Diener had argued, "Whosoever pushes North America away from Europe breaks its connection at the Bering Straits with the Asiatic continental block" (Wegener, 1924: 78). Wegener dismissed this difficulty by pointing out that Diener had made a geometrical mistake (RS2): "the objection is only met in a Mercator's map, but not on a globe, for the movement of North America consists essentially of rotation."

Wegener shored up his general case for trans-Atlantic connections by citing examples that he took to be particularly decisive. He appealed to the apparently high percentage of "identical reptiles and mammals on both sides of the Atlantic" (1924: 79), again drawing on the work of Arldt. He referred to the distribution of a number of living organisms that appeared to Arldt "to be most decisive to the question of the North Atlantic Bridge," namely, the distribution of earthworms (*Lumbricidae*), pearl mussels, garden snails, perches and other types of freshwater fish, and the common heather (1924: 81).[5] They were generally restricted to eastern North America and western Euro-Asia, and unlikely to have ever crossed the present Atlantic. Wegener believed this to be decisive evidence for the juxtaposition of Europe and America. He also cited the similar Tertiary floras of Greenland and Europe as indicative of a former connection.

Turning to the South Atlantic, Wegener referred to the small Permian reptile *Mesosaurus*, the Permian *Glossopteris* flora, and the modern sea cow, *Manatus*. *Mesosaurus* was a small freshwater swimming reptile, incapable of traversing oceans, whose remains had been found only in Brazil and southern Africa. *Glossopteris*, destined to play a large role in the mobilism debate, was a fern-like land plant, and the discovery of it in India, South America, Africa, and Australia had led to much speculation by paleo-botanists and paleoclimatologists. The distribution of *Manatus*, found only in West Africa and tropical South and Central America impressed Wegener (1924: 83–84) because today "it lives in streams and shallow warm sea-water but is unable to swim across the Atlantic Ocean."

Having established the need for intercontinental connections, Wegener favored drift, which he claimed was superior to landbridges. He used four arguments, three biological, and one geophysical (RS3). First, he argued that his drift theory offered the more adequate solution where landmass A had life forms resembling those of B but quite different from those of landmass C, even though A is geographically closer to C than it is to B. In such cases Wegener argued that originally A had been closer to B than to C, but later drift had changed their relative positions; such an alternative was not open to landbridgers. Wegener gave three examples:

The island of Juan Fernandez is, perhaps, of especial interest ... [it] does not show any affinities botanically with the closely adjacent coast of Chile, but only with Tierra del Feugo (by winds and sea-currents, I suppose!), Antarctica, New Zealand, and the Pacific islands. This fits in excellently with our idea that South America, drifting westwards, has, in recent times, approached it to such an extent, that the difference of floras becomes very startling. The theory of the submerged bridges could not begin to account for this phenomenon.

(Wegener, 1924: 84; my bracketed addition)

According to Wegener and others, the flora of the Hawaiian Islands more closely resembles that of Japan and China than North America to which they are now closer. He wrote

This appears intelligible if it is borne in mind in accordance with my theory that in the Miocene, when the North Pole lay in the Bering Straits, Hawaii had a geographical latitude of 40 to 50 degrees, and thus lay in the great westerly wind-drift which came from Japan and China. Moreover, the American coast at that time was farther removed from Hawaii than it is now.[6]

(Wegener, 1924: 84–85)

He proposed a similar solution to account for the biological similarities of Madagascar and India; he rejected the sunken continent Lemuria, preferring their initial union and later separation by drift.

Second, Wegener argued that drift was superior to landbridges where present or past biota of landmasses A and B resemble or resembled each other even though (i) they have now the same climate, but migrants would have had to cross landbridges spanning different climate zones, or (ii) A and B are presently at different latitudes and enjoy different climates. He cited the Madagascar–India linkage as an instance of (i), and the widespread land-based *Glossopteris* flora as an instance of (ii). Concerning the former, he maintained (1924: 85):

The superiority of the displacement theory is again shown, since the two parts possess in their present positions a considerable difference of latitude, and have a similar climate and shelter similar forms of life only because the equator lies between them.

He summed up the *Glossopteris* evidence in this way:

The whole of the *Glossopteris* deposits in the southern continents may be considered, not only, as already mentioned, as proof of a land connection at that period, but also as a proof of the superiority of the displacement theory as compared with that of submerged continents, for it is impossible to assume from their present position that they could all have had the same climate at every period of the earth's history.

(Wegener, 1924: 85)

Wegener's third biological argument concerned Australia's fauna, past and present. He argued that only his drift theory could explain the tri-partite division of Australian animals, established by Alfred Russel Wallace. What Wegener regarded as the oldest element "shows affinities especially with India and Ceylon, as well as with Madagascar and South Africa" (1924: 85). "This affinity originates from the time when Australia was still connected to India" (1924: 86). The second faunal element, marsupials and monotremes, are "sharply differentiated from the fauna of the Sunda Islands" (1924: 86). "This element of the fauna shows kindred relationships to that of South America ... there are also known fossils from North America and Europe, but not from Asia. Even the parasites of the Australian and South American marsupials are the same" (1924: 86). According to Wegener (1924: 88), this second element "dates from the period when Australia was still connected *via* Antarctica to South America." The third and most recent element of the Australian fauna resembles that of the Sunda Islands and the coastal regions of Southeast Asia. Wegener (1924: 88) claimed that it "indicates a rapid exchange of fauna and flora which began in recent geological time" as Australia approached the East Indies. Wegener added that drift theory not only explained the tri-partite division of Australian fauna, but it also explained why Australia's modern fauna differs so strongly from that of the Sunda Islands and Southeast Asia, as discovered by Wallace. Wallace drew a boundary between the Oriental (Southeast Asian) and the Australian fauna, which T. H. Huxley later named Wallace's Line.

Wegener (1924: 88) was pleased with this application of his drift theory, considering it "to be of quite considerable importance to the question of displacement."

This threefold division of the Australian fauna is in most beautiful agreement with the displacement theory. It is only necessary to glance at the three reconstructed maps [of the break up of Pangea] in order immediately to read from them the explanation. Even these purely biological facts show the distinct advantages which the displacement theory possesses over that of the submerged bridges ... Does one really believe that a mere land connection [between Australia and South America] is sufficient to secure the exchange of forms? And how strange it is that Australia had an interchange of forms with the immediately adjoining Sunda Islands to which it is like a foreign body from another world! No one can deny that our assumption, which reduces the distance of Australia from South America to a fraction, and on the other hand separates it from the Sunda Islands by a broad ocean basin, provides a key for the explanation of the Australian animal kingdom.

(Wegener, 1924: 88–89; my bracketed additions)

As in 1912, Wegener claimed for these manifold reasons that drift theory offered a better solution than either isthmian connections or the landbridges of the fixists. He agreed with the permanentists that broad landbridges, although they could account for the majority of the paleontological and biological data, were untenable, being incompatible with isostasy. His theory embraced isostasy. Permanentists and the isthmian-links were in compliance with isostasy, but although they could explain some disjuncts, they had no general solution to the origin of disjunct biota.

3.4 Landbridgers revise and rebut

Wegener's work received much early attention, between landbridgers and drifters who respectively condemned and praised. Here I consider the broadside attack that greeted Wegener, mainly launched by landbridgers who cited specific examples that ran counter to his theory. The major *biogeographical* response by permanentists was delayed until the 1940s, and is described later (§3.6).

At a 1923 symposium held before the Royal Society of South Africa on the distribution of life in the Southern Hemisphere and its bearing on the Wegener hypothesis, the Cambridge-trained botanist R. H. Compton was reported to regard "the botanical evidence as completely opposed to Wegener's theory" (Anonymous, 1923: 131). However, Compton, who worked at the National Botanic Gardens in Cape Town, modified what was reported of him, remarking,

The botanical evidence for the southern hemisphere is certainly not "completely opposed" to Wegener's theory: it simply does not provide any critical test of that theory, so far as I can see at present.

(Compton, 1923: 533)

The French-born entomologist L. A. Peringuey, Director of the South African Museum in Cape Town, claimed that landbridges could just as easily account for the distribution of insects, and thought that Wegener's theory "will receive little if any support from entomology" (Anonymous, 1923: 131). A general indictment of Wegener's solution came from Edward T. Berry, an American paleontologist, who claimed at the first international meeting on drift sponsored by the American Association of Petroleum Geologists (hereafter, AAPG) in 1926 that Wegener's theory "paleontologically … raises more problems than it solves" (Berry, 1928: 195).[7]

Ludwig Diels, a German paleobotanist, criticized Wegener through the late 1920s and 1930s, arguing that drift was incompatible with paleobotanical data. According to him, the flora of eastern North America was linked to that of eastern Asia rather than to Europe; the relic flora of Europe was unlike that of North America but was related to that of Asia; and the indigenous flora of Australia was tropical rather than temperate, which was as would be expected were Wegener correct. Diels (1936) also disagreed with Wegener's proposed northward movement of the Australia–New Guinea block from its former position adjacent to Antarctica to where it is today. He argued that Australia had not been attached to Antarctica because he believed that the overwhelming majority of Australia's autochthonous (that is, indigenous or ancient) flora was Malaysian rather than Antarctic and had not recently arrived from the north as Wegener contended (Diels, 1936: 192). He also contended that flora of Malaysia continued to invade, albeit episodically, Australia, and hence they have not moved relative to one another (Diels, 1939: 194).

The Norwegian zoologist Fridthjof Ökland also argued that the fossil similarities of North America and northern Europe were too meager to require them to be directly connected, and opted for landbridges; direct contact, followed by drift should have yielded greater similarities. As already noted, the German Carl Diener (1915) argued that drift was unacceptable on paleontological grounds since Wegener had to break the Bering Strait connection when uniting North America with Europe; he claimed that Wegener's reconstruction resulted in a 35° separation of Alaska and eastern Russia.[8] The British geologist Philip Lake wrote several caustic reviews of Wegener's overall theory (Lake, 1922a, 1922b, and 1923), and raised a potentially serious difficulty with Wegener's solution to the distribution of *Glossopteris* flora; potentially serious because *Glossopteris* was one of Wegener's key witnesses to drift. Lake (1923: 227) contended that drift "had not by any means simplified the problem." The basis of his criticism was the reported discovery of *Glossopteris* flora in Kashmir, northwestern Afghanistan, and, especially, in northeastern Russia and Siberia. Even if Wegener could account for the distribution of *Glossopteris* in the southern continents and India by grouping them around the South Pole, his reconstruction could not account for these Asiatic locations; as Lake remarked, "in Wegener's reconstruction all these areas lie far from the masses that he has grouped together in the south" (1923: 227). I shall return to this question later (§3.5).

Further objections came from three vehement landbridgers, the Briton J.W. Gregory, the American Charles Schuchert, and the German H. von Ihering. Gregory defended landbridges in two lengthy presidential addresses before the Geological Society of London in 1929 and 1930. He argued that Wegener was unable to account for botanical similarities between North America and eastern Asia on opposite sides of the Pacific, similarities that were just as compelling as those found on opposite sides of the Atlantic; Wegener's reconstructed continents made them harder to explain than if left unmoved.[9] Gregory postulated trans-Pacific landbridges, which received lingering support well into the early 1960s. Schuchert, a paleontologist, argued for landbridges at the 1926 symposium sponsored by the AAPG and raised several objections to Wegener's solution. He agreed with Diener that Wegener's reconstruction severed the connection between eastern Russia and Alaska, although he thought that Diener had overestimated their separation. He also supported Lake's criticism of Wegener's solution to the distribution of *Glossopteris* flora, and raised objections of his own.

Any paleontologist who reads carefully pages 98 to 106 of Wegener's book, dealing with the distribution of the Coal Measures and *Glossopteris* floras of "Permo-Carboniferous Time," will see not only the nimbleness and versatility of his mind, but as well how very easy it is for him to make all facts fit his hypothesis. Why is this? Because he generalizes from the generalizations of others, and compares unlike things, regarding the correlation of formations by geologists as dealing with "relatively trifling differences of time." In these pages he is explaining his views of the climate of Permo-Carboniferous time, and, in doing so, shoves the south pole to a place off the southeast coast of Africa, arranging the equator accordingly. Finally, to make it easy for all of us to get his views, he pictures them on a single diagram entitled "evidences of Climate in the Permo-Carboniferous." This single diagram undertakes to represent events that took place during a lapse of something like 50 million years, makes the flora of the tropical Coal Measures fit the "polar" *Glossopteris* flora of the much younger Permian, and in order that the latter may be truly polar assumes that it was tree-less, says that Antarctica then was adjacent to south eastern Africa with the south pole at the edge of it, and on this basis arranges the climate belts around it!

(Schuchert, 1928a: 134–135)

Schuchert (1928a: 138) opted for landbridges that could "explain life dispersion far more easily than ... Wegener's Pangea."

Von Ihering was dismissive, contemptuous.

To fit together the coast-lines of these two continents Africa and South America was a naive idea, which is, however, in direct opposition to all the facts proved by geological study and the geographical distribution of animals.

(von Ihering, 1931: 377)

He also argued that trans-Pacific landbridges were needed, and that the South Atlantic required intermittent landbridges as opposed to the direct connection followed by a complete break as Wegener wanted. T. Arldt, Wegener's favorite source, himself remained a fixist landbridger.

Some major figures continued to support landbridges into the 1960s. Carl Johan Fredrik Skottsberg, the doyen of Swedish botanists, did so throughout his career. At a symposium in 1959 of the Royal Society of London on the biology of the Southern Hemisphere cold temperate zone, he argued that Antarctica had been the source of many plants found now in Africa, New Zealand, Australia, South America, the Falkland Islands, and the Juan Fernandez Islands, and it was their wide distribution that had led him to posit connecting landbridges. He made it quite clear that he preferred landbridges and that mobilism was not worth discussing.

The paths of migration appear to radiate from the heart of Antarctica, but a majority of serious biogeographers refuse to believe in vertical movements of the ocean bottom sufficient to establish land connexions, and many regard Antarctica as of little consequence ... Wegener's displacement theory, which has many supporters and claims to explain everything, seems to create new, insuperable difficulties, and I shall not discuss it here.

(Skottsberg, 1960: 453)

Twenty years earlier in 1940, G. Einar Du Rietz (1895–1967), another influential Swedish botanist, made clear his preference for landbridges:

The earlier transantarctic land connections necessary for the explanation of the facts of austral and bipolar land distribution have been reconstructed either in the form of land bridges connecting the present continents in their present position ... or by epeirophoresis, i.e. continental drift, according to the theory of Wegener ... From a phytogeographic point of view this theory well suits the fact of transantarctic and Andean transtropical connection, but it seems impossible to reconcile it with the facts of an old trans-Malaysian connection so clearly demonstrated by the distribution of *Euphrasia* [a genus of figwort], as was pointed out by myself.

(Du Rietz, 1940: 254–255; my bracketed addition)

After recalling that Diels had already shown the "impossibility of reconciling the facts of Australian and Papuan distribution with the theory of Wegener," Du Rietz declared "his own experiences essentially agree[ed]" and quoted Skottsberg for further support:

Any assumption of great movements of the present continents seems only to increase the difficulties of explaining the actual facts of plant distribution, especially the many-sided phytogeographic relations of the South-west Pacific area, and most especially the combined transtropical and transantarctic relations of some of its plants (e.g., *Euphrasia*). And "it may well be stated that the composition and distribution of the present austral and subantarctic flora can be explained without the violent changes implied in Wegener's hypothesis; the vertical movements required for a reconstruction of the Scotia Arc and other land connections are not so great as to jeopardize the permanency of the Pacific during the ages, eagerly advocated by most geographers and geologists"(translated from Skottsberg, 1940a, p. 53).

(Du Rietz, 1940: 255; his Skottsberg (1940a) is my Skottsberg (1940))

Du Rietz also participated in the 1959 Royal Society symposium. During the final discussion, he simply quoted passages from his 1940 paper, making clear that in the two decades he had not changed his mind (1960).

Cornelis Gijsbert Gerrit Jan van Steenis, a leading Dutch botanist during the mid-twentieth century, argued against a drift interpretation of the unusual flora of the East Indies (§8.14). He remained a landbridger into the early 1970s. At the 1961 10th Pacific Science Congress, he insisted on the need for at least five major land-bridges, including one across the Pacific Ocean:

Minimum number of major land bridges required by plant geography. – For a satisfactory explanation of the major features of phanerogamic plant geography are needed: (1) a tropical transatlantic land bridge; (2) probably a Seychelles land bridge connecting Madagascar with Ceylon (Gondwana Land, *sensu stricto* of Suess); (3) a wide continuous land south of the Bering Strait; (4) a tropical transpacific bridge; and (5) a South Pacific temperate bridge, which could be envisaged eventually (as Hooker, Diels, and Skottsberg prefer) to consist of an extended Antarctic continent.

(van Steenis, 1963: 227)

Van Steenis rejected mobilism, quoting Diels' work:

At one time continental drift was assumed to give the solution of all biogeographical problems; but Diels (1936) has sufficiently shown that it provides no solution for salient problems and that its consequences are opposed to affinities readily understandable without drift – among others, its implication to divorce the Malaysian-Papuasian-Australian flora from that of South America. For the understanding of the tropical transpacific flora, drift is also unimaginable, since it excludes the origin of an amphi-Pacific type.

(van Steenis, 1963: 225)

No doubt mindful that only an abstract of his talk would be published in the congress proceedings, van Steenis (1962) wrote a lengthy defense of landbridges. He intensified his attack on mobilism, arguing erroneously that its paleomagnetic support was seriously flawed (see III, §1.12). His rejection of Hess's idea of seafloor spreading did not prevent him from appropriating some of Hess's ideas in support of landbridges.

Van Steenis, a late holdout, did not explicitly support mobilism until 1979, although by 1971 he seemed to be leaning ever so slightly toward it. Discussing *Nothofagus* (the southern beech), he argued that various land connections were needed to explain its distribution but he would not accept mobilism.

I was, and am still, aware that geomorphologists are not yet so generous to grant us the ancient land (cq.) [*casu quo* (in which case)] land-connections, even though it appears irrelevant from which source it might have come and by what processes it might have gone; foundered fore-continents, subsided continents, expanding or shrinking Earth, continental drift, or plate tectonics.

(van Steenis, 1971: 75)

He argued that *Nothofagus*, which today occurs in southern South America, south-east Australia, New Zealand, and some islands of Melanesia, could equally well be explained by landbridges without continental drift. Disagreeing with L. Cranwell (III, §1.12), who had worked on *Nothofagus* in New Zealand and Antarctica since the early 1960s and who had written approvingly of the paleomagnetic support for mobilism (Cranwell, 1963), van Steenis claimed:

Mrs. Cranwell's additional hypothesis, necessary to explain the fossil finds of *Nothofagus* on Antarctica under her assumption, namely a close proximity of Antarctica and New Zealand in the Upper Cretaceous and early Tertiary by which Nothofagus was able to grow in Antarctica, and then a southward drift of Antarctica by which Nothofagus got there extinct, appears to be a completely unnecessary complication.

(van Steenis, 1971: 78)

He appeared more comfortable widening warm climatic belts than moving continents. Later, however, he was unwilling to dismiss mobilism's paleomagnetic support:

I will not reject Mrs. Cranwell's reference to Irving's palaeomagnetic data, indicating that Antarctica and Australia were close together in the early Tertiary, swinging down by the Miocene, and that Antarctica and New Zealand were separated later. What I wanted to show is that *also without continental drift hypothesis* the borderland of Antarctica as it is situated today would allow for *Nothofagus* to grow under a milder climate.

(van Steenis, 1971: 78)

But his position was changing. He had come to believe that biogeographical data alone could not distinguish between continental drift, landbridges, and changes in widths of climatic zones.

Van Steenis was now willing to admit that biogeographical data were unable to specify the type of land connection required but was unwilling to accept drift simply because of its success in other fields. He did not mention what was by 1970 the overwhelming evidence in favor of seafloor spreading and plate tectonics. If he had, he would have been obliged to take mobilism as a given and use it as a starting point to unravel biogeographical problems. He obstinately refused to accept mobilism even after the corroboration of the Vine–Matthews hypothesis. He was not alone; note, for example, Alan Keast's reluctance to accept mobilism in (§9.4).

Van Steenis finally accepted mobilism in 1979, making it seem as if his earlier opposition to mobilism had not been as great as his papers indicate.

I agree … that the early disintegration of Pangaea and the later disintegration of Gondwanaland appear to be a rough guide for the early distribution of parts of the angiosperms [flowering plants]. As such, this is an extension and improvement of my "Land-bridge Theory," which was solely to indicate where botanical "contacts" were necessary and at what times they must have functioned. Its philosophy was that the genesis of the plant world was bound by land, continuous or archipelagic, which changed considerably in the vicissitudes of time and land areas, and that this inferred that the distribution of plants must be orderly and should not be explained by hypothetical sweepstake or random dispersal. I admit that I have attributed too little significance to the possibility of displacement of land carrying its flora with it, of which the effect is in essence the same as that of land-bridges. The philosophy would probably have been better understood if the theory had been called the "Land Theory of plant-geography," a less ambiguous, more neutral term which would have fully covered its purpose. The assumed botanical connections or contacts advocated by the theory stand invariably and are not affected by new geophysical theory, under which the landbridges can now largely be understood as moving land masses.

(van Steenis, 1979: 156; my bracketed addition)

Perhaps hoping to excuse his long-standing rejection of mobilism, van Steenis retrospectively blurred the fundamental distinction between landbridges and continental drift. He seemed to have forgotten how explicitly and repeatedly he had opposed mobilism.

Landbridges, their supporters argued, accounted for the distribution of similar life forms on either side of both the Pacific and Atlantic. Disjunct biota were plentiful enough to support landbridges but insufficient themselves to warrant mobilism. Throughout their rearguard action, landbridgers were assisted by geophysicists who from a theoretical standpoint continued to raise mechanism difficulties against Wegener's theory. Although Wegener thought his theory was preferable to landbridges on geophysical grounds, drift's proposed mechanism itself faced weighty geophysical objections of its own. Jeffreys raised them in his magisterial book *The Earth*, and continued to do so even after the general acceptance of plate tectonics (Jeffreys, 1924: 261; 1970: 450–459). American geophysicists reiterated and elaborated Jeffreys' objections (§4.3). Even though Wegener and other mobilists offered rebuttals, and proposed alternative models, the mechanism difficulty was considered by many as an overwhelming obstacle, and as such it was often cited by fixist landbridgers when arguing against drift as an explanation of biogeographical disjuncts.

3.5 Mobilists rally increasing support for continental drift

All was not gloom, however, for support came from Wegener himself, Alex du Toit, and paleobiogeographers, especially those working in invertebrate zoogeography and phytogeography.

Wegener's defense in the 1929 (4th) edition of *The Origin* had four major steps. He began by repeating his former arguments. Then he made a special plea to biogeographers to consider evidence from the Earth sciences when deciding between drifting continents and landbridges; he believed that if they did all that they would come to favor his theory because landbridges were incompatible with isostasy. Third, he cited the many new converts to his theory. Finally, he (1929/1966: 110) continued to maintain: "the Australian fauna will provide the most important material that biology can contribute to the overall problem of continental drift" and hoped specialists would study it from the drift point-of-view.[10]

At this point vital support came from du Toit, mobilism's most effective supporter from the Southern Hemisphere. Over the years du Toit's assessment of the importance of mobilism's solution to disjunct biota wavered and his major contribution to the mobilist controversy was his stratigraphic work; he probably added more weight to the drift case through his own field studies than any one else before World War II (see especially §6.4–§6.7).

Du Toit (1921a) first discussed drift theory in a public lecture he gave at a meeting of the South African Association for the Advancement of Science (hereafter, S_2A_3) in Durban in July 1921, arguing that paleontological similarities among continents in the Southern Hemisphere supported drift. Two years later, he received a grant from the Carnegie Institution of Washington that allowed him to spend six months in South America studying the Paleozoic and Mesozoic successions in Argentina, Uruguay, Paraguay, Bolivia, and Brazil. That work was described in *A Geological Comparison of South America with South Africa*, published in 1927. Although he favored drift as the cause of disjunct biota, he thought that it was the stratigraphic evidence, such as that for the distribution of Permo-Carboniferous glaciation, that provided stronger support.

... *geological evidence almost entirely* must decide the probability of this hypothesis, for those arguments based upon zoo-distribution are incompetent to do so, being as a rule equally, though more clumsily, explicable under orthodox views involving lengthy land connections afterward submerged by the oceans.

(du Toit, 1927b: 118)

Ten years later in his major work, *Our Wandering Continents*, he changed his position.

The strong similarities displayed by the terrestrial life of the continents, unquestionably more marked, however, during the past, forms one of the great facts and marvels of Nature. Such, indeed, constitutes the basis of our conception of former Land Connections.

(du Toit, 1937: 290)

He pointed out that to account for certain intercontinental migrations, fixists needed some very high latitude landbridges along which adverse climatic conditions would have prohibited migrations (RS2). Drift avoided this difficulty and therefore was the better solution (RS3).

During the 1920s, several European paleontologists and biogeographers favored continental drift, and Wegener welcomed their support (Wegener, 1929/1966: 97–120). The German paleontologist, Leopold von Ubisch preferred drift because (like du Toit) landbridges sometimes stretched across climatic zones that would have been inhospitable to many animals (RS2). Von Ubisch disagreed with Ökland (see §3.4), arguing that faunal correlations across the North Atlantic were intimate enough to justify drift.

The Norwegian J. Huus, the Russian W. A. Jaschnov, and the German W. Michaelsen, all zoologists, appealed respectively to the distribution of present-day sea squirts, crayfish, and earthworms in support of Wegener's solution. The German botanist E. Irmsher argued that the present and past distribution of angiosperms was best accounted for by drift, and W. Studt of the University of Hamburg and F. Koch, who employed differing analyses, claimed that drift best explained the distribution of past and present-day conifers.

After Wegener's death in 1930, support for mobilism continued through to the end of World War II, although it remained always very much a minority opinion. Phytogeographers continued to supply important support; invertebrate zoologists supported drift more than did vertebrate zoologists, some of whom strongly attacked it. Many who favored drift qualified their support by noting uncertainties in the data, by citing alternative solutions, and, although it was not in their expertise, by repeatedly mentioning the mechanism difficulty. Most phytogeographers who supported mobilism cited the distribution of Permian *Glossopteris* and Mesozoic plants in India and the southern continents as major reasons for doing so. Phytogeographers specializing in flowering plants, especially of the Tertiary, generally did not favor drift.

Perhaps the leading phytogeographer to express sympathy for continental drift was the Cambridge paleobotanist Sir Albert Charles Seward. He first did so in his Presidential Address to the Geological Society of London. After discussing the distribution of *Glossopteris*, he remarked:

> Although it would be presumption on my part to attempt to discuss the possibility of finding assistance in our endeavour to recreate Gondwanaland in the much discussed hypothesis of Wegener, I must confess some sympathy with Dr. du Toit's acceptance of the view that the present South America, South Africa, India and Australia "represent portions of the ancient continents finally torn apart ..."
>
> *(Seward, 1924: lxxxix)*

Five years later, in his 1929 Presidential Address for Section K (botany) of the British Association for the Advancement of Science (BAAS) he renewed his support while admitting, "there are, we are told, serious objections to Wegener's hypothesis: it is at any rate true that the principle of drifting continents has still to be proved tenable" (Seward, 1929). Importantly, he spent considerable time arguing that *Glossopteris* deposits were more synchronous than Schuchert had made them out to be. In *Plant Life Through the Ages* (Seward, 1931), he linked phytogeographical support for drift with that provided by Permo-Carboniferous glaciation, one of mobilism's weightiest arguments. Permo-Carboniferous glacial deposits in India and the southern continents (not as yet in Antarctica) often occur above or below strata containing *Glossopteris*; although stratigraphically separated they have similar geographical distributions. Seward argued that the climatic demands of many Permo-Carboniferous flora could not have been satisfied unless drift had occurred, and buttressed his case by appealing to the overall distribution of *Glossopteris*. Citing the Briton G. C. Simpson (3.12), then head of the British Meteorological Office and a major defender of mobilism from the standpoint of paleoclimatology, Seward stated:

> The evidence of substantial changes in climate furnished by these examples causes the palaeobotanist to adopt a favourable and hopeful attitude towards the Wegener hypothesis. The theory is attractive; it has many supporters, among them Dr. Simpson whose papers should be consulted; it has also many opponents: as Lord Acton said, in speaking of human history, "the worst use of theory is to make men insensible to fact": the facts of palaeobotany though they cannot be regarded, in any sense, as proof of the correctness of Wegener's views, may be admitted as evidence in a case which for the present must remain *sub-judice*.
>
> *(Seward, 1931: 538)*

By 1934 Seward had become decidedly pro-drift: "Vertical movements alone are insufficient; lateral displacement of continents seems to be an almost necessary assumption" (Seward and Conway, 1934: 736–737). Although they had little patience with critics such as Diels, who spoke disparagingly of the "absurd phytogeographical consequences" of Wegener's theory, they still recognized that drift faced serious difficulties.

> The Wegener hypothesis, despite the serious criticism which it has raised, appeals strongly to the imagination ... It is by no means improbable that solutions of some of the many problems of Plant Geography both past and present will be found, not in the raising of foundered continents, but through the acceptance of the mobility of the earth's crust, as a factor not merely imagined but substantiated by evidence which, it may be suggested, will eventually be provided.
>
> *(Seward and Conway, 1934: 737)*

Seward's pro-drift stance was maintained by his students Birbal Sahni, John Walton, and Ronald Good. Sahni (1891–1949) was an Indian paleobotanist.[11] He was General President of the Indian Science Congress (1940), twice President of its botany section, once President of the geology section. He was twice President of India's National Academy of Sciences, 1937–9 and 1943–4. In 1936, he became the fifth Indian to be elected Fellow of the Royal Society (London) (FRS). He was a Foreign Member of the American Academy of Arts and Sciences. A founder of the Indian Botanical Society, he was the driving force behind the creation of the Birbal Sahni Institute of Palaeobotany in 1949 at the University of Lucknow. The Prime Minister of India Pandit Nehru, who studied botany and geology at Cambridge, laid the foundation stone at the dedication ceremony. Tragically, Sahni suffered a heart attack three days after the ceremony, and died several days later.

At first Sahni was not a mobilist. In his 1926 Presidential Address to the geology section of the Indian Science Congress, he had hoped to present a rigorous test of drift by comparing fossil floras of the southern continents and India, but recognized that not enough was known about them.

> But it was soon realised that the problem was not so simple. We are not yet in possession of sufficient facts, and opinion is by no means settled with regard to the correlation of some of the plant-bearing strata. In Wegener's reconstruction one looks in vain for two regions once contiguous but now separated by the ocean, of which the fossil floras are equally well known to form a fair basis for comparison.
>
> *(Sahni, 1926: 222–223)*

He was interested in mobilism as an explanation of the distribution of *Glossopteris* among the southern continents but admitted that landbridges had merit; Sahni wanted more evidence of synchronicity among floras in all southern continents before ruling out landbridges.

He was, however, impressed with drift's explanation of the Permo-Carboniferous glaciation, and remarked, "one may perhaps say that it is to the study of the southern glaciation, rather than to that of the fossil floras, that this remarkable hypothesis owes its main support" (Sahni, 1926: 223). In 1926 he was very circumspect, but keeping an open mind, he delayed making a decision on the paleobotanical evidence.

So far as palaeobotanical facts are concerned, it does not seem as if we are yet equipped for a direct attack upon the problem. It is at any rate safe to leave one's mind open on the matter.

(Sahni, 1926: 223)

He then set himself a more limited goal, the examination of older southern hemi-spheric floras of particular interest to Indian Earth scientists.

Under the circumstances I have thought it best, instead of attempting an excursion into more speculative regions, to content myself with the task of collecting, in the form of correlation tables, the more important of the older floras of the southern hemisphere. The subject is of special significance to us here in India, although we are situated north of the equator, for it was in this country that the great problems of Gondwanaland were first raised and discussed.

(Sahni, 1926: 223)

It took him a decade to embrace mobilism, and he finally did so through a novel approach (Sahni, 1935a, 1935b, 1936, and 1937). Instead of looking for similar flora in supposedly once joined but now separated landmasses, he looked for the opposite, different flora in landmasses supposedly once apart but now joined. If two contem-poraneous fossil floras each indicative of different climates were now juxtaposed, he could argue that the crust on which each rested was formerly separated. He was looking for paleobiogeographical indicators of landmasses that were formerly separ-ated and have since collided. He found it in the Permian *Glossopteris* flora of India and the *Cathaysian* (*Gigantopteris*) flora of China and Sumatra. If India had drifted northward as suggested by Wegener and du Toit, it must have carried the beds containing *Glossopteris* along with it. As for the distribution of *Glossopteris* among the southern continents, he favored mobilism.

Sahni's argument about the former separation and present juxtaposition of differ-ent floras was clever and new. He began by making this basic distinction and then asked if drift was needed.

From a broad survey of the Late Palaeozoic botanical provinces two striking facts emerge; (1) some countries with closely related floras lie on the opposite sides of the biggest oceans of the globe; (2) others with very distinct floras, for example, the Gondwana province of India-Australia and *Gigantopteris* province of China-Sumatra, lie dovetailed with each other. Can we explain these facts without the aid of the drift theory?

(Sahni, 1936: 320)

Regarding the first, the paleobotanical record was insufficient to justify drift, even though it offered a good solution to Permo-Carboniferous glaciation.

The distribution of the early Gondwana glaciation has always impressed me as a most weighty argument in support of Wegener. But, speaking only as a palaeobotanist, I confess that my position ten years ago was that of an agnostic. It then occurred to me that it might help us to choose between the drift theory and the old theory of land-bridges if we could compare in detail the homotaxial floras of those districts which now lie on the opposite coasts of oceans, but which in Wegener's Pangea lay in contact with each other ... We still lack data from the critical areas either on the one side or on the other. Du Toit (1927) has, it is true, made an illuminating comparison of the geological features of the South African and South American coasts. This comparison has lent strong support to Wegener's theory. But from the palaeobotanical point of view the position both here and in other parts of Gondwanaland is much the same as it was ten years ago.

(Sahni, 1936: 320–321; Sahni's reference to du Toit (1927) is equivalent to my du Toit (1927b))

Like the true specialist he was, Sahni was not going to support mobilism unless the evidence from paleobotany, his area of expertise, justified it.

Sahni then turned to the second distinctive fact, the present juxtaposition of the very different yet essentially contemporaneous *Glossopteris* and *Gigantopteris* floras. *Glossopteris* flourished throughout Gondwana; *Gigantopteris* had European and North American affinities, and was quite distinct. Sahni concluded:

It will be agreed that the two floras, taken as a whole, are very distinct: they must have been evolved on separate continents, though it appears that some means of intermigration were possible at later stages.

(Sahni, 1936: 322)

He (1936: 323) then claimed "that the two floras, one essentially northern, the other southern, must have lived in different climates." In defense, he appealed to his former teacher, concluding that mobilism was "the only reasonable solution."

This brief statement regarding the climatic conditions governing the two floras may sound more like an assertion than a conclusion drawn from evidence, particularly at the time when a recent more critical examination of the evidence has threatened to shake what little faith we had in climatological value of fossil plants. But on the general question of the climatic contrast between the northern and southern floras, taken as a whole, even Professor Sir A. C. Seward, to whom this cautious attitude is largely due, and who has discussed the whole evidence in detail, agrees that "*the climate of Gondwanaland was doubtless comparatively cold well into the Permian period and much less genial than that of the northern continent*" (Seward, 1933: 258) ... If we agree that the Glossopteris flora of India and Australia must have flourished in a climate distinct from that of the Sino-Sumatran province, is it possible to believe that these two regions, dovetailed with each other as we now find them on the map, and covering so much of the same latitudes, have always occupied their present geographical positions? This is the crux of the problem. Drift is the only rational solution. I see no escape from the conclusion that the two provinces originally lay far apart, north and south of the Tethys, and have since drifted towards each other.

(Sahni, 1936: 322–323)

He proposed that India with *Glossopteris* and Southeast Asia with *Gigantopteris* were originally far apart. As they approached each other, the geosyncline, which still separated the two landmasses, became folded along its whole length, from its western

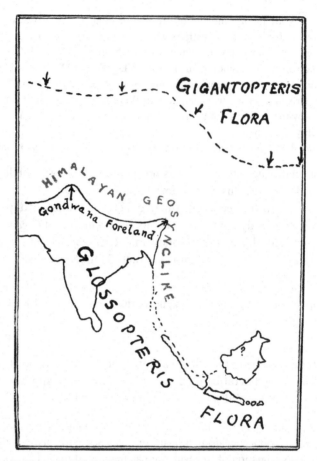

Figure 3.4 Sahni's Figure 2 (1936: 325). With the drifting of India into Asia and formation of the Himalayas, including their longitudinal extension through Burma and the Malay peninsula to Sumatra, the north and eastward spread of *Glossopteris* and south and westward dispersal of *Gigantopteris* is prohibited. The Assam and Kashmir promontories are the projections of the India block surrounded by the syntaxes of the Himalayas.

end past its knee-like bend (syntaxis) in northwest Kashmir, around the syntaxis in Assam, down through the Malay peninsula to Sumatra (Sahni, 1936: 323). The resulting Himalayan mountain belt now separates the *Glossopteris* flora of India and Australia from the *Gigantopteris* Sino-Sumatran flora. (See Figure 3.4.) By this elegant exploitation of his own specialty, Sahni had become a mobilist.

John Walton, professor of botany at the University of Glasgow, Scotland, had also been Seward's student. He favored drift as the solution of the distribution of *Glossopteris* and approved of Sahni's explanation for the juxtaposition of *Glossopteris* and *Cathaysia* flora: the absence of an acceptable mechanism bothered him, but he had the good sense to admit "we have no knowledge of forces in the Earth large enough to move continents" and "such forces and movements are beyond the range of our experience" (Walton, 1953: 184).

Ronald Good, another student of Seward's, began defending continental drift in the late thirties in his *The Geography of the Flowering Plants* (Good, 1947), and continued to do so forcefully and consistently through the early 1970s (see III, §1.12). Drift, he claimed, explained "the peculiarities of distribution to a degree of completeness far beyond that of any other theory."[12] It offered, he argued, the beginnings of a solution to the problem of the rapid development of angiosperms. He boldly denied that the absence of a known mechanism was an authentic difficulty; it was really only a recognition of ignorance.

... has all too often been accepted as final, but it must be realized that at best it is only negative. It would surely be untenable to suppose that there cannot be any such force. It is simply that no force is at present known.

(Good, 1947: 184)

Another Briton who favored drift on phytogeographical grounds was John Hutchinson, head of the African section of the Royal Botanic Gardens at Kew. His work on modern African plants led him to favor drift.

As a botanist who has studied the distribution of plants for many years, particularly African plants, I am a firm believer in Wegener's ideas ... Any botanist who has studied the floras of South Africa and Australia, or of Madagascar and India, cannot but be struck by their close relationships respectively, which may best be accounted for by the displacement theory.

(Hutchinson, 1946: 13–14)

Three paleobotanists from continental Europe, the Norwegian A. Hoeg, the Dutch W. J. Jongmans, and the Russian E. V. Wulff contributed to the discussion. Although Hoeg was not willing to endorse drift fully, he was impressed with its explanation of the distribution of *Glossopteris*.

To me personally the distribution of the *Glossopteris* flora together with the glaciations of the Carboniferous period, presents itself as the only phytogeographical problem which seems to be completely solved by means of the hypothesis of continental drift, and of which we have no better solution at present.

(Hoeg, 1937: 302)

Questioning the confident fixist arguments of Lake (§3.4), Hoeg (1937: 303) examined reported *Glossopteris* finds from Siberia in 1932, and decided that they had been incorrectly identified. Jongmans (1937: 345–362), who also corrected several mistaken identifications of *Glossopteris* from Siberia, embraced drift, agreeing with Sahni's analysis of the distribution of *Glossopteris* and *Gigantopteris* floras. Wulff, who received his degree from the University of Vienna and became curator of the herbarium at the Institute of Plant Industry in Leningrad, defended mobilism in his massive text on plant geography, rebutted phytogeographical difficulties raised against mobilism by appealing to his own fieldwork and by citing numerous European supporters (Wulff, 1950).

Support came from two American paleobotanists. In 1944 Douglas Campbell (1944: 181) of Stanford University claimed "du Toit's theory of continental drift is the only plausible explanation of the present distribution of plants, especially in the Southern Hemisphere," and continued to express support for drift in later publications. Chester Arnold, of the University of Michigan, described the strength of the paleobotanical support for mobilism as strong, but lacked the confidence in his own arguments to face down the claims by "good authority" that the forces to move continents were not there.

On the basis of present knowledge, Wegener's hypothesis or a modification of it offers the only explanation for certain distributional phenomena. On the other hand, many or perhaps the majority of geologists reject the theory as improbable or even fantastic. It is claimed on good authority that the known forces causing deformation of the crust of the earth are wholly inadequate to move continents the distances necessary to satisfy the requirements of the theory, and although it may appear to offer the answer to certain questions, there are others for which it is entirely inappropriate.

(Arnold, 1947: 399)

Henry N. Andrews, a paleobotanist at the Missouri Botanical Gardens and Washington University in St. Louis, wrote in his 1947 *Ancient Plants* that he found continental drift "the most appealing explanation" of various biotic disjuncts. After listing faunal disjuncts from opposing sides of the Atlantic, "drawn from Alfred Wegener's *The Origin of Continents and Oceans*," he turned to floral disjuncts.

In this same Atlantic region the distribution of fossils likewise speaks strongly for a former land connection. The great Carboniferous flora of North America is so similar to that of Europe that it is hardly conceivable to think of its origin as other than a continuous unit forest. Innumerable genera are found in both regions, and in some cases the species are seemingly identical, although, more than not, paleobotanists are inclined to use separate specific names simply because the specimens are found so far apart! Another distinctive upper Paleozoic group of fossil plants, known as the *Glossopteris* flora, is found in South America, South Africa, Antarctica, Australia, and India, land masses that are now separated by a broad expanse of open sea but which must have been united in one way or another in the past.

(Andrews, 1947: 249–250)

After noting that some postulate "former existence of land bridges" he (1947: 250) announced that to "the present writer the most appealing explanation lies in another direction," meaning Wegener's theory.

A few decades ago a German geologist, Alfred Wegener, ventured the view that the continents formerly existed as a great unit of land mass which split apart, with the parts gradually drifting to the positions they now occupy on the globe. On first thought this theory of continents drifting about may seem a bit fantastic, but there is a great mass of physical, geological, and biological evidence to support it.

(Andrews, 1947: 250)

Students of plants, whether of their present (plant geographers) or past (paleobotanists) distribution were more likely to support mobilism than were their counterparts among the students of present or past distribution of animals. Among zoologists and paleozoographers several continental Europeans, a few British, and fewer Americans supported drift. But Europeans favored landbridges rather than drift, while Americans, almost without exception, opted for either just a few landbridges or none at all. There was, however, an eminent French paleozoographer, L. Joleaud (1924: 325–360), who strongly supported mobilism. Initially, he believed in landbridges, but switched to mobilism in the early twenties and continued to defend it in his later work, appealing to the distributions of vertebrates and invertebrates. Joleaud proposed accordion-type continental movements allowing for openings and closings of the Atlantic.[13] But the zoogeographical support for mobilism, at least by vertebrate paleontologists, was weak. A notable exception was the British vertebrate paleontologist T. S. Westoll (1944), who had made a careful comparative study of early Permian *haplolepid–nectridian–aistropodan* faunas (primitive fish and amphibians) from Ohio, Ireland, England, and Czechoslovakia and concluded:

The case of the *Haplolepidae* can offer no decisive evidence in favor of either of these alternatives [drift or land bridge], but in the writer's opinion it is less easy to understand the distribution on the basis of "land bridges" than by accepting continental drift.

(Westoll, 1944: 110; my bracketed addition)

A decade later, Westoll had some influence on the mobilism debate during the rise of paleomagnetism (II, §4.7).

3.6 The resurgence of American permanentism: isthmian links

Permanentism was a variety of fixism. But, unlike landbridgers, permanentists rejected the idea of broad avenues of continental crust across wide oceans. They allowed only isthmian links, narrow strips of land connecting two larger land areas allowing the migration of land plants and animals. Permanentism had its main base in North America. Permanentists who were concerned with geophysical issues rejected landbridges because of their incompatibility with isostasy.

Permanentism actively promoted their ideas. Schuchert and Willis, both influential figures in American geology, presented a solution to the problem of disjunctive biota at the 1931 Annual Meeting of the GSA that greatly enhanced the status of permanentism, especially among American biogeographers and paleontologists. Schuchert, originally a landbridger, argued that narrow isthmuses were all that were needed to account for biotic disjuncts (Schuchert, 1932: 875–915). Willis (1932: 917–952) invoked several isthmian links sufficient, he thought, to account for them.

Isthmian links were made of basalt or of basalt and granite, both of which were derived from the substratum beneath ocean basins. They were short-lived. Once emplaced, they cooled and solidified, and began to sink until isostatic equilibrium was reached. Granite masses being lighter remained as islands. Eruptive masses of basalt in ocean basins may have surfaced only temporarily before they sank back into oceanic crust. If isthmuses were made of basalt and granite, they eventually became long slender peninsulas or island chains; if made entirely of basalt they eventually sank below the sea. Organisms could then have migrated along them for longer or shorter periods of time. Willis argued that because they were narrow, relatively dense, and short-lived, isthmian links were not, like landbridges, incompatible with isostasy.[14]

George Gaylord Simpson, a very influential American paleontologist, was a strong permanentist. (He is not to be confused with Sir George C. Simpson, the English meteorologist who viewed mobilism favorably (§3.12)). In papers published mainly during the 1940s, G. G. Simpson expanded the work of an earlier American paleontologist, W. D. Matthew (1915: 171–318), and developed a sophisticated permanentist scheme. Simpson, especially in his 1943 work, "Mammals and the nature of continents," launched a vigorous and effective attack against mobilists and landbridgers.[15] Du Toit (1944: 145–163) responded to Simpson's criticisms, and Chester Longwell (1944a: 218–231), who considered himself partial to neither permanentism nor mobilism, took it upon himself to reply to du Toit, Simpson being at the time in the Armed Forces.

For his purposes, Simpson could not have found a better source than Matthew, who died at the age of 59 in 1930. Matthew was a student of Henry Fairchild Osborn, who was in 1919 elected FRS and to the US NAS. Matthew, in his 1915 monograph *Climate and Evolution*, linked his view on past climate with those of T. C. Chamberlin, and proposed a fixist explanation of the disjunctive distribution of vertebrates.

Matthew's scheme was much more elaborate than the fixist solutions that Wegener had considered and rejected. He believed that terrestrial vertebrates evolved in the northern continents, the Holarctic, which had over time experienced large climate changes. As a result, less adaptable forms were forced to migrate south to gentler climes in the Southern Hemisphere. Almost all past and present-day distributions of terrestrial vertebrates, he argued, could be accounted for without resorting to former landbridges. Terrestrial vertebrate disjuncts could be the product of parallel or convergent evolution, their common ancestor now being extinct; they could also be remnants of a once widespread distribution now separated through the extinction of connecting populations. Matthew pointed to common ancestors and to extinct connecting forms in the fossil record. Where there was no common stock, he appealed to the fragmentary nature of the paleontological record, and predicted that later work would uncover them. Where separated regions had fauna with greater dissimilarities than similarities, Matthew argued that a former direct land connection was improbable because similarities would then be greater than are observed, and what few common life forms there were could have arisen by transoceanic rafting. He admitted that such rafting was very rare, but given time the likelihood was sufficient.

If then we allow that 10 such cases of natural rafts far out at sea have been reported, we may conclude that 1000 have probably occurred in three centuries and 30 000 000 during the Cenozoic. Of these rafts, only 3 000 000 will have had living mammals upon them, of these only 30 000 will have reached land, and in only 300 of these cases will the species have established a foothold.

(Matthew, 1915: 207)

Like Matthew, Simpson maintained that the distribution of mammals could best be accounted for without supposing either continental drift or extensive landbridges. He maintained "that continents were essentially stable throughout the whole time involved in mammalian history" (Simpson, 1943: 29).[16]

Simpson carefully distinguished three types of migratory routes: corridors, filter-bridges, and sweepstakes routes; three concepts that were the hallmarks of his analysis, and are still in current and fruitful use. Corridors link areas between which essentially no substantial physical barriers exist, for example, between Asia and Europe or between New Mexico and Florida today. The faunas of areas linked by corridors "will be very similar, or as far as genera or larger groups are concerned practically identical ..." (Simpson, 1940b: 147).

Filter-bridges connect regions with generally different faunas but some similar taxa. They are selective, allowing some taxa but not others to pass. Freedom to pass depends on the character and location of the bridge and the nature of the migrant. Examples are the Bering Bridge landbridge (Beringia) during most of the Cenozoic, the volcanic Isthmus of Panama for the past several million years, and the Early Eocene connection (Thulean volcanic isthmian link) between Europe and North America. He proposed two criteria for identifying filter-bridges. First, they allow for migration in both directions (1940b: 151). If, for example, marsupials from South America used a filter-bridge to get to Australia, then they could have returned later to South America. Second, "and perhaps the best criterion" (1940b: 152) of a filter-bridge "is that even though it rarely transports whole faunas, it does tend to transport integrated faunules. It does not transport all the genera of a continent, but neither does it transport one genus all by itself" (1940b: 152). He seems to have had in mind something like Willis' isthmian connections. Simpson believed that had they existed, former broad or direct continental connections as proposed by landbridgers and mobilists would have functioned as corridors rather than as filter-bridges.

There were also routes along which the "migration of single groups of mammals or of unbalanced faunas" occurred without meeting the "criteria for filter-bridge connections, and, of course, still less those for corridors" (Simpson, 1940b: 152). He labeled these routes "sweepstakes" to emphasize their indeterminate or "adventitious" character. "If a sweepstakes route exists, it depends on chance whether a given type of animal that can cross it will really do so" (1940b: 156). He cited as examples past links between the East Indies and Madagascar, between Asia and Australia, and between Alaska and Russia (Beringia). He considered sweepstakes routes the best way to account for isolated recalcitrant examples:

It has been claimed or felt, even by some adherents of Matthew's general thesis of "Climate and Evolution," that this sort of adventitious migration is dragged in when necessary to explain away the facts that contradict the main thesis ... It has not been sufficiently emphasized even by Matthew that the role of such a theory may be positive and primary, not merely negative and supplementary. Adventitious migration has indeed been used and sometimes abused simply to get inconvenient facts out of the way of a favored hypothesis, but there are instances in which adventitious migration is itself the most probable hypothesis and the most economical theory ... It is to be favored because it does explain, simply and completely, facts that the land bridge theory does not explain.

(Simpson, 1940b: 156)

He found that the percentage of "known Triassic South American reptiles (descriptions incomplete but based on latest published lists) also known from the Triassic of South Africa" was 43% for families and 0% for genera and species (Simpson, 1943: 20). Although he considered his results "preliminary but indicative," he concluded:

These figures are decidedly inconsistent with any direct union of corresponding parts of South America and Africa. The resemblance is greater than between South America and Africa today, but its small degree opposes a direct connection, even a connection by a direct bridge.

(Simpson, 1943: 22)

Simpson, contemptuous of much of the literature on disjunct biota, did not mince words in his treatment of Wegener, du Toit, Joleaud, Gregory, and von Ihering. Around 1940, he characterized the discussions as follows:

So multiple and varied are the facts and conjectures that have been published in this field that judicious selection and emphasis of them can be made to support almost any opinion not completely irrational. No one person can hope to know at first hand all the pertinent data, and a general review of the literature leaves one with the feeling that it can be taken to prove any of the dozen conflicting theories and that it therefore proves nothing.

(Simpson, 1940a: 755)

Simpson attacked mobilistic interpretations, arguing all that was required, once mistaken arguments and appeals to unreliable and incorrect data had been ferreted out, was one or other of his three types of isthmian links, each with the characteristics he had ascribed to them. He claimed that mobilists and landbridgers had overestimated the number of legitimate cases of disjunct biota, accusing them of many serious misdemeanors: they had used unreliable and inaccurate data based on inadequate taxonomical criteria; they had misidentified fossils; they had incorrectly analyzed data drawn from studies other than their own; they had neglected new fossil finds and new advances in taxonomy; they had underestimated the ability of organisms to disperse; they had overestimated the similarities and underestimated the dissimilarities between various fauna; and they had neglected to consider the possibility that many supposed disjuncts really were a consequence of parallel or convergent evolution (RS2).[17] By these serious accusations Simpson attempted to discredit drifters and landbridgers, and he met with much success. Just as Simpson, the permanentist, intended, his attack had the effect of decreasing considerably the number of disjunctive life forms seemingly requiring explanation. Using with great dexterity his three types of migratory routes, he argued that the degree of similarity between biotic disjuncts was often too low to require corridors or broad continental connections at all, and, in many instances, filter-bridges or sweepstakes routes were sufficient. In the eyes of many, his arguments presented acute difficulties for mobilists, especially if the similarities between biota could be deemed less than should be expected if continents had once been face-to-face along an entire coastline.[18]

Simpson dealt in detail with the unusual mammalian faunas of Australia. Tactically, this was clever, because Wegener had placed much weight on them. He argued that his solution was "complete" and "simple."

The faunal relationships of Australia are completely and simply explained by the view that Australia has had about its present relationships to other continents since the Cretaceous, at least, that the marsupials entered by the island route from Asia, and that rodents entered later over the same route.

(Simpson, 1943: 16)

Australian mammalian faunas fall into three groups. There are the ancient egg-laying monotremes, whose place of origin is unknown. Next are marsupials, which, Simpson believed, traveled by a sweepstakes route from Asia, although not extant there; they are related to fossil northern forms, and not to South American forms as Wegener would have them; at best, according to Simpson, Australian and South American marsupials share only one superorder and one family. Finally, there are the placentals: rodents, bats, dingos, and humans; he claimed that they used the same Asian sweepstakes route as did the marsupials, but later. As a result, except for humans, they never became overwhelmingly successful because their marsupial competitors had long been there and were well adapted.

Simpson argued confidently that his solution was the only adequate one. It alone could account for the new discovery (unknown to Wegener) that Asian placental mammals had existed prior to marsupials entering Australia. Only an early arrival of marsupials (and monotremes) by sweepstakes routes could, he believed, explain why placentals had not entered and become dominant as they had elsewhere. Simpson criticized the landbridge explanation of many of his fellow fixists; those who posited a landbridge between Australia and South America to account for Australian marsupials had grossly exaggerated the similarities between them and South American forms; had there been a land connection, similarities there would have been greater. South American placentals also are likely to have taken advantage of such a connection, and would have come to dominate Australian mammalian faunas, which they clearly have not. Simpson had no time for drift theory. Wegener had agreed with the prevailing opinion that Australian placental mammals came from Asia in post-Pleistocene time (past 10 000 years), but Simpson argued fossil placental mammals of Pleistocene age (approximately 1 million years) had been found in Australia, and there was indirect evidence that their journey had begun as early as Miocene (20 million years). Moreover, Wegener had made matters worse by maintaining the connection between Australia and Antarctica until Eocene.

Therefore the most probable interpretation of the faunal evidence leaves him only about the span of the Oligocene for the whole drift of Australia from Antarctica to its present position. If du Toit's opinion is accepted that Argentina–Antarctica–Australia were connected "down to the Oligocene at least," exceedingly little time is left for Australia's supposed shift from the Neotropical to the Oriental faunal sphere – the continent must have been speeding rather than drifting. An adjustment of the general drift theory – like so many others that have had to be made – could remove the inconsistency involved in these particular drift theories, but there is another point still harder to reconcile with any such theory: after drifting to about its present position in the Miocene, and doing so with most remarkable rapidity, Australia's drift relative to Asia would then have to stop. If it had not been about as close to Asia as it is now, the rodents could not have reached it, but if it had gone on drifting it would either be closer now than it is or would, by a radical change in the theory, have been closer in the immediate past and then would certainly have lost the degree of faunal isolation that it actually has.

(Simpson, 1943: 15–16)

Simpson ruled out continental drift.

3.7 Du Toit, Simpson, and Longwell debate

G. G. Simpson's attack against continental drift did not go unanswered. Du Toit attempted to counter many of his specific arguments about mammalian distribution to show that within the framework of the mobilism debate as a whole Simpson's work ought to be given little weight. Du Toit gave three reasons. First, retreating from his 1937 position that paleontology could offer a critical test of mobilism, he returned to his more moderate 1927 stance.

It is true that paleontologists have thus far not shown any leanings towards the new ideas and seemingly remain "orthodox" in outlook, but such is understandable for, as the writer remarked in 1927, "geological evidence almost entirely must decide the probability of this hypothesis for those arguments based upon zoo-distribution are incompetent to do so." After studying Simpson's paper and consulting those authorities cited therein that are available in South Africa, the author sees no reason for modifying that apparently sweeping pronouncement.

(du Toit, 1944: 146)

Du Toit felt that there were too many possibilities open to biogeographers, even when data were good. The fossil record was fragmentary, especially in the Southern

Hemisphere where most of the spectacular biotic disjuncts occur.[19] Du Toit agreed with some of Simpson's criticism, data often were unreliable; it was not unusual for specimens to be misidentified, and he pointedly cited the incorrect identification of *Glossopteris* flora in Russia, and misunderstandings that it had created.

One example of outstanding importance will be cited, namely the recording by Amalitsky in 1899 of typical members of the *Glossopteris* Flora ... in the Permian of Russia, an association that has never been questioned, at least in print ... the writer's examination in 1937 of several of the types identified as *Gangamopteris* and *Glossopteris* by Amalitsky, or of duplicates so labeled by him, showed that those particular specimens, while outwardly very similar, yet failed to agree in their venation with those type genera of the *Glossopteris* Flora.

(du Toit, 1944: 148)

Taxonomies were constantly being revised; there were disagreements among authorities, and there was some reluctance to ascribe identities to specimens separated by large distances even if they were in fact identical. Second, du Toit maintained that mammalian distributions had little to do with the drift question because mammals only became significant after continents had done most of their drifting.

By the writer the Hypothesis of Drift is regarded as essentially established by the Paleozoic and early Mesozoic evidence; the rest is largely a logical corollary, which in the main is consistent in outline, but has not yet been worked out in detail, for that would be a mighty task ... The true role of the mammalia will be the fixing of land bridges not in the horizontal, but in the vertical plane, through revealing the uprising and downwarpings of the crests of such linkages, themselves positioned by other, and mainly older, criteria.

(du Toit, 1944: 153–154)

Finally, he claimed that mammalian distribution was at odds with the distribution of lower vertebrates, invertebrates, and plants that was more germane to drift.

In contravention it may be questioned whether the mammalia have not created almost more biogeographic difficulties than they have solved. Even Matthew is forced to admit a discordance between the vertebrate and invertebrate evidence. Regrettably they have been robbed of much of their value through their late appearance on the scene, well after continental rupturing is deduced to have occurred, yet while such supposed movements were in full swing.

(du Toit, 1944: 146)

Longwell's response, broader in scope than du Toit's, was immediate and generally sympathetic to Simpson. He agreed with du Toit's assessment that the problems of disjunct biota and mammalian distribution were of minor importance in the overall mobilism debate. But he pointed out that du Toit had shown little restraint in evaluating them in his book, *Our Wandering Continents*. Furthermore, Longwell argued that mobilism faced obstacles in areas other than mammalian distribution, and turned to the major difficulties posed by the biotic similarities on opposite sides of the Pacific. He cited the distribution of the genus *Cornus* (dogwoods) found in

North America, southeastern Asia, on islands in the South Pacific, but absent from South America and Australia, and suggested:

The hypothesis of continental drift does not help explain the dispersal of dogwoods – rather it increases the difficulties. Obscure ways in which plants and also some animals must have been distributed during long geologic time intervals are suggested by other examples from the vast Pacific region. When these numerous problems are viewed together, emphasis that has been placed on the *Glossopteris* flora as "compelling evidence" of once-continuous lands seems dangerously near the unscientific procedure of selecting evidence to support a favored theory.

(Longwell, 1944a: 224)

Longwell did not want du Toit to forget that mobilism had its own biogeographical difficulties. He (1944b) wrote another brief note, and du Toit (1945), not wanting to allow him the last word, replied. He directed his response primarily at Longwell and at Willis (1944); their discussion is described later (§6.8).

3.8 Support for permanentism continues through the mid-1950s

G. G. Simpson's work was very influential among American zoogeographers. They were permanentists, and almost without exception were not just mildly, but vehemently opposed to continental drift. Alfred Romer, the only prominent American vertebrate zoologist to show sympathy with mobilism prior to the rise of paleomagnetism, compared early Permian reptiles and amphibians from deposits in Texas and Czechoslovakia and found them very similar. Checking against their modern-day counterparts, he showed that the degree of similarity was much greater during the Permian than at present. But he expressed himself cautiously, and was unwilling to choose between the landbridges and drift.

Consideration of Paleozoic vertebrate faunas as a whole thus leads to the conclusion that during this time North America and Europe were connected in such fashion that extremely free and relatively rapid faunal interchange was possible among the rapidly evolving vertebrate groups. Discussion of the type of connection involved is handicapped by the fact that until Middle Permian times we know almost nothing of the vertebrates of any regions except Europe and North America. It is possible that the similarities between these two continents were due merely to a condition in the Paleozoic in which vertebrate faunas of other regions, if known, would also be similar in composition ... The available evidence strongly suggests (although it does not prove) intimate and direct connection in the later Paleozoic between Europe and North America whether by apposition of the two fixed continental positions, or by a substantial North Atlantic bridge since destroyed.

(Romer, 1945: 440)

American specialists in invertebrate zoology and plant geography also followed Simpson's lead. P. J. Darlington, Jr., an entomologist at the Museum of Comparative Zoology, Harvard University, defended permanentism in the late 1940s and throughout the 1950s. I shall later show how he subsequently came to embrace mobilism in the mid-1960s (III, §1.12). At this stage however, his work on insects and other invertebrates led him to claim that they are unhelpful in addressing biogeographical questions; cold-blooded vertebrates, he believed, were more helpful.

I am an entomologist, but what I know about insects has convinced me that they, and most other invertebrates, are not yet good material for the study of zoögeography. They are in every way too little known, and their powers of dispersal are too great or too doubtful. The cold-blooded vertebrates are better known; they have left a better fossil record, although not so good a one as could be wished; and their powers of dispersal are limited and better understood.

(Darlington, 1948: 1)

Darlington disagreed with Matthew's view that the northern temperate zone has been the principal center of evolution of cold-blooded vertebrates. Instead, he placed their center in the tropics, and discussed their dispersal in terms of three filter-bridges "between North and South America," and intermittently between "Eurasia and North America" and "Asia and Australia" (Darlington, 1948: 115) without recourse to continental drift.

I agree ... that Matthews's analysis ... was enormously important in counteracting a reckless building of land-bridges and in inaugurating a new, more critical, and more logical phase in the study of animal distribution. And I agree with Matthew himself that it is not necessary to remodel the world to account for vertebrate distribution.

(Darlington, 1948: 119)

Darlington (1949) reviewed Jeannel's *La genèse des faunes terrestres: Eléments de Biogéographie*, firmly rejecting its mobilist interpretation. Eight years later, Darlington (1957) returned to biogeography, still unsympathetic but now willing to consider drift.

Nothing in zoogeography has brought forth more arguments or more demands for an open mind – the other man's mind – than the idea of continental drift. I have tried to keep my mind open on this subject and have made a new beginning by trying once more (as I have done before) to see if I can find any real signs of drift in the present distribution of animals. I can find none. So far as I can see, animal distribution now is fundamentally a product of movement of animals, not movement of land.

(Darlington, 1957: 606)

Among paleobotanists, William Darrah, Ralph Chaney, and Daniel Axelrod stand out as leading permanentists. Darrah, in his 1939 *Textbook of Paleobotany*, wrote:

The concept of Wegener that the continents have tended to drift apart from a single original land mass, "Pangea," has long been used as the basis for explaining the general distribution of Paleozoic plants ... A fundamental error in this reasoning is the belief that at no time subsequent to the Permian, did a cosmopolitan flora develop. Such great floras are characteristic of each major geologic period.

(Darrah, 1939: 326)

Chaney (1940: 486) argued that Wegener's theory condemned American Tertiary forests to freezing, and "far from finding support from Tertiary floras, must meet the

evidence that forest distribution has been controlled by existing land and sea relations throughout the Cenozoic." In this Chaney was correct; Köppen and Wegener
(1924) incorrectly placed the North Pole close to North America because, at the time,
they did not realize Earth was then in a much warmer general state. Axelrod (1952)
had no sympathy for mobilism, arguing against it until well after the vast majority of
Earth scientists had embraced plate tectonics (III, §1.12).

A sense of the general consensus of the North American Earth science community
in favor of permanentism was neatly displayed during the 1949 symposium, "The
Problem of Land Connections across the South Atlantic, with Special Reference to
the Mesozoic," held in New York by the Society for the Study of Evolution (Mayr,
1952: 85–258). Of the seventeen participants, only Romer, Wendell Camp, and
Kenneth Caster supported the idea of a trans-South Atlantic Mesozoic land connection, but they could not decide between landbridges and drift.

Camp, Curator of Experimental Botany and Horticulture at the Academy of
Natural Sciences, Philadelphia, attempted to determine whether land connections
were needed to account for the rise and diversity of angiosperms. Although their
fossil record was at the time so sparse, especially in the tropics, where, he maintained,
records were most needed, he determined general phyletic patterns of extant taxa
using them to trace major phyletic divergences (Camp, 1952: 205). He concluded that
most primitive members of the basal groups of flowering plants were distributed
along a major trend line extending from Australia through Indonesia, Malaya, India,
Africa, and South America. The main primitive members were spread either side of,
and along the whole or part of the trend line. There also were minor trend lines back
and forth between Africa and South America, between Australia and South America,
South America and North America, Africa and Europe, and extending from Southeast Asia through China and Russia into North America. Camp, unlike many
specialists, believed that angiosperms originated in the Southern Hemisphere.

I am unable to accept the Matthewsian dictum that the primitive members of a group occur at
the outer margin of a group's range because they were forced there by the later developed and
more aggressive members ... Because of this array of evidence, I [believe that] not only have
the lands crossed by this great phyletic trend line been the ancestral homes of our flowering
plant groups, but the bulk of their primary evolutionary divergences also took place in these
same areas.

(Camp, 1952: 209; my bracketed addition)

Camp addressed tangentially the competing view that angiosperms arose in Holarctica
and migrated from there to the southern continents. What he said also showed the
importance of undertaking paleobotanical investigations of flora in Africa (north of
the Gulf of Guinea) and South America (Guiana shield), and what often happened to
those who did.

This much is fact: Almost every botanical expedition to critical areas in South America and
Africa returns with additional material ever more closely linking the plants of the mountainous

areas north of the Gulf of Guinea with those of the mountains of the Guiana shield. Some have attempted to dispose of the problem by claiming that these are examples of parallel, or perhaps convergent, evolution. These instances are now so numerous and in so many groups of plants that botanists are no more willing to accept this as a tenable hypothesis than are zoologists the hypothesis that the interfertile races of man are the result of convergence, having descended from different kinds of ape-like creatures. Such speculations, which only obscure the problem, would be unnecessary if it were to be ascertained that the basic elements of our angiospermous flora already were present in the Mesozoic and that the final segmentation of Gondwana did not occur until the same time. Considerable patience and tolerance towards exploratory hypotheses will be necessary until additional paleogeographical facts are uncovered.

(Camp, 1952: 211–212)

It sounds as if Camp felt that his permanentist peers were too drift intolerant.

Camp then turned to the question of Gondwana. He favored its existence, but thought that the final decision rested with geologists. In closing, he left the question of continental drift unanswered.

I have but one conclusion pertinent to the topic of this symposium: If any significance is to be placed on the sequences of phyletic patterns among our basic groups of flowering plants it would seem evident that they did not come down out of Holarctica but originated and underwent their primary development and early dispersal on a common land mass within the Southern Hemisphere. This early dispersal within the Southern Hemisphere is tentatively placed at some time in the Mesozoic.

(Camp, 1952: 212)

As I shall explain in §6.2, Caster was one of the very few North American paleontologists sympathetic toward mobilism. His recollection of what happened when he was asked to participate in the symposium explains the awkward title of the symposium and the absence of strongly committed mobilists. He received an invitation to participate in the symposium while in Brazil, where he spent three years studying Paleozoic fossils. He underscores the strong resistance to mobilism in North America.

While still in Brazil, I had been invited by Professor Glenn L. Jepsen of Princeton to take part in "A symposium on Continental Drift" sponsored by the Society for the Study of Evolution (of which I was a founding member) to be held at the American Museum in December of 1949. He suggested that the results of my Brazilian Paleozoic studies might be a relevant paper. I agreed, and set about outlining such a paper. But about the time of my leaving Brazil I was somewhat nonplussed to get an apologetic and embarrassed letter from Jespen saying that the title of the symposium had been changed at the insistence of several invited participants who were fearful that it would be professionally hazardous for them to associate themselves with a colloquium bearing the title-words "Continental Drift," the results of which were to be printed! (The actual title as published was "The Problem of Land Connections.") Since it was obvious that the subject I had been invited to speak on did not fit under the revised title, would I have any data from my Brazilian work, say, on the Triassic, about which I might shape a paper? They still very much wanted me to participate. It happened that I did have such material, admittedly of quite secondary interest to me, but my curiosity was so piqued by the reported attitude that I wouldn't have missed the meeting for the world! So I agreed, wrote and delivered the paper that Jepsen had suggested. However, it was very much pro-Drift.

(Caster, 1981: 7)

Caster later learned who objected, and suspected they were not alone.

I later learned that Walter H. Bucher and Marshall Kay of Columbia University were chief among the fearful, although I am sure George Simpson and Ernst Mayr shared their reluctance.

(Caster, 1981: 7)[20]

Bucher and Kay were very influential North American geologists. Bucher was also very influential in the geophysical community; he was twice President of the American Geophysical Union (1948–53), and one of its senior medals carries his name; he was a structural geologist. The 1933 appearance of his *Deformation of the Earth's Crust* marked him as a major opponent of mobilism. His vehement resistance to mobilism will resonate throughout the controversy, continuing into the 1960s (§6.10, §7.3, §8.4, §8.8, §9.6; III, §1.10, §6.3). Kay, a sedimentologist, developed a sophisticated fixist account of mountain building during the 1940s and 1950s (§7.3; III, §1.14) by which time, as already described (§3.6), Simpson had from the paleontological standpoint developed an improved version of permanentism and showed his opposition to mobilism during his debate with du Toit.

Ernst Mayr was perhaps the most influential evolutionary and theoretical biologist of his generation. He adamantly opposed mobilism, and had spoken against it at the 7th Pacific Science Congress in February 1949, ten months before the New York symposium (Mayr, 1953). A specialist on birds of the southwest Pacific, he argued that Australia and New Guinea had not moved relative to the Asiatic continent. In his confident magisterial fashion, he raised four objections to the idea that the Australia–New Guinea block had only recently approached Asia.

(1) If the drift hypothesis were correct, one would expect two faunas in New Guinea, an old endemic Australian one with South American and South African relationships and a very recent Malayan one. However, as I have shown elsewhere (Mayr, 1944), the Australo-Papuan fauna is characterized by a complete gradation between the oldest, very isolated endemics and the most recent immigrants. This makes sense only if Australia were available to Asiatic immigrants throughout the Tertiary.

(2) The existence of endemic or subendemic tropical Papuan families, such as the birds of paradise, is inexplicable if late Tertiary drift is accepted. Neither is that of the Australo-Papuan groups in which the endemic Papuan element is about as strong as the Australian one.

(3) If Australo-Papua had crashed only recently into the Moluccan-Melanesian chain, one would expect a strong relationship between the Moluccan and Melanesian biota, in contrast with the Papuan one. However ... this is not the case.

(4) The frequency of endemic genera (of Australo-Papuan origin) found in the Moluccas and the Celebes region cannot be explained unless the Australo-Papuan area has been in its present position for a long period of time ... The same is indicated by the great number of genera that are rich in species both in the Malayan and in the Australo-Papuan region.

Mayr claimed that these objections appeared to rule out continental drift at least for the Cenozoic (Mayr, 1953: 13–14). But Mayr went too far: his evidence pertained only to Wegener's claim that Australia converged with Asia in the Tertiary. He (1953: 14) acknowledged that phenomena such as gymnosperm distribution favored drift, but was firm that present-day distribution of Papuan bird fauna did not support drift. He heard nothing in New York to convince him that continents had drifted during the Mesozoic; he agreed with G. G. Simpson, and believed Bucher had shown that there was no geological evidence in favor of mobilism but much against it. He (1952: 256) did admit, however, that Romer's study of Mesozoic reptiles, which he presented at the meeting (see below), "favors a trans-Atlantic African-South American connection"; but, unable to let it pass without comment, added unhelpfully, "additional discoveries in the very poorly known North American Triassic may lead to a modification of this conclusion"; as if scientists were expected to consider results that had not yet been obtained. Moreover, Mayr (1952: 258) gave no indication that he thought a landbridge, let alone continental drift, was needed to explain the affinities between African and South American Triassic reptiles. He cautioned that former hypothetical land connections should not be accepted unless consistent with all evidence, as if nature and capable scientists were in need of restraint. He apparently felt, as the authority, that he had to say something to keep things on an even keel, to prevent things from getting out of his control.

Caster felt that mobilism topics had been deliberately excluded, was troubled by the rarity of mobilists at the meeting, and later found out that one of his own colleagues at the University of Cincinnati had not been invited because her data supported drift.

When the final program was presented, I was dismayed, for to my mind, and to a very few others in attendance, it was manifestly a "stacked symposium" composed mainly of those paleontologists and biologists as speakers who were outstanding opponents of Continental Drift, or specialists in groups of organisms, especially vertebrates and higher plants, whose record would be irrelevant or indicate disjunct evolution.

While Jepsen was formulating a program I had suggested to him that my very distinguished colleague in the Botany Department at the University of Cincinnati, Dr. Margaret Fulford, be invited to present her highly relevant data on relationships of leafy liverworts, probably earliest and certainly most primitive, of terrestrial plants, which she had personally collected and studied in many parts of the Southern Hemisphere. On the advice of Harvard participants, I later learned, she was never invited because her data would be more supportive of Drift than not. Dr. Fulford had found identical or closely related leafy liverworts in places now separated by oceanic vastness, South Africa, Tierra del Fuego, Australia and old oceanic islands. The crux of her plants in this connection is that she had been able to show that, contrary to botanical opinion, her tiny leafy plants were more primitive than the thallus-forms commonly cited as most primitive of living flora; that the thalliform condition is the end-product of a reduction series in evolution. Moreover, the leafy liverwort's spores will not survive transport, either by wind or water or on the feet of birds, they must fall into wet ground, in close proximity to the sporangia. Desiccation and sunlight are speedily lethal. Thus they must have "walked" on moist land to achieve their current dispersal. (See M. Fulford, "Distribution patterns of the genera of leafy Hepaticae of South America," *Evolution*, 1951, v. 5, pp. 243–264, etc.). Such evidence would disturb the tenor of the planned biologic "death knell" to Continental Drift!

(Caster, 1981: 8)

Although Caster talked only of his Triassic data, his published paper did pretty much what he had originally been asked to do.

I gave my paper in the form desired, and heard Carl Dunbar's adverse discussion of it (in its published form his words were supplemented by Kurt Teichert's adverse criticism), but pleaded with Ernst Mayr, Editor, for time for revision, due to the short notice I had had of changed subject and my immediately previous preoccupation with fieldwork in Columbia and the lower Amazon. I must confess to a certain perverse satisfaction in being the last participant to submit a completed manuscript, and one that was so revised that I pretty much accomplished what I was first requested to do ... Among other things, in this paper I attempted to account for certain minor stratigraphic discrepancies and stratigraphically dissimilar glacial occurrences on the basis of lateral facies changes between South Africa and Brazil when the land masses were contiguous.

(Caster, 1981: 8–9; Teichert's views are discussed in §9.3)

Caster met Romer at the meeting, and they became friends.

At the meeting in New York, only Alfred Romer and I spoke favorably of Continental Drift; Al[fred] in criticism of vertebrate paleontology papers. As a result we became fast friends – despite my never accepting his personal views on the fresh-water origin of fishes!

(Caster, 1981: 9; my bracketed addition)

Caster's recollection, however, is mistaken; neither he nor Romer actually spoke favorably of mobilism at the meeting. Caster claimed that the continents had not been contiguous, but had been joined by a broad landbridge or a great craton that subsequently foundered to make the South Atlantic basin. He wrote:

To this writer there seems to be no satisfactory manner of accounting for ... relations as preserved in South Africa and South America without intimately connecting their history across the South Atlantic basin. This history requires the absence of oceanic deep where now the basin exists. This means restoring a continental, cratonic sector between the present borderlands ... It would appear that the marginal "foundering" of the Afro-American linking craton has continued throughout the post-Rhaetic. In fact, the very late Cenozoic block-fault subsidence of the whole south Brazilian coast from Bahía to Rio Grande do Sul forming the present narrow continental shelf appears to be the latest expression of this craton-sapping or foundering process. Whether or not there has been actual drifting apart of the two continents is a subsidiary detail and one towards the investigation of which this survey was not directed. Indeed, it is doubtful if the stratigraphic record and associated fossils in the relic masses of the present continents would give much reflection of continental drifting if it has occurred.

(Caster, 1952: 145)

This is hardly advocacy of continental drift. But, he did not rule out some combination of drift and cratonic foundering.

The present consensus of geophysicists seems to be that it is equally impossible (possible?) to founder as to drift areas of continental character and proportions. Certainly a bridge across the present South Atlantic basin would be a respectable piece of craton. Perchance the real explanation of the coming into existence of the South (and North?) Atlantic basin is progressive growth through the operation of both "impossible" processes.

(Caster, 1952: 145)

What about Romer? He joked about his radical position, next he might have to appear before House Committee on Un-American Activities of the US House of Representatives.

I have been reared in the tradition of continental stability and was early impressed by Matthew's demonstration (ably reinforced by Simpson in recent years) that most if not all Tertiary land bridges are not merely delusions but snares to the unwary. However, I find myself here, after consideration of the evidence, rather strongly inclined towards belief in the existence of a southern inter-continental connection between South America and South Africa in the Triassic. To my embarrassment; for in such a "leftish" position I am disturbed (like many a liberal in political circles) by the company (of bridge builders, radical continent shifters, and Gondwanaland collectivists) which this may entail.

(Romer, 1952: 250)

Romer criticized Edwin H. Colbert's earlier contribution at the symposium on Mesozoic tetrapods of South America and Africa (Colbert, 1952: 237–249). Colbert, then at the American Museum of Natural History, argued that, although the evidence slightly favored a trans-South Atlantic connection, the permanentist thesis, that the common reptiles of South America and Africa immigrated there by way of Europe and North America, was defensible. He found that among reptiles of the whole Triassic, South America shared 100% of suborders and 75% of families with Africa, and 71% suborders and 63% families with North America. Romer believed it was more realistic to confine comparison to the Middle Triassic.

To sum up, it is difficult to furnish Brazil with a Triassic reptile fauna from North America, since there is no evidence that most of the groups needed to stock the fauna were present in the north at the appropriate time. On the other hand, we have in Africa a contemporary fauna almost identical in nature with that of South America. To apply to the actually comparable Middle Triassic reptile faunas the type of treatment used by Colbert, the matter of identities stands as follows:

	Suborders	Families
South America–Africa	83%	72%
South America–North America	33%	?14%

The evidence as it now stands strongly supports the hypothesis that some type of connection existed between Africa and South America in Middle Triassic times; the burden of proof rests upon those who wish to deny its existence.

(Romer, 1952: 253)

Romer (1952: 253) also regarded *Mesosaurus* as the "strongest single piece of evidence for South American transatlantic connections" because its remains had been found only in South Africa and South America and "no one would contend that it was capable of breasting the waves of the South Atlantic." However, he was challenged on the latter point; at the very same New York meeting Carl Dunbar (1952: 155), long-time fixist from the Peabody Museum (Yale University), thought that *Mesosaurus* could have crossed the South Atlantic.

G. G. Simpson's work influenced zoogeographers throughout the world. Although in the 1920s some Australian zoologists had favored mobilism or landbridges in order to explain their fauna, after Simpson's analysis they swung to permanentism (§9.5). Many European zoologists likewise favored permanentism. For example, L.F. de Beaufort (1951), professor of zoogeography at Amsterdam, supported permanency in his *Zoogeography of the Land and Inland Waters*, and Carl H. Lindroth (1957), a prestigious Scandinavian zoologist, similarly defended permanency theory in his *The Faunal Connections Between Europe and North America*. Even Holmes, who was sympathetic to mobilism, admitted in his review of the 1949 New York symposium,

In recent years the weight of evidence has become less oppressive, and this symposium has left me with the general impression that a few land bridges or linkages by island stepping-stones would probably suffice for the biogeographical problems.

(Holmes, 1953: 671)

3.9 Questioning reliability and completeness of the biogeographical record

There were numerous instances during the paleontological sub-controversy when researchers used data or techniques that were later shown to be unreliable. As described already (§3.4), there was Lake's appeal to *Glossopteris* flora in Russia that was based on mistaken identifications. Simpson pointed to the too close similarity between Australian and South American marsupials assumed by drifters and land-bridgers (Lake, 1923). Du Toit had to agree with Simpson that much of the fossil data used by drifters was not reliable, but argued that they were not the only offenders. Even Longwell, who lamented the use by others of unreliable data, did so himself when he claimed the distribution of *Cornus* was evidence against drift, his claim being based on erroneous taxonomy, as Good pointed out at a 1950 meeting of BAAS on the status of drift theory.[21]

An early speaker warned his audience against the uncritical use of reported cases of discontinuous distribution, and it was significant that the only instance of plant geography referred to afterwards was just such a case, that of *Cornus*, which has recently been quoted in connection with continental drift.

(Good, 1950: 586)

In 1937 the American ichthyologist from Stanford University, George S. Myers (1938: 340–341) predicted, "despite the fulminations of those opposed to the theory of continental drift, there is a considerable body of weighty evidence in favor of it, and I for one, should not be surprised to see it finally prevail." However, he had this to say about ichthyological data:

In reviewing the distribution of American continental faunas in connection with some recent studies of West Indian fresh-water fishes, I have been struck with the amount of misinformation that passes as sound ichthyological evidence among zoogeographers. Part of this is the fault of the zoogeographers themselves. Papers on fish distribution written by competent ichthyologists and based on modern paleontological data are scarce, and the distributional information in two recent general textbooks on fishes . . . is scarcely to be relied on.

(Myers, 1938: 343)

Notwithstanding, Myers (1938, 1953) did not support mobilism. Rejecting Matthew's idea of the dispersal of vertebrates from a Holarctic center, he proposed that the distribution of freshwater fishes in South America and Africa is best explained by assuming late Mesozoic or early Tertiary land connection.

What I have said will, I believe, make it clear that our present knowledge of the fishes very distinctly favors a late Mesozoic or very early Tertiary South Atlantic land connection and makes a direct northern origin of at least the South American ostariophysans seem exceedingly unlikely.

(Myers, 1938: 354)

In specifying the type of land connection, Myers (1938: 352) rejected a "gigantic land bridge to account for the similarities plainly seen in the characins, cichlids, nandids, and others" of Africa and South America.

One fact alone prevents me from believing in a wide, open continental connection across the South Atlantic during the life of the present South American and African families of freshwater fishes. I have mentioned the occurrence in Africa of a most remarkable assemblage of undoubtedly old families of isospondylan teleost fishes, to say nothing of the vastly more primitive bichirs; I cannot but believe that these were probably present in Africa before the cichlids, carps, or characins. If there had been a wide Tertiary or even late Cretaceous bridge across the South Atlantic, why do we find not a single solitary representative of any of them (save again the osteoglossids) in South America? I refuse to believe that competition in South America could have killed them off; they have survived just as acute competition in Africa.

He came out in favor of Willis and Schuchert's isthmian connections.

> If we are to have a South Atlantic bridge in the late Cretaceous or earliest Tertiary, the only kind I can conceive as fitting the requirements of the fishes is one like Willis's (1932) Brazil-Guinea isthmus in the Atlantic, and from Schuchert's data, it may have still been partly in existence up to the very end of the Mesozoic. If our fishes were present in the earliest Eocene, they may have been there at the end of the Cretaceous. Such a narrow isthmian connection, with its short, swift rivers, would not provide a broad highway for all the fresh-water fishes, but it would allow to pass the same aggressive types that are now held in common by Africa and South America.
>
> *(Myers, 1938: 353)*

Myers (1953) continued to reject broad landbridges and mobilism.[22] Turning his attention to the Pacific, he declared:

> There is nothing in the distribution of the fresh-water fishes of the Pacific islands favourable either to continental drift or to J. W. Gregory's Pacific continent, and much that opposes the latter. In fact, the first evidence points directly to the conclusion that the islands of Oceania are truly oceanic, in the sense of Wallace.
>
> *(Myers, 1953: 46)*

Even though Myers believed that the data were incomplete and often unreliable, and that mobilism might turn out to be correct, he thought there were enough good data to swing the argument in favor of isthmian links.

In his Presidential Address to the geology section of the BAAS, W. N. Edwards explained the incompleteness and unreliability of the phytogeographical record and emphasized the need for more reliable techniques. As an example, he described, as others had done, the mistaken identification of *Glossopteris* in Russia.

> The supposed *Glossopteris* from Russia, originally produced by Amalitsky, who was not a botanist, as a tidbit for the participants in the Geological Congress of 1897, was not even figured until many years later, but found its way into textbooks with astonishing rapidity. When the Geological Congress again went to Russia in 1937, several visiting palaeobotanists were able at last to see the actual specimens, which turned out to be completely worthless as records of *Glossopteris*; the Russians themselves were meanwhile putting these and other somewhat similar leaves into new genera.
>
> *(Edwards, 1955: 167)*

Despite this, Edwards, like Myers, still thought that the balance of evidence favored fixism. After summarizing data on flowering plants, and citing Chaney's arguments (§3.8) based on his study of Cenozoic North American forests, Edwards made this literary allusion:

> To explain the distribution of flowering plants from the mid-Cretaceous to the present day, Chaney sees no reason to invoke the aid of continental drift. As he says, "forests under compulsion of climatic change, rather than the continents on which they live, appear to have

been the wanderers during the history of life upon the earth." Or, to put it another way, when Birnam Wood came to Dunsinane it moved by organic agency and not by crustal drift.

(Edwards, 1955: 175)

But Edwards, perhaps because he recognized difficulties with paleontological data, was not such an unyielding fixist as Chaney. Taking note of the new science of paleomagnetism, and again paraphrasing Shakespeare, he noted:

Here geophysicists who have been measuring "fossil" magnetic direction come in, and report that in such and such periods in the past the direction of the earth's magnetic axis has been greatly different from that of to-day. Assuming then that the magnetic axis coincides with the rotational axis, they proceed to plot the presumed positions of the poles at various periods. (In a sense, of course, it is not the pole which moves, but the earth itself in relation to the more or less static pole.) It is too early yet, since little has been published on the magnetic data, to say how the theory will fit the geological and palaeontological record. If it [the paleomagnetic support for polar wander] should prove to be true, then all I have said here is no more than a tale told by a palaeobotanist, signifying nothing. The conclusion, however, would rather seem to be that past history of plant communities on land does indicate a continuous arrangement of the land masses, both in relation to each other and to the equator and poles, roughly corresponding with that of the present day.

(Edwards, 1955: 175; my bracketed addition)

Edwards is to be commended; to have given an unbiased comprehensible description of the paleomagnetic support for mobilism as early as 1955 was unusual.

By the early 1950s it was evident that biogeographical data alone were insufficiently complete to bring the debate to a conclusion, and it was doubtful if they could be brought to such a state in the foreseeable future. More and better studies of *Glossopteris* and vertebrates in Africa and South America were needed; the paleobiogeographical record of icebound Antarctica was virtually unknown; more studies were needed in Asia to search there for the missing marsupial fossils desired by permanentists; early angiosperm fossils were extremely rare, and the overall botanical picture of the Southern Hemisphere was sketchy. Sahni wanted more data on *Glossopteris* before he felt he could choose between drift or landbridges. Simpson, du Toit, and Longwell all lamented the fragmentary nature of the fossil record.

These shortcomings are made apparent in the work of Sir Thomas Holland. Holland, elected FRS in 1904, spent from 1890 until 1909 in India serving as Assistant Superintendent and later Director of the Geological Survey of India (Fermor, 1949a). He saw there for himself much of the stratigraphic, paleoclimatic, and biogeographic evidence that favored mobilism. He later became Principal and Vice-Chancellor of the University of Edinburgh, where he was instrumental in appointing Holmes as Regius Professor in Geology (Lewis, 2000: 185).

In his 1941 Bruce-Preller Lecture before the Royal Society of Edinburgh and in his 1943 joint address to the Linnean and Zoological Societies of London, Holland spoke in favor of mobilism. He was impressed by Sahni's argument for the former separation of peninsular India from the rest of Asia based on his documentation of the distribution of *Glossopteris* and *Gigantopteris* floras (Holland, 1941: 156; 1944: 117). But he was also worried about the incompleteness and reliability of the biogeographical data, and thought that other independent evidence was needed to justify mobilism:

... evidence from the distribution of animals and plants is full of pitfalls and must accordingly be treated with caution, because we have very little information about the ancestry, and therefore origin, of the particular continents; for example, the present limitation of monotremes, marsupials, and struthious birds, like the emu, ostrich, and rhea, to the southern continents. These like the distribution of rainworms, scorpions, and other land forms, would have a simple explanation if we could otherwise and independently prove that the continental masses on which they now live were once in close proximity to one another.

(Holland, 1941: 152; see also 1944: 114)

The incompleteness of biogeographical data was also a common theme at the 1949 New York conference on trans-South Atlantic landbridges. For example, Theodor Just (1952: note 93, 189), a plant geographer from the Chicago Natural History Museum, stressed that data from the Southern Hemisphere were fragmentary and that, as a consequence, "conclusions based on such limited data are at best tentative rather than convincing."

To a great extent the usefulness of biogeographical data was compromised by their incompleteness (gaps in the fossil record) and the fluidity of taxonomic revisions. As a result, it was often hard to unequivocally identify truly disjunct taxa. Moreover, the ability of fossil forms to migrate was hard to assess. Two ichthyologists, Bob Schaffer, from the American Museum of Natural History, and David H. Dunkle, from the United States National Museum, argued that incompleteness makes it a matter of speculation as to whether certain alleged disjunct freshwater biota were restricted to fresh water or could have spent part of their lives in the sea and thus traveled between continents. Without relevant ecological evidence such distinctions could not be made. The dispersal abilities of such fish could not be determined, and the occurrence or non-occurrence of direct connections based on paleo-ichthyology remained up in the air. Dunkle agreed with Schaffer.

Schaffer's paper includes a summary of our present knowledge of the fossil fresh-water fishes of South America, and it appears unnecessary either to qualify or to augment any of the specific points offered by him. Rather, I propose to comment on some general paleo-ichthyological matters that bear on the broader aspects ... It is hoped that the very brevity and generality of these comments will serve to indicate the incompleteness of the data now available to the paleo-ichthyologist and to suggest the caution with which the facts must be employed biogeographically ... Our present knowledge of the Mesozoic faunas offers little insight into the origins and modes of dispersal of the primary division fresh-water fishes that previously have inhabited and now occupy that continent [of South America].

(Dunkle, 1952: 235; my bracketed addition)[23]

3.10 Permo-Carboniferous glaciation: Wegener's 1922 solution; key support for Wegener

Of all the solutions that mobilism provided during the classical stage of the drift controversy, that of the distribution of Permo-Carboniferous glaciation enjoyed the greatest advantage. But it was not without its difficulties, and it was not on its own capable of resolving the drift controversy. Also, there were many specialists who, although they believed that mobilism offered by far the best explanation of the distribution of Late Paleozoic glaciation, were for other reasons unwilling to accept mobilism. As in the biogeographic sub-controversy, participants tempered their support for mobilism by noting repeatedly that it was plagued by the lack of an acceptable mechanism.

Wegener began with a discussion of polar wandering, that is, the motion of the geographic (rotational) pole relative to the body of the Earth.

... traces of [Permo-Carboniferous] ice-action are found on all the southern continents, sometimes with such surprising clearness that the direction of movement of the ice-masses can be read from the scratches on the polished surfaces of the rocks. These traces were first and best studied in South Africa, but they were then also discovered in Brazil, Argentina, in the Falkland Islands, in Togoland, in the Congo area, again in India, and in the West, Central and East Australia. If we merely place the South Pole in the conceivably best position (50° S.45° E.) in the midst of these traces of ice, the remotest traces of land-ice in Brazil, India and Eastern Australia would possess a geographical latitude of not quite 10°, that is, a complete hemisphere of the earth was buried beneath the ice, and had therefore also the polar climate necessary for that purpose. The other hemisphere, however, the Carboniferous and Permian deposits of which are in most areas very well known, does not show sure traces of glaciation, but on the contrary at many places the remains of tropical vegetation. It need scarcely be said that this result is absurd.

(Wegener, 1924: 96–97; my bracketed addition)

Polar wandering alone failed because glaciation would have had to extend to within ten degrees of the equator, and there had to be refrigeration of almost the entire Southern Hemisphere concurrent with tropical climates in North America and Europe, even Spitsbergen. Wegener (1924: 97–98) argued "that this apparent inherent contradiction of observations among one another has absolutely crippled the development of paleoclimatology." Only his continental displacement theory sufficed, Wegener said with characteristic rhetoric.

The riddle of the Permo-Carboniferous glacial period now finds an extremely impressive solution in the displacement theory: directly those parts of the earth which bear these traces of ice-action are concentrically crowded together around South Africa, then the whole area formerly covered with ice becomes of no greater extent than that of the Pleistocene glaciation on the northern hemisphere. It is no longer merely a question of simplification which the displacement theory provides, it rather affords the first possibility of any explanation whatsoever.

(Wegener, 1924: 98)

By assembling Australia, southern Africa, South America, and India around the South Pole, he could in a unique way explain the glaciation without supposing general refrigeration in one hemisphere and tropical climate in the other.

Argand, du Toit, Holmes, and van der Gracht viewed drift's solution to the end-Paleozoic glaciation as an important reason for supporting continental drift. The Swiss geologist, Argand, professor of geology at the University of Neuchatel, Switzerland, was one of Europe's leading structural geologists, and one of the first to come out in support of Wegener's drift theory (§7.3, §8.7). In his 1924 *La Tectonique de l'Asie*, he gave a unique, detailed, and extensive mobilist account of mountain building. Argand had presented it two years earlier as his inaugural address to the XIIIth International Geological Congress in August of 1922. Carozzi (1977: xiii) describes it as "the manifesto of 'mobilism' in geology, the revolution against the paralyzing and classical 'fixism,' which for so many years had obstructed any form of meaningful progress in geology." The first edition of Wegener's *Origin* "led to a complete change in Argand's thinking. It provided him with a new doctrine – 'mobilism' – that satisfied to a much greater extent the space and time requirements of his synthesis than the concept of the general contraction of the Earth" (Carozzi, 1977: xvii). Very soon after the appearance of Wegener's first edition in 1915, Argand first publicly supported continental drift in November of 1916 at a meeting of the Neuchatel Society of Natural Sciences. Having accepted drift he swiftly recognized its explanatory power, and emphasized its explanation of Permo-Carboniferous glaciation (Carozzi, 1977: xxv).

Based on eighteen years of fieldwork, Du Toit had by 1921 come "to recognize only too clearly the magnitude and complexity" of Permo-Carboniferous glaciation, and like Argand, swiftly endorsed Wegener's overall theory. He proposed a detailed and refined mobilist reconstruction of the formation of the glaciation (du Toit, 1921a: 188).

Holmes regarded Wegener's explanation of the origin of Late Paleozoic glaciation as of utmost importance. In the third part of his lengthy 1929 review of continental drift, he wrote:

Here continental drift has undoubtedly more than one decided advantage. The opponents of drift have no way of explaining the distribution of the late Carboniferous glaciations of Gondwanaland, which accordingly continues to be the basis of Wegener's most powerful argument.

(Holmes, 1929: 340)

In a 1931 discussion on problems of the Earth's crust sponsored by the geography section of BAAS he went so far as to claim:

To my mind the evidence of the distribution of climates during the Upper Carboniferous constitutes a conclusive proof that both continental and crustal drift have operated on a very extensive scale.

(Holmes, 1931b: 450)

The Dutch geologist, van der Gracht, keynote speaker at the 1926 AAPG symposium on Wegener's theory, claimed:

One of Wegener's chief arguments for continental drift is the explanation it affords for the problem of these [Permo-Carboniferous] glaciations, and their association elsewhere with the assumedly tropical Coal Measures flora.

(van der Gracht, 1928b: 220; my bracketed addition)

Du Toit actually traced out the Permo-Carboniferous glaciation in southern Africa, but like Holmes and van der Gracht, he was not a specialist in paleoclimatology. Nevertheless, they all singled out this major climatic phenomenon as providing the strongest evidence for mobilism.

3.11 Permo-Carboniferous glaciation: fixists attack Wegener's solution and refurbish their own

Critics soon attacked, beginning in the 1920s. They raised meteorological and paleo-climatological objections to Wegener's claim that his solution was the only viable one. As a bonus, they reiterated drift's mechanism difficulty.

Arthur P. Coleman was a highly respected Canadian geologist and glaciologist, President of GSA in1916, FRS, and recipient of the Penrose and Murchison Medals. He traveled widely. In 1905, the BAAS met in South Africa, and he was convinced by the evidence for Late Paleozoic glaciation.

Some hours of scrambling and hammering under the intense African sun in lat. 27° 5′, without a drop of water, while collecting striated stones and a slab of the polished floor of slate, provided a most impressive contrast between the present and the past, for though Aug. 27 is still early spring, the heat was fully equal to that of a sunny August day in North America. The dry, wilting sun glare and perspiration make the thought of an ice sheet thousands of feet thick at that very spot most incredible, but more alluring.

(Coleman, 1926: 123)

In 1914, Coleman (1926: 104, 142), during another meeting of BAAS, saw Late Paleozoic glacial strata in Australia, and also in India on his way home. Three years later, he (1926: 142) went to South America and saw glacial beds there of similar age. He became perhaps the first person to inspect personally all the major Late Paleozoic glacial occurrences then known. He saw the ubiquitous Permian *Glossopteris* flora, occurring in strata above or interbedded with glacial strata (1926: 136, 174), and also the occurrences of *Mesosaurus* in southern Africa and Brazil. He did not doubt the former existence and wide extent of the glaciations. His doubts lay elsewhere.

On climatological grounds, however, Coleman doubted the viability of Wegener's reassembly of these Late Paleozoic strata into a single entity. He was responsible for two climatological criticisms, agreed with another, and like many at the time endorsed the mechanism objection to drift. He expressed doubt first in January of 1923 at the meeting of the Geological Section of BAAS at which Wegener's hypothesis was discussed, and repeated it in 1924, 1925, 1926, and 1933. He raised the theoretical difficulty that assembling continents around the South Pole in one huge supercontinent would have resulted in exceedingly dry conditions. Where, he asked, would the moisture come from to feed such vast glaciers?

A glance at Wegener's map, however, shows that this grouping of the continents would place the south pole and the area of glaciation in the heart of an enormous continent thousands of miles away from any source of evaporation to provide for snowfall. The supposed gigantic ice cap would lie in an arid and snowless a desert as that of Gobi in Central Asia.

(Coleman, 1933: 415)

Coleman (1925) voiced the equally potent criticism that the marine derivation of some of the glacial deposits was inconsistent with Wegener's explanation. According to Coleman, the tillites from the Southern Hemisphere were laid down close to sea level. Whereas Wegener's wholly terrestrial reconstruction indicated that many of the glaciers would have formed well above sea level in the interior of a huge continent.

... it has been proved that ice sheets on all the continents supposed to unite about the south pole reached sea level. This is true of India, South Africa, South America and Australia, and in the latter two the ice touched the sea on both sides of the continent, as proved by fossiliferous marine deposits associated with the tillites. The usual idea of a vast and lofty Gondwanaland on which an enormous ice sheet could arise must be given up; for the different glaciated areas were separated by oceans or at least by arms of the sea in which marine animals survived.

(Coleman, 1925: 602)

Another criticism, repeated by many including Coleman, was most forcefully expressed by the Englishman C. E. P. Brooks, editor of the *Meteorological Magazine* for over twenty years and in 1931 recipient of the Buchan Prize of the Royal Meteorological Society. He attacked comprehensively Wegener's solution in his *Climate Through the Ages* (Brooks, 1926, 1949). Wegener placed central and southern North America near the Permo-Carboniferous equator where Brooks cited new evidence of glaciation, and he therefore rejected Wegener's solution. He opined that there were conglomerates in Oklahoma and Kansas, and the Squantum Tillite with massive boulder and "varved" beds near Boston, that closely resembled glacial beds found in Australia and elsewhere (Brooks, 1949: 232). The presence of these varved beds was particularly harmful for Wegener because they probably resulted from seasonal melting under glacial conditions. Brooks' (1926, 1949) greatest contribution to this sub-controversy was his fixist paleogeographic account, not unlike that independently proposed by Schuchert (1926a, 1932). Brooks' solution appealed to many readers – Jeffreys was impressed.[24] Both Brooks and Schuchert utilized the standard fixist map of the Upper Carboniferous (Figure 3.5) with its extensive, horn-shaped Gondwana, Tethys, and the Volga Sea.

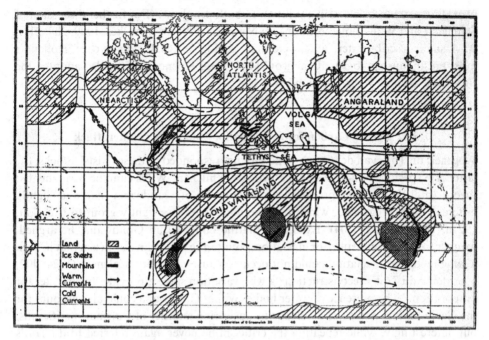

Figure 3.5 Brooks' Figure 29 (1949: 248). The geography of the Upper Carboniferous.

Brooks argued that the conditions generated by this geographical reconstruction would have brought about extensive glaciation in the Southern Hemisphere in this way: by the diversion of the whole of the equatorial ocean current into the Northern Hemisphere and by the presence of an extensive elevated continent along the equator, extending much farther into southern than into northern latitudes, which would render the Northern Hemisphere abnormally warm and shut off the southern ocean from all warm currents. Global refrigeration due to the presence of abnormally large quantities of volcanic dust could, Brooks (1949: 261) added, have triggered the entire event. Brooks argued that his solution was climatologically superior to Wegener's, and geophysically superior because it avoided mechanism difficulties.

Schuchert's account mirrored Brooks', and was presented in its most extensive version at the 1931 GSA annual meeting. The former landbridger teamed up with Willis, and at the meeting they presented a permanentist explanation not only of glaciation but also of disjunct biota. Schuchert, finally realizing the seriousness of the inconsistency between isostasy and sinking broad landbridges, encouraged Willis, an unrelenting permanentist (Willis, 1928, 1932, 1944), to see if he could locate isthmian links that would account for the disjuncts. Willis said he could, and they published side-by-side papers the following year (Schuchert, 1932; Willis, 1932). (See §7.2 for Willis' continued opposition to mobilism.) Their papers were very influential, providing permanentists with a solution to both problems. (For further discussion of Schuchert and Willis' collaboration, see Newman, 1995 and Oreskes, 1999: 208–213.) As I shall describe later, a few German geologists (§8.4) and two of Holland's most prestigious geologists, Umbgrove (1947) and Kuenen (1950), also adopted the Brooks-Schuchert-Willis solution (§8.13).

3.12 Permo-Carboniferous glaciation: mobilists counterattack

Wegener, du Toit, van der Gracht, and Holmes responded to these fixists' attacks, the latter two specifically addressing the Brooks–Schuchert–Willis climatic permanentist theory. In 1921 du Toit proposed a construction of Gondwana that was slightly different from Wegener's, which had the continents less tightly grouped (du Toit, 1921a, 1921b, 1927b: 104). Extending marine gulfs within Gondwana, he argued, would provide the moisture for glaciation, countering Coleman's criticism that in such a large landmass glaciation could not occur because of aridity. Coleman (1924: 402) considered du Toit's reconstruction "a wise variation since it provides open water for evaporation." The presence of water bodies within Gondwana could also account for the glacio-marine nature of the deposits, answering another of Coleman's criticisms, but, interestingly, while repeating the criticism he never again referred to du Toit's looser reconstruction.

Wegener (1928, 1929/1966), van der Gracht (1928b), and Holmes (1929) addressed the vexing question of Permo-Carboniferous glaciation in North America. In the report of the 1926 AAPG symposium on continental drift, Wegener admitted that if "all or any of these conglomerates are truly glacial, they would be in flagrant contradiction to" his theory, and sought to shift the burden of proof to the fixists.

We have a multitude of indications for the past geological climate in the United States, and although we admit that several details are uncertain, they yet match and give us a harmonious major picture. In the main, it is clear and simple that there is cumulative and concordant evidence that in the upper Carboniferous the United States was within the edge of the pluvial tropical belt, and during the Permian in the hot tropical desert zone.

(Wegener, 1928: 98–99)

Van der Gracht agreed, and noted its incompatibility with the notion of contemporaneous glaciation. He remarked that assignments of strata as glacial are often uncertain and offered an alternative interpretation of the conglomerates in Texas, Kansas, Oklahoma, and Colorado.

We have to be very careful with "tillites." I do not consider it proved that any of the Permo-Carboniferous conglomerates of Texas, Kansas, and Oklahoma, and particularly of Colorado, can be considered as of glacial origin. No one who is familiar with cloudbursts, particularly such as occur in deserts or along the edge of arid belts, can be surprised that great thicknesses of unsorted, coarsely clastic and partly angular material are deposited by the floods generated by such rains.

These floods are extremely violent, though of short duration. The streams are almost more mud than water, and the mixture is of such density that it will not only transport unbelievably large boulders, but also prevent any assortment of the material. It needs no ice to explain it. We see the same thing happening now in all deserts, including those of the American West.

(van der Gracht, 1928b: 220–221)

Van der Gracht also questioned whether the Squantum Tillite was truly glacial. He admitted that he himself had once incorrectly identified certain Permian conglomerates as glacial.

Even facetted and striated boulders need not be glacial, unless striation is quite common, particularly such boulders as are made of very dense and hard rocks. Such rocks, surprisingly like glacial boulders and erratics, in Permian conglomerates of northwestern Europe, with clear indications of "glacial" characteristics, are now considered as merely sheared fragments. I was once, in 1909, guilty myself of describing one of these European conglomerates as a tillite.

(van der Gracht, 1928b: 221)

He had not seen the Squantum Tillite, but the accounts of others were not univocal.

Unhappily, I am not personally familiar with the Squantum tillite, but I recently had a conversation regarding it with Laurence La Forge of the U.S. Geological Survey. This did not leave me with the impression that this deposit was assuredly of glacial origin.

(van der Gracht, 1928b: 221)

Although van der Gracht's words of caution had little influence on fixists at the AAPG symposium, they impressed Wegener. He cited van der Gracht in the 1929 edition of *The Origin*, and quoted the above passages (Wegener, 1929/1966: 134–135), but commented ruefully, "Only van Waterschoot van der Gracht appears to share my doubts" about the supposed glacial origins of the Squantum Tillite (Wegener, 1929/1966: 136). However, Holmes also shared Wegener's doubts, claiming in his review of the AAPG symposium:

One of these [inconsistencies] is provided by the position of the Squantum tillites, near Boston, Massachusetts. Wegener himself discusses this special difficulty in his contribution to the Symposium. If the beds in question are truly of glacial origin, they appear to stand in flagrant contradiction to his views, since they occur near his Permo-Carboniferous tropical belt. He pleads for an independent and impartial investigation of the problem, but he adds, not without justification, that the glacial hypothesis of these puzzling beds is also hopelessly in conflict with the adjoining palaeoclimatic evidence of the time. Neither the drift nor any other theory can be expected to explain interpretations that are mutually exclusive.

(Holmes, 1929: 341; my bracketed addition)

In the report of the AAPG symposium, van der Gracht criticized Brooks' meteorological explanation of the Permo-Carboniferous glaciation. Before doing so, he noted that Brooks evidently considers the explanation of the climatic problems by the drift theory a possibility, but the mechanical difficulties impress him so strongly that he does not consider such drift probable, and offers another explanation.

(van der Gracht, 1928b: 223)

Van der Gracht claimed that the "difficulties inherent in" ... Brooks' ... "theory, however, seem insurmountable," and then proceeded to spell them out:

He, first of all, requires in the Permo-Carboniferous a large Gondwana continent of great elevation, similar to the present highlands of Asia, which should since have sunk below the ocean. Isostasy, if at all existent (and who denies it?), precludes this possibility. Even then, Brooks' centers of glaciation so close to the equator, seem impossible without enormously more pronounced glacial conditions in the Northern Hemisphere, particularly Asia, where the paleontological facts are against it. His cold Antarctic currents in his ancestral Indian Ocean, between Africa and Australia, are assumed to have caused glacial conditions in India, *after having crossed the equator* in what should there have been a land-locked gulf, still more apt to generate warm currents than the present Caribbean and Mexican Gulf! The still doubtful glacial conditions in New England, which Brooks, however, accepts as real, would have occurred on a coast directly exposed to warm currents of the equatorial Tethys! I must confess that I prefer Köppen's and Wegener's conception, regardless of all of its problems and difficulties.[25]

(van der Gracht, 1928b: 223)

Holmes, in his review of the AAPG symposium, agreed with van der Gracht's assessment of Brooks' hypothesis, and raised further difficulties.

[Brooks has] adopted as a working hypothesis "a great plateau in the interior of Gondwanaland, rising gradually to an elevation of 10000 feet." This assumption of great height certainly eases the meteorological problem, but it has no geological justification. Moreover, it doubles the difficulty of the physical problem, for now we should have to explain, first a great thickening of the sial of Gondwanaland and then its total disappearance from the very extensive oceanic areas that now intervene between the existing southern continents.

(Holmes, 1929: 340; my bracketed addition)

Turning to the paleoclimatological difficulties, he noted that Brooks himself was unhappy with his own account of the Late Paleozoic glaciation of India. He also remarked that neither Wegener nor Brooks could account for the lack of glaciation in Antarctica during the Permo-Carboniferous, but argued that this is "certainly more damaging to his [i.e., Brooks'] attempt at a solution than it is to the less rigid hypothesis of continental drift" (Holmes, 1929: 340; my bracketed addition; unknown at the time, glacial strata were later found in Antarctica (Long, 1962b; see §7.4)).

Mobilism was helped along by Sir George Clarke Simpson (1878–1965), the most eminent British climatologist to defend drift, FRS, President of the Royal Meteorological Society (1941–2), and recipient of the Symonds Gold Medal in 1930. At the 1923 meeting of the Geological Section of BAAS, he claimed that Wegener's theory "was a wonderful one from the meteorological point of view, as it explained the marked changes of climate given by the geological record" (Simpson, 1923). But Simpson's greatest contribution to the debate came in his 1929 Alexander Pedler Lecture, *Past Climates*, in which he developed a sophisticated meteorological solution to the general problem of glaciation, even offering an explanation of interglacial intervals – something lacking in Brooks' account. He was quite explicit that his solution depended on Wegener's displacement theory being correct.

If, as I am prepared to admit, there was at one time ice in the present tropical zone and simultaneously sub-tropical vegetation in the present temperate zone, then I am forced to conclude that Wegener is right and there has been a considerable shift of the continents relative to the pole and the climatic zones.

(Simpson, 1929: 23)

Simpson restated his position the following year at a discussion sponsored by the Royal Society (London), which was held to bring together scientists from different disciplines (meteorology, geology, zoology, and paleobotany) to talk about past climates. He (1930: 299) argued that so long as Earth continues to rotate around an axis inclined approximately 66½° to the ecliptic, "there must always have been climatic zones with the equatorial zones warmer than the polar zones." Variations in the solar radiation could cause global changes in mean temperature; however, they would not eradicate climatic zones. He not only strongly supported drift as an explanation of the distribution of Permo-Carboniferous glaciation but also Köppen and Wegener's overall paleoclimatic defense of mobilism.

The large climatic variations which have caused the climate of Europe to be temperate to-day, frigid in the Pleistocene, sub-tropical in the Miocene and tropical in the Carboniferous, and have given sub-tropical if not tropical conditions in polar regions and polar conditions in equatorial regions, can only be explained by Wegener's hypothesis of the shift of the poles and the drift of the continents, the climatological consequences of which have been worked out in such great detail in Köppen and Wegener's book, "Die Klimate der geologischen Vorzeit."

(Simpson, 1930: 302)

Brooks spoke. He defended his fixist position, arguing (1930: 312–313) that Simpson had "under-estimated the possible effect of variations of land and sea, especially the effect of ocean currents." Two other participants, J. W. Gregory, and C. Tate Regan, also supported fixism. Gregory, a champion of landbridges (§3.4), agreed that the Permo-Carboniferous glaciation is hard to explain without moving continents, but sided with Brooks anyway; epicontinental seas of varying extent and terrestrial changes in elevation were sufficient.

In Gondwanaland there was a series of scattered glacial accumulations and they were not exactly simultaneous; as one of them waxed, its neighbor waned. They were all fairly near the sea, in positions where there was cold water coming from the south, and the mountains had a high winter snowfall, and a cold, cloudy summer. All those centres of glaciation may be explained by a different distribution of land and water and different elevation of the land.

(Gregory, 1930: 312)

Regan, an ichthyologist, made no bones about it.

I am therefore opposed to Wegener's hypothesis. From the distribution of fresh-water fishes I conclude that Africa had in Cretaceous times much the same climate as it has to-day, and from the distribution of marine fishes I conclude that in Eocene times the Atlantic Ocean was a wide one. Africa was not much nearer to South America, nor to the South Pole, than it is now.

(Regan, 1930: 315–316)

Mitchell, a vertebrate zoologist, voiced no opinion about drift, but argued that little definite information about past climates could be obtained from the study of fossils.

My general point is merely that we zoologists cannot infer from the presence of fossils any particular kind of climate. In fact, the inference must be reversed; if there are fossils of a particular kind of animal found under conditions where, meteorologists and geologists tell us, there was such and such a climate, we know that these animals must have been able to endure those climates.

(Mitchell, 1930: 309)

The final two speakers, Seward and H. Hamshaw Thomas, both paleobotanists, favored mobilism. Seward (1930), whose views have already been discussed (§3.5), favored mobilism, but recommended care when inferring past climates from fossil plants. Thomas, another of Seward's pro-drift students, proclaimed his support for mobilism.

In spite of all that has been said to-night, there still remain many striking anomalies in the past distribution of plants, which require explanation by some theory such as that outlined by Dr. Simpson. In my opinion Dr. Simpson's hypothesis explains some of the difficulties much more satisfactorily than any other view which has been put forward.

(Thomas, 1930: 316)

It is notable that Holland (1941, 1944) was impressed with Simpson's work. Although concerned about mobilism's solution to disjunct biota because of the variable reliability and incompleteness of biogeographical data (§3.9), he thought mobilism's solution to the distribution of Permo-Carboniferous glaciation carried much greater weight: it was mobilism's strongest defense. At the 1943 joint meeting of the Linnean and the Zoological Societies of London, he declared that the glaciations "are not to be explained away as the results of elevation or local special meteorological conditions," and turning to Simpson:

All-world refrigerations have doubtless occurred at various times in the past, as they did in the so-called Great Ice age of the Pleistocene period; but, as Sir George Simpson (1930) has stated with recognized authority, so long as the earth has been rotating on its present axis inclined to the ecliptic, the climates must always have been zoned from a warm or relatively mild equatorial belt outward in both directions to the higher latitudes, both north and south. If, then, glacial conditions actually prevailed at any time at sea-level over the tropical belt, the whole world would have been encased in ice; and that we know to be untrue for all past palaeontological time. We could hardly think of glaciation at sea-level in the tropics, and corals flourishing in our own seas at the same time. Anyone who can find a flaw in this form of reasoning from the abundant and consistent evidence will shake my faith in the drift theory. But, if the conclusion on this one point be admitted, we can search more hopefully for physiographical conditions which will satisfy the geophysicists.

(Holland, 1944: 115–116)

Holland was convinced, but fixists remained unmoved.

G. C. Simpson's contribution was the last new and important addition by a mobilist to the Permo-Carboniferous glaciation sub-controversy until the rise of paleomagnetism in the early 1950s. Wegener elaborated on his theory in the 1929 edition of his *The Origin*, du Toit's *Our Wandering Continents* appeared in 1937, Holmes defended drift in his 1944 *Principles of Geology*, but they offered no new evidence regarding the Permo-Carboniferous glaciation such as would have given a decided advantage to mobilists or pause to the fixists. Holmes nicely summed up the status of the sub-controversy from the mobilist point of view in his review of the New York 1949 symposium on "The Role of the South Atlantic Basin in Biogeography and Evolution" held by the Society for the Study of Evolution. After admitting that continental drift might not be needed to account for various intercontinental disjuncts, he appealed to meteorologists to rule on the plausibility of such vastly different Late Paleozoic climates in the North and South Hemispheres.

Meanwhile there remains the most serious enigma of all: the Permo-Carboniferous glaciations. Dunbar points out that the Paleozoic glaciations in low latitudes present "a problem still unsolved, unless we accept continental drift." But if we accept continental drift only to explain these and other still older glaciations, it becomes an *ad hoc* hypothesis. As such, it may still be justified as a stimulant to research, but it may also stand in the way of progress by distracting attention from the real problem. Can the meteorologist not come to our assistance and tell us whether or not widespread equatorial and low-latitude glaciation is possible while high latitudes for the most part enjoy a genial climate?

(Holmes, 1953: 671)

Clearly Holmes believed that mobilism offered the best explanation of Permo-Carboniferous glaciation, and put little stock in the fixist accounts of Brooks, Schuchert, and Willis.

Fixists remained unmoved. For example, Brooks, in his new (1949) edition of *Climate Through the Ages*, thought that his explanation:

gave at least as good an approximation to the facts as does Wegener's theory, and it is accordingly inferred that continental drift is not necessary to account for the distribution of past climates.

(Brooks, 1949: 10)

He also countered Wegener's attempt to discount as anomalous the alleged glacial deposits of North America, arguing that his assertion that the data were unreliable was itself biased.

According to the "drift" theory, these coal beds [in North America] represent a luxurious tropical rain-forest, and the equator is therefore drawn as nearly as possible through the middle of them. The evidences of glacial action which have been adduced from time to time in close proximity, both in space and time, to these coal beds are dismissed out of hand as not genuine. The American evidence, however, seems to be too well founded to be dealt with in his summary fashion. Thus, S. Weidmann ... describes conglomerates of Upper Carboniferous to Permian age in the Arbuckle and Wichita Mountains of Oklahoma and in Kansas, associated with all the paraphernalia of glaciation – scratched boulders, erratics, fluted and polished floors, and U-shaped valleys. Some of the boulders in marine deposits have apparently been carried by icebergs, and the author attributes the phenomena to islands in the Late Palaeozoic sea bearing local valley glaciers. J. A. Taff ... found boulders up to 20 feet across and 5 or 6 feet thick, 50 miles or more from their source, in the marine Caney shales of eastern Oklahoma. "No other competent means of their transportation than ice – presumably heavy shore ice – has been suggested." Similarly, A. P. Coleman ... considers that there is good evidence for glaciation in Oklahoma, Nova Scotia, and Alaska (Thousand Isles). As regards Nova Scotia, Coleman writes, "it is ... probable that there were moderate elevations from which, under a cool climate, glaciers spread out over the plains on which coal forests had been growing not long before."

In the Squantum tillite near Boston, Mass., there are massive conglomerates 2000 feet in thickness, which cover a considerable area. The chief interest of these beds, apart from the presence of striated boulders, lies in the associated "varve" beds – banded clays which are similar in all respects to those formed during the retreat of the Scandinavian and North American ice-sheets at the close of the Quaternary glaciation, and also similar to those formed in Australia during the Upper Carboniferous glaciation. These clays owe their banding to the seasonal variations in the rate of melting of glaciers, and are therefore incompatible with an equatorial climate. In places the banding is disturbed during the deposition of the shales, probably by the grounding of floating ice-masses.

Wegener recognizes that the Squantum tillites demand serious consideration, and the effort he makes to explain them away tacitly implies that they form a very serious objection to the "continental drift" theory. He considers [in his and Köppen's *Die Klimate der geologischen Vorzeit*] the possibility that they are real, but were formed at a very high level in the Appalachian mountain system, then young and vigorous, and agrees with the general opinion of geologists that the chances of preservation of high-level glacial deposits over a wide area would be very slight. As we have seen, there are other indications that the glaciers reached low levels. He therefore concludes that smoothed floors have not been found beneath the moraines, the remaining phenomena, although very suggestive of glacial action, could have originated in other ways. As all the other evidence indicates that Boston lay in the equatorial rain zone during the Carboniferous and in the region of hot deserts during the Permian, *"the glacial nature of these tillites is in irreconcilable opposition to the numerous climatic traces of another kind which surround it both in space and time."* He therefore says that the burden of proof that the deposits are really glacial rests with the opponents of the "drift" theory. The Squantum tillites are, however, accepted by all American geologists, while the Caney shales in the Arbuckle and Wichita Mountains were examined by an impartial observer, J. B. Woodworth, who concluded that while the striae on the boulders which he observed were probably not glacial, the transport of the boulders was almost certainly effected by floating ice.

(Brooks, 1949: 231–232; my bracketed additions)

Brooks then pointed to an analogous case of bias:

There is one other piece of evidence in connexion with the climatic zones of the Permo-Carboniferous which may be referred to, not so much for its intrinsic importance, which is small, as because it illustrates the methods too often adopted by Köppen and Wegener in dealing with items which do not quite fit their theory. Salt beds occur in Angola, formerly attributed to the Carboniferous. The authors class these as Permo-Triassic, because during the Carboniferous "Angola was too near the South Pole for salt beds to form." The drift hypothesis has certainly not reached a stage of proof in which it can be asserted that evidence which does not fit it is thereby proved false.

(Brooks, 1949: 234)

Like so many others, Brooks (1949: 223) raised the mechanism difficulty faced by mobilism, characterizing it as the "chief weakness of Wegener's theory"; it was, he maintained more severe than the geophysical difficulty Holmes had raised regarding Brooks' elevated Gondwana.

The critical assumption [of my solution] is the considerable elevation of the central parts of Gondwanaland, but this is not entirely unsupported by evidence, and is at worst not less hazardous than the extensive migrations of the continents through some fifty degrees of latitude. We find that outside Gondwanaland the only extensive glaciation occurred in Nearctis exactly where we should expect it, while on the theory of continental drift this region would lie on the equator and would be unglaciated. The main difficulty [against our theory], the non-glaciation of the Antarctic [not yet discovered], is common to both theories.

(Brooks, 1949: 260; my bracketed additions)

It is evident, however, that many Earth scientists who had thought seriously about the origin and distribution of the Permo-Carboniferous glaciation believed mobilism was a necessary part of the solution.

3.13 The geodetic sub-controversy over the westward drift of Greenland

The biogeography and paleoclimatology controversies developed from problems that had been extensively studied before Wegener's day. The origin of the geodetic sub-controversy was quite different. In Wegener's day, measuring drift geodetically hardly seemed a practical possibility, and there were few measurements. Wegener precipitated the sub-controversy by proposing that geodesists measure current motions between continents. As early as 1912, he realized that this might be possible, and hoped for definitive proof (§2.7). Geodesists undertook studies in 1922, 1928, and 1929, and Wegener thought that they had established continental drift, but his victory was short-lived. Critics thought the measurements unreliable, and measurements acquired several years later supported their charge, effectively ending the sub-controversy for half a century. Definitive measurements were not made until the 1980s (Gordon and Stein, 1992; Kearey, Klepeis, and Vine, 2009).

As previously described, Wegener (1912) initially thought of his theory as a working hypothesis, arguing that it should be so treated "until the persistence of these horizontal displacements in the present is determined by astronomical position-fixing with an exactness which eliminates any doubt" (1912b: 185). He already knew of such measurements that could be interpreted in terms of continental drift, especially those taken in 1823 and 1870, indicating that Greenland had moved westward relative to Europe.[26]

Wegener's excitement extended beyond the simple fact that these measurements had suggested westward drift of Greenland. He knew that his friend and fellow Greenland traveler J. P. Koch had made measurements of longitude there in 1907, and he considered this a good place to undertake additional studies because he believed the northernmost Atlantic had opened during the Quaternary so Greenland should, he thought, have moved rapidly enough relative to Europe to be measured reliably.[27] Wegener sent Koch an outline of his displacement theory, and encouraged him to work up his results and to compare them with older ones. Koch's measurements were made during the 1906–8 *Danmark* Expedition of which Wegener had also been a member. But Koch was unable to furnish Wegener with proof. In 1929, Wegener recalled:

Koch made a provisional calculation from the data and informed me that in fact a difference of the expected order of magnitude was shown, but that he was not able to credit the difference to a displacement of Greenland.

(Wegener, 1929/1966: 26)

Koch (1917), however, changed his mind. After further analysis, which Wegener (1929/1966: 26) regarded as "definitive," Koch argued that his results were best explained by supposing an average westward drift of Greenland relative to Europe between 1870 and 1907 at a rate of 32 meters per year. But, there was a snag. Koch's measurements, like the older ones, were based on an old method of taking lunar observations to fix longitude. Wegener acknowledged this but still thought they offered support for continental drift, commenting that his friend should receive credit for "the first discovery of the alteration of co-ordinates" (1924: 117). But to convince others, Wegener needed accurate measurements, better than could be obtained by lunar observations. He needed a more reliable method, wireless telegraphy.

In 1922 the Danish Geodetic Institute began taking measurements in Greenland using time transmissions by radio telegraphy. P. F. Jensen (1923) published his findings and, as Wegener noted (1923, 1927, 1928), they supported a continued westward drift of Greenland; Wegener summarized in his contribution to the proceedings of the 1926 AAPG symposium on continental drift.

In the summer of 1922, P. F. Jensen made new observations at Godthaab in western Greenland, on behalf of the Danish Geodetic Survey. This is the same locality where determinations had already been made in 1863, and 1882–83. These earlier observations had also been made with lunars, and consequently are subject to the same objection [i.e., the possibility of large systematic errors]. Jensen, however, has made his observations by the method of radio time signals. Consequently, they are more reliable and should have no greater discrepancy than an average error of 1/10 of a time-second. A comparison of his results with those of the earlier observations again resulted in an average drift of 20 meters yearly away from Europe. Consequently we now have a determination which is at least not exclusively based on lunar observations, and the coincidence that we continue to obtain an average of 20 meters per year makes the correctness of the earlier results decidedly probable. In order to make it possible that at some future time we may obtain a fully decisive answer regarding this question, a permanent station has been erected in a suitable locality at Kornok, in the interior of Godthaab Fjord. Jensen's observations in the summer of 1922 were the first which were made at this permanent station. Determinations of longitude, based on radio time-signals, will be periodically repeated. The next measurements are planned for the summer of 1927. If, indeed, there exists a yearly drift of about 20 meters, the amount should have come within the range of observation at the next measurement.

(Wegener 1928: 101–102; my bracketed addition)

While acknowledging that the results were not decisive, Wegener was pleased. He would have to wait until the Danish Geodetic Institute had analyzed their 1927 measurements.

When the new results became known, Wegener was so impressed that he began his fourth edition with them.

We begin the demonstration of our theory with the detection of present-day drift of the continents by repeated astronomical position finding, because only recently this method furnished the first real proof of the present-day displacement of Greenland – predicted by drift theory – because it also constitutes a good quantitative corroboration. The majority of scientists will probably consider this to be the most precise and reliable test of the theory.

(Wegener, 1929/1966: 23)

After comparing Jensen's first telegraphic results with the older lunar results, he introduced, with the Dane's permission, the latest measurement.

The determination of the longitude of Kornok has now been repeated ... using the modern impersonal micrometer which eliminates the "personal equation." This allows far greater accuracy to be achieved than was possible in Jensen's measurements ... *Comparison with Jensen's figure yields an increase in the longitude difference relative to Greenwich, i.e., in the distance of Greenland from Europe, of about 0.9 seconds (time) in five years, or about a rate of 36 m/yr.*

This increase is nine times larger than the mean error of observations, and there is no question of any systematic error in radio telegraphic time transmissions. *The result is therefore proof of a displacement of Greenland that is still in progress,* unless it should be supposed that Jensen's "personal equation" amounted to 0.9 seconds of time – a most improbable hypothesis.

As a result of this first precise astronomical proof of continental drift, which fully corroborates the predictions of drift theory in a quantitative manner, the whole discussion of the theory, in my view, is put on a new footing: interest is now transferred from the question of the basic soundness of the theory to that of the correctness or elaboration of its individual assertions.

(Wegener, 1929/1966: 29–30; Wegener's italics)

Wegener thought he had a difficulty-free solution. Successive wireless telegraphic determinations of the longitude of Greenland, taken five years apart, appeared to show that Greenland was drifting westward relative to Europe. The new studies, he thought, eliminated the difficulty that had plagued all previous studies. Moreover, the direction and rate of Greenland's displacement were consilient with his paleogeographic estimates. He claimed that continental drift had been established. Of course, there were still difficulties with mobilism's other solutions, and that was to be expected, but Wegener was confident that with an unassailable solution in hand, drifters could now concentrate on improving mobilism's other solutions, and extending its application into new areas.

Du Toit (1937) welcomed the results, and argued that they provided strong support for continental drift, but fixists were not impressed. Chester Longwell (1944a) for example, addressed the issue in his rejoinder to du Toit's attack on G. G. Simpson. After discussing the older lunar determinations, Longwell turned to the telegraphic results.

Du Toit (1937, p. 300) remarks that even if all the measurements referred to above be rejected, we still have two modern determinations at one Greenland station, with an interval of five years, which indicate westward movement at the rate of 36 meters per year, a figure "far in excess of the probable errors." The skeptic with only moderate qualifications in mathematics will answer promptly that a single example cannot be accepted as the basis of a confident conclusion, since conceivably it may involve errors several times the theoretical probable error.

(Longwell, 1944a: 227)

Longwell noted that earlier determinations between Greenwich and Washington taken in 1913 and 1926 indicated that the distance between North America and Europe had actually decreased:

... by more than 9 m., an average of about 0.7 m. per year ... These contradictory results make it appear that the possible error of such determinations is at least several meters, at old established stations using the most refined modern methods. Since conditions for observations in Greenland are far less satisfactory at best, and since all but one set of available data involve to some extent the inaccurate lunar method, objective judgment can hardly indorse du Toit's statement, "it must therefore be concluded that a positive shift of crustal matter has been instrumentally observed."

(Longwell, 1944a: 228)

Longwell then asked:

Why should we be hasty in reaching conviction in this matter? Modern methods of determining coordinates are closely accurate and will eventually reveal the truth. Jensen's Greenland station of 1922 can be reoccupied after the war, and the interval of more than 20 years should give a far better basis for judgment than any Arctic data now available ... I have no personal aversion to westward movement of Greenland – in fact I should be pleasantly excited by the vista of tremendous implications attendant on authentic confirmation of such movement. However, in this matter we are not dealing with *probabilities*. Astronomers and geodesists will eventually tell us that observations either have or have not demonstrated a shift of Greenland with relation to Europe. Until that time the only objective attitude is one of "wait and see."

(Longwell, 1944a: 228)

But Longwell was behind the times. Unbeknown to him, the Danish Geodetic Institute had already repudiated their 1922 results. In his George Darwin Lecture to the Royal Astronomical Society in London on May 14, 1937, N. E. Norlund ended the sub-controversy in this way:

I just mentioned that we have recently carried out a number of longitude determinations in Greenland, and I shall conclude with a short remark on these observations. We know that Wegener considered that his hypothesis of the displacement of continents had been supported by longitude determinations in Greenland, and from them he calculated that Greenland is moving westwards about 20 metres a year. But the old observations used by Wegener were carried out with primitive instruments and he overrated their exactitude. In 1927 and 1936 the Danish Geodetic Institute carried out longitude determinations at Kornok on the west coast of Greenland with a first-class transit instrument, both times on the same pillar. Practically speaking the two measurements gave the same result. It is most likely that the deviations of the old observations from the new ones are the result of observational errors.

(Norlund, 1937: 505–506)

Longwell was told about Norlund's paper by F. Lowe of the University of Melbourne, Australia, which prompted him to publish two notes summarizing the work and its relation to continental drift. In one (1944b) he quoted the above passage from Norlund's paper, in the other piece (1944c), he placed Norlund's finding within the overall context of the geodetic sub-controversy.

In 1927 the Danish Geodetic Institute again occupied the station [used by Jensen], and announced a value about 180 meters west of Jensen's. Wegener and du Toit [Longwell here cites du Toit, 1937, and Wegener's fourth edition of *The Origin of Continents and Oceans*] lay especial emphasis on comparative results of the two modern determinations, which they interpret as an indication that Greenland moved westward in the 5-year interval at an average annual rate of 36 meters. However, the Danish Institute repeated the observations in 1936, and obtained a longitude value essentially identical with that of 1927. In his published statement the director of the institute does not specifically mention Jensen's value of 1922, but presumably he includes it in his general opinion that "the derivations of the old observations from the new ones are the result of observation errors."

(Longwell, 1944c: 404; my bracketed additions).

Wegener's solution was certainly not difficulty-free; the 1922 results were not reliable. Jensen's "personal equation" dismissed by Wegener as unproblematic proved his undoing. Jensen's "personal equation" had to do with his telegraphing the precise instant when the particular star that he had selected to use for his measurement crossed the zenith. Because his 1922 zenith telescope did not have an automatic clicker, Jensen had to manually press the telegraph at the precise instant when he observed the star at its zenith. For later observations, the Danish Institute installed an automatic clicker that eliminated both the "personal equation" and any measurable westward movement of Greenland. The very same Institute that seemed at first to have provided Wegener with a difficulty-free solution, not only took it away, but also turned it into a possible anomaly. Of course it was easy to remove the anomaly by supposing either that Greenland had in the interim stopped drifting westward relative to Europe or that its rate of drift was less than could be measured by techniques used. These new findings destroyed Wegener's case that mobilism had been securely established.[28]

3.14 Use of research strategies in the three sub-controversies

Participants acted in accordance with the three research strategies introduced in Chapter 1 in order to improve their own solutions and to decrease the effectiveness of their opponents' solutions. Once Wegener had presented his theory, fixists raised difficulties with it (RS2), and developed new solutions of their own (RS1). Wegener, aided now by converts, counterattacked (RS1) by developing new solutions or modifying former ones to remove or circumvent the difficulties that fixists had raised; some, they argued, were insignificant being based on erroneous data (RS1). Wegener and supporters raised difficulties against fixism (RS2). Mobilists and fixists compared solutions, each believing theirs to be superior (RS3). In all three sub-controversies mobilists and fixists successfully raised difficulties against their opponents' solutions, but never succeeded in nullifying all the difficulties that had been raised against their own. The result was that throughout the classical stage of the debate nobody constructed and maintained a solution without some outstanding difficulty remaining.

In the biogeographical sub-controversy, fixists presented a solution that was as successful as that of mobilists. Mobilists fared somewhat better in the paleoclimatological sub-controversy, but they still had to remove all difficulties. Fixists countered by proposing their own paleoclimatological solution. Wegener thought he had found geodetic "proof" of Greenland's westward drift, but he was wrong. In all three sub-controversies, questions remained about the reliability of the data used either to support mobilism or to remove key difficulties.

Wegener, in his second edition of *The Origin*, launched a broadside attack against permanentists' explanation of the very widespread biotic disjuncts. He bluntly claimed that permanentists could not explain the origin of most of them (RS2), whereas continental drift could (RS3). He also raised difficulties against landbridgers (RS2), insisting that he could explain disjuncts better (RS3). Consider, for example, his discussion of the situation in which taxa from landmass *A* resembled those from *B* but were different from those of landmass *C*, although *A* is currently more distant from *B* than *C*; then landbridgers often had to suppose that the migrants had occupied or traversed regions of very different climates which would be uncongenial. In particular, Wegener argued that his explanation of the derivation of Australian fauna was much better than others (RS3). He noted that the landbridge explanation was inconsistent with isostasy (RS2), whereas his solution of partition and lateral displacement was not (RS3).

Landbridgers repeatedly raised anomaly difficulties with Wegener (RS2). Some simply claimed that he did not explain most biotic disjuncts. Others, whether they were referring to fossil or extant forms, were more specific. For example, Diels stated that the flora of eastern North America was related to Asiatic flora, and Gregory and von Ihering argued that Wegener's solution could not explain disjuncts separated by the Pacific Ocean. Diels also believed that the indigenous flora of Australia was of tropical not of temperate origin as Wegener required. To answer Wegener's argument that his mobilist solution to the distribution of *Glossopteris* was so much better than fixism, Lake, Ökland, and Schuchert incorrectly alleged that *Glossopteris* flora in north-eastern Russia and Siberia created an important anomaly difficulty for mobilism.

Landbridgers attacked mobilist solutions (RS2) while making comparisons favorable to their own (RS3). For example, Gregory, von Ihering, and Schuchert cited as a key advantage fixists' ability to explain the trans-Pacific disjuncts, which they claimed were unexplained by drift. Schuchert countered Wegener's claim that drift offered the best solution to the distribution of occurrences of *Glossopteris*, which he argued were not nearly as synchronous as Wegener had maintained; according to Schuchert, Wegener had incorrectly summarized the various studies on *Glossopteris*. The most frequently raised objection to mobilism by landbridgers and permanentists was the mechanism difficulty. Just about every biogeographer who argued against mobilism's explanation of the origin of disjunctive biota reiterated the mechanism difficulty ad nauseam.

In the 1920s, Wegener and supporters improved the problem-solving effectiveness of drift by expanding its range and removing difficulties (RS1). In the 1929 edition of *The Origin*, he cited new work in the 1920s by biogeographers who had increasingly begun to invoke continental drift to explain disjunct biota; some specialists frankly claimed that continental drift provided an excellent explanation of them; others contended that certain now widely separated flora were similar enough botanically to definitely require drift rather than landbridges. Wegener also continued to repeat the advantages and disadvantages of drift that he had documented in earlier editions of *The Origin* (RS3).

After Wegener's death, drift theory continued to command the attention of paleontologists, notably that of paleobotanists interested in *Glossopteris*. Seward, for example, countered Schuchert's criticism of the alleged synchroneity of *Glossopteris* flora (RS1). Sahni used drift to account for the juxtaposition of *Glossopteris* and *Gigantopteris* floras, which could not have flourished side-by-side and hence India and Asia could not always have been united (RS1). Other paleobotanists pointed out that the Russian specimens had been misidentified as *Glossopteris,* removing the supposed anomaly (RS1).

G. G. Simpson deftly employed all three research strategies in a sophisticated way. He distinguished dispersion of organisms via corridors, filter-bridges and sweepstakes routes (RS1). He complained about the unreliability of many paleontological studies (RS2), and that once these were set aside, the remaining disjuncts were too few to warrant either landbridges or drift (RS2). He offered a complex explanation of Australia's fauna (RS1), weakening Wegener's mobilist solution (RS3), with which he raised several anomaly difficulties (RS2); his own, he insisted, was better (RS3).

Du Toit tried to weaken Simpson's permanentist solution (RS2). Although he did not directly attack Simpson's explanation of Australian faunas, he noted that the distributions of vertebrates and invertebrates were generally at odds, and that the former, especially mammalian distribution, may be explained without appeal to drift. He argued that alleged mammalian disjuncts, even if correct, had little to do with drift because as a group they did not begin to flourish until after the main breakup of continents (RS2). Du Toit also criticized Simpson's assessment of the number and extent of biotic disjuncts. Agreeing with Simpson that many paleontological studies were unreliable, but fewer than Simpson claimed, he turned Simpson's complaint on its head by pointing out, advantageously to mobilism, that paleobotanists (as just noted) had corrected the erroneous identification of *Glossopteris* occurrences in Russia, removing an apparent anomaly difficulty (RS2).

The exchanges of Simpson and du Toit are a good illustration of the underlying structure of the disjunct debate and the use made by participants of the standard research strategies outlined in §1.13. I add only that this casting of these subcontroversies could be continued through the 1960s.

Turning now to the distribution of Permo-Carboniferous glaciation, Wegener attacked every fixist solution (RS2). He began by attacking fixist solutions that did not invoke polar wandering on grounds of their incompatibility with data from coeval strata in the Northern Hemisphere; they required periodic refrigeration over a long interval of time (Late Carboniferous and most of the Permian) throughout the Southern Hemisphere and northward to India, whereas the paleoclimatological data from the Northern Hemisphere indicated that the climate there was uniformly warm or hot, hardly a likely circumstance. Wegener raised the same objection to fixist explanations that did invoke polar wandering; these placed the Permo-Carboniferous southern pole in the center of the massive glaciation, but still required widespread refrigeration of the Southern Hemisphere contemporaneous with warm or hot climates in the other half of the world (RS2). He argued his was decidedly better than any fixist solution because it could account for this gross inter-hemispheric imbalance and retain symmetrical climate zones comparable to those at present (RS3). Argand, du Toit, van der Gracht, and Holmes agreed with Wegener that his drift theory had the advantage over the fixist competition in explaining Late Paleozoic glaciation (RS3).

Launching these counterattacks, Wegener's opponents acted in accordance with all three research strategies. There was the regular citation of their old alibi, the mechanism difficulty. Coleman argued that there would be no source of moisture for glaciation to occur in Wegener's Permo-Carboniferous supercontinent Pangea, which itself is inconsistent with the common presence of marine fossils in the strata associated with the glacial deposits (RS2). Coleman and Brooks also pointed to what were fast becoming a fixist standby, the apparently anomalous glaciation, the Squantum Tillite and conglomerates elsewhere in North America (RS2). Brooks, Schuchert, and Willis developed a new fixist meteorological model to explain the distribution of the Permo-Carboniferous glaciation (RS1) and they even began to argue that their solution was the better because it faced neither the mechanism difficulty nor the need of explaining the troublesome Squantum Tillite (RS3).

Right on cue, mobilists countered by raising difficulties with the Willis–Schuchert–Brooks meteorological solution (RS2) and by removing ones faced by their own (RS1). Wegener, Holmes, and van der Gracht argued that the interpretation of the Squantum Tillite as glacial was dubious (RS1); van der Gracht also proposed that the conglomerates in the central and western United States could have resulted from massive and short-lived cloudbursts typical of desert climates which prevailed there during the Permo-Carboniferous (RS1). Du Toit modified Wegener's solution by allowing "Mediterranean" seas to invade Gondwana, seas that could have provided sufficient moisture and habitat for marine animals (RS1). Holmes and van der Gracht sought to discredit the meteorological solution by raising a theoretical difficulty (RS2), arguing that Brooks had no geological justification for raising the altitude of Gondwana sufficiently to cause glaciation, and he could not explain its return to present elevations in a manner consistent with isostasy. The Briton G. C. Simpson amended the mobilist explanation (RS1) so that it could, unlike Brooks', explain interglacial periods (RS2), thereby making it preferable (RS3).

Finally, there is the geodetic debate. Wegener believed that continental drift was on-going, and so should be detectable geodetically. If it could, then mobilism would have a new and most persuasive solution to its credit (RS1), and fixism a fatal difficulty (RS2). Wegener's theory predicted that Greenland was drifting, but the reliability of the early geodetic results was questionable. Later, Wegener (1929/1966) claimed that new results obtained in 1928 did show Greenland drifting west relative to Europe providing direct evidence for mobilism, what in effect was a difficulty-free solution (RS1) and a potentially fatal difficulty for fixism (RS2), giving mobilism a decided advantage (RS3). Fixists wisely advised caution. Longwell, for example, still questioned the significance of the data (RS2), and advised a "wait and see" attitude. In 1936 the new longitude determinations indicated no significant drift when compared with those made in 1928. Once Longwell (1944b) found this out, he denied that the geodetic results provided any support for continental drift and his judgment stood for forty years.

3.15 Köppen and Wegener determine ancient latitudes

During the turmoil of these sub-controversies it became clear that if Wegener's jigsaw, its subsequent partition, and drift of its fragments were correct, then the latitudes of the fragments would have changed greatly relative to one another. Wegener made his initial reconstruction by matching and then disassembling the geographic, geologic, and biotic intercontinental disjuncts according to their ages. His maps gave positions of continents relative to one another, and were blind to ancient latitudes or longitudes; they gave no information about their positions relative to the geographic poles; their paleolatitudes that is. Wegener constructed his continental maps on an arbitrary grid, keeping Africa fixed in its present-day position (Figures 3.1, 3.2). He, however, did find a way to determine ancient latitudes with help from his father-in-law Wladimir Köppen, an eminent meteorologist and paleoclimatologist. Their collaboration began, I believe, while Wegener was studying the distribution of Permo-Carboniferous glaciation. Their work together culminated about four years later with *Die Klimate der Geologischen Vorzeit* (*The Climates of the Geological Past*) (Köppen and Wegener, 1924). Wegener described their collaboration in this charming way.

As the first volume was made possible by the selfless advice and cooperation of Cloos, so the second edition is marked by the no less valuable cooperation of a climatologist: its layout and logic (Ausarbeitung) happened in a daily exchange of ideas with W. Köppen and I had the satisfaction that he, initially cool and doubtful, immersed himself with increasing warmth into the ideas of drift (displacement?) theory, and finally convinced himself with great satisfaction that here the red thread in the labyrinth of paleoclimatology had been found. Several chapters were written in such close exchange of ideas with him, that the boundary of intellectual ownership can no longer be decided. His most important ideas on the subject Köppen will publish in two papers in "Petermanns Mitteilungen."
(*Wegener, 2nd edition of* The Origin, *1920; Introduction, p. iv; translation courtesy Dieter Weichert, provided by E. Irving*)

They began working together before the publication of the 1920 second edition of *The Origin*, for he included a map showing his reconstruction of the continents during the Permo-Carboniferous relative to the Geographic South Pole.[29] Wegener began by compiling the occurrences of temperature-sensitive deposits such as

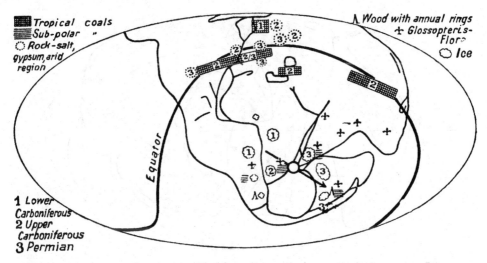

Figure 3.6 Wegener's Figure 17 (1924: 100). Wegener's positioning of the continents during the Permo-Carboniferous relative to the Permo-Carboniferous equator and South Pole. Wegener superimposed the paleoclimatically determined latitudes on his "jigsaw-fit" map of the Upper Carboniferous based on geographic, geologic, and biotic disjuncts (Figure 3.1).

tropical and sub-polar coal deposits, rock salt (gypsum in arid regions), wood with annual rings, *Glossopteris* flora, and glacial deposits. Using his reconstruction of Pangea as a basis, Wegener initially positioned the Permo-Carboniferous South Pole by centering it with respect to glacial deposits of that age in Brazil, India, eastern Australia, and southern Africa. Then, applying Köppen's model of climate zones based on Earth's present climate, he drew latitudes concentric about this and the antipodal North Pole so that they conformed to the climatic indicators (Figure 3.6). The results were consistent with his reconstruction of the continents, which he claimed was "a strong proof of the accuracy of the displacement theory."

The whole of these evidences of the climate of the Permo-Carboniferous period give such a convincing picture of the climatic zone prevailing then that I do not see how this conception of the position and direction of movement of the poles can be dismissed. In this way these evidences become a strong proof of the accuracy of the displacement theory.

(Wegener, 1924: 108)

After noting that he had deliberately chosen the Carboniferous period as his first test case primarily because it best shows "the simplification produced by the displacement theory," and that the position of the climatic belts had to be worked out for the other periods to further test his theory, he announced that he and Köppen were doing just that, and then listed pole positions in terms of present co-ordinates, with Africa stationary (Wegener, 1924: 108–109).

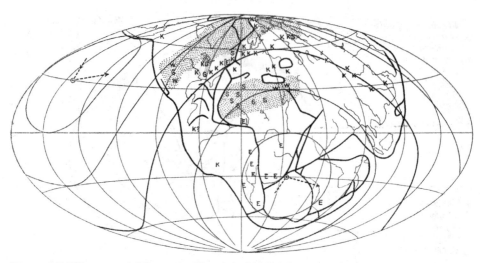

Figure 3.7 Köppen and Wegener's Figure 3 (1924: 22). Map of Pangea during the Late Carboniferous with paleoclimatically determined latitudes superimposed. Reproduced by Wegener as Figure 35 in the 4th edition of *The Origin* (Wegener, 1929/1966: 137). He labeled the Fig. "Ice, bog, and desert in the Carboniferous." E, ice traces; K, coal; G, gypsum; W, desert sandstone; stippled, arid zones. Geographic poles indicated by circles with direction of past and future movement indicated by dashed arrowed lines.

Wegener's use of paleoclimatology went beyond simply explaining the distribution of Permo-Carboniferous glaciation. Wegener acknowledged Köppen's aid in the fourth (1929) edition of *The Origin*. This was a most apt collaboration.

Since the previous edition of this book appeared, the problem of the climates of the geological past was systematically examined by W. Köppen and myself. The scope of our book was hardly inferior to that of the present work. Although our book was in essence a collection of geological and palaeontological material, subjects in which the geophysicist and climatologist are beset by difficulties and the danger of errors that the specialist can avoid, we felt nevertheless that such an investigation was justified, since palaeoclimatology can thrive only as a unification of these sciences, and the literature in this field which has appeared so far shows all too clearly that the meteorological and climatological basis for it is inadequate. Our detailed exposition will be referred to extensively in the present chapter.

(Wegener, 1929/1966: 121)

Wegener was proud of his work with Köppen. He was particularly pleased with its interdisciplinary nature, and made clear its novelty. He included two of the figures that he and Köppen had drawn on the basis of their paleoclimatic investigations. One of the figures is shown in Figure 3.7.

   Comparison of the two maps of Pangea during the Late Carboniferous (also visible at www.cambridge.org/frankel1) makes clear the great strides made between Wegener's "jigsaw-fit" and Köppen and Wegener's paleogeographic

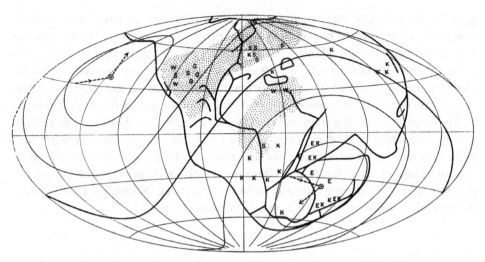

Figure 3.8   Köppen and Wegener's Figure 4 (1924: 23). Map of Pangea during the Permian with paleoclimatically determined latitudes superimposed. Reproduced by Wegener as Figure 36 in the 4th edition of the Origin (Wegener, 1929/1966: 137). He labeled the Fig. "Ice, bog, and desert in the Permian." E, ice traces; K, coal; G, gypsum; W, desert sandstone; stippled, arid zones. Geographic poles indicated by circles with direction of past and future movement indicated by dashed arrowed lines.

synthesis. Wegener's jigsaw reconstructions based on matching geographic, geologic, and biologic disjuncts did not provide paleolatitudes; they came only from applying Köppen's zonal climate model.

Although *The Climates of the Geological Past* expanded the paleoclimatological support for mobilism far beyond Wegener's original appeal to Permo-Carboniferous glaciation, it had little influence during the classical stage of the controversy. It was never translated into English, and was much less widely read than the third edition of *The Origin*. Few others besides van der Gracht (1928a) and du Toit (1937) used its findings in their defense of mobilism.

I end this brief discussion of Wegener's work in paleoclimatology and his fixing of paleolatitudes with a look ahead at the second and third stages of the mobilism controversy. Paleomagnetism, discussed in Volume II, like paleoclimatology, yields estimates of paleolatitudes, but they are numerical estimates free of assumptions inherent in Köppen's model of past climatic belts based strictly on present-day ones, assumptions that got them into trouble (Irving, 2008). Brooks (1926) correctly objected to Köppen and Wegener's insistence that climatic belts have remained basically unchanged throughout the geological past, and Wegener (1929/1966) was negligent in not acknowledging his challenge. Brooks was correct in proposing that Earth had long warm "non-glacial" periods, when the temperature differences between climatic belts were small compared to what they are now under colder "glacial" conditions (Brooks, 1949: 31). Paleomagnetism provided and continues to provide more precise and more accurate determinations of paleolatitudes than paleoclimatology. Indeed paleoclimatological determination of past latitudes during non-glacial periods ("greenhouse") is difficult, and during pan-glacial ("snowball") periods hardly possible at all. The ideas of a "greenhouse" or "snowball" Earth, although heavily dependent on paleoclimatology and its associated disciplines, require paleomagnetic determinations of paleolatitudes for their full justification.

The maps reflecting the culmination of the third stage of the mobilism controversy may be prepared with Wegener's jigsaw fit of the continents. Seafloor spreading and plate tectonics are blind to paleolatitudes and paleolongitudes, and the historical reconstruction of ocean basins and continents based on them accurately record the past movements of continents relative to one another back through the Late Triassic, but it is only paleomagnetism that pins down paleolatitudes.[30]

Notes

1 The first edition of *Die Entstehung der Kontinente und Ozeane* appeared in 1915, and was followed during Wegener's lifetime by three more editions. The second appeared in 1920, the third in 1922 and the fourth in 1929. The third and fourth editions were translated into English. The English translation of the third edition appeared in 1924, Alfred Wegener, *The Origin of Continents and Oceans* (translated by J. G. A. Skerl, London: Methuen & Co., 1924); the English translation of the fourth edition appeared in 1962, and was reprinted in 1966, Alfred Wegener, *The Origin of Continents and Oceans* (translated by John Biram, New York: Dover, 1966).

2 Wegener later underscored the importance of understanding that his description of the breakup of Pangea could have been described based on other reference systems. In the 1929, fourth edition of *The Origin* he stated:

It is important to realize the complete arbitrariness of the African reference systems we used. When, for example, Molengraff [in his contribution to the symposium on continental drift sponsored by the AAPG] stresses that the mid-Atlantic ridge shows that Africa drifted from there toward the east, I cannot discern any disagreement with drift theory in his statement. Relative to Africa, America and the mid-Atlantic ridge drifted westwards, the former at about twice the rate of the latter; relative to the ridge, America drifted westwards and Africa eastwards at about the same rate; relative to America, both the ridge and Africa migrated eastwards, the latter twice as fast as the former. On the basis of relative movements, all three statements are identical. But, once we choose Africa as the reference system, we cannot assign a displacement to this continent, by definition.

(Wegener, 1929/1966: 148; my bracketed addition)

3 Wegener also offered solutions to the origin of rift valleys and island festoons. Rift valleys form along the line of separation where landmasses split apart. He cited the East African rift valleys, in which he included the Red Sea, the Gulf of Aqaba, and the Jordan valley, as "the most beautiful example of such rifts" (1924: 166). He also suggested that the formation of rift valleys is the first step in the formation of an ocean basin, if the landmasses continue to separate. The Atlantic basin, according to Wegener, "is really an enormously widened rift" (1922: 115). Island festoons broke off from drifting continents. They generally broke off from a trailing edge of a drifting continent. Wegener mentioned this solution in his *Discovery* article, linking it to his explanation for the formation of mountains:

We have already instanced the frontal resistance which the American continental masses experience in moving through the ancient and deeply cooled bottom of the Pacific, a resistance which has led to the throwing up of the gigantic mountain chain of the Andes. Because this frontal resistance must have a much greater influence for small masses than for larger, these small masses will be left behind in the general westward movement. Thence arises the great sweep of the Antilles, left far to the east by America, and the great bend of the so-called southern Antilles between Tierra del Fuego and West Antarctica. Thence also comes the partial separation of the eastern edge of Asia in the form of chains of islands, and the separation, long ago completed, of the former Australian coastal chain which now forms New Zealand. By the same measurement Ceylon has been broken away from India, and we see evidence of it also in the bending of all the ends of continents toward the east, such as the southern end of Greenland, of Tierra del Fuego and the northern end of Graham Land.

(Wegener 1922: 118)

4 Wegener presented a short version of this argument in his *Discovery* paper, where he used the analogy of a torn drawing instead of a newspaper (1922: 115).

5 Arldt's and Wegener's views about migration of earthworms (Lumbricidae) from Europe to North America before opening of the Atlantic were incorrect. Earthworms (and perhaps some of the other listed forms) are a recent immigrant to North America, accompanying humans aboard ships (Hendrix, 1995).

6 Wegener was mistaken about the position of the geographical pole relative to North America in the Miocene. It was much closer to the present geographic pole than determined by Wegener (and Köppen). The close proximity of the Miocene pole to the present pole relative to North America was first determined paleomagnetically by Campbell and Runcorn (1956). In addition, Hospers (1954, 1955) explicitly tested Wegener and Köppen's paleoclimatically derived polar wander path and found, based on his own results from Iceland and Northern Ireland and those of others from northern England, France, and Scotland from Early Quaternary through Eocene, that all paleopoles were closer to the present geographic pole than determined by Wegener and Köppen (1924). Hospers showed that if the geographic pole had indeed wandered it had already arrived very near its current position by Miocene times. This is discussed in Volume II, §2.10. Even though paleomagnetists found that the continents had moved relative to each other roughly as Wegener had claimed, they also showed that the North Atlantic had begun to open well before Wegener (and Köppen) had claimed.

7 See Newman (1995) for an excellent discussion of the AAPG symposium. Newman was the first to point out that the proceedings do not reflect what happened at the symposium. Others, including myself, had failed to find out exactly who spoke at the symposium.

All accounts I have seen of this symposium are grossly inaccurate. They assume that the symposium booklet published in 1928 including papers by fourteen geologists, represents pretty much what happened in New York, with, of course, some editing and expanding the speeches. Far from it. Only six people actually spoke; van der Gracht, Schuchert, Chester Longwell, William Bowie, Charles P. Berkey, and Andrew C. Lawson. The other geologists whose remarks are printed in the 1928 publication sent manuscripts later to van der Gracht. Berkey and Lawson, who spoke in 1926, did not submit manuscripts. So we really do not know what happened that evening in any detail.

(Newman, 1995: 72)

8 A good synopsis of Diels, Ökland, and Diener may be found in E. V. Wulff (1950: 164–201). Wulff's Russian text was first published in 1923. He was an enthusiastic defender of Wegener's solution, and he offered a fairly complete review of the treatment of the problem over the previous decade by phytogeographers essentially from the time Wegener initiated the debate in 1912.

9 Explaining botanical disjuncts between North America and eastern Asia was a long-standing difficulty for mobilism, and was not explained until after its acceptance. These botanic disjuncts are remnants of a formerly widespread Paleogene tropical (greenhouse) flora. Magnolias, for example, have this classic disjunct distribution, documented in Hebda and

Irving (2004). This disjunct has been long recognized and was discussed by Charles Darwin and Joseph Hooker circa 1890 (Good, 1947).

10 I have considered the general reception of Wegener's theory by specialists in several of the sub-fields of Earth science in H. Frankel (1976: 304–324).

11 For details of Seward's life see Thomas (1941), M. R. Sahni (1952), and Sitholey (1950).

12 Ronald Good (1947: 328). The manuscript for this first edition was completed in 1939, but its publication was postponed until 1947 because of World War II. *The Geography of the Flowering Plants* has gone through four editions. The second appeared in 1953, the third in 1964, and the fourth in 1974. Compared to most other texts on phytogeography, Good's treatment of drift's solution was extensive in the first two editions. His treatment became even more extensive in the 1964 edition wherein he considered the new advances in paleomagnetism and marine geology in support of drift theory. By 1974, Good simply provided his readers with key references on seafloor spreading and plate tectonics.

13 Argand (§8.7) put forth his ideas in 1922, in his presentation at the 13th International Geological Congress in Brussels. The written version of Argand's talk was published two years later (Argand, 1924/1977). Argand may have influenced Joleaud regarding the Proto-Atlantic.

14 See Newman's splendid account of the Schuchert and Willis joint effort, both of whom were inveterate critics of mobilism. Newman (1995: 76–78) presents and discusses their correspondence that led up to their joint project. Oreskes (1999: 207–219) also discusses their correspondence and joint solution.

15 G. G. Simpson's more important papers on the problem are 1940a, 1940b, 1947, and 1952.

16 Much of this account is taken from Frankel (1981). Laporte (1985) gave a similar account. See also Frankel (1984b) for further discussion of controversy in biogeography after the development of plate tectonics.

17 In order to appreciate the severity of G. G. Simpson's criticisms of mobilism, here are two examples:

> Here may be exemplified the tendencies of both drift and trans-oceanic continent theories greatly to exaggerate the affinities of the groups they discuss. For instance du Toit ... speaks of the "diprodonts" of Australia and South America as belonging to "identical or allied species." This is so far from the truth that even the zoologists most convinced of the reality of the relationship never place the Australian and South American forms in the same families. Most students place them in different superfamilies or suborders.

> Similarly, Wegener ... referring to another supposed connection, quotes from Arldt. What he, Wegener, says are percentage figures for "identical" mammals and reptiles on the two sides of the Atlantic [are not as he says.] [Referring] directly to Arldt ... shows that these figures (which incidentally, were unreliable to begin with) are for families or subfamilies, which gives them implications decisively different from those carried by Wegener's word "identical" (or "identisch" in German). Such looseness of thought or method amounts to egregious misrepresentation and it abounds in the literature on this perplexing topic.
>
> *(Simpson, 1943: 3–14; my bracketed additions)*

Simpson was equally harsh on Joleaud and J. W. Gregory.

18 Simpson credited du Toit with realizing that similarities should be greater for drift than for landbridges, but then argued that du Toit had failed to recognize that mobilism received less support from the fossil evidence than did landbridges.

19 As a matter of fact, du Toit had admitted the fragmentary nature of the paleontological data in South Africa at the 1923 meeting in South Africa on the distribution of life forms in the Southern Hemisphere and its relation to drift theory (see Anonymous, 1923). Moreover, in light of these general difficulties with the paleontological data, du Toit suggested that Simpson's quantitative comparison of the similarities between fauna might be misleading, especially in cases when only a small number of specimens were known.

True to his word, he cited the small number of specimens in Simpson's comparison of Triassic reptiles of Brazil with those of South Africa, even though this was one of the strongest zoological arguments in support of drift.

20 Caster, wishing to acknowledge Kay's later work in support of mobilism, separated him from the others but, I add, exaggerated its importance.

Incidentally, Marshall Kay more than made amends later, when he became one of the earliest ardent supporters of Plate Tectonics, and for which much of his splendid work in the northeastern Appalachians was prime documentation.

(Caster, 1981: 7)

21 The Dutch geologist J. H. F. Umbgrove, an adamant fixist, repeated Longwell's misuse of the distribution of *Cornus* as evidence against mobilism, going so far as to reproduce Longwell's figure of its distribution and quoting extensively from him.

The distribution of the Gondwana-flora, including plant remains like *Glossopteris* and *Gangamopteris* has often been considered as in favour of the theory of continental drift. However, Longwell has given an example of similar problems presented by the distribution of modern plants belonging to the primitive dogwoods of the genus *Cornus*. The distribution of the genus, containing about 40 species and its near allies (fig. 144) [from Longwell and reproduced beside the following quoted passage from Longwell], "poses greater problems than that of the *Glossopteris* flora, since it involves greater oceanic distances and more formidable climatic barriers. Moreover, the hypothesis of continental drift does not help explain the dispersal of dogwoods – rather it increases the difficulties. Obscure ways in which plants and also some animals must have been distributed during long geologic time intervals are suggested by other examples from the vast Pacific region. When these numerous problems are viewed together, emphasis that has been placed on the *Glossopteris* flora as 'compelling evidence' of once continuous lands seems dangerously near the unscientific procedure of selecting evidence to support a favored theory." Umbgrove's views are discussed in §8.13.

(Umbgrove, 1947: 232–233; my bracketed addition)

22 Myers' argument is generally consistent with what is currently known about the early stages of the opening of the South Atlantic. It opened earlier than Wegener had proposed and probably was crossed by volcanic (hotspot ridges) which many have intermittently acted as isthmian links between South America and Africa.

23 Launcelot Harrison and G. E. Nicholls, two mobilists who studied the origin and composition of Australia's fauna (§9.4), also raised doubts about the reliability of biogeographical data.

24 A year after the publication of Brooks' *Climate Through the Ages,* Jeffreys pronounced that the "main lines of Brooks' theory seem extremely plausible," and claimed that Brooks' solution to the origin of Permo-Carboniferous glaciation was "very plausible."

The Permo-Carboniferous glaciation presents greater difficulties [to the paleoclimatologist than the recent glaciations]. At this period there was ice in comparatively low southern latitudes, combined with an extensive development of coal and glaciation almost side by side in North America. A reconstruction of the distribution of land and sea at the time, already inferred by geologists on purely geological grounds, leads Brooks to a very plausible explanation.

(Jeffreys, 1927: 279; my bracketed additions)

25 The Squantum Tillite continued to play a role in the mobilism debate, and the question of whether it is glacial continued to plague researchers. (See III, §1.2 for details.) There is still some disagreement about whether it is glacial; however, it has now been accurately dated to be Late-Precambrian, not Late Paleozoic. As a result, even if caused by glaciation, its presence is no longer a difficulty for mobilism.

26 I shall concentrate almost exclusively on the geodetic attempt to show that Greenland continues to drift westward relative to Europe, because this provided mobilists with their

strongest support. Geodesists and astronomers, however, attempted to determine longitudinal shifts between landmasses. Included among their attempts are Europe and North America, Madagascar and Europe, Europe and Honolulu, eastern Asia, Australia, and Indochina. (See, for example, Chapter 3 in the fourth edition of Wegener's *The Origin*.)

There was also a joint attempt by the International Union of Geodesy and Geophysics (IUGG) and the International Astronomical Union (IAU) to set up a world longitude-net by the aid of radio signals to test continental drift. Newman (1995: 69–70) and Oreskes (1999: 231–236) have discussed this attempt. The idea was put forth in 1922 at the joint meeting of these organizations. The results were inconclusive (Oreskes, 1999: 236). Bowie (1925a, 1925b), then the doyen of US geodesists, was solidly behind the project. Oreskes and Newman wondered why Bowie, who was adamantly opposed to continental drift, supported the project. Both, I believe, correctly claimed that Bowie wanted to promote his science. Oreskes also examines but rejects the idea that Bowie wanted to refute continental drift, although he realized that negative results would not refute mobilism. Although I agree that such results would not have refuted drift, they could have shown that a worldwide effort had failed to find support for mobilism and could have set upper limits to drift. Wegener already had been touting the early geodetic results from Greenland, and a negative finding from the IUGG-IAU team would have weakened his case. In short, Bowie wanted to show that drift was worth refuting, to disprove drift, and also to promote geodesy.

Oreskes (1999: 260) also argued that Bowie later changed his mind about mobilism, and recommended continued geodetic testing of it. According to Oreskes, Bowie, who had based his rejection of mobilism on its incompatibility with Pratt's model of isostasy, no longer rejected mobilism once he realized, *circa* 1936, that new evidence supported Airy's model. Thus Oreskes (1999: 260) thinks, "Ten years after he had called continental drift impossible, William Bowie now suggested that geodesists would be the ones to prove it." Bowie actually said:

The Wegener Hypothesis has received a great deal of attention in recent years and deservedly so. It is based on the idea that we have perfect isostasy. This implies that the sial which lies under the continents and islands has residual rigidity while the sima is plastic. Many students of the Earth's crust feel that the Wegener Hypothesis does violence to certain mechanical principles, but in any event, it is something that should be looked into. The geodesists have cooperated with the astronomers in two great world-wide longitude-campaigns which were designed to furnish the basis for determining whether continents and islands are wandering over the face of the Earth. Nothing has been found up to this time that would substantiate the Wegener Hypothesis, but we cannot tell what the future may disclose.

(Bowie, 1936: 20)

Unlike Oreskes, who also quotes most of the same passage, I do not think that Bowie substantially changed his mind about continental drift. He certainly does not claim to support "the Wegener Hypothesis" in the above. More importantly, some of Bowie's objections that he had raised against mobilism ten years before at the AAPG symposium were not based on its incompatibility with Pratt's model of isostasy. Oreskes (1999: 172) correctly argued that given Pratt's isostasy the sialic continents could not be strong enough to transmit horizontal stresses over great distances because they had to have been weak enough to accommodate isostatic adjustments within crustal prisms. But Bowie had other "impossibility arguments" against mobilism. Indeed, even given a weak sial, Bowie still thought that the forces Wegener thought responsible for continental drift were not large enough to break apart the continents.

The only forces which may be operating to any extent on continental masses to make them change their positions geographically seem to be the equatorward forces exerted on a floating body due to the change in direction of gravity with elevation, and the tidal forces. There are other forces that have been suggested by Wegener and others as possible causes of continental drift. All of these forces, including the two just mentioned, are exceedingly small; in fact they are so small that it does not seem possible that they could tear continental masses apart and make them drift over the earth.

(Bowie, 1928: 180)

Also, Wegener required a sima of no strength. But, even given isostasy, Bowie argued that sima was not without strength: strong oceanic earthquakes proved it to be so.

In order that the continents might wander as postulated by Wegener, the sima would have to be entirely lacking in residual rigidity or strength. It might be rigid to stress acting for short periods of time but to stresses operating through many years it would have to yield like tar or sealing wax. But how can the sima be entirely lacking in residual rigidity in view of the fact that many of the most violent earthquakes occur under the oceans, presumably in sima? An earthquake is generally understood to result from the rupturing of rock due to the stress differences which may accumulate in the crust. In order that there may be a break of crustal rock it must have residual rigidity sufficient to accumulate stresses beyond the elastic limit of the material. It would seem to me that the occurrence of earthquakes under the oceans is an argument against a sima with no residual rigidity, but if the continents are to move through the sima under small forces assumed by Wegener, the sima cannot have any residual rigidity whatsoever.

(Bowie, 1928: 183)

Oreskes (1999: 171) also quotes from a 1928 letter Bowie wrote to Schuchert in which he dismissed Holmes' early ideas that convection could cause continental drift.

In addition, Bowie in his 1935 paper, entitled "The origin of continents and oceans," might have agreed with Wegener about the choice of a good title, but he certainly did not agree with him about continental drift. Bowie adopted Osmond Fisher and George Darwin's idea that the Moon had been created from sialic material that had spun off from the Earth because of tidal forces. Originally, the Earth was completely covered with a sialic crust. But the birth of the Moon left a huge gap in the sialic cover, exposing the underlying sima in a cavity that became the Pacific basin. Bowie then reasoned:

It is rather interesting to note that the two coasts of the Atlantic are so nearly alike that they have the appearance of the shores of a great river. Is it not possible that North and South America could have been torn away from the crustal material that forms Europe and Africa just before the disruption occurred?

(Bowie, 1935: 447)

Thus Bowie thought he was able to explain the congruency of the peri-Atlantic continental margins, but he made it quite clear that he did not have anything in mind comparable to Wegener's post-Paleozoic breakup of the continents.

Some other hypothesis may be advanced to account for oceans and continents, but the one which we may label the Darwin-Fisher hypothesis certainly is the most probable one that has been advanced up to the present time. The question may be asked, "When did all this occur?" The answer must be, "A long time ago." According to the best geological evidence the sedimentary age of the earth began about one billion six hundred million years ago. It is reasonably certain that the continents and ocean basins were formed prior to that time, for without erosion there could be no sediments. In order to have sediments we must have running water. Running water carries material from high areas to low ones.

(Bowie, 1935: 447–448)

The continents and oceans had formed before sediments were deposited, which was well before continental drift was supposed to have occurred. If Bowie had changed his mind only one year later, I would think that he would have explained why the Darwin–Fisher view was no longer tenable. I see no reason to think that Bowie changed his mind about mobilism, and thought that geodesists would prove continental drift. I think he believed that their test would be negative.

27 Wegener erroneously believed in a very late opening of the northernmost Atlantic because of the matching of Quaternary terminal moraines (see, for example, Wegener's Figure 19 showing the boundaries of the Quaternary inland ice that he used partly to base his estimate of when the northernmost Atlantic opened) (Wegener, 1929/1966: 77).

28 Although the sub-controversy ended, geodesists continued to fix latitudes of Greenland relating them to the question of continental drift when appropriate. Longwell kept track of their findings. S. Warren Carey invited Longwell to present an honorary paper at Carey's 1956 symposium on continental drift. Longwell reviewed the latest findings by the Danish Geodetic Institute.

The latest report of the Institute (Sinding, 1955) makes the summary statement, "As to possible westerly drift of Greenland it can be stated that the observations by E. Nielson at Qornoq 1948 showing a difference of $0^s.02$ against Gabel – Jorgensen's determination in 1927 do not indicate a variation in the longitude of Qornoq during the period in question." The slight difference between the two results is attributed to "personal equation".

(Longwell, 1958: 3)

29 Marvin reproduced Wegener's 1920 figure as her Figure 29B (Marvin, 1973: 72). She also discussed how Wegener was only able to fix the continents to the geographic poles over time, and thus discuss polar wandering, through his appeal to paleoclimatology (Marvin, 1973: 79).

30 Even if hotspots are fixed relative to the mantle, whether or not hot spots are fixed relative to the geographic poles requires the use of paleoclimatology.

4

The mechanism sub-controversy: 1921–1951

4.1 Introduction

An important and long-lasting sub-controversy that occurred during the classical stage of the mobilism debate was concerned with providing an acceptable mechanism that could cause continental drift. This sub-controversy became particularly important when the mechanism proposed by Wegener was shown in the 1920s to be incorrect and beyond repair, placing mobilists, in the eyes of many, in the untenable position of having to defend an impossible theory. Consequently, during the 1920s and 1930s, several Earth scientists, in particular Daly, Joly, van der Gracht, and Holmes developed new mobilist solutions to the mechanism problem. This chapter examines the evolution of the mechanism sub-controversy, and the new solutions offered by Daly, Joly, and van der Gracht. Because of the superiority and greater importance of Holmes' solution, it will be examined separately in Chapter 5.

Throughout the controversy, both sides employed the standard research strategies; fixists raised difficulties with mobilist solutions (RS1); mobilists countered by proposing alternative solutions (RS2), and arguing that despite the mechanism difficulty the overall problem-solving effectiveness of mobilism was greater than fixism (RS3). I begin by describing Wegener's proposed mechanism, and the criticism of it. I shall then examine mobilists' responses to that criticism, particularly those of Wegener, Daly, Joly, and van der Gracht. Daly, a well-traveled and learned man, was the most prominent North American geologist to support continental drift. Initially attracted to mobilism but dissatisfied with Wegener's proposed mechanism, he offered an alternative and defended it throughout the 1920s and early 1930s. Joly, a well-respected Irish geophysicist, offered lukewarm support for mobilism during the 1920s. At the 1926 AAPG symposium, the Dutch geologist van der Gracht recast Joly's work as a mobilist theory, which he championed as a solution to the mechanism problem.

4.2 Wegener's 1922 mechanism

Wegener devoted the last chapter of the third edition of *The Origin* to the mechanism of continental drift, arguing that tidal force and *Polflucht* accounted for the westward and equatorward movement of continents, respectively. Because these forces are

159

small, and because he viewed the continents as plowing through the seafloor and its substratum (sima), he had to present and defend the view that the latter had little strength; in the long-term it behaved as a fluid.

Wegener began the final section of *The Origin* with a chapter devoted to Earth's viscosity. He (1924: 120–127) identified several phenomena which suggested that the substratum lacks long-term strength and behaves as a highly viscous fluid when acted on by forces acting over long periods of time. He cited the tendency of large-scale features of the Earth's crust to be in isostatic equilibrium, which indicated that the crust floats on a substratum that behaves as a liquid over long intervals. Continental drift and polar wandering both require flow within the Earth, as does the oblateness of the Earth with its slightly flattened poles. Of course, his appeal to continental drift and polar wandering did not carry much weight with those who did not believe in them.

Wegener was fully aware that many geophysicists did not share his idea of flow within the Earth. They believed, Wegener (1924: 128) noted, that the high velocity of earthquake waves and the Earth's ability to recover from tidal forces and oscillations of its poles proved that the Earth is "about two to three times as rigid at room-temperature as steel . . ." but they did not disprove that Earth behaves as a viscous fluid when subject to small forces over long periods. There is no contradiction between an Earth that is rigid over the short term but viscous over the long term because (i) the Earth is large and (ii) there is much time, a "point which has been quite insufficiently appreciated in the previous literature, but which is of greatest importance in geophysics" (Wegener, 1924: 129–130). Regarding (i), he argued that the interior of a steel body the size of Earth would flow because of high pressures and temperatures at depth.

We know that steel loses its rigidity under pressures as we can mechanically produce, and becomes plastic. We cannot erect an indefinitely high column of steel without reaching a limit at which the foot of this column begins to flow. If we imagine an entire continental margin of steel, its uppermost portion would certainly remain rigid, but the lower layers would become plastic under the pressure of the mass lying above it, and would flow out laterally.

(Wegener, 1924: 131)

The effect of (ii) was even greater. The phenomena indicative of a rigid Earth, "quickly alternating tides, or still more earthquake waves, and perhaps also the oscillations of the poles" are responses to short-term forces (Wegener, 1924: 132). Rigidity is a measure of resistance to short-term forces; strength, a measure of its resistance to long-term forces. Consequently, these short-term phenomena indicate nothing about Earth's long-term strength; paradoxical as it sounds, a rigid body can lack strength over the long term. As he had in 1912, Wegener cited pitch as an example.

Pitch, for example, behaves as an absolutely solid body when subjected to blows and percussion, but, given time, it begins to flow under the influence of gravity; a piece of cork cannot be forced through a sheet of pitch, but after a lengthy period its slight buoyancy is sufficient to allow it to rise slowly through the pitch from the bottom of the vessel.

(Wegener, 1924: 132)

Wegener then narrowed his attention from the whole Earth to the ocean floor and underlying substratum, what is now called the oceanic crust and upper mantle. He needed to establish at least the possibility that the simatic ocean floor and substratum can yield to the sialic continents when acted on by long-term forces even though the simatic ocean floor and substratum are more rigid than the sialic continents. He drew another analogy, comparing the continents to candle-wax or tallow and the ocean floor and underlying substratum to sealing-wax, and noted the general confusion over viscosity, solidity, and rigidity. Citing James Clerk Maxwell's *Theory of Heat*, he claimed:

Maxwell calls a body soft if it reacts quickly to an impulse, but only after a certain limit of force has been exceeded; on the other hand, he calls it "a viscous fluid" when it reacts to an infinitely small impulse, although infinitely slowly. "When this continuous alteration of form is only produced by stresses exceeding a certain value, the substance is called a solid, however soft it may be. When the very smallest stresses, if continued long enough, will cause a constantly increasing change of form, the body must be regarded as a viscous fluid, however hard it may be. Thus a tallow candle is much softer than a stick of sealing-wax; but if the candle and the stick of sealing-wax are laid horizontally between two supports, the sealing-wax will in a few weeks in summer bend with its own weight, while the candle remains straight. The candle is therefore a soft solid, and the sealing-wax a very viscous fluid."

(Wegener, 1924: 133; the quotation is from Maxwell's Theory of Heat*)*

Although Wegener (1924: 133–134) recognized that sima and sial were composite materials and therefore unlike Maxwell's homogeneous sealing-wax and tallow, he maintained that the different responses sima and sial exhibited to long- and short-term forces were comparable, and the analogy valid.

Sima is more rigid than sial; sial is softer than sima; so sima is more resistant than sial to short-term forces. But sial is stronger than sima; sial is more resistant to long-term forces than sima. Oceanic rock, made of sima, behaves like a viscous fluid, with little or no long-term strength. So he claimed that although sima is more resistant than sial to short-term forces, it is more yielding than sial to long-term forces.

Wegener concluded his characterization of sial and sima with a brief discussion of how temperature affects them. After reviewing the little available data, he favored the idea that both sima and sial probably reach their melting point at the base of continents. But he noted (1924: 136) that it "must certainly not be thought that this melting temperature is at the same depth all over the world, and that the depth is constant at all times," adding that "radioactive transformations perhaps play a part" in altering the melting depth. However, Wegener (1924: 137) claimed that temperature was not paramount, and that displacement would probably occur even below the melting temperature of sial and sima.

Having developed the model of long-term relatively strengthless substratum and seafloor, Wegener identified the forces he thought could cause drift: *Polflucht* caused equatorward movements; tidal forces, westward movements. R. Eötvös, a Hungarian geophysicist, recognized the existence of *Polflucht* during the first decade of the twentieth century. *Polflucht* is a differential gravitational force caused by Earth's shape and rotation. Because Earth is a rotating oblate spheroid with slightly flattened poles and bulging equator, Wegener argued that continents in higher latitudes tend through centrifugal force to move to lower latitudes as the Earth rotates, displacing sima as they go. As Earth rotates in an eastward direction, the tidal force, brought about by the gravitational attraction of the Sun and Moon, retards Earth's rotation, acting more strongly on the upper layers, and moving continents westward relative to oceanic sima.

Although Wegener believed his mechanism was viable, he felt that it should be treated only as a working hypothesis. In his *Discovery* paper he wrote:

If the standpoint of the displacement theory be taken up, numerous problems immediately present themselves, of which the most important is perhaps the nature of the forces which give rise to displacements. Here no final conclusion can be reached, but the problem has been so far examined by the theoretical physicists and geophysicists as to leave no doubt as to the possibility of such a force existing.

(Wegener, 1922: 118)

4.3 Wegener's mechanism attacked: 1921 through 1926

If Wegener or anyone else can throw light on these baffling problems, he is entitled to a hearing. However, certain demands are made of this new and romantic speculation before it is admitted into the respectable circle of geological theories. It must meet the test of established scientific principles, and *it must not create more problems than it pretends to solve.*
(Chester Longwell, remark at the 1926 AAPG symposium Longwell, 1928: 146; emphasis added)

This lofty statement by Longwell reflected the general attitude in the 1920s and 1930s to Wegener's mechanism; it did not conform to "established scientific principles"; it was not yet respectable; romantic, yes, but a little scatter-brained perhaps and not quite believable. This attitude, at times patronizing, and the great weight that was solemnly and repeatedly assigned to the wrongness of Wegener's and all subsequent attempts to provide a mechanism for continental drift, reoccurred throughout the classical stage of the controversy. It is one of the grand ironies of the mobilism debate, as will become clear in later volumes, that when, in mid-century, difficulty-free solutions were achieved, they were based on physical measurements of displacement that neither provided for nor required a solution to the mechanism problem.

Although the vehemence and content of the initial attacks against Wegener's mechanisms varied somewhat, by the end of the 1920s they had coalesced and become a severe obstacle. His mechanism was considered impossible because of its dependence on a highly dubious, if not outright mistaken assumption that the oceanic crust and the substratum beneath the continents are devoid of strength. Wegener argued that the relatively weak tidal and *Polflucht* forces acting over long periods could cause continental displacement. Critics argued that even though estimates of the strength of the seafloor and substratum are uncertain, both are stronger than continental crust, and both are far too strong to yield to such minute forces. Moreover, there are no other known adequate forces to cause continents to plow their way through the seafloor. Wegener never claimed that his mechanism was necessarily correct, yet critics argued that he had not demonstrated even the possibility of there being a mechanism for drift. Continental drift as envisioned by Wegener was, to them, impossible.

In addition to this central impediment (external theoretical difficulty) there were, critics argued, other obstacles (internal theoretical difficulties). Some said that the relative continental displacements invoked by Wegener were mutually inconsistent, or inconsistent with the relative magnitudes of *Polflucht* and tidal forces. At least one critic claimed that large drifting continents would break apart into small landmasses as they plowed along through the ocean floor. Others took the opposite view, arguing that continents, after initial separation, would not fracture while drifting, as Wegener had supposed, to produce island festoons and oceanic ridges. Other critics raised an inconsistency between his mountain building and mechanism solutions. Wegener maintained that sima is weaker than sial, and yet mountains form on the leading edge of drifting continents as sial crumbles upon itself because of the resistance of sima; sima could not be both stronger and weaker than sial. Finally, some argued that even if tidal or *Polflucht* forces were sufficient to cause continental drift, both forces were definitely too weak to cause mountain building.

Attacks against Wegener's mechanism began in the early 1920s. In 1921, German geophysicists Epstein and Schweydar raised doubts about the viscosity of the simatic oceanic crust and whether it would yield to the weak *Polflucht* force. The American geodesist Lambert (1921), after examining *Polflucht* force, agreed with them. In 1922, Jeffreys, then a young British geophysicist, inaugurating his life-long disapproval of continental drift, critically evaluated Wegener's mechanism and declared that sima was not devoid of strength. Daly, a Canadian-born geologist from Harvard, first objected to Wegener's mechanism in 1923. Daly thought the notion of continental drift had merit and was worth pursuing, but he was not impressed with Wegener's ideas as to its cause. He, like Jeffreys, forcefully argued that sima was not devoid of strength. Lambert reconsidered Wegener's hypothesis in 1923, arguing that *Polflucht* and tidal forces were probably too weak to cause sima to yield before the sialic continents. Daly and Jeffreys continued to expand and intensify their attacks on Wegener's mechanism throughout the 1920s. In 1926, many participants at the AAPG symposium repeated earlier criticisms, often strengthened them, and raised new ones. Most critics, Daly excepted, were ardent fixists; even Longwell, who was the most even-handed, raised difficulties (§3.7, §3.9, §6.7, §6.8; II, §6.16).

I shall concentrate on the criticisms of Lambert, Jeffreys, Daly, Longwell, and Holmes, who also joined the chorus. Lambert's attack was representative of those that appeared prior to the 1924 English translation of *The Origin*, and was similar to Epstein's and Schweydar's. Jeffreys was repeatedly invited to lecture on the geophysical aspects of mobilism, and criticized Wegener, Joly, and Holmes' theories. Daly, the most eminent North American Earth scientist to support in print the general notion of continental drift before the 1960s, proposed an alternative mechanism, which he developed, in part, because of the deficiencies he perceived in Wegener's. Similarly, Holmes, who developed the most viable mechanism, did so, in part, to avoid the difficulties facing Wegener's solution. Longwell is important because he was viewed by mobilists as one of the few fair-minded fixists.

Jeffreys launched his first offensive at a meeting of the Geological Society of London, held on September 11, 1922, and from then on, throughout his long career, was in the vanguard of the attack on mobilism. He argued correctly that oceanic crust and upper mantle were not devoid of strength; that the forces to which Wegener appealed (*Polflucht* and tidal) were far too small to account for the formation of mountains, and that the northward movement of India clear across the equator was particularly problematic. Wright summarized Jeffreys' remarks.

Dr. Harold Jeffreys stated that the rotational force which could be invoked to explain the movements of the continents was very small and quite insufficient to produce the crumpling up of the Pacific ranges. The ocean floors also presented a difficulty, for being composed of basaltic rock, they would be less radio-active and therefore stronger than the continental crust. The withdrawal of India northward and its gathering up into the Himalayan folds were moreover not easily accounted for.

(Wright, 1923a: 31)

Lambert, like Jeffreys, an applied mathematician and geophysicist at the United States Coast and Geodetic Survey, first turned his attention to continental drift in 1920, when he was asked by Daly to determine if Wegener's mechanism, in particular the *Polflucht* force, was sufficient. Without identifying the "well-known" geologist who asked him to investigate the matter, Lambert stated:

If you will pardon a bit of personal reminiscence, I will tell you what suggested the apparently fantastic problem of the huge rolling sphere. [i.e., Lambert's theoretical posing of the question about whether *Polflucht* or the Eötvös force could produce an equatorwards movement of the continents.] A well-known geologist told me that he had been convinced by geologic evidence that a part or a whole of each continental block has shifted its position in past time by moving towards the equator and in so doing has probably twisted in direction with reference to the meridian. He had been led to think along these lines by two articles [which Lambert identified as Taylor (1910) and Wegener (1912b)] published quite independently of each other, in which the authors evolved the hypothesis of continental creep in explanation of mountain building. If we accept this idea, the inevitable question is: what forces caused this motion of the continental masses.

(Lambert, 1921: 136–137; my bracketed additions)

Daly later identified himself as the "well-known" geologist.

Wegener also assumes the "Polflucht" and, to explain it, relies on the projection of a continent above the suboceanic crust. Three years ago the present writer requested Mr. Lambert to calculate the value of this force; the result has been published and proves the insignificance of the force.

(Daly, 1923b: 369)

Lambert also was clearly uneasy with Wegener's appeal to *Polflucht* as the cause of the equatorwards movement of continents.

All this is quite speculative of course; it is based on the hypothesis of floating continental masses and on the assumption of a sustaining magma that would, of course, be a viscous liquid, but viscous in the sense of the classical theory of viscosity. According to the classical theory, a liquid, no matter how viscous, will give way before a force, no matter how small, provided sufficient time be allowed for the latter to act in. The peculiarities of the field of force of gravity will give us minute forces, as we have seen, and geologists will doubtless allow us aeons of time for the action of the forces, but the viscosity of the liquid may be of a different nature from that postulated by the classical theory, so that the force acting might have to exceed a certain limiting amount before the liquid would give way before it, no matter how long the small force in question might act. The question of viscosity is a troublesome one, for the classical theory does not adequately explain observed facts and our present knowledge does not allow us to be very dogmatic. The equator-ward force is present, but whether it has had in geologic history an appreciable influence on the position and configuration of our continents is a question for geologists to determine.

(Lambert, 1921: 138)

Although Lambert was unwilling in 1921 to dismiss *Polflucht* as inadequate, he thought that the force was minute even where its effect is greatest (1/3 000 000 part of gravity at 45° latitude), and that it could only have the desired effect if the continents were moving through a region of little strength. Within two years, his opinion about the inadequacy of *Polflucht* had hardened.

Both Lambert and Daly criticized Wegener's mechanism at a meeting in April 1923 of the Washington Academy of Sciences, jointly sponsored by the Geological and Philosophical Societies of Washington. Its purpose was to discuss the Taylor–Wegener hypothesis. Taylor, Daly, and Lambert were asked to speak. Both Daly and Lambert ignored Taylor's hypothesis, addressing almost exclusively Wegener's theory.

Lambert examined Wegener's analysis of a crust without strength, extended his discussion to include tidal force, and argued that both it and *Polflucht* were wholly inadequate to cause drift. According to him (1923: 449), the Earth adjusts itself to tidal forces "by plastic flow or by ruptures here and there, the continents remaining in the same general relative position with respect to their surroundings." Tidal force could not cause a westward displacement of sial (continents) relative to sima (oceanic crust) because its action is not long-term. He rejected *Polflucht*, and his skepticism of Wegener's mechanism now matched Daly's.

There is, however, a small residual equatorward force that acts on an object floating on the earth's surface . . . This force is invoked by Wegener as an explanation of the equatorward movement of a continental block of "sial" floating in "sima." The force is so small (about 1/1 000 000 of gravity) that it seems inadequate to overcome the resistance of the "sima." In rebuttal to this objection it is urged that a very small force acting through geologic ages might produce considerable effects, since, in the yielding of a viscous liquid, time is the all-important element rather than the magnitude of the force. This argument assumes that so-called solids like the "sima" are really extremely viscous liquids. There is, however, a real distinction between soft solids and viscous liquids, as was pointed out long ago by Clerk Maxwell, and as has been more recently verified . . . It seems far more probable that the "sima" is a solid with a yield-point well above the stresses due to this extremely minute equatorward force than that it is a viscous liquid; if the "sima" is a true solid, the force in question would be ineffective in producing equatorward displacements of the continents.

(Lambert, 1923: 449)

Lambert had shifted his stance, now believing that Wegener's appeal to isostasy as a justification for viewing continents as floating on a substratum of no strength was mistaken.

The fact that the higher portions of the earth's crust are lighter than the deeper-lying portions and the hypothesis of isostasy based on this fact both suggest the conception of floating continental blocks, of which Wegener has made such free use. But this whole hypothesis of a floating crust is rather a convenient simile than an adequate statement of all the facts and must not be pressed too far. On the hypothesis of isostasy the stresses in the crust are not hydrostatic (that is, such as occur in flotation) until the depth of compensation is reached. The assumption of an absolutely rigid continental block floating in a liquid is therefore an unsatisfactory basis for calculating the stresses involved.

(Lambert, 1923: 449–450)

In conclusion he predicted correctly the future attitude of geophysicists toward Wegener's theory, at least until the rise of paleomagnetism.

The hypothesis of continental migration is a serious attempt to coordinate and explain facts that need explanation, but the suggested mechanical explanations of the migration are unconvincing. Till some more adequate explanation is offered, mathematicians and physicists are likely to doubt the validity of the hypothesis.

(Lambert, 1923: 450)

At the Washington meeting Daly (1923a, 1925) emphasized several difficulties with Wegener's mechanism. He cited evidence that the oceanic crust was not devoid of strength but was stronger than continental crust. He echoed Jeffreys' skepticism about the inadequacy of Wegener's theory to force India across the equator, and agreed that *Polflucht* and tidal forces were inadequate.

Wegener's assumption of practically no strength in the sub-oceanic crust, the sialic, continental part having notable strength, is basal to his whole reasoning and yet appears to be quite indefensible. Believing the sima, cold or hot, to be essentially fluid, he could permit himself to think that rotational and tidal forces are adequate to cause the *Polflucht* and *Westwandering* of floating continents. Lambert, Epstein, and Schweydar have proved the insignificance of these forces. Wegener lays down as a principle that, during the westward migration, the larger continental blocks should outstrip the smaller. Yet he considers the less massive Americas to have moved faster than Eurasia-Africa which was also left behind by the long but narrow fragment represented in the mid-Atlantic swell. This is but one of several inconsistencies in his reasoning . . . With the earth's axis in the positions given in the third edition of his book, one has trouble in accounting for many facts, including . . . the postulated movement of India (which actually had to climb the *Polflucht* slope). Wegener assumes peninsular India to have been endowed with energy out of all proportion to its mass, but offers no reason for the peculiarity of this particular fragment of the Paleozoic continent.

(Daly, 1923a: 447–448)

Jeffreys expanded his attack in the first edition of *The Earth* (1924); Appendix C, entitled "The Hypothesis of the Indefinite Deformability of the Earth by Small Stresses," was primarily devoted to a critique of Wegener's mechanism. He argued that rocks in the outer layers including the oceanic crust had finite strength. Because the seafloor and underlying substratum have cooled several hundreds of degrees below their point of solidification, they should be viewed as solids, possessing finite strength. He also cited geodetic evidence in support of an upper layer of finite strength, noting (1924: 260–261), "the crust in the continents is strong enough to hold up Mount Everest, and in the ocean" to "hold down the Tuscarora Deep," and concluded:

the whole weight of the continents will not be enough to produce permanent deformation in the upper layers; much less will the small fraction of it that acts tangentially, on account of their asymmetry of the earth's figure. We cannot therefore accept hypotheses of the widespread migration of continents, unless forces enormously greater than any yet suggested are shown to be available.

(Jeffreys, 1924: 261)

According to Jeffreys, the ocean floors and their underlying upper mantle are not devoid of strength, and any force that can displace the continents through the ocean floor must be sufficient to overcome its resistance. Neither tidal nor *Polflucht* forces are sufficient. Both produce less stress than does the weight of continents, which in turn is too weak to fracture Earth's upper layers. Even if ocean floors and their underlying mantle are devoid of strength, neither of these forces could account for the origin of mountains; their formation would require much larger forces, forces sufficient to overcome the weight of mountain ranges. Jeffreys concluded with a warning.

The assumption that earth can be deformed indefinitely by small forces, provided only that they act long enough, is therefore a very dangerous one, and liable to lead to serious errors.

(Jeffreys, 1924: 261)

The next major assault on Wegener's mechanism was in New York City at the 1926 AAPG symposium on continental drift; its proceedings were not published until 1928, and included papers not given there. Authors of most of the published papers had a field day attacking Wegener. Chester Longwell's criticisms were largely derivative but are worth quoting at length because of their completeness and clarity.

Wegener's forces are too weak, and tidal effects are chiefly elastic and therefore cannot explain slow plastic yielding.

The tidal forces, due to the attraction of the moon and sun, have been named as the cause of drift from east to west. This claim can be considered in more than a qualitative way, as these forces, and their effect on the solid body of the earth, have been evaluated by careful calculation and experiment. Michelson and Gale found that the difference in phase between observed and computed body tides is very small, thus demonstrating that the solid earth yields chiefly in an elastic way to the tidal stress. Schweydar, who is on the whole favorable to Wegener, also states that the tidal deformation is elastic in character, and therefore cannot be called on to explain slow plastic yielding. Wegener refuses to accept this verdict as final, presumably because there is no other known force acting consistently in an east-west direction in all latitudes. At best, however, the force available for his purpose must be infinitesimally small. The total tidal force is calculated to be one eleven-millionth as large as the value of gravity. Certainly the small rhythmic response in the solid earth is for the most part elastic, and any plastic yielding must be almost if not quite inappreciable.[1]

(Longwell, 1928: 146–147)

Longwell, relying on Lambert, emphasized the inadequacy of *Polflucht*. He also argued that if continents drifted toward the equator, they would break apart along east-west trending faults as they moved.[2] He further plausibly argued that because *Polflucht* is stronger than tidal forces, the equatorward drift of the continents, contrary to Wegener, should be greater than their westward drift.

Suppose that the mass [moved by *Polflucht*] is a continent – will it be urged through the sima? The force is very small, although it has about four times the value of the tidal force, or one three-millionth the value of gravity in latitude 45°, where it is a maximum. It would seem, therefore, that a continent in mid-latitudes would show much stronger drift toward the equator than toward the west. However, this is not true of North America, whose supposed westward drift has had a relatively small southward component. Further, it would seem that as the equatorward force is differential, being strongest at 45° and decreasing northward, it should tend to pull the continental plates apart – assuming that it has any effect in causing movement – with the formation of east-west rifts. There is no such major effect in any of the continents and yet on deductive grounds it should be expected more logically than north-south rifts, which Wegener assumes are caused by the small and uniformly distributed tidal stresses.

(Longwell, 1928: 149; my bracketed addition)[3,4]

Longwell also raised the difficulty about the strength of oceanic crust, carefully distinguishing rigidity from strength.

The sima is not fractured or ruptured – it merely yields as does the water before a floating raft. Therefore it must have, throughout, the properties of a viscous liquid. But if we should admit these properties for the deeper zone, where high temperature is a factor, can we admit the conception for the upper zone also? Whatever may be the depth at which the sima becomes distinctly weaker, certainly the upper zone consists of crystalline basic rock – basalt and its relatives, diabase and gabbro – known to be among the strongest of rocks, I do not have in mind their rigidity merely, but their strength, which, according to laboratory data, is comparable to that of average granite. The crystalline zone, according to our conceptions of the outer crust must be at least 30 miles deep. [Here Longwell cited Daly for support.] Are we to believe that this material, superior in strength, is displaced like a liquid by the floating sial? I shall not, as some others have done, say emphatically that it is not so. I shall merely say that in light of all our knowledge in geophysics and geology the conception is improbable in the highest degree.

(Longwell, 1928: 150; my bracketed addition)

Longwell then turned to the inconsistency between Wegener's accounts of mountain building and drifting continents, and, although he disagreed somewhat with Jeffreys, he agreed with him that Wegener's forces were too weak to produce mountain ranges.[5]

Another of Wegener's ideas calls for a geophysical test. The continents in their drifting are supposed to encounter resistance on their forward sides, with the result that the rocks are deformed to make mountain structures, like those of the Andes and the Rockies. If this supposition is correct, the propelling forces not only do the work of transporting the immense continental masses; they do the enormous work of thrusting and folding the strong continental rocks. The inconsistency of Wegener's argument is obvious. In order to conceive that the continents may be propelled at all, we must assume that the sima is devoid of strength with relation to secular forces. Therefore there could be no resistance, no back thrust, to fold the mountains. In any case the discrepancy between the work performed in making mountains and the forces assumed for the task is very great. Jeffreys has computed that the forces charged with moving the Americas westward amount to 1/100 000 dyne per square centimeter; the force necessary to make the Rocky Mountains is 1 000 000 000 dynes per square centimeter. Jeffreys' conclusion may be criticized . . . Even so, the intensity and large scale of mountain deformation imply powerful forces. The puny secular stresses . . . appear to be wholly inadequate for the task.

(Longwell, 1928: 150–151)[6]

He concluded on a cautionary note.

It is obvious that the results of geophysical examination, so far as they go, are generally unfavorable to the displacement hypothesis, but they are not conclusive. In fact the geophysicists turn the problem back to us, with the statement that geologists alone can determine whether the geophysical forces have had "in geologic history an appreciable influence on the position and configuration of our continents." [The quotation was from Lambert.] Let us therefore re-examine some of our own evidence, to see whether it is compelling. If it is, then physical geologists should be content to accept the *fact* of displacement, and leave the explanation to the future.

(Longwell, 1928: 152; my bracketed addition)

Longwell found geological evidence insufficient, but this was no reason not to treat continental drift seriously; it should be treated as a working hypothesis.

If the doctrine of continental displacement is accepted as a working hypothesis, to be tested and tried fairly along with others, it may be productive of valuable results. Members of this

Association [AAPG] may have the opportunity to make important contributions bearing on the problem of displacement, for much of the critical area in South America and Africa will be explored and mapped in the search for petroleum. If this hypothesis is kept in mind and the workers are not blinded either with the zeal of the advocate or with the prejudice of the unbeliever, we shall finally have a sound basis for geological tests.

(Longwell, 1928: 157; my bracketed addition)

Most authors at the AAPG meeting were harshly critical, but, in Longwell, mobilists seem to have gained a sympathetic critic. Readers are referred to Newman's exemplary account of the meeting (Newman, 1995).

4.4 Van der Gracht modifies Wegener's mechanism

Mobilists responded to these attacks by modifying Wegener's mechanism or by replacing it. Wegener used the first approach in the 1929 edition of *The Origin*, but he also acknowledged that mechanisms developed by others showed promise. Van der Gracht utilized both approaches; Daly, Joly, and Holmes the second. All of them thought that Wegener's general theory had merit on other grounds and should not be set aside because his proposed mechanism was faulty. I deal with the first three here and with Holmes in the next chapter.

Van der Gracht, the organizer of the New York AAPG symposium, gave a lengthy account of continental drift in his opening address (1928a), and in his closing response (1928b) he discussed the difficulties raised by other speakers during the symposium. In his opening address he made several attempts to alter Wegener's mechanism to minimize difficulties that had been raised. For example, he tried to reconcile Wegener's mechanism and his own solution to the origin of mountains by amending Wegener's model for the crust and upper mantle (RS1). After agreeing with his assessment of the relative strength and rigidity of sima and sial, he disagreed with Wegener's assertion that thin simatic oceanic crust is slightly more rigid than the upper simatic mantle underlying continental and ocean crust. He considered that the oceanic crust is rigid enough to crumble and fold the advancing sialic continents, but is not strong enough to prevent drift. He also maintained that both the oceanic and continental crust crumble and fold, but because sima has less strength, isostatic adjustments tend to smooth out folds.

We can perfectly well press a chisel of soft beeswax into a block of hard pitch, provided we push in our chisel slowly enough. That is what happens both in isostatic adjustment and at the front of a continent floating forward into sima. If it went slowly enough, it possibly would not need to be compensated at the front at all. Compression, however, exists. It is probably caused by the great rigidity of the upper sima crust, the ocean floor, as compared to its deeper layers. The sima, of course, is also deformed, but it does not keep its deformation permanently as the sial does. It behaves like pitch. Figures modeled from pitch will not endure, but again flow apart, while figures modeled from beeswax will endure; they have been proved to have endured for centuries.

(van der Gracht, 1928b: 200)

4.5 Daly's early attitude toward mobilism

Reginald Aldworth Daly (1871–1957) was born in Canada, the youngest of four sons and five daughters.[7] He attended Victoria College, Ontario, where Coleman, who argued against Wegener's solution to the origin of the Permo-Carboniferous glaciation (§3.12, §6.8), introduced him to geology. Daly received A.B. and S.B. degrees in 1891 and 1892. He then went to Harvard to pursue a career in geology, receiving his Ph.D. in 1896. He taught at Massachusetts Institute of Technology, continued his education in Europe, and engaged in fieldwork throughout the world. He returned to Harvard in 1912 where he became Sturgis Hooper Professor of Geology, a position he held until retirement in 1942. He remained professionally active during retirement, publishing his last paper in 1951. He published over 150 papers and seven books. Daly received honorary degrees from the universities of Toronto (1923), Heidelberg (1936), Chicago (1941), and Harvard (1942). He was President of the GSA (1932), and received its Penrose Medal two years later. He was awarded the Wollaston Medal by the Geological Society of London and the Bowie Medal from the AGU. He was elected to the NAS in 1925.

It is no accident that Daly was honored by the AGU as well as by prestigious geological societies. Trained as a geologist, he believed that geologists and geophysicists should attend to each other's data and arguments. Near the end of his career, he wrote, "the geophysicist has to become a geologist, just as a geologist, to be worthy of the name, does his best to master methods and results of geophysical study" (Daly, 1951: 23). Daly did his best, keeping abreast of geophysical studies of the Earth's interior, and his proclivity to promote and utilize geophysical data led to his postulation of an upper substratum of little strength, for which he later adopted the term "aesthenosphere," from Barrell (1914). This idea of an aesthenosphere of little strength was central to his version of mobilism, and it remains central to modern mechanisms. He did not shrink from speculation. Unlike many North American geologists, he sought solutions to fundamental problems even though they were speculative because he believed they would direct future research fruitfully. Throughout his long career, he offered explanations of igneous rocks, mountain belts, glaciers, coral reefs, and submarine canyons. He regarded many of his solutions as highly speculative, including his ideas on mobilism.

Daly became a mobilist in the early 1920s. He spoke extemporaneously in favor of mobilism at the GSA's December 1922 annual meeting (Newman, 1995).[8] Four months later, he argued in favor of mobilism (1923a). He defended mobilism through the remainder of the 1920s, and proposed a new mechanism, which he called (1929: 271) "the *down-sliding, or landslide, hypothesis*" (1923b, 1923c, 1925). He forcefully and extensively defended mobilism in both editions of his *Our Mobile Earth* (1926, 1929). Daly proposed his down-sliding hypothesis (which is somewhat akin to what would now be called "slab-pull") as an explanation of continental drift, mountain belts, and ocean basins. After *Our Mobile Earth* appeared, his support for mobilism diminished, and he began (1933, 1938) to separate his down-sliding hypothesis from extensive continental drift. Although he denied neither the possibility of large-scale continental drift nor that his hypothesis could explain it, he did claim that his down-sliding hypothesis could solve the problem of mountain building even if continents had drifted only hundreds or even tens, rather than thousands of miles. By this hesitant approach, Daly may have hoped to gain support for his down-sliding hypothesis as an explanation for mountain belts even among those who rejected mobilism.

Daly attempted to blunt fixist attacks by replacing Wegener's solutions to the origin of mountain ranges and (at least initially) the mechanism for continental displacement with his own (RS1). He thought that although his new theory circumvented the difficulties that Wegener's had encountered, it was unworthy of full acceptance because it faced its own shortcomings. Despite these reservations, Daly argued that his mobilist theory offered a better solution to the origin of mountains than competing fixist and other mobilist theories (RS3). He also argued that his theory of mountain building explained more than Wegener's did; for example, the formation of mountain ranges prior to the breakup of Pangea (RS3), a fundamental problem Wegener had not addressed.

4.6 Daly's mobilist theory presented in *Our Mobile Earth*

Eppur si muove! [And yet it moves]
> *(Daly placed this apocryphal claim of Galileo's on the title page of* Our Mobile Earth*)*

Although Daly had laid the groundwork for his down-sliding hypothesis in early chapters, it was not fully presented until the last two, "The Origin of Mountain Ranges" and "Evolution of the Face of the Earth." Central was his model, very different from Wegener's, of the crust and substratum. He made (1926: 110) three novel proposals: that basaltic oceanic crust is slightly denser than continental, granitic crust; that suboceanic basaltic crust is denser than the substratum immediately beneath continents; and that the latter is made of glassy not crystalline basalt. He viewed the glassy basaltic substratum as a worldwide reservoir of upwelling and eruptive basalt which lacks strength; this was Daly's aesthenosphere. Unlike Wegener, he believed both continental and oceanic crust are rigid and strong.

The gravitational instability of crust and substratum resulted in continental migration and formation of mountains.

The general hypothesis is, then, to be presented in the following form: The continents appear to have slid down-hill, to have been pulled down, over the earth's body, by mere gravity; mountain structures appear to be the products of enormous, slow *landslides* (crust-slides). Each chain has been folded at the foot of a crust-block of continental dimensions which was not quite level, but slightly tilted. Because the continent was tilted, it pressed against a geosynclinal prism at the foot of the slope. The pressure was equal to a fraction of the whole weight of the huge continental block, with the thickness of the earth's crust, taken to be about 40 miles. In an analogous way, a ladder, inclined against a wall at a low angle, exerts at the foot of the ladder horizontal pressure, which is a fraction of the weight of the ladder.

(Daly, 1929: 263)

If the continent becomes tilted or "de-leveled," a region of tension forms where it is elevated or domed, and a region of compression, or geosyncline, arises where it is depressed. Suppose a massive continent becomes domed in its center and depressed along its coastlines. Eroding sediments collect in geosynclines along continental margins, forming a prism. If the pressure due to the domed continent's weight becomes sufficient to overcome the strength of the crust, the crust breaks, and severed continental blocks slide toward the marginal geosynclines compressing and folding the prism of sediments in them.

When, therefore, the horizontal pressure reaches a certain intensity and the geosynclinal prism of sediments has reached sufficient thickness, the crust will break at the geosyncline. The dome is no longer supported from the sides. Tension pulls are generated at and near the center of the dome. There the crust becomes torn apart. The crust of the dome is torn into blocks or plates of the sizes of the existing continents. These blocks tend to slide slowly toward the geosynclinal.

(Daly, 1929: 268; notice, as Marvin (1973) did, his use of "plates")

The crust slides downhill over the slippery glassy substratum, its tip breaks off, sinks into the hotter substratum, and melts. Because heating and melting cause expansion, the folded geosynclinal prism rises, forming mountain belts. Meanwhile, upwelling basalt floods into the gap between separating continents, producing new oceanic crust. If separating continents migrate thousands of miles, then new ocean basins form in their wake, a vague but prescient notion, reminiscent of what is now called seafloor spreading. In this way he hoped to avoid the theoretical difficulties faced by Wegener's mechanism.

To explain the tilting of continents, Daly proposed that the Earth's crust had been domed at the poles and equator, and furrowed in the mid-latitudes.

. . . the Canadian-Greenland Shield stood up, high and dry, during most of geological time, while the United States was largely submerged. Similarly, the great Russo-Scandinavian and Siberian shields have tended to stand higher than the regions to the south. All three of these shields are high-latitude elements, clustered around the North Pole. They are relics of a great North Pole dome of high ground . . . Similarly; Africa-Arabia and peninsular India have as stubbornly remained high ground. They are relics of a long crust-dome with axis near the equator. The Brazilian Shield is another relic of this equatorial belt of high ground. Just as mid-latitude United States was generally low and usually covered with transgressing seas, so a broad east-west, mid-latitude belt of Eurasia, from France to China, tended to be submerged under the sea.

(Daly, 1929: 264)

Daly noted (1929: 264) that the intra-continental furrowed regions had often been covered by transgressive seas, and "were the sites of the principal east-west geosynclines and of the future Alpine-Himalayan, Antillean, Argentinean, and South African systems of mountains." He further postulated a former unification of continents into a supercontinent, and argued:

On the other hand, the crust of the land hemisphere, as *one whole*, clearly stood high above the level of the crust under the Pacific, or primitive, Ocean. There is good reason to believe, in fact, that the ancient land as a whole was slightly domed, so that part of the earth's crust had on every side a general slope toward the Pacific . . . Geosynclines were developed . . . along the border of the Pacific.

(Daly, 1929: 265)

He suggested three reasons for this worldwide pattern of doming and furrowing regions.

First, the earth is distorted because it contracts. Second, the earth is distorted because its speed of rotation is changing. Third, the earth is distorted because its lands are being eroded.

(Daly, 1929: 265)

Although he thought that classical contraction theory could not explain mountain building, he still maintained (1929: 266) Earth's contraction was primarily responsible for the doming. He also speculated that the decrease in the rate of Earth's rotation enhances doming and furrowing.

At present one can do little better than guess as to what shape the slowing-down earth would assume. My own guess is this: that the equatorial belt would continue to stand relatively high when compared with the mid-latitude belts, which would also stand lower than the polar regions; in other words, that the original continental land would be furrowed in mid-latitudes, both in the northern and southern hemispheres.

(Daly, 1929: 274–275)

His remarks on erosion were brief. Erosion "would perpetuate, and perhaps increase, the warping set up by the other two processes," and sediment would collect in the geosynclines and force down the leading edge of a down-sliding continent (Daly, 1929: 274).

4.7 Daly's defense of continental drift and his down-sliding hypothesis

Some geologists, especially European geologists, saw at once how the new hypothesis [of continental drift] explains automatically not only mountains but also a dozen other mysteries in their science. But there is a *difficulty*. Neither Taylor nor Wegener has shown *why* the continents should move. They have not discovered the force which did the gigantic work of overcoming the resistances to continental migration. Nor have they evaluated those resistances. For these reasons geologists are going slow in placing such mobility of continents among the accepted principles of science.

This conservatism is justifiable until some one has discovered the force available for the movement of continents. To offer the general reader a new suggestion on this fundamental question is a decidedly bold step. However, after prolonged study of the subject, I have come to the conclusion that the suggestion now to be presented is in principle inescapable, if the continents are not securely anchored and have bodily migrated. The supplementary hypothesis really implies a restatement of the idea of continental migration.

(Daly, 1929: 263; my bracketed addition; emphasis added)

Daly began by listing nine geological phenomena (or "sets of facts" as he called them) that were explained by continental drift and particularly by his down-sliding hypothesis. The first three, the seventh, and the ninth of these phenomena related to his ideas on down-sliding; the remainder echoed Wegener.

First, he appealed to the distribution of mountain chains (RS1). Those surrounding the Pacific were formed on the leading edge of sliding continents, the Americas to the west, Eurasia eastward. Older mountain chains, such as the Appalachian and Hercynian, formed before the breakup of the original supercontinent by intracontinental sliding into east–west trending deep furrows. The younger Alpine and Himalayan mountains formed as Eurasia slid southward.

Second, Daly argued (1929: 277) that his hypothesis explained the elevation of mountains and plateaus (RS1). Mountains are high because of thermal expansion of their roots and of foundered crust blocks. He also suggested that if the foundered block is not directly below but adjacent to the original geosynclinal prism of sediments, the melted block transforms into raised plains and plateaus such as the Great Plains of America adjacent to the Rockies, and the high plateaus of central Asia adjacent to the Himalayas.

Third, he claimed that his hypothesis accounted for the formation of new ocean basins (RS1). New seafloor formed as upwelling basalt arose in the space created in the lee of sliding continents.

Is it too extreme to suggest that North America slid westward, crumpling rock beds into the structural turmoil of our Pacific Mountain System; that South America slid westward, folding the Andean strata on the way; that Eurasia slid eastward, crushing together the festooned arcs of eastern Asia; that the great rigid block of Australasia slid eastward, folding the rocks "down-stream" into the mountains of New Guinea, Fiji, and New Zealand? Was the sliding so prolonged as to open the new, wide basins of the Atlantic and Indian Oceans at the cost of the area of the primitive Pacific Ocean? Similarly, was the Arctic basin in part formed by southward sliding of both North America and Eurasia, where the east-west Mediterranean Zone of mountains of the northern hemisphere was created; in other part by the sliding of America and Eurasia toward the central Pacific?

(Daly, 1929: 278)[9]

Fourth, following Wegener, he argued that his down-sliding hypothesis offered a solution for the former existence of landmasses east of North and South America, west of Europe and Africa, and between Africa and Australia (RS1). He dispensed with landbridges and foundering landmasses, restating the inconsistency between them and isostasy (RS2).

Before the new idea of continental migration was offered to science, it was thought by geologists that these lands had foundered, block by block, under the ocean. This older explanation of the Atlantic and Indian ocean-basins is fatally affected by difficulties arising from the physics of the case. We shall not enter on a discussion of these difficulties . . . However, gigantic founderings of the kind are not necessary assumptions, if the two young ocean-basins have been opened by the migration of the continents.

(Daly, 1929: 279)

Daly also proposed a solution to the origin of the Mid-Atlantic Ridge (RS1), suggesting that it represents

a long strip of the original continent, a strip left behind when that continent was torn into fragments, which slid away, respectively to westward and to eastward.

(Daly, 1929: 280)

He (1929: 280) considered that Wegener's maps of continental positions during the Upper Carboniferous, Eocene, and Older Quaternary illuminated "the general principle of the extensive horizontal displacements of land masses," even though inaccurate in "certain details" (RS1).

Fifth, Daly argued, just as Wegener had, that Pacific and Atlantic types of coastlines represent leading and trailing edges of drifting continents (RS1).

Sixth, Daly, echoing Wegener, turned to the marginal congruencies and geological similarities between the Americas and Europe and Africa, considering them as good as could be expected (RS1). He then turned to the geological disjuncts, first in the north and then in the south (RS1).

. . . one can hardly fail to be impressed with the remarkable matching of the rock structures on the two sides of the Atlantic. Thus, the Appalachians of Newfoundland and Nova Scotia correspond in composition, structure, and date of formation with the Hercynian mountain-mass of Britain and Brittany. The old rocks of the Scottish Highlands are identical in habit and structure with the equally old rocks of Labrador. The famous Old Red Sandstone of England has its mirror image in the Old Red Sandstone of Nova Scotia. The coal-beds of Britain and Belgium match well with the coal-beds of Pennsylvania.

(Daly, 1929: 284–285)

Although he thought the disjuncts between the geology of South Africa and Argentina were "more striking" than between other regions with a "dozen peculiar features of the one region . . . repeated in the other," he discussed only the more widespread Permo-Carboniferous glacial evidence, offering Australia as another region with such glacial deposits. Daly argued, albeit cautiously, that drift offered *an explanation of these disjuncts* (RS1).

Striated pavements [caused by glacial migration over country rock] of the same kind [which appear in South Africa] are found in South Australia, where a large glacier, contemporaneous with that of South Africa, molded the rocks . . . Similar phenomena appear in Argentina, also on the large scale. Geologists who know them well are now doubting that the South African and South American deposits and markings were made by two separate ice-caps. Some of these geologists are sympathetic with the hypothesis of a forceful tearing apart of the two continents, and therewith the glaciated tract, since the old glaciation.

(Daly, 1929: 287; my bracketed additions)

Daly's seventh "set of facts" pertained to the arcuate shape of Asian mountain belts and island arcs. Drawing on the work of Taylor and Suess, he argued that his down-sliding hypothesis with its postulated southward and eastward movement of Asia explains the arcuate shape of the Himalayas and island arcs along the Pacific coast of Asia. Arguing that his hypothesis differed from Taylor's, he nevertheless admitted that he had grafted Taylor's solution onto his (RS1).

The hypothesis now being presented is somewhat different, as it involves continental move-ment directly toward the central Pacific, as well as movement along the meridians; but the fact which is now of primary importance is that discerned by Taylor, who, sixteen years ago, emphasized the necessity of postulating crustal creep, bodily movements of continents, in order to explain the arcuate forms of the high mountain chains.

(Daly, 1929: 288)

Eighth, Daly (1929: 289) next turned to oceanic trenches, which he described as troughs or foredeeps, comparing them to depressed intra-continental regions found on continents in front of mountain belts where two continents have slid into each other (RS1). He (1929: 317) singled out the "Geosyncline of Punjab" the trough-like structure in India just south of the Himalayas. This intra-continental foredeep was formed as Eurasia was "crushed into the Afro-Indian remnant" of the original super-continent.

His ninth argument in favor of his down-sliding hypothesis was that it explained the origin and upheaval of mountain belts (RS1). It was, he claimed, the best available because it alone explained the puzzling fact that folding within mountain belts occurs long before their elevation (RS3).

As already stated, the landslide hypothesis explains one of the long-standing mysteries of geology, the upheaval of mountainous belts and the surrounding regions long after folding has ceased. No other explanation of mountains accounts so well for this fundamental fact of geology.

(Daly, 1929: 289)

Daly's final chapter gave "a speculative outline of Earth's history in its broader aspects" in which he further defended his landslide hypothesis (Daly, 1929: 300). He described how Earth had formed from gaseous solar material that had been drawn away from the Sun as a result of its near collision with a passing star, a view developed earlier, albeit differently, by T. C. Chamberlin, Jeans, and Jeffreys (see Brush, 1996a, 1996b). Daly adopted Jeffreys' view that the newly formed gaseous Earth quickly condensed. With further cooling, Earth formed a double-layered, solid crystalline crust. The upper, less dense layer was granitic and concentrated in one hemisphere. The lower, denser layer was basaltic. To explain the hemispheric asymmetry of the granitic crust, he listed three possibilities, which he thought no better than reasonable guesses. The first two, originally proposed by G. H. Darwin and Poincaré, involved early fission of the Moon from Earth, both of which faced a "serious difficulty" (Daly, 1929: 308). The third, which he preferred, was that the asymmetry had been inherited from an original asymmetry created during the condensation phase, and preserved because condensation was rapid. Once created, the less dense granitic hemisphere stood higher than the basaltic hemisphere, and such adjustments could have "led to the formation of a single continent, a land hemisphere; and a single ocean basin, a water hemisphere" (Daly, 1929: 310).

In order to extend the reach of his down-sliding hypothesis he offered it as an explanation of two additional problems, the first being the folding in ancient basement complexes, extensive in all continents.

If such develeling of the crust became sufficiently advanced, the higher crust would tend to slide into the opposite hemisphere. If Asia and America slid toward the Pacific in relatively modern times, it is logical to inquire if the earliest high-standing continent did break up and slide toward the antipodal region, crumbling the rocks as it moved. Dare we think of extensive sliding, so that the whole of the granitic crust was ultimately concentrated in one hemisphere? If it was, then it must have been greatly deformed, greatly mountain-belt. In fact, we can imagine more than one such oscillation of the earth's rocky structure each adding complexity of the crustal structure.

(Daly, 1929: 310–311)

According to Daly, the rocks of the original continent began to deform as the latter slid downhill towards the original ocean basin. This deformation caused the extensive folding found in basement complexes and, as the process repeated itself, deformation would intensify. Daly then suggested that these early down-slidings of successive ancient continents explain why the continents, through crustal thickening, became dry land.

The hypothesis . . . explains the very existence of dry land. The continental surfaces now stand so high, because the granitic crust was folded and rafted together, and therefore thickened; and, secondly, because the rocks thus aggregated are specially light rocks. Today each continent floats higher than the denser, heavier basaltic crust under the ocean. The sliding hypothesis has no small advantage, as it gives some kind of answer to the primary question: Why is there any dry land at all? I know of no other answer that appears valid.

(Daly, 1929: 311)

Daly himself never accepted his own down-sliding hypothesis, nor did he ever argue that it ought to be so accepted, but he believed that it offered the best available solutions to the origin of mountain belts and the folding of ancient basement complexes, and that in the 1930s it was the only explanation of how continental crust thickened and became dry land. But he warned readers that his solutions to the latter two problems were plagued by empirical difficulties, because they depended on data that were fragmentary and of variable quality. Moreover, he acknowledged that they depended on speculative cosmogonical theories, which were themselves plagued by theoretical difficulties. Although Daly preferred his solution to the origin of mountains to any of the alternatives, and thought his mechanism superior to Wegener's, he clearly stated that it remained a matter of speculation as to whether the amount of doming required for down-sliding actually had occurred.

To sum up. The problem of this chapter is the manner in which the great mountain chains of the globe have been developed. The forces responsible for their folded structure were directed horizontally, and have been speculatively attributed to the down-sliding of continents. Before any of the epochs of mountain-building, the continental surface and the continental crust were not as level as now. They were warped, in the form of huge, gently sloping domes and basins. That is a fact, proved by geologists. Speculation enters when it is held that the crust in any one of these domes was sufficiently warped, thrown out of level, to compel its breaking up into fragments sliding down the flanks of the dome and crumpling the rocks ahead into mountain structures.

(Daly, 1929: 290)

He believed his ideas on the cause of continental drift were an improvement on Wegener's and Taylor's, and that his theory of mountain building was preferable to that of fixists or other mobilists. However, he saw problems ahead.

4.8 The reception of Daly's down-sliding hypothesis

Daly's speculations caused little stir, but were not completely ignored. Although some were mildly favorable, and at least one recommended that it deserved serious study, nobody entertained it as a working hypothesis. It is hard to avoid the conclusion that reviewers, at least of his *Our Mobile Earth*, were reluctant to criticize.[10] He was highly respected.[11] Schuchert and Longwell, the two major fixists who discussed Daly's view, seemed not to want to embarrass him; neither looked favorably on his theory but found ways to praise him for his efforts: his mechanism was an improvement on those of Wegener and Taylor, and they commended him for his willingness to speculate. Schuchert even used Daly's speculations as pretext for presenting his own highly speculative fixist theory. Mobilists, although often welcoming Daly's support for mobilism, ignored the details of his down-sliding hypothesis, and either offered no assessment of its merits or gently criticized it, suggesting that it merited serious study which they promptly left to others.

Fixists respectfully allowed Daly his speculations and little more. Longwell reviewed *Our Mobile Earth*, and briefly discussed Daly's mechanism at the 1926 AAPG symposium. He characterized the book as an "admirable contribution," noted that it was written for intelligent laymen rather than professional geologists and geophysicists, and placed it within the corpus of Daly's speculative works. Longwell noted Daly's long interest in the problem, viewed his drift mechanism as superior to Wegener's, but acknowledged that, like Wegener's, it "may be criticized without difficulty."

Conservative geologists may feel that the book would be better, if the hypothetical discussions in the last chapters had been omitted. However, this material is very properly included, as it is the product of the author's serious thinking through many years. Daly is one of the few American geologists who have been strongly attracted by the doctrine of continental displacement. As the original proponents of this hypothesis have not suggested any force that appears adequate to do the work of moving continents and making mountains, Daly devotes his attention chiefly to this phase of the problem. The conception of continental *sliding* on the flanks of great bulges or domes is proposed in place of the continental *drifting* favored by Wegener. Either of these speculations may be criticized without difficulty. It may be said for the Daly hypothesis that it recognizes the ordinary principles of mechanics, and that it makes use of older helpful conceptions such as, the contraction theory.

(Longwell, 1927: 525)

Longwell raised a difficulty against Daly's hypothesis at the 1926 AAPG symposium.

Daly, aware of the work involved and the magnitude of forces required, seeks to substitute *sliding* for drifting, assuming that broad domes or bulges form at the earth's surface, and on the flanks of these domes the continental masses slide downward, moving over hot basaltic glass as over a lubricated floor. This is an interesting and stimulating speculation; but Daly has not yet demonstrated that the doming can actually occur on the scale he assumes, or that the gradient will be sufficient to give the desired effect. At present, therefore, his suggestion must be regarded as purely speculative.

(Longwell, 1928: 151)

Schuchert's review was in a similar vein. Beginning with praise for book and author, he placed Daly's work within the speculative realm.

This book, "boldly planned on an endlessly difficult theme," is primarily intended for the general reader; even so, working geologists and especially teachers of the earth sciences will find "Our Mobile Earth" not only replete with facts but also interesting, and inspirational to deep thinking . . . It goes easily from the known to the unknown, and finally becomes highly speculative, and why not, since "science progresses through systematic guessing in the good sense of the word"?

(Schuchert, 1926b: 624)

He devoted a third of his review to Daly's hypothesis. Rather than criticize Daly directly, Schuchert used Daly's own words to relegate his mechanism to a realm "forever invisible to human eyes."

It has long been known that Daly sees much of value in the Taylor-Wegener hypothesis of continental drift, which is now being widely discussed. This hypothesis, he holds, "must be seriously entertained as the true basis for a sound theory of mountain building. . . . Taylor and Wegener believe that the mountain chains of the globe were formed by the horizontal crushing of geosynclinal prisms which lay in front of *slowly moving, migrating continents*." Parts of the continents are bowed or raised up and accordingly gravitate downhill, as it were, over the potentially mobile substratum, and so by their sliding force crumple together the weaker places in the crust. As he says, "The continents appear to have slid downhill, to have been pulled down, over the earth's body, by mere gravity; mountain structures appear to be the product of enormous, slow *landslides*."

"Our solid earth, apparently so stable, inert and finished, is changing, mobile and still evolving . . . And the secret of it all – the secret of the earthquake, the secret of the 'temple of fire,' the secret of the ocean-basin, the secret of the highland – is in the heart of the earth, forever invisible to human eyes".

(Schuchert, 1926b: 624)

Invisible to the human eye, yes, but not, as will become evident in Volumes II and III, to the paleomagnetist's magnetometer, the seismologist's seismometer, or the marine geologists deep-sea corers. Indeed, Daly's intuitions were remarkable. Plates do slide downhill and gravity is the main driving force. They slide down on inclined planes because as they age the lithosphere thickens and becomes denser, so they just plunge downward beneath ocean trenches (Oliver and Isacks, 1967; Isacks, Oliver, and Sykes, 1968; McKenzie, 1969; Elsasser, 1971).[12]

Schuchert also considered Daly's mechanism in the final section of his contribution to the AAPG symposium volume, where, using Daly as his authority and guide, and unable to pass up the opportunity of taking a shot at Wegener's Pangea, he himself entered the realm of high speculation.

Daly's new book, *Our Mobile Earth*, sounds the keynote for the attempt to save the germ of truth in the displacement theory and reconcile it with the facts that geology already has at hand. Following his lead, the writer has set down below the sequence of earth development that he finds necessary to fit our determined geologic chronology, on the one hand, and the known development and distribution of ancient faunas, on the other. He realizes that this plan, as well as that of Wegener, presents at least one difficulty for which no solution is yet in sight, namely, the breaking down of the land bridges between continents and of the many border-lands, but he is confident that the geophysicists will in time find the way in which this was accomplished. In any event, it seems less insurmountable than the many inaccuracies and "imaginings" that stand against the theory of Pangea.

(Schuchert, 1928a: 142)

Schuchert even followed Daly in allowing for great mobility during Archeozoic and Proterozoic, speculating about sialic doming and large-scale movement with the eventual formation of a sialsphere containing the continents positioned where they are now but connected by wide landbridges. So Schuchert, despite his criticism of Daly, used his mobile Earth to provide the paleogeography he needed before the collapse of his cherished landbridges. If speculation allowed Daly continents sliding under gravity, it could allow Schuchert his fixist continents and collapsing landbridges.

Granted these or similar conditions, progressive geology meets the knowledge of orthodox geology and grants it the permanency of the earth's greater features, a knowledge on which the biogeographer of ancient life has built his paleogeography; he must have all this – the long and intricate migration routes – to explain the evolution that the migrating hordes have undergone. A Pangea, the postulated single continent that began to rift in the Carboniferous and split and wandered apart after Jurassic time, will never explain the life of the seas and lands as seen by the paleontologist.

(Schuchert, 1928a: 143)

Schuchert was not softening on mobilism.

Mobilists also treated Daly with respect, boasted about his shift to mobilism, but distanced themselves from his down-sliding hypothesis. Wegener (1929: 53) welcomed Daly's support, but discussed neither his theory of mountain building nor his drift mechanism. Mentioning him only twice, he remarked, "It is especially noteworthy that Daly's book (*Our Mobile Earth*, London, 1926) is based altogether on the drift theory." Van der Gracht, like Longwell and Schuchert, discussed Daly's version of mobilism at the New York AAPG symposium; characterizing *Our Mobile Earth* as "remarkable," he offered this assessment.

Remarks – we may or may not favor this explanation as an improvement. Possibly there is more truth in this view, when applied to movements of smaller amplitude, than for the entire Pangea as a whole. It might, for instance, be used as a partial explanation for the mechanics of great overthrust sheets as we know them in the Alps and Himalayas, which are also difficult to explain mechanically. It may be a cause for certain movements of drift in the interior of continents, such as the "creep of Asia," which Suess already accepted, and which was again so clearly set forth in 1922 by Emile Argand in his *Tectonics of Eurasia* and was also considered by him a result of "the plasticity of Asia."

(van der Gracht, 1928a: 39)

Holmes and du Toit were more critical. Holmes made short work of Daly's mechanism in 1931 in a paper that contained the first detailed account of his own mobilist theory of mantle convection (see Chapter 5).

Daly has suggested for discussion a hypothesis of continental creep due to the sliding of sial blocks on a lubricating zone of glassy basalt [in his *Our Mobile Earth*]. A bulging of the polar and equatorial regions with a depression between, towards which the continents migrate, is pre-supposed. No explanation is offered for the initiation of so unstable a deformation of the globe.

(Holmes, 1931a: 562; my bracketed addition)[13]

Du Toit devoted a paragraph to Daly's mechanism in his introduction of his (1937) *Our Wandering Continents*.

R. A. Daly's "Downsliding Hypothesis" (1923) or "Landslide Hypothesis" (1926) views the older continental masses as moving under the influence of gravity by reason of an initial tilting produced through the combined actions of erosion, contraction of the Earth and changes in its speed of rotation, the crust tending to fold or fracture along the weaker oceanic borders. Sliding is supposed to take place upon a deep-seated stratum of basaltic glass. There are, however, *some serious difficulties connected with the detailed application of this theory*, particularly on the colossal scale postulated, though the general idea would seem correct. Daly's views on the physical basis of Drift are worthy of careful study.

(du Toit, 1937: 22; emphasis added)

Du Toit neither detailed the "serious difficulties connected with the detailed application of" Daly's mechanism, nor did he attempt a "careful study" of it.

4.9 Joly's thermal cycles and his ambivalence about mobilism

John Joly (1857–1933), born in Ireland, was an undergraduate at Trinity College, Dublin, where in 1882 he gained first place and special certificates in a variety of subjects including engineering, mechanics and experimental physics, geology, chemistry, and mineralogy at the Engineering Degree Examination.[14] He promptly became assistant to the professor of engineering at Trinity College and finally professor of geology in 1897, remaining at Trinity until his death. Joly became a major figure in British Earth science. He was elected to the Royal Dublin Society in 1881, being its president from 1929 to 1932, and to the Royal Society (London) 1882. He received the Royal Medal (1910) from The Royal Society (London), the Boyle Medal (1911) from the Royal Dublin Society, and the Murchison Medal (1923) from the Geological Society of London.

He had broad interests: as a student, they were mineralogy and petrology, later they ranged from Schiaparelli's "canals" on Mars, to sunspots, and the physiology of color vision. With the discovery in 1903 by Curie and Laborde that radium maintains a temperature above its surroundings, Joly quickly realized that radioactive heat, particularly that generated by radium and thorium, had a major effect on Earth's heat budget. This realization and his development of its implications were, I believe, his major contributions. He was a pioneer of the use of radioactivity to measure the age of the Earth.

Joly developed his theory of thermal cycles during the 1920s, presenting it in papers and lectures, and in the two editions (1925 and 1930) of his book *The Surface-History of the Earth* (hereafter, *The Surface-History*). It sought to explain the origin of mountains, flood basalts, rift valleys, island arcs, the transgression of oceanic waters onto continents and their regression, and the periodicity of mountain building. Radioactivity and isostasy were central to his theory of thermal cycles.

He contended that continents and ocean floors, granitic and basaltic respectively, float in isostatic equilibrium on a basaltic substratum, and that radioactivity in the substratum and in Earth's deeper interior generates heat, which accumulates, more being produced than is lost through conduction. By "substratum" he meant what nowadays is called the mantle. As the temperature reaches the melting point of basalt, the substratum liquefies and convection begins, hotter material rising into the upper substratum. Liquefaction decreases the density of the substratum, continents sink, water from the oceans spills onto them forming shallow seas. Meanwhile, westerly directed tidal and precessional forces act preferentially on the liquefied substratum and crust causing them to rotate westward relative to the main body of the Earth. The weak geosynclinal areas of continents sink even further down into the substratum and fill with sediments, and being lower are more affected by eastward-directed rotational forces. These forces, Joly argued, acted on the western edge of continents, squeezing sediments and underlying rocks together and upwelling basalt intrudes into them. Heat escapes by conduction from the substratum, primarily through ocean floors which may partially liquefy. The substratum begins to cool and re-solidify, and its density increases. Continents ride higher, and the compressed sediments and intruded basalt are elevated into mountains. Then, according to Joly, the whole cycle is repeated: mountains erode, sediments collect in geosynclines, radioactive heat liquefies the basaltic substratum, continents sink, the substratum loses heat and re-solidifies, the continents rise, and so on.

Joly (1923a, 1923b) realized he could tinker with his theory of thermal cycles so as to include the drift of continents, and as a bonus perhaps provide a means of displacing them. Also, he thought he could use continental drift to remove a difficulty, the origin of east–west fold belts, from his own theory of mountain building. However, his commitment wavered. With the publication of the first edition of *The Surface-History* in 1925, he was ready to abandon all thought of embracing drift, being unsure that he could solve the mechanism problem. He also had found a way other than drift to avoid the difficulty faced by his own theory of mountain building. He was unenthusiastic about drift in his 1928 contribution to the proceedings of the 1926 AAPG symposium, and just kept drift alive as a possibility in the second edition of *The Surface-History* (1930).

Joly's first discussion of continental drift appeared in January 1923, approximately six months after Wegener's first publication in English appeared in *Discovery*, and a year before the appearance of the English translation of the third edition of his *The Origin*. Acknowledging that "Wegener has brought forward much evidence in favor of continental movements," Joly, like everyone else, raised the mechanism difficulty:

But I do not think he has discovered any adequate source of the motion. The *polefluchtkraft* is too feeble; it is purely meridional in direction and is inconsistent with the existing distribution of land. It is probably ineffective.

(Joly, 1923a: 79–80)

He offered his theory of thermal cycles as a solution. Uniquely, he proposed that some continents drift differentially eastward rather than westward as Wegener would have them.

The fact is Wegener works out the theory on the basis of a westerly drift of the continents. In so doing I think he is in error. An adequate force appears available provided an easterly drift is postulated; and so far as I can see the theory grows in probability when examined from this new point of view.

(Joly, 1923a: 80)

According to Joly, the lower-riding continents move eastward relative to the ocean floors and to the higher-riding continents, because the effect of westward directed tidal forces decreases with depth. He appealed to G. H. Darwin and Hayford.

According to Sir George Darwin, the tidal effects of sun and moon acting on "a stiff yet viscous planet . . . must produce a retardation of the surface crust relative to the interior. He states that this is speculative as regards the earth; but this was written twenty-five years ago. The great fact of isostatic compensation of the continents, proving their flotation in a viscous magma, was not then supported by such strong evidence as Hayford and others have since adduced. I assume that differential motion exists (or formerly existed) and that the floating continents possess a slightly less rotational velocity than the deeper parts, of the underlying magma, the velocity of which continues to increase downwards until a more rigid interior is reached.

(Joly, 1923a: 80)

Low-riding continents move eastward relative to liquefied oceanic crust and to higher-riding continents.

According to this view, America did not leave Europe and Africa but was left behind by them. Their increased easterly velocity was, possibly, ascribable to the great Laramide submergence of Southern Europe, South Asia, and North Africa . . . In a similar manner New Zealand left Australia: the force in this case being plainly referable to isostatic compensation demanded by the lofty ranges of New Zealand. So also Ceylon was torn from Peninsular India; the fracture line of the eastern Asiatic coast was produced, etc.

(Joly, 1923a: 80)

He argued that such differential continental drift could have produced the east–west trending Himalayas "by direct pressure between land masses," that is by what would now be called continent–continent collision; Joly explained east–west mountain building by invoking continental drift (Joly, 1923a: 80).

Joly's strongest statement in support of continental drift came two months later during a lecture on March 7, 1923, to the Royal Dublin Society, and published the next month. After summarizing his theory of thermal cycles and how they caused mountains, he remarked how most of them trend north–south, which he thought his theory explained.

Such great ranges as the Cordilleras of North and South America rose up out of troughs of sediments in this manner. They were specially favourably oriented to receive the easterly pressure of the underlying magma, and, correspondingly, they are in many respects the greatest mountain developments of the globe.

(Joly, 1923b: 605)

Other mountain systems, such as the Himalayas, trend east–west, and his theory of thermal cycles alone could not account for them.

However, while it seems easy to understand that the formation of mountain ranges directed more or less north and south might arise in this manner, it is more difficult to imagine chains of mountains like the Himalayas or like the Pyrenees originating in the west-to-east force arising from tidal or precessional effects.

(Joly 1923b: 605)

He (1923b: 605) then introduced "the consideration of the possibility of the continents having shifted their relative positions during geological time." He remarked that Wegener's drift theory offered an explanation of what are here called biotic disjuncts, and was consistent with isostasy, adding that his theory of thermal cycles provided drift with a mechanism.

Many are now weighing evidence for and against such extraordinary possibilities as to whether the Atlantic Ocean is not a comparatively recent innovation; whether New Zealand was not recently detached from Australia, and India from the eastern shores of Africa, and so on. Before this interesting question arose biologists and geologists generally got out of their difficulties by assuming the former existence of land connexions or "bridges" which subsequently "foundered" and disappeared. Now, according to the present explanation of the surface movements of the earth, the foundering of such "bridges" would be difficult to realise; for they are of lower density than the basaltic magma upon which they at one time floated. So that it becomes very difficult to imagine the former existence of these bridges. Not only is this the case, but also the present theory certainly suggests that differential movements of the continents might quite possibly have taken place.

(Joly, 1923b: 605)

Joly had three reasons to fold continental drift into his own theory of thermal cycles – doing so removed, he argued, the difficulty that his theory had in explaining the east–west trending mountains in Asia and Europe, it provided a mechanism, and it was consistent with Wegener's explanation of biotic disjuncts. Nevertheless, he made it clear that continental drift was neither a consequence nor a requirement of his theory, allowing himself leeway to jettison mobilism if need be, but retain his own theory.

I do not mean to convey that these supposed great movements necessarily arise out of our theory, but it is at least remarkable that a theory which appears to explain much – and on a basis which can claim to be more than merely hypothetical – should offer what may be regarded as a *vera causa* for continental drifting, if other considerations require it. The continents during times of revolution became acted upon by forces tending to move them towards the east; and, what is even more relevant, these forces must of necessity be different in intensity from one continent to another.

(Joly, 1923b: 605)

Joly then justified his two claims, first regarding the mechanism problem.

Another consideration in favor of continental drifting must be taken into account. The continents become acted upon by these forces only during the period of magmatic fluidity. We saw that this fluidity is ultimately lost, mainly in consequence of heat escaping through the ocean floor; this floor being probably more or less melted away during the process. It may be that the reduction in thickness of the ocean floor is carried so far as to remove what is really the main obstacle to differential continental movement – the existence of a strong and rigid ocean floor, holding the continents immovably fixed to one another.

(Joly, 1923b: 605–606)

Drift also explained the east–west trending mountain belts.

We return for a moment to the problem of the elevation of such ranges as the Himalayas, which trend more or less east and west. We are now prepared for the possibility that the explanation of these events was due to a certain small amount of continental movement. It is a fact that tidal and precessional forces are greatest in equatorial regions. May it not have been that the great continent of Africa, experiencing the effects of this, rotated just a little, its southern extremity moving eastward; and so also for Peninsular India; so also for the Spanish Peninsula? A small turning movement, crushing the ancient geosynclines, would suffice. For, after all, the greatest mountains are but very tiny wrinkles upon the surface of this huge world.

(Joly, 1923b: 606)

Remarkably, no sooner had he achieved this mobilistic synthesis than he began to retreat from it. As an editor of *Philosophical Magazine*, Joly arranged for publication of "The Movements of the Earth's Surface Crust" (1923c) in which he gave a non-drift dependent account of the origin of east–west trending mountain belts, and in which he began to doubt that liquefaction of the basaltic substratum and oceanic crust would reduce their rigidity sufficiently to allow deeper seated continents to move through them.

This led him to his new explanation of east–west trending mountain belts. He argued that liquefaction of the substratum increases its volume by 10%, which causes a 65 km increase of the Earth's circumference. This affects only the ocean basins, and he estimated an expansion of 50 km for the Pacific and 15 km for the Atlantic. With this expansion, compressional mountains would form in the geosynclines marginal to large oceans.

With the subsequent solidification of the substratum, the downward movement of the ocean floor gradually begins and the continents become pressed-upon, on every side, by the enlarged and strengthened oceanic floors. Even if buckling of the floor results, the lateral pressure must continue, for the rigidity of the cooled basalt is considerable. This effect will give rise, pre-eminently, to very intense horizontal thrusts (although of limited range) on the continental land masses. The pressure must long continue and may be adequate to find out the weak places of the land and to fold and shear the sediments in the upper few miles of the geosynclinal accumulations, seconding the intervention of other forces to be presently discussed.

(Joly, 1923c: 1181)

The wider the ocean, the greater the lateral compression, and the larger the adjacent mountains.

In the case of the greater world-wide movements when the rise and fall of the entire crust takes place the wider the oceanic reach the greater must be the orogenic effects. Here we recall the significant remark that the greatest mountain ranges face the widest oceans. There are no mountains around the Atlantic comparable with the Cordilleras. The Himalayan fold-mountains range perpendicularly to one of the greatest oceanic reaches of the Globe.

(Joly, 1923c: 1183)

Joly no longer needed continental drift to account for the Himalayas; he had found another way to remove his former difficulty, a way that he thought to be a direct consequence of his thermal cycles.

Hedging his bet, Joly still allowed for the possibility of very limited continental drift, which he associated as before with the formation of the Alpine–Himalayan mountain belt. He no longer required drift, but limited drift may, notwithstanding, have occurred.

These conditions [of a fluid upper substratum and oceanic crust and tidal action] involve differential effects upon the floating land-masses. For the more deeply compensated continental features experience the west-to-east magmatic drive more than shallower land-masses . . . Such rotational effects may have affected mountain development. Comparatively small movements would suffice: thus, if the African continent and Peninsular India were, even to a small extent, affected in this way, by the more active equatorial currents upon their southern extension, much of the Eurasian folding would find explanation. We have seen that at the height of revolution the ocean floor is, probably, much disturbed as well as reduced in thickness, and its resistance thereby diminished.

(Joly, 1923c: 1185; my bracketed addition)

Devoting the final section of his *Philosophical Magazine* paper entirely to continental drift, Joly now seriously questioned whether the ocean floor can, during thermal cycling, lose enough rigidity to allow for differential drift of continents.

The question of the possibility of continental movements having occurred in past times arises naturally in connexion with the foregoing views. The main point at issue, however, is one which we do not seem able to determine – how far, during times of revolution, the ocean floor may have been reduced in rigidity and thickness. For it is obviously hard to imagine that in times when a thick and rigid ocean floor binds the continents together – any forces with which we are acquainted can have attained such intensity as to rupture the connecting crust. However, during the height of a great revolution, under the operation of convective currents – probably in part superheated – assailing the basaltic ocean floor, almost anything may have happened so far as the stability of the floor is concerned. It may have been fractured or softened to an extent permitting of differential forces – already referred to as arising from tidal or precessional sources – to occasion slow drifting from west-to-east. For such must be the direction of continental movement relative to the earth's surface.

Under such conditions Eurasia may have drifted from America in Laramide times. New Zealand, in virtue of its great loading of mountain ranges, may have parted from Australia; the Festoon Islands may have been carried eastward from Asia; Peninsular India may have parted from the African Continent, etc., etc.

But as I have intended to convey, while the possibility of continental drift appears to be involved potentially in the present views, the former existence of all the necessary conditions does not seem to be demonstrable. Final support for the theory of continental drift must come from evidence which is outside the scope of this paper to discuss.

(Joly, 1923c: 1188)

In a lecture he gave on May 2, 1923, before the Geological Society of London, Joly had become even less enamored with continental drift. Two years later, Joly added the text of this lecture at the end of *The Surface-History* in the hope that "its perusal will enable the reader to obtain a clear retrospective view" of his theory of thermal cycles. He limited his discussion of continental drift to one paragraph, which contained nothing about Eurasia moving eastward relative to the Americas, no appeal to continental drift as a factor in the formation of east-to-west trending mountain belts, only an increased concern about the rigidity of the oceanic crust.

The thesis that the continents may have shifted during geological time gains some support from the foregoing history of terrestrial tectonics. For we observe that in the course of this history the ocean floor has been repeatedly assailed by thermal effects tending to reduce its thickness and rigidity, and that the continents were exposed to long-continued magmatic forces acting from west to east. If the ocean floor were in times of Revolution so far reduced in rigidity as to yield to those forces continental movements in an easterly direction would probably have taken place. We cannot now evaluate the intensity attained by the several factors concerned, and so we must leave to other sources of evidence the final verdict in this interesting theory. An affirmative answer would not be out of harmony with possibilities arising out of the present views.

(Joly, 1925: 186–187)

He did not dismiss continental drift, it was not necessary to his theory of thermal cycles, but drift "would not be out of harmony" with it.

Through the early 1920s, Joly's support for continental drift lessened and, at best, was lukewarm. In the 1925 first edition of *The Surface-History*, his support had declined further. Again, although he (1925: 170) mentioned that a small eastward rotation of Africa and peninsular India relative to Eurasia might have affected the growth of Eurasian mountain belts, he no longer thought he needed such rotations to account for them. Moreover, as an avowed supporter of isostasy, he (1925: 135) even allowed for the possibility of some foundering of landbridges, at least to some extent, as an explanation for biotic disjuncts. He still did not dismiss mobilism entirely, devoting a short appendix to it (Joly, 1925: 172). Drift was a possibility, but he now agreed with critics that tidal or precessional forces were insufficient to propel the continents through the ocean floor even when the substratum is liquefied. But if the ocean floor becomes "greatly attenuated," if it becomes so thin, or is rifted to such an extent that it almost disappears, then differential continental drift may occur.

Joly later considered continental drift in his short contribution to the proceedings of the 1926 AAPG symposium. He did not emphasize the merits of continental drift and its relation to his theory of thermal cycles. Continental drift, he remarked "is not improbable during periods of fluid substratum," adding that it was his own theory that gave it some legitimacy (Joly, 1928a: 88–89). There was no renewal of commitment to continental drift, if anything it had dwindled further.

4.10 The Joly–van der Gracht mechanism

Although several Earth scientists discussed Joly's theory of thermal cycles as a mechanism for continental drift, only van der Gracht took it seriously. Those who did, did so in terms of van der Gracht's modified version.

Van der Gracht (1928b) devoted much of his seventy-five-page essay, the final paper in the proceedings of the AAPG conference, to showing why he thought continental drift was a likely consequence of Joly's theory. Although he believed that Joly's theory was worth entertaining on its own merits, he thought that combining it with drift enhanced the credibility of both, because drift solved problems that Joly's theory did not even address, and Joly offered a more plausible mechanism than Wegener.

He agreed with Joly that tidal and precessional forces tended to rotate the Earth's crust westward relative to the interior, causing a general westward movement of continents. He also agreed with Joly that those continents that sink further into the liquefied substratum would be more affected by easterly rotations; they would be less affected by westerly rotations than continents that float higher in the substratum. Where he disagreed with Joly, and this was the heart of the matter, was whether or not the ocean floor could become weak enough to allow for *differential* continental drift.

As I said before, Joly himself does not feel much inclined to accept Wegener's differential drift. He discusses it in his last book, but evidently considers it rather improbable, at the best a remote possibility. His objection is based upon the resistance of the ocean floor. Would the westward drive of Joly's magmatic tide and of Schweydar's precessional retardation suffice to overcome the resistance of the ocean floor? I think that the effect of the resistance of the sima, even at times of complete, or nearly complete, solidification, is exaggerated, and that its viscous yielding against a *steady push over an immense period of time is underestimated*. Perhaps drift comes to rest during *complete* solidification, but this is only a comparatively short climax of the long cycle, during most of which the sima should remain very near its melting point, and partially fluid in thinner layers . . . just below the sial crust. Below large continents such local fluid patches may never completely solidify, if so, they re-form immediately (volcanism and batholiths). Perhaps the ocean floor, combined with a highly resistant substratum, would indeed be an effective barrier against *rapid differential* drift during Joly's periods of full solidity of the sima, but his *mere skin* of solid sima of 10 kilometers or less below the oceans, during his times of fluidity, would certainly not appear to be a barrier to a magmatic drive, gripping continental sial floats of a thickness ranging from 30 to 100 kilometers, with local protuberances to double his depth. What we speak of here is *differential* drift causing relative displacements of continents as to the oceanic crust, and of one continent as to the other; for instance, of North America *relative* to Europe, and South America *relative* to Africa, opening and widening the Atlantic Ocean. These relative displacements would be only part of a *general* absolute westward drift of the entire crust, involving both sial and sima ocean floor, which we cannot ascertain.

(van der Gracht, 1928b: 74–75)

Van der Gracht repeated his belief that the seriousness of the mechanism difficulty had been overestimated by mobilism's critics. He was not convinced that the strength (residual rigidity) of the sima was sufficient to prohibit displacement of the continents through the ocean floor, and it is understandable that he wanted to combine Joly's thermal cycles and continental drift. He seems to have understood and appreciated the importance that Joly's theory had for continental drift far better than Joly did himself.

4.11 Fixists reject the Joly–van der Gracht mechanism

Fixists, Berry, Gregory, and Longwell, dismissed the Joly–van der Gracht mechanism in their contributions to the AAPG New York symposium. Berry, then an emeritus professor of paleontology at Johns Hopkins University, who believed that drift's explanation of biotic disjuncts "raises more distributional problems than it solves" (Berry, 1928: 195), remarked:

Of Joly's theory, based upon radioactive changes in the earth's crust, I can only say that it is highly interesting. It cannot be proved or disproved at the present time; therefore, it can hardly be cited in confirmation of Wegener's hypothesis.

(Berry, 1928: 196)

Gregory, a landbridger, thought the Joly–van der Gracht theory was a "reinforcement of the Wegener theory by Joly's periodic melting of the sima by radioactive heat" and was "an important contribution to it," but still raised a difficulty (RS2).

The efficacy of this source of heat is, however, open to doubt. Van Waterschoot van der Gracht lays much stress on the need for outlet for the radioactive heat; effective outlets other than the melting of the sima appear available.

(Gregory, 1928: 94)

Although Longwell found the joint theory interesting, he noted that Joly himself had raised a difficulty about sima's rigidity (RS2).

Joly's hypothesis, outlined by Van Waterschoot van der Gracht, has much in its favor from a strictly qualitative standpoint. It accounts admirably for the rhythmic advance and retreat of seas and for the large facts of igneous geology; and it provides a periodic mobility not conceivable under any other assumption. Granting Joly's principal postulates, it is not difficult to conceive that the continents suffer some horizontal displacement at times; enough, perhaps, to relieve part of our embarrassment in seeking to explain the very great shortening in mountain zones. It does not seem probable, however, that wide separation of continental masses could ever occur. The strong floor of the ocean must remain, even at the most liquid stage, as an obstruction to large differential displacement. The liquid zone would greatly facilitate slipping between the crust and the interior; but as Joly points out, any considerable movement would be likely to occur as a slipping of the entire outer shell over the interior rather than differential displacement of the crust. Further development of the Joly hypothesis will be watched with interest.

(Longwell, 1928: 151–152)

These criticisms by fixist contributors to the AAPG proceedings pale by comparison with the attack launched by Jeffreys. Beginning in 1926, a year after the publication of the first edition of Joly's *The Surface-History*, Jeffreys raised serious difficulties with Joly's theory of thermal cycles (RS2), and Joly promptly responded (RS1). Jeffreys ignored Joly's flirtation with mobilism, which remained outside their discussions, but his theory of thermal cycles gave them plenty to disagree about, and increasingly nasty exchanges ensued.

Jeffreys raised unreliability and anomaly difficulties. Joly had estimated the average thickness of continental crust to be about 30 km. According to Jeffreys, Joly's estimate was based on outdated and unreliable data. Newer and more reliable data indicated that the average thickness of continental crust was closer to 15 km, and Joly's theory required twice that because a thinner crust would make less likely the buildup of sufficient radiogenic heat to melt the upper substratum.

Jeffreys raised three theoretical difficulties. He argued that Joly's model was inconsistent with the theory of thermal conduction. Joly claimed that radiogenic heat causes the substratum to undergo cyclic melting and re-solidification. According to Jeffreys:

The basic rocks below the granitic layer also contain radioactive matter, and it is upon the heat generated in them that Prof. Joly relies for his main object, which is to explain periodic variation of temperature within the crust. Now there are many soluble problems involving the steady supply of heat to a conducting solid with an outer boundary at a constant temperature, but they all have a common property, namely, that the temperature distribution tends asymptotically to a steady state: there is no possibility of periodicity.

(Jeffreys, 1926b: 924)

Once maximum temperature is reached, it remains constant, regardless of whether the substratum is solid or liquid. There would be no cyclical changes.

Jeffreys further argued that Joly's idea that the blanketing effect of continents prohibited dissipation of heat was inconsistent with the theory of heat.

This contradiction appears to arise from the statements on pp. 102–3 of Prof. Joly's book. "If (the substratum) is solid, the radioactive heat continually being developed . . . beneath the continents is almost entirely accumulated." "The blanketing effects of continental radioactivity . . . effectively block its escape by conductivity." It seems to be supposed that heating a region prevents the transmission of heat through that region; which is not the case. The equation of heat conduction being linear, the effects of all sources of heat are purely additive; radioactivity within the granitic layer can do absolutely nothing to prevent the transmission to the surface of the heat developed below, and only ensures that the temperatures below that layer become higher than they would in its absence.

(Jeffreys, 1926b: 927)

Jeffreys denied that tidal force could bring about a westward rotation of the entire Earth's crust relative to its interior even with a liquefied substratum.

There is a further difficulty in making the crust revolve at the required rate. Tidal friction is appealed to for this purpose; but it is well-known that perfect fluidity on an extensive scale in a crust region is likely to abolish the ocean tides altogether . . . There is therefore no force tending to move the ocean along the boundary, and there can be no ocean tides. The magma layer being itself a perfect fluid, there can be no dissipation of energy within it, and there is therefore no tidal friction.

(Jeffreys, 1926b: 928)

Joly responded immediately. He countered Jeffreys' charge about over-estimating the thickness of the sialic crust by noting (1926: 933) "that so careful a writer as Dr. A. Holmes" favors an estimate more in keeping with his rather than Jeffreys' estimate, and argued that even if Jeffreys' estimate were correct, it would not matter because the substratum produces enough radioactive heat to cause its own liquefaction.

Turning to the alleged inconsistency with the theory of heat conduction, Joly said that Jeffreys had not understood his theory.

Dr. Jeffreys next proceeds to combat the view "that periodic variation of temperature" arise within the crust; which he describes as my "main object." Now this is not at all my main object. If Dr. Jeffreys had stated that my main object was to show that there is periodic accumulation and discharge of heat from within the crust he would be correct. But the heat is latent. There is but little variation in temperature. Throughout the rest of his argument he appears to labour under this initial error, and to assume that the problem he undertakes to discuss is concerned mainly with a periodic variation of temperature gradient; that is to say, the problem is one of thermal conductivity.

(Joly, 1926: 934)

Joly seemed to be claiming that the theory of thermal conduction does not apply to his theory of thermal cycles. What is at issue, according to Joly, is not simple conduction and periodic variation in temperature. Heat accumulates as latent heat, liquefies the substratum and initiates convection.

"There is," he says, "no possibility of periodicity." This is true of the mathematical theory of heat conduction. The periodicity which I have referred to in all my writings upon this subject arises from the gravitative instability which affects the ocean-floor when an underlying substratum assumes, in consequence of accumulated latent heat, the fluid state, and becomes of less density than the solid floor. The consequences arising out of this condition allow a very great transfer of latent heat by convection from the lower to the upper parts of the magma, so that the upward heat-flux during the fluid phase is very much greater than the mean, while the upward heat-flux during the solid phase is very much less than the mean.

(Joly, 1926: 934)

Joly also argued that both of Jeffreys' attempts to show that his theory of thermal cycles clashed severely with theories of heat transfer were largely irrelevant because the conditions, which arise upon heating of the substratum, are too complex for idealized theories of heat transfer to be applied. Here, for example, is his rebuttal to Jeffreys' objection that continents would not block the escape of heat from the substratum.

From my book (pp. 102–103) I am quoted as follows: – "If the substratum is solid, the radioactive heat continually being developed beneath the continents is almost entirely accumulated." "The blanketing effects of continental radioactivity . . . effectively block its escape by conductivity." Surely the meaning is plain enough; and is expounded many times throughout the book! Nevertheless, here is Dr. Jeffreys' comment: – "It seems to be supposed that heating a region prevents the transmission of heat through that region; which is not the case." We seem to have here the mathematical point of view with a vengeance.

(Joly, 1926: 935)

Joly also attempted to answer the difficulty Jeffreys raised about the inadequacy of tidal forces. After noting Jeffreys' analysis of the conditions needed for tidal force to be effective, he argued that such conditions are inherent in his theory; "while Dr. Jeffreys' conclusion leads us to a closer recognition of all the conditions, it raises no difficulties in the physical basis of the theory" (1926: 938).

Jeffreys renewed his attack in a paper that appeared in the *Geological Magazine*, summarizing his dissatisfaction with Joly's theory in this way:

In any case the onus of proof that the theory of magmatic cycles explains anything at all rests with its supporters, at least if the word "explain" is understood in the sense given to it by most physicists. A theory explains a fact if it starts from a set of stated hypotheses and shows how the fact in question follows from them. But the theory of magmatic cycles considers itself at liberty to use the observed facts to enable it to infer its own consequences; and any fact can be explained on any theory if that is permitted.

(Jeffreys, 1926b: 521)

As far as Jeffreys was concerned, Joly's theory disregarded what was known about the physics of heat transfer.

A year later Joly responded to the alleged inconsistencies between his theory and theories of heat transfer. He cited several cyclical phenomena that are brought about by a constant heat supply and compared them to the behavior of the substratum as envisioned in his theory of thermal cycles; a geyser, for instance, is fueled by a continuous supply of heat and faithfully supplies periodic ejection of hot water and steam.

The phenomena in the geyser and in the substratum are plainly analogous and thermally alike. In the case of the geyser, as in the case of the substratum, there is insufficient loss of heat "from an outer boundary" to control thermal accumulation and consequent instability. Hence it, ultimately, develops cyclic changes analogous to those affecting the substratum. They differ mainly in the nature of the change of state. The refrigerator, too, differs in character. In the one case it is mainly the atmosphere, in the other mainly the ocean which carries away the heat. The like principles affect both. Both appear to be impossible according to Dr. Jeffreys' statement that cyclic events cannot arise out of a steady thermal supply.

(Joly, 1927: 342–343)

He ended by turning to a broader issue, warning of the danger in applying mathematical physics to geotectonics.

It would appear that Dr. Jeffreys' simplifying assumption – although for mathematical reasons possibly unavoidable – must result in depriving his final conclusion of the significance which he attaches to it. The limitations he deduces have, in fact, been introduced by his own assumptions, which have led him to omit important factors concerned in the genesis of a general shift of the outer crust over the inner parts of the surface of the earth.

(Joly, 1927: 348)

Their debate continued until 1929 when Jeffreys ended it by summarizing his attack on Joly's theory in an appendix to the second edition of *The Earth*. Their heated exchanges in 1928 were published together in the *Philosophical Magazine*. Both continued their point by point debate. Both continued to debate the role of mathematics in geophysics, and to display increasing mutual disdain.

Jeffreys restated his former objections and dismissed Joly's rebuttals. He agreed that some systems underwent cyclical changes under a constant heat supply, but they were not analogous to the substratum. They were exceptions.

Prof. Joly's new paper (Aug. 1927) offers a series of periodic phenomena due to a steady supply of heat, and considers that they provide an answer to my statement that periodic phenomena cannot be produced in this way. Perhaps, I ought to have added a list of exceptional cases where the proposition might not be expected to hold, as is done at great length in works of pure mathematics. But in physics the discovery of these is usually left to the reader. After all, the whole science of engineering is devoted to constructing systems that will, if suitably treated, behave in an extraordinary way. But in the present problem we are entitled to suppose that the system has not been so designed. Prof. Joly has indeed done a service in mentioning the four examples he does, because they are admirable examples of the kind of exception that proves the rule. But as the conditions of convection in a layer of fluid of greater horizontal extent are by this time fairly well understood, and as the basaltic layer, if fused, would be such a layer, we may well consider the actual case first.

(Jeffreys, 1928: 211–212)

In geysers, for example, liquid is restricted from escaping until considerable pressure accumulates within the enclosing chamber. However, there is no such restriction in the substratum and crust, where the area is large compared to its depth. According to Jeffreys, a better analogy

with the conditions within the earth's crust would indeed have been the boiling of the domestic kettle or, still better, the melting of a tray of paraffin wax by heating below.

(Jeffreys, 1928: 214)

After dismissing Joly's examples, Jeffreys broadened his attack, showing his disdain for Joly and for his cavalier attitude toward mathematical physics.

This investigation will necessarily involve a considerable amount of mathematics; at least the equations of fluid motion and of heat conduction must appear in it and be solved. But until it is carried out, there is no reason to believe that the consequences would be anything like those asserted by Professor Joly. The onus of proof in a question of theoretical physics is on the advocate of a theory, not on its opponents; and it ought in this case to have been carried out before Professor Joly proceeded to publication in an important physical journal and followed it up with an ambitious book. If the problems to be solved are in any way more complex than I have considered so much the more complex must the problem be that Professor Joly has to solve before his theory will command the respect of physicists. The reason for this situation is simple: inadequately tested theories can be constructed more rapidly than people in the habit of taking reasonable care can point out the mistakes. I am sorry to have to point this out, but the number of suggestions that have been made in public and in private by non-physicists, to the effect that it is the duty of a mathematical physicist to drop any work he may consider more important in order to examine some theory of the type under consideration – with the certainty that the results will be accepted only if they are favorable – have made it necessary. I add that, since I have already devoted more attention to Professor Joly's theory than I consider it worth, I do not propose to take any part in further discussion of it unless he produces more satisfactory arguments than he has done hitherto.

(Jeffreys, 1928: 214)

Jeffreys was fed up with global theories espoused by geologists like Joly who, he felt, either knew little mathematical physics or disregarded what they knew. Joly, obviously upset with Jeffrey's attack, began his rejoinder with the following:

My paper entitled "Dr. Jeffreys and the Earth's Thermal History" (*Phil. Mag.* August 1927) has stirred Dr. Jeffreys into a rejoinder revealing so much irritation that I feel a real difficulty in pursuing the matter any further lest quite unintentionally I should say anything which might prove to be a source of further annoyance.

(Joly, 1928: 215)

Agreeing with Jeffreys that geysers were constrained, Joly claimed that analogous constraints were present within the substratum.

In the case of the geyser the medium is fluid, and constraints hindering free circulation are required while the energy necessary to attain the boiling-point at all levels is accumulating. In the case of the substratum the medium is solid to start with, and the corresponding constraint is in the rigidity of the working substance which determines quiescence while the energy necessary to attain the melting-point at all levels is accumulating. The difference in the character of the constraints, which arises necessarily out of the differing mobility of the media, possesses no fundamental significance. As thermal energy accumulates similar conditions of instability arise in both cases – i.e., a medium charged with energy proper to the pressure at all levels and necessarily changing state if any movement occurs to bring any part of it to a higher level.

(Joly, 1928: 216)

Of course, Jeffreys disagreed. In the second edition of *The Earth*, published in 1929 (p. 324), he restated his objections to Joly's theory, and dismissed Joly's replies, including those in his 1928 article, as "a series of evasions and irrelevancies."

4.12 Mobilists show little sympathy for the Joly–van der Gracht mechanism

Van der Gracht was the only major mobilist to develop a hypothesis extensively based on Joly's hypothesis of thermal cycles. Holmes welcomed Joly's appeal to radioactivity, and during 1925 through 1927 developed what he considered an improved version of Joly's thermal cycles. At this time, Holmes was not a mobilist, although he found Wegener's hypothesis intriguing. He became a mobilist in 1928, when he developed his own mechanism and became critical of Joly's theory.[15] Du Toit is the only other mobilist who made use of Joly's theory. He first praised it in 1924.

Among the few who have approached the problem with a perceptive mind can be mentioned Joly, whose lucid explanation of crustal diastrophism as the logical outcome of radio-activity in the crust would seemingly provide the urgently needed key to the problem under discussion.
(du Toit, 1924: 73)

Then, in his book, he looked to radioactivity as a means for supplying much of the energy needed to displace continents, but in developing his mechanism he borrowed more from Holmes than Joly (du Toit, 1937: 22–23, 329). Daly also discussed Joly's theory. Agreeing with Jeffreys' and Holmes' objections, he raised additional difficulties (Daly, 1933: 224–226). Although he labeled Joly's attempt "brilliant," he did not think it merited further investigation (Daly, 1933: 226). Wegener had little to say about Joly. In the 1929 edition of *The Origin*, he may have been put off by Joly's ambivalent attitude toward mobilism, but commented rather vaguely that he had provided "new evidence in its favor by considering radioactive heat generation" (Wegener, 1929/1966: 53). The Dutch geologist Gustaaf Adolf Frederik Molengraaff (1860–1942), an early mobilist sympathizer, thought van der Gracht's extension of Joly's mechanism deserved "full attention" but was "highly speculative" (Molengraaff, 1928: 92). He (1928: 90–91) raised three difficulties (RS2). First, he doubted that there is enough radioactive heat to melt the substratum beneath continents. Second, he questioned the assumption that such heat would accumulate and remain in the sub-continental sima because it could escape through continents. Third, Molengraaff pointed out that even if the joint theory could account for a general westward drift of all the continents, it could not explain the differential drift of the continents surrounding the Atlantic; he believed that the Mid-Atlantic Ridge represented a rift valley fixed to the interior, along which the Americas separated from Africa and Europe, with the Americas drifting westward and Africa and Europe drifting eastward.

It is worth pausing for a moment to introduce Molengraaff and examine his explanation of the origin of the Mid-Atlantic Ridge. Molengraaff took part in an 1884–1885 expedition to the Dutch East Indies. He returned there in 1894, and in 1910–1911 led an expedition to Timor. He also worked in southern Africa. Appointed "state geologist" of the Transvaal Republic in 1897, he discovered the Bushveld complex while mapping there. An early convert to mobilism (Molengraaff, 1916), he was particularly impressed with its solution to problems concerned with the Dutch East Indies and South Africa (§8.14). As I shall later show, this was in marked contrast to Dutch opinion in the 1930s (§8.14). Molengraaff returned to the Transvaal in 1927 as a guide on the Shaler Memorial Expedition, and must have enjoyed discussing mobilism with fellow participants du Toit and Daly.

Molengraaff appears to have been the first to suggest that the Mid-Atlantic Ridge was entirely simatic and formed as volcanic material arose through the giant fracture caused by the separation of the Americas from Europe and Africa. As he (1926: 91) noted, he first presented this idea in 1916, just one year after the publication of the first edition of Wegener's *Origin*, when he also reviewed Darwin's explanation of the origin of coral reefs and atolls. Darwin had maintained that coral reefs and atolls continued to build as the ocean floor subsided. Although most agreed that the reefs and atolls subsided as they grew, ocean floor subsidence was questioned by supporters of isostasy. Molengraaff (1916) suggested that reefs and atolls were found on the summits of volcanic islands, which were out of isostatic equilibrium when formed. As isostatic equilibrium was restored, they subsided slowly with reef-growth keeping pace. Molengraaff extended this general explanation to the Mid-Atlantic Ridge which, unlike Wegener (and Daly), he thought was sima. Sympathetic to mobilism, he made the important suggestion as I have just said that separation of the Americas from Europe and Africa caused a huge fracture in the ocean floor through which massive amounts of volcanic material was discharged. The ridge, originally higher when first formed, is slowly sinking as isostatic equilibrium is attained.

Perhaps we may see in this remarkable mid-Atlantic ridge the final result of volcanic activity along an enormous fracture of the same extent, where from numerous fissures and vents volcanic material was discharged. Thus a volcanic mountain chain and cones being formed, which nowadays subside through yielding under the influence of gravity and nearly all have sunk back to a level approaching the average level of the deep submarine ridge. Here and there a few islands, whose volcanic activity lasted longer or has existed to this day, still rise above the sea . . . The cause for the extrusion of such enormous masses of volcanic material might perhaps be sought in the disruption of the American continent from the European-African one with which it formerly cohered. This disruption was assumed by Pickering and Taylor and a plea for it is again brought fore by Wegener on page 68 of his paper quoted before. On this supposition the mid-Atlantic ridge would in my opinion indicate the place where the first fissure occurred and the sima was first laid bare. From this it would follow logically that the ridge itself must consist entirely of sima and not of sial, as Wegener assumes on page 69.

(Molengraaff, 1916: 626–627; the reference to Wegener is to the first edition of The Origin)

Later, Molengraaff (1928: 91) restated his solution, compared the Mid-Atlantic Ridge to the East African Rift Valley and argued that recent gravity studies supported his analysis.

To my mind the mid-Atlantic ridge is nothing but the cicatrix of the former rent or fracture, along which the disruption of the American continent from the European-African continent took place. America drifted from the rent on which the volcanic mid-Atlantic ridge has been built up in a westerly direction, but Africa drifted toward its present position in an easterly direction. It has since then been the site of volcanic activity, and this activity is not yet completely exhausted. If so, this mid-Atlantic fracture is strictly comparable to the great rift-valley (Ost-Afrikanische graben) in East Africa . . . If this supposition is correct one must expect to find the mid-Atlantic ridge to be composed entirely of effusive volcanic material of relatively high specific gravity. The latest measurements of gravity, rather recently made in the Atlantic Ocean above the mid-Atlantic ridge on board a submarine by Vening Meinesz, have proved that the mid-Atlantic ridge shows a positive anomaly of gravity (excess of gravity), which remarkable fact gives support to my suggestion on the nature and origin of this ridge.

Molengraaff was not anticipating seafloor spreading. He claimed that freshly rising sima formed the Mid-Atlantic Ridge, which then slowly subsided. He did not claim that newly formed sima spread sideways creating new seafloor. Formation of the ridge was a byproduct of the initial phase of continental drift. But he certainly long anticipated Heezen and Tharp's comparison of the Mid-Atlantic Ridge with the East African Rift Valley (III, §6.7).

The decline in interest in and demise of Joly's theory was probably caused by Jeffreys' sustained attack. Holmes, after becoming a mobilist, and Daly, for example, cited and agreed with Jeffreys' criticisms. Moreover, Holmes and Daly had their own mechanisms, and Holmes' enjoyed a better reception than Joly's (§5.5). Moreover, mobilists such as van der Gracht and du Toit, who both used parts of Joly's theory, were no doubt discouraged by his vacillating, often tepid, attitude toward mobilism. Joly's major contribution to the mobilism debate for which he is mainly remembered was, as Wegener pointed out, that he brought radioactively generated heat to the forefront of the discussion. It was this that Holmes adopted, as I shall describe in the next chapter.

Notes

1 R. T. Chamberlin (1928: 87), citing Lambert, claimed that tidal forces were too weak to cause continental drift. Bowie (1928: 180) did also. Edward Berry (1928: 195) raised the following inconsistency with Wegener's appeal to tidal forces (and *Polflucht*) and his more rapid westward movement of the Americas compared to Eurasia.

   If centrifugal and tidal forces are the causes of the westward drift and rotation of the Americas, these same forces should cause Eurasia to move more rapidly, since the latter's mass is very much greater; hence, the Atlantic rift ought never to have occurred.
   (Berry, 1928: 195)

2 Willis also argued that the continents would not survive their postulated journeys through the seafloor. Willis' difficulty differed from Longwell's, pertaining as it did to the postulated westward movement of continents.

   By hypothesis the floating mass of South America has moved forward several thousand kilometers with sufficient momentum to crush up its frontal margin. But the compression in front must, according to the principles that govern the movement of a floating body, be related to an equivalent tension behind. The mass of the supporting medium that is displaced by the advance is the same as that which must be drawn in behind to fill the vacuum that is left . . . We may imagine the potential vacuum to have developed to the degree that the tension became equal to the tensile resistance of one or the other of the two masses affected by it, namely the sima, which remained behind, and sial, that was moving forward. We have just deduced the fact that the latter must be weaker in resisting compression, since it is the continental margin that is crushed up. It would naturally also be weaker in resistance to tension than the sima and therefore would be the one which would yield as the tension developed to the breaking strain . . . If this be admitted we should expect that the eastern side of the moving continent would have suffered normal faulting to a degree at least compatible with the uplifts on the western side. But that is not the fact. On the contrary the mountains and plateaus of Brazil demonstrate by their physiographic expression as well as by their elevation that the eastern side of the continent has been subjected to compression simultaneously with the similar compression of the western margin (Willis, 1928: 79–80).Willis, noting that Wegener, in his discussion of how Madagascar, the Antilles, and the festoons of the East India Archipelago were formed, had recognized that continents should break apart at their trailing edge, raised a theoretical "inconsistency" difficulty: Wegener recognized that continents would break apart in some cases but not in others – "these apparently detached islands seem to support his theory. So also does the integrity of South America. Unfortunately the facts are not consistent one with another."

(Willis, 1928: 80)

3 Longwell may have been the first to use the term "plates."
4 Chamberlin (1928: 87), again referring to Lambert, claimed that *Polflucht* was inadequate to move continents. Bowie (1928: 181) cited both Lambert and Schweydar for support of the insignificance of this force.
5 In the second edition of *The Earth*, Jeffreys concisely explained this difficulty concerning the inconsistency between Wegener's mechanisms for drifting continents and the origin of mountains.

But reference must be made to the assertion that the situation of the Rocky Mountains and the Andes is what would be expected on the theory of continental drift. On the contrary, it is one of the most definite pieces of evidence against it. Either the materials of the ocean floor are stronger than those of the continents, or they are weaker. If they are stronger they will not give way to let the continents move through them; if they are weaker, the continents would advance, if at all, without being fractured, and no mountains would be formed.

(Jeffreys, 1924: 322)

6 Many other participants commented on the apparent inconsistency of Wegener's mechanisms for continental drift and mountain building, Willis, for example, said:

When one body floating in another moves forward one or the other will be displaced. In the ordinary case, as for instance if a raft moves forward on the surface of the water, it is the water which is displaced, for the water is the more yielding of the two. It is not easy to find a simile for the relations assumed by Wegener, which imply that the moving body is the one that yields. Perhaps a leaden chisel thrust into steel will do as well as another. The point is that if the cordillera has been forced as by the advancing margin of the continent, then the latter is the yielding mass. It is clear that the sial must under these assumptions be weaker than the sima, but the reverse is the essential postulate stated by Wegener, who compares the sial to "solid wax" and the sima to flowing sealing wax. He [Wegener] himself saw the difficulty, but evades it.

(Willis, 1928: 78; my bracketed addition)

Bowie made the same point.

According to Wegener, the sial drifting through the sima meets such resistances as to crumple up the front of the moving mass to form mountain systems. This, it seems to me, is an impossible condition, for if the sial can move through the sima there can be no bulwark opposed to the sial which would distort the mass. We cannot have a sima which has both the quality of a material of no residual rigidity and at the same time has a rigidity far in excess of that of the drifting sial. This is a contradiction in Wegener's hypothesis that has probably created the strongest antagonism to it.

(Bowie, 1928: 184–185)

Chamberlin (1928: 86), Singewald (1928: 191), and also Termier in his 1924 critique of Wegener's theory (§8.8) pointed to this inconsistency.

7 Biographical information is from Birch (1960) and Billings (1959).

8 Although it is difficult to date precisely when Daly was first attracted to mobilism, he was sufficiently impressed with it in 1920 to have asked Lambert to evaluate Wegener's appeal to *Polflucht*; three years before his first published espousal of mobilism. Alex du Toit may also have influenced Daly during this period. Daly was a member of the Shaler Memorial Expedition to the South Atlantic and South Africa. Du Toit was included in the expedition, and they did fieldwork together in South Africa during the first half of 1922. Du Toit, whose fieldwork greatly impressed Daly, already had begun to support mobilism during the previous year. However, du Toit did not have to convince Daly that mobilism was worthy of investigation because he already had asked Lambert to examine Wegener's solution to the mechanism problem before leaving for South Africa. Moreover, Daly and du Toit were attracted to mobilism for quite different reasons. Daly wanted to use it to develop his solution to the origin of mountain belts and recognized that he needed to develop a new solution to the mechanism of continental drift; du Toit was attracted to mobilism primarily because of the solutions it offered to problems concerned with the South African geology – the origin of the Permo-Carboniferous glaciation, disjunctive distributions and geological similarities between South Africa and other regions. Molengraaff, who became a mobilist in 1916, also served as a guide on the expedition (Molengraaff, 1916). (See §4.12, and §8.13 for discussion of his work.)

9 Daly's hypothesized movement of the continents is different from both Wegener and Taylor. He has movement of continents toward his ancient mid-latitude furrows and furrows on the margin of the original continent. Thus, he has neither a general westward drift nor an equatorward drift. Nor does he suppose polar wandering. Daly noted that his down-sliding hypothesis generally differed from his predecessors, and added that it could be possible even without accepting lengthy migration of continents.

Neither Taylor nor Wegener has described the evolution of the continents and oceans exactly along the lines of the foregoing questions [i.e., the ones quoted in the above passage], but both authors of the general hypothesis agree in believing continents to have migrated much more than 1000 miles. Other geologists are more cautious and assume maximum migration of no more than a few tens or scores of miles. It is still too soon for any one to come to a definite conclusion on the subject, but there seems to be no compelling evidence against the hypothesis that the Arctic, Atlantic, and Indian basins have been opened up by the horizontal movement of continents. Meantime, I shall continue to present the general conception in terms of the more generous estimate of the extent to which continents have moved.

(Daly, 1929: 278–279)

Holmes' mobilism theory also differed substantially from others. Both he and Daly believed that the differences were important, especially their proposed mechanisms; the various versions of mobilism shared the same hard core beliefs (see Frankel, 1979).

10 Daly was harshly criticized at the GSA's 1922 annual meeting when he supported mobilism. Newman (1995: 66) quotes the caustic remarks by W. H. Hobbs, a permanentist, and A. P. Coleman (see §3.12 for a discussion of Coleman's views).

11 Neither Birch (1960) nor Billings (1959) mentioned Daly's mobilist theory in their memorials of him. Because their memorials were written before the triumph of mobilism, I suspect that they did not want to taint Daly's memory with such mention. Daly put a great deal of effort into his defense of mobilism, and omitting any reference to this, gave distorted accounts of his work. Like Daly, Birch and Billings spent most of their careers at Harvard University. Marvin (2001), who was a graduate student at Harvard during the 1940s, also remarked about the absence of any reference to Daly's mobilism in Birch's and Billings' memorials and speculated that their omission was "presumably to shield his [Daly's] professional reputation."

Marvin also recounted a poignant remark by Daly that she happened to overhear. Putting the remark in context, she also noted that Birch was rumored to be upset with Daly's continued support of mobilism.

Another eminent professor who opposed drift was Francis Birch, who succeeded Daly in 1942 as the Sturgis Hopper Professor of Geology. Oreskes scarcely mentions Birch, but he was one of the first investigators to conduct high-pressure experiments on rocks in the program set up by Daly and directed by Percy W. Bridgman. His results persuaded Birch that Jeffreys was correct: rocks at depth lack plasticity and therefore cannot sustain either convection currents or continental drift. Rumors circulated that Birch was upset with Daly for continuing to support drift. One day in 1945 I was in the office of petrology professor, Esper S. Larsen, when Daly entered, clearly distressed: "Here I am an old man," he said, "and I may have to give up everything I have believed in and worked on for half of my life." I left the room and heard nothing more. I cannot be certain that Daly was referring to his theory of gravity-sliding, but I find it hard to imagine what else it might have been.

(Marvin, 2001: 213–214)

Perhaps Birch also did not want to blemish Harvard's reputation by pointing out that one of Harvard's most eminent Earth scientists was a mobilist.

12 Marvin (1973: 177) also remarks that Daly anticipated subduction with his down-sliding hypothesis.

13 Of course, Daly did offer an explanation. He attributed the tremendous deformation to contraction, decrease in the Earth's rate of rotation, and erosion. Consequently, Holmes must have come to think so little of these factors that he believed they did not even count as a possible explanation. Indeed, he must have because by this time he had come to reject contractionism. Curiously, Holmes actually cited Daly's 1923 article, "Decrease of the Earth's Rotational Velocity and its Geological Effects," in the list of references at the end of his 1931 work, but did not refer to the article in the body of the text. Was this simply an oversight? Did he think that Daly's explanation was not worth refuting? Or, did he think that a careful refutation was not needed and would only serve to embarrass Daly?

14 This brief biographical sketch is drawn from Dixon's (1934) obituary of Joly.

15 Holmes' attack on Joly's theory will be discussed in §5.3 in the context of the development of his ideas on subcrustal convection.

5

Arthur Holmes and his Theory of Substratum Convection: 1915–1955

5.1 Introduction[1]

The aim of this chapter is to explore Holmes' contributions to mobilism and their reception by both mobilists and fixists. Accounts of his work on determining the age of Earth and geological eras and periods may be found in Brush (1996b), and in Lewis' (2000) lively biography.

Initially, Holmes was a fixist and favored thermal contraction as the cause of orogenesis. Later he abandoned contractionism and became a mobilist, based on his novel account of substratum (now called the mantle) convection, which is activated by radiogenic heat, and which creates new ocean basins, causes continents to drift, and plays a central role in the formation of mountain belts. Holmes first developed these ideas in the late 1920s and early 1930s, with revisions in the mid-1940s. His model of Earth behavior is recognizably modern. Although not without its difficulties, his was a plausible solution, and the mechanism he proposed could not be dismissed as readily as the earlier ones of Wegener, Taylor, Daly, and Joly. I shall pay special attention to the arguments he used in favor of his own theory and against rival ones, both mobilist and fixist, and I shall document his use of the standard research strategies. The arch-fixist, Jeffreys, thought that Holmes' substratum convection offered a better drift mechanism than its predecessors, and grudgingly admitted that he had moved mobilism from the impossible to the highly unlikely: henceforth continental drift could not on mechanical grounds reasonably be considered as entirely impossible. I shall follow their debate. Holmes' was a step forward that some later workers did not acknowledge.

Mobilists' reaction to Holmes' hypothesis of substratum convection was mixed. Du Toit more or less adopted it, but Daly rejected it. In the debate that followed many rejected it regardless of their attitude toward mobilism; some were willing to allow it but remained fixists. I shall argue that Holmes, although highly inclined to believe that substratum convection occurred, was not adamant about it, recognizing that important difficulties remained.

There were three stages in the development of his work. In the late 1920s Holmes became disenchanted with contractionism because it could not explain the huge role that igneous activity plays in the geological record. In the early 1930s he used Joly's ideas of radioactive heating in the substratum and A. K. Bull's thoughts about convection to develop his own ideas on substratum convection. In the later 1930s and 1940s, although Jeffreys (confirmed by Hales and Pekeris) had to concede that substratum convection, albeit highly improbable, was not impossible, Holmes' enthusiasm for mobilism waned as, under the banner of permanentism, new and ever more sophisticated objections to mobilism were formulated, and as his own contribution was little recognized. By the early 1950s, general belief in mobilism was, perhaps, at its lowest ebb, and Holmes' attitude seemed to have mirrored this.

5.2 Holmes' scientific career

Arthur Holmes (1890–1965) was one of the most distinguished Earth scientists of the twentieth century. He had a thorough understanding of geology, geophysics, and physics, which enabled him to play a central role in the development of the Phanerozoic radiometric geological timescale, and to work on many other fundamental problems in geology such as the age of the Earth. His *Principles of Physical Geology* (hereafter, *Principles*) was a successful and fair-minded geology textbook, one of the very few that provided a balanced account of the mobilism controversy up to the 1960s.

Holmes wrote over fifty of his papers on radiometric dating. Fewer than ten dealt with continental drift. Holmes' hallmark was his ability to confidently develop global theories based on a wide range of evidence, while at the same time recognizing that he might be wrong. He was willing to revise or abandon theories in light of new evidence. His knowledge was broad and deep. It is no accident that he was highly respected by such different scientists as du Toit and Jeffreys.[2] Although his mathematical abilities could not match Jeffreys', Holmes was one of the few geologists Jeffreys respected for his understanding of geophysics. Jeffreys was one of Holmes' proposers for the Royal Society (London). Though not so productive as du Toit as a field geologist, he nevertheless had much field experience, and was widely read in world geology. Daly also had great respect for Holmes.

He was an only child, born in 1890, near Newcastle-upon-Tyne, England. He attended Gateshead Higher Grade School, where he was introduced to physics and geology by reading Kelvin and Suess.[3] Holmes was awarded a scholarship to the Royal College of Science, London, renamed Imperial College in 1911. He entered the College in 1907, where he became associated with R. J. Strutt (later, Lord Rayleigh). Strutt had recently discovered that radium is widely distributed throughout many rocks, which raised questions about Earth's thermal history. Strutt also realized that radioactive decay could provide an absolute timescale. During Holmes' first year, he took courses in mathematics, mechanics, chemistry, and physics. He also attended lectures in geology from the newly arrived William W. Watts. He switched to geology. He enjoyed Watts' lectures, but also decided to become a geologist partly because (according to his second wife, Doris Holmes née Reynolds, herself a petrologist) "he thought that too many good physicists were being trained at Imperial College at the time."[4] Needing to begin his own research, Holmes was invited by Strutt to work on radiometric dating, and he dated several rocks (Lewis, 2000). But, short of funds, in 1911 he took a job with Memba Minerals Limited in Portuguese East Africa (now Mozambique), where he contracted malaria complicated by blackwater fever.

After about eight months, he returned to Imperial College, where he remained until 1920. During his first year back at Imperial College (1912), he wrote *The Age of the Earth* (1913), and later *The Nomenclature of Petrology* (1920). While still at Imperial College he began a series of papers investigating the effect of radioactivity on Earth's thermal history. In 1920 he took a job with an oil company in Burma (Myanmar). He returned to academe in 1925 as a reader in geology at the University of Durham, where he remained until 1943, and it was there that he developed his mobilism theory, presenting it first in 1928. During his final years at Durham, he wrote the first edition (1944) of his *Principles*. In 1943, he was appointed to the Regius Chair of Geology at Edinburgh. Because of deteriorating health, he gave up the Chair in 1956. Holmes died in 1965, one year before the confirmation of the Vine–Matthews hypothesis and seafloor spreading, but not before the appearance of new work in paleomagnetism and marine geology supportive of mobilism, which he described in the second edition of *Principles* (1965). Although Holmes' enthusiasm for mobilism wavered at times, he never abandoned it. (See III, §2.16 for his later work.)

Holmes was elected FRS in 1942, one of the many scientific societies of which he became a member. In 1956 he received the Wollaston and Penrose Medals, the highest awards of the Geological Society of London and GSA; the Fourmarier Medal of the Royal Academy of Belgium, 1957, and in 1964, the Vetlesen Prize from Columbia University, New York.

5.3 Holmes before becoming a mobilist

Holmes began thinking about Earth's thermal history while a student and it remained a major interest. He wrote five papers entitled "Radioactivity and the Earth's thermal history." The first two appeared in 1915, the third in 1916, and the others in 1925. In the first three he defended contractionism, working comfortably within it. In the last two, written almost a decade later, he rejected it.

In 1915 and 1916, Holmes, seeking to reconcile radioactivity and contractionism, assumed that as Earth cooled from a molten state, radioactivity did not produce enough heat to interfere with overall cooling as contractionism required. As he put it in 1925, referring to his earlier papers:

As a working hypothesis it was further assumed that the total amount of the radioactive elements in the Earth was limited by the conditions that the Earth had in fact cooled down and was not, as seemed to be a possibility, growing hotter.

(Holmes, 1925a: 504)

He assumed that radioactivity decreases downward exponentially, and calculated a temperature–depth curve extending 300 km below the continental crust based on his radiometrically estimated age of Earth. This was favorably received by Jeffreys, who later calculated his own curve based on Holmes' work.

With this in hand, Holmes, in 1916, argued that there was a weak zone underlying the level of isostatic equilibrium, and boldly resolved "to carry the investigation still further by calculating the level of no strain and discussing the possibilities of mountain building by compression, and igneous activity in general" (1925a: 505). But, as Holmes recalled, Jeffreys had already examined and developed an improved contractionist theory of mountain building.

Fortunately, Dr. Jeffreys had become interested in the work at this stage ... His results definitely rejuvenated the thermal contraction theory of mountain building. Whereas on the Kelvin theory of cooling the level of no strain was not far below the present surface and the amount of contraction available was woefully inadequate, the new theory gave the level of no strain at over 100 kms. below the surface, and a total amount of compression of apparently the same order of magnitude as that actually implied in the crumpled and overthrust formations of existing mountain ranges.

(Holmes, 1925a: 505)

Note that Jeffreys had not considered the question of igneous activity.

But in 1925 Holmes found himself in a peculiar position; Jeffreys and other contractionists praised his explanation of how radiogenic heat could be accommodated within contractionism, but he had now come to believe that contractionism should be rejected and some of his previous work revised.

Although Holmes clearly regarded Jeffreys' account of mountain building as a decided improvement over earlier versions, he realized that it was not without difficulties, but he did not think them sufficient to reject contractionism. In 1925, Holmes wrote:

My own opinion was that the theory explained about a third or a quarter of the folding and overthrusting that has demonstrably occurred during the earth's known history. However, the original radiothermal theory of the earth's heat could not be held to stand or fall by its success or failure in explaining tectonic structures, for many other processes besides thermal contraction might conceivably have been involved.

(Holmes, 1925a: 506)

Holmes believed it possible to remove the difficulties facing Jeffreys' contractionist theory by combining it with other orogenic processes to produce orogenic belts.

But Holmes also believed that contractionism faced one really outstanding difficulty, namely, it could not account for the vast quantity of igneous rocks that are not associated with mountain belts or large sedimentary deposits.

With regard to igneous activity the theory [of thermal contraction] has been much less happy [as compared to its solution to the origin of mountain belts] ... Where the amount of radioactive matter in any column of the crust is greater than the continental average already arrived at the temperature will rise and favour the growth of magmas. Such excesses of heat are particularly likely to accumulate beneath geosynclines after thick masses of sediments have been deposited; and also in the depths above which compressed mountain ranges have arisen ... But where, instead of sedimentation, denudation has been long-lived; where stresses have been tensional rather than compressional; and where epirogenic movements have been in control; the genesis of magmas by superheating due to radio-thermal energy clearly cannot be invoked on the theory under discussion. Yet plateau lavas and associated types of igneous activity have broken out with the utmost vigour in just such regions as appear to be the most improbable. There can be no doubt that here the theory breaks down completely, and the object of this paper is to recognize the fact.

(Holmes, 1925a: 506–507; my bracketed additions)

The temperature–depth curves of Holmes, Jeffreys, and others, calculated using the constraints of thermal contractionism, did not provide enough heat to produce plateau lavas in places far removed from orogenic belts and deep sedimentary basins. Although Holmes saw a possible way of forming these plateau lavas through fusion of peridotite far below Earth's surface, he thought that was inconsistent with other aspects of contractionism. This, Holmes recognized with his wide knowledge of igneous rocks and their distribution, was the Achilles' heel of contractionism (RS2).

In conclusion, it appears to me that the vague possibility that igneous activity may begin with the fusion of peridotite hundreds of kilometres below the surface, while it cannot be positively excluded, is one that raises so many difficulties at every stage that it cannot be reasonably invoked as a practical explanation of vulcanism. Moreover, if it is invoked, then the contraction theory of mountain building already developed becomes not only partly but hopelessly inadequate.

(Holmes, 1925a: 514)

Holmes had an imaginative and far-reaching alternative: it was that radiogenic heat produced the vast expanses and thickness of plateau basalts. This would require Holmes to reject his ten-year-old temperature–depth curve, and to develop one unrestrained by contractionism, in effect to give up contractionism, be it his own or Jeffreys' version.

We are therefore left with only one known way out of the impasse: there must be more uranium and thorium in deep-seated rocks than has hitherto been thought possible on the theory of a cooling earth. This conclusion was rejected ten years ago, and has since been rejected by other workers in the same field [e.g., Jeffreys], partly because it was thought necessarily to imply that the earth must be getting hotter, and partly because the difficulties in the way of explaining vulcanism were not, and have not been, fully realized. However, since it is an obvious truism that a theory to be tenable must not contradict observational facts, we must give up the older theory and start afresh with different assumptions, guided as far as possible by such scanty data as are available and relevant.

(Holmes, 1925a: 514–515; my bracketed addition)

Holmes was sure that he had to reject contractionism, and that he had to suppose that radioactivity was not restricted to Earth's crust but existed in the substratum, but initially he was unsure how to proceed.

Unfortunately, once the straightforward conception of a continuously cooling Earth is abandoned, the possibilities become hard to visualize, unwieldy in their complex interrelations, and difficult to check except by the most complete details of geological history, and of the physical properties of material under familiar conditions. Nevertheless an attempt must be made, and if mistakes are involved at first, then at least their recognition and correction in the future will mark a beginning of sound progress.

(Holmes, 1925a: 515)

During the next several years Holmes sharpened his attack on contractionism. In the last of his five papers, he continued to argue that the amount of radioactivity beneath Earth's crust was more than permitted by contractionism. Here, "instead of deducing the distribution of rock-types in depth from their radioactive contents," as he had as a contractionist, he attempted

to determine the downward distribution independently, so that the radioactive effects may be deduced without reference to any limiting hypothesis of a steadily cooling earth.

(Holmes, 1925b: 529)

In 1927, he raised several further difficulties with contractionism. Citing passages from Jeffreys' new book *The Earth* (1924), Holmes claimed (1927: 263) that the questions of the origin of magma, the origin and distribution of mountains in time and space, the alteration of marine transgressions and regressions, and the formation of geosynclines all raised severe obstacles to contractionism.[5]

Although, as I have said, in 1927 Holmes was not sure how to proceed, he was certain that *"The loss of heat from the earth must be a discontinuous process"* (1927: 276). He also believed (1927: 276) that the large quantity of heat generated from radioactive processes "is accumulated for a time in the formation of magmas, and afterwards discharged by the ascent of the magmas to higher levels." Because Joly's theory of thermal cycles was based on radiogenic heat playing a central role, and because he believed that loss of heat is discontinuous, it is not surprising that Holmes (1926) was at first sympathetic to Joly's theory, which at this stage he came to prefer to Jeffreys' thermal contractionism. He (1926) even amended Joly's theory, incorporating his own account of magma generation into it, and offered a partial defense of it against some of the objections raised by Jeffreys; perhaps he was attracted to Joly's theory because it allowed for continental drift. Although he soon abandoned Joly's for his own newly minted theory, he still commended Joly for now linking continental drift with the removal of excess heat from the substratum.

Though his mechanism for bringing about the loss of excess heat is unsatisfactory, Joly has boldly faced the situation that arises if an excess of heat is generated in the substratum, and has clearly realized that some form of crustal drift is absolutely necessary to permit the discharge of such heat.

(Holmes, 1931a: 565)

Holmes' publications written before he developed his mobilist theory contain two remarks about continental drift, in his favorable reviews of both Joly's 1923 "Movements of the Earth's Crust," and his Halley Lecture of 1924. In these, Joly was more sympathetic to continental drift than in his book *The Surface-History* (1925). Holmes stated:

It will be clear ... that if the processes described by Prof. Joly be accepted in principle *it is no longer possible to resist Wegener's intriguing* displacement theory on the grounds that the continental blocks are at the present day embedded in a rigid substratum. Besides this apparently all-sufficient line of attack, other adverse criticism has been a matter of comparatively unessential detail. But the insuperable mechanical difficulties that seemed to stand in the way of the alleged continental drift are now seen to be temporary and intermittent. Prof. Joly points out in his first paper that during the climax of a period of fusion almost anything may have happened so far as the stability of the ocean floor is implicated. It seems inevitable that under the influence of magmatic currents the ocean floor would be not only fractured but also engulfed to such an extent that it would no longer be able to present a continuous resistance to bodily movement of the continental blocks. The probability of at least occasional periods of drifting is therefore intimately involved in Prof. Joly's views.

(Holmes, 1925c: 532; emphasis added)[6]

Evidently by 1925 Holmes, sparked apparently by Joly's work, had become interested in continental drift. He was generally familiar with criticisms that had been made of it, the mechanism difficulty being the principal one, and he thought that Joly had offered a possible solution. A year later, in an attempt to improve Joly's theory, Holmes speculated:

Indeed, had radioactivity not been already discovered it is likely that geologists would have been driven to postulate the self-heating properties which led to the evolution of magma. Feeling no doubt that "The earth is rude, silent and incomprehensible – at first," Argand wrote in the prelude of his luminous work, "Nous ne pretendons pas reduire la tectonique a la physique: c'est l'affaire a l'avenir." [We do not intend to reduce tectonics to physics: this is an affair for the future.] With the advent of the theory of magmatic cycles that future is definitely begun, for already tectonics have been physically liberated from the deadening conception of fixity, against which – on grounds very different from those of Wegener – Argand has revealed a world of eloquent testimony.

(Holmes, 1926: 308)

Holmes was reading works on continental drift, including Argand's *La Tectonique de l'Asie*, which described a mobilist process as the cause of mountains but had left its mechanism as a future task for physicists and geophysicists (§8.7). Holmes believed that Joly's theory opened up the mechanism problem by its appeal to radioactivity. From 1925 through 1927 he was favorably inclined toward Joly's theory, and amended it with his own account of how plateau basalts originated.

So by 1925, Holmes had rejected contractionism because it could not explain the origin of plateau basalts, which he proposed were the results of discontinuous loss of radiogenic heat. Two years later, he argued that contractionism could not explain geosynclines, origin of mountains, and the alternation of marine transgressions and regressions. He recognized continental drift as *intriguing*, and finally rejected contractionism.[7] He thought that the lack of a known mechanism was mobilism's most serious difficulty.

5.4 Holmes develops his mobilistic theory, 1928–1931

Holmes' first public defense of continental drift was on January 12, 1928, in an address given to the Glasgow Geological Society. A short summary was published that year, and a full and amended version in 1931 (Holmes, 1928a, 1931a). He also wrote two reviews of the published papers from the 1926 AAPG symposium on continental drift, the shorter appeared in *Nature* (Holmes, 1928b), the longer, in three parts, in the *Mining Magazine* (Holmes, 1929).

When developing his mobilistic theory, Holmes began by comparing mobilism and competing theories. He believed that mobilism offered solutions to a wide range of problems but, Joly's theory excepted, the mechanisms proposed so far were plagued by serious difficulties (RS3). In 1927 Holmes defended Joly's hypothesis by arguing that many of the difficulties that had been raised against it, including Jeffreys', were not serious enough to merit its rejection. A year later he changed his mind, restated earlier difficulties, raised new ones, and argued that Joly's theory should be rejected (RS2, RS3). Finally, he saw mantle convection currents driven by radioactive heat as the fundamental global tectonic process. I shall later assess his changing attitude toward his own new mobilist theory.

Holmes either came across or remembered a paper he had formerly read, which contained a proposal that mountain belts were a product of convection currents fueled by heat produced by radioactive decay of materials within the Earth's substratum. The author was Bull and his paper was published in 1921 in *Geological Magazine*, a journal that Holmes knew well. From the beginning Holmes acknowledged his debt to Bull.

Admitting that the continents have drifted, there seems no escape from the deduction that slow but overwhelmingly powerful currents must have been generated in the underworld at various times in the earth's history. In Joly's hypothesis movements due to expansion and contraction are taken fully into consideration. But the only lateral movements considered are those due to tidal actions leading to a westerly lag of the whole outer crust. There remains to be considered a happy suggestion which we owe to A. J. Bull, namely that convection currents may be set up in the lower layer as a result of differential heating by radioactivity.

(Holmes, 1928a: 238)

Bull made no reference to continental drift, although later he did (Bull, 1927).[8] Nonetheless, it is easy to understand why in 1925 Holmes would have found such a hypothesis attractive; Bull postulated radiogenic heat to fuel his tectonic mechanism, claiming that heat loss is discontinuous. Holmes, as I shall soon explain, agreed.

5.4.1 Holmes weighs the advantages and disadvantages of continental drift

Holmes appreciated mobilism's explanatory range. In his first pro-drift paper (1928a: 237), he characterized as a "straightforward and consistent reading of the rocks" drift's ability to account for the congruence of continental margins, geological disjuncts, the origin of mountains, and the origin and the distribution of Permo-Carboniferous glacial strata.

In his two reviews of the AAPG symposium volume, Holmes (1928b, 1929) noted that some authors had raised difficulties, which he attempted to mitigate by citing replies by van der Gracht and Wegener, by appealing to the work of du Toit, Argand, or others not at the symposium. He was at great pains to set the record straight. He also proposed his own solutions to these difficulties. The following illustrate his approach.

Noting the theoretical difficulty that Schuchert, Longwell, and White had raised about the implausibility of Pangea remaining intact until the Mesozoic, he appealed to van der Gracht:

> There is, of course, neither proof nor probability that there was ever a single "Pangea," and it is reasonably suggested by van der Gracht that there may have been a pre-Carboniferous "Atlantic" which was closed up by the Caledonian diastrophism. He is careful, however, to commend Wegener for not leading us into a discussion of remote periods, regarding the palaeogeography of which our evidence is still lamentably meagre.
>
> *(Holmes, 1928b: 431; also see 1929: 346)*

By citing van der Gracht and Argand, Holmes answered difficulties raised against Wegener's assertion of the congruence of continental margins.

> Several authors are concerned to prove that the opposing shore lines of the Atlantic do not fit so closely as Wegener supposes. Van der Gracht rightly lays no stress on the validity of geographical pattern as an argument, for surely if drift has occurred it is mechanically impossible that the sial blocks should have moved without internal and peripheral distortion. Argand's conception of varying plasticity is a valuable corrective to the exactly fitting coast lines of Wegener's too dogmatic maps.
>
> *(1928b: 431; also see Holmes, 1929: 206–207)*

After listing criticisms of Wegener's intercontinental comparisons, Holmes cited recent studies, based on more reliable data than Wegener had, and which now supported him. For example, he referred to Bailey's 1928 Presidential Address to Section C of the BAAS (§8.12) which countered Schuchert's claim that the Hercynian orogenic belt of Europe did not mesh with the Appalachians in North America.[9]

> Schuchert presents a valuable summary of the geological similarities and differences between the opposing Atlantic lands. He admits that Wegener is correct in connecting the Caledonian trends of Britain with those of Newfoundland, but denies that the Hercynian trends of Europe connect with the Appalachians. Against this we may refer to Mr. E. B. Bailey's statement of the comparison in NATURE of Nov. 5, 1927. Mr. Bailey is by no means one of Wegener's sponsors, yet he says, "It is as if the Atlantic did not exist or, in other words, as if Wegener, after all, were a true prophet."
>
> *(Holmes, 1928b: 431; also see 1929: 207–208)*

In reviewing the geological similarities between Africa and South America, Holmes let du Toit (§6.6) serve "as our guide" (1929: 208).

Reference should be made to du Toit's book [*A Geological Comparison of South America with South Africa*, sponsored by the Carnegie Institution of Washington, and published in 1927] for a presentation of a most remarkable series of parallels (stratigraphical, palaeontological, tectonic, magmatic, and climatic) in the geological history of the opposing lands of the South Atlantic, the whole assemblage of data pointing persuasively – unless Nature be a misleading witness – to the probability of a formerly closer union.

(Holmes, 1929: 208; my bracketed addition. Also see Holmes, 1928b: 432)

Holmes noted occurrences of the rare mineral thorianite as evidence for mobilism (RS1).

Among radioactive minerals possibly the most significant coincidence of occurrence is displayed by thorianite. The name immediately suggests Ceylon, and according to Wegener's reconstruction of Gondwanaland, Ceylon formerly lay to the east of Madagascar. Thorianite, as it happens, is known to occur only in Ceylon and in one other place. That place is Madagascar!

(Holmes, 1929: 286–287)

He raised difficulties with landbridge solutions (RS2). Focusing on Gregory's endorsement of the landbridge hypothesis (§3.4), he noted its inconsistency with isostasy.

In Professor J. W. Gregory's contribution to the Symposium he does not positively object to the drift hypothesis, but he maintains his long-held opinion that the main cause of the present distribution of land and sea is to be found in uplifts and subsidences due to the shrinking of the earth, the latter process being only in part a consequence of cooling. This is, of course, the view of the older orthodox geology, and it should be clearly realized that those who hold it must be prepared to face geophysical difficulties just as serious as any with which the advocates of continental drift can be confronted.

(Holmes, 1929: 287)

Holmes also criticized Brooks' fixist solution to the origin of the Permo-Carboniferous glaciation (§3.12) (RS2).

[Brooks] adopted as a working hypothesis "a great plateau in the interior of Gondwanaland, rising gradually to an elevation of 10 000 feet." This assumption of great height certainly eases the meteorological problem, but it has no geological justification. Moreover, it doubles the difficulty of the physical problem, for now we should have to explain, first a great thickening of the sial of Gondwanaland and then its total disappearance from the very extensive oceanic areas that now intervene between the existing southern continents.

(Holmes, 1929: 340; my bracketed addition)

He (1929: 341) echoed the consensus of other mobilists critical of the proposed presence of glaciation in North America during the Permo-Carboniferous (§3.13).

Finally, Holmes separated himself from Wegener. Sensible as ever, he wanted to shift the focus of the debate from proving Wegener wrong to assessing the likelihood of drift, to sifting the evidence for it. He agreed with critics that Wegener often used doubtful data and methods.

There is a general agreement that Wegener's methods ... are to be condemned. His plausible selection of data, frequently erroneous age determinations, faulty analysis of causes and devious reasoning have undoubtedly had the effect of weakening his case. There is, indeed, a distinct danger that the easy disproof of large sections of the Wegener hypotheses may be mistaken for a demonstration of the impossibility of continental drift as a geological process. The important issue is now not so much to prove Wegener wrong as to decide whether or not continental drift has occurred, and if so, how and when.

(Holmes, 1928b: 431)[10]

After this optimistic assessment of mobilism's ability to solve many varied problems, he noted a major stumbling block, the lack of an acceptable cause. Holmes first noted that mechanism had always been a major stumbling block for mobilism.

There can be no doubt that the reluctance on the part of many geologists to accept the straightforward testimony of the rocks in favor of continental drift is due to the fact that no gravitational or other force adequate to move continental blocks in the required directions has been recognized.

(Holmes, 1929: 344–345; see also 1928a: 237)

He reviewed and agreed with Jeffreys' attack on Wegener's mechanism (RS2). He also criticized Daly (RS2).

Daly has suggested for discussion a hypothesis of continental creep due to the sliding of sial blocks on a lubricating zone of glassy basalt [in his *Our Mobile Earth*]. A bulging of the polar and equatorial regions with a depression between, towards which the continents migrate, is pre-supposed. No explanation is offered for the initiation of so unstable a deformation of the globe.

(Holmes, 1931a: 562; my bracketed addition)

Holmes, formerly a supporter of Joly's theory of thermal cycles, now criticized it (RS2).

This particular mechanism advocated by Joly for bringing about alternative accumulation and discharge of heat appears to be unsatisfactory by itself because (a) it would lead to abnormal heating of ocean water (this follows if there is no light layer over the Pacific floor); (b) it does not explain the great amplitude of movement implied in Alpine structures; (c) it is inconsistent with the simultaneous occurrence of tension in the Urals with compression of the Caledonian mountains; (d) it does not account for the apparent drift of continental masses radially outwards towards the Pacific and the Tethys in both the northern and the southern hemi-sphere; (e) it has not yet been stated in terms satisfactory to mathematical physicists.

(Holmes, 1928a: 238; also see 1931a: 564–565)

Point (a) had been raised by MacCarthy in 1926, but Holmes (1926: 309) dismissed it, arguing that the temperature of the oceans would not be raised appreciably because heat from the rising magma would be absorbed by "a light surface layer which can float in the magma." However, Holmes had now come to believe that the Pacific floor was devoid of an upper sialic layer. He also criticized van der Gracht's version of Joly's theory, claiming that it could not provide continental displacements in directions required by geological evidence. Behind objection (e) was Harold Jeffreys, who was not happy with Joly's theory; in stating it Holmes wanted to make plain that his theory, unlike Joly's and those of other mobilists, could be stated in terms satisfactory to Jeffreys. Jeffreys was a friend and fellow Northumbrian with whom he differed amicably, but who also had to be humored because he was *the* preeminent mathematical geophysicist. Closing his review of the status of Wegener's theory, Holmes made this comment:

It is thought that all these difficulties may be avoided by a development of the conception of large-scale convection currents in stiff material from which more mobile portions are squeezed out as magmas.

(Holmes, 1928a: 238)

5.4.2 Holmes sets out his mobilism theory

Holmes began by setting out a model of Earth's crust and substratum based on a review of the available petrological, seismological, geochemical, and thermal data. Dividing the solid Earth into lower, intermediate, and upper layers, he distinguished crust from substratum on thermal grounds.

The term *crust* is used for the upper and intermediate layers and any part of the lower layer that is crystalline, while the term *substratum* is adopted for the underlying thermally "fluid" or glassy part of the lower layer.

(Holmes, 1931a: 569)

Holmes' substratum comprised most of what we now call "mantle." He characterized the upper layer as sialic – granite, granodiorite, and diorite – and, following Jeffreys, estimated its thickness to range from 10 to 12 km or slightly more. He (1931a: 558) identified the intermediate layer as metamorphic rock – amphibolite passing downward into granulite – and maintained that both could transform into very dense eclogite if subjected to sufficiently high pressure. He noted that Jeffreys and others estimated the crust's thickness to be 20 to 25 km. He identified the lower layer as either eclogite or some form of peridotite. He (1931a: 569) thought it "probable that the top of the lower layer may be crystalline," and noted that seismological evidence does not prove that the substratum is crystalline, but demonstrates only that, in the short term, it is highly rigid. Over the long term it is viscous. Thus, according to Holmes, the substratum, except perhaps near its top, is fluid or glassy, and lacking in strength. Using the thermal definition of crust and substratum, Holmes estimated that the crust (essentially what is now called the lithosphere) was approximately 60 km thick. The substratum extended downwards to the metallic core, which begins 2900 km from the Earth's center. Holmes (1931a: 570) identified two layers of oceanic crust. Judging the overall thickness of oceanic crust to be the same as continental crust, he thought they shared a common bottom layer. He tentatively identified the top layer of oceanic crust as gabbro passing down into amphibolite. He maintained that the Atlantic crust was covered with "patches of sial lying above the 'gabbro-amphibolite' layer," the Indian crust had a few such patches, the Pacific none.

Turning to more familiar ground, Holmes argued that the substratum is a viscous fluid because its radioactive content keeps it hot. A heat-generating substratum was fundamental for Holmes, and in support he referred to his own previous work, and raised a further difficulty with Jeffreys' opposition to it. Both Holmes and Jeffreys believed that convection would proceed in the substratum until radioactive material had been completely transferred to the crust. Afterwards, the temperature gradient would become too small to support convection, and it would crystallize. But they strongly disagreed about the time of complete removal. Jeffreys thought that it had happened early in Earth's history, Holmes that removal was still incomplete. Holmes contended that Jeffreys' claim that removal was now complete (RS2) clashed with what was known about magmatic differentiation and the radioactive content of peridotites and basalts of different ages – if Jeffreys were correct, peridotites of great age should be more radioactive than young peridotite.

All the experience of geochemistry is strongly opposed to such a hypothetical possibility [i.e., complete removal of all radioactive material from the substratum]. No process of magmatic differentiation is known that could bring it about. It is likely that the material of the upper layer, together with the oceans and the atmosphere, was differentiated from the general body of the earth while the latter was still a relatively mobile fluid. The process of "gaseous transfer" so implied would be very effective in leading to a marked concentration of radio-elements in the first-formed crust. But the known radioactivity of basalts and peridotites of widely different ages shows that the suggested process was far from removing all the radio-elements from the substratum. Yet clearly, if we admit only a slight radioactivity in the substratum there must be generated within it an excess of heat which would maintain fluidity.

(Holmes, 1931a: 573–574; my bracketed addition)

I suspect that Holmes relished presenting these criticisms, letting Jeffreys know that he was not well versed in "all the experience of geochemistry" and much of geology.

Holmes now had all he needed for his convection hypothesis, namely, radioactive material within the substratum, warming it and making it behave as a liquid over the long term, and a means of removing excess heat by processes such as igneous activity, continental drift, or Joly's thermal cycles. Having rejected the third, he argued for drift and the discharge of heat by igneous activity.

For the accumulated heat of 200 million years to escape through . . . the oceans it would be necessary for one third of the whole of the ocean floors (taken at 60 km. thick) to be engulfed and heated up to 1000 °C and replaced by magma which cooled down to form new ocean floors at 300 °C. A process competent to bring about this result on the scale indicated would be some form of continental drift involving the sinking of old ocean floors in front of the advancing continents and the formation of new ocean floors behind them. We may therefore conclude that (a) if the crust of the Earth makes good the loss by surface, and (b) if the substratum has only 1/700 of the heat-generating capacity of plateau basalt, then (c) the substratum cannot yet have cooled sufficiently to have crystallized, but must still be in the stage of convective circulation, and (d) to avoid permanent heating-up, some process such as continental drift is necessary to make possible the discharge of heat.

(Holmes, 1931a: 574)

Showing his familiarity with geophysics, the geologist Holmes reviewed the estimates of the substratum's temperature, viscosity, and strength to see if they fell within ranges expected for convection. Estimating the critical temperature gradient required for convection to be about 3 °C per km, he argued that it could be easily exceeded by radiogenic heat. Noting that too high a viscosity would prevent convection, he cited Jeffreys' 1927 opinion "that the viscosity of the substratum cannot by itself prevent convection" (Holmes, 1931a: 575). Turning to the third question, Holmes claimed that the substratum was fluid and lacked strength as indicated by the tendency of continents to be in isostatic equilibrium. However, Jeffreys believed that the latter only put a limit on the strength of the substratum; it was not devoid of strength. Holmes restated Jeffreys' criticism.

Jeffreys, however, infers from the fact that the equator of the geoid is not exactly circular that deep-seated stress-differences must be resisted by the substratum, and that the latter must therefore have at least a little strength.

(Holmes, 1931a: 576)

He then offered an alternative (RS1), turning Jeffreys' argument on its head.

The bulge of the equator corresponds to an excess of material about the longitude of Central Africa. Such an upward bulge of Africa is undoubtedly genuine, but it has developed since the Cretaceous from a state of almost perfect peneplanation. Thus Central Africa has been affected – and still is – by a process which has actively uplifted the region faster than isostasy could restore equilibrium . . . So far from proving the existence of deep-seated strength, the development of the bulge proves that some dynamic process is operating in the substratum sufficiently actively to maintain a slight departure in advance of complete compensation.

(Holmes, 1931a: 576)

According to Holmes, it is convection currents in the substratum that cause drift and allow for the discharge of excess radiogenic heat commonly expressed as igneous activity. Holmes illustrated the general behavior of sub-continental convection currents with the two diagrams shown in Figure 5.1.

Holmes argued that forceful convection currents arise beneath continents because the more radioactive continental rock is hotter than oceanic rock. These ascending currents diverge, move toward the continental edges, and "produce a stretched region," a region under tension, which eventually becomes new ocean floor. Much of the rising heat escapes through the new ocean floor.

When the ascending currents turn over, the opposing shears and resulting flowage in the crust would produce a stretched region, or a disruptive basin which would subside between the main blocks. If the latter could be carried apart on the backs of the currents, the intervening geosyncline would develop into a new oceanic region. The formation of a new ocean would involve the discharge of a great deal of excess heat.

(Holmes, 1931a: 579)

Turning to what happens just beyond the leading edge of drifting continents where the convection currents moving below continents meet weaker sub-oceanic convection currents and both descend, he envisioned:

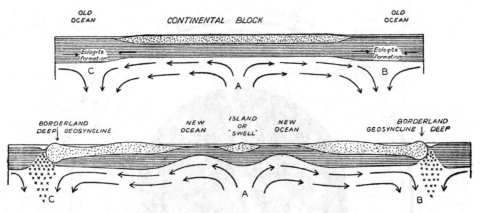

Figure 5.1 Holmes' Figures 2 and 3 (1931a: 579). Sub-continental currents are generated within the substratum. They ascend at A beneath the middle of a continental block, where they separate, and move along the horizontal towards the periphery of the continent. They descend at B and C, where they meet weaker convection currents arising under oceans. The upper granitic layer is dotted; the intermediate layer (primarily amphibolite) is line-shaded; descending eclogite is represented by the x's at B and C; the deeper substratum is unshaded.

The crust above the zone of contact will be thrown into powerful compression and the amphibolite layer will tend to be thickened by accumulation of material flowing in from two directions. The observed effects of dynamic metamorphism at high temperature and differential pressure on such material lead us to expect that recrystallisation into the high-pressure facies, *eclogite*, will here take place on a large scale.

(Holmes, 1931a: 580)

Because eclogite is denser than amphibolite, it would sink as it cooled.

In order to elevate his idea of mantle convection into a mobilist theory, Holmes needed a mechanism; he needed to show that his convection currents could move the continents in the direction required by drift reconstructions. Wegener's forces were too weak. Daly had no way to elevate the continents high enough to make them slide. Joly's theory, even if it allowed for a sufficient decrease in the rigidity of the ocean floors, had not been stated in a way acceptable to mathematical physicists. To overcome these difficulties (R1), Holmes boldly claimed that convection currents, fueled by radiogenic heat, could move continents horizontally.

But even if there is a force sufficient to displace continents horizontally, they would be unable to plow through the ocean floor because oceanic rock is more rigid than continental. Holmes countered this by having continents over-ride ocean floor, which as it sinks is transformed into the denser eclogite, weighing it down.

Other possible consequences of eclogite formation will be suggested later; meanwhile we may notice that it provides a mechanism for "engineering" continental drift … Each part of the continental block would be enabled to move forward, partly by the fracturing and foundering of the belt of ocean floor weighed down with eclogite immediately in front, and partly by over-riding the ocean floor along thrust planes lubricated by magmatic injections from below.

(Holmes, 1931a: 580)

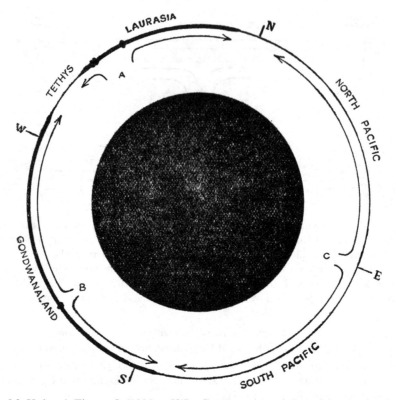

Figure 5.2 Holmes' Figure 5 (1931a: 586). Currents A and B, arising under Laurasia and Gondwana, were responsible for their breakup. Current C, which followed the planetary circulation, arose in the Pacific and descended at the approach of the other current systems.

Continents over-ride ocean floor along thrust planes lubricated by magmatic injections, and ocean floor makes way by falling further into the substratum pulled down by higher density eclogite. Cooling and densification speeds the descent of ocean floor whereas lighter continental rock floats at the surface.

    In an attempt to show that his convection currents could move continents in the manner required by drift reconstructions, Holmes (1931a: 577) suggested (erroneously as it turned out) that their general circulation is "like that of the planetary systems of winds." Currents rise at the equator and sink at the poles. The second major factor controlling the direction of flow concerned the supposition that there would be strong sub-continental currents because there is more radiogenic heating within continental crust; it was these that Holmes considered responsible for the breakup of Laurasia and Gondwana. Holmes, like du Toit (§6.5), preferred a separate Laurasia and Gondwana, not a single Pangea, like Wegener. In all, he proposed three major convection systems (Figure 5.2).

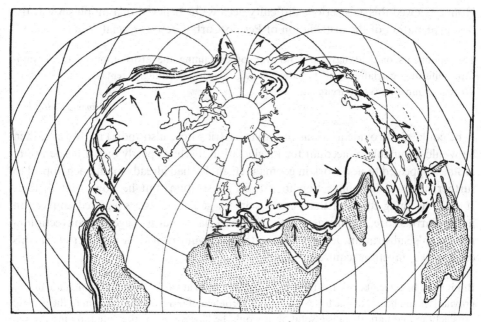

Scale: 1 : 170 × 10⁶.

Figure 5.3 Holmes' Figure 6 (1931: 589). Laurasia fragments and North America drifts westward relative to Europe. Mountain belts form on the leading edge of North America, and island arcs form along the Pacific edge of Asia as it moves toward the Pacific. The southward movement of Eurasia combined with the northern movement of Africa and India led to the shrinking of the ancient Tethys into the modern Mediterranean and the formation of the Alps and Himalayas. (Figure 7 illustrates the breakup of Gondwana with the separation of South America, Africa, and Australia.)

Holmes provided two figures (Figure 5.3) illustrating the continental movements brought about by his proposed convective systems, movements that qualitatively were in keeping with drift reconstructions.

Although Holmes thought that these convection systems explained most of the displacements required by the mobilist theory, he recognized that they did not offer an explanation of certain northward movements; namely Permo-Carboniferous glaciation necessitated a northerly movement of Africa and also India because drift theory required that southern Africa and India had then been situated near the South Pole; also the paleoclimatological data indicated that Britain had been tropical at that time, and had since moved northwards.

There still remains, however, one very serious difficulty ... I indicated that in addition to the radially outward movements of Laurasia and Gondwanaland there appeared to have been a general drift of the whole crust over the interior with a marked northerly component on the African side. Convection currents may explain the former; they cannot, unfortunately, have much bearing on the latter.

(Holmes, 1929: 347)

Holmes viewed this as a serious difficulty because he believed that mobilism was the best explanation of the distribution of Permo-Carboniferous glacial strata.

> The opponents of drift have also no way of explaining the distribution of the Permo-Carboniferous glaciations of Gondwanaland, which accordingly remains the basis of Wegener's most powerful argument.
>
> *(Holmes, 1928b: 433)*

If he could not provide a means of moving Africa (and also India) northward, then he would be unable to account for its glaciation. Perhaps partly out of desperation, Holmes appealed to new work in geomagnetism. He had already made such appeal to further his case for convection in a fluid substratum, but he now went farther. It was one thing to claim that recent work in geomagnetism suggested convection in the Earth's core, but it was more tenuous to claim that it suggested convection in the Earth's substratum, let alone hope that it would provide a rotation of Earth's crust in the direction required.

> But it must be remembered that there are other processes at work besides those that are due to gravitation and heat. No one has yet solved the problem of terrestrial magnetism to the general satisfaction, and until there is a solution it would be hazardous to speculate too far as to the possibilities of forces that may be set up by the interaction of magnetic and electric fields. Meanwhile, until these are adequately explored – and they are undoubtedly of the kind called for to solve this final riddle – no one can say that the crust may not be able to move relative to the poles.
>
> *(Holmes, 1929: 347)*

Two years later, Holmes announced his own solution. He proposed that the whole crust moved northward, which he said "goes far to justify ... tentative acceptance" of his mobilism.[11]

> But this [that South Africa lay near the South Pole] being granted, we must envisage not only a radially outward drift of each great land mass towards the Pacific and the Tethys, but also a general drift, possibly involving the whole of the crust, with a northerly component on the African side sufficient to carry the South African region from the South Polar circle, and Britain from the tropics, to their present positions ... Here, then, we have a severe test of the convection hypothesis. Can it provide a northerly drift for the African hemisphere? In the figure reproduced on the previous page [Holmes' Figure 5] the combined effects of the two types of circulation so far considered are represented for the time at which the movements leading to our present geography began. A quantitative treatment is at present impracticable, but a roughly qualitative assessment is possible on the assumption that the sub-continental circulation becomes stronger than the planetary circulation. In Gondwanaland the chief ascending current would most probably develop near the Cape Mountains and their continuations. Thus the planetary circulation would be completely reversed beneath Gondwanaland. In Laurasia the chief ascending currents ... and the planetary circulation would thereby be strengthened in the north but reversed in the south. The resultant forces ... would clearly be in

the direction of rotating the whole crust towards the north on the land hemisphere and towards the south on the Pacific side. The success of the hypothesis in providing a possible mechanism for this long-puzzling movement goes far to justify its tentative acceptance.

(Holmes, 1931a: 585; my bracketed additions)

This enabled Holmes to call attention to what he believed was the chief advantage of his theory, its plausible mechanism (RS3).

5.4.3 Holmes explains mountain belts

Holmes also argued that his theory of convection currents removed the incompatibility that Jeffreys, Bowie, Longwell, and Willis had raised between Wegener's mechanisms for drift and the origin of mountains (RS1).

It has been frequently pointed out that, if the material of the ocean floor is weaker than that of the advancing continent, then mountain building will not take place; while if it is stronger, continental advance becomes impossible. With the mechanism here suggested this *impasse* does not arise. Mountain building occurs as a result of rock-flowage set in operation by the underlying current; it will generally occur provided that the horizontal component of flowage is greater from behind than in front.

(Holmes, 1931a: 581)

To explain how Holmes avoided the difficulty, his solution to the origin of mountains is outlined. Like Daly's, Holmes' theory is intertwined with his drift mechanism. However, Holmes separated mountain formation from the advance of continents. Continental blocks move forward because of fracturing and foundering of the ocean floor weighed down by eclogite ("sinkers" they are now sometimes called) immediately in front of them. Oceanic deeps form with the downward flexure of the ocean floor before advancing continents.

The hypothesis of eclogite formation is supported by the fact that it provides a reasonable explanation of oceanic deeps. It is consistent with their depths, as judged from the requirements of isostasy, and with their occurrence in front of the active orogenic arcs bordering the Pacific on the Asiatic and Australasiatic side.

(Holmes, 1931a: 580)

Mountain belts and island arcs form at advancing continental margins. Holmes used his Figure 3 (my Figure 5.1) to illustrate how the advancing sialic layer initially becomes thickened while over-riding the sinking oceanic floor.

The upper or sialic layer of the continental margin will also be thickened by the differential flowage of its levels towards the obstructing ocean floor. Here thickening of the crust and mountain building will occur, and the mountain roots, unable to sink, will begin to fuse and give rise to igneous activity of the Circum-Pacific type with basalt-andesite-rhyolite volcanoes.

(Holmes, 1931a: 581)

This thickening of the sialic layer at its leading edge gives rise to island arcs and geosynclines, which form behind the island arcs. The geosynclines, which Holmes

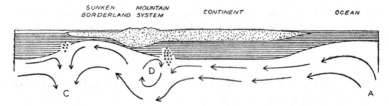

Figure 5.4 Holmes' Figure 4 (1931a: 582). Mountain system is established, and borderland sinks. Opposing currents (D) form beneath mountain roots. The upper granitic layer is dotted; the intermediate layer (primarily amphibolite) is line-shaded; descending eclogite is represented by x's where currents converge and descend.

(1931a: 582) referred to as "*interior* geosynclines," fill with sediment, and continue to subside and folding begins.

By the accumulation of sediments from the continent behind and the borderland or mountain arc in front, further subsidence of the geosyncline will occur. Thus the base of the crust will be kept hot, and may even grow hotter, and the process of geosyncline development will continue. It should be noted that this region, though tending to subside, will nevertheless be under compression whenever rock-flowage in the underlying sial is more rapid than that in front. Thus folding may proceed concomitantly with subsidence and accumulation of sediments.

(Holmes: 1931a: 582)

Mountain formation ensues. Referring to his Figure 4 (reproduced as Figure 5.4), Holmes claimed:

Sooner or later a time will come when the buried crustal material beneath the infilled geosyncline will become hotter and weaker than the material behind and in front. Moreover, the grip of the underlying current will have weakened as a result of higher temperature and lowered viscosity. At this stage thickening of the geosynclinal belt due to compression will become more effective than thinning by magmatic corrosion; mountain building will have set in (Fig. 4).

(Holmes, 1931a: 582)

Meanwhile the sialic mountain arc is stretched and thinned into a sunken borderland by convecting currents accelerated by the heat produced at the infilled and compressed geosyncline (see Figure 5.4).

He viewed the formation of island arcs, such as those along the east coast of Asia, as an early stage in the genesis of mature mountain belts such as the Andes and the Cordillera of western North America.

The long history of the Cordilleran geosyncline of Western America, and its subsequent orogenesis, supplies a spectacular example of the process theoretically deduced. The Mesozoic and Tertiary history of Alaska tells the same story of long-continued subsidence interrupted and ultimately overcome by the effects of compression ... The hypothesis suggests that western America represents a more mature stage of development than eastern and south-eastern Asia.

(Holmes, 1931a: 582–583)

This was Holmes' solution to the apparent incompatibility between mountain building and continental drift. He rejected the idea of sial crumbling against sima. Instead, sialic geosynclines, filled with sediments and weakened by rising temperature, are crushed between unweakened sialic crust. According to Holmes, mountain belts do not form at the leading edge of a drifting continent as they drive their way through the simatic ocean floor, instead they rise out of interior geosynclines located between the leading edge and main part of the drifting continent.

Thus Holmes clearly avoided the difficulties that he and others had raised against previous theories, and provided, through thermal substratum convection, the means to move continents driven by radioactivity roughly in the manner required by drift reconstructions. Unlike Joly's theory of thermal cycles, it did not lead to abnormal heating of the Pacific Ocean, it was presented in a manner that was not immediately vulnerable to criticisms of mathematical physicists, and it allowed for synchronous tension and compression. Unlike Jeffreys' thermal contraction, it offered a geologically realistic solution to the origin of mountains, geosynclines, and plateau basalts.

Holmes took a characteristically modest view of his hypothesis, recognizing its speculative nature.

It is not to be expected that the first presentation of a far reaching hypothesis and its manifold applications can be wholly free from errors. The workings of the inner earth reveal themselves only indirectly, and their actual nature in space and time is certain to be far more complex than can be visualized.

(Holmes, 1931a: 600)

Certain issues remained to be investigated, and he hoped that they would be because his hypothesis was, he thought, "a working hypothesis of unusual promise."

So far the treatment has been almost entirely qualitative and therefore it inevitably stands in need of criticism and quantitative revision. The hydrodynamics of the substratum and its behavior as a heat engine need to be attacked on sound physical lines. The capacity of substratum currents to promote magmatic corrosion, transport and crystallization, and to produce migrating sub-crustal wave forms, calls for detailed treatment. The full bearings of the hypothesis on petrogenesis have yet to be investigated. Meanwhile its general geological success seems to justify its tentative adoption as a working hypothesis of unusual promise.

(Holmes, 1931a: 600)

5.5 Reception of Holmes' hypothesis of substratum convection

The reception by mobilists of Holmes' solution to the mechanism problem was mixed. Du Toit welcomed it, incorporating it as a component in his synthesis in *Our Wandering Continents* (1937). Daly was impressed with Holmes as a scientist, but he rejected his hypothesis and, insofar as he remained a mobilist, continued to support his own continental down-sliding. The Dutch geologist B. G. Escher adopted Holmes' hypothesis to explain several features of the Dutch East Indies (§8.13). Adamant fixists ignored Holmes' theory, except for Jeffreys who, throughout the 1930s, debated with Holmes the relative merits of their ideas. Jeffreys found Holmes' idea of substratum convection the best mobilists had to offer. He went so far as to encourage Hales and Pekeris, two mathematical geophysicists, to investigate the possibilities of such convection. Holmes' theory was also discussed by Vening Meinesz and Griggs; both were attracted to the idea of convection currents quite independently of whether it could serve as a mechanism for moving continents. Vening Meinesz, the Dutch geophysicist best known for his pioneering work on mapping gravity anomalies at sea, also postulated convection within the substratum but with a different pattern; he rejected continental drift, but supported the idea of polar wandering. Griggs, while a member of Harvard's Society of Fellows and the Lowell Institute, developed a convection theory of mountain building, but he did not support continental drift.

I shall begin by reviewing a symposium held at the BAAS 1931 annual meeting where Holmes' substratum convection was first publicly discussed. Both Holmes and Jeffreys were there, and they and several others discussed it. I shall trace Holmes' (1933) and Jeffreys' (1935) continuing debate over the merits of Holmes' convection hypothesis, and then turn to the work of the aforementioned geophysicists, examine the reactions of Daly and du Toit during the 1930s, and discuss Holmes' mobilism as presented in the first edition (1944) of his *Principles*. Finally, I shall describe how Holmes' convection fared with several other participants in the mobilism debate, and assess the status of convection theories twenty years later, just prior to the revival of mobilist ideas in the 1950s and 1960s.

The geography section of the British Association sponsored a session on problems of Earth's crust. There were fifteen speakers. Six argued against continental drift, including Jeffreys, Hicks, secretary of the Geographical Society, and Gregory, a confirmed landbridger (§3.4). Five supported drift, including Holmes, Poole who had worked with Joly, and G. C. Simpson, the British meteorologist (§3.13). Davies, a British geomorphologist, and another speaker remained neutral, while the Norwegian Werenskiold and the final participant made no general comment about drift. Six speakers commented specifically about Holmes' substratum convection hypothesis: Holmes, Jeffreys, Hinks, Poole, Davies, and Werenskiold.

Here are the positive comments. Poole (1931: 444) noted that Holmes had made a convincing case for the convection current theory, but added that mobilist theories, including his own amended version of Joly's theory, had advantages and disadvantages. Holmes himself reviewed and defended his theory, and presented a brief account of his critique of contractionism. He also expanded his solution to the origin of ocean trenches by noting their association with the negative gravity anomalies recently discovered by Vening Meinesz (RS1). Holmes (1931a) already had argued that ocean deeps form where converging convection currents descend. As amphibolite metamorphosed into the denser eclogite in descending material, the region would sink forming oceanic trenches; however, if a region has a negative gravity anomaly, then its density is less than would be expected, and it should rise and attain isostatic equilibrium. But the seafloor of oceanic trenches is sinking, not rising, and some active process, identified by Holmes as converging downward directed convection currents, must be forcing it down.[12]

Hinks, Werenskiold, and Jeffreys were less complimentary (RS2). Hinks questioned whether convection currents could hold down trenches against isostatic forces. Drawing an analogy with the annealing of optical glass, he argued:

To explain in particular Dr. Vening Meinesz's long strip of negative anomalies in the Netherlands Indies Holmes invokes two opposing subcrustal currents which approach the strip and turn down beneath it, compressing and exerting a downward drag. This idea of currents that can exert a drag upon the crust above is evidently going to be used extensively, and it seems important to visualize the process. Such currents must have at most an extremely small velocity: may one compare them to the currents that redistribute the material in a block of optical glass in the process of annealing? The block is kept for weeks at a temperature much below melting-point, during which time inequalities of density are averaged out. Currents of a sort must be flowing, but is it possible that they can produce any drag upon the supports? Currents are generally effective in proportion to some rather high power of the velocity. Is it conceivable that there can be subcrustal currents so intense, so uniformly directed, and so continuous in action, that they could produce the drag required to maintain Meinesz's long belt of negative anomalies against the forces trying to compensate them?

(Hinks, 1931: 536)

Although Werenskiold (1931) agreed that Holmes' convection currents would produce sufficient crustal drag to cause continental drift, he thought it highly unlikely that they occur at all. Werenskiold thought that seismological data indicated density differences within the substratum, which implied that it consisted of several concentric layers of different rock-types. He therefore questioned Holmes' assumption that the substratum is sufficiently homogeneous to allow convection to extend from its base to its top; convection might exist, but would be too small-scale to shift continents.

Jeffreys thought Holmes' theory a decided improvement over previous mobilist theories.

I have examined Professor Holmes's theory of subcrustal currents to some extent, and have not found any test that appears decisive for or against. So far as I can see there is nothing inherently impossible in it, but the association of conditions that would be required to make it work would be rather in the nature of a fluke.

(Jeffreys, 1931: 453)

He raised three difficulties (RS2). The first two led him to believe that Holmes' characterization of the substratum as rigid, yet highly viscous, and lacking strength was probably incorrect. He contended that there were, at best, insufficient data to determine whether such material could exist under the high pressures found within the substratum.

He [Holmes] assumes that radioactivity is uniformly distributed in the rocky shell, from about 40 km. to 3000 km. down, and that the state of the material is neither a true liquid nor an ordinary solid, but what I have called a liquevitreous solid, possessing rigidity, but a finite viscosity and no permanent strength. Whether this state is possible at high pressures in the actual material is not known; I think the evidence is against it.

(Jeffreys, 1931: 453; my bracketed addition)

Jeffreys wrote with caution. He did not agree with Holmes, but he was not a person to be trifled with. According to Jeffreys' theory of thermal contraction, the substratum had solidified very early in Earth's history, and, unlike Holmes, he believed that hardly any radioactive material remained there. Jeffreys could not understand why Holmes thought that radioactive material remained in the substratum because his own study of Finland granites indicated an early complete separation.

I cannot understand Holmes's objection to the idea of strong upward concentration of radioactive matter. He himself inferred from geological evidence that in the Finland granites the radioactive constituents moved upwards when the granite was fused. If the Earth was once fluid there could be no solidification so long as the radioactivity was uniformly distributed; so long as it remained fluid the radioactive matter would rise until solidification became possible. I cannot see why Holmes will not admit the generality of his own conclusions.

(Jeffreys, 1931: 452)

He, however, appeared to agree with Holmes that if substratum convection really did occur, it would produce enough drag to move continents.

But granting as a working hypothesis that it may be so, and that the supply of heat is greater than could be conducted away at the adiabatic gradient, the state of equilibrium would be unstable, and vertical currents would be generated, rising in some places and descending in others. Matter on reaching the top would travel horizontally to a descending current, and on the way would produce a viscous drag on the bottom of the crust. Holmes appeals to this drag to overcome the strength of the crust and produce displacements of the continents. The currents would be slow, but the viscosity would be high, and the drag might be great.

(Jeffreys, 1931: 453)

Jeffreys' third difficulty was theoretical; large-scale convection was improbable and even if it occurred, it would not remain stable long enough to cause drift.

For the theory to succeed would require the currents to be in the same direction over regions of continental dimensions; that is, the instability developed must correspond to the lowest mode of disturbance and no other. If the supply of heat was greater than could be carried off by the currents in the lowest mode, we should get irregular motions of small extent, resembling those that occur in the boiling of a kettle, and these would be no use for producing widespread geological effects. So Holmes's theory requires that the rate of supply of heat should fall within two limits, which in a deep region like the Earth's rocky portion would probably be very close together; if it was too low there would be no convection currents at all, while if it was too high the currents would be of the wrong kind. I think therefore that while the theory merits further examination, its validity would be a remarkable accident. The positions of the currents would be very easily disturbed, so I should say it was unlikely they could be in the same places for whole geological periods.

(Jeffreys, 1931: 453)

At the meeting, Holmes offered no response to Hinks and Werenskiold, nor did he respond to Jeffreys' contention about the unlikelihood of large convection currents remaining stable long enough to cause drift. He did, however, argue that his Finland study created greater difficulties for Jeffreys' theory than for his own (RS1, RS2, and RS3), indeed his study showed that the transference of radioactive material from the substratum to the crust falls far short of that required by Jeffreys' contraction.

Dr. Jeffreys ... says he cannot understand ... [why I believe that radioactive material has not completely moved from the substratum to the crust] since I have always a certain amount of upward concentration. The evidence for the latter however is also evidence against the upward concentration being complete. If we start with an average earth-magma in which crystallization is proceeding, the radioactive elements will distribute themselves in certain proportions between the liquid and crystal phases. What the contraction hypothesis requires is that practically no radioactive matter should go into the early formed crystals. Let us look at the facts. The first-formed crystals from basaltic magma with $U = 2.2 \times 10^{-6}$ gm./gm. consist of olivine with $U = 0.7$ in the same units ... The peridotites formed by the accumulation of the early crystals from basaltic magma have $U = 1.5$. Similarly for the Finland granites. Refusion has certainly brought about upward concentration, but only to the extent of separating an initial $U = 4.4$ into 6.2 above and 2.4 below. In widely different materials like stony and iron meteorites the uranium contents are 1.7 and 0.07 respectively, the partition-ratio being about 25 to 1. This represents the "cleanest" separation known (apart, of course, from the local crystallization of uranium minerals), but the ratio is far from approaching the 100 000 or 1 000 000 to 1 required by the contraction hypothesis.

(Holmes, 1931b: 541; my bracketed addition)

The separation of radioactive material in the Finland granites is only 2.58 to 1, which falls far short of that required by Jeffreys. The Finland granites present a difficulty not for convection but for contraction theory.

What about his own theory? Holmes admitted that the Finland granites and other evidence, such as the comparison between stony and iron meteorites, were not wholly compatible with his theory; they did not provide enough separation for him, but he argued again that the shortcoming was less for his theory than for Jeffreys'.

It is only fair however to point out that the convection hypothesis has to face what is quantitatively the same difficulty. I have already shown that if the substratum had only 1/700 of the radioactivity of basalt, the heat generated within the substratum would not be able to escape without some form of continental drift. A greater amount of radioactivity would imply an embarrassing amount of internal heat. Roughly the convection hypothesis requires that the substratum should carry only 1/100 to 1/1 000 of the radioactive matter to be expected on the evidence of investigated materials, whereas for the contraction hypothesis even this amount would be a thousand times or so too high.

(Holmes, 1931b: 541)

Holmes ended hopefully.

As a possible way out of this difficulty I would suggest that the outer concentration of the radioactive elements may be a consequence of the separation not of crystal phases from liquid, but of liquid phases from gas.

(Holmes, 1931b: 541)

He admitted that his theory was speculative. However, even if his theory required a greater transference of radioactive material than presently indicated, contractionism would require even more.

My own opinion is that the distribution of radioactive matter required by the contraction hypothesis is farther removed from these probabilities than that required by the convection hypothesis.

(Holmes, 1931b: 542)

Holmes, invited to the United States by Daly, presented a comparison of his own, Jeffreys', and Joly's theories in a lecture to the Washington Academy of Sciences and the Geological Society of Washington in April 1932. He decided to take on Jeffreys.[13] Holmes attempted to remove Jeffreys' difficulty concerning the improbability of convection currents of sufficient size and stability to cause drift (RS2). But Holmes did not disagree with Jeffreys. The required currents were unlikely, but he turned this into a virtue; after all the very existence of Earth itself is unlikely.

Jeffreys ... can find nothing inherently impossible in the hypothesis, but he thinks that "its validity would be a remarkable accident." I agree; but then I think the earth is no less a remarkable accident. It is impossible to be a geologist without realising that – in the dim light of the knowledge we have so far gained – the earth we live on is a strange and most improbable planet.

(Holmes, 1933: 194)

Holmes considered the geological history of the Earth unlikely in light of what was currently known about the Universe. But it was not as unlikely as the existence of a geologist who did not believe that Earth's existence was unlikely.

Jeffreys next discussed Holmes in *Earthquakes and Mountains*, first published in 1935. Jeffreys regarded this little book as a summary of *The Earth*, but written for a more general audience, by which he meant those with limited mathematical abilities, a group Jeffreys thought included most geologists. Unperturbed by Holmes' implication that he did not know much geology, Jeffreys declared, "My own work has ... become increasingly concerned with the borderland between geophysics and geology" (Jeffreys, 1935: v).

Jeffreys subtly restructured and reinforced his previous attack. He returned to Finland granites, ignoring Holmes' 1933 rebuttal.

Holmes, in some of his recent work, supposes that these convection currents in the lower layer still exist and are maintained by radioactivity. For some reason that I have not succeeded in understanding why he refuses to admit that the tendency of radioactive matter to rise to the top, which he himself showed in the Finland granites, has a general application.

(Jeffreys, 1935: 169–170)

Jeffreys categorized Holmes' mantle convection as of little more than academic interest.

The interest of his theory, in my opinion, is that it shows what kind of assumptions are necessary if we insist on avoiding the view that radioactivity is sufficiently concentrated to the top to permit the earth to cool. For this reason a full quantitative examination of Holmes's theory would be worthwhile; but it seems to be beyond the range of our present knowledge of hydrodynamics, and may require also a much more complete theory of viscosity at high temperatures and pressures than now exists.

(Jeffreys, 1935: 170)

Jeffreys (1935: 170) added that there are "a few obvious difficulties" with Holmes' mobilism. After repeating his objection that the amount of heat required to produce convection currents falls within a very small range, and appealing to geology, he raised a new difficulty with Holmes' theory (RS2), one that he believed contraction-ism avoided (RS3).

Nor can I see anything in the theory [offered by Holmes] to explain the observed intermittence of mountain formation. The stresses it implies act all the time, unlike those involved in the contraction theory, which are relieved at every yield; thus they would apparently either produce mountains all of the time or not at all.

(Jeffreys, 1935: 171; my bracketed addition)

The Geological Society of London and the Royal Astronomical Society (hereafter, RAS) held a joint meeting on continental drift in January 1935 (Anonymous, 1935). Astonishingly, in the published comments of eight speakers (W. B. Wright, F. Dyson, A. Smith Woodward, Dr. J. de Graaff Hunter, O. T. Jones, Harold Jeffreys, James Jeans, and R. Stoneley), Holmes' mobilism was not even mentioned. Except perhaps for Wright, the only speaker who was not FRS, nobody was favorably disposed toward mobilism, no one championed it. Of all the meetings specially convened on continental drift in Britain, this was the least hospitable. Wright discussed modifications and additions by Joly and Argand. Dyson hoped that astronomers could help determine if Wegener's theory is correct by continuing to measure any changes in the relative position of continents, but cautioned that it would take twenty to thirty years before reliable results could be obtained. Woodward claimed that continental drift's success in explaining faunal disjuncts had been greatly exaggerated; he gave counter-examples, and invoked parallel evolution to explain some of the faunal similarities in places now remote from one another. Jones, who would later befriend Runcorn and supervise some of his students while Runcorn was away, was also not favorably inclined toward mobilism; citing Gregory, he claimed there was abundant faunal evidence that the Atlantic Ocean had separated the Americas from Europe and Africa throughout the Lower Paleozoic, but without mention of Argand's Proto-Atlantic, which he could have used to explain it. Stoneley simply noted that Antonio Snider had suggested continental drift many years before Wegener. Jeans, asked if he could suggest a reason for the formation of a single primeval continent such as Wegener's Pangea, suggested two possibilities, but said nothing about its fragmentation. De Graaff Hunter argued that Eötvös force was the strongest force available, but was still too weak to cause continental drift. Jeffreys agreed with de Graaff Hunter about the Eötvös force, and added that if it actually had caused drift, then all the continents would be nestled along the equator. Interestingly, he also claimed that if tidal force could move continents, then Earth would have to behave as a fluid to tidal forces, which would have eliminated oceanic tides. Even Jeffreys did not discuss Holmes' substratum convection. These participants reiterated the easy ways to attack mobilism. They dismissed Wegener's mechanism and ignored Holmes' alternative. It was a remarkably unprogressive meeting.

The last speaker, E. W. MacBride, FRS, an embryologist and zoologist, was not on the schedule, but he could not contain himself. Perhaps wanting to liven up what must have been a rather boring day, he produced this amusing and elegant comment, ridiculing the use of the migrations of modern animals as evidence for continental drift.[14]

Professor E. W. MacBride said that he had not intended to intervene in the discussion – he had come as a visitor to learn. There were, however, certain zoological facts which seemed to him impossible to explain unless land-masses had been displaced from their original positions. These facts related to the migrations of animals. First, in British Columbia there lived a species of plover. This bird every autumn migrated to the Hawaiian islands, flying over 2300 miles of sea without a break, and returned in the spring. The young birds hatched in the summer accomplished this flight. Secondly, at intervals of 10–15 years extraordinary increases in the population of lemmings (forest rats inhabiting northern Scandinavia) took place. The lemmings invaded the pastures and arable lands, pressing westwards. The survivors met their end by plunging into the Atlantic Ocean, swimming westward. Thirdly, similar periodical increases in the population of springboks, a small antelope inhabiting South Africa, took place. In this case also the survivors plunged into the Atlantic Ocean. Lastly, eels inhabiting European rivers when mature passed into the sea and swam 3000 miles to a spot south of Bermuda, where they spawned. The larvae took three years to reach the home rivers.

(Anonymous, 1935: x)

The migratory habits of Atlantic eels had been suggested as support for mobilism (Wegener, 1929/1966: 104).

5.6 Work on convection currents during the 1930s

Several others worked on substratum convection during the 1930s, the more important being Vening Meinesz (1934a, b, c), Pekeris (1935), Hales (1936), and Griggs
(1939); none supported continental drift. The Dutch geophysicist Felix Andries
Vening Meinesz maintained that convection causes the formation of island arcs
and associated oceanic trenches, and can even bring about polar wandering, which
he thought of as the movement of the whole crust relative to the Earth's interior. The
Lithuanian-born Pekeris and the South African Hales took up Jeffreys' suggestion
that they investigate Holmes' idea of large-scale convection. Both acknowledged
Jeffreys' help, offered mathematically advanced models of convection systems, and
showed that convection currents could exert enough drag on the crust to bring about
drift. Although neither claimed that such currents actually exist, they blunted
Jeffreys' charge that Holmes' convection currents were highly unlikely. The US geophysicist Griggs proposed convection cells within the substratum to explain mountain
building, but neither endorsed Holmes' convection theory nor stated that it served as a
mechanism for drift.

Vening Meinesz (1934a, b, c) first argued in favor of convection in 1934, noting
(1934a: 37) that Holmes and another worker had led him "to take up the study of the
convection-hypothesis." He had attended Holmes' April 1932 lecture (§5.5). But as
Allwardt (1990: 64) noted, quoting from a letter by Doris Holmes to Donald
B. McIntyre, he did not favor Holmes' theory.[15] Although Holmes' theory inspired
him to examine substratum convection, Vening Meinesz did not, from the beginning,
favor it; his view (1934a: 37) was "somewhat different from what Holmes" had
suggested.

Both Vening Meinesz and Holmes maintained that currents rise under continents
and sink beneath oceans, but the resemblance otherwise was superficial. Vening
Meinesz made six points. (i) Continents contain more radioactive material than
ocean floors, probably brought about by once rising currents beneath continents
and sinking ones beneath oceans, which probably persist to this day. (ii) Currents
occur in the upper part of the substratum, their lower limit being at 1200 km, the
depth of a seismological discontinuity indicative of a change in density. (iii) Prior to
the formation of continents the pattern of convection was haphazard, afterwards it
adopted the general pattern in (i). (iv) In light of (i) and (iii), he formulated this in
response to Jeffreys' contention that stable large-scale convection currents are highly
unlikely:

This consideration [of how large-scale convection becomes stabilized] meets, in the opinion of
the writer, a difficulty that Jeffreys has brought forward with regard to the hypothesis of
convection-currents. Although Jeffreys does not believe them to be impossible, he thinks that
they are so improbable that they could be called a freak ...

(*Vening Meinesz, 1934a: 40; my bracketed addition*)

(v) He linked convection with his gravity work over island arcs and oceanic trenches, in which he had discovered strong negative gravity anomalies flanked on both sides by positive anomalies. He proposed that horizontal currents create a viscous drag on the crust, causing it to buckle. At trenches, there is a central downward buckle flanked by two less pronounced upward bulges, causing the central negative and lateral positive anomalies respectively. Vening Meinesz used downbuckling to explain the origin of oceanic trenches and island arcs. (vi) He thought that convection currents caused the crust to buckle but not to fracture and cause large-scale relative displacements of fragments. Instead, he maintained that the crustal drag brought about by convection rotates the entire crust relative to Earth's interior and its axis of rotation, which is a form of polar wandering.

With Vening Meinesz's model, the poles may wander, but there is no significant horizontal movement of crustal blocks, no continental drift.

The probability of these considerable tangential forces working on the crust makes it also likely that the convection-currents will bring about shifts of the whole crust with regard to the core of the Earth. This would imply movements of the poles with regard to the crust. So these movements, which many geologists have wanted to admit for explaining climatic changes, that otherwise appear inexplicable are by no means impossible in light of the convection-hypothesis.

(Vening Meinesz, 1934a: 44–45)

Thus Vening Meinesz did not endorse mobilism theory, and his convection-current hypothesis contained little or nothing that was supportive of Holmes.[16]

In 1935 Pekeris took up Jeffreys' suggestion to examine the hydrodynamical implications of Holmes' theory. Pekeris, basing his calculations on the best available estimates for the temperature, viscosity, and strength, claimed that convection could occur in the substratum if it were homogeneous enough. Convection would proceed with a temperature difference of just a few tens of degrees, even if the substratum were as strong as Jeffreys had claimed; the substratum could have strength and still convect (1935: 348). In one of his models, convection currents could extend from the core boundary to the bottom of the crust with maximum velocities of the order of 5 cm/year, and would exert enough drag upon the crust to overcome its estimated strength. Pekeris envisioned that the crust would be pushed upwards under warmer continental regions where convection currents ascend, and pulled downwards under colder oceanic regions where currents descend. He did not claim that convection actually took place, only that it was possible.

While it would appear from the above discussion that convection currents in the shell of the earth are not an impossibility, it is not our purpose here to attempt to establish their existence. We shall in the following *assume* that the shell is a plastic fluid possessing a finite coefficient of viscosity, and shall investigate the nature of the circulation which is caused by a given zonal temperature perturbation in it.

(Pekeris, 1935: 348)

Pekeris did not positively endorse Holmes' hypothesis. However, he began what Jeffreys said needed to be done, namely, to determine if large-scale convection could occur within the substratum, and according to Pekeris, it could. Pekeris even offered Holmes a partial answer to the assertion by Werenskiold (§5.5) that convection currents that did not extend from the bottom to the top of the substratum would not exert sufficient drag on the crust to move it.

Hales, who spent 1931 through 1933 at Cambridge completing the Mathematical Tripos, and "taking a sequence of courses in geophysics, most given by Harold Jeffreys," introduced his study with a reference to Holmes and recalled Jeffreys' suggestion:

it is necessary to investigate whether the theory of convection currents in the outer shell [substratum] is consistent with the fact that gravity anomalies show that there are deep-seated stress differences of the order of 5×10^7 dynes/cm^2 [Jeffreys' estimate of the strength of the crust].

(Hales, 1936: 372; my bracketed additions)

Hales concluded that such convection currents were possible, and made two observations. First, it did not appear that the temperature gradient needed for convection necessarily required radioactive material in the substratum.

Holmes supposed that the currents were maintained by a deep-seated layer of radioactivity. Jeffreys has pointed out that it is inconsistent with the fact, shown by Holmes for the Finland granites, that there is a tendency for radioactive material to rise to the top. It seems doubtful whether the assumption of a deep-seated layer of radioactivity is necessary. If the conductivity in the core is greatly in excess of that in the shell [substratum], then even with the lower temperature gradient the heat brought to the lower surface of the shell by conduction would be more than could be carried away by conduction through the shell. This supply of heat would therefore maintain the convection currents in the shell. The conductivity of the core is probably sufficiently large for this to be possible.

(Hales, 1936: 379; my bracketed addition)

Thus he offered Holmes an answer to Jeffreys' assertion that complete separation of radioactive material from the substratum was needed. Hales also answered Jeffreys' charge about the improbability of large-scale convection remaining stable long enough to cause drift.

Jeffreys has also suggested that since the pattern of the instability is always changing there would be no uniformity of stress over continental areas and for geological periods of time. The time necessary for an appreciable change of pattern will be large compared with the time for a complete cycle which is of the order of 10^{18} years, *i.e.*, is far greater than a geological period of time.

(Hales, 1936: 379)

Neither Holmes nor his supporters, however, utilized Hales' suggestion. However, Hales should not be regarded as a supporter of Holmes, because he, like Pekeris, never argued that convection actually occurred. Hales (1936: 379) showed only that convection currents of sufficient magnitude to overcome the strength of the crust, as determined by Jeffreys from an examination of gravity anomalies, are possible. At this stage of his career and for some time later, Hales was not a mobilist; during the early 1950s he disagreed with the paleobotanist Edna Plumstead, a supporter of drift, when both were at the University of Witwatersrand (§6.3), but became sympathetic toward mobilism during the second half of the 1950s because of the new paleomagnetic evidence which he himself was in part responsible for obtaining from South Africa (II, §5.4). Nonetheless, Hales was greatly influenced by Jeffreys throughout his entire career.[17]

Griggs at Harvard University developed a convection theory to account for the formation of mountains based on laboratory experiments. He first presented his theory at the 1938 GSA Annual Meeting, and published his first paper on it the following year (Griggs, 1939). Building on work on rock-deformation under high pressures by Percy Williams Bridgman (1946 Nobel Laureate in Physics for work on high-pressure physics) and by others at Harvard's Lowell Institute, Griggs discussed convection as a form of solid-state creep under great pressure; importantly he did not require a fluid substratum. Griggs' work, as Allwardt (1990) has shown, helped provide geophysicists with a new conceptual framework for investigating the behavior of rocks under very high pressure that eventually lessened the importance of the question, central to the discussion of Jeffreys and Holmes, about the fluidity of the substratum. Griggs, the first American to present a sophisticated theory of convection currents within the "solid" Earth, noted that his work was based partly on the work of Holmes, Vening Meinesz, Pekeris, and Hales. But Griggs endorsed neither Holmes' substratum convection nor continental drift, and applied his theory of convection currents only to mountain building. He was not a mobilist.

After reviewing the work of his predecessors, and summarizing the difficulties with theories of mountain building such as thermal contraction, Griggs described a laboratory model he had constructed that allowed him to simulate the interaction between thermal convection in the substratum and the crust. In constructing his model, Griggs heeded the advice of Bridgman and King Hubbert. Bridgman had become Hollis Professor of Mathematics and Natural Philosophy at Harvard University, and was immersed in high-pressure physics. King Hubbert (see Doel, 1999), who later became well known for his 1956 prediction that oil production in the United States would peak between 1966 and 1971, was also investigating the behavior of rocks under high pressure. Hubbert (1937) and Bridgman (1922) stressed, as German workers had done earlier,[18] that for laboratory models to simulate geological processes, they must be scaled to reflect accurately the spatial and temporal dimensions and relevant properties of Earth's interior, and Griggs took special care to scale his experiments correctly. He used slowly rotating drums to simulate thermal convection cells, which he filmed.

Central to Griggs' theory was his notion of a convection cycle. A full cycle lasted about 550 million years, and was divided into four phases: slowly accelerating currents (25 million years duration), rapidly accelerating currents (5 to 10 million years), decelerating currents (25 million years), and a long final quiescent phase (500 million years). He also proposed that convection currents extend from the top of the core to the bottom of the crust, although he thought his theory would not be harmed if, following Vening Meinesz and one of Pekeris' models, convection currents originated at a depth of 1200 km, the 20° seismic discontinuity.

Griggs correlated these phases with phases of the mountain-building cycles inferred from geology. During the first phase the crust is compressed and begins to buckle downward, dragged down by convection currents. Downbuckling creates a geosyncline which collects sediments. During the second phase (rapidly accelerating convection) the crust is forced down further into the substratum, sediments of the crust and the geosynclinal region are compressed and folded, and mountain roots are formed. The downbuckled light sialic crust, held down by the rapidly accelerating convection currents, produces negative gravity anomalies associated with island arcs. During the third phase of decelerating convection, downward forces decrease, the formerly downbuckled region rises seeking isostatic equilibrium, and elevated mountains develop.

Although Griggs followed Holmes in employing convection currents to account for the formation of mountains, island arcs, deep ocean trenches and their associated negative gravity anomalies, he did not suggest that convection causes continental displacements. Indeed, Griggs' distribution of convection current cycles in time and space implied that continents had not changed their relative positions. He rejected Holmes' and Pekeris' placement of rising convection currents beneath continents. Griggs suggested, prophetically, that convection currents arise under oceans, and sink at the periphery of continents where island arcs form. During the current cycle, rising convection currents in the Pacific basin and perhaps in the Indian basin, respectively, have formed the circum-Pacific mountains and Himalayan–Alpine mountain belt; during the previous cycle, they had arisen in the central Atlantic Ocean, forming the Appalachian and Hercynian mountains. Griggs explicitly contrasted his geographical placement of convection currents with that of Holmes and Pekeris.

Holmes (1932) and Pekeris (1936) have suggested that the blanketing effect of the continents with their high radioactive content will cause sufficiently excess temperature under the continents to initiate rising currents there. These currents would act to spread the continents and to form mountains peripheral to them. Holmes published maps showing the hypothetical effect of this action on continental structures.

It seems conceivable to the writer that the temperature differences within the substratum inherited from the preceding convection cycle may be of more importance in localizing the cells than the blanketing effect of the continents. This opens the attractive possibility of a convection cell covering the whole Pacific basin, comprising sinking peripheral currents localizing the circum-Pacific mountains and rising currents in the center. Such an interpretation would partially explain the sweeping of the Pacific basin clear of continental material, in the manner demonstrated by the model. A minor cell might be suggested with its center in the southwest Indian Ocean, accounting for the Himalayan-Alpine bifurcation.

If this be assumed, then one may carry the speculation further and suppose that the previous cycle occurred as far from this location as possible – namely, about the central Atlantic Ocean. This location is nearly central to the Appalachians, Hercynian mountains, and the Post-Carboniferous mountains of Brazil and Africa.

> *(Griggs, 1939: 647; Griggs seems to have got the dates wrong on his references to Holmes and Griggs, for they should be Holmes (1933), and Pekeris (1935))*

Griggs supported the idea of convection currents as a theory of mountain building, but not as a basis for continental drift.

For Griggs' mountain building, convection model to work, continents had to remain fixed (or move exceedingly slowly) relative to an advancing convection current. On his theory, fixed continents surround ocean basins. Convection turns downward at continental edges. By contrast, on Holmes' convection theory, continents split apart, riding on the backs of convection currents beneath them, and mountains form on the leading edge of advancing continents; convection caused the breakup of the supercontinents, Laurasia and Gondwana, and formation of the Atlantic. Griggs believed the Atlantic basin already existed before the previous convection cycle had begun; convection did not split apart the continents and form the Atlantic basin. Convection occurred, as Griggs put it, "about the central Atlantic Ocean" and formed the Appalachian and Hercynian fold belts; Griggs' Atlantic Ocean existed before these belts were formed.

As I understand him, Griggs never intended his convection hypothesis to provide a mechanism for continental drift, and neither Holmes nor Gutenberg, who favored mobilism (see III, §2.3 for Gutenberg), regarded Griggs as a supporter of continental drift. He did not link convection with continental drift. He was not, as I have said, a mobilist. To think otherwise would give an erroneous impression about the support for mobilism among North American geophysicists. I believe the manner in which Griggs presented his theory may have led some to believe incorrectly that he supported mobilism, and it is worth explaining why I think this is so.

In the passage quoted above, Griggs stated that a convection cell covering the whole Pacific basin "would partially explain the sweeping of the Pacific basin clear of continental material, in the manner demonstrated by the model." He also wrote:

This indication that a singly active cell may sweep off the superjacent continental crust opens wide avenues for speculation as to the formation of the circum-Pacific mountains and indeed as to the primary segregation of the continental masses themselves. Here is a possible deformation force which could effectively counteract the tendency of erosion to distribute the continental material uniformly over the surface of the globe.

> *(Griggs, 1939: 643)*

What did Griggs have in mind by "sweeping of the Pacific basin clear of continental material," by "sweep off the superjacent continental crust," and by "primary segregation of the continental masses themselves?" What does he mean by continental crust? It is, as he explained in the last sentence, material on the ocean floor derived by erosion of continental sedimentary rocks. It is not continental crust in the accepted sense. As Griggs recalled almost fifty years later:

Dr. Day was president of the GSA when I gave my paper on convection currents in the mantle, complete with working model and a movie which showed the lithosphere plunging under the continents creating massive subduction, *meanwhile sweeping the sedimentary layer from the oceans.*

(Griggs, 1974: 1343; emphasis added)

Griggs was not postulating or seeking to explain continental drift. He sought only to explain the origin of mountains. Detritus from the continent, which had settled on the ocean floor, was being returned to the continent whence it came, and this became mountains. Finally, what did Griggs mean by "primary segregation of the continental masses themselves?" Again, he had in mind (1939: 643) the plastering of continentally derived sediments back onto continents, which "could effectively counteract the tendency of erosion to distribute the continental material uniformly over the surface of the globe."[19]

Griggs did not claim that his theory of mountain building was in any sense complete, only that he thought it worth serious attention, and was more attractive than any other at the time.

The hypothesis of orogeny developed from these observations is attractive because it seems to satisfy better than any other the three fundamental conditions of a mountain-building mechanism. (1) Provision of an adequate compressional force. (2) Local provision of sufficient contraction for orogenesis. (3) Explanation of the intermittent nature of orogenic processes and the threefold character of the mountain-building cycle.

Evidence sufficient to establish any theory of this kind can hardly be found within a short time, and has never been put forward in support of any previous theory. This difficulty seems inherent in the very nature of the problem. Accordingly the present theory, necessarily founded on insufficient evidence is here presented because it can be effectively tested only by use and the critical study of others.

(Griggs, 1939: 649)

The above brief examination of the major attempts to develop convection theories during the 1930s shows that Holmes' theory received little direct support.[20] Hales and Pekeris did not argue that convection in the substratum actually occurs; only that it was possible. Vening Meinesz and Griggs were more willing than Hales to believe that convection occurs; neither supported continental drift; both suggested convection patterns different from that of Holmes and from each other. Thus, although Holmes' theory helped promote interest in convection, it did not gain support for continental drift among non-mobilists, or even among those who believed that substratum convection was possible or even likely. This must have been very discouraging for Holmes.

5.7 Reception of Holmes' substratum convection by mobilists Daly and du Toit

As noted above, Holmes' convection hypothesis was not even mentioned at the remarkable 1935 RAS London meeting on mobilism. Furthermore, his work attracted only four minor comments at a large symposium on the structure of the Atlantic Ocean and the theory of continental drift held in Germany in 1939, the proceedings of which filled 700 pages in the prestigious *Geologische Rundschau* (1939) (§8.4). However, it would be a mistake to conclude that Holmes' hypothesis was completely ignored, and that it did not help to promote interest in convection currents. It certainly caught the attention of Daly and du Toit, who, along with Holmes, and perhaps also Gutenberg (III, §2.3), were the most influential proponents of mobilism from the time of Wegener's death (1930) through the mid-1950s.[21]

Daly and Holmes maintained a long correspondence (Allwardt, 1990: 83, 106; Lewis, 2000: 105–106). Holmes considered Daly a personal friend. Daly invited Holmes to Harvard in 1932 to give the Lowell Lectures, and Holmes stayed at Daly's home. Any discussions, however, which they surely must have had, failed to reconcile their differences. Holmes, who had criticized Daly's down-sliding hypothesis in 1931 because it did not provide for the formation of domed regions sufficient for down-sliding to occur, never wrote again on Daly's theory. Daly raised several difficulties with Holmes' theory (RS2) in the second and greatly revised edition (1933) of his *Igneous Rocks and the Depths of the Earth*, which appeared the year after Holmes had delivered the Lowell Lectures. Daly raised these difficulties before the appearance of works by Pekeris, Hales, Vening Meinesz, and Griggs.

After summarizing Holmes' theory, Daly (1933: 226) claimed that it "faces difficulties." There was, Daly remarked, Werenskiold's criticism of Holmes' assumption that the substratum is sufficiently homogeneous to permit non-layered or single-tier convection. He pointed to seismological studies indicating marked density discontinuities in the substratum, and noted (1933: 226) that the degree of chemical homogeneity of a substratum, required for single-tier convection, would be "irreconcilable with the value of the earth's moment of inertia." Consequently, even if convection occurred, the cells would not be large enough to move continents. He argued that it was more reasonable to suppose greater concentration of radioactive material in the upper than in the lower substratum, so temperature would increase upwards, the opposite of that required for convection. He argued that convection could not occur in the substratum unless it is devoid of strength, and if it is devoid of strength, then convecting currents would not produce sufficient drag upon the crust to either raise mountains or drift continents.

The explanation of mountain chains by direct drag of thermal-convection currents is subject to doubt. In the first place, can the assumed horizontal current exert enough viscous pull on a continental block so that the rocks on the "prow" side shall be crumpled and thrust? All mountain building appears to have been an exceedingly slow process in absolute measure. During one of these prolonged periods one would expect the elastic coupling between crust and substratum to be completely broken by viscous displacements in the substratum. Unless, then, the material of this shell did retain some undecayed rigidity or what amounts to strength, it could not force the crust to do the great work of orogeny. Yet thermal convection is impossible in a medium that has finite strength.

(Daly, 1933: 254–255)

Daly also argued that if single-tier convection in the substratum were to occur, it would lead to tremendous uplift (even more than that required for Daly's own theory) and fusion of the crust.

If with Holmes we assume convection in a shell bottomed by the 2900-kilometer discontinuity, and rise of the deep material all or nearly all of the way to the crust, we encounter another setback. For the temperature at the depth of 2900 kilometers is probably not much lower than half of the 10 000° [C] calculated by Holmes. The expansion of this exceedingly hot material in the rising branch of the current should be enough to lift the crust in that sector 10 or more kilometers, and the ultimate temperature near the surface would much exceed that of mere fusion of the crust. Neither consequence matches the geological record. Nor is there any evidence that the opposite tendencies in the sector occupied by the diving branch of the postulated current have left their mark on the globe.

(Daly, 1933: 255; my bracketed addition)

After offering these and other criticisms of Holmes' theory, Daly praised (1933: 227) Holmes for his "courageous attempt." He noted that Holmes' theory "is not destined for ready acceptance," but that the idea of convection in the substratum could prove quite useful for his (Daly's) hypothesis, if single-tiered convection were replaced with "delayed, tandem" convection, which restricts convection currents to regions having the same density that do not extend throughout the whole depth of the substratum. Daly went further, he invoked tandem convection upon re-considering his own down-sliding theory. Daly, admitting (1933: 234–238) that the amount of crustal sliding might be "quite limited in distance of travel," proposed that tandem convection might help to bring about the uplift of the crust required for continents to slide downhill.

Alex du Toit, the other major mobilist to consider Holmes' hypothesis in any detail, adopted its major aspects, although in the final chapter of *Our Wandering Continents* he amended it and combined it with other elements. After favorably introducing Holmes' hypothesis, he explained:

The writer, nevertheless, feels that, in the form hitherto stated, *convection by itself is not wholly competent to account for continental drift in the full meaning of the term,* and that some modification of the theory is accordingly demanded.

(du Toit, 1937: 322)

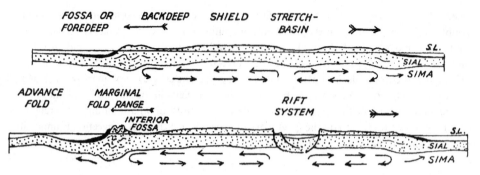

Figure 5.5 Du Toit's Figure 48 (1937: 325). His caption read as follows: Hypothetical stages in the fracturing of a continental block. Black, orogenic sediments; large arrows, direction of stretching.

Du Toit (1937: 325) proceeded to reproduce Holmes' Figures 2 and 3 (my Figure 5.1), and provided two of his own very similar figures (Figure 5.5).

Although there are differences between the two hypotheses, and the details of du Toit's are difficult to understand, it is clear, I think, that he followed Holmes. Du Toit argued that convection currents could provide sufficient force to move the continents, and postulated that currents arise under continents and sink under peripheral geosynclines. Du Toit was further indebted to Holmes because he invoked the reversible pressure-induced phase change of basalt to denser eclogite; he reasoned that, during drift, downward pressure created at a continent's leading edge transforms the basalt beneath into eclogite, and that erosion lightens the rear of the continent. As a result, the continent slides downhill as convection currents propel it through the substratum. Du Toit did not entirely adopt Holmes' mechanism, but he nevertheless was its only major supporter.

5.8 Holmes reconsiders his substratum convection hypothesis, 1944

Holmes returned to his convection theory a decade later as he began working on *Principles*, a book written primarily as an introductory text for undergraduates, less technical than his earlier writing. He characterized his hypothesis as an explanation for mountain building, and argued that it was superior to time-honored thermal contraction. In his final chapter he gave a general defense of mobilism with the addition of fulsome praise of du Toit's fieldwork, especially his careful comparison of the geology of Africa and South America. He invoked convection as the cause for continental drift.

He now incorporated the new work of Griggs, going so far as to reproduce his figure of his 1939 mechanical model, which simulated the action of subcrustal currents upon the crust to form mountains (1944/1945: 412). Following Griggs, he (1945: 410) divided a convective cycle into three active stages, and matched them with the three stages of the orogenic cycle.[22]

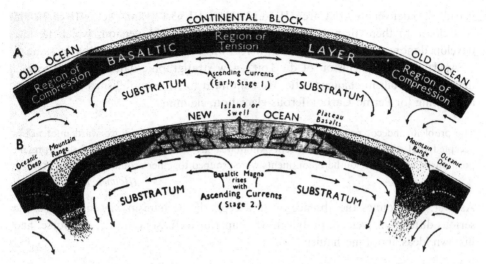

Figure 5.6 Holmes' Figure 262 (1944/1945: 506). His caption read as follows: A purely hypothetical mechanism for "engineering" continental drift. In A sub-crustal currents are in the early part of the convection cycle ... In B the currents have become sufficiently vigorous ... to drag the two halves of the original continent apart, with consequent mountain building in front where the currents descend, and ocean floor development on the site of the gap where the currents ascend.

He also changed his description of how new seafloor was formed between separating continents, adding two new illustrations (1945: 506; see Figure 5.6) of his convection theory, which he offered as "a purely hypothetical mechanism for 'engineering' continental drift."

The difference between this and earlier versions may be seen by comparing diagram B in Figure 5.6 with Figure 2 from his 1931 paper (my Figure 5.1). In his new version, ascending convection currents bring basalt up and into the gap created by the tearing apart of old crust.

Most of the basaltic magma ... would naturally rise with the ascending currents of the main convectional systems until it reached the torn and outstretched crust of the disruptive basins left behind the advancing continents or in the heart of the Pacific. There it would escape through innumerable fissures, spreading out as sheet-like intrusions within the crust, and as submarine lava flows over its surface. Thus, in a general way it is possible to understand how the gaps rent in the crust come to be healed again; and healed, moreover, with exactly the right sort of material to restore the basaltic layer.

(Holmes, 1944/1945: 508)

In his new version, rising basalt definitely intrudes into the gaps created by the torn and stretched crust creating new seafloor; it also covers old seafloor, and caps recently created seafloor. As he did in 1928, Holmes invoked thinning of the seafloor, but now he proposed the creation of new basaltic seafloor. This new version predated by over a decade the later versions of Carey (II, §6.12), Hess (III, §3.14), and Dietz (III, §4.9), and was closer to them than to his own previous 1928/1931 version, which he had developed more than a dozen years earlier. It was closer to them, but not the same.[23]

Apart from his inclusion of du Toit's new stratigraphic data, Holmes' general defense of mobilism changed little. He continued to stress mobilism's advantage in accounting for Permo-Carboniferous glaciation, claiming:

> The problem, indeed, remains an insoluble enigma, unless the straightforward inference is accepted that all the continents except Antarctica lay well to the south of their present positions, and that the southern continents were grouped together around the South Pole.
>
> *(Holmes, 1944/1945: 503)*

As ever, for Holmes the absence of an acceptable mechanism was mobilism's one serious difficulty. Even so, mobilism was superior to fixism, and he, of course, had his own ideas about mechanism.

> The only serious argument advanced against the validity of the above solution is that it merely exchanges one embarrassing problem for another – the difficulty of explaining how continental drift on so stupendous a scale could have been brought about. By itself this consideration might be a reason for sitting on the fence, but the real antithesis is not so simple. If one rejects continental drift and accepts the possibility that Central Africa could have been glaciated while Britain had a tropical climate, one must also admit the necessity for land bridges, which have since subsided to oceanic depths. The continental drift solution has the advantage that it reduces two baffling problems to one, while at the same time it removes many other less intractable difficulties.
>
> *(Holmes, 1944/1945: 504)*

Despite the absence of an acceptable cause, mobilism had the greater problem-solving effectiveness.

In *Principles*, Holmes gave a short defense of his convection hypothesis. He thought it the best available mechanism and noted its advantages over Taylor's and Wegener's. He did not mention Daly's down-sliding; he did not say that du Toit had adopted his idea of thermal convection. He responded to some, but not all, of the criticisms that had been made of his theory. He did not mention any of the differences between his own theory and those proposed by Vening Meinesz or Griggs, and he said nothing about the works of Pekeris or Hales, although he must have known about them. Perhaps he did not consider them appropriate in an introductory text.[24]

Holmes, however, did attempt to answer Jeffreys' contention about the improbability of large-scale convection.[25] He did so by considering why continental drift should not have begun until Mesozoic time.

> Incidentally, it should be noticed that the coalescence of the usual chaotic or small convective systems into three gigantic ones involves a coincidence that can rarely have happened in the earth's history, and one that is just as likely to have come about during the Mesozoic era as at

any other time. The often-asked question: How is it that Pangea did not begin to break up and unfold until Mesozoic time? thus ceases to have any significance. If continental drift could have been caused by the gravitational forces invoked by Wegener, then it should have occurred once and for all early in the earth's history, since those forces have always been in operation. If convection currents are necessary, continental drift may have accompanied all the greater paroxysms of mountain building in former ages, but, if so, it would usually have been on no more than a limited scale. That there was a quite exceptional integration of effort in Mesozoic and Tertiary times is forcibly suggested by eruptions of plateau basalts and building of mountains on a scale for which it would be hard to find a parallel in any earlier age.

(Holmes, 1944/1945: 505–506)

Not only was Jeffreys answered, but Holmes felt he had an additional reason for preferring his to Wegener's mechanism.

Holmes also discussed the difficulty he had raised in 1929 about the need to shift the whole crust relative to the Earth's interior in order to account for the northward drifting of both Africa and Britain. At the time he admitted that he had no solution, but two years later in 1931, he gave one (§5.4) that he now seemed less comfortable with.

To go beyond the above indication that a mechanism for continental drift is by no means inconceivable would at this present time be unwise. *Many difficulties still remain unsolved.* In particular, it must not be overlooked that a successful process must also provide a general drift of the crust over the interior: a drift with a northerly component on the African side sufficient to carry Africa over the Equator, and Britain from the late Carboniferous tropics to its present position. The northward push of Africa and India, of which the Alpine system and high plateau of Tibet are spectacular witnesses, could not have been sufficient by itself to shove Europe and Asia so far to the north. To achieve this the aid of exceptionally powerful sub-Laurasian currents directed towards the Pacific is required. The total northward components might then overbalance the southward components, and a general drift of the crust would be superimposed on the normal radial directions of drift.

(Holmes, 1944/1945: 508; emphasis added)

Holmes seems to have become doubtful whether further work on his highly speculative theory in the near future would be helpful. Even after a decade, "many difficulties still remain unsolved." He was not nearly as sanguine as he had been about the time needed to test his theory. He had become rather pessimistic about the prospects for an all-embracing mobilism theory.

It must be clearly realized, however, that purely speculative ideas of this kind, specially invented to match the requirements, can have no scientific value until they acquire support from independent evidence. The detailed complexity of convection systems, and the endless variety of their interactions and kaleidoscopic transformations, are so incalculable that many generations of work, geological, experimental, and mathematical, may well be necessary before the hypothesis can be adequately tested. Meanwhile, it would be futile to indulge in the early expectation of an all-embracing theory which would satisfactorily correlate all the varied phenomena for which the earth's internal behaviour is responsible. The words of John

Woodward, written in 1695 about ore deposits, are equally applicable to-day in relation to continental drift and convection currents: "Here," he declared, "is such a vast variety of phenomena and these many of them so delusive that 'tis very hard to escape imposition and mistake."

(Holmes, 1944/1945: 508–509)

Why had he become so pessimistic? As I have previously argued (Frankel, 1978), and Doris Holmes agreed, it was because little new data about the substratum and the seafloor needed to test Holmes' hypothesis had been forthcoming during the intervening years.[26] In addition, Holmes knew, as he implied in the above passage, that the new work of Vening Meinesz, Pekeris, Hales, and Griggs suggested different possible patterns of convection. Many key questions were not even close to being answered and would remain so for years to come.

Furthermore, as Allwardt (1990) has argued, Griggs, Bridgman, and others at Harvard's Lowell Institute had begun to develop new concepts in solid-state physics and to envisage Earth's substratum as a solid that can flow under stress. Holmes certainly was pleased with Griggs' endorsement of convection, but much of his own theoretical work on the conditions needed for convection had been rendered irrelevant by Griggs' work. Moreover, seismological studies were indicating density changes in the substratum, which created a new difficulty for Holmes' single-tier convection. It is therefore understandable that Holmes should have become somewhat pessimistic about the prospects of testing his convection theory in the near future. He had answered Jeffreys' criticism, and had shown that mobilism was not physically impossible. The kind of work that he and Jeffreys had hoped their disagreement would spawn was far from complete and the only results so far had only shown how complex and involved the problem of substratum convection was. Indeed, his 1944/5 assessment above more closely echoed Griggs' than his optimistic earlier 1931 one. Of course, it would be a mistake to conclude that Holmes was no longer inclined toward mobilism in 1944. He still considered it superior to fixism, but he recognized that formidable obstacles lay ahead.

5.9 Reception of Holmes' 1944 presentation of his convection hypothesis

Although Holmes' convection theory was not ignored during the remainder of the 1940s, it did not lead to any significant research. Substratum convection remained highly speculative. Progress on it was at a standstill. I shall now briefly examine three reviews in British journals of Holmes' *Principles*, and two symposia held in 1950, on continental drift and convection. All praised Holmes generally, but none claimed that his theory of convection merited serious attention. Two questioned the usefulness of its inclusion in an introductory textbook. Interestingly, none of the reviewers rejected Holmes' theory outright, but wished he had addressed more of its difficulties. Another made clear that many disagreed with Holmes.

W. B. R. King (1889–1963), then Woodwardian Professor of Geology at the University of Cambridge, and future President of the Geological Society of London, praised the book as clearly written, and found (1945: 46) its first two parts to be an "ideal elementary textbook." However, he was not as pleased with Part III in which Holmes discussed continental drift and convection. Although King expressed no opinion about the merits of Holmes' theory, he was concerned that introductory students might be misled into thinking Holmes' speculative theory a matter of fact.

It is certainly not suggested that Part III should not have been written, but whereas Parts I and II can be wholeheartedly recommended as a textbook to every elementary student of geology it may be felt that a beginner may be led away by Professor Holmes's clear and attractive presentation and fail to notice that he does not claim that all the opinions expressed in this part are of the same proved or accepted value as the old established views given in the earlier parts.

(W. B. R. King, 1945: 47)

In contrast S. E. Hollingworth (1899–1966), Yates-Goldsmid Professor of Geology at University College London, had high praise and, unlike King, welcomed Holmes' speculative approach (Hollingworth, 1944: 122). It was, he wrote, a "real achievement in the portrayal of geology as a science that is full of vitality with intriguing problems awaiting solution." He also praised Holmes' presentation of his theory, but wished that Holmes had discussed the difficulty it faced over whether the substratum contained layers of different density.

Professor Holmes gives a masterly exposition of the hypothesis of convection currents in the fluid subcrustal regions, as applied to mountain building and, in the last chapter, to the theory of continental drift. More information on such questions as the reconciling of a density-zoned earth with deep-seated convection currents would have been welcome.

(Hollingworth, 1944: 122)

H. H. Read (1889–1970), professor of geology at Imperial College, University of London, also praised Holmes' book, claiming (1944: 721) that it "rises well above any recent work in English in its own field." However, Read expressed reservation about Holmes' inclusion of mobilism. He worried that readers might be misled, and noted that Holmes' view was not widely believed.

The last chapter discusses the theory of continental drift. This section differs in tone from the previous one, in that it is now admitted that there are a great many problems still unsolved; but Professor Holmes appears to suggest that solutions for most of these will eventually be found along the lines proposed by him. Thus, the notion of internal convection currents, elaborated by the author in recent years is applied to account for mountain-building and for volcanic activity, and is used as the mechanism for continental drift. It is, of course, right and proper for a geologist or anyone else to have a profound belief in his own proposals, but this is no guarantee that the belief will be widely held by others. It would have been better and fairer for the general reader if some of the many other speculations on these topics had been mentioned, if not discussed.

(Read, 1944: 721)

This guarded, slightly negative attitude should not be interpreted as fixist, because Read himself was sympathetic toward mobilism. Read's own textbook, published in 1949, was pro-mobilist, and Read even claimed (1949: 167) that the convection current theory "is admittedly attractive," but "must be judged as non-proven," which, of course, was just what Holmes had said all along.

Daly, in his review, expressed his great respect for Holmes, and welcomed the inclusion of continental drift and his convection theory.

For this boldness he will doubtless be chastised by some "tender-minded" geologists, but, in view of the widespread popular and professional interest in this hypothesis, it is at least a question whether the relevant ideas of Taylor, Wegener, du Toit, and others should be omitted from even an elementary book on physical geology. While tolerant of the hypothesis himself, Professor Holmes has wisely guarded his readers against putting final faith in it. He writes (p. 508): "It must be clearly realized, however, that purely speculative ideas of this kind, specially invented to match the requirements, can have no scientific value until they acquire support from independent evidence." Like many other passages, this last chapter well illustrates our author's teaching, that the student should keep an open mind on all the greater problems about the earth.

(Daly, 1945: 572–573)

Nonetheless, Daly (1945: 573–574) did criticize; he restated the difficulty of having single-tier convection in what appeared to be a multi-layered substratum, and he reiterated his own idea of tandem convection as preferable to Holmes' single-tier.

Holmes' theory received hardly any attention at the 1950 symposium on continental drift held in Birmingham at the annual BAAS meeting, jointly sponsored by the Sections of Geology, Zoology, Geography, and Botany. Of the six participants whose comments were published, two (R. Good and H. E. Hinton) were mobilists, two (Jeffreys and J. H. F. Umbgrove) were fixists, S. W. Wooldridge adopted a fixist position, and J. R. F. Joyce took a neutral position. Only Hinton (1951) and Wooldridge (1951) mentioned Holmes' theory.

Umbgrove (1951), an influential Dutch geologist, was strongly opposed to mobilism. Drawing on his critique of mobilism in the second edition of his *The Pulse of the Earth* (1947), he attacked Wegener's drift mechanism (RS2). He began by arguing that sima was as strong, if not stronger, than sial.

The basic assumption in explaining the mechanism of continental drift appears to have satisfied Wegener and many of his followers. Yet, it is a well-known fact that basaltic material (sima) has a higher melting point, approximately 1300°C., than granite (or sial), approximately 700°C. This means that the crystalline crust of the pool of sima must be as strong as or even stronger than the continental masses. Yet this crust must disappear somehow in front of an advancing continent.

(Umbgrove, 1951: 67)

Claiming that the idea of a rigid sima is supported by seismic and gravimetric evidence, and giving as an example the presence of positive gravity anomalies over the Hawaiian Islands, he (1951: 67) declared that the anomalies are best explained by a downbending of the ocean crust, which requires a crust with substantial rigidity, rendering drift impossible.

Appealing to sedimentation on the seafloor, Umbgrove raised two anomalies (RS2), turning first to the Pacific.

Suppose the American continents to have drifted towards the Pacific. We have to consider the fact that the ocean floor is blanketed by some 4 Km. of deep-sea sediments. Would not they have been squeezed and piled up like ice in front of a square-nosed ship? Taking the westward drift of South America at 3000 Km., 1500 Km.^{3} of deep-sea oozes would have been piled up in front of each kilometre of the western coast. After complete isostatic compensation at least 1000 Km.^{3} would still emerge out of the sima. Such a mass would form a plateau 200 Km. broad at sea level. Actually the continental slope is one of the steepest of the world and is fronted by deep-sea troughs.

(Umbgrove, 1951: 69)[27]

Moving to the Atlantic, he wondered why, if continental drift had occurred, the Atlantic seafloor was, it seems (erroneously as we now know), overlain by a thick layer of sediments.

And what would be the result of drift in the Atlantic Ocean? If the Atlantic originated in comparatively recent times as a result of continental drift we certainly would expect to find much thinner strata of deep-sea sediments. The facts, however, are not in accordance with this hypothesis. Petterson, the leader of the Albatross Expedition, mentions that depth-charge soundings of the thickness of the carpet of sediment gave a maximum value of nearly 10 000 ft. in the Atlantic, representing a time-span of 300–400 million years, possibly still more.

(Umbgrove, 1951: 69)

He then turned to the problem of biotic disjuncts, and argued that the existence of oceanic ridges provided support for isthmian links.

The most prominent problems of biogeography have been explained by various able and outstanding biologists and palaeontologists without the acceptance of drifting continents. Yet some biologists and palaeontologists may think there are facts which force them to accept trans-oceanic land-connections in the geological past. What can geological science offer them in this respect? Among the features which are liable to change in one or perhaps several respects are the details of the submarine relief. For example the "rejuvenation" of the relief, which is so strikingly illustrated by the Mid-Atlantic Rise and Carlsberg System. These movements remind one at once of the "isthmian links" of Schuchert and Willis. I consequently incline towards the view that no essential objection can prevent us from accepting the idea of isthmian links as postulated by Schuchert and Willis. The existence of land-connections during certain given periods cannot be proved geologically but such land bridges need not be discarded *a priori* as mere products of the imagination.

(Umbgrove, 1951: 70)

To close his presentation, Umbgrove mentioned that he favored substratum convection without mentioning Holmes. Umbgrove's critique of mechanism was outdated, even antiquated; it applied to Wegener's but not to Holmes'. Umbgrove knew this. In *The Pulse of the Earth*, which was published in 1947, three years before the 1951 meeting in Birmingham, he tentatively invoked large-scale periodic convection as the underlying cause of periodic geological changes.

Most readers will be acquainted with the interesting suggestion which Joly and Holmes put forwards some years ago, when they attempted to derive periodic phenomena from radioactive processes, and will probably also have read the opinions of such authors as Schwinner, Escher, Holmes, Pekeris and Vening Meinesz on convection currents. Some may have come across Griggs' publication in which this author tries to show that convection currents may in fact occur intermittently, and will have admired the ingenious model by which he sought to demonstrate the activity of cyclic convection currents.

(Umbgrove, 1947: 324–325; for Escher's views on convection see §8.13)

Somewhat surprisingly, Umbgrove did, however, thank Holmes, noting (1947: xxii): "Professor Holmes even kindly undertook the onerous task of reading the page proofs." I wonder if they ever discussed Holmes' convection theory. (Umbgrove's views are further discussed in §8.13.)

At the 1950 Birmingham meeting, Jeffreys directed almost all of his attention to the mechanism difficulty. Although he mentioned convection theories and dismissed them because, he claimed, they required an Earth of no strength, he did not specifically mention Holmes' convection; his criticisms were actually directed at Wegener's mechanism, already a lost cause. Jeffreys had enough of continental drift, and thought that no more time should be wasted on it until a mechanism was developed.

Considering that the theory has been advocated for 35 years, it is remarkable that nobody has suggested one of the right amount or even in the right direction ... I seriously suggest that no more time should be spent on discussion of this theory until a mechanism for it is produced; what it has done and continues to do, is to distract attention from the serious problems of geophysics. This is the fourth time that I have taken part in a public discussion of this theory. In each previous one a distinguished biologist or geologist has presented the case for drift, and has been followed by equally distinguished ones who have pointed out facts that it would render more difficult to explain.[28]

(Jeffreys, 1951: 80)

So much for Holmes' mechanism; it was no longer even worth mentioning.

H. E. Hinton, a zoologist and biogeographer at the University of Bristol, who favored mobilism over fixism, revisited the debate between du Toit and Gaylord Simpson (§3.7). He characterized Simpson's "attack on Wegener's theory from the field of zoogeography" as "one of the best informed" (1951: 76). He also praised du Toit's reply, and actually thought that Simpson's data supported mobilism. Simpson had argued that the Triassic reptiles, so abundant in South Africa, were not as well represented in South America as would be expected if they had been together. Hilton reexamined Simpson's data, and argued "that not only do his figures show the exact opposite of what he claims, but that they constitute one of the best zoogeographical proofs of the displacement theory" (1951: 77). Hinton mentioned Holmes' theory, but being primarily interested in biogeography, did not single it out as any better than Wegener's or du Toit's.

S. W. Wooldridge mentioned and even quoted Holmes, but what he chose to quote and how he used it, show that he thought it was time to forget Holmes' theory. Referring to Holmes' comment about the many years it would take to adequately test his hypothesis of substratum convection, Wooldridge thought it would be more profitable to look instead at vertical movements.

Continental drift and related topics were the subject of a discussion sponsored by Section E at the London meeting of 1931. The position as then set forth has undergone no very radical change in the years between. Then as now no demonstrably adequate mechanism for such movement was in sight, though Prof. Holmes had begun his persuasive advocacy, in quantitative terms, of the hypothesis of sub-crustal convection. As he himself has since written, "many generations of work, geological, experimental and mathematical may well be necessary before the hypothesis can be adequately tested." Meanwhile it is idle to overlook the evidence that great vertical movements have, in fact, taken place.

(Wooldridge, 1951: 82)

Wooldridge's seeming preference for fixism at this meeting is strange, because he and R. S. Morgan had favored mobilism in 1937 in the first edition of their textbook on geomorphology, and he again favored mobilism in the 1959 revised edition (Wooldridge and Morgan, 1937, 1959).

Appealing to new bathymetric data, J. R. F. Joyce raised difficulties with the positioning of the Scotia Arc in the reconstructions offered by Wegener, du Toit, and King. Joyce offered his own solution, but expressed no great enthusiasm for it.

The primary claim that can be made for the proposed reconstruction is that of morphological completeness, for first there is a pleasing geometrical fit; secondly, the fold girdle to East Antarctica has been restored; thirdly, the "sialic vacuum" of the sea thought to exist between Australia and Asia in Pangaean times has been destroyed, and lastly the difficulty of disposing of two arcuate features in the Pangaean scheme has been overcome. The suggestion is therefore made that if Pangaea did in fact exist then this new organisation of continental masses and island arcs at the opening of the Palaeozoic era is more in accord with the known data.

(Joyce 1951: 87)

5.10 Geophysicists' attitude toward convection around 1950

... when I gave my paper on convection currents in the mantle, complete with a working model and a movie which showed the lithosphere plunging under the continents creating massive subduction, meanwhile sweeping the sedimentary layer from the oceans. Harry Hess presided, and endeavored to get favorable discussion of these then controversial ideas, but circumstances prevented him. Andy Lawson, sitting in the front row got up and squeaked, "I may be gullible. I may be gullible! But I'm not gullible enough to swallow this poppy-cock." After this long tirade, before Harry could do anything, Bailey Willis, who was sitting in the second row got up, turned to face the audience and said, "All you here today bear witness – for the first time in twenty years, I find myself in complete agreement with Andy Lawson."

(From David Griggs' (1974: 1342–1343) acceptance speech upon receiving the Arthur L. Day Medal from the Geological Society of America)

This was probably the most hostile reception Griggs ever received during his prestigious career, and shows that convection, whether as a cause of continental drift or mountain building, was not held in much regard by many Earth scientists in 1939. Contractionism in its various versions was then in vogue and remained so into the 1960s. Vening Meinesz, Griggs, Holmes, Bull, and several others, including Harry Hess, who favored some sort of substratum convection, were a small minority.

In order to gain an understanding of the status of Holmes' and other convection theories at the beginning of the 1950s, I shall summarize events at a meeting which took place in Hershey, Pennsylvania, in September 1950 as part of a colloquium on plastic flow and deformation within the Earth. It was jointly sponsored by the International Union of Geodesy and Geophysics (hereafter, IUGG) and the International Union of Theoretical and Applied Mechanics, and brought together leading geophysicists and geophysically minded geologists. The special editorial committee was chaired by Beno Gutenberg, the German-born seismologist and geophysicist who emigrated to the United States in 1930. Next to Daly, Gutenberg was the most renowned mobilist working in the United States during the first half of the twentieth century (III, §2.2, §2.3). The meeting was arranged around four themes: inelastic deformation of material, formation of folded belts and geosynclines, gravity anomalies and postglacial uplift, and convection currents. Holmes was not present. Griggs (UCLA), Hess (Princeton University), Vening Meinesz (Royal Netherlands Meteorological Institute), Francis Birch (Harvard University), Umbgrove (Institute of Mines, Delft, Holland), Cox (Free University of Brussels), E. H. Vestine (Carnegie Institution of Washington), and Bullard (National Physical Laboratory, Teddington, UK) gave presentations to the session on convection.

It is worth noting that participants spoke of mantle convection not of convection in the substratum, "mantle" having replaced "substratum." Griggs even used "mantle" instead of "substratum" when referring to his former work (§5.6).

Published remarks did not include any reference to Holmes' mantle convection, nor did Griggs mention Holmes in his summary of modern theories of convection. He cited Vening Meinesz, Pekeris, and himself; perhaps he thought Holmes' theory was no longer modern (Griggs, 1951: 527). Vening Meinesz (1951: 532) cited Griggs, and discussed his own theory of mantle convection. Bullard, who also spoke at the session on the formation of folded belts and geosynclines, mentioned Vening Meinesz, Griggs, and Hess (Bullard, 1951: 520). None of the participants argued for convection as a mechanism for drift; none of their published remarks included any direct reference to drift. Gutenberg, himself a mobilist, summarized the Hershey meeting; he noted (1951: 541) that even though two participants had mentioned geodetic support for possible movement of Greenland relative to Europe, others "pointed out that during recent observations with more accurate methods and means no movements of the continents have been observed." Everyone agreed that mantle convection was highly speculative. Hess (1951: 530), who strongly favored convection to account for island arcs, admitted: "At present it is not possible to prove that convection currents exist in the mantle nor is it possible to show that they cannot exist." Participants used convection to account for the origin of mountains, island arcs, deep focus earthquakes or negative gravity anomalies, but not continental drift. None claimed that support for convection had substantially improved since Griggs' pre-war work. If anything, support for single-step convection had probably taken a downturn. Birch (1951), a rising star among American geophysicists, introduced new seismological evidence for a major discontinuity within Earth's mantle, which, as Griggs acknowledged, highlighted the difficulty Daly and others had raised against single-tier convection.

A principal uncertainty in convection hypotheses is the unknown composition of the Earth's mantle. In order that thermal convection be important in its mechanical effect on the Earth's crust, the outer several hundred kilometers of the Earth's mantle must be free from stable density stratification ... An inherent density discontinuity ... of one percent will serve as an important bar to convection. To date, seismological observations have not established the existence of such a density discontinuity short of 2900-km depth. Francis Birch, however, will present an interpretation of seismic evidence which favors inhomogeneity of the outer mantle, and thus poses a problem to advocates of convection hypotheses.

(Griggs, 1951: 528)

That the session was held at all shows that mantle convection continued to be a viable alternative to contractionism as a cause of mountain building. The majority were favorably inclined to contractionism, but convection had its supporters. Neither contractionism nor convection merited full acceptance, and both were viewed as speculative. But the very fact that Holmes' theory was not specifically discussed shows that it had become of little interest to technically well-versed members of the generation that succeeded his. Indeed, the absence of any but the briefest mention of continental drift shows how unimportant it was for the geophysicists and geophysically minded geologists assembled at the Hershey meeting in the fall of 1951.

5.11 Holmes' attitude toward mobilism in the early 1950s

To end on a personal note: I should confess that, despite appearances to the contrary, I have never succeeded in freeing myself from a nagging prejudice against continental drift; in my geological bones, so to speak, I feel the hypothesis to be a fantastic one. But this is not science, and in reaction I have been deliberately careful not to ignore the very formidable body of evidence that has seemed to make continental drift an inescapable inference. In recent years the weight of evidence has become less oppressive, and this symposium has left me with the general impression that a few land bridges or linkages by island stepping-stones would probably suffice for the biogeographical problems.

(Holmes, 1953: 671)

Holmes wrote this in his review of the 1949 New York symposium on Mesozoic landbridges across the Atlantic Ocean (§3.8). Most participants argued that the various trans-Altantic disjuncts, structural, biotic, and stratigraphic could be explained without resorting to either mobilism or landbridges, and Holmes agreed that they probably were correct. He noted that the overall evidence for mobilism in "recent years" has "become less oppressive," which was very much in keeping with general opinion in the early 1950s when support for continental drift was at perhaps its lowest ebb for almost three decades. He wrote, "it is now safe to say that India cannot have been where Wegener [had] placed it relative to Africa" (1953: 671). There is little doubt that Holmes' enthusiasm for mobilism had waned since writing his first edition of *Principles*.

Although Holmes had come to think that mobilism might not be needed to account for disjuncts, and that the "formidable body of evidence" in its support was not as "oppressive" as before, he still believed there remained much evidence in its favor. He took issue with two of the symposiasts, Walter Bucher and Maurice Ewing, both adamant fixists from Columbia University. Bucher, a long time foe of continental drift, who has already been introduced (§3.8), argued extensively against it in his 1933 *Deformation of the Earth's Crust* (§8.8), and continued attacking mobilism for years to come (§6.9, §7.3, §8.4, §9.6; III, §1.10, §6.3). Ewing, who will play a major role during later stages of the overall debate, also was solidly against mobilism (see especially III, Chapter 6). Ewing, a physicist by training with little background in geology, was strongly influenced by Bucher. Ewing (1952) described recent discoveries about the ocean floor that indicated great differences between continental and oceanic crust, which he argued supported fixism. Holmes questioned Ewing's conclusion, and pointed to the apparent lack of sediments on the ocean floor, a matter that was soon to become critically important.

Among other general contributions, Dr. M. Ewing presents a useful summary (pp. 87–91) of geophysical evidence (up to 1950) and concludes that both seismic and gravity data show the great similarity of the Atlantic crustal structure to that of the Pacific, and their great dissimilarity to the continental crust. So far as such evidence goes, it is consistent with the hypothesis of the permanence of continents and ocean basins. But the Atlantic data are still too sparse to justify regional conclusions. Moreover, the discovery near the surface of crustal layers having the elastic properties of simatic rocks raises the question of what has happened to the many kilometers of sediment that must have accumulated over the original ocean floor, if the latter has existed for ~3×10^9 years. Have they been deeply buried by repeated coverings of basic lavas and sills, and metamorphosed beyond seismic recognition? If so, their radioactive contents might account for the unexpectedly high heat flow encountered through recently tested parts of the ocean floor. If not, then the seismic evidence recorded by Ewing could be interpreted as against the "permanence" and "land bridge" hypothesis and in favor of "drift."
(Holmes, 1953: 670)

Holmes was harsher on Bucher, who ignored the possibility of mantle convection and evidence supportive of mobilism.

Prof. W. H. Bucher sets out to deal with the general problems of "Continental Drift versus Land Bridges" (pp. 93–103), but his treatment is largely confined to the thesis that the topography, depressions, mountain ranges and structural relations of the ocean floor cannot be harmonized with Wegener's assumption that the simatic crust is or has been so weak that sialic bodies could drift through it. The criteria are not strictly so decisive as Bucher claims, since he fails to consider fully the effects of metamorphism and altogether ignores the hypothesis of sub-crustal convection currents; nor does he even mention any of the evidence which favors the "drift" hypothesis.

(Holmes, 1953: 670)

Although Holmes had come to disagree with Wegener's placement of India next to Africa (a matter in which he has since been proved correct), there was still the excellent fit between Africa and South America. Holmes and others had attempted to obtain radiometric dates for Pre-Cambrian rocks from South America, Africa, and India. Indeed, it was this work that led him to disagree with Wegener's positioning of India. Holmes, who had obtained age determinations of South African rocks, wondered whether radiometric age determinations from South American rocks would agree with those from South Africa. At the time measurements were inconclusive.

A remarkably good "tectonic fit" between parts of South America and Africa can be recognized, but so far no correlation by age is possible, because the South American age determinations have not been correlated with the tectonics and have not been checked by isotopic analyses. No doubt this temporary frustration will be overcome, and the matter will be decided one way or the other. At least we know how the problem can be settled.

(Holmes, 1953: 671)

Holmes still was impressed with mobilism's superior solution to the problem of the origin and distribution of Permo-Carboniferous glaciation. However, his commitment to continental drift had weakened, and he even questioned the rationality of continuing to work on it.

Meanwhile there remains the most serious enigma of all: the Permo-Carboniferous glaciations. Dunbar points out that the Paleozoic glaciations in low latitudes present "a problem still unsolved, unless we accept continental drift." But if we accept continental drift only to explain these and other still older glaciations, it becomes an *ad hoc* hypothesis. As such, it may still be justified as a stimulant to research, but it may also stand in the way of progress by distracting attention from the real problem. Can the meteorologist not come to our assistance and tell us whether or not widespread equatorial and low-latitude glaciation is possible while high latitudes for the most part enjoy a genial climate?

(Holmes, 1953: 671)

5.12 Significance of Holmes' convection hypothesis

By the 1950s, Holmes' solution to the mechanism problem had two major and somewhat paradoxical consequences. First, on a strictly logical basis, *it had removed the burden initially placed on mobilists by Wegener himself to provide a solution to the mechanism problem that was not deemed physically impossible.* The highly critical reception, almost since its birth, of Wegener's mechanism had put mobilists on the defensive. Joly's work, even with van der Gracht's help, did not remedy the situation. Jeffreys had relegated Joly's theory to the realm of the impossible. Daly's hypothesis did not help; Holmes deemed it physically impossible. However, once Jeffreys proclaimed that Holmes' hypothesis was possible, although he thought highly unlikely, and Pekeris and Hales agreed, fixists could not in all fairness justifiably claim that every version of mobilism was impossible. I am not saying that some did not make that claim, just that they were not being reasonable when they did so.

Second, *Holmes' solution to the mechanism problem did little to change the minds of mobilists and fixists; his work failed to diminish the importance that Earth scientists generally attached to the mechanism sub-controversy.* Despite the fact that Holmes provided mobilists with a possible mechanism, he had failed to generate much active interest in it among mobilists, and he did not provide a good reason from them to agree among themselves about the best way to settle the mechanism question. Neither Vening Meinesz nor Griggs, the other two most important proponents of convection, supported Holmes or mobilism. Holmes' carefully crafted and plausible convection theory was, in general, ignored; it is very telling that in the 1930s through the 1950s, it was ignored at meetings on mobilism or convection such as the 1935 London meeting of the RAS (§5.5), the 1939 meeting in Germany on the structure of the Atlantic Ocean and continental drift (§8.4), the 1949 New York meeting about Mesozoic landbridges across the Atlantic Ocean (§3.8), and the 1950 international meetings in Birmingham (§5.9) and in Hershey, Pennsylvania (§5.10). Too little was known about Earth's interior and solid-state physics to develop physical tests of Holmes' or any other mantle convection hypothesis, and thus command broad interest. Mobilists neither devoted much attention to Holmes' hypothesis of mantle convection, nor did they develop alternative new solutions to the mechanism problem. As Holmes made clear at the time, he had not given mobilists a difficulty-free solution to the mechanism problem; although he did give them a physically plausible solution, it was as yet untestable. Because it was untestable, and spawned no testable corollaries or auxiliary hypotheses, it can in retrospect be seen as premature.

There have been several accounts of the legacy of Arthur Holmes (Dunham, 1966; Allwardt, 1990; Lewis, 2000). As Lewis shows, he made enormous contributions to dating Earth. I have been mainly concerned with his theory of convection in the substratum, and its application to tectonics, a substratum that must be liquid. All that changed with Griggs' work and his introduction of solid-state creep. I do not think, however, that this overshadows Holmes' role in disproving contraction as the major driver of tectonics; his demonstration that contractionism fails utterly to explain the widespread occurrence of massive igneous activity, which is a major feature of the geological record, is surely one of the arguments that finally discredited contractionism.

5.13 Appeal to historical precedent: another manifestation of standard research strategy one

"There are more things in heaven and earth, Horatio, than are dreamt of in your philosophy."

(William Shakespeare, Hamlet, *Act I, Scene V)*

Before Holmes provided mobilists with what physically was perhaps an unlikely but a not impossible mechanism, many believed that continental drift could not happen. A small minority argued that it would be a mistake to reject mobilism solely because it appeared to be physically impossible because other theories, once deemed impossible, have turned out to be correct. Many used this argument; du Toit proposed it in his 1924 Presidential Address to the S_2A_3 (§6.5), as King (1953) did in his defense of mobilism, taking aim at North American Earth scientists (§6.9).

Van der Gracht, in his opening address at the 1926 AAPG symposium, compared his audience's denial of mobilism with that of many geologists regarding extensive overthrusting in the Alps. He argued that continental drift should still be taken seriously even though symposiasts may judge it impossible.

The problem of continental drift has raised considerable and spirited discussion in geological circles. Many authorities, entitled to all respect, advocate it; others are undecided but favorably inclined; still others do not favor it, and some are violently opposed. The whole controversy reminds me vividly of the discussions during my student days on the problem of sheet-overthrusting in the Alps. As now in the discussion of continental drift, so there was then much opposition, in which no less an authority than Albert Heim took a leading part before his conversion to the new idea. Its mere possibility was then as firmly denied, as is now the possibility of continental drift. The facts have since proved beyond any doubt that these sheets exist, not only in the Alps, but universally. Still their detailed mechanism, their "possibility," remains almost as much a riddle as it was then. The possibility has only been demonstrated by fact, not explained.

(van der Gracht, 1928b: 3–4)

He (1928b: 5) strengthened the analogy by arguing that current knowledge of Earth's interior was as limited now as knowledge of mountain building had been when Alpine geologists originally proposed huge overthrusts.

On another occasion, Rastall (1929: 448), who taught at Cambridge, repeated the argument, noting that van der Gracht already had used it. He drew analogies between previous assessments about Earth's age and the existence of ice ages. Eloquently putting the case in general terms, he wrote:

The discussion of the theory of continental drift has now reached such a point that it is believed that the setting forth of a few useful analogies may possess some value. It is hardly too much to say that the present status of the controversy is that geological evidence continues to accumulate showing that lateral movement of continental masses has taken place, as is indeed admitted, either directly or tacitly by many of the opponents of the Wegenerian theory, while mathematicians and cosmogonists continue to reiterate that such is impossible. Now the real meaning of this attitude is that the mathematicians and astronomers have not yet discovered a cause, which is not by any means the same thing as proving that there is no possible cause, a philosophical distinction which is very commonly disregarded.

(Rastall, 1929: 447)

After describing three such instances, Rastall concluded:

Here, then, we have three instances in which mathematical theory has either led to completely wrong results, or failed to produce an explanation, or has not yet tackled one of the fundamental points of tectonic geology. All of these cases lead to "impossibilities" which nevertheless have occurred. Is it then too much to ask for a suspension of mathematical judgment on the problem of continental drift, until geology and other sciences of observation have shown whether such drift did or did not take place?

(Rastall, 1929: 450)

In his 1929 edition of *The Origin*, Wegener appealed to historical precedent.

The determination and proof of relative continental displacements ... have proceeded purely empirically ... This is the inductive method, one which the natural sciences are forced to employ in the vast majority of cases. The formulation of the laws of falling bodies and of the planetary orbits was first determined purely inductively, by observation; only then did Newton appear and show how to derive these laws deductively from the one formula of universal gravitation.

The Newton of drift theory has not yet appeared. His absence need cause no anxiety; the theory is still young and still often treated with suspicion. In the long run, one cannot blame a theoretician for hesitating to spend time and trouble on explaining a law about whose validity no unanimity prevails. It is probable, at any rate, that the complete solution of the problem of the driving forces will still be a long time coming ...

(Wegener, 1929: 167)

Likewise, Wooldridge and Morgan also appealed to historical precedent in both the first (1937) and second (1959) editions of their co-authored textbook. But by the time of the 1950 Birmingham meeting, Wooldridge had changed his mind and sided with fixists (§5.9). Citing Jeffreys, he claimed that recent findings in seismology and geomorphology, his own specialty, had again made the idea of sinking landbridges viable. He also thought that even though Holmes had persuasively advocated subcrustal convection, no demonstrably adequate mechanism for mobilism had been given. In 1959, however, Wooldridge changed his mind yet again, and switched back to mobilism. Although Wooldridge and Morgan thought that Holmes' solution to the mechanism problem looked promising, they did not think the mechanism problem had been solved. Comparing current views about mobilism and past views about the Alps, they, following Rastall, argued that the lack of a proven mechanism was no good reason for rejecting mobilism.

We cannot claim to have recognized the force with certainty, though suggestions made by Holmes point to a likely line of inquiry. Meanwhile, the essential question at issue is whether drift has occurred. If its reality could be demonstrated our present inability to explain it would not really constitute a criticism of the theory. A cogent parallel has been drawn between the present position and that formerly existing in regard to folded mountain ranges such as the Alps. The structures portrayed by geologists in such regions were formerly dismissed as mechanically impossible. Now that virtual unanimity has been reached as to their existence we are little nearer to a complete understanding of the mechanics of their origin – but this evidently does not prove that they are impossible.

(Wooldridge and Morgan, 1959: 41)

Before Jeffreys and other geophysicists proclaimed Holmes' hypothesis physically extremely unlikely but not utterly impossible, these defensive responses were the best that mobilists could offer. Holmes himself, however, essentially provided them with a much stronger reply: "Mr. Jeffreys, mobilism is not impossible!" and once this response became known, mobilists no longer had to support a view deemed by the majority to be physically impossible. However, difficulties still plagued Holmes' theory, and appeals to historical precedent continued. In addition, given the vastness of unknowns regarding Earth's interior, even Holmes had perhaps come to feel that it would be fruitless to keep working on the mechanism problem; it would be a long time before the question of what drives the continents could be answered, before mechanism theories could be properly tested. Of course, as it turned out the foolproof way to remove the charge that his theory was impossible was to show by observation that drift actually happened. This was yet to come. By the early 1950s, mobilists still had no empirically based solution that was even remotely difficulty-free.

5.14 Difficulty-free solutions, theory choice, and the classical stage of the mobilist debate

I have proposed that a difficulty-free solution to a first-stage problem warrants acceptance, and is unreasonable to reject (§1.12). Neither mobilists nor fixists presented even a single difficulty-free solution during the first stage of the controversy. As a result, and as I shall continue to document in this volume, neither fixists nor mobilists were obliged to change their minds or were shown to be decisively unreasonable. Also, it was possible quite reasonably to sit on the fence, and many did. The geodetic solution that Wegener had proposed to the problem of the westward drift of Greenland relative to Europe turned out to be based on unreliable data. Reasonable, unanswered objections were raised against both mobilist and fixist solutions to biotic and geologic disjuncts. Mobilism's most general solution, its explanation of Permo-Carboniferous glaciations, still faced well-recognized difficulties even after the many attempts by mobilists to remove them. At the beginning of the 1950s, neither mobilists nor fixists had difficulty-free solutions.

It is beyond question that almost everyone, be they mobilist or fixist, regarded the alleged physical impossibility of the process of continental drift as the strong reason for its rejection. Even committed mobilists recognized this as a formidable obstacle to their cause, and perhaps because of it, many may have lost heart, or simply adopted a wait-and-see attitude, going on to address other problems. As developments in the late 1960s showed (Volume IV), it is one of the great ironies of the mobilism debate that the absence of an acceptable mechanism ceased quite suddenly to be seen as an obstacle to mobilism.

In closing this chapter, I want to raise the following hypothetical question: What if Holmes or someone else had offered a difficulty-free solution to the mechanism problem, would this have led to the collapse of the fixist's case and the general acceptance of mobilism? I think not. Suppose, for a moment, that in the 1930s and 1940s, there had been good reason to believe that large-scale mantle convection was likely. Most fixists, I believe, still would not have seen any compelling reason to believe that it had occurred, simply because no quantitative physical measurements of such vast displacements had been made. Vening Meinesz and Griggs favored mantle convection, but they did *not* favor continental drift. *The apparent impossibility of a process was almost universally regarded as a reason for rejecting drift, but the possibility of such a process was insufficient reason for drift's acceptance.* Perhaps many fixists who ignored Holmes' convection theory decided that it was not worth taking seriously because, at best, it explained how something that they felt all along was highly unlikely might have happened. So long as continental drift lacked a physical demonstration of displacements, it did not have a difficulty-free solution to a first stage problem: it was unworthy of acceptance. Thus, as I proposed in §1.12, having a difficulty-free solution to a second stage problem is not sufficient for acceptance; to show that something definitely could have happened, is not sufficient to show that it did happen.

In 1950, there were two possible scenarios that could, I believe, have led to the development of either a mobilist or fixist theory worthy of acceptance. First, the development of a difficulty-free solution to one of the first stage problems that had already been solved, a problem for which a solution had already been offered but not accepted. This did not happen. Second, a difficulty-free solution could have been developed to an entirely new first stage problem. This is what in fact happened. Indeed it happened four times (see §1.12). Paleomagnetists were the first to develop a difficulty-free solution. I shall argue in Volume II that their pro-mobilist explanation of intercontinental scattered paleopoles or widely divergent APW paths, achieved difficulty-free status by 1959, and by 1964 had been fully confirmed. However, not many, outside paleomagnetism, recognized its difficulty-free status, and most continued to reject mobilism. In Volume IV, I shall show the development of the other three, the Vine–Matthews hypothesis, the notion of ridge-ridge transform faults, and plate tectonics.

Notes

1 Part of the ensuing discussion of Holmes' shift from contractionism to continental drift and the presentation of his 1928/31 hypothesis of substratum (mantle) convection has been drawn from my previous paper, "Arthur Holmes and Continental Drift" (1978). Alan Allwardt (1990) in his dissertation, *The Roles of Arthur Holmes and Harry Hess in the*

Development of Modern Global Theories, has greatly extended my analysis. Allwardt provides a detailed analysis of the evolution of Holmes' ideas from contractionism to continental drift, the evolution of Holmes' theory of convection, and its reception and further defense by Holmes. Consequently, I have incorporated some of Allwardt's analysis here. My debt to Allwardt is extensive, especially in regard to certain aspects of Holmes' defense and how others who also hypothesized convection within the Earth's interior received his version of convection currents. I have also been aided by Naomi Oreskes' account of Holmes' mobilism (see Oreskes, 1988, 1999). In my earlier account I failed to place enough emphasis on Holmes' account of mountain building in his defense of his mobilism. I think Oreskes got it right. Oreskes and I, however, disagree about many aspects of Holmes' hypothesis, its defense, and its reception. We also disagree about Jeffreys' importance (Oreskes, 1999).

2 According to Doris Holmes, Jeffreys and Holmes became friends while Holmes was a student at Imperial College. She also commented on her husband's opinion of Jeffreys as a mathematician, physicist, and geologist.

Harold Jeffreys and Arthur were students together at Imperial College. Arthur respected Jeffreys as a mathematician and physicist but always said that he knew no geology.
(letter from Doris Holmes to author, February 4, 1979)

Although Holmes and Jeffreys certainly got to know each other while they met in London, Jeffreys was never a student at Imperial College. He was a student at Cambridge while Holmes was a student at Imperial College. Nevertheless, it is clear that Holmes and Jeffreys maintained a long-standing friendship. Doris Holmes noted that her husband and Jeffreys used to exchange rude postcards.

I was always astonished at the rude postcards they sent to each other, the sort of thing that only contemporary students would do in later life.
(letter from Doris Holmes to author, February 4, 1979)

3 This brief synopsis of Holmes' life and career is primarily drawn from four sources: C. Lewis' *Dating Game* (1980); K. C. Dunham's "Arthur Holmes" (1966); and Doris Reynolds' "Memorial of Arthur Holmes" (1968). Lewis (2000) is the first biography of Holmes, and is an insightful examination of his life.

4 Letter from Doris Holmes to author, February 4, 1979.

5 It was during this period that Holmes and Jeffreys debated the severity of the difficulties Holmes had raised against Jeffreys' theory of thermal contraction. The key articles are Jeffreys (1926b, 1927), Holmes (1925a, b, 1927). However, I shall not follow their entire debate, but restrict my attention to their exchanges about Holmes' mobilism. For details, see Allwardt (1990).

6 When I wrote my previous analysis of Holmes (Frankel, 1978), I outright missed this comment by Holmes contained in this short review of Joly's work. Allwardt, in his more complete analysis, cites it (Allwardt, 1990: 35).

7 Allwardt (1990: 34) was informed by Doris Holmes:

From the time of the publication of Wegener's book (1915) Arthur Holmes was convinced of the reality of continental drift. He merely added drift to his own conviction that convection currents were inevitable and saw that they fitted.

Allwardt questions this recollection of Doris Holmes. So do I. What bothers us is her claim that Arthur Holmes was convinced of the reality of continental drift as far back as 1915 with his reading of the first edition of Wegener's book. Holmes doesn't even mention continental drift in his publications until 1925. In 1915 and 1916 he argued that the discovery of radioactivity did not create a theoretical difficulty for contractionism. Holmes' own published remarks indicate he was only intrigued by continental drift. Moreover, as Allwardt points out, Doris Reynolds did not know Holmes during this period. In fact, Reynolds and Holmes married in 1939, six years after she joined the department at

Durham. Continental drift certainly attracted Holmes. Perhaps, he privately entertained the idea of continental drift. Allwardt (1990: 34) concludes:

we should likely take Mrs. Holmes's statement at face value and at least admit
the possibility that Arthur Holmes had mixed scientific loyalties in 1915. If so, it would
not be the last time he found himself torn over the subject of continental drift ...

Allwardt is more generous than I am. I think that Doris Holmes' simply overstated Holmes' early attitude toward continental drift. I'll stick with Holmes' own assessment that he was intrigued by it. However, it may be and probably is true that he thought about continental drift and read much of the literature on it. Moreover, he certainly thought that the mechanism difficulty was severe, and only mentioned continental drift when discussing Joly's possible way out of the mechanism difficulty.

8 However, Bull (1927) did discuss continental drift in his next paper on convection currents and mountain building. He spoke favorably in its support, citing its solutions to problems in paleontology and paleoclimatology, and noted difficulties with previous mobilist solutions to the mechanism problem. Here he even mentioned Daly's theory, raising the same difficulty that Holmes later offered.

In order, however, for the block of crust to be tilted sufficiently to slide under the force of gravity, there would probably be required very considerable elevation of the tension regions
(Bull, 1927: 155)

Bull next suggested (1927: 155) that his convection current hypothesis may provide "a suitable force to move large portions of the crust," but noted that "little evidence can be adduced in support of this." Nevertheless, Bull thought that his convection current hypothesis might provide mobilists with a solution to their mechanism problem.

Thus, we may wonder whether Holmes came up with the link between convection and mobilism on his own, or found it in Bull. There is no question that Bull inspired Holmes. But, did Bull provide him with an explicit link? I don't know. Holmes presented his ideas on mantle convection and how continental drift might relate to it in January of 1928. Holmes noted Bull's "happy suggestion," but did not provide a specific reference. Holmes' first reference to Bull's 1927 article does not appear until 1929, at least a year after Holmes had come up with his hypothesis (see Holmes, 1929: 347). In addition, Bull, who was president of the Geologists' Association, presented his paper at his Presidential Address, and published it in the association's proceedings. But this is not an important question. Regardless of whether Holmes made the link between convection and mobilism or got it from reading Bull's 1927 paper, Holmes developed it extensively, garnered evidence in its favor, and defended it throughout the ensuing years. Holmes, of course, credited Bull with the idea. But, he correctly noted:

The possibility that currents might be set up by differential radioactive heating
was recognized by Bull in 1921, though in 1927 he considered that little evidence could
be adduced in its support.

(Holmes, 1931: 565)

Besides Bull, there were others who had mentioned the idea of convection currents as providing a mechanism for continental drift. Holmes discovered some of them either directly, or from reading the AAPG symposium on continental drift. However, because he did not mention how they served as an inspiration, it most surely was only Bull who played that role.

9 Holmes' choice of Bailey was particularly wise. Bailey was highly respected, and was not viewed by fixists as a diehard mobilist. Nevertheless, Bailey (1927, 1928a, b, 1929) was impressed with the match ups between mountain belts on both sides of the North Atlantic.

10 I owe this particular point to Allwardt (1990: 35) for he first recognized this move by Holmes. This is not to say, however, that Allwardt would classify it as an instance of one of the standard research strategies that I've identified.

11 We now know this to be so, although the rates of northward movement vary among continents. See, for example, Blackett, Clegg, and Stubbs (1960).

12 Davies, for some reason unbeknownst to me, claimed that physicists working in geomagnetism would find Holmes' appeal to convection currents useful in attempting to discover the origin of secular variation (Davies, 1931: 537).

13 Allwardt (1990) presents a splendid analysis of the debate between Holmes and Jeffreys over both Jeffreys' thermal contraction and Holmes' mobilism. However, I shall discuss only Holmes' rebuttal to Jeffreys' attack against Holmes' mobilism.

14 MacBride (1938) supported continental drift, arguing in its favor primarily because of its explanation of the distribution of Permo-Carboniferous glaciation.

15 Donald B. McIntyre, one of Holmes' first graduate students at Edinburgh, enjoyed a long-standing professional relationship with Doris Holmes. Both shared similar views about granitization. McIntyre, an excellent source for Allwardt, supplied him with the following excerpt from a letter McIntyre received from Doris Holmes in 1975.

[Arthur Holmes] first met [Vening] Meinesz when he lectured to the Washington Academy of Sciences [April 7, 1932; see Holmes, 1933], and Vening Meinesz told him that his idea was like lifting oneself up by one's boot strings.

(Allwardt, 1990: 64)

16 Vening Meinesz (1934d) also argued that the convection currents that downbuckle the crust in the West Indies would, if they moved the Americas at all, move them toward Europe and Africa. Although he presented the argument in terms of Wegener's theory, it would apply to Holmes' and other versions of mobilism.

Considering this result in relation to Wegener's theory of continental drift, there seems to be a discrepancy. According to this theory the substratum in the oceans is not covered by a rigid crust capable of transmitting stress and so a sinking current in the Atlantic ought to bring about a tendency of the adjoining continents towards each other instead of away from each other. So, for both theories in mutual agreement, we should have to suppose that the direction of the currents has changed since the Atlantic Ocean came into being. We should have to expect that geodetic measurements would nowadays show a narrowing of the Atlantic instead of a broadening. This whole conception does not appear to fit in with the arguments of the defenders of this theory.

(Vening Meinesz, 1934c: 195–196)

17 Hales later recalled that the lectures he took from Jeffreys "(as might be expected) covered a very wide range of topics in geophysical hydrodynamics and elasticity. I found them most stimulating, and indeed most of my activities since then have their origin in those lectures, with the significant exception of paleomagnetism" (Hales, 1986: 1).

18 Greene (1998) raises an interesting point regarding Wegener's use of scaled-down models that use forces and materials that accurately depict what they are supposed to model. Wegener did just that in his experiments to produce impact craters. Greene also explains that King Hubbert's advice about proper scaling had already appeared in German-language periodicals as far back as 1912, that Hubbert noted it had been said, and that he lamented the fact that it seemed to have been forgotten (Greene, 1998: 120).

19 There is another possibility, which Griggs never explicitly stated. Suppose continental material completely covered the Earth early in its history. A series of convective cycles might tend to consolidate continental material and lead to the "primary segregation of the continental masses themselves" from ocean basins. But if this is what Griggs had in mind, such convection would have occurred very early in the Earth's history, and did not constitute Wegenerian continental drift. Even Harold Jeffreys thought that very early convection might have occurred, and Vening Meinesz also proposed such an idea before he began to support continental drift.

20 Even T. H. Holland, who was pro-drift and later helped Holmes secure his position at Edinburgh University (§3.9), was not very disposed toward Holmes' convection theory. At his 1941 Bruce-Preller Lecture before the Royal Society of Edinburgh, Holland declared:

> Favourable conditions for segregation [of basalts] also account for the concentration of radioactive minerals in the outermost layer; but unless these conditions were perfect, there may well be still enough radioactive material at depths sufficient to justify a theory of convection currents, like that proposed by Holmes (1928–1929) to account for continental drift as one way in which the accumulated heat has been intermittently dissipated.
>
> *(Holland, 1941: 164; my bracketed addition)*

21 Wegener, I believe, was unaware of Holmes' hypothesis of convection currents (as he may have been of Brooks' (1926) work on changes in climate zones (§3.13)) when writing the 1929 edition of *Die Entstehung der Kontinente und Ozeane*. His latest reference to Holmes is to a paper published in December 1927. Wegener included references to several works that were published in 1928, including the proceedings of the AAPG symposium. However, he did not refer to Holmes (1928a), the short summary of the presentation of his hypothesis to the Glasgow Geological Society, to Holmes (1929b), or to Holmes' short review in *Nature* of the AAPG symposium. There are no references to works that were published in 1929. However, Wegener was aware of several other works whose authors put forth the idea of convection as the cause of continental drift. Wegener mentioned works by O. Ampferer (Wegener, 1929: 59–60), who published an article in 1925, R. Schwinner (Wegener, 1929: 59–60 and 178), who published an article in 1919, and G. Kirsch (Wegener, 1929: 60 and 178), who wrote a book on geology and radioactivity in 1928. He introduced them in this way:

> At this point we should mention the authors who attribute the phenomena of the outermost crust to "undercurrents," such as Ampferer [69] and Schwinner [69], among others. According to Ampferer, undercurrents have dragged America westwards; Schwinner believes that there are convection currents in the liquid layer caused by non-uniform output of heat, and that these currents draw the crust along and compress it at areas where they take a downward path. In connection with the excess radioactive heat production in continental blocks, Kirsch [70] has made extensive use of the idea of thermal convection currents in the fluid layer. He assumes heat was generated beneath them ... this led to circulation of the fluid substratum, which flowed downwards due to increased thermal loss, while rising under the center of the continental region. The continental platform was finally broken up by the friction and the fragments were separated in all directions by the current.
>
> *(Wegener, 1929: 59–60; citations refer to Ampferer, 1925,*
> *Schwinner, 1919, and Kirsch, 1928)*

Returning to their ideas at a later point in his 1929 book Wegener concluded:

> If the theoretical basis of the ideas should prove adequate to support them, they [convection currents] could in any case be considered as contributory factors in the formation of the surface of the earth; it is still not possible at present to survey the theoretical background.
>
> *(Wegener, 1929: 178–179; my bracketed addition).*

Wegener, as noted already (§2.7), mentioned internal currents as a possible cause of mobilism in his *Petermanns* paper (1912b).

22 All references to Holmes' first edition of *Principles* are to the 1945 American version.

23 Although I do not know what importance Holmes placed on this change he made, it is certainly interesting in light of Hess's and Dietz's later versions of seafloor spreading. Holmes' 1944 version is similar to Hess and Dietz – actually, it is closer to Dietz than to Hess. Oreskes (1999: 270–271) makes precisely the above point. But there was a major difference between Holmes of 1944, and both Hess and Dietz, which turned out to be very significant. The Vine–Matthews hypothesis is not a "corollary" of Holmes' view. With Hess and Dietz, new seafloor is created only within the narrow region at the center of ridge

axes, a necessary condition for the Vine–Matthews hypothesis. Holmes (1944/1945: 508), however, envisioned the rising basalt "escaping from innumerable fissures, spreading out as sheet-like intrusions within the crust, and as submarine lava flows over its surface." Thus, there would be places where older seafloor would be considerably closer to ridge axes than newer seafloor, and the well-defined zebra pattern of normally and reversely magnetized stripes of seafloor would not develop as well as it did. Dyke injections along ridge axes appear to be injected within a zone whose width is only a few hundred meters (Fowler, 2000: 299).

24 There is no question that Holmes was aware of their work. Allwardt refers to Donald B. McIntyre's lecture notes of April 11, 1944, from Holmes' undergraduate course in "Advanced Physical Geology," which recount Holmes' discussion of Pekeris' calculation of how the amount of crustal drag of convection upon the crust would be sufficient to rift a continent. Citing McIntyre, Allwardt notes:

> However, Holmes never published these calculations – and not because he lacked the opportunity: the course that McIntyre took in 1944 was essentially a preview of Holmes' forthcoming textbook, *Principles of Physical Geology* (Donald B. McIntyre, oral communication, 1984).
>
> *(Allwardt, 1990: 113)*

25 Wegener's idea that drift began with the breakup of Pangea in the Mesozoic caused considerable difficulty for mobilists. Holmes was clearly bothered by it. Argand (§8.7) took exception to it, and proposed a proto-Atlantic whose closing helped to form Pangea. The first physical evidence that continents had drifted independently of each other before the formation of Pangea came from Irving and Green's (1958) paleomagnetic work in Australia (II, §5.3).

26 Doris Holmes, agreeing with my former assessment, wrote the following:

> You are right that Arthur's suggestions were ignored at the time of publication because there was no way of testing them physically at the time, and because physicists had known of no way in which the continents could move they denied the very possibility of drift.
>
> *(letter from Doris Holmes to author, February 4, 1979)*

27 Kuenen (1950) raised the same difficulty against mobilism (§8.13).

28 The three other times Jeffreys had earlier discussed mobilism at public meetings before were in 1922 at a meeting of the Geological Society of London (§4.3), in 1931 at a meeting of the geography section of the BAAS (§5.5) and in 1935 at the joint meeting of the Geological Society of London and the RAS (Britain) (§5.5).

6

Regionalism and the reception of mobilism: South Africa, India, and South America from the 1920s through the early 1950s

6.1 Introduction

Drift is better observed and studied in the Southern Hemisphere. There are several very clear fragments of Gondwanaland and they each preserve better evidence of their Palaeozoic relations than anything in Laurasia. The simplest example is the Karroo system and its equivalents. As a former student of mine wrote from South America, "Doc, it's exactly like Africa, and one can always know what series will lie below that one is standing on, and how the fossils will change." See one place, see one continent and you have seen the lot. When the American (W. E. Long) described a Palaeozoic tillite in the Transantarctic Mountains, Adie (Yes, he was my student, as a matter of fact his ice axe is six feet from me in my study and my Antarctic boots beside it.) chuckled and said, "You've got the Dwyka tillite." Long protested against any such glib identification, so Adie arranged his visit to South Africa where he stood in wonder at the things he first knew in Antarctica. I've seen it also in southern Australia, in Brazil, and in Kashmir.

(Lester King, second December 29, 1981 letter to author)[1]

In Chapter 1 (§1.14, §1.15), I gave a general account of how the reception of mobilism in different regions of the world depended, in large part, on how well it explained local or regional phenomena. *Regionalists'* attitudes toward mobilism were largely shaped by their study of their own regions, and *specialists'* attitudes were largely shaped by their specialty. Regionalists may be contrasted with *globalists*, specialists with *generalists*. Globalists and generalists were rare during the classical phase of the mobilism debate, and Earth scientists who were both, like Wegener and Holmes, were rare.

Most geologists from North America and Australia who participated in the controversy were opposed to mobilism. Responses from continental Europe and the British Isles were more varied. Workers on the geology of India, southern Africa or eastern South America were inclined to support mobilism. Mobilism better explained the distribution of Permo-Carboniferous glaciation, the strong similarities in the Late Paleozoic and early Mesozoic sedimentation history of the three regions, and the congruency between the facing margins of Africa and South America. For the times, mobilism also provided a reasonable explanation of the Himalayas and Andes. By contrast, those who worked on the origin of mountains in North America or Australia were much more likely to support fixism; fixist solutions to the origin of mountains worked particularly well in such regions with an ancient stable craton bordered by increasingly younger orogenic belts. Feeling more confident in their own work than in the work of others, and being very much more familiar with their own geology than of Earth as a whole, most participants in the fixism–mobilism debate opted for explanations that best fitted the geology of their own region. Making geological observations is an interpretative activity and fieldworkers are often unwittingly influenced by the theories they support. Readers in one region often doubted reports and interpretations from other regions that were not in agreement with their own theories and interpretations.

Besides being distrustful of field reports from elsewhere, regionalists did not look seriously at alternative views, views that explained much more successfully than their own the geology of regions elsewhere. Regionalists were reluctant to entertain seriously the possibility that there might be virtue in alternative ideas developed elsewhere, reluctant to consider whether some phenomena in their own region might be explained better by those ideas or to redo their own fieldwork to see if they had missed anything that might comply with them, or even to go elsewhere to do fieldwork in places where the alternative theory explained the geology well. Finally, and I believe most importantly, regionalists did not properly evaluate their regional theories within a global context. When considering the question, "Is my fixist theory, which nicely explains the geology of my region, the best global theory?" regionalists gave little weight to intercontinental similarities. In contrast, mobilists gained much of their support from their attempts to explain intercontinental similarities. Regionalist attitudes were generally not helpful to mobilism.

The next three chapters elaborate this idea that theory preference depends, in part, on region, and the specific regional problems addressed. In this chapter, I examine the reception of mobilism in South Africa, India, and South America, three regions where it was favored by most locals who participated in the controversy, and begin by comparing two autobiographical essays. One is by Ken Caster (1981), the American who has already been mentioned (§3.8); the other is by Edna Plumstead (1982), a South African stratigrapher and paleobotanist, who, in 1952, made the important discovery of the fructifications of *Glossopteris*, the ubiquitous Permian plant of Gondwana. Both essays were written in response to questions I asked them about their involvement with mobilism, and they make an informative contrast. Both authors had long careers, and were sympathetic to mobilism. Caster lived in North America where mobilism was generally rejected. Plumstead's homeland was very different; South Africa had two of the most vigorous defenders of mobilism (Alex du Toit and Lester King), one prominent geologist mildly opposed (Sidney Haughton), one prominent geophysicist initially against mobilism (Anton Hales), and a few minor fixists.

After describing mobilism's reception in South Africa, I shall examine the attitudes of geologists who devoted their professional lives to understanding the geology of India, and discuss a symposium, held in 1937 at the 24th Indian Science Congress and entitled "Wegener's Theory of Continental Drift with Reference to India and Adjacent Countries," at which ten scientists, including Birbal Sahni, spoke. In contrast to most drift-related symposia held elsewhere, twice as many preferred mobilism to fixism because it solved Indian problems. I shall then describe the attitudes of the two leading Indian geologists, D. N. Wadia and M. S. Krishnan. I shall also describe how mobilism was favorably received by geologists in Brazil and Argentina.

6.2 Ken Caster and his attitude toward continental drift

It might seem strange to begin this chapter with a North American. I do so to illuminate an important almost self-evident feature of regionalism: although it could cripple the development of global solutions by workers who stayed in one place, it could be cured by exposure to an "elsewhere," to another and very different geology. Scientific tourism, however, was insufficient; only a deep professional commitment was an effective antidote. Caster is a good example of this, as are Darlington (§3.8; II, §1.12), Long (§7.4, §7.5), Hamilton (§7.6–§7.9), and Opdyke (II, §5.13–§5.15), whom we shall encounter later.

In his 1981 essay, Caster recalled that he first became interested in continental drift in 1927, one year after the AAPG symposium on continental drift. He was born in 1908, and grew up in Ithaca, New York, where he attended Cornell University. Continental drift was not often discussed in his classes, but was talked about among students. He thought of himself as somewhat unconventional when it came to ideas.

I first became enamoured of the concept of Continental Drift during my junior year at Cornell University (1927) . . . for I received no mention of the theory in either Heinrich Reis' introductory course or in G. D. Harris' Historical Geology, the latter being "Suessian" and very much a traditionalist land-bridger of the permanent oceans.

Harris was to become my major professor in the Graduate School, when invertebrate paleontology became my field. Harris was one of the world's outstanding students of Tertiary Mollusca and I was one of his last students before retirement. For Tertiary invertebrate paleontologists the status quo of our present oceans and continents are axiomatic, and theories of ancient drifting continents had little appeal . . . My interests, unlike most of his former students, lay in Paleozoic relationships of strata and fossils, and present geographic realities were less satisfactory paradigms. Among my graduate peers in paleontology, most of whom were Tertiary students, the Drift concept was a matter of less than cogent speculation. But drifting was very much on the minds of other graduate students in the department, and especially those who had been exposed to Professor C. M. Nevin's courses. Nevin was a sedimentologist and structural geologist, out of Penn State, and newest senior member of the geology faculty. In addition to new courses in structural geology and sedimentation, he offered a course in "Major Problems in Geology" which, among other topics, encouraged discussion of Wegener's hypothesis, for which the balance of the geology faculty, all much older, had little use . . . Among advanced biology students, especially in Vertebrate Zoology, there was also debate about "Wegenerism." Being by nature unorthodox and heretical, I was inclined toward what was still, after 17 years, in America, at least, a new idea. My copy of Skerl's 1924 translation was bought in 1929, and went with me into the field as diversionary reading during three years of study of the late Devonian of the Oil Region of western Pennsylvania, 1929–1932.

(Caster, 1981: 1–2)

Given Caster's interest in drift, he was lucky that Nevin was on the faculty at Cornell. Nevin (1892–1975), unlike most North American structural geologists, had an enlightened attitude toward continental drift, giving a straightforward evaluation of it in his *Principles of Structural Geology* (1949). Following Wegener, he introduced drift as a promising compromise between permanency and landbridge theories. Permanency could not explain biogeographic and other intercontinental disjuncts, but landbridges were inconsistent with isostasy. However, he raised difficulties with drift (RS2): Why did it happen only once? "The question is reasonable, and the answer is not forthcoming" (Nevin, 1949: 291). For the Americas, he found the link between mobilism and mountain building problematic.

If the continents were together when sediments were being deposited in the Appalachian geosyncline, there would have been plenty of land material. However, this reasoning does not apply to the western source of sediments which compose the Andes Mountains. An ocean basin would lie in that direction no matter how the continents were grouped or drifted. This is very disturbing because it touches the general theme of permanence of continents, and theory of continental drift furnishes no solution.

(Nevin, 1949: 291)

And, like so many others, he raised the mechanism difficulty, commenting that the "defect in the theory becomes no less grave even if we remember that the forces which cause mountain deformation and major vertical uplifts are also complete mysteries" (Nevin, 1949: 291). He (1949: 290) accused Wegener "of special pleading whereby he glossed over, or completely ignored many of the difficulties and at the same time presented his viewpoint in the best possible light." He (1949: 290) claimed, "The fact of the matter is that geologists are not as yet in a position either to build up or tear down the theories of continental drifting," and argued:

Whatever may be the final outcome of the many theories of continental drifting that have been proposed, the fact remains that the conception of large horizontal movements of the crust is valid. In any discussion of the tectonics of the crust no one should ignore the possibility of both inter- and intra-continental drifting.

(Nevin, 1949: 291)

Caster took a course in paleobotany from Loren Petry, and met Arnold (§3.5), a Ph.D. student under Petry. Both Petry and Arnold were pro-drift. Caster also was impressed with du Toit's work.

Somehow, now obscure, a copy of Alexander du Toit's (1927[a]) "Fossil Flora of the Upper Karroo Beds" came to me. I had become much interested in botany, and paleobotany especially, through Professor Loren Petry at Cornell and Chester Arnold, who was doing a doctorate in paleobotany with him, and was my botany laboratory instructor. Both were favorably inclined toward continental drifting. I believe I studied paleobotany with Petry the first time he offered such a course . . . [Du Toit's] close comparison with the Brazilian coeval flora impressed me as being important new "Wegenerian" data, as did his preprint (1927) of a geological map of a part of South America . . .

(Caster, 1981: 3; my bracketed additions)

Caster also mentioned his reaction to the AAPG symposium and to Schuchert's 1932 paper in which his former landbridges became Willis-like isthmian connections.

It was in these formative years that the A.A.P.G. symposium (1928) on The Theory of Continental Drift – a generally favorable presentation, and "shocker" to the earth science establishment, was held. It engendered much debate among us Cornell proto-geologists. But so did Schuchert's (1932) rebuttal (Gondwana Land-Bridges), which most of us would not "buy."

(Caster, 1981: 2)

It is no surprise that Caster did not accept Schuchert and Willis' isthmian connections (§3.6); however, it is surprising that he had a favorable impression of the AAPG symposium (§4.3) because just about everyone who has reviewed the symposium remarked on its anti-mobilist tenor (see Newman, 1995). It must have been van der Gracht's extensive mobilist-leaning introductory and concluding comments (§4.4), and the fact that the AAPG had even held a symposium on such a widely unpopular topic, that gave Caster this impression.

In 1936 Caster took a job at the University of Cincinnati with support from Bucher, who already was becoming a major force against drift (Caster, 1977: 651). He continued to read du Toit, whom he had met in Washington at a meeting. Du Toit impressed him.

Meanwhile . . . Du Toit's masterful (1937) *Our Wandering Continents* had appeared. I had met Alexander du Toit at the Washington International Geological Congress in 1934 and we had become fast friends when he found me taken by the Drift Hypothesis. Probably it was acquaintance with the man that firmed up my already established predilection for Continental Drift.

(Caster, 1981: 3–4)[2]

Caster also met King, and they became friends.

I believe it was at the Pan-American Scientific Congress in Washington (1940) that I first met Lester King of Durban, South Africa; then, or shortly thereafter, when he lectured in Cincinnati. It was another case of instant rapport. He too was an ardent advocate of Continental Drift, but with a new, physiographic, approach.

(Caster, 1981: 3–4)

After reading the debate between du Toit and Simpson, Caster (1981: 4) recalled, "I became ever more eager personally to see the southern world for myself, as well as to accumulate relevant literature." The opportunity arose in 1945, and he spent three years in South America.

In 1945 my chance came, thanks to the American State Department and the John Simon Guggenheim Foundation: I was invited as Visiting Professor at the University of Sao Paulo,

Brazil. This was another aftermath of a friendship made at the Pan-American Scientific Congress where I met the Director of the Servico Geologico do Brasil, Dr. Glycon de Paiva. My expected one-year Visiting Professor visit to Sao Paulo extended to three years, when I replaced the Head of the Geology Department upon arrival in Brazil.

(Caster, 1981: 4)

Caster's field studies and examination of museum fossils increased his appreciation of the similarities between the Paleozoic fauna of southern Brazil and Argentina and those of South Africa.

This long stay, with adequate financial support, gave me opportunity to visit the widespread Paleozoic, and especially Devonian, fossiliferous sites in Brazil . . . I was able to collect and make field studies in the southern Brazilian states of Rio Grande do Sul, Santa Catarina and Paraná – all readily accessible; also by expedition to the deep interior of Goiás and Matto Grosso, and especially on the Rio Araguaia. In the northeastern state of Maranhão I was instrumental in adjudging a totally new fossiliferous Devonian area, and during the boreal summer, after leaving Brazil, I had the opportunity, thanks to the Brazilian National Petroleum Council, to examine the remote outcrops of the Paleozoic on the lower Amazon and its north and south tributaries. Evermore the oneness of the Austral Devonian fauna, especially the southern expressions of it in south Brazil, Argentina and South Africa became manifest.

(Caster, 1981: 5)

Caster thought that the Geology Department at São Paulo should have a copy of du Toit's *Geological Comparison of South America and South Africa*. Du Toit urged him to translate an updated edition into Portuguese, and offered to send notes he had made in preparation for a new edition. J. Frenguelli, an Argentine mobilist, also offered Caster notes intended for the upcoming Portuguese edition.

When I was just arrived in Brazil I tried unsuccessfully to locate a copy of du Toit's 1927 "Geological Comparison of South America and South Africa." Not a copy was to be found in any library in São Paulo or Rio de Janeiro. Having had my personal copy sent from the States, I wrote to du Toit asking for a complimentary copy, if possible, for the Geology Department library. His reply was negative, but he suggested that perhaps I would be interested in a mass of notes he had been accumulating for a revision that now he would never make. Considering the rarity of the work, and its very great relevance to Brazilian geology, had I considered a Portuguese translation, for which his notes could be made available? While these ideas were in the air I met Dr. Joaquin Frenguelli, Director of the La Plata Museum of Argentina at a science congress in Petrópolis. He was much addicted to Continental Drift, and enthusiastic at the idea of a Portuguese rendition of du Toit's works. He also offered to give us a mass of relevant notes he had accumulated on further Argentine documentation.

(Caster, 1981: 5–6)

Caster's colleague at São Paulo, J. M. Mendes, helped him with the translation. Unfortunately Mendes left on sabbatical for the United States, and Caster went to the Andes for a year on a Guggenheim Fellowship before the galleys appeared. Caster recalled what happened.

The handsome volume came out in 1948. This edition was immediately suppressed, however, by the University when Mendes and I discovered that it contained an unacceptable number of typographical errors and downright text changes. An acceptable publication was achieved later by the Brazilian Institute of Geography and Statistics in 1952 (Servico Grafico do Instituto Brasileiro de Geografia e Estatistica, Rio de Janeiro, 179 pp., 12 pls., 1 mp). This is du Toit's last word on the subject of Continental Drift, for soon after publication of the annotated translation he died.

(Caster, 1981: 6)

Before leaving South America, Caster was invited to speak at the 1949 New York meeting on Mesozoic South Atlantic landbridges, where he argued for a land connection across the South Atlantic and against continental drift (§3.8). A short while after, he and his co-author, O. D. von Engeln, favored mobilism in their 1952 textbook, *Geology*. They (1952: 533–534) discussed the Permo-Carboniferous glaciation of Gondwana. In their reconstruction of Gondwana they placed Africa and South America in direct contact, attached Australia to Antarctica which touched Africa, and rested India and Madagascar against Africa's east coast. They also raised two difficulties against foundering landbridges (RS2).

The two great objections to the former existence of so vast a land mass (in addition to those now present) are (1) that its later foundering is difficult to account for isostatically and (2) that, if it did sink, the enormous displacement of ocean water entailed by its submergence would have produced a giant inundation of the remaining continental areas.

(von Engeln and Caster, 1952: 534)

Their next section, "Theory of continental drift," began with a brief summary of its development, and noted that it avoided the difficulties of foundering landbridges (RS3).

To escape some of the difficulties involved in the disposal of most of a hypothetical Gondwana by a vertical foundering continent, and to find a plausible explanation for a variety of phenomena that are otherwise extremely difficult to account for, a *hypothesis of drifting continents* has been proposed and vigorously supported. This hypothesis had its inception from simple map inspection. It was noted that the eastern shoulder of South America could be rather snugly fitted into the armpit of Western Africa if the two continents were shoved together across the intervening Atlantic Ocean. Once this idea was grasped, it was noted that numerous similar fittings, after the manner of resembling a jigsaw puzzle, could be made. Across the North Atlantic many parts of Eastern North America and Western Europe would, if brought together, interlock without doing violence to structural conditions along the line of junction. Madagascar could go nicely back into the outline of Eastern Africa across the Mozambique Channel.

(von Engeln and Caster, 1952: 535)

They described the drifting apart of Gondwana's fragments.

The drifting-continent hypothesis does not require a shift in the position of the axis of rotation of the earth. Instead, it merely infers that a thin outer shell of the crust of the earth has slipped around on an interior core. In the course of such slipping or drifting, a formerly continuous land mass might be torn asunder. Detached marginal fragments could then set about independent voyaging on the surface of the earth at different rates and in different directions. The Gondwana dissevered after the manner would now be represented by widely separated pieces.

(von Engeln and Caster, 1952: 535–536)

They noted, as others had done, that continental drift avoided the need to suppose extensive equatorial glaciation (RS2), arguing that it was the better solution (RS3).

Most important, perhaps, Gondwanaland intact, with the continental glaciers, could then, in Permian time, have been situated in the high latitudes of the Southern Hemisphere. Thus the difficulty of a continental glacierization centered over the equatorial regions is overcome.

(von Engeln and Caster, 1952: 536)

Although for them drift was plausible, they did not accept it outright.

All that may be said with assurance in regard to these interpretations [of the glaciation in terms of drift] is that they provide a plausible explanation for a challenging assemblage of geologic phenomena.

(von Engeln and Caster, 1952: 536; my bracketed addition)

Thus Caster, by the early 1950s, had finally expressed guarded support for mobilism.

Ken Caster showed that North American Earth scientists were not beyond redemption. Extended exposure to other places, where mobilism explained well the phenomena that interested them, could prompt them to change their minds about the relative merits of fixism and mobilism. Regionalism worked both ways. It worked to restrict belief with exposure to just one region; it worked to expand belief with exposure to other regions, especially if one was primed as Caster was for conversion; at Cornell, Petry favored mobilism, and Nevin was willing to entertain it, and then he met du Toit and King. Ready to go and understand an "elsewhere" where support for mobilism was strong, his successful Guggenheim application allowed him to work in Brazil for three years, and this experience later led him to support mobilism. As I noted above and shall describe later, Caster was not the only American to have such a life-changing experience.

6.3 Edna Plumstead and her support for continental drift

"Africa forms the Key".

(Alex du Toit, from the title page of his 1937 Our Wandering Continents*)*

Edna Plumstead (1903–89) née Janisch was born in Cape Town (Maguire, 1990). Youthful experiences led to her work in paleobotany and in Antarctica.

I realise now that two aspects of my early youth had an important influence on my future career of plant zones and Antarctica. My first seven years were spent in the Cape Peninsula, where I learned to love the wild flowers of the area which is now the type area of the S.W. Cape flora, and which is an important feature of the modern flora of S. America and Australia. In 1911, at the time of Union [of South Africa] my father was transferred to Johannesburg but we spent every summer holiday at the Cape, climbing the mountains and searching for the wild flowers. During my teens we used to read aloud in the evenings, all the books then appearing on Capt. Scott's Antarctic Expedition [1910–1913] which ended so tragically and where, on the last section of their return from the pole, Dr. Wilson found and collected specimens of coal and fossil plants (later identified as *Glossopteris*), which he and Capt. Scott carried to their last resting place and which were subsequently described as the most important scientific results of the whole expedition.

(Plumstead, 1982: 1; my bracketed addition)

Plumstead attended Witwatersrand University, where she was introduced to continental drift.

My undergraduate years 1920–1923 were spent at the University of the Witwatersrand, where I included Geology in my first year and found it so fascinating that I changed to a Science degree choosing Geology and Geography as major subjects (3 years) and various one year courses including Botany, Chemistry etc., but Geology under Prof. R. B. Young remained my main interest. In 1924 I took an Honours B.Sc. degree in Geology under Prof. Young, in which I made my first direct contact with Continental Drift. The course included Palaeontology and I became aware that early life on the earth had not been universal but that a zonal distribution was apparent and that a southern floral zone which differed from the three northern zones had been detected in S. America, Africa, Australia and also in India from the late Palaeozoic onwards. We also heard of the work of Feistmantel in the latter part of the nineteenth century which had led to the suggestion of a land, "Lemuria," including India, Madagascar and Africa and subsequently of a hypothetical continent called Gondwanaland based mainly on the southern distribution of a Glossopteris fossil flora.

(Plumstead, 1982: 2)

Unlike Caster, Plumstead needed to travel only thirty miles from her university to examine an occurrence of key importance for the theory of continental drift, *Glossopteris* at its type locality, the spot from which the original specimens described by Seward had been obtained (§3.5).

Up to this time my only contact with fossils was an annual one day excursion for 1st year students of geology, to Vereeniging, 30 miles south of Johannesburg, which was the nearest coal area where fossil plants could be found, and at which a local resident, Dr. Leslie, had made a large collection. He sent a number of plant fossils to Prof. Seward of Cambridge University, to identify. These became type fossils of the southern Glossopteris flora. This excursion for 1st year students was repeated throughout my teaching years and I became fairly familiar with the plant assemblage.

(Plumstead, 1982: 2)

Upon graduation, she was appointed junior lecturer in the Department of Geology, and worked on a thesis on "Phosphate Deposits in the Northern Transvaal," which she completed for her M.Sc. degree in 1925. She was awarded both the Geo. Corstophine Medal by the Geological Society of South Africa for geological research of outstanding merit, and an overseas postgraduate scholarship from Witwatersrand University which made it possible for her to study at Cambridge University.

Before describing her experiences at Cambridge, Plumstead mentioned two significant papers on continental drift that were widely discussed and important for her own development.

During 1925 General Smuts had delivered his Presidential Address to the S_2A_3 (South African Association for the Advancement of Science), in the course of which he dealt extensively with Continental Drift, making reference to du Toit's popular evening lecture to the same society in 1921, "Land Connections between the other continents and S. Africa in the past." I was not at that time a member of the S_2A_3 and did not read either lecture until later but the topic was discussed in all geological communities ... Smuts, as a keen botanist, made important comments on the Cape flora and its associations in S. America and in Australia. This brought the subject to the notice of South Africans like myself, who were familiar with their own flora.

(Plumstead, 1982: 2–3)

In his 1921 address, du Toit had embraced mobilism, especially as an explanation of Permo-Carboniferous glaciation and biotic and other intercontinental disjuncts (§3.5, §3.11, §3.12, §6.5). Smuts said that his purpose was not to defend Wegener's theory, but to show why it was important for South African science; his positive attitude toward continental drift was evident.

Jan Christiaan Smuts (1870–1950) was a man of amazing achievements. Born in Cape Colony, he went to Victoria College, which later became the University of Stellenbosch, where he received first class honors in arts and sciences. He then studied law at Christ's College, Cambridge, again obtaining a first. By the time he gave his Presidential Address to the 1924 S_2A_3 annual meeting, he had been a full-time soldier in the Second Boer War (1899–1902) fighting the British, a statesman who played a major role in the formation of the Union of South Africa, Air Minister in Lloyd George's cabinet in the United Kingdom during World War I, and Prime Minister of The Republic of South Africa from 1921 to 1924. He was the founder of holism, and was in the process of writing *Holism and Evolution*, which appeared in 1926. He also was an avid amateur botanist, well versed in the theory of evolution. Smuts told his audience:

I wish to speak on the place of science in South Africa, science from the South African point of view, its special sphere and role in this sub-continent as distinguished from its universal sphere and role.

(Smuts, 1925: 2)

What special role in science could South African play? It could test Wegener's theory. Smuts understood the value of scientific ideas, and the importance of South Africa to Wegener's theory.

Within the last five years a great impetus has been given to this way of looking at Africa by the Wegener theory, or rather hypothesis, for it is, perhaps, not yet more than an hypothesis. Now let me say at once that I am not to-night going to argue the correctness or otherwise of the Wegener hypothesis. I disclaim any competence or desire to do so. The Wegener hypothesis as an explanation of great groups of problems may be right or it may be wrong. Its profound significance to us is not so much the particular solution it propounds as the attention it focuses on those problems . . . for us in South Africa it has a special interest in its account of the origin and distribution of continents in the Southern Hemisphere . . . For us in this part of the world, the most interesting feature of the scheme is that in it Africa assumes a central position among the continents; it becomes, in fact, the great "divide" among the continents of the Southern Hemisphere. It appears as the mother-continent from which South America on the one side and Madagascar, India, Australasia and their surrounding areas on the other, have split off and drifted away, have calved off, so to speak. The evidence for all this is strong; but it may well be that the evidence is yet insufficient to account for the whole Wegener hypothesis . . . But, even so, it may be right in assigning to the African continent a central determining position in respect of many of the great unsolved problems of Geographical Distribution, and in making that position the key which science will have to use in ever increasing measure if it wishes to unlock the door to future advances.

(Smuts, 1925: 4)

Smuts told his audience to go forth and test Wegener's theory, and to work with scientists from other countries in the Southern Hemisphere instead of looking back to Europe; Wegener's theory offered the best of reasons to stop thinking like a colonial.

One important line of research which [Wegener's hypothesis] suggests to us is the East-West aspect in addition to the hitherto prevalent North-South line of orientation. Hitherto, as I have said, it is the European affiliations which have guided our thought and our research; we have looked to the North for explanations as well as our origins. In future, on the lines of Wegener's speculations we shall look more to East and West – to our affiliations with South America, India and Madagascar and Australasia for the great connections which can explain the problems of our past and present.

(Smuts, 1925: 5; my bracketed addition)

Smuts sought to ensure that South Africa would occupy the central position among southern continents.

We shall look upon Southern Africa as the centre of the Southern Hemisphere and correlate all the relative scientific problems of this hemisphere from that new point of view. The grouping of the southern continents and lands and the intimate connections and interdependence of their scientific problems will be our new point of departure. It may prove a most suggestive and creative point of view for science in general. This new aspect will establish new contacts, and it is generally such new contacts which prove fruitful and reactive for scientific progress. Our workers, in following up the new clues, may reach solutions which will have a far-reaching value for universal science.

(Smuts, 1925: 5)

Up to now, Smuts' talk could be viewed as that of a scientifically knowledgeable politician wanting his country to make its own mark in science, and to free itself from European intellectual domination. But he had a deeper understanding of Wegener's theory, and asked his audience "to bear with me if I now proceed to indicate in a general and cursory way some of the lines of scientific work which may be usefully followed up from this point of view" (Smuts, 1925: 5). He divided his lines of inquiry into geological, botanical, zoological, and climatological. He urged that the geological similarities between South Africa and South America, India, and Australia be further investigated, and added a few specifics of his own.

Several of our formations at the Cape seem to be continued or paralleled by identical or similar formations in India and South America. A proper correlation of the geological systems may lead to most interesting results, and may also throw great light on the past of the three continents. We may be enabled thereby to explain just why they are practically the sole producers of the world's diamonds; why the diamond fields of South-West Africa are situated on the one edge of the Atlantic and those of Brazil on the other; why the coal fields of these three countries and of Australia are confined to the eastern portions of each of these land masses; and why the curious and ancient banded ironstones are so widely spread in South Africa, Brazil, peninsular India and Western Australia, though absent from Europe. The results of such a comparative study for the Southern Hemisphere might be most valuable and might settle many of the problems which still agitate science as to the past of the earth. And correlation of the several geologies of the Southern Hemisphere would decidedly throw a new light on all of them.

(Smuts, 1925: 6)

Regarding zoology, he again stressed comparative studies of the southern continents, and specifically suggested "the present and past distribution of our scorpions and land mollusca in South Africa and in the lands of the Southern Hemisphere generally would give valuable evidence in regard to our zoological origins" (Smuts, 1925: 10).

As for studies of climate, he stressed the Permo-Carboniferous glaciation.

Great ice-ages are known to have occurred far back at the beginnings of geological time before the present sedimentary formations were laid down. To pass to the other extreme, Europe during the Permo-Carboniferous period, when the coal measures were mostly laid down, possessed the climate of a subtropical rain-forest, and, at a much later date, the Magnolia and similar tropical plants flourished in Greenland and Spitzbergen. At the time Europe was mostly covered by shallow seas, and its tropical climate was balanced by a cold, dry climate which existed in the contemporaneous Gondwanaland of the Southern Hemisphere. The Glossopteris flora of the latter was the vegetation of a cold, dry climate. And the glaciation of many parts of Gondwanaland of which evidence is visible over a large part of South Africa shows that great ice-masses must have covered its high table-land. Much other evidence points to the fact that the ancient Africa which formed the centre of Gondwanaland was, on the whole, a cold and arid country.

(Smuts, 1925: 11)

Du Toit could not have said it better. In fact, Smuts (1925: 8) paid homage to the great South African geologist, remarking, "In his lecture before this Association in 1921, Dr. A. L. du Toit gave a most illuminating account of Gondwanaland, its rise, decline, and fall, to which reference may be made for full details." In timely fashion du Toit would soon publish his 1927 *A Geological Comparison of South America with South Africa*, just the sort of comparative study Smuts wished for (§6.6).

Smuts was at his best when he made use of his knowledge of botany, of Wegener, and of Darwin. He suggested that angiosperms might have originated in Gondwana, not in the northern continent.

We have two distinct floras in South Africa; the one, the South African flora, which covers most of sub-tropical Africa and is clearly of tropical origin; the other, a temperate flora, found only in the south-west of the Cape Province on the seaward side of the first great mountain barrier, with outliers extending to the north along the mountain systems into the tropics. The two floras are, apparently, quite different and distinct and are engaged in a mortal conflict with each other, in which the temperate or Cape flora is slowly losing ground.

(Smuts, 1925: 6)

Smuts then inquired about their origin, and doubted that both were derived from European flora.

. . . we meet again with what I may call the European fallacy, or the fallacy of the European origin. The current idea among botanists is that Northern Europe is the source and the north temperate flora of Europe is the origin of both our South African and Cape floras. The north temperate flora of Europe is supposed to have been driven south by the onset of the last great Ice Age in Europe and in the much cooler climate of the tropics at the time to have migrated southward along the eastern mountain systems of Africa until Southern Africa was reached.

(Smuts, 1925: 7)

Instead he suggested that both floras came from the ancient lands of the Southern Hemisphere "which are covered by the Wegener hypothesis." He raised a difficulty against their northern origin (RS2).

Even according to our present knowledge, the African floras do not seem to fit in well with the current view of their origin. Apart from the Cape flora in the extreme south, and the Mediter-ranean temperate flora in the extreme north, the African flora – better known as the Tropical African flora or the Palaeotropical African flora – covers the rest of the continent. In this flora an element predominates which is peculiar to this part of the world but is more or less closely related to the floras of India, Madagascar, Australasia and South America. In other words, the special Tropical African flora has peculiar affiliations with the floras of certain countries in the Southern Hemisphere. The current view of the northern origin may therefore not be the last word so far as Botany is concerned.

(Smuts 1925: 7)

Wendell Camp (§3.8) would make the same argument twenty-five years later.

Smuts turned to Darwin. Remembering that one of the current supporters of the European-origin fallacy credited Darwin with having been the first to suggest the southward migration of the European flora during the Great Ice Age, Smuts noted that Darwin himself had been uneasy about the idea and had suggested another possibility.

Moreover, if Darwin was the author of this theory of the northern origin, it is clear from his letters that an element of doubt remained in his mind on this most important point . . . as late as 1881 we find him writing to his friend Hooker: – "Nothing is more extraordinary in the history of the Vegetable Kingdom, as it seems to me than the apparently very sudden and abrupt development of the higher plants [i.e., angiosperms]. I have sometimes speculated whether there did not exist somewhere during the long ages an extremely isolated continent, perhaps near the South Pole."

(Smuts, 1925: 7–8; my bracketed addition)

Setting the historical record straight, Smuts noted further that there had been substantial advances in paleobotany since Darwin's day, and he appealed to none other than Professor Seward, citing his opinion about the former existence of Gondwana, and arguments based on the widespread occurrence of Permo-Carboniferous glaciations and *Glossopteris*. Smuts then made the obvious connection between Darwin's guess and Wegener's hypothesis.

Finally, Darwin's wonderful guess has found expression in the great southern continent of Wegener's hypothesis, a continent which existed in Permo-Carboniferous times, of which Africa was the centre and South America, Madagascar, India and Australia, as well as Antarctica, were integral portions, a continent which must have become disrupted somewhere in the Mesozoic period, and parts of which have gradually drifted away to the positions which they to-day occupy in the Southern Hemisphere.

(Smuts, 1925: 8)

Smuts asked South African paleobotanists to consider the possibility that angiosperms might have originated in Gondwana. There was much to do, and South Africa and other southern continents were plausible places to look.

Indeed, in the palaeobotany of the Southern Hemisphere we are only at the beginnings; and who knows whether further discoveries in this largely virgin field of research may not yet give point and substance to Darwin's surmise that the existence far back in the long ages of an extremely isolated southern continent is somehow to be linked with the mysterious origin of Flowering Plants. I say this, not in order to express any particular opinion as to the northern or southern origin of the Angiosperms, but to point to the necessity of further research in the fossil botany of southern Africa. Some of the greatest problems of botany, of geographical distribution, and of the past of the earth will have to wait for their solution until palaeobotany has made much further advances in South Africa and the Southern Hemisphere generally.

(Smuts, 1924: 9)

Then, like a good politician or scientist, Smuts made a plea for funds.

In this connection, a great opportunity lies before science in South Africa, and I trust a step will be taken by the establishment of a Chair of Paleobotany at one or other of our South African Universities. It will be a small step, but its significance will be great and its results may be far-reaching.

(Smuts, 1925: 9)

Smuts impressed Plumstead. He was familiar with the Cape flora which she had grown up with and loved so much. Much of her later work led her to conclusions similar to those of Smuts. Plumstead also believed that Smut's appeal for funding led to the establishment of the Bernard Price Institutes for Palaeontology and Geophysics, where she later worked.

In the course of stressing the importance of Palaeontology in interpreting modern plant and animal distribution he appealed to S. African benefactors to finance a Chair of Palaeobotany at one of the Universities. This plea must have influenced Dr. Bernard Price who a few years later established the Bernard Price Institute for Palaeontology and the Bernard Price Institute for Geophysics at the University of the Witwatersrand, from which some outstanding contributions to S. African science have been made.

(Plumstead, 1982: 3)

Plumstead agreed with Smuts about the European-origin fallacy. Referring to it as a contradiction, she recalled:

I realized that this [the similarity of the Cape Flora with floras of South America and Australia] was a contradiction of the view generally held in the N. Hemisphere that flowering plants had originated in Greenland and had gradually migrated southwards via the 3 extensions of continents to Cape Horn, the Cape Province and Australia, developing en route their peculiar characteristics. This view had already seemed absurd and I welcomed General Smuts' contradiction which automatically implied a close relationship between the southern continents, thus supporting a Gondwanaland concept.

(Plumstead, 1982: 3; my bracketed addition)

Years later she did as Smuts had directed. She discovered (1952) the fruiting bodies of *Glossopteris*, which she argued were the ancestors of angiosperms (1956a and 1961). Before her work, many had thought that *Glossopteris* reproduced from spores (Sahni, 1938). Others found that associated with *Glossopteris* leaves were what appeared to be winged pollen grains resembling those of gymnosperms, from which they argued that *Glossopteris* were their ancestors (Virkki, 1937, 1939; Sahni, 1938). Plumstead later (see below) made a new argument in favor of mobilism based on her identification of *Glossopteris* specimens from Antarctica (1961).

According to Plumstead, her involvement in the controversy began in earnest at Cambridge where she chose to work on the microscopy of coals.

I was determined to study at Cambridge University and chose as my research subject "The microscopic study of South African Coals," being a new science which I could justifiably claim could be best studied in England. During 1926 I lectured and also, during vacations, I visited a number of S. African collieries, collecting samples of coal. My application was accepted and in December 1926 I departed for Newnham College, Cambridge as a Research scholar. My studies necessitated a full course in Palaeobotany and since Prof. Seward had retired, Dr. Hugh Hamshaw Thomas, who had been appointed my supervisor, gave the lectures.

(Plumstead, 1982: 3–4)

Hamshaw Thomas (1885–1962) was a student of Seward's. His major area of research was on *Caytonia*. He discovered its fructification and argued that angiosperms arose from it (Thomas, 1925). He also wrote on *Glossopteris* (Thomas, 1952,

1958). Although he (1930) was favorably disposed toward mobilism (§3.13), he published nothing substantial on it. Plumstead continued with her narrative.

In addition to the lectures and laboratory work in Cambridge, I visited Manchester, where my slides were made, and Sheffield University, where the first coal microscopy had been developed. I studied all available slides of British coals as well as a collection of Australian coal slides in the Nat. Hist. Museum, and found the latter to be far more like my own than the English ones.

(Plumstead, 1982: 3–4)

She also recalled an evening she spent in the company of Harold Jeffreys.

Perhaps even more important for the future were the daily contacts with the other geological research students from many universities in Britain and from the Empire. We discussed all modern developments including Continental Drift. We were eligible to be members of the senior geological club of the University to which all earth scientist lecturers belonged, and Dr. Harold Jeffreys was one of these. I remember one evening meeting when Continental Drift was a topic that Dr. Hamshaw Thomas and I supported – from our knowledge of southern fossil floras, while Dr. Jeffreys, as a physicist, opposed it.

(Plumstead, 1982: 4)[3]

Unfortunately, Plumstead had to leave Cambridge before obtaining her degree.

This happy and fruitful period came to an abrupt end in January 1928 when I received a cable from the University of Witwatersrand requesting me to return at once because a critical staff shortage had arisen in the geology department. I felt I should not desert my alma mater and thought I would be able to return to Cambridge later but the occasion did not recur and since a minimum of two years in residence was part of the Cambridge conditions for the conferment of their doctorate I had to abandon it.

(Plumstead, 1982: 4)

She recalled that she became, with help from du Toit and Seward, fully convinced of drift during an international conference held in South Africa in 1929.

On my return to Johannesburg I was a full lecturer and was elected on to the council of the Geological Society, which was preparing for the 1929 International Conference in South Africa. It was to be a joint meeting of the International Geological Congress of the I.U.G.S. (4 yearly) and also a visit from the British Association for the Advancement of Science. It was therefore of considerable importance as leading scientists from the whole world were attending it, and I was able to meet both geologists and biologists. I was one of the guides on the goldfields excursions and also attended a long post congress excursion through Bloemfontein and Natal to Zululand, which was largely palaeontological. Alex du Toit was on this excursion also and I got to know him much better for we lived in a train and moved from place to place, so that discussions were possible.

Probably for many delegates the most important meeting during the congress was the debate on Continent Drift in which, Dr. du Toit, Prof. Seward and several Americans took part. I listened fascinated and from then on was ardently pro-drift but there was still no acceptable suggestion of a mechanism. Du Toit was busy with his book "Our Wandering Continents" which appeared in 1937 but his visits to other areas had only increased his belief.

(Plumstead, 1982: 5)

Unfortunately Plumstead either forgot or neglected to identify the "several Americans" who took part. Perhaps she did not care who they were. Haughton (1962) also recalled the meeting, but he said nothing about the debate and its importance for South African geology. He mentioned two Canadians and one American who might have participated in the debate.

One outstanding [accomplishment] ... was the holding of the XVth International Geological Congress in the Union [of South Africa] in 1929. Although the numbers attending this Congress were far fewer than have attended more recent Congresses – such as those in London, Mexico and Copenhagen – a tremendous amount of preliminary planning with respect to accommodation, excursions and the preparation of guide-books was necessary ... Many of the visitors whom we were so glad to see at that time, men of world-wide reputation such as Coleman, Daly, Bailey Willis. . .

(Haughton, 1962: xi; my bracketed additions)

Plumstead continued teaching and became "completely absorbed in University work but read all available literature on the subject of Drift, but was not directly involved" (Plumstead, 1982: 5). Then in 1934, at the age of 31, her life changed drastically.

At the end of 1934 I resigned to marry Edric Plumstead, a mining engineer, and I became, for the next 12 years, heavily involved in domestic activities and mine life. We had five children and my only academic work was to produce two more editions (enlarged) of the students' geological mapping book.

(Plumstead, 1982: 5)

She returned to the University of Witwatersrand in 1947 as Senior Lecturer; the geology department asked her to return because of the huge influx of ex-servicemen after World War II.

It was at Witwatersrand that Plumstead discovered the reproduction organs of *Glossopteris* in 1950 at Vereeniging, the type location where she had collected *Glossopteris* many years before. Here is what happened.

During the war a new industry, The Vereeniging Brick and Tile Co. had started excavating Karroo shales at Vereeniging and I had heard that a number of good plant fossils had been uncovered and I decided to explore them. When the first general influx of students had abated a new course for Coal Mining Engineers was starting and I was in charge of the Geological section of their work. I planned to take the students to the new quarries to introduce them to plant fossils; it was on this visit that I found the very first fructifications attached to *Glossopteris* leaves. On the same day we visited a friend of one of my students, a Mr. S. F. le Roux, a cabinetmaker who had made a hobby of collecting fossil plants. In his collection were several peculiar "fructs" of which he had sand-papered the surrounding rock and left them like medallions. I was tremendously excited by this discovery and explained to him the importance of attachment and thereafter he increased his efforts and soon found a few more species attached to *Glossopteris* leaves.

(Plumstead: 1982: 5–6)

Plumstead and le Roux built up a large collection. Smuts' appeal, which led to funding of the Institute, had borne more fruit. Seward had his Dr. Leslie, Plumstead, her Mr. Roux. Plumstead returned the favor.

Mr. le Roux knew no botany but longed to study. He had passed his matriculation but had then branched into cabinet making. He was over 40 years of age but was prepared to return to studies if he could be financed. I was able to procure a scholarship grant from one of the industrial companies in Vereeniging and he started a 3-year grind of botany and geology, and graduated. During weekends whenever possible I went to the quarries with him and together we built up the large collection now in the B.P.I. (paleontology) as the S. F. le Roux collection.

(Plumstead: 1982: 6)

She realized that it was she who would have to describe the collection, and spent two years doing so.

Meanwhile I had undertaken the arduous work of describing the new finds which I knew to be of world-wide importance. Unfortunately Dr. du Toit had died before the first discovery and Prof. Birbal Sahni of India had died shortly afterwards and there was no one else available.

(Plumstead, 1982: 6)

Plumstead related the "Leaf" classification of *Glossopteris* (Glossopteris means "tongue-shaped fern") to the proper botanical classification based on reproductive organs. She (1952) eventually identified five species of *Glossopteris*, each with characteristic fructification and leaf. She isolated a sixth species, but its fructification had no attached leaf. She grouped them into two genera, *Scutum* and *Lanceolatus*. In her 1952 paper she still thought *Glossopteris* was a pteridosperm ("seed" fern). Two years later, she found evidence that some species of *Scutum* had bisexual fructifications, which led her to the belief that they were forerunners of angiosperms. She was less sure about *Lanceolatus*.

In 1952 fructifications borne of five species of *Glossopteris* were described in the Transactions of the Geological Society of South Africa as belonging to two genera named *Scutum* and *Lanceolatus* . . . New material has been found since 1952 in which 4 species of the first genus, *Scutum*, are preserved in the early pollination or "flora" stage which show the organs to have been bisexual. Pollen grains have been obtained from the microsporophylls of two of these species. No pollen bearing organs have been found on *Lanceolatus* but the evidence suggests that this genus too was bisexual. These fructifications are so unlike any previously known that the plants can no longer be classified as pteridosperms. They are considered, however, to have so many angiospermous characteristics that the author believes these two members of the Glossopteris flora may well be Permian forerunners of the Angiosperms. Their very rapid diversification and distribution over the Southern Hemisphere in lower Permian times is regarded as due to their superior organisation and anticipated the comparable dominance of true Angiosperms in the Cretaceous period.

(Plumstead, 1956a: 51)

Plumstead (1956b) found more forms of fructification, this time attached to leaves that had not been identified as *Glossopteris*. The leaves were similar to *Glossopteris* but were classified as *Gangamopteris*, which was the second most common fossil leaf present in the *Glossopteris* flora. She (1956b) also identified one more genera, and argued that the three genera should be placed in a distinct class, *Glossopteridae*. She later (1958a) identified three more genera of this class.

Plumstead (1958b) also attempted to reconstruct what Glossopteridae looked like in life, its general structure and growth habits, and she explained why she wanted to do so.

This paper is in reply to a challenge. In February, 1956, whilst making a tour of fossil plant collections, I was privileged to see for the first time, in the South African Museum, at Cape Town, the magnificent diorama created by Dr. L. D. Boonstra of Karroo life in the Permian period. The reptiles were of course accurately realistic, but the plant life which surrounded them included Gangamopteris leaves growing in clumps directly from the ground, as Agapan-thus lilies and Irises do, and Glossopteris on low scrubby bush, rather like the Karroo bush of to-day. Loose leaves lay on the ground beneath, awaiting fossilisation.

(Plumstead, 1958b: 82)

I suspect that she was particularly pleased to undertake the task of reconstructing appearances of the *Glossopteris* flora to add a South African touch to the dioramas, which were based on opinions of reptilian specialists from the Northern Hemisphere, including the fixist Colbert (§3.8).

The reconstructions had been based on artists' impressions included in two comparatively recent papers by von Huene in 1931 and Colbert under the auspices of the American Museum of Natural History in 1948. In both, of course, the authors were concerned, like Dr. Boonstra, primarily with reptilian reconstruction; but the plant life they pictured represents the only attempt to envisage the dominant plants of the Southern Hemisphere in the Permian period. These reconstructions differed so radically from mine on the subject that I felt it encumbent upon me to sum up all the available evidence of the habit and habitat of growth of the Glossopteridae. The evidence presented here is in part material and new and, in possibly the greater part, circumstantial; but I believe it to be relevant and that it may help to penetrate a little the curtain of ignorance which still surrounds these plants.

(Plumstead, 1958b: 82)

She (1958b: 51) claimed that new specimens from Wankie, Southern Rhodesia (now Zimbabwe), indicated that *Glossopteris* formed with large clusters, or short shoots, of leaves "growing, at well-spaced intervals, on thick, rather fibrous stems." She argued:

Glossopteridae were mainly deciduous, woody plants of arborescent habit and that the leaves, flowers and fruits grew as short shoots, at fairly wide intervals from the wood stem, and also terminally, and that they represent a new experiment in plant evolution in Paleozoic times.

(Plumstead, 1958b: 1)

Plumstead was pleased and gratified to examine the fossil plants collected on the Trans-Antarctic Commonwealth Expedition, 1955–8, gratified because she found strong support for mobilism.

My greatest satisfaction was the invitation by the Antarctic Committee in London of the Trans-Antarctic Commonwealth Expedition to describe all the plant fossils collected by the geologists of Sir V. Fuchs's Weddell Sea Station, and a few weeks later by the New Zealand party under Sir E. Hilary of the Ross Sea area. This gave me the unique opportunity to study a complete section across the Continent, and to find that there was complete agreement of Antarctica with other Gondwana continents from Devonian through Permo-Carboniferous, and Triassic to Jurassic periods. See Transactions Antarctic Expedition, No. 9, published in 1962 (MS was in Britain for nearly 2 years before publication). [Plumstead, 1962] I produced charts to show the relationship of every fossil plant fragment with each of the other Gondwana continents and the resemblance provided dramatic support of continental drift and of the similarity of a single climatic zone passing through all the now so widely scattered Glossopteris zones.

(Plumstead, 1982: 7; my bracketed addition)

Before her full-blown descriptions of the Antarctic material appeared, Plumstead (1961) wrote a vigorous defense of continental drift. In it she linked her own discovery of Glossopteridae fructifications in South Africa and the later discoveries of them in India, South America, Australia, and Antarctica, with her hypothesis that angiosperms evolved from Glossopteridae. She noted the large differences between the flora of the Southern and Northern Hemispheres. Her paper fulfilled the "commands" of General Smuts to test Wegener's hypothesis and to search out the origin of angiosperms without committing the "European fallacy" (Smuts, 1925). In fact, Plumstead referred to Smuts' talk in her paper, describing it as "remarkable" (Plumstead, 1961: 178). As had Smuts, she also quoted from Darwin's letter to Hooker explaining that he "sometimes speculated whether there did not exist somewhere during long ages an extremely isolated continent, perhaps near the South Pole," that might have been the source of flowering plants. She summarized previous thinking about the ancestry of flowering plants, saying that although many favored an Arctic origin, much had been learned recently about the history of Southern Hemisphere plants, and that it now looks "as though Darwin with his uncanny foresight, may have been right" (Plumstead, 1961: 176). Smuts would have been pleased.

In the section of her 1961 paper entitled "The Case for Continental Drift," she argued that the great similarities among the ancient flora of Africa, South America, peninsular India, Australia, and Antarctica supported the existence of Gondwana, and then asked two rhetorical questions:

Moreover, if the position of the continents has not changed how could the same genera and even species of plants have migrated east, west and south across the wide oceans which now separate them but still have failed to reach North America, Europe and Asia along easy routes? If the climatic zones were the same, how can we explain the fact that the fossil plants of peninsular India are closely comparable, and often identical with those of Australia, Africa south of the Sahara, Argentina and Brazil but above all, with those found in the heart of Antarctica at 86°S where today no vascular plant could live? It might be possible to explain an occasional, individual occurrence, and in fact a very few instances of supposed *Glossopteris* in Asia in late Permian times, have been recorded – but the presence of whole and almost identical floral assemblages in such widely separated areas defies explanation by any means other than that the continents themselves have moved and were once together.

(Plumstead, 1961: 178)

Smuts would have been delighted further with her concluding remark.

The story has been barely outlined and much of it requires factual proof for as yet we know practically nothing of the Cretaceous floral and faunal life of Africa apart from coastal deposits; but the possibilities are far reaching and the biologists of Africa have an enviable field for research.

(Plumstead, 1961: 181)

Edna Plumstead made a great contribution to the understanding of *Glossopteris*, providing key evidence confirming the mobilist interpretation of the trans-Gondwana Glossopterid disjuncts. She recognized the opportunity that presented itself that day in the early 1920s when she saw the fossilized fruiting body of a *Glossopteris* specimen at Vereeniging, thirty miles from her university. She took the opportunity and developed it over many decades. She was an authentic, trustworthy voice for mobilism. She spoke with enthusiasm but without polemics. She received encouragement; Wegener's hypothesis, as Smuts emphasized, offered South African Earth scientists lots of research projects of global significance. She did not have to abandon an anti-mobilist training, and she did not, in order to become a mobilist, have to study in a region other than her homeland where she learnt her geology.

Du Toit, whom I consider below, had less need of Smuts' encouragement, and devoted much of his career to testing, applying, and modifying Wegener's hypothesis. In fact, du Toit did so much that some of his contemporaries might have felt that there was little more to do in South Africa in defense of mobilism. But mobilism did more than provide research opportunities; it offered solutions to key problems in South African geology, paleobotany, paleozoology, and paleoclimatology. The very strong preference for mobilism among some South African Earth scientists makes sense because it offered them rich research opportunities and a global framework for solving problems.

6.4 Alex du Toit: his life and accomplishments

. . . du Toit . . . with his unequalled first-hand acquaintance with the stratigraphy of the Gondwana System and the Carboniferous Glaciation, as well as many other geological features in the Southern Hemisphere and India, was, so to speak, impelled to become an active protagonist of Continental Drift.

(Gevers, 1949: 101)

Du Toit spent over twenty years as a field geologist in South Africa, and he became the leading expert on the Karroo System. He announced his support for continental drift in 1921, his fieldwork having convinced him that drift satisfactorily solved the central problems of South African geology. In 1923, he spent five months mapping Karroo equivalents in Brazil, Argentina, and Uruguay, and his advocacy became more assured. With the publication of *Our Wandering Continents* in 1937, du Toit's case for mobilism had become both global and general, including detailed discussion of the geology of all continents and appeal to work in different fields of geology.

Du Toit was one of the kindest of men and I owe him thanks for help of many kinds. Du Toit was very highly respected for his practical knowledge of almost every branch of geology, drift included. An Englishman would call him the doyen of South African geology.

(King, 1981, letter to author)

T. W. Gevers (1949), a mobilist South African geologist and old friend of du Toit's (they first met in 1923), wrote an excellent account of his life and work.[4] Haughton (§6.11) provided a briefer account for the Royal Society London (Haughton, 1949). Oreskes (1999) has added to our understanding of du Toit through her archival work. I shall draw on all of them.

Alex du Toit, the oldest of four siblings, was born near Cape Town in 1878. He earned his B.A. degree at the University of Cape of Good Hope, and traveled to Glasgow, Scotland, where, in 1898, he obtained a diploma in mining engineering at the Royal Technical College. He spent 1900–1 studying geology at the Royal College of Science, London. Returning to Glasgow, he became a lecturer in mining and mine-surveying at the Royal Technical College and in geology at the University of Glasgow, but remained there only one year, returning to Cape Town in 1903 as an assistant geologist with the Geological Commission of the Cape of Good Hope. In 1912, two years after the formation of the Union of South Africa, du Toit joined the newly created Geological Survey of South Africa, where he remained until transferred to the Department of Irrigation in 1920. He resigned from governmental service in 1927, and became Consulting Geologist to De Beers Company. Du Toit retired from De Beers in 1941. He died in 1948, aged seventy.

Du Toit was the first South African geologist elected (1943) to the Royal Society of London. The Geological Society of London awarded him the Wollaston Fund in 1919 and the Murchison Medal in 1923. The S_2A_3 awarded him the South Africa Medal in 1930, and he received the Draper Memorial Medal from the Geological Society of South Africa in 1933 and its Jubilee Medal in 1945. He was twice President of the Geological Society of South Africa, President of the South African Archaeological Society, President of the S_2A_3, and, at the time of his death, President Elect of the Royal Society of South Africa.

Du Toit became a major leader in the continental drift debate because of his work in South Africa, especially his first-hand experience of the Karroo System, because he took the opportunity to do fieldwork in regions of Gondwana other than his own, and because he was particularly interested in understanding the origins of what he observed rather than simply recording it. Like Daly and Holmes, du Toit was out to test hypotheses.

As a field geologist, Du Toit was accurate, thorough and efficient. Gevers (1949: 24) estimated that before he left the Geological Survey in 1920 du Toit had "mapped in detail over 50 000 square miles of country, mostly in the Cape and Natal." Du Toit himself emphasized the importance of firsthand observation in his (1912) first single-authored book, *Physical Geography for South African Schools*. As Gevers noted:

> Du Toit with his well-known prowess as a field geologist, always underlined the necessity for carrying out field studies. Thus he says in the Preface to his *Physical Geography for South African Schools*, first published in 1912: "Too much stress cannot be laid on the importance of actual observational work out of doors to supplement the teaching of textbooks, while, wherever possible, photographs, sketches and plans of instructive features and areas should be taken."
>
> *(Gevers, 1949: 36)*

He became an expert on the Karroo System.

> Already in 1918, when President of the Geological Society of South Africa, he had given in his Presidential Address that unique summary of the correlation and phasal variations of the various groups of the Karroo System, entitled *The Zones of the Karroo System and their Distribution*. This comprehensive paper at once established him as *the* specialist in Karroo stratigraphy in South Africa and was the direct cause of the universal sympathy and aid he received in other countries of the Southern Hemisphere, when later he extended his investigation of the equivalents of the Karroo System outside the borders of Southern Africa.
>
> *(Gevers, 1949: 25)*

His knowledge of the Karroo System prepared him to recognize and appreciate its equivalents elsewhere in Gondwana, and readied him to embrace and defend mobilism. During the first half of 1914, several years before du Toit publicly declared his support for mobilism, he went to Australia where he saw for himself the Australian equivalent of the Karroo System.

Gevers recalled what happened when he had used Du Toit's geological maps as a guide during his student days. He not only learned how good du Toit's maps were, but also heard that R. A. Daly had proclaimed du Toit "the world's greatest field geologist."

> At that state I had met du Toit only once, very briefly as a student in Cape Town. Gazing up at the towering mountains and looking down into deep gorges and ravines, I was not only overawed with the general ruggedness of the landscape, but very soon also with the accuracy of du Toit's mapping. Had I then known under what circumstances and in how short a time he had done it all, I would have been even more impressed . . . Later when I had occasion to go over other of du Toit's maps in Pondoland, East Griqualand, Natal and Zululand, I found that this flair for rapid observation and accurate recording of detail was an unfailing characteristic

of the man. So when that no less a person than Professor R. A. Daly of Harvard University on the occasion of his visit to South Africa with the Shaler Memorial Expedition in 1922, had enthusiastically referred to du Toit as the "world's greatest field geologist," I could only express wholehearted approval.

(Gevers, 1949: 8)

Daly formed his opinion when he, F. E. Wright of the Carnegie Institution of Washington, and the Dutch geologist G. A. F. Molengraaff worked together during the Shaler Memorial Expedition studying the Bushveld Igneous Complex. Du Toit was their host, and spent several weeks in February and March 1922 with them in the field (Oreskes, 1999: 93). Daly and Wright, who also thought highly of du Toit's fieldwork, helped secure funds for him from the Carnegie Institution of Washington to undertake a comparative study of the Karroo equivalents in South America in 1923 (Oreskes, 1999: 157–163). This work led to his 1927 *A Geological Comparison of South America with South Africa*. Du Toit recalled the help he had received from Wright and Daly in securing funds.

Following the sympathetic representation by Dr. Fred. E. Wright, of the Geophysical Laboratory of the Carnegie Institution of Washington, and Dr. Reginald A. Daly, of Harvard University, the president (Dr. John C. Merriam) and trustees of the Carnegie Institution of Washington, appreciative of the scientific importance of such a mission, most generously offered a grant in aid, and, making use of my six months of leave from official duties, fortunately then available, I was enabled to cross the Atlantic and spend five months in Brazil, Uruguay, and Argentina.

(du Toit, 1927b: 2)

Du Toit was an ideal person for the task; he knew the Karroo intimately and had the capacity to map large areas accurately and efficiently.

6.5 Du Toit's early defense of continental drift

Du Toit's support for mobilism evolved through three stages. It began with his fieldwork on the Karroo System, his visit to Australia, and his review of work on Karroo equivalents in these and other Gondwana fragments; these are described in this section. He then investigated the Karroo equivalents in South America, which strengthened support for the former juxtaposition of Africa and South America. In the final stage, he expanded his global case to the Northern Hemisphere and to topics other than stratigraphy, and presented his own global version of continental drift, becoming in the 1930s and 1940s its most avid defender.

It was not until seven years after returning from Australia in 1914 that du Toit announced (on July 15, 1921) at the 19th Annual Meeting of the S_2A_3 that he was a mobilist. He proposed the existence of two primordial continents, Gondwana in the south and Laurasia in the north. This he maintained throughout his career in contrast to Wegener's single Pangea.

That the continents might have originated by the actual tearing apart of one or more much larger masses is no new doctrine. Although generally dismissed as fantastic, it has been very ably championed recently by Wegener, and, when the hypothesis is studied in detail, the evidence in its support is found to mount up so remarkably as to become almost overwhelming . . .

(du Toit, 1921a: 126)

As Wegener had done, he argued for former land connections between Gondwana's fragments to explain the wide distribution of Carboniferous glaciation (§3.11). Like Wegener, he favored continental drift, because it explained the glaciation better than sunken continents or landbridges (RS2, RS3).

The foregoing represents Gondwanaland as generally conceived, and, although its southern limits are problematical, it is obvious that in size it would have rivaled Eurasia. With a restoration on these lines the geologist is confronted by several difficulties, the most formidable of which – one that has hitherto been insuperable – arising out of the extraordinary location of the ice-centres, namely, in the temperate girdle, and their peculiar attitudes to one another. For explanation I am advancing in all seriousness the view, revolutionary and heretical as it will appear to orthodox geologists, that Gondwanaland was a much smaller continent than as usually conceived, that its centre lay somewhat further to the south, that the Carboniferous ice-sheet was an almost continuous mass, and that the land fragments still preserved represent portions of the ancient continent forcibly torn apart, subsequently modified in outline by erosion, deposition, etc., and now separated by vast stretches of ocean.

(du Toit, 1921a: 125–126)

He reemphasized the superiority of drift's solution to the Late Paleozoic glaciation problem and, as additional evidence, cited geological and biological disjuncts characteristic of Gondwana's fragments (§3.5, §3.7).

It will forthwith be realised that by thus supposing the several units to have been spaced much closer together in the past, the numerous remarkable lithological and palaeontological resemblances between them become more explicable, while the difficulties that beset migration become much reduced. Moreover, the areas known to have been capped by ice become roughly grouped around the South Pole, not far from, if not well within, the present northern limits of drift-ice, and a serious stumbling block to the interpretation of the Carboniferous Ice Age is thereby removed (see Fig. 1).

(du Toit, 1921a: 126; du Toit's Fig. 1 is reproduced in Figure 6.1)

He discussed the rise and spread of *Glossopteris* flora (1921a: 127); the distribution of the "primitive free-swimming little *Mesosaurus*, the earliest known reptile of the Southern Hemisphere" in the Cape, South-West Africa (Namibia), Brazil, and Uruguay (1921a: 128); the desert conditions during the Triassic in South Africa, southern Brazil, and India (1921a: 130); the huge contemporaneous outpourings of basalt, the Rajmahal Traps in India (mistakenly thought at the time to be Jurassic, now known to be Cretaceous), eastern South America, and Africa associated with the break-up (1921a: 131); and the extension of the Southern Cape Ranges westwards into Argentina, Northern Chile, and Bolivia (1921a: 132) (RS1). In support of continental drift, he noted similarities of shape and especially geological features between opposite sides of the Atlantic (RS1).

Figure 6.1 Du Toit's Gondwana (1921a: 127). His caption read as follows: Hypothetical restoration of Gondwana at the close of the Carboniferous Epoch. The ruled areas are those known to have been covered by ice, the arrows indicating the direction of movement of the latter.

Realizing the magnitude of the changes that might have occurred along the coasts, one would not put too much weight upon the extraordinary resemblances in outline between the opposite coastlines of Africa and South America, were it not that the geological peculiarities of the two areas are so amazingly similar; not only does this apply to the South but also to the North Atlantic.

(du Toit, 1921a: 135)

Du Toit briefly turned to mechanisms. Without mentioning them specifically, he followed Wegener, appealing to *Polflucht* force and a fluid-like substratum.

Our hypothesis starts with the assumption that the Continent was first of all severed by great tear-lines and that these fragments then started to move apart, just as though driven asunder by the centrifugal forces set up through spinning around the polar axis. The action is to be conceived as a slipping of the outer part of the crust upon its yielding foundation and the phenomenon could indeed be compared to the gradual opening out of cracks developed in a sloping asphalt pavement.

(du Toit, 1921a: 135)

He ended his 1921 presentation by suggesting that mobilism might explain Tertiary mountain building. Sediments accumulated around the periphery of Gondwana from the Carboniferous to the Tertiary, and as it fragmented, these were compressed and formed mountain belts along the advancing margins of the drifting fragments. Thus were formed the Andes in the west, passing through Venezuela and becoming the

Atlas-South European-Iranian-Himalayan folds on the north, the Malay-Polynesia-New Zealand crumblings on the east, and the West Antarctic belt on the south, joining with Patagonia and thus completing the circle.

(du Toit, 1921a: 136)

Between the drifting fragments, regions of tension developed, forming the Red Sea, Arabian Gulf, and the "remarkable system of trough-faulting extending down through Eastern Africa – the great Rift Valley" (du Toit, 1921a: 136).

Du Toit (1921b) defended mobilism in a second paper which began with a summary of his fieldwork on the Carboniferous glaciation of South Africa.

Eighteen years of fieldwork have indeed caused me to realise only too clearly both the magnitude and the complexity of the problem, and only the unlikelihood of being able to continue this investigation has led to the setting down of the following outlines of this revolutionary period of earth history.

(du Toit, 1921b: 188)

He then went on to describe equivalents in other fragments of Gondwana, and he argued:

the solution follows in the simplest manner with the perception of the idea that the several sections represent portions of Gondwanaland actually disrupted and forcibly torn apart from one another subsequently to the Triassic. This revolutionary idea of continental and oceanic evolution ably championed by Wegener, was recently applied to this problem by the writer [du Toit, 1921a], and not only has a most remarkable mass of evidence in its favour hard to account for otherwise but has proved extraordinarily stimulating in its application and indeed can be looked upon as a master-key to the Past.

(du Toit, 1921b: 219; my bracketed addition)

The areas covered by glaciation, assuming drift, were much closer to the pole than they would have been otherwise, and he compared its extent to that of Pleistocene glaciation in the Northern Hemisphere.

Under the hypothesis adopted all the areas known to have been glaciated, even Peninsular India, can be brought together inside the forty-fifth parallel south, which we may note is close to the existing northerly limit of drift ice, so that the Carboniferous glaciation of this continent on these assumptions would present no greater difficulties than that of the Northern Hemisphere in the Pleistocene.

(du Toit, 1921b: 222–223)

He had adopted continental drift because it offered the best solution to one of the greatest problems of South African geology.

A year later, du Toit (1922: 6) said of himself that he was "accepting provisionally" the "daring hypotheses of Taylor and Wegener," but did not explain why.

Du Toit devoted his 1924 Presidential Address to the geology section of the S_2A_3 to the role South Africa should play in shaping general geological thought. Foreshadowing what Smuts (1925) would say the following year, du Toit lamented the dominance of European geology and argued that geology would have evolved differently had it originated elsewhere.

Of all the natural sciences the one that has been most influenced by environment in its development is Geology. The succession among the stratified rocks, the fossil faunas of the various groups, the division of the stratigraphical column into systems, series and stages, the periods of major crustal deformation, etc., have been based almost entirely upon European observation and experience, whence it follows that the entire fabric of Geology bears the indelible impress of Europe . . . With its beginnings thus shaped and with its superstructure added to mainly by workers in that quarter of the globe, the European aspect has become unduly emphasized, with a tendency – if one may be pardoned for saying so – towards a somewhat narrow outlook on what are proving to be really world-wide problems. Such was, however, inevitable, since certain phenomena of the highest importance were either absent or poorly represented in Europe; *indeed had this science taken its origin in any of the other continents, it would doubtless have evolved along different lines.* Geology, unfortunately, as the history of the subject shows, has from its beginning been strongly dogmatic; even to-day geological "unorthodoxy" is looked upon rather coldly, while in certain places the doctrine of "authority" has not altogether been disposed of.

(du Toit, 1924: 52; emphasis added)

How would geology have developed had it begun in South Africa? He did not answer directly but he did identify the contribution of South Africa's geology to the drift debate as its most important.

The influence of South Africa in helping to shape geological ideas and sometimes to introduce an absolutely new viewpoint, has been by no manner of means inconsiderable, though insufficiently recognized . . . There is hardly a branch of geology that has not beneficially been touched and advanced by investigation in South Africa, although a great deal of such valuable work, as is only natural, has come to be overlooked abroad, or the bearing thereof insufficiently appreciated. Within the several branches of the subject it is in problems of petrogenesis, of former glacial climates, but more particularly of the distribution of the continental masses in the past, that South Africa has been supplying so splendid a field of investigation . . . Some of these problems are now, however, becoming larger, and not only drawing in for their solution the regions to the north of the Zambesi, but are involving the other parts of the Southern Hemisphere as well.

(du Toit, 1924: 53)

He seems to have sensed already the adverse effects that regionalism had had on the reception of continental drift by northern Earth scientists. Determining the past position of continents required working in the Southern Hemisphere. Imagine if geology had begun in South Africa, and then spread throughout the Southern

Hemisphere and peninsular India before being studied in the Northern Hemisphere? Continental drift may have been orthodoxy from the start!

In his Presidential Address, he considered two difficulties that had been raised against mobilism, first the goodness of fit of continental margins (RS2).

It must not be supposed that the existing land masses ought to fit one another after the manner of the parts of a jig-saw puzzle; a more proper analogy would be to the breaking up of a tabular ice-floe with the melting away of the several portions along their edges during the drifting apart of the fragments.

(du Toit, 1924: 72)

The details of continental margins were not immutable. Then there was the mechanism difficulty. Du Toit suggested that Joly's appeal to radioactivity was promising, but argued that the absence of an acceptable mechanism should not be viewed as a serious impediment; science abounded in examples of something first judged impossible being later shown to have occurred (see §5.13 for other appeals to historical precedent).

That certain physicists are denying the possibility of such continental disruption is of small moment, for we can recall their unyielding attitude prior to the discovery of radium, when faced by the demands of the geologists and biologists for a greater age for the earth.

(du Toit, 1924: 72)

Du Toit noted imperfections in Wegener's presentation.

. . . it has to be admitted that the hypothesis as presented by Wegener introduces numerous improbabilities as well as errors, but it must nevertheless be recognised that the geological facts have not yet been properly marshalled and presented.

(du Toit, 1924: 72)

Of course, du Toit and other Gondwana geologists were well placed to correct Wegener's errors and thereby strengthen mobilism's factual support; in fact, du Toit (1924: 72) as a result of his work in South America had been able to do just that.

Even to outline the data which go to support this daring speculation is impossible, but I may remark that a recent opportunity of studying the geology of South America to this end has greatly strengthened in my opinion the evidence that collectively points to the former union of these two continents and to a much closer relationship geographically during the early Mesozoic.

6.6 Du Toit compares geology of South America and Africa

Du Toit left Cape Town for South America on June 12, 1923, reaching Rio de Janeiro two weeks later. He (1927b: 4–5) worked in Brazil for nearly six weeks, in Uruguay for another six, in Argentina for nearly two months, crossed the Andes to Valparaiso, Chile, and returned to Cape Town on December 10, 1923. In Brazil he visited diamond mines, saw Triassic traps (now dated Early Cretaceous) and sediments, and a full Devonian to Cretaceous sedimentary succession of the Santa

Catharina System – equivalent of the Karroo and Gondwana Systems in South Africa. In Uruguay, he "confirmed the existence of Carboniferous glacial beds," in Argentina he studied various horizons in the Santa Catharina System, and the closely matching Silurian to Carboniferous "equivalents of the 'Cape Fold Ranges'" in the Sierra de la Ventana (1927b: 5, 19). At the time there were some geological maps of Brazil and Uruguay, but "no general map of Argentina," and "much of Paraguay, Bolivia and adjacent parts of Brazil" had not been geologically explored (1927b: 3).

Du Toit made his pre-drift geological restoration of South America and Africa (Figure 6.2), remarking on the numerous matching geological disjuncts along opposite margins, and arguing that their variety, current separation, and common changes over time testified to their former union.

While the general geological resemblance between those portions of the two continents that face the South Atlantic basin has long been perceived, the outcome of these present studies is essentially to emphasize this geological parallelism, which is nothing less than extraordinary, considering the enormous stretches of ocean parting these two land-masses. Such points of resemblance have now become so numerous as collectively almost to exceed the bounds of coincidence, while they are, moreover, confined not to one limited region nor to one epoch, but implicate vast territories in the respective land-masses and embrace times ranging from pre-Devonian almost to the Tertiary. Furthermore, these so-called "coincidences" are of a stratigraphical, lithological, palaeontological, tectonic, volcanic and climatic nature.

(du Toit, 1927b: 109)

Facies variations within either continent, he claimed, were often greater than variations between their present near-coastal occurrences.

Of prime importance, moreover, is that evidence obtainable from the study of the phasal variations displayed by particular formations when traced within their respective continents. In illustration, let us consider the case of two equivalent formations, the one in South America beginning on or near the Atlantic coast at A and extending westward to A′ and the other in Africa starting similarly near the coast at B and stretching eastward to B′. Then it can be affirmed that more than one such instance can be designated, where the change of facies in the distance AA′ and BB′ is *greater* than that found in AB, although the full width of the Atlantic intervenes between A and B. In other words, these particular formations along the two opposed shores tend to resemble one another more closely than either one or both of their actual and visible extensions within the respective continents.

(du Toit, 1927b: 109)

He continued to underscore the importance of drift's solution to Gondwana's Permo-Carboniferous glaciation; it was better than the alternatives (RS3) because it

brings into close association *away from the equator* all the areas glaciated in the "Permo-Carboniferous" and thus succeeds in eliminating an outstanding difficulty in necessitating somewhere or other a refrigeration in the subtropics or tropics, a trouble inherent under any other hypothesis, even in those postulating a wide movement of the South Pole.

(du Toit, 1927b: 119)

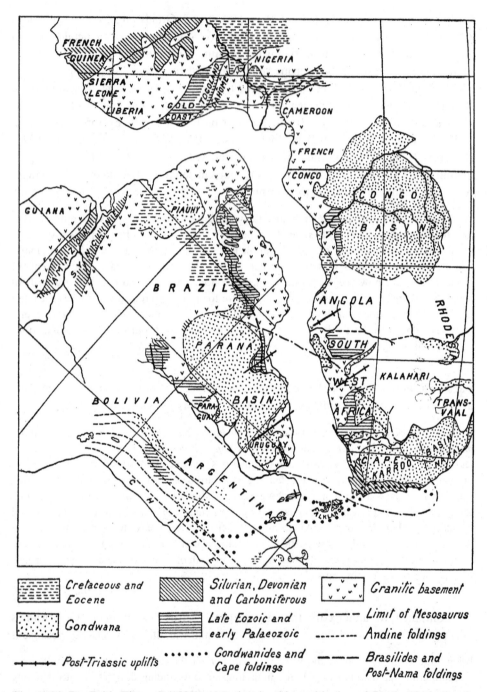

Figure 6.2 Du Toit's Figure 7 (1927b: 116) showing his positioning of South Africa, South America, and the Falkland Islands, which he placed along the folded belt linking the Cape with Argentina, slightly nearer to Africa than to South America.

He argued that his recent fieldwork had demonstrated that the South American glaciation had begun from "an ice-center situation out in the present Atlantic" (1927b: 102), which was also supported by the "north-south directed groovings beneath tillite in the Falklands" (1927b: 103). He also proposed:

It would appear that glaciation commenced in the middle or early upper Carboniferous in the Southern Hemisphere in the Seaham areas [of New South Wales, Australia], extended to Argentina, then to South Africa, to the west of Australia, Tasmania, and India during the upper Carboniferous, and recurred during the early Permian in New South Wales and perhaps still later in Bolivia and the eastern Congo.

(du Toit, 1927b: 103, my bracketed addition)

He further argued:

Under the displacement hypothesis, not only would all the areas affected be brought together in the most simple manner, within an oval perhaps no larger than the African continent, but the non-synchronism of glaciation in the several parts would find their explanation as the natural outcome of an extensive active ice-cap, the margins of which progressively advanced in one direction and retreated in the other, so that with time the sheet had as a whole *migrated over* the surface, glaciating various sections in turn, exactly as seems to have been the case with Pleistocene Ice Age in the Northern Hemisphere.

(du Toit, 1927b: 103–104)

He added (1927b: 104) that his solution of migrating glaciation "would perhaps meet Coleman's well-founded criticism of the excessive size of the single ice-cap under this supposition" (RS2) (see §3.11 and §3.13 for more on the glaciation debate).

Du Toit emphasized the good morphological fit between South America and Africa when a globe is used. The fit is further supported by the Mid-Atlantic Ridge situated midway between them, and the regular manner in which it mirrored their margins indicated a common origin.

Of prime importance is the extraordinarily close correspondence in the outlines of the opposed shores of the two continents, as has been pointed out and discussed by others long before Wegener, and which is particularly marked when comparison is made not with maps, but on the face of a terrestrial globe. Next is the presence of the central Atlantic rise beneath the ocean, with its surprisingly symmetrical position nearly midway between the Old World and the New. Interpreted mathematically, the great regularity of these three features, extending through the entire length of the South Atlantic, would betoken an enormously high probability that such features had owed their origin to one and the same set of tectonic forces at a relatively late geological period. Upon this rise are, furthermore, aligned certain of the volcanic islands of the southern Atlantic.

(du Toit, 1927b: 110)

Although he thought the stratigraphic evidence more important than the paleontological, he discussed the distribution of *Glossopteris* (1927b: 119 and elsewhere), *Mesosaurus* (1927b: 79 and elsewhere) and other disjunctive distributions (see §3.5 and §3.7 for du Toit's changing attitude toward mobilism's paleontological support).

He argued, as he had in 1921, that there had been two primordial continents, not one as Wegener would have it. He agreed with Wegener that the fragmentation of Gondwana created mountains that formed on the leading edges of drifting fragments, and that other mountains formed where drifting fragments of Gondwana collided with Laurasia.

In the hypothesis as formulated by Wegener, the New World is regarded as having parted from the Old up the full length of the Atlantic from south to north at quite a late stage in the Tertiary, against which supposition many objections could be raised. I have tentatively suggested [1921a] the radically different conception of two such parent continents, that to the south being Gondwanaland, which, by impinging on each other, gave rise to the enormously extended orogenic structure made by the Atlas, Alps, Carpathians, Caucasus, Himalayas, etc. . . .

(du Toit, 1927b: 119–120; my bracketed addition)

Du Toit, who later more or less adopted Holmes' substratum convection as continental drift's mechanism (§5.7), reminded readers that landbridge theories also had their own mechanism difficulty (RS3).

The fact that the many eminent scientists have cast doubt upon its geophysical possibility should not be permitted to cloud the issue any more than the existence of former "land bridges" should be denied because of cogent objections based upon the doctrine of isostasy.

(du Toit, 1921: 110)

Returning to South America and Africa, he went on to explain:

the intention is merely to set forth some of the data regarding Africa and South America and to state the conclusions to be drawn therefrom, that are distinctly awkward of explanation under the current and orthodox view of "land bridges," but which, on the contrary, appreciably favor the "hypothesis of continental disruption."

(du Toit, 1927b: 110)

Despite the close similarities between African and South American geology, he did not argue that mobilism should be accepted solely because of them. Recognizing the vastness of the task ahead, he urged others to undertake comparative studies of the geology of North America and Europe, and once again appealed for cooperation among specialists in physics and various branches of the Earth sciences.

These pages have of necessity dealt in very summary fashion with this fascinating subject, more particularly as related to those territories beyond the respective opposed regions known personally to the writer. It is therefore to be hoped that scientists acquainted with those particular outside regions may be induced to set down their observations on those areas, supporting or else refuting the presumptions here put forward on behalf of this hypothesis, since a strict and impartial criticism is indeed required if this riddle in early history is to be deciphered. The cooperation of geologists, palaeontologists, zoologists, botanists, and physicists is urgently needed, the discussion advanced here dealing admittedly with only a very limited, albeit extremely important, section of the globe, from which, under the eclectic hypothesis of crustal instability, corroborative or destructive evidence might reasonably be expected. By virtue of their enormous lengths of opposed coast-line and extraordinary geological parallelisms, the two most favored continents from which evidence is to be drawn would appear to be Africa and South America, and, should the details herein set forth appear to be worthy of serious consideration, it is to be hoped that more detailed investigations may shortly be instituted elsewhere on the two sides of the Atlantic for the purpose of clearing up some of the many crucial questions that must be regarded as *sub judice*.

(du Toit, 1927b: 120–121)

In the years to come North American and most European geologists did not heed his request for comparative stratigraphic studies of the two continents; their interest lay mainly elsewhere, and in the lean years of the 1930s research funds were not made available to support workers of du Toit's vigor and acumen in the field, to personally carry out such a comparison.[5]

6.7 Du Toit's *Our Wandering Continents*

A decade later, du Toit (1937) dedicated his mature and most complete account of mobilism, *Our Wandering Continents*, "to the memory of Alfred Wegener for his distinguished services in connection with the geology of Our Earth." New elements were added to his earlier defense of mobilism, among them his recognition of the trans-Gondwana Samfrau Geosyncline, his extensive updating of evidence from the Southern Hemisphere, his detailing of evidence from the Northern Hemisphere in support of mobilism, and, following Köppen and Wegener (1924), his use of paleoclimatological data in support of mobilism. As inscribed on the title page, he still maintained "Africa forms the Key," which he justified by masterly marshaling the stratigraphic evidence from there and from the Southern Hemisphere generally, seeking to correct what he believed to be the overemphasis that other mobilists had placed on tectonics and on the Northern Hemispheric evidence.[6] He was a spokesman for the Southern Hemisphere.

Taylor, Argand and Staub have each over emphasized the tectonic aspect, while concentrating upon the Northern Hemisphere, wherefore the stratigraphical viewpoint will be given its rightful place and fuller attention paid to the Southern Hemisphere, for which the evidence is, as it happens, clearer and less equivocal.

(du Toit, 1937: vii; see §8.7 for Argand, and §2.11 for Taylor)

True to his word, and based on his analysis of the Gondwana stratigraphy, du Toit recognized the late Paleozoic *Samfrau* Geosyncline, which extended from South America across South Africa to Australia.

This feature, which seems to have played so vital a role during the evolution of Gondwana, and which was already in existence in the Ordovician, can conveniently be called the "Samfrau" Geosyncline – a contraction of the words "South America-South Africa-Australia." Its eastern portion has indeed for long been known as the "Tasman" Geosyncline, to the outer side of which belong New Zealand, New Caledonia and other islands.

(du Toit, 1937: 63)

He alleged that the same sequence of Silurian through Cretaceous strata occurred in Argentina, the Falklands, the Cape region of South Africa, and eastern Australia (Figure 6.3). The Samfrau skirted Antarctica, but, owing to the limited stratigraphic evidence from there at the time, the continuation of Samfrau through West Antarctica was largely conjectural.

Emphasis must be laid at this point on the extraordinarily similar sequence of events in South America, South Africa and Australia, stage by stage, from the Silurian up to the early Cretaceous as shown in the attached table [my Figure 6.3]. Taken in conjunction with a mass of supporting evidence, it favours the idea of a major geosyncline directly connecting these countries – traversing Bolivia, north and central Argentina, Cape, Weddell Sea, passing east of King Edward VII Land [West Antarctica] and through Edsel Ford Land [West Antarctica], crossing Tasmania and the eastern part of Australia to New Guinea, its inner margin advancing or retreating from time to time.

(du Toit, 1937: 62; my bracketed additions)

Adding the Samfrau Geosyncline to his criteria for reassembling Gondwana, he presented his reconstruction as shown in Figure 6.4.

Du Toit also discussed extensively mobilism's support in the Northern Hemisphere, devoting two chapters to the history of Laurasia. Although he continued to emphasize that the roots of his theory were in Africa and the Southern Hemisphere, he now adopted a global perspective in *Our Wandering Continents*.

This attempt by the writer to explain the elaborate architecture of the Globe has necessitated a critical revision of Geological Principles as well as a review of the geology of the whole Earth.

(du Toit, 1937: vii)

He added several themes, most not new. He thought it unfortunate that many knew only Wegener's theory, and not other versions of mobilism that avoided some of the difficulties that had in the past 25 years been raised against it (1937: 1). Du Toit argued that orthodoxy in geology favored fixism partly because most Earth scientists worked in the Northern Hemisphere where the evidence for mobilism was weaker than in the Southern (1937: 2–4). The lack of comparative intercontinental studies obscured both mobilism's strengths and fixism's weaknesses (1937: 4). Mobilism solved more problems than fixism (1937: 9–10), and he emphasized the superiority of its explanation of Permo-Carboniferous glaciation (1937: 72–76). He adopted Holmes' substratum convection and added his own idea of a sub-crustal paramorphic zone created by the reversible transformation of basalt to eclogite (§5.4). But he continued to argue that it was legitimate to believe in continental drift without there being a generally acceptable mechanism (1937: 6).

Du Toit (1937: 9) listed ten facts that are "very imperfectly or not at all accounted for" by fixism (RS2). Here they are, in his own words.

(1) That each of the continental blocks shows youthful marginal foldings along but a section of its borders, and an enlargement in that direction despite the resulting compression, making the structure thereof characteristically *asymmetrical*;

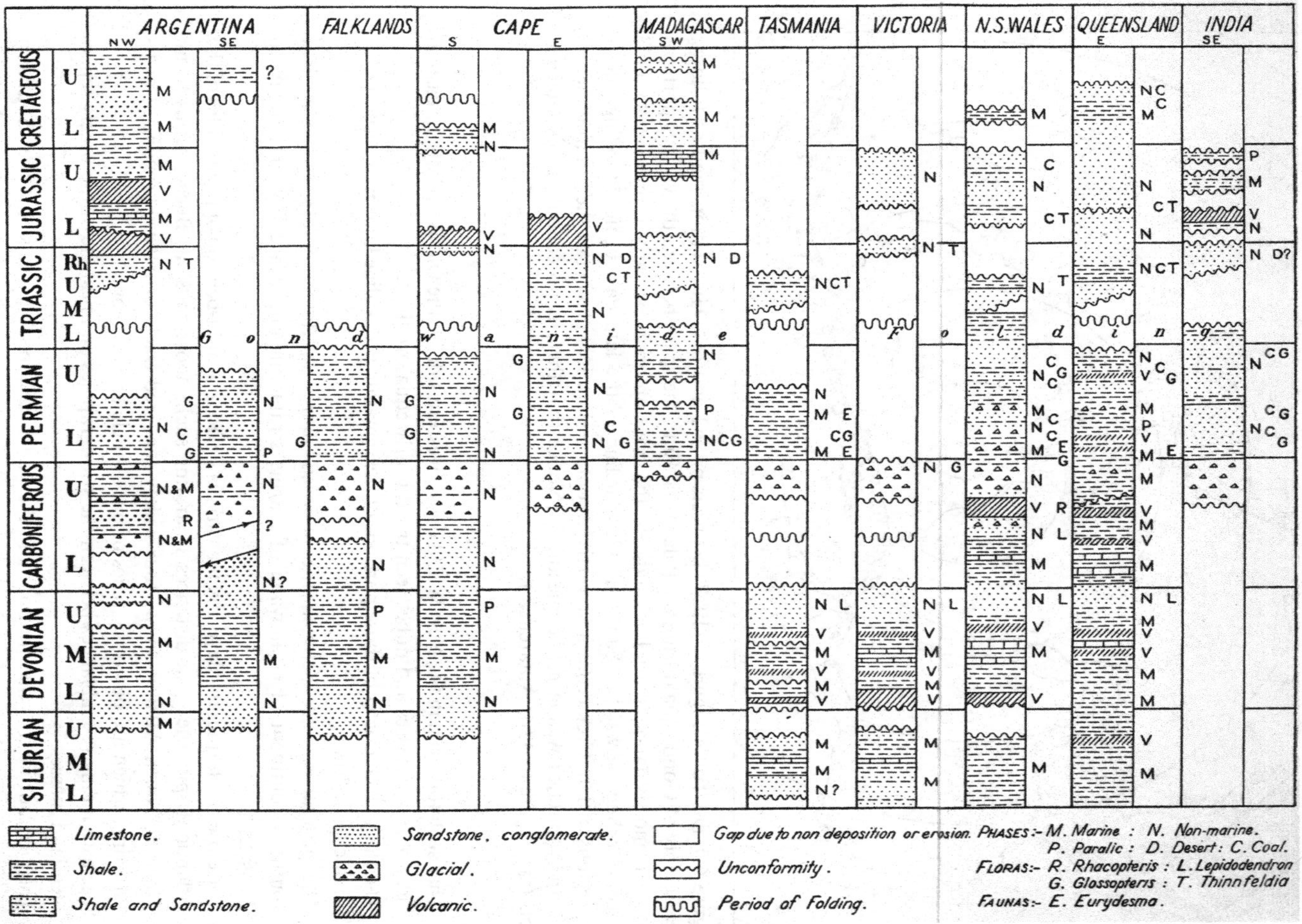

Figure 6.3 Du Toit's attached table (1937: opposite p. 61). Du Toit used this diagram to display the parallel phases of the Samfrau Geosyncline from fragments of Gondwana.

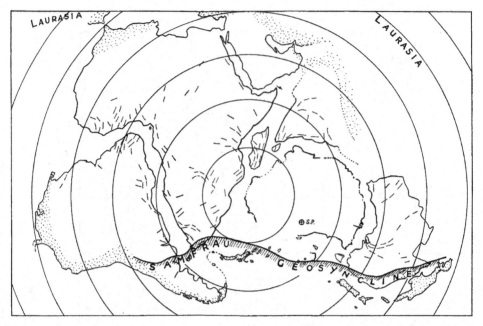

Figure 6.4 Du Toit's Figure 7 (1937: between pp. 62 and 63). His caption read as follows: Reassembly of Gondwana during the Palaeozoic Era. The space between the various portions was then mostly land. Short lines indicate the pre-Cambrian or early Cambrian "grain." Diagonal ruling shows the "Samfrau" Geosyncline of the late Palaeozoic. Stippling marks out regions of late Cretaceous and Tertiary compression. His placing of the Samfrau across western Antarctica was largely conjectural. (Notice that there is no column in his attached table (my Figure 6.3) for Antarctica.)

(2) That such folds run in the New World on the western side, in the Old World on the eastern side, of those masses and in between trend nearly equatorially;

(3) The elevated character of those portions of the remaining coast-line known or suspected to be due to faulting;

(4) The general absence between the opposed shores in the remnants of the ancient southern continent of Gondwana of marine strata ranging in age from Devonian to Triassic;

(5) The close similarities displayed by the various Palaeozoic fold-belts on opposite sides of the Atlantic despite the great distances separating such "loose-ends" as Bucher terms them. Thus the North Atlantic shows four widely sundered tracts of intense (Taconian) over-thrusting, which are obviously relics of but a single tectonic zone;

(6) The astounding resemblances between the stratigraphies and past life of the widely separated land-masses;

(7) The extraordinary distribution of the Carboniferous glacials in the Southern Hemisphere (including India) that are today situated in the anticyclonic belt (with highest snow-line), two of them having the ice-movement directed *away from the equator*;

(8) The rift-valley systems of Africa and other lands and their unique physiography;

(9) The enormous extent of terrestrial deposits formed between the Carboniferous and the Triassic; and

(10) The discovery of peculiar faunas in isolated and remote positions, as for example certain marine Permian mollusca in Brazil characteristic of New South Wales.

He (1937: 9) added that the above "are but a few of the numerous problems ordinarily only explainable in a rather nebulous fashion, which find their ready solution under our Hypothesis" (RS1, RS3).

Du Toit also raised a general difficulty with fixism not shared by mobilism (RS2, RS3).

A serious objection arises from the concept of the general fixity of the continents, or more properly of their more stable parts or "shields." The farther apart they lay, the less the chance of their evolutionary history having been able to follow identical lines. Now the outstanding feature of geological analysis is the wonderfully similar histories – stratigraphical, tectonic, climatic, biological and eruptive – of particular pairs of the land-masses at various periods, which is finely displayed between Greenland and Scandinavia, South America and Southern Africa, East Africa and India, and South Africa and Australia. That they should have reacted so similarly while at so great a distance apart one from the other is improbable, accepting orthodox views, whereas under the Displacement Hypothesis such is not only reasonable but inevitable. To use a homely analogy one could readily picture a number of bathers standing in shallow water holding hands and plunging up and down in unison, but scarcely if they were out of sight of one another.

(du Toit, 1937: 10)

Wegener had his torn newspaper; du Toit, his synchronized swimmers. Wegener stressed the number of similarities that matched in space at a moment in time; du Toit, the many similarities that matched spatially and over a span of time, a notion captured by his recognition of the similar sequences from Silurian to Cretaceous across Gondwana and in his reconstructed Samfrau Geosyncline.

Despite du Toit's strong words about mobilism being "inevitable," he admitted that there were serious obstacles, but they certainly were no worse than fixism's.

To sum up, it will be seen that current theories rest upon foundations which are far from secure, that they fail to explain with conviction many of the larger features of the globe, and that they demand diastrophic changes on a far greater scale. It is freely admitted that our Hypothesis of former continental rapprochement is faced by not a few difficulties, though, it would seem, not any greater than those striking at the basis of orthodox theories. The trenchant criticisms of the Displacement Hypothesis are hence a long way from being justified.

(du Toit, 1937: 10)

He asked only that mobilism and fixism be treated equally. He noted that van der Gracht had pleaded with fixists for more tolerance. Fixists had treated mobilism harshly, and he intended to respond in kind.

> He [van der Gracht] has also put in a plea for more tolerance, though the writer's viewpoint is that, the most effective defence being in attack, a good deal is, on the contrary, to be gained from an exposure of some of the weaknesses and fallacies of current or orthodox Geology, by which is meant that collective body of opinion which does not admit of extensive horizontal displacements of the continents. Such deficiencies are by no means insignificant, as will briefly be set forth below. Only a few of the critics have as yet been prepared to concede with readiness the point that current views might require some appreciable revision in the light of the new thesis, even though the latter be ultimately rejected.
>
> *(du Toit, 1937: 2; my bracketed addition)*

Launching his attack, du Toit identified fifteen opponents of mobilism, and it is no surprise that all but one were from the Northern Hemisphere. Seven were from Europe (W. Sörgel, A. Penck, P. Lake, P. Termier, E. Krenkel, P. D. Kreichgauer, and L. Kober), six from North America (A. Keith, H. S. Washington, R. T. Chamberlin, C. Schuchert, B. Willis, and G. V. Douglas), one from Brazil (Betim P. Leme), and lastly, B. Gutenberg, which was very strange indeed as Gutenberg sympathized with mobilism, and had even developed his own version (III, §2.3).

He marshaled arguments in favor of mobilism into seven broad categories: physiographic, stratigraphic, tectonic, volcanic, paleoclimatic, paleontologic, and geodetic. He thought (1937: 51) "it . . . essential" that they be applied "not only singly, but so far as possible collectively." He (1937: 53) considered the first five arguments as "possessing great weight." Physiographic, which included the congruence of opposed margins, was to be "used with discretion" (1937: 51). Stratigraphic included the equivalence of strata on opposing coasts established "with due regard to their mode of origin, lithology, facies, attitude, metamorphism, fossils, etc.," which du Toit claimed was "a most important criterion" (1937: 52). Tectonic included the crossing of fold-systems of varying ages, and he (1937: 161) singled out that of Caledonides and Hercynides on both sides of the Atlantic, appealing to the work of Suess (§8.11) and Bailey (§8.12). Volcanic arguments included the "synchronous intrusion of batholiths in equivalent fold systems – those of the Appalachian-Hercynian orogeny" and matching "petrographical provinces with similar eruptive suites of varying ages" found in Brazil and West Africa (du Toit, 1937: 52–53).

Du Toit devoted an entire chapter "Past Climates and the Poles," to paleoclimatic arguments, some "possessing great weight." As I shall show below, he believed them to provide the most effective demonstration of continental drift. Paleoclimatic evidence was derived from (du Toit, 1937: 53; my bracketed addition):

(a) Strata denoting a special environment, particularly extreme climatic types such as tillite, varved shale, laterite, "evaporite" (salt, gypsum, etc.), aeolian sandstone, coal, coral limestone, banded ironstone, etc. . . .

(b) Distribution of the above with reference to the climatic girdles of the past as indicating relative polar shift.

(c) Glacial deposits with special reference to ice-centres and ice-movements – [in] Gondwana.

He drew heavily on Köppen and Wegener's 1924 book *Die Klimate der geologischen vorzeit* with its extensive analysis of past climatic zones that testified to displacement of continents relative to the geographic poles. He began by dismissing Schuchert's (1928a) attack on Köppen and Wegener's paleogeographic reconstructions. Schuchert had to ignore or greatly deform climatic zones in order to keep continents fixed; Köppen and Wegener, and du Toit moved continents and kept climatic zones unchanged.

Köppen and Wegener's continental groupings in relation to the polar axes of past geological periods come in for adverse comments, but the orthodox and rigid alternative proposed by Schuchert (his Fig. 20) – one which doubtless makes appeal to the conservatively minded – is to the writer meteorologically mysterious.

(du Toit, 1937: 32)

By keeping continents fixed, Schuchert had made meteorology mysterious, which I take him to mean, inexplicable.

Using his favorite example, the climate of the Permo-Carboniferous, du Toit repeated Wegener's allegation (§3.10) that fixism had crippled the development of paleoclimatology.

Huge areas in the Southern Hemisphere far removed from the present South Pole were, for instance, heavily ice-capped in the Permo-Carboniferous, although equivalent refrigeration is almost unrepresented in the Northern Hemisphere; well-developed floras indicative of milder environments are known from high latitudes, during the Permian from close to the south Pole, during the Jurassic from Graham Land and during the Cretaceous from Greenland; in the Triassic aridity affected extensive regions in both hemispheres, and so on.

Such climatic incongruities form one of the major problems of geology, since they are difficult or impossible to account for satisfactorily under current theories with the lands and poles fixed as at present, and that, too, despite the invoking of pretentious land-bridges, ocean currents and winds. Orthodoxy is indeed endeavoring to defend a wholly untenable position. As Wegener has caustically remarked, the general failure to admit the hypothesis of drift has absolutely crippled the science of Palaeoclimatology.

(du Toit, 1937: 270)

The starting point for paleoclimatology was that there had been latitudinal climate zones in the past as there are today, the very thing which fixists had neglected to do.

The Earth's surface is today divided into the seven climatic girdles: the low-pressure moist equatorial zone, the two bordering high-pressure zones with much lower rainfall – in which are located almost all the desert regions – the two temperate zones and the two polar circles.

(du Toit, 1937: 271)

He was willing to allow for general changes in width of climate zones in time, but there always had been a decrease in mean temperature from equator to poles.

He then defined the task at hand, and introduced the idea of wandering of the geographic pole.

Our essential object is to trace out beneath these almost-world-wide and obscuring undulations of climate the earth's underlying zonal pattern and to deduce therefrom the approximate course of the equator of the time, and in that way the positions of corresponding poles. Throughout this discussion, the term "pole wandering" or "polar shift" will explicitly refer to the creeping of the thin and distorting crust over its rigid core that has been rotating upon a fixed axis and not to a change in the direction of that axis in stellar space.

(du Toit, 1937: 272)

Du Toit wanted to make clear that he was not suggesting an actual shifting of the polar axis relative to the main body of the Earth, a point on which he disagreed with Wegener.

The apparent changes in the positions of the poles at various epochs are interpreted by Wegener as due to actual shifting of the polar axis, the feasibility of which has been stoutly denied by most physicists. Such incongruity, which incidentally has lost Wegener considerable support, is easily met by other hypotheses, which picture the axis of rotation as fixed while the crust slides around over the denser core.

(du Toit, 1937: 19)

Following Köppen and Wegener (1924), du Toit (1937: 273–275) identified six indicators of past climatic zones: glacial deposits indicative of polar zones; coals indicative of either tropical or temperate wet climates; fossil soils including laterites, bauxites and kaolinites formed in the equatorial zone, and pale sandstone formations indicating high latitudes; salt and gypsum, aeolian sandstones, wind-faceted pebbles, and superficially silicified rocks indicative of deserts; coral limestones indicating the equatorial zone; and animal and plant remains, which, du Toit cautioned, must be used with care as climate indicators.

By these means du Toit proceeded to determine the position of paleogeographic poles through time insisting that this be done separately for Laurasia and Gondwana because

They must be viewed as separate and in certain respects as independent masses belonging to the Northern and Southern Hemispheres, and their displacements, with reference to their corresponding poles, as having been not necessarily equal nor simultaneous.

(du Toit, 1937: 275)

He reasoned that by observing the changes in the climate of landmasses, he could determine their changes in latitude and the movement of the pole relative to each, and that the differences in these changes would reflect their relative movements. This is analogous to the paleomagnetic test for continental drift carried out twenty years later (II, §3.3).

But du Toit thought it premature to determine, as Köppen and Wegener (1924) had tried to do

[The] geographical co-ordinates of places to within a few degrees for the pre-Tertiary epochs . . . It cannot be too firmly impressed that this Climatic Method is not sufficiently sensitive to do more than establish the major polar shifts; the minor movement can only be deduced from orogenic considerations.

(du Toit, 1937: 276; my bracketed addition)

Notwithstanding, he did reproduce three of Köppen and Wegener's maps, which showed the past positions of the continents and paleogeographic grids during the Carboniferous and Permian (p. 278), Triassic (p. 282), and Cretaceous Periods (p. 284), and noted:

While our polar restorations follow in a general way those of Köppen and Wegener, they depart materially therefrom, partly through the somewhat different fitting together of the lands and the different manner of their partition, but, above all, through the independency of horizontal movement postulated for the two great continents.

(du Toit, 1937: 276)

In tracking the movement of Laurasia relative to the geographic North Pole, du Toit found a slight difference in the movement of North America and Eurasia, which indicated, he argued, their relative displacement.

A general movement can therefore be deduced for North America and Eurasia at first, north, then north-east, thereafter north again and finally east, modified to some extent by the continued *divergence of the two continents.*

(du Toit, 1937: 283)

He then turned to the movement of Gondwana's fragments relative to the South Pole. He deduced that relative to Laurasia it drifted northwestwards and thereafter northwards with an anticlockwise rotation from the early Carboniferous through the Jurassic, and added:

In the early Cretaceous Africa tore itself free [from Gondwana], renewed its northerly drift into warmer climes, came again into conflict with Eurasia and proceeded to push up the Atlas-Persian ranges.

(du Toit, 1937: 287; my bracketed addition)

Continuing his paleoclimatic-based analysis he argued:

Upon the breaking away of the "wings" made by South America, India and Australasia, the Antarctic nucleus continued on its southerly drift through the Tertiary and thus came to acquire its present co-polar position. The radical dispersal of the three fragments, on the contrary, took them back into progressively warmer latitudes, their climatic histories thereafter proving consistent with the main idea expounded above.

(du Toit, 1937: 288)

In closing, du Toit (1937: 288) emphasized that pole positions deduced from paleo-climates "of Laurasia and Gondwana have been arrived at *separately and quite independently*" yet are "mutually consistent" (RS1).

Even making allowances for the many uncertainties in this problem, it will still have to be admitted that the positions of the present continents must have experienced radical changes in their distances from the poles through geological time. Since polar movement, in the sense of a changed axis of rotation, is generally denied by geophysicists, it assuredly follows that those lands must have suffered displacement across the surface of the globe with reference to that fixed axis. Indeed, from the mid-Palaeozoic onwards the lands must have crept northwards for thousands of kilometers to account for their deduced climatic vicissitudes. *Such, indeed, constitutes the most telling demonstration of the reality of Continental Drift.*

(du Toit, 1937: 289)

6.8 The reception of *Our Wandering Continents*

Although du Toit was greatly respected by many Earth scientists, I do not think that *Our Wandering Continents* caused very many fixists to change their minds. As Holland, a strong supporter of mobilism (§3.9, §3.13), pointed out in his 1941 Bruce-Preller Lecture to the Royal Society of Edinburgh, many readers questioned du Toit's objectivity, thinking he argued too aggressively, with too great a conviction, that drift had occurred.

The distinguished South African geologist, A. L. du Toit, published three years ago a comprehensive work entitled *Our Wandering Continents*, in which he is frankly aggressive to the so-called orthodox naturalists, assuming openly that the most effective form of defense is attack (du Toit, p. 2). As a consequence of this attitude, both critics and supporters naturally tend to dispute or to hold in abeyance many of du Toit's arguments which cannot be independently verified, or would be regarded as justifiable accessory evidence if the drift theory were definitely established as a "fact" in geological history.

(Holland, 1941: 151)

Also, I suspect that not very many studied his book carefully. It is not an easy read. The work certainly did not affect the views of Longwell, R. T. Chamberlin, Schuchert, and Willis. Longwell (1938) and Chamberlin (1938) reviewed du Toit's *Our Wandering Continents*. Schuchert and Willis responded privately. Chamberlin (1928), who had objected so strongly to Wegener's mobilism at the AAPG symposium, thought du Toit's new defense of mobilism an improvement, and commended him for gathering much diverse evidence.

The author of this interesting and attractive volume has elaborated the hypothesis of continental drift in much detail by presenting an impressive array of geological observations which seem to him in accord with this hypothesis and to be better explained by it than by any other hypothesis. The pertinent observations and facts have been gathered by a painstaking study of the geology of the whole globe in its many aspects. Historical, stratigraphic, biologic, climatic, tectonic, and geophysical evidence has all been brought to bear on the problem. In fact, the focusing of so many different lines of evidence upon the main questions in hand may well be regarded as the chief contribution of this new study. Irrespective of the conclusions expressed, the assembled information itself and the skillful handling of it make the volume worthy of careful reading by any student of larger earth problems.

(Chamberlin, 1938: 791)

But Chamberlin was not satisfied. Unable to resist sarcasm and taking cheap shots, he declared that du Toit moved continents about with abandon; he moved them where and when he wanted.

> Certain advantages of the hypothesis are at once apparent. If a land mass can be drifted for a desired distance and can, in addition to drifting, be rotated either clockwise or counterclockwise at the wish of the manipulator, the possibilities of producing a given fit, pattern, or arrangement with other similarly movable land masses must be very considerable. Such maneuvering has few restrictions.
>
> *(Chamberlin, 1938: 791)*

Chamberlin also thought that du Toit had not demonstrated consilience between continental reconstructions based on different lines of evidence, and he also did not think that mobilism was necessary to explain intercontinental similarities.

> But is it necessary? The reviewer is not convinced that broadly corresponding stratigraphic successions in two separate regions, more or less contemporaneous lava outpourings on a large scale, comparable glacial deposits, or related faunas necessarily indicate the close proximity of the areas which the author assumes to be required for their more or less parallel development.
>
> *(Chamberlin, 1938: 791)*

Chamberlin gave no alternative view. Instead, he questioned du Toit's objectivity, albeit not without an attempt at humor.

> The author's conviction that the displacement hypothesis is far superior to any alternative current hypothesis is strongly expressed. As a manifestation of the writer's enthusiasm, some readers will doubtless note how often works cited for interpretations or conclusions in harmony with the continental drift hypothesis are characterized as "brilliant," "illuminating," or "able," whereas some of those opposed suffer considerably in contrast. But any derogatory implications must be taken in good spirit by opponents as return compliments for their own severe criticism of the hypothesis of drifting continents. Dr. du Toit's new book is a stimulating contribution, whether one accepts his conclusions or not.
>
> *(Chamberlin, 1938: 792)*

Longwell was impressed with du Toit's marshaling of evidence, which he saw as a continuation of his earlier work.

> Du Toit is a well known champion of the continental-drift hypothesis. More than ten years ago he made extensive field studies in South America and found striking similarities to his own South Africa . . . He has continued to study the problem widely, in the field and in the literature . . . he summarizes an impressive quantity of stratigraphic and structural information from the literature of the world. Armed with this material, he makes a vigorous attack on the viewpoint of "orthodox" geology and decides that the evidence now available overwhelmingly favors the concept of large-scale horizontal movements of all the continental masses.
>
> *(Longwell, 1938: 704)*

Longwell commended du Toit's painstaking work, and remarked that it was only through such work that mobilism could be adequately tested, but his words would not have inspired many to actually read the book.

Compared with Wegener's well known volume the present work is less entertaining to the general reader, because it presupposes considerable knowledge of geology, and the recitation of field evidence necessarily involves much tedious detail. However, it is only by the most thorough study of geologic facts that we can hope ever to give the hypothesis an adequate test.

(Longwell, 1938: 705)

So much for encouraging geologists to bother attempting to test mobilism! Longwell in his measured way raised a difficulty with du Toit's proposed mechanism (RS2).

The mechanism by which continental masses have been propelled horizontally presents a difficult problem to proponents of the hypothesis. Du Toit attempts to combine into an eclectic scheme several earlier suggestions, among them those of the centrifugal force of the earth's rotation and convection currents below the continental plates resulting from radioactive heating. To these the author adds his "paramorphic principle," according to which the minerals that are characteristic of the outer part of the earth's crust become changed, if carried below a critical depth, to other minerals that have greater density and smaller volume. This principle agrees with the view of petrologists that the heavy rock eclogite is the high-pressure equivalent of basalt. However, Du Toit goes further and assumes that complete inversion to the lower-density forms takes place if the pressure is sufficiently reduced through long-continued erosion . . . Unfortunately, in his quantitative estimate of paramorphic expansion the author arrives at a result about four times too great under the assumed conditions (p. 239). Du Toit's concept of paramorphism is attractive, but it is extremely doubtful whether inversion to low-density minerals takes place on the scale he suggests.

(Longwell, 1938: 705)

But Longwell did not like du Toit's rhetoric for a reason that must have surprised du Toit.

Perhaps the chief criticism of Du Toit's presentation is that it betrays somewhat too clearly the viewpoint of the zealous advocate. Most geologists now have an open-minded interest in the concept of continental drift and will weigh fairly any new factual evidence relating to it.

(Longwell, 1938: 705)

Longwell painted far too rosy a picture of geologists' open-mindedness; the rigid fixists Schuchert and Willis remained unaffected by du Toit's arguments. Both wrote to du Toit appreciative of his enthusiasm (Oreskes, 1999: 294), and thought his presentation of mobilism the best yet. Schuchert died in 1942. Willis, Longwell, and du Toit continued to disagree through the early 1940s.

The pro-drift paleobotanist Seward also reviewed *Our Wandering Continents*. He praised du Toit as a geologist, but found the book difficult to read.

The question is: how far has the author been successful? He has accumulated an enormous mass of information which, despite lack of coherence, will be of great value to readers who desire to follow the devious trails blazed by many adventurers and to study in true perspective the bearing of conflicting hypotheses and the formidable array of facts upon one of the most baffling and fascinating problems of earth-history. Readers are assisted by a useful glossary which might have been still more helpful if Greek and Latin derivations had been given. The author, with the best intentions, has not succeeded in producing a book that can be read and understood with reasonable ease. The reviewer has the highest admiration of Dr. du Toit's ability and enviable familiarity with the contributions of authors interested in a bewildering variety of subjects; he is also definitely in sympathy with the hypothesis usually associated with the name of Wegener, and he frankly admits that some of the difficulty he has experienced in piecing together the several sections of the volume into a connected whole are due to his own ignorance. The impression left after reading the book is that the author might have presented a more convincing case had he devoted a larger proportion of space to general summaries, written in a form intelligible to the layman. In short one feels that Dr. du Toit has yielded to the temptation of using the contents of note-books as separate entities, leaving readers to make their own synthesis, instead of co-ordinating the results of his amazingly laborious researches, and presenting his considered opinions in a series of essays. This criticism is made by one who shares many of the author's views and in the hope that the great service already rendered may be made still greater and more effective by a less controversial presentation in simpler language.

(Seward, 1938: 319–320)

Seward (1938: 322–323) then provided an excellent chapter summary of the book, offered a few gentle objections, and concluded his review by once again praising the author, and faulting the book.

In the conclusion he tells us that his aim has been to set forth "in an elementary fashion certain individual ideas of Earth Structure and Evolution." He has put forward "a provisional attempt to interpret the host of phenomena that do not appear to be adequately, if at all, explained under current theories of geology." What is elementary to Dr. du Toit will not be so regarded by many of his readers. One may criticize his method of presentation, his impatience with obstinate opponents, and his disinclination to admit that even the orthodox are seekers after truth. Truth, it has been said, is the hypothesis which works best. Dr. du Toit has brought together a most impressive collection of facts and has expressed his views in no uncertain language: he has made a notable contribution to knowledge which will not only stimulate geologists, palaeontologists, and others to examine afresh and with new ideas their beliefs and disbeliefs, but will convince a fair proportion of readers that the conception of wandering continents is not mere fancy, but fancy that is just one fact the more.

(Seward, 1938: 322–323)

Perhaps some readers became mobilists after studying du Toit's book, but I wonder how many actually studied the book, especially if they were not already sympathetic to mobilism. Looking ahead, there were at least two US readers, Warren Hamilton and Bill Long, who were deeply affected by reading *Our Wandering Continents*; after reading it, Hamilton, one of the few North American geologists who favored mobilism during the 1950s, became sympathetic to mobilism (§7.1). Long, who was leaning toward mobilism once he started to believe he had discovered evidence of Permo-Carboniferous glaciation in Antarctica in the late 1950s, became a mobilist after reading *Our Wandering Continents* (§7.5). If the erudite Seward had trouble understanding some chapters, how difficult it must have been for others. Like many classics, it may have had an honored place on the bookshelf, where it remained little read.

6.9 Du Toit's later contributions to mobilism

Just before World War II, at a conference in Germany on mobilism, du Toit (1939a) described the origins of the Atlantic and Arctic Oceans and their relevance to surrounding continents. He also wrote a companion work (1940) on the evolution of the Pacific for the Sixth Pacific Science Congress. Coleman (1938) launched his final criticism of drift's explanation of the Permo-Carboniferous glaciation, and du Toit (1939b) responded. Du Toit (1944) also responded to G. G. Simpson's (1943) attack on drift's explanation of biotic disjuncts (§3.7).

I begin with the last exchange of the du Toit–Coleman debate. Coleman (1938: 999) reiterated his opinion that Permo-Carboniferous glaciation could not have formed on a dry supercontinent that Gondwana likely was, and the supposedly Permo-Carboniferous Squantum Tillite near Boston was inconsistent with a low-latitude of North America at that time required by drift. He also reiterated the lack of mechanism (RS2).

May an old-fashioned geologist ask the advocates of drifting continents how they account for the extraordinary performances they so readily assume? On the Labrador coast one may see icebergs pushing southward, thrusting aside the ice-floes, or sometimes ridging them up in front, but one knows that the greater part of the berg is sunk in the arctic current, so that the motion is accounted for.

(Coleman, 1938: 998)

Coleman questioned the analogy; continents were not icebergs, and Earth's substratum was not water.

Our iceberg-like continents are solidly frozen into the sea bottom crust, supposedly miles in thickness of strong basalt. Are there currents in the supposedly plastic basic substratum in which the bulbs of the continents are enclosed? Or what forces push them in one direction rather than another? To say they "drift" is, of course, begging the whole question. Permanently enclosed rock-masses cannot drift – they must be pulled or pushed. What power do the advocates of the "drift of continents" suppose dragged India from the Antarctic regions thousands of miles north, over the bulge of the equator, to leave it in the Northern Hemisphere?

(Coleman, 1938: 998)

Coleman also wanted to see evidence of drift on the seafloor (RS2).

Where are the heaped-up ridges of rock which it thrust on each side, and where is the scar it left after its passage? Again why did Gondwanaland explode and send South America in one direction and South Africa in another and leave Antarctica where it is?

(Coleman, 1938: 998)

This old-fashioned geologist had had enough of continental drift, and ended with this remarkable rhetorical outburst.

These causal driftings of massive blocks of the earth's solid crust should have some reasonable explanation before being used to account for the distribution of plants or animals. The Gondwana plants are mainly ferns, equisetums and club-mosses, especially ferns – all cryptogams the spores of which could easily be transported by the wind. A gale would quickly carry them hundreds of miles. Why send continents crashing through the solid earth's crust to effect their distribution?

(Coleman, 1938: 999)

Du Toit was restrained. He had earlier (§3.13) countered Coleman's "lack of moisture" objection with "bays and gulfs penetrating into" Gondwana. He now appealed to new research on the extent of Pleistocene glaciation in the Northern Hemisphere, which showed that the center of the Pleistocene glaciation had been further from the sea than that of the Permo-Carboniferous glaciation on his reconstructed Gondwana (1939b: 243). Du Toit, however, did not respond to Coleman's other objections. Instead, he turned the tables on him, asking how he, Coleman, would explain Gondwana's Permo-Carboniferous glaciation. He first claimed that Coleman had no answer.

Coleman has deplored not being able to explain the distribution of past glacial deposits, which suggests that he must have overlooked some vital factor. Others too have failed to interpret such glacial phenomena in terms of existing geographical and meteorological environments.

(du Toit, 1939b: 243)

Coleman should develop a fixist solution to the glaciation problem instead of criticizing mobilism.

Instead of contending that the Permo-Carboniferous glacials could not have been developed upon a single land-mass as he does, I feel it would be more helpful if Prof. Coleman could show how they could have been laid down at all in the several widely-parted areas in the positions and latitudes which they occupy to-day.

(du Toit, 1939b: 243)

Coleman was unable to accept du Toit's challenge; he died on February 27, 1939, sixteen days after du Toit's paper appeared.

Du Toit (1945) also responded to Willis (1944) and to Longwell (1944a, b), who had substituted during World War II for G. G. Simpson in his debate with du Toit (§3.7). Both had raised several difficulties with mobilism. Longwell remained skeptical of continental drift, but thought that it should not be dismissed; Willis rejected drift in his "Continental drift, *ein märchen* [fairytale]."

In the April number of the *American Journal of Science*, my well-balanced colleague, Chester R. Longwell, who leans neither backward nor forward in scientific discussion, analyzes the evidence for continental drift, adopting what seems to be a most cautious attitude of skepticism. I would like to join him on the fence, but I cannot. I confess that my reason refuses to consider "continental drift" possible. This position is not assumed on impulse. It is one established by 20 years of study of the problem of former continental connections as presented by Wegener, Taylor, Schuchert, du Toit, and others, with a definite purpose of giving due consideration to every hypothesis which may explain the proven facts. But when conclusive negative evidence regarding any hypothesis is available, that hypothesis should, in my judgment, be placed in the discard, since further discussion of it merely incumbers the literature and befogs the minds of fellow students.

(Willis, 1944: 509)

Although du Toit disagreed with most everything Willis said, he concurred with the diehard fixist that Longwell's skeptical stance was no longer viable.

Longwell's protest in regard to insufficient objectivity will, so far as the writer is concerned, be conceded. Nevertheless the differences between the doctrines of "fixed" and "moving" continents are fundamental and the acceptance of the one must largely exclude the other. My attitude, however, is that synthesis has advanced sufficiently to justify the rejection of the current idea of fixed continents and all the principles that go therewith. Consequently there would seem no particular gain in taking up a neutral or waiting attitude; the Wegener Hypothesis – or some variant thereof – must be either rejected or supported, and in so doing one's viewpoint is bound to become colored by the decision taken.

(du Toit, 1945: 404)

Du Toit and Longwell joined forces attacking Willis' criticism of mobilism's mechanism, a restatement of what he had said at the 1926 AAPG symposium (see Endnote 2, Chapter 5).

It is my firm conviction that the laws of mechanics and dynamics govern terrestrial structures in their development and movements, whatever the magnitude of the masses and forces involved. Continents present no exception . . . Now, it is a well-established principle of mechanics that any floating object moving through air, water, or a viscous medium creates behind it a suction of the same order as the pressure developed in front of it. This law applies equally to airplanes, ships, rafts, and drifting continents (if there are any). The pressure which could raise the Andes must, therefore, have been approximately equaled by the suction and tension in the rear. Sections of the continent must have been sucked off. They should now remain as islands in the Atlantic; but there are none such.

(Willis, 1944: 509)

Du Toit became impatient.

Dr. Willis, like the majority, rejects drift on mechanical grounds. But do the simplified dynamics of rigid bodies favored by him apply inexorably to "living evolving continents" with circulatory systems evinced by perpetual transfers of materials horizontally or vertically, by changes in size, shape, internal and external stress, flotation height, internal temperature, etc. and not always gradually, but often rhythmically? Why indeed should their crumpling edges be assumed to offer enormous resistance to movement? What degree of opposition is met when a red-hot poker is pushed into pitch or solder? Again, why should a fracture in the rear of a moving continent produce an equivalent vacuum? How comes the supposedly tough crust to have been so readily diked, and whence originated the forces that have repeatedly snapped it in such diking?

(du Toit, 1945: 406)

Willis thought that du Toit begged the question, claiming that he and other mobilists assumed that drift had occurred and then supposed that "since continents did drift there must have been some competent mechanism of some kind" (Willis, 1944: 510). Du Toit disagreed.

So far from "begging the question," as Willis asserts [1944: 510] the writer has discussed at length [*Our Wandering Continents*, Chapter XVII] various factors that would or could have contributed to Drift. Too much has been made of the impossibility of sliding because adequate forces have not yet been disclosed by mathematical analysis. All such calculations have been based on assumptions that, considering the complexity of the problem, are regrettably limited in number and in certain cases extremely doubtful.

(du Toit, 1945: 407; my bracketed additions)

Longwell also was displeased with Willis' treatment of mobilism's lack of a mechanism.

I must say – in a modification of Professor Willis' own phraseology – that I should like to join him in believing the vexed question of continental drift can be settled now, once for all. In my view, however, the issue is not as simple as he represents it. No better test than the one he proposes could be chosen to illustrate the difficulty of passing summary judgment. The principle of mechanics that demands suction in the rear of a floating object applies if the object is driven by an independent force through the medium; the principle has no relation to objects that float passively on a broad current in the medium itself. Thus the test suggests an objection to the particular mechanism proposed by Wegener, a mechanism which in any case fails much more convincingly under quantitative evaluation of the forces he invoked (Jeffreys, 1929). The test has no validity, however, in an appraisal of convection currents as a possible mechanism of continental displacement (Holmes, 1933) – a suggestion that has received serious consideration from competent geophysicists. This hypothesis of convection currents may be attacked in its turn, by use of tests other than the suction principle. However, even if convection could be ruled out, other suggested answers would confront the test which Professor Willis views as conclusive.

(Longwell, 1944b: 514; Longwell's references to Jeffreys
and Holmes are the same as mine)

Willis was wrong and had not kept up to date. Jeffreys had already raised much better objections to continental drift, and Holmes' substratum convection theory, Longwell claimed, was invulnerable to Willis' criticism.[7]

Du Toit in his replies to Willis and Longwell restated how mobilism better explained the paleoclimatogical evidence. Longwell agreed that the evidence of Permo-Carboniferous glaciation was strong, but noted climatologists disagreed about whether mobilism was required or not. He cited the 1930 Royal Society of London meeting where G. C. Simpson, Brooks, and others debated the issue, and opinions differed with two for mobilism, two for fixism, and one neutral (§3.13).

The most striking unit in du Toit's "Samfrau" belt consists of late Paleozoic glacial deposits; and without question the eloquent testimony that widespread ice caps reached to low latitudes in the southern continents furnishes the most cogent argument that has been offered in favor of continental drift. The marvelous displays of this evidence, particularly in South Africa, serve to create in the geologist a humble attitude and a feeling of tolerance toward all attempts to explain this great enigma. Students of climatology do not agree on the significance of the evidence, however; G. C. Simpson interprets it in favor of continental drift, but Brooks points out other possibilities (G. C. Simpson *et al.*, 1930).

(Longwell, 1944a: 225; Simpson reference is equivalent to my Simpson, 1930)

Du Toit, I suspect, was disappointed by the fact that hardly anyone paid any attention to Köppen and Wegener's use of climatic zonation as a basic principle in analyzing past climates.[8] He thought that mobilism's solution to the origin and distribution of Permo-Carboniferous glaciation was just part, albeit the most telling, of the application of latitudinal zonation to the problem of past climates.

Few apparently, apart from Wegener, have troubled to study the evidence of past climatic girdles as deduced from the particular sedimentary facies around the Earth, though that constitutes one of the most important of criteria for Drift. Such however, involves much searching through voluminous literature for the basic facts. More direct is "polar environment" as indicated by the presence of ancient tillites. Now, represented in Africa are at least two such glacial formations (of which a few could well prove equivalent), some with wide distribution. Either geologists are suffering from delusions when interpreting those formations or else the Pole of those times wandered somewhere in or near that continent, and I am still conservative enough to reject the first explanation. With fixed continents or poles Palaeoclimatology just does not make sense.

(du Toit, 1944: 405–406)

It would be another decade before geologists began to take the tectonic implications of paleoclimatology seriously, brought about by the revitalization of paleoclimatology by paleomagnetists from a new standpoint who reinvigorated the case for mobilism in the mid-1950s (II, §3.13, §5.13; III, §1.3, §1.7, §1.15).

Du Toit also attempted to develop a method for determining mobilism's probability as a function of the number, degree, and variety of similarities on opposing margins of formerly connected landmasses (Oreskes, 1999). He wrote Daly about the idea. Oreskes (1998: endnote 59: 364) discovered an unpublished manuscript of du Toit's, "On the mathematical probability of drift," which was written *circa* 1944–5. Oreskes (1999: 298), who also examined du Toit's personal annotated copy of *Our Wandering Continents*, argues that du Toit "was evidently planning a revised edition of his book, in which he would" demonstrate that the mathematical probability of mobilism was very high. He died in 1948 before he could write it.

6.10 Lester King

It's largely a matter of *where* a man works whether he sees the *fact* of drift.
(Lester King, first December 29, 1981 letter to author; King's emphasis)

After du Toit died, Lester King (1907–89) became the leading mobilist in South Africa, and one of its important defenders worldwide (see Twidale (1992) and Maud (1989) for details of his life). Born in London, he and his parents emigrated to New Zealand when he was two. In 1925, he entered Victoria University College, Wellington, and trained as and became a schoolmaster. While teaching, he studied geology on a part-time basis, and obtained a B.Sc. degree in 1928. He obtained an M.Sc. (with First Class Honors) in 1930. King studied primarily under Charles A. Cotton, and, like him, became a geomorphologist. King lectured in geology at Victoria University

College from 1930 to 1934. With the growing economic depression his position lapsed. He wrote around explaining his interest in securing a position. S. J. Sand, of Stellenbosch University, told the authorities at the Natal University College at Pietermaritzburg about King, and in 1935 he was appointed lecturer in geology and geography. He obtained his Ph.D. from the University of South Africa in 1936, and in 1939 the University of New Zealand awarded him a D.Sc. degree. In 1949 he became the founding head of the Department of Geology at the University of Natal, Durban, where he remained until he retired in 1974.

King was elected president of the South African Geographical Society in 1943 and president of the Geological Society of South Africa three years later. He received the Dumont Medal of the Geological Society of Belgium in 1958, the Draper Memorial Medal of the Geological Society of South Africa, and in 1965 the Founders Medal of the Royal Geographical Society, London, for his "geomorphological explorations in the Southern Hemisphere."

By the 1950s, King had become a leading mobilist (King, 1950, 1953, 1958a, b, c, d, 1961, 1965). Gaining a reputation as mobilism's formidable and eloquent defender, he gave spirited lectures in its favor in North America while a distinguished lecturer for the AAPG in 1951–2, in Tasmania at a symposium arranged by S. W. Carey in 1956 (II, §6.16), and in Britain in an address before the Geological Society of London in 1957. While on tour in North America, King debated Bucher at Columbia. It was a debate made famous by those who heard it and lived through the eventual triumph of mobilism. When asked about this "famous" debate, King recalled:

The occasion of debate was probably at Columbia in 1951. The geology society asked if I would debate against W. B. Bucher. When the debaters retired, the students took an opinion poll, 90% were for drift. In later years I met many of those students. They all said with solemn voices, "I was present at the debate."

(King, 1981 letter to author)

John O. Wheeler of the Geological Society of Canada, then a graduate student at Columbia, was one of those students. Although Wheeler did not mention a vote, his memories coincide with King's. Wheeler, however, is sure that the debate occurred on Halloween, and is fairly confident that King won.

The debate was held on October 31st, 1951 during one of the Department of Geology's "Journal Club" meetings. The reason I know the date is that I mentioned it in a letter to my fiancée which I retrieved recently from our "family archives." The passage from the letter follows: "Tonight (Oct 31/51) we had grand debate over 'continental drift' with Dr. Bucher of Columbia and Dr. King of University of Natal taking sides. Afterwards we had a good discussion and, all round, it was one of the best meetings we've had in many a moon!"

I am sorry to say that after 50 years my recollections of the event are rather dim. My overriding memory is that Lester King was a clear winner and that continental drift had very likely taken place in the past.

Other things that vaguely come to mind are that King used stratigraphic correlations between South Africa and South America, proposed earlier by the great South African geologist Alex du Toit, as one of his main arguments that the two continents were once joined. He also referred to the Paleozoic glaciations in South America, Africa, India, and Australia – parts of the former Gondwanaland continent. He particularly stressed the direction of ice-flow features, preserved in the Paleozoic rocks of the Karroo Basin of South Africa, which indicated ice flow from beyond the southeast coast of Africa, presumably from parts of Gondwanaland in Western Australia. King also used evidence from the correlations of erosion surfaces developed from processes of peneplanation that were recognized in South Africa and in South America.

Another factor in the debate that favored Dr. King was that he presented his arguments and supporting data in a lively and convincing manner. Dr. Bucher, however, read his paper – one he had given at some earlier time that, I believe, dealt with the Bering Sea land-bridge and its role in facilitating the migrations to and from Asia of various fauna, including humans. Dr. Bucher, of course, put forward a fixist view of Earth history – much of which is outlined in his book "Deformation of the Earth's Crust," published in 1933. Dr. Bucher's presentation was lackluster compared to Dr. King's. This was unusual because his lectures in his structural geology course were interesting and stimulating. Indeed, his exposition of the tectonic evolution of the Alps was absolutely riveting. My feeling is that Dr. Bucher was either too busy with other matters or that he did not take the debate that seriously.

Judging by the passage I quoted earlier from the letter to my fiancée the discussion must have been lively and interesting but I can recall no specifics other than it was concluded that continental drift must have taken place in the past but nobody knew what processes caused the drifting.

(Wheeler, November 11, 2002 letter to author)

Hugh Gabrielse, another graduate student at Columbia, also heard the debate. He agreed with Wheeler; King definitely won the debate, and appealed to Gondwana's geology in arguing for mobilism.

The Bucher-King debate was a fairly one-sided affair. Lester King presented data that was well accepted by South African geologists concerning the probability that continents had shifted their relative positions at least since Permian time. He pointed out the coincidence of Gondwanan fossils in India, South Africa and South America and the directions of ice movements on the three continents during the Permian. His fit of South America, when restored, to Africa was of course rather convincing as it had been many years before to Alfred Wegener and, later, Du Toit. Bucher, on the other hand, seemed to have dusted off a talk that he had given previously pretty well expressing the general North American view that the continents had remained fixed since their formation. He had held these views since publication of his book "The Deformation of the Earth's Crust" in 1933. His analysis of ice age phenomena, geosyncline formation and geosynclinal deformation suggested to him that continental drift did not explain these criteria any better than a fixist hypothesis. In particular he had difficulty in seeing how continental drift could explain deformation of intra cratonic geosynclines.

(Gabrielse, October 8, 2002 letter to author). (See III, §1.2 for R. H. Dott's recollection of the debate and IV, §7.2 for Jack Oliver's view.)

King presented a written defense of mobilism, specifically aimed at North American geologists, in his 1953 paper "Necessity of Continental Drift." He described how the reception of mobilism varied from region to region, and attributed it to the comparative abilities of mobilism and fixism to solve regional problems. The paper reads like the opening set of remarks in a debate, not surprising perhaps in light of his recent debate with Bucher. King (1953) referred explicitly and implicitly to Bucher's (1933) attack on mobilism. I suspect that the paper was a written version of what he talked about during his AAPG lecture tour, and probably included many points he made during their debate.

King began aggressively, noting that few North American geologists knew much about the evidence for continental drift and that few were in a position to make an informed judgment about it.

Lecturing recently to various groups of the American Association of Petroleum Geologists upon the subject of "Continental Geomorphology," involving the possibility of intercontinental correlation of cyclic erosional surfaces, the writer found much greater interest evinced in the hypothesis of continental drift than had been expected. The subject is apparently little taught in American colleges and many in the audiences professed ignorance and a desire for information on which they could base a judgment for themselves. A new spirit seemed to be abroad of impartial inquiry and broad outlook.

(King, 1953: 2163)

Readers of this chapter could conclude that King's belief about a new spirit of impartial inquiry and broad outlook was something of an illusion.

King accented the great divide along regional lines attributing it to different regional geologies.

In the past there has been remarkable cleavage of opinion among geologists on the subject of drift, a large number of Gondwana geologists resident in India, Africa, and South America expressing a profound belief in the hypotheses (e.g., du Toit, Robert, Guimares, Leonardos, Fermor, and Windhausen), whereas a number of equally distinguished Laurasian geologists have held strenuously to the opposite view (e.g., Schuchert, Bailey Willis, and J. W. Gregory). The basis for such a cleavage of opinion is doubtless to be found largely in the respective environments in which the two groups of savants have worked. Thus Gondwana, being apparently broken into many fragments, making many discontinuities between the respective parts (Fig. 1), presents geologists there with a set of peculiar and typical problems to which their attention is naturally directed; whereas Laurasia, being in two major pieces only, has not presented observers with a like number of allied problems, and the interest of these, in all but a few cases, has remained purely academic.

(King, 1953: 2163; King's Fig. 1 (p. 2164) was his latest reconstruction of Gondwanaland).[9]

He then turned to the need to explain trans-Atlantic geological disjuncts, noting that both drifters and anti-drifters agreed at least on the need to postulate lands beyond the boundaries of present continents to provide detritus for sedimentary basins now at continental edges.

Problems of this type [e.g., explaining the formation of the Great Karroo basin] are, of course, not peculiar to Africa; North America has likewise its enigmas concerning the derivation of great quantities of sediment apparently from beyond the present boundaries of the continental mass. Americans likewise have explained these phenomena by postulating hypothetical lands of which there are three in particular: Cascadia in the northwest, Appalachia in the east, and Llanoria in the south. These were originally deemed to have been broad and fairly high, and to have shed vast quantities of sediment onto the main disc of North America. Of late years there has been some retreat from this view, Llanoria being rejected altogether by some authors and Cascadia reduced to a narrow mountain range or island arc more or less at the edge of the continental shelf. Appalachia still seems to be generally accepted.

(King, 1953: 2165–2166; my bracketed addition)

These hypothetical forelands from which sediments were deposited in geosynclines and then uplifted to form mountains had been proposed by many North American geologists, including Willis (1907), Schuchert (1928b), and Bucher (1933).[10] King's mention of island arcs acknowledged Marshall Kay's solution to the origin of mountains (§7.3).

   Although fixist solutions typically supported by North Americans may be adequate to explain their geology, they could not, King argued, explain South African geology (RS2).

Even if narrow borderlands and island arcs on the continental shelf are adequate for explanation of North American geology they do not even approach a solution of the African problem which is on a much broader scale, reaching from side to side of the continent.

(King, 1953: 2166)

He (1953: 2166) also noted that sinking former sialic landmasses was inconsistent with isostasy (RS2), and added that mobilism did not require changes in the Earth's radii.

The mechanics of continental drift, on the other hand, requires two fundamental crustal levels of sial and sima, maintains always the principle of isostasy, and does not involve any such radical physical changes as shortening and lengthening of global radii.

(King, 1953: 2166)

When referring to "shortening and lengthening of global radii," King surely had Bucher in mind because he (1933: 219–223) had supposed both in his account of mountain formation.

   Turning to mechanism, it would, he wrote, be a mistake to reject mobilism, as Rastall and others had done (§5.13), merely because it appears to be physically impossible because other theories, once deemed impossible, turned out correct (RS1). King appealed to biological evolution.

If the demonstration of a mechanism was vital to acceptance of the idea of drift, the present immature state of the geophysical discipline might be a great impediment. But proof or disproof of drift does not lie in the production of a satisfactory mechanism; the principle (like that of biologic evolution) may well be established on other premises, while explanation of its mode of operation awaits further geophysical data.

(King, 1953: 2167)[11]

Marshaling the Southern Hemisphere support for mobilism, he admitted at the outset that some had used mobilism to explain too much, implying that he would not make the same mistake. He challenged those who rejected mobilism on stratigraphic grounds, reviewed du Toit's analysis of the stratigraphic similarities between South Africa and South America, discussed its updating by Caster and Mendes (1948), and mentioned the stratigraphic similarities among the Falkland Islands, South Africa and South America, and India.

No man is entitled to reject the theory of continental drift on stratigraphic grounds until he can disprove the truly astounding analogies that du Toit has adduced on opposite sides of the South Atlantic Ocean . . . When many other correlations, as between East Africa and India, are studied the correspondences become truly astonishing, especially as more and more detailed information becomes available. So close, indeed, are many of these correspondences that they are not susceptible of explanation under any other hypothesis than that of former contiguity of the now sundered lands.

(King, 1953: 2169)

King thought mobilism had the advantage as an overall problem-solver. Once again, he had Bucher in mind, believing he had failed to appreciate the variety and magnitude of the similarities.

Bucher (1933), for instance, has pointed to sundry clear resemblances between sequences and fossils in the Alps and in the Himalaya, regions 4000 miles apart, and has argued from this that the correspondences between South Africa and South America may have no more significance than that beds were laid down in the same geosyncline. But this viewpoint fails utterly to appreciate the many types of correspondences between the two southern continents. These correspondences involve not only marine but past terrestrial sequences that show much more varied facies, they involve numerous corresponding mountain structures (including crossing of axes) that can not be accommodated in any other way than by drift . . . they involve igneous action, volcanic hypabyssal, and plutonic, of many ages; and include metamorphic sequences. Nor are the correspondences found only along narrow geosynclinal belts, but down the whole length of the opposed coasts from West Africa to the Cape, from the Guianas to Patagonia, at right angles to many, if not most, of the structures for a distance of 3000 miles. Not before the Cretaceous period do marine formations follow the existing coastal outlines of eastern South America and the western coast of Africa.

(King, 1953: 2173)

Until North American geologists became immersed in Gondwana's geology, and undertook their own comparative studies, much as he had done, they would be unable, King argued, to overcome their regional biases.

As for the objection of drift's uniqueness and lateness, he first argued that it could have occurred more than once (RS1).

This does not mean that continental drift had not occurred before, Laurasia and Gondwana, the northern and southern super-continents, respectively, may or may not have been free to drift – we do not pursue the subject here – but their fragmentation into the present continents and the subsequent dispersal of those masses is an unique event in geologic history and is dated as mid-Mesozoic by clear geological evidence on the world's coasts.

(King, 1953: 2173)

There was, in any case, he argued, nothing inherently strange about drift occurring only once, adding that drift did not require accretion of continents.

Philosophically, the concept of two primordial continental masses, Laurasia in the north and Gondwana in the south, of similar size and originally in opposed, semi-polar positions, is satisfying. It is easier to derive as a primitive state on the young earth than almost any other pattern. It involves little more than differentiation and convection current in the cooling earth. Under the drift hypothesis all terrestrial sial was originally concentrated in these two oval, primitive masses, a neatness of distribution that appeals to the mind. There is an economy of material that imposes a rigid check upon later distribution of landmasses for the quantity of sial can not be added to significantly in later geologic times. Continents can not "grow."

(King, 1953: 2173–2174)

Hearing King's rhetoric it is understandable why students who heard the debate with Bucher thought that he had prevailed.

King also suggested a "crucial" test for mobilism. It had to do with whether the findings of the recently completed Norwegian–British–Swedish Expedition of 1949–52 to Queen Maud Land (Antarctica) would reveal a sequence of rocks that matched those from southeast Africa. Whalers had told him about seeing "a range of Black Rock Mountains in the coastal hinterland" when they visited Seal Bay, off Queen Maud Land (King, 1950: 353). Because ranges of black rocks are often basalt, King thought that the black rocks of the Antarctic range might correspond with the Drakensberg basalt.

On the reassembly of Gondwana which accompanies this note (Fig 1) part of Antarctica is seen apposed to southeast Africa. We have already noted that the geology of this part of Africa is incomplete and that a continuation of it must exist beyond the present borders of the continent. Under the hypothesis of drift some of the missing part of Africa should appear in Queen Maud Land, Antarctica. Literally, Queen Maud Land should consist of rocks similar in nature and sequence to parts of the South African succession. There should be a gneissose basement; possibly followed by some lesser thickness of terrestrial Karroo sediments forming the eastward extension of the great Karroo basin . . . which now appears truncated through nearly 10° of latitude along the eastern slope of South Africa; and the whole should be topped off with flows of plateau basalt of the Drakensberg type. Our colors are now nailed to the mast.

(King, 1953: 2170)

The results soon appeared, and he (1958c) discussed them five years later, getting most of what he wanted.

Still within the Lower Permian of Natal and eastern Transvaal are the arkosic and gritty coal measures of the Middle Ecca derived from northeasterly and easterly sources (du Toit). The reconstruction suggests that these sources lay in the Queen Maud Land sector of Antarctica where Roots (1952, 1953) has discovered just such a thick sequence of unfossiliferous mixed sandstones overlying granitic basement.

(King, 1958c: 54)

But, the predicted overlying plateau basalts were missing. King acknowledged their absence.

The reconstruction, of course, suggests that a continuation of the irruption should be sought in Queen Maud Land of the Antarctic, where a continuation of the Karroo sediments has already been demonstrated upon the work of the Norwegian-British-Swedish Expedition in 1948–1950. The specimens recovered by geologists of the expedition did not include basalt, but basic plutons analogous to those of Griqualand East [in southern Africa] were identified.

(King, 1958c: 64; my bracketed addition)

It was not the expected basalts but dolerites, their grey-black hypabyssal equivalent. But to maintain hope that lavas would be found he did add that cropstones from penguins of Queen Maud Land included basalt.

Mobilism, King emphasized, was not footloose. It had rigid controls, based on fieldwork. Those who thought otherwise should do some fieldwork, which, for him, meant comparative intercontinental studies.

The more the philosophy of drift is examined with its many and rigid controls, the more attractive it becomes, especially when compared with other hypotheses of continental genesis. The clear perceptions that arise from it tempt one ever farther and farther afield. But we must focus our main point once more. The hypothesis of continental drift is not to be proved by idle, armchair theorizing, but stands or falls on one thing – hard work in the field.

(King, 1953: 2174)

King was proud that he had done fieldwork on seven continents. He became a geologist in New Zealand, a region where mobilism was not in vogue. He saw evidence for mobilism in intercontinental similarities among South Africa, India, South America, and the Falkland Islands. He wanted North American geologists to go and see it too. Over the next decade Long (§7.5) and Hamilton (§7.7) did just that, and became advocates of mobilism.

6.11 Other South African mobilists

Other South African mobilists were L. J. Krige, F. Dixey, Raymond J. Adie, and Heno Martin. Krige (1926, 1930), who was Smut's brother-in-law, stressed the importance of the role that South Africans could play.

For South Africa, Wegener's theory is of special interest. It furnishes an explanation of the occurrence of the glacial deposits of the Southern Hemisphere, and India, and these are exceptionally well represented in S. Africa. Furthermore the geology of the west coast of Africa and of the east coast of South America seem destined to prove or disprove the disruption theory. It is to be hoped that South Africans will take a leading part in the solution of this problem, and especially the geological formations of the Southern Hemisphere.

(Krige, 1926: 214)

He drew heavily on Daly's (1925) down-sliding hypothesis (§4.7), and thought it might supply drift's mechanism. He devoted his 1930 Presidential Address to the Geological Society of South Africa to the mechanism difficulty. He had an excellent understanding of Joly's theory, and of Holmes' initial sympathy with it (§4.9).

Dixey argued that many geological features of central and southern Africa are best explained by continental drift.

> The physiographical development of southern and part of central Africa, and also the probability that the main Tertiary movement of Africa, and of other parts of Gondwanaland now bounded by "fault-line coasts," are . . . discussed in relation to the Theory of Continental Drift, to which they would appear to lend appreciable support.
>
> *(Dixey, 1939: 169)*

Adie, a South African and student of King's, knew the geology of the Falkland Islands well, and correlated their rock sequences with those of the south coast of South Africa. Throughout his career in South Africa and later in the UK, he provided steady support of continental drift. He (1952: 409) argued, correctly as it turned out, for positioning the islands "some 250 kilometers east of the Eastern Cape Province of South Africa." He hoped that in recognizing the striking similarities between the "Falkland Islands and the South African east coast" that he had given "additional strength to the displacement hypothesis." It was Adie who would later explain to Long (§7.5) that he had had the good luck to have discovered the equivalent of the Dwyka Tillite in Antarctica, and he later helped convince Drummond Matthews that continental drift had probably occurred (IV, §2.6).

Henno Martin was an avid mobilist. A former student of Hans Cloos (§8.3) in Germany, he emigrated to South Africa before World War II. He "declared for drift, after inspecting the geology of Brazil" (King, first December 29, 1981 letter to author). Martin (1961) in his 7th Alex du Toit Lecture updated du Toit's 1927 comparison of Africa and South America, summarizing what had been done since then. He singled out the work of R. Maack (1934, 1953), a German geologist who had lived and worked in Brazil for many years. Martin also mentioned the 1952 Portuguese translation of du Toit's treatise that Caster had discussed.

> Thirty-four years ago, in 1927, A. L. du Toit published his "Geological Comparison of South America with South Africa" (Du Toit, 1927b). This treatise had a considerable influence on the course of the geological investigations in Argentina and Brazil. The best testimonial to the outstanding quality of this work is the fact that it was, twenty years after its publication, translated in Portuguese and republished with copious footnotes by the Geological Survey of Brazil (Du Toit, 1952), a singular honor accorded to a publication based upon a reconnaissance trip. The similarities of the stratigraphy of large parts of the two Continents, established by this study, are so striking, that they have ever since formed one of the corner stones of the hypothesis of Continental Drift. It is therefore very desirable that the evidence should be re-evaluated from time to time. R. Maack (1934) had added new data which were used by Du Toit in his book "Our Wandering Continents" in 1937.
>
> *(Martin, 1961: 1; his du Toit (1927b) is the same as my du Toit (1927b))*

Martin then summarized many studies in Brazil, including his own, in Argentina, and in South Africa.

It was stated in the beginning that no conclusive evidence for or against continental drift can be expected at this stage, that it should however be possible to see whether the more detailed knowledge has tended to increase or decrease the similarities between the two sides of the Atlantic Ocean. There is not the slightest doubt that from the Silurian to the Cretaceous every correction of the stratigraphy has increased the similarities. The stratigraphic and lithologic columns for this period of some 200 m.y. have become almost identical. I do not think that, for a comparable length of time, a similar likeness between parts of any other two continents can be found. It is probably not even possible to match parts of a single continent – excepting perhaps in a geosynclinal belt – as closely if separated by a distance of more than about 2000 miles (3000 km).

It is true that every single similarity can be explained without recourse to continental displacement. But the astonishing similarity of the whole succession does certainly favor the hypothesis of Continental Drift or Global Expansion, especially if the independent fit of the continental margins and the doubt cast by the directions of ice movement on the former existence of the Atlantic Ocean are also taken into consideration.

*(Martin, 1961: 43–44; Earth expansion theories are discussed
in Volume II, Chapter 6, and Volume III, Chapter 6)*

Martin, a stratigrapher, placed great emphasis on these stratigraphic and lithologic disjuncts persisting over such a long time, but he gave less emphasis on biotic ones.

The paleontological evidence, on the other hand has remained as doubtful as it has always been. The endemic nature of the fauna of the Paraná Basin indicates that it was not connected with the Great Karroo Basin. This conclusion is supported by the directions of ice-flow, which strongly speak for a larger elevated area to the east of the Paraná Basin. Under these circumstances the absence of identical species of aquatic animals has no great weight as an argument against a continental connection. Fossils of terrestrial animals of approximately the same age have only been found in East Africa and Southern Brazil, at two localities which are, even if the Atlantic is closed, so far from one another that identical species are not necessarily to be expected. Palaeobotany strongly suggests a connection, but contiguity of the continents is not the only possible connection.

(Martin, 1961: 44)

Perhaps Plumstead would have disagreed with the last sentence, but Martin thought that *Glossopteris* could have spread by floating across oceans. Dunbar had suggested as much at the 1949 New York meeting on land connections across the South Atlantic.

During recent years Virkki (1937) has found peculiar double-winged spores in such abundance in association with the leaves of Glossopteris and Gangamopteris that the Indian paleobotanists believe they belong to these plants (Sahni, 1938). Such winged spores were presumably specialized for transportation by the wind. But whether these plants were distributed by spores or by seeds, when we consider the endless adaptations of animals and plants, it appears not improbable that this peculiar and highly specialized tribe had developed some device that permitted their transportation across open water. The coconut and other modern plants of the pacific islands at least show us that such adaptations are possible. Until we know more about the manner of reproduction within the Glossopteris flora it is hardly necessary to drift continents . . .

(Dunbar, 1952: 155)

Martin thought that although the evidence for drift based on new descriptions of the disjuncts between Africa and South America had greatly increased since du Toit's 1927 work, he did not think drift had yet been proved. Echoing what Smuts (§6.3) said just over thirty-five years before, he urged Earth scientists from Gondwana to work together.

A final proof for or against this challenging theory can perhaps only be found in the depths of the Atlantic Ocean. So, what we on the Southern Hemisphere need, is not a "Mohole" but a "Gohole" – Go for Gondwana – to establish the age of the oldest sediments in the South Atlantic. Once the techniques for drilling in the deep-sea have been developed, this should not be an impossible task, if tackled as a co-operative effort of the Gondwana Countries.

(Martin, 1961: 44)

Thus Plumstead, Adie, King, Martin, and to a lesser extent Dixey, like du Toit, appealed to mobilism to solve problems that arose out of their work in South Africa or comparative work in South Africa and South America. Krige was sufficiently attracted by mobilism because of its success in solving problems of South African geology to attempt to remove the mechanism difficulty. These workers all preferred mobilism because it was better suited to the geology of their region of birth or adoption. Caster, who was neither of these, eventually returned to the United States after his work in South America; while writing positively about continental drift giving it a fair hearing, he never embraced it with quite their enthusiasm. Adie, King (III, §1.4), and Plumstead (III, §1.12) would later appeal to paleomagnetism in support of mobilist interpretations.

6.12 South African fixists

In her autobiographical essay, Plumstead mentioned three South African fixists: S. H. Haughton, Anton Hales, and B. B. Brock. Hales, an eminent geophysicist, worked with Jeffreys at Cambridge University in the 1930s, studying the feasibility of mantle convection (§5.6). He later became Director of the Bernard Price Institute for Geophysics, where he helped initiate paleomagnetic research during the second half of the 1950s (II, §5.4), and as a result of which he became sympathetic to continental drift. Plumstead recalled that Hales was

my chief opponent at this time. We had many friendly arguments, but as a physicist, he needed proof of movement acceptable to his profession. He showed signs of wavering when he returned from an annual Conference in the U.S.A., where palaeomagnetism was . . . discussed, and he put it into practice in S. Africa. At first he had no success, for the very high incidence of lightning on the Witwatersrand apparently masked the true position. Later he took specimens from greater depths and capitulated most magnanimously in a public lecture. I cannot now remember the exact date of the lecture, but he gave all his objections to the theory first and then said, "That is what I used to think but now . . ."

(Plumstead, 1982: 11)

King's remark about Hales is more pointed, and reveals his suspicion of geophysicists who said drift was impossible. King saw "the fact of drift" every day, and he resented geophysicists telling him otherwise.

Hales was a mathematician who looked down on geology. When he left S. Africa he declared that geologists would have to give up drift ideas. Geophysics would settle the matter. Did it?

(King, first December 29, 1981 letter to author)

King was wrong about Hales. By 1957 Hales (II, §5.4) had become sympathetic toward mobilism through paleomagnetic work (Graham and Hales, 1957).

B. B. Brock (1904–72) was a geologist with heretical views – "remarkable" views according to K. C. Dunham (1983: 256). Plumstead said of him:

Dr. B. B. Brock (Consulting Geologist of the large Anglo American Mining Co.) was hotly opposed to the drift theory, and advocated "Vertical Tectonics." He would not even allow friendly discussion, although we were friends and neighbors.

(Plumstead, 1982: 10)

Brock (see Pretorius, 1975) was born in Canada in 1904. He received his Bachelor of Applied Science in geological engineering from the University of British Columbia, and his Ph.D. from the University of Wisconsin. He emigrated to Northern Rhodesia (now Zambia) in 1934, and worked for several mining companies, becoming Consulting Geologist for the Anglo American in 1952. In his Presidential Address to the Geological Society of South Africa in 1957–8, he opposed continental drift (Brock, 1957), and continued to be adamantly opposed to seafloor spreading and plate tectonics. His (1972) posthumously published book, *A Global Approach to Geology*, is a monument to his heterodoxy.

Without question, however, the most eminent geologist in South Africa to oppose mobilism was Sidney H. Haughton. Born British in 1888, he read geology at Cambridge. In 1911 he accepted an appointment at the South African Museum, Cape Town, where he remained until becoming a senior geologist at the Geological Survey of the Union of South Africa. He was Director of the Survey from 1934 through 1947 when he resigned to become Honorary Director of the Bernard Price Institute for Palaeontological Research. He remained active for many years, and was elected to the Royal Society (London) in 1961, at the age of seventy-three.

In his summary of "Fifty years of geology in parts of Africa," Haughton noted:

Two sets of theories in particular have occupied the minds of geologists, in South Africa and elsewhere, for a number of years and have caused much ink to flow – the ink sometimes, being mixed with an excessive amount of gall. One set concerns the mode of origin of the gold and uranium in the Witwatersrand blankets [auriferous conglomerates]; the other is the theory of continental drift. Their formulation has been, in each case, of immense value in that it has stimulated thought and induced a search for new facts (which, unfortunately, is often stated to be "a search for new evidence" as if the postulant were a suspect criminal on trial).

(Haughton, 1962: xx; my bracketed addition)

Certainly, Haughton put drift on trial.

Trained as a general geologist, he devoted much of his career to collecting and understanding the early history of sub-Sahara reptiles. Haughton (1953) argued that the overall distribution of non-marine reptiles did not support drift, echoing the majority view at the 1949 New York meeting on Atlantic landbridges (§3.8).

The history of these animals during Upper Carboniferous, Permian and Triassic times, therefore, does not demand the existence of a Gondwanaland as visualised either by the early exponents of the idea or by the later supporters of continental drift. Rather does it favour a configuration of the land masses during those periods which approximates to that of to-day save for the existence of a bridge joining North America to western Europe by means of which the Upper Carboniferous cotylosaurs and pelycosaurs could cross from one area to the other . . . The supposition of a land-bridge between southern Africa and Brazil in Triassic times is at best not proven, and future discoveries in North America of the animals that left such abundant footprints in the Triassic Moenkopi formation of North America may prove that it was from the north that the almost cosmopolitan rhynchosaurs and their associates reached the Parana basin.

(Haughton, 1953: 30)

What about *Glossopteridae*? Like Dunbar and Martin, he thought that their chances of "dispersing across water barriers is much greater than for land-dwelling vertebrates" (Haughton, 1953: 30). After his second year as an undergraduate at Cambridge, Seward told him that he had done best in botany, but he still specialized in geology in his final year (Dunham, 1983: 246). Had he become a botanist, he might have favored drift.

Of the three, Brock, Hales, and Haughton, the third's rejection of mobilism might seem counter to my thesis that theory preference depends on the relative effectiveness of competing theories to solve problems in the regions scientists study. But specialists in terrestrial vertebrate paleontology who favored mobilism were rare; they were something of an exception. Westoll (§3.5) and Harrison (§9.5) are the only two I have found who, before the rise of paleomagnetism, stated in print that they preferred drift to landbridges; Romer might have, but he (§3.8) showed no preference for either in his presentation at the 1949 New York meeting. Brock's views were peculiar, and Hales was a geophysicist and therefore probably little influenced by geological evidence and regional problems; he was initially influenced by Jeffreys and was critical of drift because it did not have a viable mechanism, but became sympathetic to mobilism because of his paleomagnetic work in South Africa. However, as King noted, only a few geologists in South Africa were interested in broad issues; most were economic geologists working in the mining industry, and drift did not impinge on their professional lives; most did not participate in the mobilist controversy.

6.13 Favorable reception of mobilism among Indian geologists

Mobilism offered the best solutions to an impressive list of problems in Indian geology: its Permo-Carboniferous glaciation, the puzzling juxtaposition of *Glossopteris* and *Gigantopteris* floras in India and China respectively, the formation of the Himalayas, and the geological and paleontological similarities between India and other parts of Gondwana.[12] If regionalism were the main determinant of attitude to continental drift, then the proportion of those scientists who voiced an opinion and who preferred mobilism to fixism should have been higher in India than almost anywhere else.

In order to determine this I shall examine contributions to the symposium on "Wegener's Theory of Continental Drift with Reference to India and Adjacent Countries" that was part of the 24th Indian Science Congress held in Hyderabad in 1937. Ahead of time, Birbal Sahni (1936) referred to the upcoming symposium and noted the global and regional interest in Wegener's theory.

At the forthcoming session of the India Science Congress, to be held in January 1937 at Hyderabad-Deccan, there is to be a joint meeting of the sections of geology, botany and zoology with a view to discuss Wegener's theory of continental drift. Although this subject has been widely discussed in scientific circles the last ten or twelve years it still claims the attention of men of science all over the world. This must be due, partly at least, to the regional appeal of much of the evidence, apart from the intrinsic interest of this many-sided problem.

(Sahni, 1936: 322)

Of nine presentations, four (B. Sahni, S. P. Agharkar, C. S. Fox, and A. K. Dey) definitely favored continental drift, one (L. Rama Rao) rejected it, one (W. D. West) favored it supplemented by certain landbridges, one (P. Evans) was neutral or slightly inclined toward it, one (S. L. Hora) slightly favored landbridges over drift, and one (D. H. Wiseman and R. B. Seymour Sewell) made no mention of it. Thus, except for Sewell and Wiseman, who described the findings of the recent John Murray Expedition to the Indian Ocean, every participant had spent most or all of their careers studying the geology, zoology, or botany of India. At the 1937 Hyderabad symposium, the majority favored continental drift.

What advantages did mobilism have over fixism as a solution to Indian problems? I begin with Wiseman and Sewell, and their discussion of what they had found in the Indian Ocean – Sewell led the expedition. All the rocks they had retrieved from the Carlsberg Ridge and deeper areas of the Arabian Sea were basalt, leading them (1937: 509) to reject the idea of sunken "continental" landbridges: "There is little or no indication that any older continental mass or land isthmus, such as the hypothetical continent of Gondwanaland, or the isthmus of Lemuria, ever existed . . ." Describing how the topography of the floor of the Arabian Sea resembled that of the early Tertiary African Rift Valley, and that minor earthquake belts run down both the African Rift Valley and the Carlsberg Ridge, they suggested that the Carlsberg Ridge, like the African Rift Valley, formed during the early Tertiary. They said nothing about continental drift. Sewell was not yet a mobilist, although he later came to favor it (§8.12), and here I did not count him in the above survey.

What about the presentations favoring mobilism? Sahni explained his new mobilist explanation of the juxtaposition of the very different *Glossopteris* and *Gigantopteris* flora as described in §3.5. He then related how

The main facts of the Gondwana glaciation have always strongly urged me in favour of Wegener. But speaking, as I should, only as a palaeobotanist, I confess that my position until recently was that of an agnostic. Latterly I have felt myself drifting gradually towards Wegener's idea of continental displacements.

(Sahni, 1937: 502)

He also noted that recent work by Evans and Wadia had established the structural continuity of the Burmese and Himalayan mountains around the Assam syntaxis, which supported his explanation of the juxtaposition of these two very distinct floras (Sahni, 1937: 503–504). He wondered about the feasibility of determining geodetically if India was still moving northward relative to Asia; it would offer direct proof, but he was not too sanguine about the chance of getting reliable data.

If as some geologists believe, the Himalayan uplift is still in progress, this fact may provide indirect evidence that the continental displacement is still going on. Accurate readings of latitudes and longitudes in the region between Szechwan and Celebes on the east, and in Hazara and Afghanistan on the west, if continued over a long enough period, may yield direct proof of the suggested pivotal movements [of the Asian block] round the Assam and Kashmir promontories. Possibly the suggested northward displacement of the Australian block may also be checked directly by observations of latitude. But it appears that the practical difficulties in the way of reliable observations in the critical regions are still too great to inspire hopes for the near future.

(Sahni, 1937: 505; my bracketed addition)

(Figure 3.4, §3.5, shows the Assam and Kashmir promontories of the India block respectively projecting against the eastern and western knee-like bends of the Himalayas.)

Sahni commented unfavorably on Gutenberg's version of continental drift (III, §2.3), which supposed that continents drifted as a consequence of the spreading of a huge sialic block that underlay the Atlantic and part of the Indian Oceans, and which was akin to Taylor's theory (§2.12). Sahni thought instead that Gutenberg's sialic ocean floors might represent sunken landbridges. Finally, he noted that the Murray Expedition had shown that the physical features of the Arabia Sea were too recent to have any relevance to the drift question; ironic, because Vine and Matthews based their hypothesis on magnetic surveys over the Carlsberg Ridge in the Arabian Sea (IV, §2.12).

Shamkar Pandurang Agharkar (1884–1960), a paleobotanist from University College, Calcutta University, examined post-Triassic Indian flora. He claimed, "the occurrence of the Palaeo-African element in India is better explained by Wegener's theory of continental drift than by any other hypothesis," reasoning that "the Palaeo-African element of the Indian flora reached India towards the end of the Jurassic and the Cretaceous period when there was [according to Wegener] a direct land connection between Peninsular India and Africa through Madagascar" (Agharkar: 1937, 508; my bracketed addition).

A. K. Dey, a paleontologist at the Indian Geological Survey, who received his Ph.D. at the University of London, spoke about the distribution of invertebrate marine organisms and concluded:

The distribution of marine fauna in the Middle and Upper Mesozoic seas of the East African continent and India, therefore, provides clear evidence of free inter-communication between the Eastern Tethys and the India ocean across the present day Arabian Sea. This conclusion supports Wegener. For according to Wegener, the Arabian Sea has been created since the Mesozoic time by the drifting of India towards the northeast.

(Dey, 1937: 519)

British born Cyril S. Fox (1886–1951) served as Director of the Geological Survey of India. He was well-versed in geophysics. In the mid-1930s, he rejected mobilism because of mechanism difficulties, claiming (1935: 11) that continental drift cannot "be taken seriously as a working hypothesis," because the upper mantle is insufficiently weak to allow it. But his position soon changed, and he began to entertain drift as a working hypothesis because it offered a possible explanation of these two apparently contradictory features of Indian geology.

The uplift of an ocean floor with its deposits of marine sediments into a mountain range such as the Himalaya is good evidence for believing in the instability of lands and seas. The geological history of the Indian peninsula, however, shows that it has remained a land region since the middle Paleozoic era . . . and is evidence in support of the belief in the remarkable permanency of continents. Thus two adjacent regions in India supply contradictory evidence in regard to an important geological question and it is to be concluded that the truth must be in some theory which permits both possibilities. This nice compromise appears to be best attained, according to some geologists, in an hypothesis which involves the lateral displacement or so-called drift of the continents.

(Fox, 1937: 512)

He noted:

The general hypothesis of continental drift helps to explain so many features of geography and geology and, in particular, gives a complete solution to the mystery of the great Ice Age of Gondwanaland – one of the unsolved problems of Indian palaeogeography – that it requires the most careful consideration.

(Fox, 1937: 512)

Fox (1937: 514) also dismissed the idea that something like the Himalayan uplift could have caused the Permo-Carboniferous glaciation, for even if locally significant, it could not have caused glaciation on the trans-continental scale observed.

Fox also referred to an abortive geodetic attempt to determine if India and Asia were moving toward each other.

In the hope of finding evidence for the southward drift of Asia, an examination of the records of latitude observations in India through the past hundred years was made but only yielded contradictory results. The differences in latitude which were noted were thought to be due to errors of observation rather than to actual changes in latitude.

(Fox, 1937: 512)

For him, the lack of a viable mechanism remained the main hindrance to the acceptance of drift.

If it could be established that the available motive forces were sufficient, or that they periodically become able, to cause the continents to drift, not necessarily towards the equator or westward, but in the direction of least resistance, then the theory would be widely accepted. It is thus worth while to examine the available evidence and see what may have been overlooked, or whether any new data can be made to give an answer to the question – Do the continents "creep" with respect to the plane of the equator and in process of time change their latitude and longitude?

(Fox, 1937: 512–513)

He was impressed with Joly's mechanism (§4.9), but thought it inadequate in its original form, and hoped it could be amended.

The viscosity of the sub-crustal layer is perhaps the most rational objection to continental drift where the determination of a measurable rate of movement is desired. It seems impossible to ascertain any definite changes of latitude or of longitude free of error of observation as the movements must be exceedingly slow if there is no fluid sub-crustal layer for horizontal slipping. The question is whether the present solid condition of a sub-crust is permanent or whether there is any direct evidence to show the possibility of its periodic melting. This probability of periodic melting is so well-established that it led Professor Joly to develop his splendid hypothesis to explain the periodicity of volcanic eruptions – roughly every 40 million years – and thus account for the periodic regional earth (crustal) movements ... Such an hypothesis would also permit continental drift. Unfortunately the available data shows that the radio-active elements are less than Professor Joly assumed and that the heat generated on the smaller proportion works out to be insufficient. Here again we are dealing with calculations on data which may be found in need of modification.

(Fox, 1937, 515)

Fox returned to drift's ability to solve problems in Indian geology, especially its solution to the origin of the Himalayas.

From a geological point of view the general evidence appears to favour the travel of the granitic continental masses irregularly across the surface of the earth, the crust of which seems to be unequally loaded. There seems to have been a large southern continent, Gondwanaland, in Palaeozoic times which broke up and drifted northward as, among other fragments, India and Africa towards and over the equator. There also appears to have been a northern continent, Angaraland, which drifted south-westward as Europe-Asia drifted against the northern ends of India and Africa. While shearing has split the continent of Africa from the Zambesi to Akba the marine geosynclinal between Asia and India suffered intense compression which resulted in the upheaval of the Himalaya and other mountain chains. Different movements have naturally occurred due to the opposing continental masses – Asia and India meeting obliquely. The Indian peninsula has been severely fractured in various directions by the thrust from the Asiatic mass, and local irregularities such as the rise of the Assam range and the mountains of Burma and the North-Western region of India have appeared.

(Fox, 1937: 516)[13]

On balance Fox was mildly sympathetic toward drift generally, and was enthusiastic about its ability to solve Indian problems. The absence of an acceptable mechanism worried him, and he thought that Joly's hypothesis, if modified, showed promise.[14]

The only speaker who was adamantly against drift was L. Rama Rao (1896–1974), a prominent geologist from the University of Bangalore. His specialty was micro-paleontology, particularly the study of algae (Radhakrishna, 1996). He helped found the Geological Society of India in 1958 and served as its journal's first editor. The society later named its Gold Medal, awarded for contributions to Indian stratigraphy and paleontology, after him.

Rao began with this summary.

When Wegener put forward his theory of continental drift, it was generally welcomed as it once seemed to offer a possible solution for several geological and biological problems which had till then baffled all attempts at explanation. But as a result of a closer and a more careful examination of the theory in light of observations that have been accumulating during the last few years, it is coming to be increasingly realized that this hypothesis is not quite so acceptable. In discussing the validity of this theory, we must naturally focus our attention on the history of the earth during the late Palaeozoic and early Mesozoic periods. Any conclusions based entirely on Tertiary and post Tertiary phenomena will not be useful.

(Rao, 1937: 511)

More specifically, and citing the conclusions of others, Rao criticized the alleged geological similarities of the Northern Hemisphere continents, and in the Southern Hemisphere, the significance of the distribution of *Glossopteris* and Permo-Carboniferous glaciation.

In the Northern Hemisphere, recent studies from various aspects of the geology of the countries on either side of the Atlantic by Washington, Chamberlin, Gregory, Mrs. Reid and others have shown that the features on the two sides do not fit in as Wegener imagines. In the Southern Hemisphere with which we are more concerned in the present discussion, the theory has been examined in its bearing on two problems: (1) the distribution of the *Glossopteris* flora and (2) the Permo-Carboniferous glaciation. From one or the other of these points of view the theory has been critically examined within the last 15 years by several leading geologists and palaeontologists, and none of them find any evidence in support.

(Rao, 1937: 511)

Familiar with the literature critical of drift, he offered no new analyses. He gave no references, but was influenced evidently by the 1926 AAPG symposium (§4.3).

Rao then argued that Sahni's use of India's drift to explain the juxtaposition of *Glossopteris* and *Gigantopteris* floras was insufficient to counterbalance the difficulties that mobilism faced on other questions; he hoped for future alternative explanations less drastic than continental drift.

The remarkable case of two originally distinct floral provinces (the *Glossopteris* flora of India and Australia and the *Gigantopteris* floras of China and Sumatra) now seem in close juxtaposition and even dovetailed with each other to which Dr. Sahni has drawn attention, no doubt suggest a movement of these land areas towards each other and therefore seems to support the general idea of a continental drift. But in view of the fact that the theory has been tried and found wanting in the solution of the more major problems of geological structure, paleoclimates, and former distribution of faunas and floras, it seems doubtful if we have still to involve the aid of this theory for explaining this occurrence. No other explanation may just now be possible; but in course of time, with a more detailed knowledge of these two contrasted floras and a better understanding of the factors controlling life distribution, it is quite probable that we may discover an alternative explanation without involving such drastic and large scale movements of land as are contemplated in Wegener's theory.

(Rao, 1937: 511)

He paid homage to India's doyen of botany, but did not follow him. He seemed conflicted like Romer, Caster, and Camp were a dozen years later at the 1949 New York symposium on trans-Atlantic land connections, but with his sentiments interchanged. Reared in a country where, as he acknowledged, mobilism found much support, he could not rid himself of fixism, and just hoped that mobilism would eventually go away.

I now turn to the presentations of William Dixon West (1901–94) who presided over the Hyderabad symposium. After graduating from Cambridge University in 1923, he joined the Geological Survey of India, and became its Director from 1945 until independence in 1947. He received the Lyell Medal from the Geological Society of London in 1951, and in 1983 the Wadia Medal from the Geological Society of India (Radhakrishna, 1999). West (1937: 502) "opened the discussion with a brief review of the theory in general and of the difficulties raised, and referred to alternative solutions." He favored continental displacement, but thought that some trans-oceanic landbridges were also needed. Drift, he argued, as Sahni had done, offered a solution to the origin of the Himalayas and the juxtaposition of *Glossopteris* and *Gigantopteris* floras. West took issue with Rao because he had failed to mention du Toit's work, especially his 1927 comparison of South America and South Africa.

Prof. Rama Rao, in drawing attention to the geological evidence adduced by various writers as to the dissimilarity of the geological structure on either side of the Atlantic, had ignored the much more important researches of A. L. du Toit, who had shown that the similarity between South America and Africa, if the two continents were brought to within about 250–500 miles of each other, was so striking as to be beyond the possibility of coincidence.

(West, 1937: 520)

But he did not think that India had drifted as much as Wegener had supposed. He noted the similarity between certain Deccan trap rocks and those of Madagascar, but rejected juxtaposing them.

It is well known that Kathiawar [in Western India] was a special focus of igneous activity during the Deccan trap eruptions, and produced special rock types not found elsewhere in the Deccan trap. The extensive types include very basic varieties rich in fresh olivine and augite, designated ankaramite and oceanite. These have been analysed. Precisely similar rock types, and of the same age, occur in north Madagascar, and have been described and analysed by Lacroix. Two alternative conclusions can be drawn. Either Madagascar was once close to Kathiawar, and partook of its peculiar Deccan trap rock types; or similar conditions of differentiation prevailed in the two areas at the same time, and produced the same rock types. The latter is the more probable, but the possibility of the former should not be entirely excluded.

(West, 1937: 510; my bracketed addition)

West summarized the presentations and discussions at the Hyderabad symposium and opted for compromise.

It seemed likely that in the formation of the Himalayas there had been a movement northwards of India underthrusting Asia, rather than the reverse, and this was in agreement with the direction of movement postulated by Wegener. At the same time it would be very rash, with our present knowledge, to conclude with Wegener that India was once adjacent to South Africa and Madagascar. Two rival theories had to be considered, continental movement or the sinking of land-bridges. As regards the formation of the Arabian Sea, the speaker [i.e., West] was inclined to think that both these types of movement had been operative, and it was probable that, as in the case of so many rival theories, the true solution would be found to lie in between.

(West, 1937: 520; my bracketed addition)

Percy Evans, an oil geologist, worked in Assam where the Digboi oil field was located. He favored Sahni's solution to the juxtaposition of *Glossopteris* and *Gigantopteris* flora.

The evidence collected mainly by geologists of the Burmah and Assam Oil companies strongly supports the suggestions of Prof. Sahni and others that structurally and stratigraphically there is a definite continuity around the head of the Assam Valley.

(Evans, 1937: 507)

Evans then estimated the lateral movement during the formation of the mountains that looped around the Assam promontory as 400 miles, which, he noted, was much less than the northward drift of India.

It does not seem possible to condemn Wegener's hypothesis merely on account of the magnitude of the postulated drift when there is this evidence of late Tertiary lateral movements of several hundred miles, although this is, of course, much less than the total movement required by the hypothesis.

(Evans, 1937: 510)

Evans then turned to the work of Wade (1934, 1935), another British petroleum geologist, who had searched for oil in Papua New Guinea (§8.12), and favored Gutenberg's version of mobilism (III, §2.3). He found one of Wade's arguments particularly interesting.

Wade has shown that Gutenberg's hypothesis receives support from the curious distribution of oil. In certain regions oil-forming conditions have recurred at intervals since Cambrian times; in some other countries it is limited to the Tertiary beds, and in yet others it is all but absent. Oil formation and preservation require certain conditions of sedimentation and folding which are most likely to be present in the equatorial regions, and Gutenberg's hypothesis postulates that North and Central America have travelled along a path which has kept them near the equator for a very large part of geological time – so greatly favouring the formation of oil. By contrast, South America, Africa and Australia remained for a long period too far from warm equatorial conditions for sufficient organic matter to be available and were also unfavourably placed for folding and sedimentation. The occurrence of oil in

India and Burma and neighbouring countries appears to fit in with Gutenberg's hypothesis. Wade has suggested that the subject deserves further consideration, particularly from petroleum geologists.

(Evans, 1937: 510–511)

Evans was clearly intrigued by drift, and agreed with Wade's view that it "deserves further consideration, particularly from petroleum geologists." At times he seemed mildly disposed towards drift, but I classify him as undecided.

S. L. Hora (1896–1955), a paleontologist from Calcutta University, believed the distribution of Indian freshwater fish required a strong India–Africa connection.

The facts adduced by the writer concerning the origin and the geographical distribution of the Indian freshwater fishes negate the theory of the permanence of oceans and continents, for they postulate the existence of a land connection between India and Africa. As to whether this connection was in the form of a "land-bridge" between the two continents, or the two continents were juxtaposed at some remote period but later drifted apart, it is very difficult to decide.

(Hora, 1937: 507)

But his preference was for landbridges.

In the discussion that followed Dr. Hora explained that though his studies did not support or refute Wegener's hypothesis they postulated a continuous land connection between India and Africa as late as the Middle Tertiary so as to permit the migration of the modern bony fishes of the Oriental Region to Africa. According to Wegener's theory India and Africa began to drift apart in the Jurassic period, but this contention is not borne out by the distribution of fishes. If the continuity of land during the Tertiary age is conceded, then it seems more probable that the formation of the Arabian Sea must have resulted from a subsidence of land connections between India and Africa during late Tertiary times rather than by drifting apart of the two continents in the manner and time envisaged by Wegener.

(Hora, 1937: 520)[15]

This examination of work in India during the classical phase of the mobilism debate demonstrates that geologists who preferred mobilism did so because it offered better explanations of Indian geology than provided by fixism. Sahni, Evans, Dey, West, and even Rao were impressed with mobilism's solution to the juxtaposition of *Glossopteris* and *Gigantopteris* floras. Sahni, Dey, and Fox thought highly of mobilism's solution to Permo-Carboniferous glaciation. Dey and Sahni preferred mobilism's solution to the distribution of *Glossopteris* throughout Gondwana. West, Evans, and Fox highlighted mobilism's solution to the origin of the Himalayas. Evans discussed mobilism's relevance to the worldwide distribution of petroleum, and to occurrences in Indian and Burma. Agharkar, sticking entirely to his own expertise, argued for mobilism because it better explained the distribution and nature of India's Mesozoic flora. Fox, who worried much about the absence of an acceptable mechanism, at first rejected mobilism but later gave it his guarded support.

6.14 L. L. Fermor supports mobilism

In 1898 Lewis Leigh Fermor (1880–1954), later Sir Lewis, entered the Royal School of Mines (London). Two years later, he received an associateship in metallurgy, and planned to continue studying at London University. However, following the advice of a teacher, he applied for and was offered a post in the Indian Geological Survey. He arrived in India in 1902. He was Director of the Survey from 1930 until 1935. He left India the following year, spent the next years in Kenya, South Africa, and Malaya, and then returned to England. He was elected FRS in 1935. A founding member of the Society of Economic Geologists, he served as its Vice-President in 1932. He was President (1951) of the Institute of Mining and Metallurgy, and Vice-President of the Mineralogical Society (1943–6) and of the Geological Society of London (1945–7). Crookshank and Auden (1956) give details of his life.

After leaving India, Fermor wrote three pro-drift papers (1944, 1949b, 1951). He brought to the task his firsthand knowledge, appreciation, and understanding of Indian geology, his knowledge of mineral deposits found only in Gondwana, his preference for drifting continents over vanishing landbridges, and his interest in mineralogically based mechanisms for drift.

Fermor (1949b) presented his most interesting defense of drift in a lecture to the Rhodesia Scientific Association in August 1947. He began by referring to Blanford's 1857 discovery of Late Paleozoic glaciation in India. Blanford's work was probably of special interest to him because it was he who, in 1902, interviewed and recommended Fermor to the Indian Geological Survey (Crookshank and Auden, 1956: 102). After describing the Indian "Gondwana formation," sequences of sandstones and shales, and coal beds with *Glossopteris* overlying Carboniferous glacial beds, Fermor (1949b: 12) pointed to similar sequences in the other Gondwana fragments, and emphasized how different they were from "the better-known land of Europe and North America."

Drawing on his own studies of Precambrian mineral deposits in India and elsewhere, Fermor cited three features of their occurrence in Gondwana.

In Gondwanaland we have not only the Gondwana formation, which in Africa is known as the Karroo formation, but underlying the Gondwana formation is the old crystalline Archaean basement of granites, gneisses and schists, such as are found over a considerable portion of Rhodesia. Although granites, gneisses and schists are found in other parts of the world, yet Gondwanaland has peculiarities in respect of its *mineral deposits* that distinguish it from the remainder of the earth's crust. The first is that all the really deep *gold* mines in the world are in Gondwanaland; starting from west to east, we have the St. John del Rey gold mine at Ouro Preto in Brazil, the deep mines of the Rand in the Transvaal, and the Kolar goldfield in Mysore, in all of which depths of 9000 feet and upwards from the earth's surface have been reached. In addition, no less than 99.98% of the diamonds of the world come from Gondwanaland – from Brazil, West Africa, the Congo, South Africa and India. Finally, as a third peculiarity, we may mention that of the seven countries yielding over 100 000 tons of manganese-ore annually, no less than five are in Gondwanaland, namely, Brazil, the Gold Coast, Sinai, South Africa, and Peninsular India.

(Fermor, 1949b: 12)

Having established, at least to his own satisfaction, the fragmentation of Gondwana, Fermor (1949b: 12) noted that foundering continents face "the most incredible difficulties," while continental drift is "apparently equally difficult to swallow," and confessed his preference for drifting continents. Continents could be reconstructed into Gondwana on three different bases, the congruency of their margins, the distribution of Late Paleozoic glacial strata, and the trend lines of their Precambrian basements. (Fermor's own work was on India's Archean basement rocks.) Fermor preferred drift because the three independently based reconstructions agreed, something that foundering continents could not explain (RS1, RS3).

As to the mechanism of drift, Fermor turned to garnet, a semi-precious stone he knew particularly well from his own researches in India.

Therefore, accepting Gondwanaland and its break-up by wandering or flotation or displacement, whichever phrase you prefer, the question we must now discuss is how is this physically possible? It is now that we have to discuss the horse, the motive power which causes continents to move, and this involves a discussion of the mineral garnet. Garnet is doubtless known to all of you as an inexpensive gem-stone . . .

(Fermor, 1949b: 13–14)

What was Fermor's "horse"? Fermor argued that huge basaltic lava flows such as the Deccan traps suggest the presence beneath them of a "basaltic" shell whose upper part is gabbro.

Basalt itself, however, is a surface lava, which, if it consolidates below the surface slowly and under considerable pressure, takes the form of a completely crystalline rock known as gabbro, so that the upper portion at least of this basaltic shell must really be gabbro.

(Fermor, 1949b: 14)

The lower part of the "basaltic" shell is made, Fermor argued, of eclogite of which garnet is a major constituent. He appealed to a reversible phase transformation between eclogite and gabbro, to which Holmes had earlier called attention (§5.4). There are, Fermor claimed,

Very strong reasons for supposing that the lower section of the basaltic shell is not gabbro but a rock known as eclogite. The difference . . . is that eclogite is full of garnets, which are absent from basalt and gabbro. The chemical composition is the same, but eclogite is much denser and occupies about 14 per cent. less space . . . It is possible to write down chemical equations showing the passage of the minerals composing basalt or gabbro into eclogite, and it is found that this passage results in not only a reduction of volume (the 14 per cent. mentioned) but also the absorption of heat; therefore, in accordance with a well-known principle of physics, in which a system tends to take that form in which the external energy is smallest, it follows that under the high temperatures prevailing at depths in the earth's crust and the high pressure of the superincumbent rocks, basalt or gabbro must be compressed into eclogite, and therefore the lower portion of the basaltic shell must be in the form of eclogite.

(Fermor, 1949b: 14–15)

Fermor (1949b: 15) claimed that this reversible transformation explained "the isostatic balance of high mountain ranges against low-lying plains," and was the cause of continental drift. He cited the Himalayas and the Gangetic plains, which led into his mechanism.

The second point that we can explain is the isostatic balance of high mountain ranges against low-lying plains. How are the high Himalaya supported as compared with the lower Gangetic valley or plains? This can be explained by the roots of the Himalaya being lighter than the sub-crust of the Gangetic plains, a result effected by the passage of eclogite into gabbro under the Himalaya. As the Himalaya rises, eclogite expands into gabbro; as the Ganges valley sinks, gabbro is compressed into eclogite and thus the balance is maintained.

(Fermor, 1949b: 15)

He goes on:

If we wish to move continents, what is the mechanism that permits of the adjustment of the crust to all the stress and strains involved? Obviously a reversible reaction operating at the irregular surface or junction between the overlying gabbro and the underlying eclogite could provide the mechanism, since exactly such changes in one direction or the other can occur as are needed by the circumstances of the movement. The garnet-bearing rock eclogite provides the possibility of the apparatus or mechanism, and since the peculiar properties of eclogite depend fundamentally upon those of the contained garnets, we can treat this mineral as the "horse" that helps the continental "cart" to move.

(Fermor, 1949b: 16)

leaving his readers to fill in details.

Fermor returned to the issue of continental drift in his 1951 Presidential Address to the Institution of Mining and Metallurgy. He concentrated on the differences between the mineral deposits of Gondwana and elsewhere, referred to his explanation of the cause of continental drift, and looked to Antarctica for further support for drift by making a prediction about its mineral deposits, adding to earlier predictions by du Toit and Wade.

But if we can have the courage to adopt the hypothesis of continental drift, which students of Antarctic geology seem to favour, and accept the corollary that Antarctica was once in close apposition to the other fragments of Gondwanaland in some such manner as is shown in du Toit's reconstruction then we may perhaps dare to be more selective in our prediction . . . In view of the presence of uranium deposits in all these neighbouring countries, one is bound, on this reconstruction, to expect uranium in Antarctica. In addition the proximity of the north-west and north coast of Antarctica (the Enderby Quadrant), as thus placed, to Madagascar, Ceylon, and India, with the Ruby Mines tract of Burma in the offing, compels one to envisage the possible occurrence in that part of Antarctica of numerous kinds of gemstones, whilst in the west end of Antarctica one might expect diamonds. And as du Toit remarks the further discovery of rare mineral such as monazite or thorianite in the Enderby Quadrant should cause no surprise. Treating East Antarctica as a whole, gold should also be found, as well as deposits of iron-ore and manganese-ore. Other minerals on the list might also be revealed by our lifting of the mantle of snow and ice. But there are two minerals we should not expect to find. One is bauxite, for we have no evidence that Antarctica was ever far enough north to experience the tropical climatic conditions that seem to be necessary for lateritization. The other is petroleum. We should not expect this in the Gondwanaland section of Antarctica. On the other hand it seems possible that the conditions in the orogenic belt that crosses West Antarctica may have been suitable for the accumulation of petroleum, as already noted.

(Fermor, 1951: 465)

Fermor's work in India had given him the "courage to adopt the hypothesis of continental drift." He sensed that current students of Antarctic geology were favoring drift.

Fermor later played an advisory role in the mobilism controversy, for it was to Sir Lewis that E. Irving in October 1951 went to ask about rocks in India suitable for paleomagnetic studies when he was planning his paleomagnetic test of India's displacement (II, §1.18).

6.15 The differing views of D. N. Wadia and M. S. Krishnan

Two of India's leading geologists during the first two-thirds of the twentieth century, Dadrashaw Nosherwan Wadia (1883–1969) and Maharajpuram Sitaram Krishnan (1898–1970), did not take part in the 1937 Hyderabad symposium on drift. But an account of Indian attitudes to mobilism would be incomplete if they were not mentioned. Wadia rejected permanentism but, I believe, showed no particular preference for continental drift or landbridges. Krishnan was a mobilist.

Wadia was the fourth of nine children. He attended Baroda College, University of Bombay, obtaining a B.Sc. in 1903 in botany and zoology, another B.Sc. in botany and geology in 1905, and an M.A. degree in biology and geology in 1906. A year later he was appointed professor of geology at Prince of Wales College, Jammu, Kashmir. In 1912 he joined the Geological Survey of India. Although not the first Indian to be recruited as a scientific officer by the Survey, he was the first whose degree was not from a European university. He remained with the Survey until 1938, when upon retirement he became Government Mineralogist of Ceylon, a position he held until 1944. Wadia returned to mainland India in 1945, and continued working in various capacities for the Indian government until his death in 1969. In 1962 he became the first geologist to be made a National Professor by the Government of India. He was elected to the Royal Society (London) in 1957. He received the Lyell Medal from the Geological Society of London (1943), the Joyakishan Medal from the Indian Association for the Advancement of Science (1944), and the Khaitan Gold Medal from the Asiatic Society. (See Stubblefield (1970) and West (1965) for details.)

Wadia's most important and original work was on the Himalayas, particularly his demonstration of its geological continuity around the Assam and Kashmir promontories of the India block. West (1965: 2) speculates that Wadia's years in Jammu amongst the mountains of Kashmir provided opportunity and inspiration for his Himalayan work, and briefly described Wadia's identification of the bends, putting them in historical perspective.

This work led up to the publication in 1931 of his well-known paper on "The syntaxis of the North-West Himalaya: its rocks, tectonics and orogeny," wherein he put forward an hypothesis to explain the remarkable fact that the Himalaya, after having pursued for 1500 miles, from Assam to Kashmir, a north-westerly direction, suddenly turns from this direction to a south-westerly course within 18 minutes of longitude. The great Austrian geologist Eduard Suess had explained this as due to the meeting of two distinct mountain systems, the Himalaya and the Hindu Kush. But Wadia's detailed fieldwork on both sides of this bend . . . showed that the [westerly] mountains of Hazara belong to the Himalaya, and are but a continuation of the Himalayan geosynclinal belt . . . In a subsequent paper . . . Wadia considered also the structure at the eastern end of the chain, where it had once been thought that the Himalayan axis extends eastward into China. Though detailed surveys in this rather inaccessible area had not yet been made, Wadia suggested that the chain turns sharply on itself . . . pursuing a south-westerly course and southerly course through Assam and Burma. Here again it was postulated by him that this acute knee-bend was caused by the Assam wedge acting as a pivot around which the Himalayan folds were molded.

(West, 1965: 2–3; my bracketed addition)

Wadia identified the eastern syntaxis in 1936, and Sahni (1937) promptly appealed to it to support his explanation of the juxtaposition of *Glossopteris* and *Gigantopteris* floras, the non-mobilist unintentionally assisting him become a mobilist.

Wadia's major book, *Geology of India*, went through four editions (1919, 1939, 1953, 1961). It is almost entirely descriptive. Throughout his book he remained content to describe geological occurrences and the speculations of others, but with little analysis of these speculations and adding few of his own. He seemed detached from the mobilism debate, not wanting to get involved in it. Wadia, however, did discuss the concept of Gondwana, and he believed that it once existed as an entity.

The Ancient Gondwanaland – Rocks of later age than Vindhyan [Pre-Cambrian] in the Peninsula of India belong to a most characteristic system of land-deposits, which range in age from the Upper Carboniferous, through the greater part of the Mesozoic era, up to the end of the Jurassic . . . This enormous system of continental deposits . . . is distinguished in the geology of India as the Gondwana system . . . Investigations in other parts of the world, viz. in South Africa, Madagascar, Australia, and even South America, have brought to light a parallel group of continental formations, exhibiting much the same physical as well as organic characters. From the above circumstances, which in itself is competent evidence, as well as from the additional proofs that are furnished by important palaeontological discoveries in the Jurassic and Cretaceous systems of India, Africa and Patagonia, it is argued by many eminent geologists that land-connection existed between these distant regions across what is now the Indian Ocean, either through one continuous southern continent, or through a series of land-bridges or isthmian links, which extended from South America to India, and united within the same borders the Malay Archipelago and Australia. The presence of land connections in the southern world for a long succession of ages, which permitted an unrestricted migration of its animal and plant inhabitants within its confines, is indicated by another very telling circumstance. It is the effect of such a continent on the character and distribution of the living fauna and flora of India and Africa of the present day. Zoologists have traced unmistakable affinities between the living lower verte-brate fauna of India and that of Central Africa and Madagascar, relationships which could never have subsisted if the two regions had always been apart, and each pursued its own independent course of evolution . . . The northern frontier of this continent was approxi-mately co-extensive with the central chain of the Himalayas and was washed by the waters of the Tethys.

(Wadia, 1939: 123–124; my bracketed addition)

Wadia thought the evidence for Gondwana was strong.

The evidence from which the above conclusion regarding an Indo-Africa land connection is drawn, is so weighty and so many-sided that the differences of opinion that exist among geologists appertain only to the mode of continuity of the land and details of its geography, the main conclusion being accepted as one of the settled facts in the geology of this part of the world . . . The term Gondwana system has been consequently extended to include all these formations, while the name of Gondwanaland is given to this Mesozoic Indo-African-American continent or archipelago.

(Wadia, 1939: 124)

Regarding Gondwana's demise, Wadia showed no preference for fragmentation and dispersal (drift) or landbridges.

The Gondwanaland, called into existence by the great crust-movements at the beginning of this epoch, persisted as a very prominent feature in ancient geography till the commencement of the Cainozoic age, when, collaterally with other physical revolutions in India, large segments of it drifted away, or subsided, permanently, under the ocean, to form what are now the Bay of Bengal, the Arabian Sea, etc., thus isolating the Peninsula of India.

(Wadia, 1939: 124)

Perhaps Wadia did not care. Perhaps he feared that if he did take sides, readers would think his account of Indian geology prejudiced. He discussed India's Upper Carboniferous glaciation without mentioning drift.

The Gondwana system is of interest in bearing the marks of several changes of climate in its rocks. The boulder-beds at its base tell us of the cold of a Glacial Age at the commencement of the period, an inference that is corroborated, and at the same time much extended in its application, by the presence of boulder-beds at the same horizon in such widely separated sites as Hazara, Simla, Salt-Range, Rajputana, Central Provinces and Orissa. This Upper Carbon-iferous glacial epoch is a well-established fact not only in India, but in other parts of Gondwanaland, e.g. in Australia and South Africa.

(Wadia, 1939: 126)

He thought the Himalayas arose from a geosyncline, a view that could be fixist or mobilist depending on the causes of its compression and subsequent uplift – again no mention of drift.

> The Himalayan zone is, according to this view, a geosynclinal tract squeezed between the two large continental masses of Eurasia and Gondwanaland. This subject is, however, one of the unsettled problems of modern geology, and one which is yet *sub judice*, and is, therefore, beyond the scope of this book.
>
> *(Wadia, 1939: 151)*

His association of geosynclines, the elevation of the Himalayas, and contraction could indicate that he did not think drift caused their uplift. It is hard to tell. He did not seem very enthusiastic about contraction either, being "beyond the scope of this book."

In the third edition (1953) of his *Geology of India*, Wadia included this brief description of the mobilist controversy.

> Both supporters of Wegener's theory of Continental Drift and its opponents have looked for evidence in support of their respective views in the later geological history of the different units of Gondwanaland. The separation of the now discrete units of the once continuous southern continent of the Palaeozoic (Pangea) was brought about, according to one view, by the drifting away (i.e., north-easterly drift) of India from Africa; and by the fragmentation and foundering of large segments of the land under the oceans, according to the other.
>
> *(Wadia, 1953: 178)*

Wadia even cited du Toit, Schuchert, and Willis; his own opinion on drift he kept secret.

> Palaeontological facts clearly show that the Indian Mesozoic systems . . . are more closely related to those of Madagascar and South Africa than to Europe. Only at the end of the Cretaceous does the fauna . . . show relationships to Sind, Persia and further west. In an important paper on the geographical relations of Gondwanaland, the eminent American geologists, Schuchert and Bailey Willis, present geological and biogeographic evidence which strongly supports the existence of land-bridges or isthmuses of the nature of Cordilleras, rather than a continuous land-mass, connecting Brazil, Africa and India, from the pre-Cambrian to the end of the Cretaceous. A. L. du Toit, on the other hand, supports the hypothesis of continental drift in a paper on the geological comparison of the sedimentary sequence in South Africa.
>
> *(Wadia, 1953: 178; Wadia cited du Toit's* Our Wandering
> Continents, *Schuchert (1932), and Willis (1932) in his bibliography)*

As far as I can tell, he remained firmly on the fence. In 1964, at age 81, Wadia presided over the 22nd International Geological Congress in New Delhi. In his opening address before the general assembly, having no reason not to express his view, he remained uncommitted, merely reporting that both "the proponents of continental drift theory and their opponents are finding many sustaining facts and data in support of their respective hypotheses" (Wadia, 1965: 35–36). Considering again the origin of the Himalayas he remained silent on drift, only remarking that the direction of thrusting – from or towards Asia – remained undetermined. Mobilists maintained that the thrusting was from India to Asia (Wadia, 1965: 36). Stubblefield (1970: 547–548) also wondered about Wadia's attitude toward drift and its application to the origin of the Himalayas. Strong well-documented paleomagnetic evidence in favor of the northward drift of India was available by the late 1950s (II, §5.2), but Wadia did not acknowledge it. For many geologists in India, as well as mobilists elsewhere, his lack of commitment on this central factor in Indian geology must have been disappointing.

M. S. Krishnan was born in a village near Tanjore (now Thanjavur) in southern India in 1898.[16] He received his Ph.D. in geology in 1925, and was appointed to a senior post in the Geological Survey of India, "unthinkable in the then British dominated Geological Survey of India" (Sankaran, 1998: 1084). In 1951, twenty-six years later, he became the first Indian Director of the Survey. He was known for his precise geological mapping. Earlier in his career, he taught at several colleges in India including Presidency College, Madras (now Chennai) (1920–1), Forest College, Dehra Dun (1928–30), and Presidency College, Calcutta (now Kolkata) (1933–5). Through his teaching he realized the need for a new comprehensive book on Indian stratigraphy. He told this to Fox who suggested that he write one. He did, *Geology of India and Burma*, which went through four editions (1943, 1949, 1958, 1960). Krishnan devoted much of his life to improving geological studies in India. He was Director of Indian Bureau of Mines, New Delhi (1948–51); Director of Indian Bureau of Mines, Dhanbad (1957–8); Head of the Geology and Geophysics Department, Andhra University (1958–60); and Director, National Geophysical Research Institute, Hyderabad (1961–3). In January 1970 he received the Padma Bhushan Prize, the third highest civilian prize awarded by the government of India. He died seven months later on August 24, his birthday.

Sankaran (1998) noted that Krishnan was open to new ideas. Certainly he welcomed continental drift, arguing that it explained many features of Indian geology, and offering mobilistic narratives of what happened to India during its drift history.

Krishnan, like Wadia, first established Gondwana as an entity. He attributed the formation of the Hercynian Orogeny and the redistribution of the land and sea in the Carboniferous to a common albeit unidentified cause.

After the deposition of the Vindhyan [Precambrian] rocks and their uplift into land, there was a great hiatus in the stratigraphical history of the Peninsula. At the end of the Palaeozoic Era, i.e., towards the Upper Carboniferous, a new series of changes took place over the surface of the globe, which brought about a redistribution of the land and sea and which was responsible for the mountain-building movements called the *Hercynian*. At this time there existed a great Southern Continent or a series of land masses which were connected closely enough to permit the free distribution of terrestrial fauna and flora. This southern continent included India, parts of the Malay Archipelago, Australia, South America, South Africa and Madagascar, which was probably at that time very close together. This southern continent called Gondwanaland, shows evidence of the prevalence of the same climatic conditions and the same type of deposits. The sedimentary era initiated at this time began with a glacial climate for we find the deposits commencing with a glacial boulder-bed which has been recognised in all the above-mentioned lands. The bulk of the deposits which followed the glacial conditions were laid down as a thick series of fluviatile or lacustrine deposits with intercalated plant remains which now form coal seams . . . The plant remains embedded in these sediments have close affinities in all the lands mentioned, and comprise *Glossopteris, Gangamopteris, Neuropteridium,* etc. This floral assemblage, called the *Glossopteris* flora, is very characteristic of the deposits of the lower part of this system. The amphibian and reptilian fauna of this era are strikingly similar and point to unrestricted intermigration.

(Krishnan, 1949: 241–242; my bracketed addition)

Up to this point he showed no special preference for fragmentation and drift or for landbridges.

This southern Gondwana continent seems to have persisted through the great part of the Mesozoic era and was broken up probably at the end of the Cretaceous, either by the sinking of certain connecting areas or by the drifting apart of the component parts. The close faunal and floral affinities of the Gondwana strata in India and Africa, for instance are reflected in their present day distribution and relationship.

(Krishnan, 1949: 242)

But Krishnan's enthusiasm for drift becomes evident in his section on "Palaeogeography of the Gondwana Era." He admitted that mobilism faced difficulties, but nonetheless favored it. He described the fragmentation of Gondwana in terms of drift, and related it to Himalayan uplift.

The similarity of the Gondwana deposits is so great that it has been suggested by Wegener that South Africa, Madagascar, India, Australia, Antarctica and South America formed parts of a continent which probably lay in the Indian Ocean around what is now South Africa. India then lay alongside South Africa and Madagascar, and Australia to the east of India. South America was joined to South Africa, Argentina curving round the Cape of Good Hope. The southern part of this continent was Antarctica. Australia seems to have drifted apart in the Jurassic when the Bay of Bengal moved or took its present shape. India began to drift northward or northeastward perhaps in the late Cretaceous. The different phases of the Himalayan upheaval may be looked upon as active phases of this drift and underthrust of India into the Tethyan region. South America is supposed to have drifted westward from Africa. These drifts may have been accompanied by the breaking off of some portions of the crust which foundered into the undercrust. Though there are gaps in our knowledge and certain details difficult of explanation, yet Wegener's conception of continental drift gives a remarkably interesting explanation of the geology and the subsequent history of Gondwanaland. For discussion of the various phases of this subject, the works of Wegener, Du Toit, Fox and others may be consulted.

(Krishnan, 1949: 283)[17]

He also accepted Sahni's mobilist explanation for the present juxtaposition of India's *Glossopteris* and China's and Indochina's *Gigantopteris* floras.

[These] floras are quite distinct from each other though India now adjoins China and Indo-China. The extensive changes which took place in late Mesozoic and early Tertiary eras caused India to drift and wedge itself into the Chinese regions. The line of separation is well marked along the Burma Malay arc. Just as Gondwanaland now lies scattered over the whole of the Southern Hemisphere, the land of the European flora lies on both sides of the Atlantic and the land of the Gigantopteris flora on both sides of the Pacific. These facts are best explained by the theory of continental drift associated with the name of Alfred Wegener.

(Krishnan, 1949: 281; my bracketed addition)

Krishnan continued to support mobilism in later editions. In the fourth edition (1960), he repeated many of the passages quoted above. But there were changes, notably the expansion of sections on the Gondwana System in regions other than India. He described the intermittent severe glaciation in the Late Carboniferous and Permian, the formation of coal, and the presence of *Glossopteris* flora throughout Gondwana during warmer intervals; he emphasized the similarities between South America and South Africa.

There is a considerable mass of data which leads to the conclusion that South America and South Africa were contiguous or were part of one land mass, as has been pointed out by Du Toit and more recently by Caster (see paper in *Bull. Amer. Mus. Nat. Hist, 99*, part 3, 1952). The remarkable similarity and even identity of numerous species in the flora points to the unrestricted migration over lands which were close to each other and were not separated by any large water barriers. Though the reptile and fish faunas do not indicate such closeness of relation as the flora, there is, nevertheless, a considerable amount of similarity between them. Lester King has pointed out that the meagre data from Antarctica indicate that the gigantic escarpment seen in Queen Maud Land is strongly reminiscent of the Natal Drakensberg and that the former region was probably a continuation of the Karroo basin of Africa.

*(Krishnan, 1960: 321–322; his references to King were to his 1950
and 1953 papers; the Caster reference is the same as mine)*

His discussion of areas other than India, and his appeal to Caster and King indicate that he was abreast of the literature. That he read Caster indicates that he was aware of the negative reaction of others to drift at the 1949 New York meeting, but was unaffected by it (§3.8). Krishnan also argued that the great outpourings of lavas throughout much of Gondwana during the Jurassic found a ready explanation in continental drift.

The great continent of Gondwanaland apparently began splitting up in the Jurassic or early Cretaceous. The vast outpourings of lavas in the Jurassic were a manifestation of the tension to which the crust was subject, resulting later in the separation and drift of the continents.

(Krishnan, 1960: 322)

He (1960: 326) also argued that the high ash content of all Gondwana coals indicated their former proximity; "even the best seams contain not less than 5 or 6 per cent ash."

In his address before the first International Oceanographic Congress (1959), Krishnan described the creation of the Indian Ocean and gave a timetable for the fragmentation of Gondwana. Paying little attention to new developments in oceanography, he appealed, as Wegener and du Toit had done but now with much expanded evidence, to sedimentary, geological, and faunal disjuncts among Gondwana's fragments.

The site of the present Indian Ocean was occupied by Gondwanaland in Permo-Carboniferous times. During the Permian, when Tethys was formed after the Hercynian revolution, a shallow epicontinental sea transgressed southward from Baluchistan [western Pakistan] and Persia over Arabia, Eritrea and East Africa. In and around this arm were deposited continental Karroo formations with a few marine intrusions in East Africa and western Madagascar. This became a fossiliferous marine realm in early Jurassic, probably by separation of Madagascar from East Africa. Australia was also separated from the rest of Gondwanaland in Permo-Triassic times as coal-bearing and marine Permian formations are found on its northwestern coast, followed by marine Jurassic and Cretaceous strata. The very close resemblance of the Jurassic fauna of Madagascar and Cutch (Western India) indicates that they were close together. Similarly, the close resemblance of the Archean geology of South India, Ceylon, Madagascar and East Africa is an argument in support of these units forming part of a land mass in Pre-Cambrian times and later, until they drifted apart.

The eastern coasts of S. Africa, of Madagascar and of India took shape towards the end of the Jurassic or early in the Cretaceous by a southerly extension of the Mozambique rift and by the severing of Madagascar from India. The west coast of India has some Tertiary deposits. After the Miocene transgression and an uplift, this coast was finally faulted down in Plio-Pleistoncene times, the fault extending into the Persian Gulf region.

Thus the Indian Ocean began to form in the East African and in the west Australian regions in the Permian. It seems to have become wider and deeper, with the separation of Madagascar – India from Africa and of Australia from the rest, in the Triassic or early Jurassic. The dismemberment of Gondwanaland was complete by early Cretaceous when the different units drifted apart. The final shape was attained during the Tertiary, at the time of the great lava flows of Deccan (India) and the formation of the Himalayan mountains system with their extension westward into Persia and Syria, and eastward into Indonesia.

(Krishnan, 1959: 34; my bracketed addition)

Thus the two most influential mid-twentieth-century Indian geologists were not opposed to continental drift. Krishnan favored it strongly, and described its ability to explain the geology of India and disjuncts between the geology of India and other Gondwana fragments. Wadia recognized Gondwana as an entity, but neither opposed nor favored mobilism; he amiably sat on the fence.

6.16 Favorable reception of mobilism in South America

Rare among South American geologists was the Brazilian A. Betim P. Leme, who (1929) favored fixism. According to du Toit (1937: 32), Leme ruined the fit between South America and Africa by overlapping them instead of fitting them together along their margins; du Toit also noted Maack's (1934) objections to Leme's interpretation. Maack (1953) elaborated on du Toit's comparison between South America and Africa. King (1953: 2163) referred to Guimaries, Leonardos, and Windhausen as "expressing a profound belief in the hypothesis." The Brazilians A. I. de Oliveira and A. H. Leonardos (1940) endorsed mobilism in their *Geologia do Brasil*. Djalma Guimarães (1894–1973), another Brazilian, continued to favor mobilism in his own 1964 *Geologia do Brasil*.

Windhausen, an oil geologist who worked in Patagonia in the 1930s and later through-out Argentina, supported mobilism in his 1931 *Geología Argentina*. J. C. Mendes, the Brazilian geologist who helped Castor translate du Toit's *Geological Comparison of South America with South Africa*, argued in favor of mobilism (§6.2) (Caster and Mendes, 1948) as did F. F. Almeida (1953) (Martin, 1961: 17). Joaquin Freguelli, Director of the La Plata Museum of Argentina, was a mobilist (Caster, 1981). During the 1950s and first half of the 1960s, the Brazilian geologists Almeida (1953, 1954) and J. J. Bigarella and R. Salamuni (1961) in their studies of paleowinds provided evidence in support of mobilism (III, §1.9). Bigarella, anxious to assist in the paleomagnetic test of continental drift, advised Creer on collecting suitable samples (II, §5.5). This cursory survey makes clear their overwhelmingly mobilist outlook; however, a study of the reception of mobilism among South American geologists is needed.

6.17 Summary

Of the thirty-five workers from South Africa, peninsular India, and Brazil and Argentina whose opinions I have reviewed in this chapter, twenty-nine favored mobilism; they are Adie, Dixey, Gevers, King, Krige, Martin, Plumstead, Smuts, du Toit, and Hales (after 1956) from South Africa; Agharkar, Dey, Evans, Fox, Krishnan, Sahni, and West from India; Almeida, De Almedia, Bigarella, Freguelli, Guimares, Leonardos, Maack, Mendes, de Oliveira, Salamuni, and Windhausen from Brazil and Argentina; and Wade, who worked in various places in the Southern Hemisphere. Six favored fixism; they are Brock, Hales (before 1956), and Haughton from South Africa, Hora and Rao from India, and Leme from Brazil. One, Wadia from India, had no preference. Fixists were outnumbered four to one. This strong preference for mobilism arose from mobilism's ability to solve the problems of regional geology that were important to them, and which they spent a substantial part of their careers attempting to solve.

Notes

1 W. E. Long discovered Permo-Carboniferous glaciation in Antarctica when he was a member of the US team working in Antarctica during the continuation of the International Geophysical Year Antarctic project (§7.4).

2 The Washington meeting of the International Geological Congress was actually in 1933.

3 I do not know what club she had in mind by the "senior geological club." The long established geological club at Cambridge is the Sedgwick Club, which is run by undergraduates, and Ted Irving and Fred Vine both served as presidents. Faculty and graduate students as well as undergraduates regularly attend meetings.

4 T. W. Gevers was born in 1900 in Natal. He received an M.A. degree in geology at the University of Cape Town, and D. Phil. at Munich in 1926. He worked for the Geological Survey of the Union of South Africa. In 1935 he was appointed Chair of Geology at the University of Witwatersrand. Gevers was awarded the Draper Memorial Medal by the Geological Society of South Africa in 1944. He had great respect for du Toit. In his acceptance speech for the Draper Memorial Medal he said:

There is one name, however, which I feel in duty bound to record here, that of
Dr. A. L. du Toit, whose work has made the name of South Africa ring throughout
the scientific world, and whose work and example will forever be an inspiration to
all South African geologists.

(Gevers, 1944: lii)

5 I owe this point to Anthony Hallam (September 9, 2004, email to author).

6 Du Toit, however, did not ignore tectonic arguments. Indeed, he largely agreed with
Argand's explanation of the Alps, accepting, for example, his anticlockwise rotation of
Spain (du Toit, 1937: 56, 179).

7 Du Toit and Longwell were not the only ones to severely criticize Willis' mechanism
objection. Years later, the geophysicist Edward Bullard (1975a: 6) quoted the same
passage from Willis (1944) to show that some of the arguments against mobilism's
mechanism indicated the "almost complete ignorance of mechanics of several of the
contestants."

8 Besides du Toit, I have found three early mobilists who made use of Köppen and
Wegener (1924) in their own work. van der Gracht (1928a) recommended that Köppen
and Wegener's work be studied, and reproduced eleven of their diagrams; R. Staub
(1928) used the idea of latitudinal zonation in his pro-mobilist analysis of the Alps; and
Simpson (1930) spoke favorably about their work, and stressed the importance of
latitudinal zonation in his own work. Among fixists, Köppen and Wegener (1924) got
little attention. At the AAPG symposium, Schuchert (1928a) and David White (1928)
criticized them. As already pointed out, it is unfortunate that the book was never
translated into English.

9 Djalma Guimarães (also known as Guimares) was a Brazilian. He later wrote a book on
the geology of Brazil (Guimarães, 1964). Othon Henry Leonardos, also Brazilian,
co-authored a book with A. I. de Oliveira (1940), also called *Geologia do Brasil*. A second
edition appeared in 1943. A. Windhausen was an oil geologist who lived and worked in
Argentina. Du Toit (1937: 216) briefly discusses Windhausen's hypothesis about the
opening of the South Atlantic and the origin of the South Atlantic Rise. Windhausen
defended mobilism in his 1931 *Geología Argentina*. Sir Lewis Leigh Fermor is discussed
in §6.14.

10 Bucher (1933) discussed the origin of the term "Llanoria."

In the case of the Ouachita Mountains, the region from which the active thrust must have
come lies buried beneath the Cretaceous cover. H. D. Miser ["Llanoria, the Paleozoic Land
Area in Louisiana and Eastern Texas." *Am. Jour. Sci.*, 2(1921): 61–89] has summed up the
information that proves the presence of an active source of sediment south of the Ouachita
folds. By analogy to "Appalachia" he has called this land "Llanoria" … In place of the
name "Llano" first employed by Bailey Willis, "A Theory of Continental Structure Applied
to North America," *Bull. Geol. Soc. American*, Vol. 18, 1907, pp. 394–5.

(Bucher, 1933: 157–158; the bracketed reference is a footnote in Bucher, 1933: 157)

11 During the discussion following his (1958c) presentation to the Geological Society of
London, King said:

The driftist is no more obliged to adduce a mechanism to prove the fact of drift than the use
of an electrical appliance is obliged to define the nature and mechanism of electricity.
Both drift and electricity are to be known from their effects.

(King, 1958d: 76)

12 When the work about to be described was carried out, the countries now known as
Bangladesh, India, and Pakistan were one country, India. It is therefore historically correct
to refer to the collective as "India" and I shall do so in this section.

13 One wonders if Fox actually said or meant to say "from the Zambesi to Akaba" instead of
"from the Zambesi to Akba [Algeria]." Akaba is situated at the northern end of the Gulf
of Akaba at the head of the Red Sea.

14 I find it curious that Fox does not mention Holmes' theory, which offered a better drift mechanism than Joly's. Perhaps Fox did not know about it. Again, it shows Holmes' lack of influence at the time.

15 Wegener (1929/1966: 20) actually supposed that India separated from Australia in the early Jurassic and from Madagascar at the transition from Cretaceous to Tertiary. Nevertheless, this would still be too early for Hora.

16 This account of Krishnan's life is drawn entirely from Sankaran (1998).

17 Krishnan cited no specific works of du Toit or Wegener. Although he did not provide a specific reference to Fox's work in this passage, Fox is mentioned in the text more often than is any other author. He singled out Fox for his help, in the Preface to his *Geology of India and Burma*. Fox was Director of the Geological Survey while Krishnan worked on the first (1943) edition.

7

Regional reception of mobilism in North America: 1920s through the 1950s

7.1 Introduction

As just described in the last chapter, workers from South Africa, India, and South America generally preferred moving rather than fixed continents because of drift's success in explaining their geology. In contrast, continental drift was poorly received by North American workers, and I shall argue that this was because at the time they thought fixism could well explain its geology, they felt little need to look elsewhere to address their problems; they were turned inwards. This regionalism was, I believe, the major factor controlling their preference for fixed continents.

I shall first briefly describe previous studies of the reception of mobilism in North America. I shall not, as these studies have done, attempt a general review, but will concentrate on a single figure, Marshall Kay, whose views epitomize those of the geological establishment during the last twenty years of the classical stage of the mobilism debate. I shall identify the common ground held by him and several other influential, establishment figures – mandarins, as Newman (1995) called them.

I shall continue my discussion of the reception of mobilism in North America by examining the work of William Long and Warren Hamilton. Long, whom Lester King singled out in the opening quotation of the previous chapter, was originally a fixist but became a mobilist in 1962 after making the first discovery of the Permo-Carboniferous glaciation in Antarctica. Long's journey from fixism to mobilism well illustrates the effect of regionalism. He had to rid himself of prejudice against drift, step outside the framework of North American geology, and remedy his ignorance of Southern Hemispheric geology before he understood that he had actually walked on these ancient tillites in Antarctica. Hamilton was at first a closet mobilist. He read du Toit's *Our Wandering Continents* in 1949, which convinced him of continental drift. But he kept his belief to himself until he found his own evidence of it in his own work; ten years in North America provided none, and then, like Long, he found it in Antarctica, and began arguing in favor of mobilism. His application and extension of mobilism continued; in 1989 he was elected to the NAS and received the Penrose Medal, GSA's senior award. Hamilton's initial failure to find evidence for mobilism in North America and his success in doing so in Antarctica further illustrates the effect of regionalism.

7.2 Previous studies on the reception of mobilism in North America

Was regionalism the only reason most North American participants in the debate rejected continental drift during the 1920s through the 1950s? Certainly not! Newman (1995) and Oreskes (1999) showed that some of these geologists (Willis especially) offered both good and bad reasons for dismissing mobilism. Newman (1995) focused on T. C. Chamberlin, Willis, Schuchert, and Bowie. He attributed the intransigence, as he calls it, of Americans toward continental drift partly to their adoption of permanentism, developed by America's own J. W. Dana (1813–95), and to T. C. Chamberlin's doctrine of multiple working hypotheses. Newman underscored T. C. Chamberlin's strong influence on several generations of US geologists, including his son, R. T. Chamberlin, Willis, and Schuchert who, in turn, along with Bowie and several other adamant opponents of mobilism, such as Berry and Reid, he ranked among the mandarins of American geology. Newman also claimed, and I agree, that mobilism's lack of a mechanism and regionalism also contributed to drift's rejection by US scientists.

Focusing on the same individuals as Newman, Oreskes (1999) attributed mobilism's rejection by US scientists to five factors: the general acceptance of the permanency of continents and oceans, which goes back to Dana; the general adoption of T. C. Chamberlin's (1890, 1897) method of multiple working hypotheses, which commonly led to the rejection by US geologists of theories (unless they were their own) that were not developed in accordance with it; the strict adherence by American geologists to the principle of uniformitarianism, which clashed with Wegener's one time and very late drifting of the continents; and the adoption as fact of Pratt's model of isostasy, especially by Bowie, which disallowed large-scale horizontal displacement of continents. (See Chapter 3, note 25 for a different account of the importance Bowie's adoption of Pratt's model of isostasy played in his rejection of mobilism.) Oreskes (1999) also explicitly rejected the regionalist thesis, at least in the way that I have defined it, as a factor in the strongly negative reception of continental drift by US Earth scientists (see Chapter 1, note 2).

T. C. Chamberlin certainly was a giant among US geologists; he, Willis, Schuchert, Bowie, Berry, and others who rejected mobilism *were* among the senior mandarins of the US geological community. They were members of the NAS, and presidents of major geological societies. They probably influenced many North American geologists not to take mobilism seriously. North American geologists generally knew little about the geology of the Southern Hemisphere. Oreskes is right to claim that mobilism was rejected by US geologists, in part, because of their adherence to key parts of Dana's permanentism. But why did permanentism last so long as the reigning view? Was it social pressure to conform? Was it pressure to adhere to a certain methodology? Was it the passing down of Dana's wisdom to later generations? Or, was it because Dana's permanentism was at the time better equipped than mobilism to solve key problems of North American geology, especially through its later renovation by Kay and others in response to new data?

One methodological objection that fixists raised against mobilism was its apparent clash with uniformitarianism. In particular, they objected to Wegener's theory, in part, because he claimed that drift had happened only once and very late in Earth's history. Uniformitarianism held that processes repeated themselves throughout the Earth's history, or, if they occurred only once, it was very early when it was easier to envision a very different world. Schuchert (1928: 106–108) voiced such an objection in his paper from the AAPG volume as he did in his correspondence (Oreskes, 1988: 201–202). Singewald (1928: 192–193), White (1928: 187), and R. T. Chamberlin elaborated on the same theme. Chamberlin remarked:

Wegener's hypothesis is no general theory of earth behavior or earth deformation. It describes simply one supposed breaking up of a consolidated land mass and the migration of the different fragments. That was assigned to very late geological history. What was happening throughout most of geological time? Why did the continents all remain coalesced in a single mass all through geological time, only to become fragmented very recently? The geological record is rhythmic, with cycles following cycles, and gives little suggestion of any single great event apart by itself such as Wegener postulates.

(Chamberlin, 1928: 83–84)

Mobilists took this objection seriously. Van der Gracht (1928b: 203–204) responded in the AAPG symposium volume; Wegener, he argued, was just being responsible in not speculating about what had occurred earlier in the Earth's history, continental drift could have occurred earlier, and it probably did.

Dr. Schuchert and Dr. Chamberlin attack Wegener for confining his arguments and his analysis of continental drift to late geological history. Far from following them in this, I think that we should commend Wegener's method for not leading us into a discussion of conditions in very remote geological periods, regarding which we have so few facts and so little reliable evidence of a wide regional nature. In his writings, as well as in his book on geological climate, published in co-operation with Dr. Köppen, Wegener stresses the necessity of confining a somewhat detailed analysis to periods beginning with the late Paleozoic. This does not mean, as particularly Dr. Chamberlin seems to assume, that adherents of a drift theory should believe that drift was confined to very late geological history, and that all the continents had remained coalesced in one single mass all through geological time, only to break up into fragments very recently. Dr. Singewald raises the same objection, and so does Dr. White . . . *Old Paleozoic or earlier continental drift, however, remains just as possible, but it simply was not discussed in detail by Wegener, because the relevant facts are too little known* . . . I am personally inclined to believe that an originally universal sial was rolled up in very remote pre-Cambrian time into nuclei, which later coalesced into one single Pangea, but I do not at all contend, and also do not think that Wegener believes that this ancestral continent could have remained without drift and rifting of considerable extent until after the Jurassic.

Argand, as already noted, also argued that continental drift had occurred in the distant past. Du Toit agreed, but like van der Gracht, argued that there were not enough reliable data to speculate about pre-mid-Paleozoic drift.

> To conclude, Continental Sliding is deemed to be essentially the outcome of cyclonic disturbances that have affected the more or less "permanent" convective circulation in the sub-crust. It is furthermore regarded not as a special process applying to a particular period or region, but as an inherent property of the crust and as having operated throughout geological time as well as over the entire globe, only for the epochs prior to about the mid-Palaeozoic the evidence is as yet too fragmentary for its proper decipherment.
>
> *(Du Toit, 1937: 331)*

Thus it is a mistake to claim that it was only North Americans or only fixists who noted the clash between Wegener's mobilism and uniformitarianism.

Newman (1995) and Oreskes (1999) both argued that Wegener's mobilism was rejected by the elite of American geology partly because he had failed to abide by Chamberlin's method of multiple working hypotheses. Chamberlin (1897) directed his methodological principle of multiple working hypotheses especially at students; recognizing that they often fall in love with their own hypotheses, he thought they would become more objective if they were encouraged to compare their own hypotheses with alternatives. Chamberlin's method was a heuristic that was supposed to help young scientists develop and evaluate their hypotheses.[1] This criticism of the American geological establishment was really aimed at Wegener's methods, not his theory. The great irony is that Wegener had provided them with another hypothesis, but rather than seriously entertain it as the method of multiple working hypotheses obliged them to do, they dismissed it as unworthy of consideration because he had not developed it in an appropriate manner. Self-awareness does not seem to have been a strong characteristic of his American critics.

Oreskes linked Chamberlin's method of multiple working hypotheses with induction and careful fieldwork, essentially an empirical approach. Because Wegener, she argued, used something much more akin to the hypothetico-deductive model in testing his theory, rather than building his theory inductively, US Earth scientists rejected his theory as methodologically unsound. However, Oreskes (1999) also showed that US Earth scientists rejected du Toit's mobilism as readily as they had rejected Wegener's theory, even though du Toit corrected many of Wegener's mistakes, was an accurate and energetic fieldworker, and greatly strengthened the observational support in favor of mobilism. Longwell criticized du Toit for arguing too strongly; R. T. Chamberlin attacked du Toit because he seemed to characterize only pro-drift studies as brilliant and illuminating. There were, however, legitimate difficulties that could be raised against du Toit's theory, and Longwell raised one about du Toit's mechanism. If US geologists were permanentists primarily because of tradition, and if their only reason for rejecting drift was the manner in which Wegener and du Toit developed and presented their theories, then the overall rejection of du Toit by US geologists was clearly unreasonable. But, they were not, and some of their objections were very reasonable and no different from difficulties fixists outside of North America raised.

Newman believes, and I agree with him, that the mandarins of Earth science in the United States rejected mobilism, in part, because they took its mechanism difficulty seriously. Indeed, it is hard to read through the papers in the proceedings of the AAPG symposium and not conclude that they did so (§4.3). During the 1920s, Lambert (1921, 1923), Berry (1922, 1928), Reid (1922), Bowie (1928), R. T. Chamberlin (1928), Longwell (1928), and Willis (1928) criticized mobilism because there was no acceptable mechanism. Willis (1944) said it again, and Longwell (1944b) criticized him; appealing instead to Jeffreys' objection, which he claimed was better than Willis'; Longwell also noted that Willis' objection did not apply to Holmes' mantle convection, but added that Holmes' mechanism had its own difficulties (§6.9). Moreover, Daly (1923a, 1923b, 1925), the most important North American mobilist during the classical stage of the controversy, rejected Wegener's and Taylor's mechanisms, and developed his own. Gutenberg (III, §2.3) also fretted about mechanism. When Griggs proposed mantle convection as a solution to mountain building (but not continental drift), Lawson and Willis (§5.10) thought his solution absurd. In the 1950s, Birch (§5.10) argued against convection, and MacDonald joined him in the 1960s (III, §2.7).

Oreskes (1999) claims incorrectly, I believe, that the mechanism difficulties of continental drift had little to do with its rejection in the United States. One reason that Oreskes gave was that Jeffreys had little effect in North America. However, North Americans did not need Jeffreys to raise difficulties with drift's mechanism; they did so themselves. Marvin (2001) also disagreed with Oreskes. She correctly, I believe, argued that Jeffreys did influence North American Earth scientists. She recounted how each new edition of Jeffreys' *The Earth*

appeared in university libraries and classrooms throughout America where it served to reassure large numbers of geologists and their students that, in the absence of an adequate causal mechanism, they should not take continental drift seriously. Two more generations of like-minded geophysicists followed Jeffreys . . .

(Oreskes, 2001: 211)

Again, I agree with Marvin. Marvin mentions Birch and MacDonald as two US Earth scientists who raised mechanism difficulties against mobilism, and who were strongly influenced by Jeffreys. Oreskes (2001) replies that Jeffreys had little influence in shaping the anti-mobilist attitudes of US Earth scientists. However, even if Jeffreys did not influence US Earth scientists during the 1920s and 1930s, they raised some of the same mechanism difficulties that Jeffreys had.

There were good and bad reasons for favoring fixism over mobilism. Newman and Oreskes have provided examples of the latter. Willis was probably the worst offender. Newman and Oreskes (1988: 201–202) quoted passages from Willis' letters, and documented his bullying Schuchert during their joint effort to develop a combined solution to the origin of Permo-Carboniferous glaciation and biotic disjuncts. Both showed that Willis was an arrogant, self-righteous bully. Willis complained that mobilism was not made in America. Greene (1999) recounted how Willis' nephew, who accompanied Willis to China, later confessed that he and another member of the expedition seriously considered murdering Willis. But Willis and Schuchert did develop an explanation of the Permo-Carboniferous glaciation and biotic disjuncts. Although their explanation was far from diffi-culty-free, it gave fixists a counter to what was then mobilism's best arguments and it was widely referenced.

Both Newman and Oreskes concentrated almost exclusively on mobilism's reception in North America. Doing so tends to obscure the fact that a region's geology had a profound impact on the attitude toward continental drift of those who studied in that region. The reception of mobilism is evidence of this. Mobilism was better received than fixism in those regions where mobilism better explained the region's geology than did fixism; where fixism (mobilism) did a better job, it was better received than mobilism (fixism).

7.3 Permanence of ocean basins, continental accretion, geosynclines: the North American experience, Marshall Kay and others

Relatively few North American geologists have become advocates of continental drift; nor do the theories have as great an impact on their problems as in other continents . . . The writer's *knowledge* and *experience* has been principally with North America and north-western Europe . . .

(Kay 1952b: 283; emphasis added)

Kay (1904–75) was a leading North American sedimentologist and structural geolo-gist, who in 1975 received the GSA's Penrose Medal, its highest award. He obtained his B.S. (1924) and M.S. (1925) from the University of Iowa, where his father taught geology. Moving to Columbia University, he obtained his Ph.D. in 1929 and stayed the rest of his career. He retired from teaching in 1973, but remained professionally active. He was an adamant fixist until the confirmation of seafloor spreading in the 1960s whereupon he relented and began remodeling his understanding of North American and European stratigraphy along mobilist lines.

Kay (1952a) argued against mobilism at the 1949 New York conference on Mesozoic trans-Atlantic land connections and in his contribution (1952b) to the volume honoring Birbal Sahni. Kay's conception of the overall structure and evolution of North America was derived from Dana and was shared by most geologists who lived and worked there. Robert Dott Jr., Kay's student, has discussed the development, influence, and importance of Kay's ideas (Dott, 1974, 1977, 1979, 1985) and placed Kay's views within their historical context.

Kay for many years was a prominent advocate of permanency of ocean basins and gradual accretion by orogenesis of fixed continents . . . the brain child originally of Dana and Suess in the late nineteenth century. But he was in excellent company, for this was the ruling hypothesis in North America (especially) until about 1965.

(Dott, 1977: 4)

Kay believed that North America consists of an ancient, central and stable Precambrian craton, which is generally sialic in composition, and is surrounded by younger mountain belts; the craton was the original continent. Kay, like Dana, Willis, Schuchert, and many other prominent North American geologists, believed that the mountain belts arose from geosynclines – elongated depressions that bordered the craton that became filled with sediment, deformed, and uplifted. In this way new mountains were added to the growing continent. The process then repeated itself. Faill (1985) described well the centrality that the idea of geosynclines and their evolution held in North American geology, and the changes that it underwent to account for the origin of mountains, especially the Appalachians.[2]

Dana's tectonic views [late nineteenth century]. . . were based on the concept of paired geosyncline-geanticline positioned along the continent-oceanic boundary . . . Both the geanticline and the geosyncline remained dominant in Appalachian tectonics during the ensuing decades. The primary strength of the "Appalachia" geanticline concept lay in its providing a southeastern source of sediment for the geosyncline. This aspect of geanticlines led to the extension of the concept to other orogenic zones around North America and the development of the "borderlands" idea. By the early part of this century, Charles Schuchert (1923) had proposed 11 geanticlines (borderlands) in various locations around the North American continent . . . By the mid-twentieth century, the geanticline idea itself had foundered. Geologic mapping in the crystalline terranes, particularly in New England, provided the main snags . . . In addition, the early seismic studies of the North Atlantic continental shelf revealed a continent-ocean transition that in no way could accommodate large crustal blocks that simply subside and vanish as some lost continent "Atlantis" . . . The geosyncline concept fared somewhat better. In the Appalachians, it was the dominating tectonic idea until the 1970s . . . Supplanting the passive "Appalachia" borderland, Kay proposed . . . [an] active zone along the length of the Appalachians dominated by extensive volcanic activity . . . He also demonstrated the close similarity of the . . . belt with present day island arcs . . .

(Faill, 1985: 26–27; my bracketed addition)

This is how Kay summarized his views in 1949 (published in 1952).

It long has been known that the central part of North America, the hedreocraton, was relatively stable, subsiding little as compared to geosynclinal belts, along its margins, which came to be severely deformed during times of orogeny. A geosyncline is a surface deformed downward during the deposition of its contained sediments; classification of geosynclines has been summarized (Kay, 1947) and a comprehensive treatise is soon to be published (Kay, 1951). The geosynclines that margined the North American continent during the Palaeozoic were of several kinds, having distinctive relations in space and time. In the early Palaeozoic, the hedreocraton was separated by flexures from more deeply subsiding linear . . . belts. Non-volcanic subsiding zones along the hedreocraton gained sediments considerably derived by erosion from the area of the hedreocraton. Beyond these zones, sediments have thick

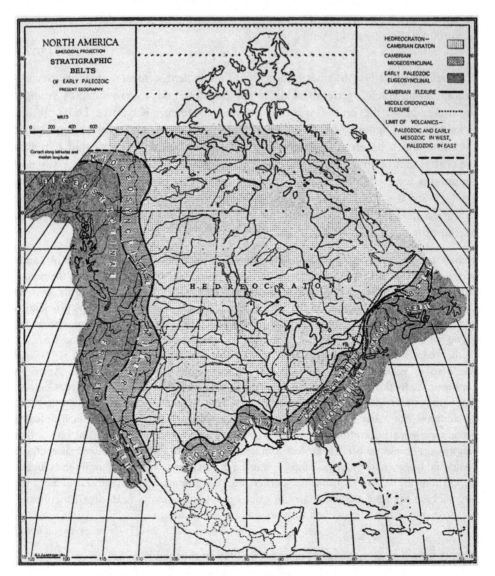

Figure 7.1 Kay's map of North America showing the North American hedreocraton and surrounding orogenic belts during the Paleozoic. The stippled inside ring is the non-volcanic miogeosyncline; the shaded marginal ring is the volcanic eugeosyncline. (An almost identical map appears in Kay, 1951: 9.)

associated sequences of lavas and volcanic detritus, as well as coarse debris that must have been derived from rapidly rising linear welts. The former, the non-volcanic linear geosynclines, have been designated miogeosynclines, and the latter, volcanic-bearing eugeosynclines (Stille, 1941).

(Kay, 1952b: 281–282)

Referring to the map (Figure 7.1), Kay compared his views with Dana's and Schuchert's.

The accompanying map of North America shows the present position of several belts of early Palaeozoic rock types; their original geographic relations differed and moved through later folding and thrusting . . . The theory of marginal volcanic troughs and island arcs (Kay, 1942; 1944) postulated that the margin of a continent had been geographically similar to modern arcs through part of their histories. The prevailing hypothesis [about North America] in the later nineteenth century when knowledge was principally of the eastern half of the continent was that the crystalline rocks . . . were Proteozoic (that is, pre-Paleozoic), and that Palaeozoic seas lapped against islands of these ancient "Archean" rocks; thus the hypothesis of Archean protaxes of Dana. By the close of the century, many of the intrusions and much of the metamorphic rock along the continental margins had been found to be Palaeozoic and Mesozoic. But some of the marginal sediments seemed to have come from lands in the regions of the present shores; thus the hypothesis of borderlands of Schuchert [which also served as Schuchert's sialic land-bridges], postulating great lands of pre-Paleozoic gneiss and granites that persisted through most of the Palaeozoic era, an hypothesis still presented in current American text-books of historical geology. But the theory of marginal island arcs seems a more satisfactory explanation of present information.

(Kay, 1952b: 281; my bracketed additions)

Kay replaced Schuchert's passive sialic borderlands with active island arcs and described the evolution of North America in terms of accretion of mountain belts which had evolved from bordering island arcs. The borderlands Schuchert proposed lay outside geosynclines; Kay's volcanic island arcs arose within them, and it was they, not Schuchert's borderlands, which provided the sediments that filled the deeper geosynclines, i.e., the eugeosynclines. Kay defended his innovation by appealing to new stratigraphic work, partly his own, which he linked with the work of Harry Hess, another rising star in North American geology, who had begun to argue that island arcs evolve into mountains. Kay acknowledged Hess's influence (see III, §3.4 for discussion of Hess's view).

The eugeosynclinal belts from their stratigraphic record must have been deeply subsiding belts having raised welts within that supplied much of their sediment, and volcanic islands that were the sites of their volcanism . . . This interpretation of the stratigraphy alone strongly suggests modern island arcs, but it has been found that ultrabasic intrusions are limited to the eugeosynclinal belts, just as they are now distributed through island arcs (Hess, 1939).

(Kay, 1952b: 282)

He saw a similar situation in the Urals, which gave him confidence in his notion of paired eugeosynclines with volcanic and non-volcanic miogeosynclines. The Urals are not currently a continental margin, but he saw enough there to remind him of North America.

In my graduate course in stratigraphy the volcanic rocks that I discussed were seen to be concentrated in belts along the two margins of the American continent, each volcanic geosynclinal belt contrasting with nonvolcanic ones toward the interior. The lava and radiolarian cherts at Nizhni-Serghinsk in the eastern Urals that I saw on an International Geological

Congress trip in 1937 . . . further confirmed the contrast between volcanic and nonvolcanic belts. With publication of the Russian congress proceedings, I learned that Hess (1939) had observed serpentine belts in positions similar to those in which these volcanic rocks had been reported. Thus, the specific observations in many localities on several continents led to generalization.

(Kay, 1974: 377)

Kay's comparison of the Appalachians and the Urals might seem contrary to my thesis on regionalism, but I do not think so, for he was generalizing from North America, fitting the world into his own region of expertise. Had he grown up in South Africa and traveled to North America and Europe, he would have found no similarities with his home country but he might have been impressed with the similarities on either side of the Atlantic, and thus favored mobilism. But that was not his experience.

Kay (1952b) described the formation of the Appalachians.

In North America, the eastern eugeosynclinal belt had become so deformed and intruded by the mid-Paleozoic time that it ceased being dominantly subsiding . . . Extensive folding and thrusting with plutonic intrusion affected parts in the late Ordovician, and orogenies and accompanying intrusions continued in late Devonian and late Carboniferous.

(Kay, 1952b: 282)

The mountains of North America's western margin were younger.

In the west, the belt remained one like modern island arcs until the late Mesozoic, welts yielding detritus to the eugeosynclines in the late Palaeozoic, and orogenies and intrusions being particularly intense in late Jurassic and latest Cretaceous.

(Kay, 1952b: 282)

In eastern North America mountain building continued into the late Paleozoic, in the west, into the Tertiary.

Kay's belief that North America grew by accreting peripheral mountain belts explained the fact that North America had a centrally located craton made of very old rocks surrounded by zones of younger rocks. This belief was shared by most North American geologists; some offered somewhat different explanations for the Appalachians, but almost always there were geosynclines, permanent oceans, and continental accretion. This model was created by Hall and Dana in the 1850s through the 1870s in the eastern United States, and had prevailed throughout North America since then. Finding a solution to the origin of the Appalachians was a longstanding and central problem for American geologists, and theories that did not explain the Appalachians were not of prime interest to them. It seems likely to me that Taylor's mobilist theory was poorly received in the United States, in part because it sought an explanation of the origin of Tertiary mountains only (§2.12), but gave none for the Appalachians, a central preoccupation of American geologists at the time: Taylor's theory was of little use to them. Nevin, Caster's teacher in structural geology at Cornell, who was somewhat sympathetic towards continental drift, also despaired at attempts to link continental drift with the building of North American mountain belts (§6.2).

Turning now to Kay's argument against mobilism, he admitted that ancient faunal identities between regions currently separated suggest former geographic juxtaposition, but they do not require it. He mentioned similarities among Lower Paleozoic faunas (trilobites) of Europe (Sweden, Poland, and Wales) and North America (Newfoundland, New Brunswick), adding that they are "quite suggestive that the two continents were once in close proximity" (Kay, 1952a: 159). But he noted that similar Paleozoic trilobites had been found in Argentina and Colombia, although nobody was suggesting that these regions were once adjacent to Europe and North America: similar fauna do not necessarily require geographic juxtaposition. Close proximity should not be invoked if it creates a severe difficulty, and if there is a better alternative (RS2 and RS3).

Kay objected to drift's account of mountain building and the formation of island arcs, especially as applied to North America. If mountains form at the leading edge of a drifting continent, mobilists must claim that when the Appalachians formed North America drifted eastward, and then changed direction and moved westward when the Tertiary mountains in the west formed; if island arcs form at the trailing edge of a continent, then North America would have had to have drifted away from Europe when the ancestral island arcs of the Appalachians formed, and toward Europe when island arcs formed off western North America. North America would have had to have drifted toward and then away from Europe, and its drifting toward Europe would have had to have occurred well before it drifted away from it. To Kay this "seem[ed] fantastic."

With present understanding of island arcs, some of the explanations given in the original hypothesis of continental drift (Wegener, 1924) seem fantastic. The arcs were thought to have formed in the lee of drifting continents, and the mountain chains to have risen along the margins of advancing blocks. Both coasts of North America seem to have had island arcs prior to the postulated time of drift. The hypothesis would require multiple advance and retreating movement on both sides of North America and along western South America. And the development would have proceeded from at least early Paleozoic time, rather than have the short span originally attributed to it. The development of both coasts has been essentially similar, though stages have been somewhat retarded along the Pacific as compared to the Atlantic. Could North America have drifted from Europe without resulting in contrasting history on the two sides of the continent? The arcuate pattern of the festoons may better be compared to the subsiding bands between the welling areas of convection in a boiling fluid than to the sinuous eddying of drift.

(Kay, 1952a: 161)

Drift theory could not explain the Appalachians unless North America had drifted toward Europe before it drifted away; Wegener's theory of continental drift only broke apart Pangea, it did not assemble it. By contrast, his own theory of mountain building could account for the origin of both the Appalachians and western Tertiary Cordillera. It could also account for the stratigraphic similarities between North America and Europe (Kay, 1952a: 160). In any case, as he already had argued, similar faunas do not require continental drift.

Of course, there is a wonderful irony in Kay's *reductio ad adsurdum* against mobilism. With the acceptance of mobilism, it has become clear that Laurasia and Gondwana came together in the Paleozoic forming Pangea and the Appalachians. But, Émile Argand (1924/1977), who had said much about the formation of mountains, invoked a Proto-Atlantic approximately thirty years before Kay's work precisely because he wanted to explain the origin of the Early Paleozoic Caledonian orogenic belts in Scandinavia and Scotland, and their extension in North America.

According to the mobilistic concept, the formation of the Proto-Atlantic becomes simply an odd outline of the Atlantic that was generated by the traction and thinning of a very old continent, both associated with drifts. The geosynclinal condition has been reached, as is well known; whether this has also been the case for the oceanic condition is not known, but fundamentally it does not matter; nor is it important to know whether the geosyncline was simple branching, or multiple. The Caledonian folding, a product of reverse drifts, of the nearing of systems that had been separated, reestablished the welding. It will probably never be known if it did reestablish it over the entire length: the present Atlantic conceals too many things. The folding of this great Caledonian branch is only the scar of an ancient wound, subsequently reopened with great consequences as shown by the gaping space of the present-day Atlantic. One of the advantages of mobilism in this matter is to reduce to the same concept the explanation of the former and the present Atlantic by using the intersection of the plans instead of being hindered by them.

(Argand, 1924/1977: 139–140)

Argand's Proto-Atlantic was Kay's absurdity. Argand had even provided a geosyncline for Kay!

The mention of Argand calls to mind another matter about the Appalachians and geosynclines. Argand grew up in the Alps where huge recumbent folds and overthrusts suggested substantial horizontal displacement. For Argand, weaned on Suess and the Alpine nappists, the adoption of Wegener was an extension of thought, not a basic change in point of view. But, the Appalachians are different (Faill, 1985). Huge overthrusts and recumbent folds are there, but their presence was not obvious to most because those parts that had not been eroded away had been blanketed with sediment. The Alps are much younger, but even there it took years before they were mapped and demonstrated to exist (Greene, 1982). North American geologists mapped folds in the Appalachians, but they were more symmetrical than Alpine folds, they did not seem to require nearly as much crustal shortening, and seemed to them to have resulted from vertical, and only minimal horizontal movements. Willis, Schuchert, Longwell, and R. T. Chamberlin, for example, agreed that large-scale folding had occurred, and Willis (1893) "reproduced" them in laboratory experiments. Moreover, North American geologists recognized the importance of horizontal compression in forming folds, but they saw it as an expression of contractionism. Faill (1985) describes this difference between "vertical" and "horizontal tectonics" and notes that Appalachian geologists had their counterparts in Europe.

In Europe, the geosyncline was introduced at the turn of the century and adopted by a number of (but by no means all) geologists. The concept was incorporated into formal classifications during the ensuing decades, forming a "fixist" school of thought in which vertical crustal movements dominated and horizontal displacements were minimal. This was in distinct contrast to the more mobile school that developed somewhat earlier, an outlook epitomized by the horizontal tectonics of Albert Heim . . . and the nappes of Marcel Bertrand . . . One of the proponents of the "fixist" school of thought, Leopold Kober, subdivided the continents into stable regions, termed Karatogens (later called kratons, or cratons), which were surrounded by tectonic zones, the orogens, in which the geosynclines developed (Kober, 1921). The geosyncline was also central to the thinking of Hans Stille, who advocated recurrent orogenic episodes, driven by the Earth's contraction, with repeated uplifts, subsidences, and unconformities (Willis 1918) . . . Stille later classified the various mobile belts, dividing the orthogeosynclines (geosynclines) into a non-volcanic miogeosyncline portion, and a volcanic-rich eugeosyncline (Stille, 1941) . . . in the Appalachians, Marshall Kay elaborated on Stille's classification.

(Faill, 1985: 26–27)

Kober and Stille both argued against Wegener. Carozzi relates Kober's fixist approach to his work on the Eastern Alps of Austria, which, like the Appalachians, hide their huge overthrusts.

As their Swiss colleagues, the Austrian geologists were facing another piling-up of overthrusts in the Eastern Alps. These overthrusts were, however, not so clearly displayed; they had a tectonic style of their own and had been generally less studied because of the immense scale. The geological thinking in Austria was unfortunately dominated by Leopold Kober from Vienna. He was a fixist and the proponent of the idea that mountain ranges form closed circular rings around old indurated continental masses which increase in size with time. Between these masses, the orogenic belts were compressed by contraction processes . . . .

(Carozzi, 1985: 130)

Kober would have been at home in the Appalachians conversing with North American geologists (see §8.7, §8.3, §8.5 for Argand, Stille, Kober, and other European tectonicists).

Kay did not speak for all North American fixists, nor did they all agree with his particular account of mountain building, but his views about continental accretion and the structure and evolution of North America and his stress on vertical tectonics were shared by most North American geologists. They believed such views to be a straightforward reading of the rocks. Therefore, regionalism goes a long way in explaining why fixism worked well in North America, and why mobilism was not needed or wanted there.

Turning to other North American geologists of the classical stage, Walter Bucher (1933) and Andrew Lawson (1932) believed in the buildup of continents by accretion. Bucher wrote:

North America. In 1849 Dana wrote, all continents "have had their laws of growth involving consequent features as much as organic structures." In his Presidential Address before the American Association for the Advancement of Science in 1855 he dwelt on the triangle, the simplest of mathematical figures. He visualized it as a block wedged between the oceans, wrinkled along its margins. Since this presentation, American geologists have been accustomed to think of continental growth, especially that of North America, as due to continental accretion.

(Bucher, 1933: 379)

He epitomized the prevailing view. He thought that Europe and Australia had also grown by accretion (Bucher, 1933: 377–379, 381–382). J. Tuzo Wilson, from the University of Toronto (1949, 1952a, b, 1959), until the end of the 1950s believed ardently in permanent continents and oceans and the accretion of continents around ancient nuclei (IV, §1.3). Later he made important contributions to mobilism (IV, especially, §1.9, §1.12). The Appalachians did not have examples of spectacular unidirectional horizontal movements. North American geologists did not live among the marked structural, stratigraphic, floral, and paleoclimatic similarities and contrasts that surrounded geologists working on the Gondwana System in South Africa (where it is called the Karroo System), Brazil and Argentina (where it is called the Santa Catarina System), and the type location of peninsular India. Most North American geologists who participated in the mobilist versus fixist debate vigorously championed permanentism until the triumph of seafloor spreading in the mid-1960s. Dana's geanticlines may have been replaced by Schuchert's borderlands, and Schuchert's borderlands by either his and Willis' isthmian links or Kay's island arcs, but geosynclines, ocean permanency, and the growth of continents by accretion worked well for North America, and remained rigidly fixed within North American thought. Moreover, there was the work of the American geodesists; work that American geologists, except for old-time landbridgers like Schuchert, appreciated. A. P. Coleman from the University of Toronto expressed precisely this point. He is particularly interesting because he was a distinguished glaciologist who appreciated the huge extent of the Permo-Carboniferous glaciation in Gondwana, having visited the classical exposure, yet remained a fixist. What he said in his 1915 Presidential Address before the GSA would have fit right in to the 1949 New York meeting (§3.8).

The essential permanence of continents and oceans has been firmly held by many geologists, notably Dana among the older ones, and seems reasonable; but there are other geologists, especially paleontologists, as well as zoologist and botanists, who display recklessness in rearranging land and sea. The trend of a mountain range, or the convenience of a running bird, or of a marsupial afraid to wet its feet, seems sufficient warrant for hoisting up any sea-bottom to connect continent with continent. A Gondwana Land arises in place of an Indian

Ocean and sweeps across to South America, so that a spore-bearing plant can follow up an ice age; or an Atlantis ties New England to Old England to help out the migrations of a shallow-water fauna; or a "Lost Land of Agulhas" joins South Africa and India . . . It is curious to find these revolutionary suggestions made at a time when geodesists are demonstrating that the earth's crust over large areas, and perhaps everywhere, approaches a state of isostatic equilibrium . . . The distribution of plants and animals should be arranged for by other means than by the wholesale elevation of ocean beds to make dry-land bridges for them. W. D. Matthew's excellent paper on climate and evolution suggests ways in which this may be done more economically.

(Coleman, 1916: 190–192)

What about drift's interpretation of the vast Permo-Carboniferous glaciation? Coleman attacked it repeatedly, and he even used North America's own Squantum Tillite to do so (§3.12, §6.6).

   Wegener took a different path. He was not a geologist, and never had to unlearn the classical teachings of geological education. He was not tied to any particular school of thought. He himself collected no data of key importance to the mobilism controversy, and he had no personal commitment to any particular line of evidence. He was not a regionalist. Why did North American geologists not follow Wegener? Because they believed that the rocks they knew said, "No to continental drift." But how could they get away with being regionally so restrictive? There are several reasons. Nobody had yet formulated a line of critical evidence that remained intact after extended criticism with all difficulties being answered; nobody had a difficulty-free solution. With influential figures such as Matthew and G. G. Simpson (§3.6–§3.8) devising fixist explanations of biotic disjuncts, Coleman (§3.12, §6.6) raising difficulties with drift's solution to Permo-Carboniferous glaciation, Brooks and Schuchert providing a fixist climatic explanation of that glaciation (§3.12), and with Schuchert and Willis (§3.6, §3.12) constructing a paleogeography that seemed to explain both biotic disjuncts and Permo-Carboniferous glaciation, North American geologists had what they regarded as overwhelmingly good reasons for being fixists, reasons that kept them afloat until the triumph of seafloor spreading in the mid-1960s.

7.4 Antarctica breaks the chains of North American regionalism: the experience of William Long

By 1960 I had taken off in the life of an explorer. New lands and new ideas fascinated me. My training had so far been of the conservative North American variety. But, Antarctica, certainly was not North America. I liked simple explanations. Continental movement, with all of the evidence described by du Toit and others made good sense to me. I became a believer, but not a protagonist.

At the time, 1960–61, I had studied and written considerably about the rocks of the Ohio Range. I recognized them as part of the Gondwana rocks. At the time my position would have been that a large continent most likely broke apart, but it might be possible that some other origin of the similar sequences could have taken place.

(Long, April 21, 2002 letter to author)

Recalling his first encounter with what he then did not recognize as the Permo-Carboniferous glaciation of Antarctica, Long described how he ended up at the foot of what later was named Mt. Glossopteris where he discovered the tillite, and what happened during his first ascent.

Let's start back in IGY [International Geophysical Year] at the Byrd Station over-snow traverse of 1958–59. The party consisted of six scientists driving three Tucker Snow Cats, each cat pulling a 1-ton military sled with supplies and fuel. The triangular route from Byrd Station would cover about 1200 miles of unknown ice cap. Traverse objectives included depth of ice measurements, yearly snow accumulation, geographic description of all features encountered.

As chief glaciologist of the team, my responsibilities, with an assistant glaciologist, were to gather snow accumulation data. Prior to departure from Byrd Station in 1958 several of us participated in an aerial reconnaissance of the route to be taken during the coming Antarctic summer.

The R4D (DC-3) aircraft carried us along the northern side of a vague area marked on Antarctic charts as the "Horlick Mountains" as observed and roughly located and mapped by aircraft during the Byrd expeditions of the 1940s. We discovered that the Horlick Mountains were broken by the interior ice sheet into three ranges and some smaller rock exposure features. These individual mountain features now carry the names Wisconsin Range, Ohio Range, Thiel Mountains and Long Hills. The Ohio Range particularly attracted my attention because of an intriguing display of sedimentary rocks [resting] on an obvious non-conformity. Also, this range appeared reasonably accessible from our traverse route.

The over-snow approach to the Ohio Range seriously tested the strength of the machines and scientists. Rough snow, the hardness of concrete, pounded the Snow Cats until all three machines suffered major structural failure; a broken frame one, a broken spring the second, and a broken 5th wheel on the third. Fortunately our mechanic (Jack Long, my brother) met the challenge and commenced welding and repairing. Navy support planes airlifted needed parts and welding equipment to make the repairs possible. *This major breakdown and delay occurred directly in front of the Ohio Range*. What a marvelous opportunity!

One year earlier, in 1957, I had received my B.Sc. in Geology from Mackay School of Mines at the University of Nevada Reno (after ten years of study and effort including military service). Stratigraphy, sedimentation and glacial geology particularly fascinated me during those undergraduate years . . . Therefore, the concept of describing the obviously excellent exposures of sedimentary strata of the Ohio Range proved an irresistible opportunity.

Four of our party, Charles Bentley, Jack Long, Fred Darling & I, shouldered packs and crossed the two miles of flat ice surface to reach the cliffs at the base of the range and began to ascend the slopes of the mountain and carefully inspect the nature of the rocks. Within the first few hundred feet of rock exposure we found fossil brachiopods in the basal sandstone unit. Higher on the slopes of [later named] Discovery Ridge, we found tracks in shale (formerly mud). Also, we had ascended over about a thousand feet of a rock with fine-grained matrix and scattered large clasts or stones of various lithologies. High on the mountain we found bed after bed of coal and many sandstone beds with fossil wood fragments and abundant leaf fossils Glossopteris.

Collections and note-taking required some time during the ascent and finally the four of us reached the summit of the mountain, [later named] Mt. Glossopteris, approximately 4000 feet above the level of our camp [which was at about 6000 feet at 85° S]. We planted the broken length of a 4 × 4 redwood post I had carried for the purpose on the summit. Then Charlie and Jack said they would carry some specimens and descend with no more interruption to return to the Snow Cats and camp. Fred & I would take the time necessary to make careful notes, elevation readings, and more collections during a slower return to camp. The extended time allowed fog and clouds to gradually close in on the foot of the mountains and our camp and Snow Cats disappeared in the mists. Eighteen hours after leaving the Snow Cats we returned. The time, effort, lack of food and anxiety caused by fatigue and fog that obscured the route for the last two miles across the ice cap took its toll. Fred was coughing up blood and hardly able to walk. I fortunately sensed the direction to our camp and went ahead to summon assistance for Fred. Soon all had returned. We had stumbled on to a "missing link" of Southern Hemisphere stratigraphic geological puzzle.

But I did not yet realize the great significance of the day's observations and collections.

(Long, January 30, 2002 letter to author; my bracketed additions)

Why did Long not realize during his ascent of over about "a thousand feet of a rock with fine-grained matrix and scattered large clasts of stones of various lithologies" that there was a possibility that he had walked over a glacial tillite, over Du Toit's and King's predicted Antarctic equivalent of the South African Dwyka Tillite, the last major piece of the Gondwana jigsaw? He did not realize because his geological education left him ignorant of what to expect, of one of the most distinctive features of Southern Hemisphere geology, the Permo-Carboniferous Gondwana tillites, and their implications for continental drift. Although he was amply prepared to survive in Antarctica, physically able to climb its mountains, and to survey living glaciers, he did not for over a year begin to fully appreciate what lay under his feet.

Long was born in 1930.[3] Raised in California, he became an expert mountaineer, taking part in many expeditions that were sponsored by the Sierra Club. In 1954 he was a member of the California Himalayan Expedition to Makalu. He also knew how to survive in severe climates, for he had served in the Air Force as a survival instructor from 1951 until 1954.

Upon deciding that he wanted to be a geologist, he went to the Mackay School of Mines, University of Nevada, Reno, where he studied under Richard Larson, a student of Kay's. What he learned and did not learn in the early 1950s is of special interest.

Prior to ascending Mt. Glossopteris, I had no knowledge of Gondwana stratigraphy. I had heard of "Continental Drift" but only as a radical idea of little significance. My professor and mentor from the University of Nevada, Dr. Richard Larson, had been a student of Marshall Kay and therefore taught much about geosynclines and their relationship to mountain building and continental accretion. My rather extensive education in microscopic identification of rocks never included a tillite. In retrospect it seems as though tillites and continental drift were unstated, forbidden subjects.

However, I did learn about graywackes and their importance in geosynclinal sequences; and observed a number of them microscopically. In thin-section a tillite remarkably resembles a graywacke; because one looks at the matrix of the tillite not the larger clasts. (A non-observance of the forest due to tree-inspection sort of thing).

(Long, April 21, 2002 letter to author)

Reflecting on his education and that of other North America geologists during the 1940s and 1950s, Long wrote:

Most geology taught in those years seemed to have originated from studies of the geology of the East Coast of North America. Even the strange geological conglomeration of the Pacific Coast seemed to be ignored. And for good reason, the units of the coastal ranges of the western North American continent do not fit into a stable continental condition. The geologists and professors of that era were brilliant, fine geologists, perhaps with limited vision and worldwide experience. They certainly seemed lacking in southern hemispheric knowledge. Their knowledge of North American geology, in fine detail was most impressive.

(Long, April 21, 2002 letter to author)

With his geology degree, work on glaciers, mountaineering and survival skills, Long was a perfect choice to work as glaciologist during the IGY (1957–9) at the Byrd Station. He spent two summers and one winter in Antarctica, from about November 1957 until February 1959. During the summers, he would go on over-snow expeditions; during the winter he remained at Byrd Station, measuring temperature and snow accumulation. During the first summer, he participated in an over-snow expedition exploring the Sentinel Range and Ellsworth Mountains, which are about 550 km from the South Pole. During the next summer, he and three companions began an expedition in November 1958 to the Horlick Mountains. Their vehicles broke down one month later at the foot of Mt. Glossopteris – so named later by James Schopf, the paleobotanist and coal geologist who identified the fossil plants Long and his colleagues collected on the mountain.

7.5 Long returns from Antarctica and becomes a mobilist

Long returned to the United States, took a position at The Ohio State University, and began working on his M.S. in geology.

The return to United States after one and a half years on the Antarctic ice sheet greeted me with some serious financial and life-goal impacts. Two months after reaching U.S. soil, the Arctic Society of North America informed me that my pay had stopped as of the day my foot reached American soil. In the meantime my debts continued and suddenly I found myself in debt. Working a few months as a Fuller Brush salesman succeeded to relieve the debt load for my family and by that time I had decided to accept a position at the newly formed Ohio State University Institute of Polar Studies [hereafter, IPS], as a research associate. Also I enrolled as a graduate student for the coming fall semester of 1959.

(Long, January 30, 2002 letter to author; my bracketed addition)

During his first eighteen months at IPS, he spent considerable time processing the weekly glacial data he had collected over the winter at the Byrd Station, and wrote up an initial report of what he had seen on what was later called Mt. Glossopteris of the later named Ohio Range, itself part of the Horlick Mountains where he initially failed to identify the tillite (January 30, 2002 letter to author). This was entirely understandable in light of his earlier training under Larson at the University of Nevada, and later his thesis reviewers at Ohio State did not object to his interpretation.

Ohio State reviewers seemed happy enough with the "greywacke" interpretation. Very little comment was offered one way or the other. OSU reviews indicated no interest in relating the stratigraphy and geology of the Horlick Mountains to Gondwana; most likely because they tacitly assumed Antarctic geology formed independently of other continents.

(Long, May 30, 2002 letter to author)

Long (January 30, 2002 letter to author) also learned about an upcoming symposium to be held in Buenos Aires in November, 1959 "dedicated to Antarctic geology." He thought it was "an ideal place to present the Mt. Glossopteris findings." His paper was accepted, and he read it at the meeting. The response was very different from that in Columbus, Ohio. He met Dr. Raymond Adie, who, as King later wrote, told Long, "You've got the Dwyka tillite." This is how Long remembered it.

Dr. Ray Adie of the [British] Falkland Island Dependencies Survey played a major role in the organization and running of the 1st Symposium of Antarctic Geology at Buenos Aires, Argentina; and at that symposium I first met Ray Adie. Dr. Adie looked more like a bouncer from a British pub or a retired boxer. He appeared tough as nails, presented his ideas very directly, yet maintained a vocabulary and speaking skills of the well-educated Englishman. [Adie is actually South African, and spoke with his native accent which can be mistaken for certain English ones.]

Upon hearing my presentation and viewing Mt. Glossopteris slides, Dr. Adie discussed with me several times the potential significance of the stratigraphic geology I had described. He seemed quite certain that the unit I referred to as a "graywacke" with scattered large boulders would prove to be the equivalent of the Dwyka Tillite. And he suggested, in fact, the unit *was* tillite; probably the tillite many geologists predicted should exist somewhere in Antarctica.

(Long, January 30, 2002 letter to author; my bracketed additions)

Long was not ready to accept Adie's interpretation. He was pleased that he might have made a significant discovery, and that it gave strong support for continental drift, but knew little about the geology of the Southern Hemisphere.

His [Adie's] ideas pleased me and seemed very logical, yet my training, experience and education seemed to create a requirement for more information. I wanted to believe that the section I had described was part of the great Gondwana continent, but I felt serious reservations. Dr. Adie recommended that I inspect the geology of South Africa, if at all possible. The idea certainly appealed and that "seed" in the back of my mind would become reality.

(Long, January 30, 2002 letter to author; my bracketed addition)

The "seed" became reality several months later. Long took one of Dr. Richard Gold-thwait's courses in glacial geology and glacial stratigraphy during the spring 1960 semester – Ohio State was on a tri-semester schedule. Goldthwait was Director of IPS.

During graduate school, in Dr. Richard Goldthwait's Glacial Geology and Glacial Strati-graphy classes, the strong realization hit me solidly and hard during one of "Doc G's" lectures. I went directly up to him after class and said that I had been walking all over tillites in Antarctica and needed to return and describe the rocks of the "Buckeye Range" (Buckeye Range seemed a good name to me but the US Board On Geographic Names deemed Ohio Range more fitting than a nick name. They were correct in my opinion.).

(Long, January 30, 2002 letter to author)[4]

Goldthwait expressed little excitement when Long told him he had walked "all over tillites in Antarctica." Nor, as far as Long remembers, did he ever express an opinion about continental drift.

Goldthwait was a glacial stratigrapher and glacial geologist primarily. He did not express strong opinions about other fields. Yet, his work all over the world had certainly provided him much more appreciation of Southern Hemisphere ideas and theories. I do not remember him making a statement, pro or con, related to continental drift. He was aware of the Gondwana stratigraphy but likely held to more conservative explanations such as "coincidental similarities." Doc G certainly believed in tillites as he knew that glaciers and glaciations operated in past geologic time.

(Long, January 30, 2002 letter to author)

Notwithstanding, Goldthwait encouraged Long to apply to the National Science Foundation (hereafter, NSF) for funding to return to Discovery Ridge and Mt. Glossopteris, and he submitted a proposal in late 1960 or early 1961, asking to make a brief stop in South Africa. Long recalled (April 21, 2002 letter to author) that the "South African tour rationale was to inspect, and learn from field observation, the nature of South African Gondwana geology in order to compare these sequences with those of Antarctica." He was funded.

Long actually became a mobilist about the time he applied for funding from NSF; he became a mobilist before returning to Antarctica and before seeing the Dwyka Tillite in South Africa. The reason being, he read Wegener, King, and du Toit. Schopf, a paleobotanist who knew both Edna Plumstead and *Glossopteris*, gave Long a copy of *Our Wandering Continents*. Schopf had much to do with Long's re-education. He also advised Long to write to Plumstead.

Other persons [besides Goldthwait] and personalities at Ohio State influenced my interpretations. Dr. James Schopf (his son is Dr. William Schopf of UCLA) was a USGS coal geologist and paleobotanist with a laboratory at Ohio State. Dr. Schopf worked tirelessly and relentlessly with all sorts of minutia related to coal and coal-bearing strata. He made his lab available to graduate students for various sorts of work, either for him or for the student's own research. Few students associated with Dr. Schopf and his coal laboratory, probably because Dr. Schopf's concentration on his work caused him to be a bit on the cantankerous side when it came to interpersonal relationships. For some reason these less than comfortable aspects of his personality did not bother me. In fact, I admired his dedication and directness. He reminded me of my father, with whom I worked well and admired.

(Long, April 21, 2002 letter to author; my bracketed addition)

I read du Toit's *Our Wandering Continents* in 1961. The book was a gift from . . . Dr. James Schopf. Dr. Schopf had been in contact with Edna Plumstead and it was he who suggested that I contact her for information about *Glossopteris* and related fossils. This information played a part in my finally touring in South Africa. My first stop was to visit Edna Plumstead.

(Long, April 21, 2002 letter to author)

Now convinced that he had discovered the old tillite, Long (1962a) gave a paper "Permo-Carboniferous Glaciation in Antarctica" at the GSA December 1961 Denver meeting. His talk did not cause much of a stir. Nobody in the audience remarked about the glaciation being a key prediction of continental drift. Long did discuss his finding with Dott. Robert Dott Jr., who became one of North America's leading sedimentologists, had recently argued that turbidity currents rather than glaciers had caused the Squantum Formation (III, §1.2).

I remember no serious argument, but I have vague recollection of some strong suggestions that the unit I called a "tillite" could be something else. Dott attended that presentation and I think that that is the time Dott discussed the turbidite possibility for the tillite.

(Long, May 30, 2002 to author)

Long (1962b) submitted a paper to *Science* in October 1961, which described the sequence of the Ohio Range (his Buckeye Range). Immediately above the basement was sandstone-shale layer (the Horlick Formation) rock, which, he (1962b: 320) argued, was Devonian based on Schopf's identification of "a favorably preserved early Devonian spore assemblage." Next came the "Buckeye" Tillite.

Overlying the Horlick Formation is a unit about 800 feet thick composed of bluish-gray silty, clayey matrix which includes pebbles, cobbles, and boulders of mixed lithology in an unsorted arrangement. The unit rests on a striated and grooved pavement that truncates the Horlick Formation and parts of the basement rock. The lithology of included pebbles shows about 70 percent sedimentary rocks, 23 percent igneous rocks, and 7 percent metamorphic rocks, all of uncertain source. More than 90 percent of the pebbles counted are subrounded to angular with about 40 percent subangular. About 10 percent are striated; an example of a striated cobble is shown as the cover photograph of this issue. Evidently these beds are of glacial origin and correspond to the well-documented Permocarboniferous tillites in Gondwana deposits of the Southern Hemisphere.

(Long, 1962b: 320)

He was able to determine that glacial movement had been from west to east, and the presence of "sandstone beds within the tillite" indicated intervals of glacial advance and retreat. Citing King (1958c), Long (1962b: 320) argued that the Buckeye Tillite was equivalent to the Permo-Carboniferous Gondwana tillites of South America, Africa, India, and Australia. He ruled out Devonian glaciation because "relations with units above the tillite are similar to those in South Africa." The unit immediately above the tillite was of carbonaceous shale, which was overlain by Permian strata containing coal and *Glossopteris* leaves – just what should be expected if the Buckeye and Dwyka Tillites were equivalent. Long also discussed the top unit. It was a diabase sill, which he thought was probably Jurassic, based on its similarity with other sills in Antarctica and other fragments of Gondwana.[5]

Although Long had come to believe in continental drift before submitting his paper, he omitted any reference to it. Less than a year after submitting his paper, however, he and his co-author, George A. Doumani, a fellow graduate student at IPS, wrote a paper for *Scientific American* about their work in Antarctica in which they argued strongly in favor of continental drift.[6] The paper appeared in September. Doumani and Long (1962) wrote like longtime drifters. They presented drift's solutions to the origin and distribution of Permo-Carboniferous glaciation and biotic disjuncts. They showed a reconstruction of Gondwana which included du Toit's "Samfrau" geosyncline, and raised difficulties with both landbridge and permanency theories. And they also appealed to the new support offered by paleomagnetism.

To the independently compelling fossil evidence the latest paleomagnetic studies lend great force. Since primary considerations require that the Magnetic Pole be located somewhere in the vicinity of the pole of the earth's rotation, the odd directions pointed by the fossil magnetism of the rocks around the world are more readily explained as evidence of continental wandering.

(Doumani and Long, 1962: 184)

Their concluding statement (1962: 184), "The seventh continent holds a master key to the earth's ancient history," sounds like du Toit writing about Africa.

Long finally made it to South Africa in summer 1962. With NSF funding, he combined his further investigation of the Ohio Range with a tour of South Africa's Gondwana geology. None other than Edna Plumstead, whom he later described as "a most dedicated paleontologist and gracious lady," was his guide. She showed him the Dwyka Tillite, and he was amazed at its similarity to what he had seen on his first ascent of Mt. Glossopteris (Long, January 30, 2002 letter to author).

Long's journey from the descriptive "graywacke" to the much more informative "tillite," from the University of Nevada where he was taught the fixity of continents and oceans by a former student of Marshall Kay's and learned little or nothing about Southern Hemispheric geology, through his work in Antarctica, his self-education at the IPS reading *Our Wandering Continents* and becoming a mobilist, his amiable meeting with Adie and his seeing the Karroo System in South Africa and what Lester King described as "the reality of drift" under the guidance of the "gracious" Edna Plumstead, shows the baleful legacy of permanentism in North America and the lack of knowledge of Southern Hemispheric geology on which it was sustained. Long's struggle is a prime example of the power that regional geology had at the time in shaping workers' allegiance to fixism. He was indeed lucky to have had Schopf as an ally at The Ohio State University.

Long's story had an instructive codicil. He submitted a paper to the first international symposium on Antarctic geology sponsored by the Scientific Committee on Antarctic Research, an interdisciplinary body of the International Council for Science, which was held in Cape Town in September 1963. Even though his paper was published in the proceedings (Long, 1964), he was prevented from attending. Long recounted what had happened.

I was not present at the Symposium on Antarctic Geology, 1963. Art Mirsky attended that symposium. You will note that my paper was presented at the symposium, I was not there. A. Mirsky read my paper at the meeting. He was to represent Ohio State's latest geological work in Antarctica and present the findings of my most recent expedition to the Nilsen Mountains; an expedition that brought handsome funding to OSU (Ohio State University) and continued to expand the knowledge and extent of the rock units of the Ohio Range. Mirsky was the deputy director (Dr. Goldthwait was director) of the Institute of Polar Studies and one of my supervisors. My application to attend the symposium had been denied by Mirsky and Goldthwait. (I certainly was too passive. I should have attended in spite of the denial.)

(Long, April 21, 2002 letter to author)[7]

Art Mirsky, Deputy Director of IPS and one of Long's supervisors, was not, according to Long, pleased with his switch to mobilism. Long argued with him, and with other graduate students. Mirsky did go to the meeting, and read his own paper (1964a) at the symposium, arguing in favor of retaining the old stratigraphic term "Beacon" to refer to Antarctica's *Glossopteris* bearing sedimentary unit – Long named the latter unit the "Mount Glossopteris Formation" which made clear its similarity to equivalent formations in other Gondwana fragments. During the closing session, King (1964) discussed the importance of Antarctica in establishing continental drift, and praised Long for his discovery of the Buckeye Tillite. Mirsky then addressed the symposiasts.

I am from the Northern Hemisphere, and during this week I have been told from time to time that individuals from the Northern Hemisphere do not understand the Southern Hemisphere problem, because if we did we would agree with the protagonists of continental drift. I am pleased to note that some people from the Southern Hemisphere do not agree either.

Regarding stratigraphic evidence, it does not seem fair to analyse Antarctic stratigraphy, as compared with that of South Africa, on the basis that both were once parts of a single landmass. I believe that Antarctic stratigraphy should be considered first as a domestic problem, and when more is known international relations can be brought in.

(Mirsky, 1964b: 734)

His remarks epitomize the regionalist outlook: domestic not international, a sure way to miss evidence of mobilism, a sure way to insulate oneself from it.

Mirsky, when I asked him almost forty years later, was unable to identify any anti-mobilist symposiasts from the Southern Hemisphere. But he did remember, however, the strongly pro-drift stance of most at the meeting, and their "almost-arrogant" attitude. He himself gave his support to the standard fixist rhetoric of the time, notably hiding behind their standard mobilism difficulty.

I do recall, however, that I (and I believe those other geologists) was not so much against the idea of continental drift as I was skeptical about the mechanism that would enable continents to *actively* move and of what I saw at the time as a kind of circular reasoning. That is, the kinds of evidence used to support continental drift (stratigraphic, paleontologic, and structural) could be explained in other ways, and to invoke one of the ways, continental drift, to explain the distribution of the evidence was not convincing . . . Moreover, I recall that I objected to the almost-arrogant "we Southern Hemisphere geologists know it all" attitude, so I got the urge to sound more hardline anti-drift than my actual "maybe it is so, but I am not yet convinced" tone. In fact, when I returned to my university office in the Northern Hemisphere, I got some chuckles out of relating the events.

(Mirsky, May 7, 2002 email to author)

Mirsky did not make a "big deal" about what he said at the meeting. As he put it in a second email:

Finally, concerning the incident(s) of the "chuckles," this was another low-key event. As you can imagine, whenever anyone returns to the office after attending a major conference such as the week-long Symposium on Antarctic Geology, the returnee is routinely asked by one or more colleagues "how was it?" And "how did your talk go?" And "what did you learn?" and so on. During my response to such questions, the "chuckles" came when . . . I related my observation that the Southern-Hemisphere geologists appeared to me to be somewhat arrogant regarding their support for "drift" with what they considered obvious evidence, and so I needled them with my apparent hard-line opposition rather than my actually more moderate, questioning view. Of course, it never occurred to me that my "game" might come back in another context decades later.

(Mirsky, May 20, 2002 email to author)

Upon seeing Mirsky's retrospective remarks about the Cape Town meeting and ensuing discussions of the meeting with his IPS colleagues, Long recalled his supervisor's resistance to mobilism. Even after forty years the resentment remained.

Your quote from Mirsky's discussion finally has enlightened me regarding resistance to my ideas of the relationship of the geology of the Transantarctic Mountain with the rest of

Gondwana, and the separation of the smaller continents from the super-continent. Mirsky's tactics and action designed to muffle and subdue my ideas seems downright nasty at this point. Most of the students in stratigraphic geology at The Ohio State University argued with me regarding my interpretations of the stratigraphy of the Nilsen Mountains and its relationship to the other described sections of the Transantarctic Mountains. Explanations involving continental drift and widespread tillites seemed too simple and easy and did not fit the geosynclinal models we all had learned about. Mirsky performed other hostile, harmful actions and seemed to influence most decisions of the director. Rather than battle, I found an interesting position with Tenneco Oil Company and left the IPS.

(Long, April 21, 2002 letter to author)

Mirsky's recollection regarding the acceptance of continental drift at IPS differed from Long's. According to Mirsky, most researchers there favored continental drift. When forty years later I asked Mirsky about his disputing mobilism with Long, he took issue with my use of the word "disputes."

About my getting into disputes with anyone at the IPS, first let me say that "disputes" is the wrong word, that "discussions" would be more accurate. Also, continental drift was not a front-line topic at the IPS; the individual research projects were much more specific, much more locally contained, and drift for most of the reports would not even come into the discussion. As I recall most of the researchers at IPS, if they were asked, were supporters of continental drift unlike the general US geological community at that time. I was one of the few who, as I said in my previous e-mail, wanted to believe but I was not yet convinced by what I saw as circular reasoning . . . My discussions with others at IPS about the "truth" of continental drift occurred from time to time as a response, perhaps when something came up (an article, a comment in a visitors talk, and so on). So my discussions generally were mild, unassuming, non-memorable, which could have been with anyone at IPS. As you noted, Bill Long, who led two expeditions to the Horlick Mountains in the Transantarctic Mountains and was responsible for the stratigraphic studies, supported continental drift. As a stratigrapher myself and titular supervisor of Bill's research, I am sure that I must have had a discussion or two with him about the drift controversy, but I do not recall any specific time when such a discussion might have taken place. Again, the drift problem was not a primary part of our research programs and, therefore, discussions were not all that frequent.

(Mirsky, May 20, 2002 email to author)

What was insignificant to a Ph.D. supervisor might be quite significant to his Ph.D. student, who was making an important scientific contribution by his discovery of a major missing piece in the continental drift jigsaw puzzle. Regardless, Long did not think that he received much support for his ideas at IPS, and this led, at least in part, to his leaving the institute. But, if Mirsky's colleagues chuckled when he related what had been said at the Cape Town meeting that was held in September, 1963, it would surprise me if most at IPS "were supporters of continental drift unlike the general US geological community at the time."

7.6 Antarctica again breaks the chains of North American regionalism: the experience of Warren Hamilton

I became a closet drifter in graduate school when I read "Our Wandering Continents," and an active "drifter" when my first Antarctic season (1958, in the Dry Valleys region of the Trans-antarctic Mountains) showed me that Du Toit's reconstruction correctly predicted the geology.

(Hamilton, July 23, 2002 email to author)

Warren Hamilton was a closet drifter before he left for Antarctica in 1958. Reading du Toit's *Our Wandering Continents* in 1949 while a graduate student at the University of California, Los Angeles (UCLA), convinced him that mobilism was substantially correct. But he published nothing on mobilism and kept his views to himself until returning from Antarctica where he had discovered similarities between its geology and other fragments of Gondwana that supported du Toit's assembly. Hamilton, unlike Long, did not become a mobilist because of his work in Antarctica; however, like Long, he became an *active* mobilist because of it. After his (1960) initial work in Antarctica, Hamilton (1963a, 1963b, 1965, 1967) further developed his mobilist interpretation of its geology and place within Gondwana. But Hamilton (1961) also began to apply mobilism to North America, initially offering a mobilist explanation of the origin of the Gulf of California, and later applying it elsewhere in western North America. In the early 1960s he was the most active North American mobilist who developed his ideas independently of contemporaneous advances in paleomagnetism and oceanography.

I shall argue that regionalism partially explains Hamilton's journey from passive to active acceptance of mobilism. Antarctica had what North America lacked; its geology in its situation and time was very different and called for mobilism. Hamilton's teachers were fixists. Du Toit's *Our Wandering Continents* convinced him of mobilism's fundamental soundness. He himself failed at first to find new evidence of it in North America; only work in Antarctica gave him the new evidence he wanted. Then returning to North America, he reinterpreted important aspects of its geology in terms of mobilism.

Warren Hamilton was born in 1925. Upon graduation from high school in 1943 Hamilton enrolled at UCLA, where he enlisted in the Naval Reserve Officers Training Corps (NROTC), which in wartime meant active duty. Hamilton spent seven semesters at UCLA, graduating in three years. The extensive naval requirements precluded a regular major, and Hamilton graduated in June 1945 with a US Navy commission, and a B.A. degree with minors in Navy, geology, and English. He recalled "no awareness of continental drift" as an undergraduate (Hamilton, July 24, 2002 email to author).

He wanted to go on to graduate school, but he was a commissioned Naval officer, and spent the next year "as a deck and gunnery officer on a carrier in the Caribbean and western Atlantic" before being released to inactive duty when he returned to

UCLA in 1946 as a graduate student in geology, where he took mostly the under-graduate courses that he had missed. During summer 1947, he did fieldwork in the Front Ranges of the Rocky Mountains in northern British Columbia with a party from Kansas University. He spent the next year at the University of Southern California where he obtained his M.Sc. in January 1949 for work on the "Geology of the Dessa Dawn Mountains, British Columbia," drawing on his 1947 fieldwork. He returned to UCLA and received his Ph.D. in 1951 on "Granitic rocks of the Huntington Lake area, Fresno County, California."

Among Hamilton's teachers at UCLA, he was most influenced by James Gilluly "for instilling the attitude that nothing must be believed just because it is written" (Hamilton, August 25, 2002 email to author).

Gilluly was a terrible, wooden lecturer, but an excellent teacher of seminars. Given, say, 8 students, he would hand out, as reading assignments, apparently random papers to 2 sets of 4 of us; and they would turn out to be loaded so that half of us came prepared next time to defend some viewpoint, the other half to argue the contrary. He instilled the invaluable attitude that one should be skeptical of everything he reads (though he was thin skinned, and did not want this applied to his own work) and should question assumptions.

(Hamilton, July 24, 2002 email to author)

His teachers were fixists. Even Gilluly was not prepared to question this dogma.

I don't recall drift as ever a topic. Jim [Gilluly] was dogmatically anti-drift in those days. . . . I remember very little discussion of continental drift in graduate school. "Permanence of continents and oceans" was dogma that did not need challenging – a topic ignored rather than refuted.

(Hamilton, July 24, 2002 email to author; my bracketed addition)

Others also influenced him.

David Griggs, for beating in awareness (upon which I acted inadequately for years there-after) that physics provides vast constraints for geology. John Crowell (who was at UCLA only while I was finishing my thesis: I took no coursework with him) who was documenting major motions on big strike-slip faults in southern California (I don't recall his being a drifter otherwise, though certainly he was such by the early 1960s). Cordell Durrell, for petrography (an invaluable tool; but not for concepts).

(Hamilton, August 25, 2002 email to author)

Griggs did not talk about continental drift (Hamilton, July 24, 2002 email to author). Hamilton was aware of Griggs' 1939 paper in which he proposed convection as the cause of orogenesis (§5.6), but does not recall him discussing it in class.

Hamilton found his introduction to mobilism elsewhere. He read du Toit. He does not remember why he read it, but he was impressed.

I read Du Toit's geologically very sound "Our wandering continents" circa 1949 and was much impressed by it, though I don't recall how or why I got into it. Several of the graduate students were sympathetic, but I have no recollection of discussions; others were indifferent, and the faculty was hostile.

(Hamilton, July 24, 2002 email to author)

He did not remember in any detail why he was so impressed with du Toit, but thought that it had to do with geological similarities among Gondwana's fragments, the fit of opposing continental margins, and Permo-Carboniferous glaciation.

Much of what you ask regarding my early inspiration from Du Toit is hopelessly lost in the foggy past. But I was certainly much impressed by the matching of orogens where truncated at high angles and separated by oceans across which they can be matched and the continental margins fitted together, and by the late Paleozoic Gondwana glacials and paleoclimatic indicators.

(Hamilton, August 6, 2002 email to author)

Hamilton also read Wegener's *Origin*, but in German. He preferred du Toit.

I also read Wegener in graduate school. He had priority, but he was sloppy and generalized, not meticulous and detailed like du Toit. Further – if my recollection is correct – I then had available only German editions, and my German was, and is, crummy.

(Hamilton, August 30, 2002 email to author)

He did not read Argand or Holmes, or any other major mobilists identified by du Toit in *Our Wandering Continents*. Hamilton knew that Holmes was a mobilist, but did not know about his convection hypothesis. He did not know about Holmes' *Principles of Physical Geology* or at least did not see it while in graduate school (Hamilton, July 24, 2002 email to author).

Hamilton's teachers discussed mountain building. Although he did not learn about Holmes' mantle convection hypothesis, he was introduced to Vening Meinesz's tectogene or downbuckling hypothesis (§8.13) and Marshall Kay's geosynclines (§7.3).

I recall, and that dimly, only vague discussions of causes of orogeny. Vening Meinesz' tectogene (which only much later did I recognize as a figment of his absurdly bad assumptions). Mostly collapsing geosynclines, Kay's and others'. Nothing coherent, nothing that provided real explanations.

(Hamilton, July 24, 2002 email to author)

Hamilton left UCLA for the University of Oklahoma, where he remained one year as an assistant professor. In 1952 he joined the USGS where he remained for forty-four years as a research scientist, from time to time accepting a visiting professorship or lectureship at universities such as Scripps Institution of Oceanography (1968), California Institute of Technology (1973), University of California, San Diego (1979 and 1990), Yale (1980), University of Amsterdam (1981), Louisiana State University (1985), and McMaster University (1990).

By the time he left on his critical, self-enlightening journey to Antarctica in 1958, he had published over fifteen papers and ten abstracts. He had written on

Precambrian rocks of the Wichita and Arbuckle Mountains in Oklahoma, metamorphic rocks of the Blue Ridge front in North Carolina, the stratigraphy of the Great Smoky Mountains in North Carolina and Tennessee, and on metamorphism and thrust faulting in the Riggins quadrangle, Idaho. But he had written nothing on continental drift, and if he read anything else on the subject, it left no impression in his papers. Nevertheless, he was seeking evidence of drift.

I don't recall what else I had read before going to Antarctica in 1958, but I was by then actively looking for evidence for drift (although I had yet published nothing on the subject, and had taken no public stands).

(Hamilton, August 24, 2002 email to author)

But he found none, or at least none he recognized at the time as strong evidence of drift.

I had no data of my own that I then recognized (as I did later for the same early data) to be strong evidence for drift, so I saw no occasion to write on the subject. I would have been discussing drift with friends in bull sessions, but have no specific recollections thereof.

(Hamilton, August 25, 2002 email to author)

My early North American work – Canadian Rockies, Sierra Nevada, Oklahoma, Great Smoky Mountains, Central Idaho – was all with topics that I did not then recognize as bearing on drift (I do now, of course). So I just didn't have occasion to go into it.

(Hamilton, August 30, 2002 email to author)

Hamilton's failure to find strong evidence for continental drift was not because he was not looking for it, but because the geology of continental North America did not at the time seem to require drift. Recall Lester King's words, "It's largely a matter of *where* a man works whether he sees the *fact* of drift." By his own account Hamilton believed that drift was probably true, but this was based on what he had read in du Toit, not what he had seen in North America. Had he been sent by the USGS to study late Paleozoic rocks not in the United States but in Brazil and Argentina like Caster, he probably would have quickly recognized strong evidence for drift.

Although Hamilton up to that time had not found strong support for mobilism, he was taught about the relationship between stratigraphy and tectonics.

My first USGS job was in the Great Smoky Mountains with Phil King. Phil was no drifter, but he was probably the most important early geologist anywhere in recognizing that stratigraphic framework controls structural style. I was ever after looking for such relationships, which served me well when plate tectonics came along and, suddenly, we had potential explanations for frameworks. He had no mobilistic concepts in mind when he compiled his Tectonic Map of N America, but the geologic units he discriminated there are precisely those needed for plate-tectonic analysis.

(Hamilton, July 24, 2002 email to author)

These lessons helped him in Antarctica, where his ability to infer orogenic structure from stratigraphic framework helped him to recognize the Early Paleozoic orogenic belt that extends across Antarctica into South Australia. Du Toit's Samfrau was not incorrect, but was bordered by a prominent Early Paleozoic orogenic belt.

7.7 Hamilton finds new evidence of continental drift in Antarctica

It was not in search of evidence of mobilism that Hamilton volunteered to work in Antarctica during IGY.

I went to Antarctica in 1958 on short notice. The USGS had been asked by NSF to provide 2 people for fieldwork for the [IGY] program-to-be. Someone else had long been signed up and then belatedly backed out. My then supervisor asked me if I would like to go, and I immediately jumped at the chance without further consideration. Drift was not on my mind when I jumped: it was a chance for adventure and travel. I invited Phil Hayes to go along; an excellent unflappable companion. But I was already a philosophical drifter, aware of the position "my" area held in Du Toit's reconstruction, so by the time the travel came I was drift-oriented. Antarctica was an opportunity, not an epiphany.

(Hamilton, July 24, 2002 email to author; my bracketed addition)

But once he decided to go, he realized that he had an excellent chance to find strong support of mobilism. He recalled that he had little time to read on Antarctica and drift.

The Antarctic trip came up on very short notice: an IGY attack on the continent was escalated and USGS was only belatedly asked to participate. I was invited to go only about a month before actually leaving, and more than a week of that went for related travel to Washington. I looked up prior reports (Old Scott expedition, etc.) on the region and probably glanced at du Toit to get his Samfrau geosyncline in my memory, but I had no time for general reading . . . before going.

(Hamilton, 30 August 2002 email to author)

He later summarized what he found, and its relation to du Toit's reconstruction of Gondwana.

Re Antarctica, Du Toit, and continental drift: I saw the Australia-Africa-South Africa continuity first in the basement (pre-Devonian) rocks of the Transantarctic Mountains (which I named; a most pedestrian label for one of Earth's major mountain systems), which obviously were not part of an Archean craton (as previously assumed) and instead exposed a Cambrian orogen. The dating was already partly available for those who looked, and more came in the next few years. Also, in the broad progression across Antarctica from Indian Ocean to Scotia Sea – Pre-Cambrian shield/Cambrian orogen/later Paleozoic orogen/Mesozoic orogen. My early interpretations of course raveled away in detail as information multiplied, but the broad progression remains valid. Beacon and diabase had long been known; no surprise there.

(Hamilton, 24 July 2002 email to author)[8]

Du Toit (see Figure 6.3) had traced the Samfrau Geosyncline across Gondwana by correlating Silurian to mid-Mesozoic strata of South America and South Africa with those of eastern Australia (the Tasman Geosyncline), but the geology of Antarctica was little known at the time. He believed that the Samfrau "was already in existence in the Ordovician" (du Toit, 1937: 63). Hamilton (1960: B379) recognized older assemblages in Antarctica that extended back into the Cambrian and "presumably, late Precambrian" that he correlated with the Adelaide Geosyncline of South Australia which lay to the west of the Tasman Geosyncline of New South Wales.

Hamilton wrote short reports in 1959 (with Hayes) and in 1960, and gave two important presentations, the first (1963a) to the Society of Economic Paleontologists and Mineralogists in 1960 and the second (1963b) to the AAPG in April 1961. These will be referred to as the SEPM and AAPG papers, and are now discussed along with the short reports. Hamilton began SEPM by remarking:

Much of the evidence for continental drift has come from the Southern Hemisphere, and particularly from Du Toit's (1937) analysis of its late Paleozoic stratigraphy and paleogeography. The present paper is concerned especially with the older Paleozoic history of Antarctica . . . and comparable features of Australia and South Africa.

(Hamilton, 1963a: 88)

This new evidence grew out of Hamilton's surprising discovery that the mountain chain which extends across Antarctica from the Ross Sea to the Weddell Sea was not, as formerly believed, a horst, an uplifted Precambrian block between two parallel faults, but was instead, as he wrote in his short report, a belt of crystalline rocks "metamorphosed and intruded by batholiths during Cambrian time" (Hamilton, 1960: B380). He first referred to the chain as the "trans-continental mountains" (Hamilton and Hayes, 1959: 575). A year later (1960: B379) he changed this to "the trans-Antarctic mountains" and still later (1964a: 676) finally settled on the "Trans-Antarctic Mountains." Placing the newly identified Early Paleozoic mountain range within a new framework of Antarctic tectonics, he claimed in his SEPM paper that

long-held assumptions regarding the tectonic pattern of Antarctica are at best oversimplified. Conventional explanations suggest Antarctica to be a great shield of old Precambrian rocks, flanked on the Pacific side by mountains of the circum-Pacific system of Mesozoic and Cenozoic orogeny: but although shield and circum-Pacific elements are indeed present, it now appears certain that belts of major Paleozoic orogeny lie between them.

(Hamilton, 1963a: 74–75)

Hamilton found that those sections of the Transantarctic Mountains that he had surveyed are partly composed of a distinctive granite. In his short report, his first publication written from a mobilist standpoint, he wrote:

The dominant rocks of this part of the chain belong to a composite batholith of plutons of various types of granitic rocks, the characteristic type being coarse-grained pink quartz monzonite ranging to granite . . . Many granodiorites, quartz monzonites, and granites known from the Antarctic batholith are characterized by lightly colored potassic feldspar which is in well-shaped crystals but is not generally phenocrystic. Plagioclase is only obscurely zoned. As most other batholiths contain potassic feldspar that is more commonly white than colored and more commonly anhedral than not, and as plagioclase in most other batholiths is commonly zoned, these granitic rocks are distinctive.

(Hamilton, 1960: B379)

He then noted that this particular type of "pink" granite is also found in Australia's Adelaide Geosyncline. Strata of the Adelaide Geosyncline, which also had formed at the edge of a Precambrian shield, were deformed in the Early Paleozoic and intruded by similar granites.

East Antarctica can thus be interpreted to be largely formed of a Precambrian shield, along whose Ross Sea-Weddell Sea side a geosyncline was filled by early Cambrian and, presumably, late Precambrian sediments. The contents of this geosyncline were metamorphosed and intruded by batholiths during Cambrian time. This inferred pattern is strikingly similar to that of Australia, where the Precambrian shield of western Australia gives way in South Australia to the Adelaide geosyncline of Lower Cambrian and upper Precambrian sediments, and that to a Cambrian batholith characterized by granitic rocks (for example, the Murray Bridge granite) which are remarkably similar to those of the trans-Antarctic mountains.

(Hamilton, 1960: B379–B380)

He strengthened his argument by noting a paleontological similarity between the Transantarctic Mountains and Adelaide geosyncline.

Cobbles and erratics containing Early Cambrian pleosponges, both metamorphosed and nonmetamorphosed, have been found in four places in or near the trans-Antarctic mountains, in positions consistent with derivation from ice-buried geosynclinal materials along the inland side of the mountains. Similar fossils characterize the Adelaide geosyncline of South Australia.

(Hamilton, 1960: B379)

Hamilton then referred to du Toit's 1937 reconstruction of Gondwana (Figure 6.4), noting these new similarities were unknown to him.

It may be significant that the position of the Cambrian (?) orogen of Antarctica accords with Du Toit's (1937) theoretical reassembly of the Southern-Hemisphere continents, and forms a bridge between the Cambrian batholiths of Australia and South Africa; Du Toit was not aware of the correlations between these granites, nor of their petrologic kinship. Many other features of Antarctic geology also fit well with Du Toit's reconstruction.

(Hamilton, 1960: B380)

When Hamilton left Antarctica after his first season there, he was not certain about the petrologic kinship between his and the Australian granites and wanted to find out.

Knowing that granites have provinciality and that the Transantarctic Mountain granites I had seen were distinctive, I radioed USGS Washington from Antarctica for, and received authorization to detour to Australia on the way home to sample granites. I flew to Sydney, and rented a car for about a week. I spent a day or so in Canberra, making contact with a few BMR [Bureau of Mineral Resources] geologists (about whom I previously knew nothing), and drove around, mostly in SE New South Wales, sampling granites (none of which looked like my Antarctic ones) [They were in fact all of Upper Paleozoic age from the Tasman Geosyncline].

(Hamilton, September 10, 2002 email to author; my bracketed additions)[9]

Unable to collect the older granites of the Adelaide Geosyncline further west, the ones he really needed, he arranged for A. H. Whittle of the Department of Mines of South Australia to send some. To complete the comparison he also wanted granites from South Africa and from other places in the Transantarctic Mountains. Subsequent chemical analysis showed that granites from these three areas resembled one another, and differed from the younger Late Paleozoic granites he had collected in New South Wales.

I junketed to the several small collections in the US of (then undated) interior West Antarctica rocks and mooched small chunks of granites for chemical analysis (and published these data, purely descriptive, in a short paper), and mooched by mail samples of Cambrian granites from far-southern S Africa and from the Adelaide "geosyncline" of S Australia. Whittle sent me older granites from the Adelaide region (which I saw on my own a few years later). They are very different. The S African and Adelaide rocks were like my Trans Mountain ones (and unlike those of SE Australia). My analytical work was limited to their Antarctic counterparts. I got chemical analyses of some of the Adelaide granites and for similar granites from southernmost Africa, both of which resemble the Transantarctic granites with which I worked, but don't think I ever published anything on them.

(Hamilton, September 10, 2002 email to author)

Once Hamilton learned of the resemblance among the granites of these three areas, he knew he had strong support for continental drift. "So I was immediately embarked on a course as a 'drifter' from my first Antarctic season" (Hamilton, July 24, 2002 email to author).

Unable to find strong support for mobilism in North America, Hamilton had found it in the granites of the Transantarctic Mountains, the Cape Fold Belt and the Adelaide Geosyncline of South Australia. He also had had to embark on a comparative study of more than one region, which is what du Toit (§6.6), King (§6.10), Martin (§6.11), and Plumstead (§6.3) had done, and as paleomagnetists in the mid-1950s had begun to do (II, §3.10, §3.13, §5.2–§5.9).

In the SEPM paper, Hamilton reviewed the recent radiometric dating of Antarctica's ancient shield which further established its Precambrian age, and the widespread occurrence of charnockites.

... typical of the Precambrian shield regions of Australia, Africa, and India. Charnockites of these and other continents are virtually restricted to old Precambrian terranes: whatever the details of their origins, they are exposed only in regions of complex geologic history and deep erosion.

(Hamilton, 1963a: 78)

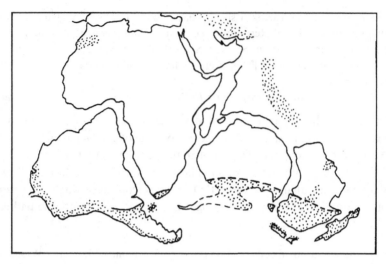

Figure 7.2 Hamilton's Figure 8 (1963a: 91) showing outlines of the present southern landmasses as part of Gondwana-Land as postulated by du Toit (1937, Figure 7; my Figure 6.4). Stippled areas are Paleozoic geosynclines.

He discussed again the peculiar granites of the Transantarctic Mountains, and their presence in Australia and South Africa. He expanded his comments about Lower Cambrian pleosponges there and in the Adelaide Geosyncline. He reported the identification by various experts (Caster among them) of a fossil collected by Long and noted their occurrence elsewhere in Gondwana.

Long (1959) found about 1500 feet of dark shales and sandstones, and other sedimentary rocks, between crystalline basement rocks and typical Beacon sandstone in the Horlick Mountains. Brachiopods which he collected near the base of sedimentary sections have been examined by G. A. Cooper, A. J. Boucot, P. E. Could and T. J. Dutro (written communications), and K. E. Caster (oral communication) who found them to be Lower Devonian terebratuloids similar to those of New Zealand, of the Falkland Islands, and of the Table Mountain sandstone of South Africa.

(Hamilton, 1963a: 81)

Hamilton argued that the similarities between the Late Precambrian through mid-Paleozoic geology of Antarctica, Australia, South Africa, and South America could be best explained by drift. Referring to his first introduction to mobilism, he argued:

Figure 8 [reproduced here as Figure 7.2] illustrates the reconstruction before drift of the southern continents suggested by Du Toit (1937). The new interpretation of Antarctic tectonics summarized in this paper is consistent with Du Toit's reconstruction, as the probable Paleozoic orogens of Antarctica lie where they are required by the drift theory . . . Drift is not required by the new interpretation, but it certainly provides new coincidental relationships which are much more easily explained with drift than without. Australia, South Africa, South America, and probably Antarctica all have middle and late Paleozoic geosynclinal rocks in the positions required by the drift theory; Australia, Antarctica, South Africa, and perhaps South America have terranes of older geosynclinal rocks that were metamorphosed and intruded by granites during early Paleozoic time, and the distribution of these rocks also can be explained by the drift theory.

(Hamilton, 1963a: 90–91; my bracketed addition)

He added, however, that one important element of Gondwana strata had not been found in Antarctica.

At the base of the Karroo is the Dwyka tillite of Late Carboniferous age, deposited from ice which radiated from centers farther north in southern Africa. Above the tillite are Permian and Triassic clastic rocks which bear the Glossopteris flora. These rocks are intruded by huge sills of Lower Jurassic (?) diabase; basalts of the same age cap parts of the Karroo system. In Angola, the eastern Congo, the Falkland Islands, South America, India and Australia, sediments of glacial origin similarly lie beneath strata with the *Glossopteris* assemblage, but no such glacial deposits have been found in Antarctica.

(Hamilton, 1963a: 90)

Long had found *Glossopteris* flora and the diabase sills, but he had not yet met Adie and realized he had actually walked on the "missing" tillite (§7.4).

Hamilton (1963b) quickly presented at the April 1961 AAPG meeting a fuller defense of mobilism based on the new work in Antarctica. Long had by now recognized the Buckeye Tillite, and Hamilton cited his abstract of the paper Long would present at the December 1961 GSA meeting announcing his discovery.

In East Antarctica, virtually all of the crystalline rocks are known to be pre-Permian, for in many regions the Beacon Sandstone . . . overlies them. The Beacon, which consists largely of clean, cross-bedded quartoze to arkosic sands, is a few hundred to a few thousand feet thick, carries a sparse Gondwana flora of Glossopteris and allied plants (for example see Seward, 1914 and Long, 1959), and in most places is intruded by thick sills of quartz diabase. The sandstones and sills share many features with correlative "Gondwana" rocks in South America, South Africa, Australia, and India, and Long (1961) has found at the base of the Beacon glacial sediments such as characterize each of those other regions, also.

(Hamilton, 1963b: 7)

What impressed Hamilton was that the new data which du Toit knew nothing about were so beautifully explained by his reassembly.

The granites and orogenic-belt approach provided new data, wholly unknown to du Toit but in accord with his implicit predictions. Adelaide, Transantarctic, and Cape granites are similar, but strikingly different from southeast Australia-northern Victoria Land-West Antarctica . . . The presence of Gondwana strata and fossils in the Beacon sandstone had been known since early in the century, and of course much more came later, so while I mentioned that, I was not adding anything dramatic.

(Hamilton, August 30, 2002 email to author)

7.8 Hamilton explains the origin of the Gulf of California in terms of mobilism

There was no connection between my Antarctic and Gulf of California speculations except that I was thinking mobilistically. The Gulf of California bit followed logically from the showing by Hill and Dibblee, and Crowell, that the San Andreas had mega-offset, for the fault went into the Gulf.

(Hamilton, July 24, 2002 email to author)

With Hamilton's success in finding new support for mobilism from his own and others' Antarctic work, he began thinking about how to apply it to North American geology. He read more pro-mobilist works, and found Carey's mobilist views interesting and helpful. (Carey, whose work is discussed in Volume II, §6.5–§6.15, was a vehement defender of mobilism during the 1950s.)

I first met Sam Carey, but only casually, at a drift meeting, I think the 1960 SEPM one. I got a copy of his 1958 Symposium in 1960 by wheedling a surplus copy from the USGS Denver Library. (There are few copies of it in the US.) The only time I spent hours, as opposed to minutes, with Carey was during my visit to the University of Tasmania in October 1963. I saw him again when I was a speaker (expressing my intellectual debt to him) at a post-retirement bash for him at the University of Tasmania in February 1977.

(Hamilton, August 30, 2002 email to author)

Hamilton was now fully primed to apply mobilism to North American geology. To begin, he needed a place where geological evidence indicated extensive horizontal movement.

I was previously aware, from work by John Crowell [1952,1960], and Mason Hill and Tom Dibblee [1953], that the San Andreas system was a megafault (though it was then regarded as much older than we learned later). This made opening of the Gulf an obvious corollary. Conversely, my awareness of the San Andreas geology made me receptive to other forms of mobilism, though I read du Toit before I knew much about the San Andreas.

(Hamilton, August 30, 2002 email to author; my bracketed additions)

Hamilton also knew that Wegener and Carey had offered mobilistic explanations of the origin of the Gulf of California. Puzzled by Wegener's postulation of a southward movement of Baja California, he dismissed it.

Wegener (1924, p. 184–185) suggested that the Gulf was opened by southward drift of Baja California; perhaps he misunderstood both the location and sense of displacement of the San Andreas fault system.

(Hamilton, 1961: 1308)

Although Carey had not worked out any of the details, Hamilton was sympathetic to his proposal.

Carey (1958, Fig 24) drew the Gulf of California as having originated by the drift away of Baja California from the mainland, apparently as a sialic raft; he did not suggest that the San Andreas fault might have affected either ocean or mainland outside the Gulf. His suggestion was not appreciably elaborated in the text, appearing chiefly as part of a map illustrating a possible restoration of the Caribbean region according to a jigsaw-puzzle scheme of continental drift.

(Hamilton, 1961: 1308)

Fixists considered the Gulf of California to be a sunken continental block. Hamilton argued that it had formed as a result of northwestward movement of the Mexican peninsula of Baja California and southwest California along the San Andreas fault accompanied by pulling away from mainland Mexico. He defended his solution by appealing to the close fit of Baja California with mainland Mexico and noted the offset of similar geological features along opposite sides of the San Andreas fault further north. He also argued that seismological studies by Gutenberg and Richter (1954) and Benioff (1959), which had recorded a concentration of earthquakes in the Gulf of California, extended the San Andreas fault system along the Gulf's length. Being familiar with the geology of California, he argued that his solution explained the contrasting tectonics of coastal California north and south of Los Angeles; the former formed under compression, the latter, under tension.

Recent work has demonstrated the probability of a strike-slip displacement of about 350 miles on the right-lateral faults of the San Andreas system; these faults trend longitudinally into the deep and seismically active Gulf of California, which has an oceanic crustal structure and probably thence into the Pacific basin. The fault displacement is on the order of the distance between the tip of Baja California and the continental-margin bulge of Jalisco to the southeast. Baja California has the proper shape and, so far as is known, geology to have been part of the mainland, since rifted away by oblique tension across the San Andreas fault system. The striking contrast in tectonic styles of coastal California north and south of Los Angeles can be explained in these terms.

(Hamilton, 1961: 1307)

For Hamilton, foundering of continental crust as an origin of the Gulf of California was out of the question (RS2). "Crustal structure of most of the Gulf is of oceanic type, so that an origin by structural depression of continental rocks is not possible" (Hamilton, 1961: 1307).

Like many mobilists before him, Hamilton felt obliged to provide a mechanism. He again returned to du Toit, and adopted his idea of sub-crustal flow with its density-phase transformation of basalt to eclogite (§5.7, §6.7). As Hamilton later confessed, "My mechanism of course was nutty, and shows how little attention I was then paying to geophysics" (Hamilton, July 24, 2002 email to author).

7.9 Regionalism and Warren Hamilton

The influence of regionalism and his escape from it goes a long way in explaining Hamilton's journey to mobilism, his reading of du Toit, his initial inability to find strong support for mobilism in North America, his subsequent discovery of support in

Antarctica, and his emergence as an active and vocal mobilist. If Hamilton had worked in a southern continent or peninsular India when he first read du Toit, he would have found it easy to become an active mobilist. But he had been educated and had worked initially in North America, where fixism worked fairly well, and so found no strong evidence that gave mobilism the advantage. With his work in Antarctica, his new-found experience in relating stratigraphy and tectonics, and his knowledge of petrology, Antarctica provided him with what North America could not offer. Well acquainted with du Toit's arguments about the fragmentation of Gondwana, Hamilton, unlike Long, anticipated support for mobilism before seeing the continent for himself.

Returning to North America, he looked for somewhere to apply mobilism and turned to the origin of the Gulf of California. He could not have made a better decision, being, perhaps, the most likely place in North America to do so. He recognized the stratigraphic similarities between Baja California and the Mexican mainland, the jigsaw fit of their facing margins, the movement along the adjacent San Andreas fault system, and perhaps the lesser apparent differences between the tectonic styles of coastal California north and south of Los Angeles.

Within a few years Hamilton had become a staunch and active mobilist. He was ready in 1962 to argue (1964) against Axelrod's paleobotanical attack on continental drift (III, §1.12). Extending northwards his mobilistic interpretation of the origin of the Gulf of California, Hamilton and W. Bradley Myers (1966) began explaining the Cenozoic tectonics of the western United States in mobilistic terms. He also began appealing to paleomagnetic support in favor of mobilism (III, §1.13), first in his attack on Axelrod, then in his mobilistic interpretations of Antarctica (Hamilton, 1965) and western North America (Hamilton and Myers, 1966). He developed mobilist interpretations for the formation of the Scotia and Caribbean arcs (Hamilton, 1966), and marshaled paleoclimatic support for mobilism (Hamilton, 1968). With the advent of plate tectonics, he modified and expanded some of his mobilistic interpretations of the tectonics of North America, putting it in terms of plate tectonics (Hamilton, 1969). He argued that the Urals were formed by Paleozoic collision of the Russian and Siberian platforms (Hamilton, 1970). In the 1970s and 1980s he worked on the tectonics of Indonesia (see Hamilton, 1988 for references) and in later years has turned to the history of Archean tectonics and the dynamics of the mantle (Hamilton, 1998, 2002).

From this record of Hamilton's very active and successful pursuit of mobilism in the early and mid 1960s, it might seem that nothing inhibited him from publishing papers in its support. This would be wrong. Journals rejected some of Hamilton's papers, and others were either blocked or rejected by internal reviewers at USGS. Hamilton's paper on the Gulf of California, which was published in the *GSA Bulletin*, had earlier been rejected.

Very large strike-slip offset on the San Andreas fault was demonstrated in a GSA paper by Hill and Dibblee (1953). My [paper on the origin of the Gulf of California] built on this to link the

San Andreas fault to oblique opening of the Gulf of California. (My manuscript had previously been rejected, as foolish speculation, by the *Bulletin of the American Association of Petroleum Geologists*; indeed, my proposed mechanism was foolish.)

(Hamilton, 1988: 1504; my bracketed addition)

The review process of USGS was cumbersome and slow.

The USGS had, throughout my tenure, an in-house manuscript-review system that had to be completed before submission of a paper to an outside journal. Each manuscript was sent to two USGS technical reviewers (abstracts to only one); often they were very helpful, but they could, at their option, operate anywhere between passing it without comment, and holding it up for many months and perhaps trying to block it. After revision, and possible recycling to the same or new reviewers, and clearance by a supervisor, a paper went on to editors for text, figures, and geologic nomenclature; this was helpful, but added delays. Then it went to some manager who scrutinized it for anything that might resemble policy. (In a 1960 paper, I noted that I did not include on a map some dubious faults that had been compiled from a bad photogeologic source on the Geological map of Oklahoma; the compiler of that map, Hugh Miser, was Associate Director and final manuscript arbiter for some years, and after that manuscript went to him he personally stalled any manuscript of mine for 3 months before passing it.) After all this, a paper went into the line for USGS in-house publication (which could be years long), or could legally be submitted to start over at the bottom of a journal's review chain. Journals also could be backlogged a year or more after they finally accepted something, and society monographs could take 3 years to get out. So publication was much more leisurely than now, and papers commonly were written long before they actually appeared. The "submission" date on a journal article also could be long after the paper started into the USGS mill. Sometimes late revisions could be made, but often not.

(Hamilton, August 6, 2002 email to author)

Hamilton's pro-drift papers often met with resistance from fixist USGS reviews.

After that [first] season [in Antarctica at the end of 1958], I wrote drift into drafts of more papers than it was published in. Conservative reviewers or supervisors would grumble that this was irrelevant, and out it would go. So the absence of drift (including paleomagnetism) in any published paper from 1959 through most of the 1960s can mean either that I didn't write any therein, or that a reviewer insisted it come out.

(Hamilton, August 25, 2002 email to author; my bracketed additions)

Hamilton wrote a lengthy monograph entitled "Continental drift" that was delayed by USGS reviews and finally never published. The paper went through several versions, the last written in 1966. Failing to get past the reviewers, Hamilton finally gave up.

"Continental drift," intended for a USGS Professional Paper. This is as far as I got with the paper that was serially abused by nothing-moves reviewers. (Earlier versions and reviewers' comments on both them and this version are things I remember discarding in 1996.) Pages of this retained version all bear dates-of-typing of 1966. Total perhaps 350 typed pages, without references list, which I recall dimly as stacks of index cards. Fattest section is Permian climatology – a mix of lithologic indicators, terrestrial paleobotany, and marine paleontology. Lesser sections on general climatology and paleoclimatologic principles, Cenozoic paleoclimatology (I skipped the Mesozoic), biogeography, and paleomagnetism.

(Hamilton, August 23, 2002 email to author)

Hamilton had another paper that was held up for almost a year before it was published.

My paper "blocked for a year by a branch chief [of USGS] and some senior geologists in Menlo [Park]" was published as Warren Hamilton, 1969, "Mesozoic California and the underflow of Pacific mantle," *Geol. Soc. Amer., Bull., 80*, 2409–2430. (It actually was less than a year, but not much, and it did get improved in the interim.) It was the first broad exposition of western US tectonics in terms of plate tectonics; enormously influential. But Menlo had the Franciscan Friars, to whom the Franciscan mélange (my accretionary wedge, in terminology that did not yet then exist) was a stratigraphic unit, and Paul Bateman and friends, to whom the Sierra Nevada (my magmatic arc) was a product of crumpling of a geosyncline, and here I was in distant Colorado telling them what their rocks meant.

(Hamilton, August 23, 2002 email to author; my bracketed additions, and paper is cited as Hamilton, 1969, in the bibliography)

To publish mobilistic papers about Antarctic geology met with some resistance, but to publish ones in major North American journals was much harder. Hamilton, perhaps more than any other North American geologist during the early 1960s, experienced this resistance. His views were opposed to those of most reviewers and editors of major journals in North America, and he worked for a conservative organization whose in-house reviewers were fixists, and whose geological experience was almost exclusively North American. He worked for an organization that had produced the Cox and Doell 1960 *Review of Paleomagnetism* that had systematically underplayed the pro-drift evidence of the new paleomagnetic work (II, §8.3–§8.11), which surely must have strengthened the hands of these fixist reviewers. Hamilton's experience is a poignant example of the effect of regionalism on the mobilism debate.

7.10 North American regionalism: a summary

North American geologists preferred fixism to mobilism because fixism appeared to better explain their geology. They believed that their continent consists of a central and stable Precambrian craton surrounded by younger mountain belts. The craton was the original continent; new mountains arose from bordering geosynclines, and were accreted around it. They thought mobilism was not needed to explain North American geology. The Appalachians did not appear to have undergone very large unidirectional horizontal movements. North American geologists did not live among the more obvious intercontinental structural, stratigraphic, floral, and paleoclimatic similarities that surrounded geologists working in South Africa, Brazil, Argentina, and peninsular India. This fixist view was exemplified by Kay, and also by other key figures such as Schuchert, Willis, Coleman, Bucher, and Wilson, and they truly dominated North American geological opinion up to the mid-1960s.

Long's and Hamilton's studies in Antarctica reflect regionalism's effects. They showed that Antarctica when compared to other Gondwana fragments had what North America lacked: what was then recognized as solid regional support for mobilism. Long, through his initial ignorance of mobilism and Southern Hemisphere geology as a result of his regionalistic education, did not at first realize that he had discovered the predicted Permo-Carboniferous glaciation of Antarctica; stumbling over the Late Paleozoic greywackes with boulders in them he did not recognize them for what they were. Adie, who had studied with King, had to explain to him that they were tillites. Long, much to his credit, quickly agreed. Hamilton, a closet drifter, was unable to find new convincing evidence for mobilism while working in North America. Finding what he needed in Antarctica and other Gondwana fragments, he argued for mobilism, returned to North America, interpreted its geology in mobilist terms, and became a leading land-based mobilist. Long, like Opdyke as I shall later describe (II, §4.2), had to leave North America to become a mobilist; Hamilton had to leave North America to become an active mobilist.[10]

Notes

1 I think it is actually somewhat unclear as to whether Chamberlin was proposing a method of discovery, a method of justification, or both. If his method was only one of discovery, then the criticism of Wegener's theory was absurd. If his method was a bit of both, which I think it was, then the criticism of Wegener was unfair because he typically compared mobilism's solution with fixist alternatives, and argued that his solution faced fewer difficulties. Oreskes (1988: 138–142) claims that Karl Grove Gilbert (1896) first came up with the method of multiple working hypotheses, but Chamberlin supplied its name and popularized it. She quite rightly points out that Gilbert's method was much more sophisticated, and claims (1988: 141) that Gilbert actually "articulated the hypothetico-deductive method of science." Gilbert certainly separated the process of formulating a hypothesis from testing it.

When the investigator, having under consideration a fact or group of facts whose origin or cause is unknown, seeks to discover their origin, his first set is to make a guess. In other words, he frames a hypothesis or invents a tentative theory. Then he proceeds to test the hypothesis, and in planning a test he reasons in this way: if the phenomenon was really produced in the hypothetic manner, then it should possess, in addition to the features already observed, certain other specific features, and the discovery of these will serve to verify the hypothesis.

(Gilbert, 1896: 1)

Gilbert proposed "hypotheses are suggested through analogy." Gilbert, as Oreskes (1988) explains, also thought that it was a good idea to work with several hypotheses, developing tests for each, and comparing their success. Indeed, Wegener's development and testing of continental drift fit well enough Gilbert's views. He looks at the bathymetric map, sees the matching continental margins, then comes up with the idea of continental drift. He thinks of further implications of moving continents, and starts reading about geology. He also begins to think of the continents as being like ice floes. Both float in their supporting media. Ice floes move through the ocean; continents move through the sima.

2 Also see Aldrich (1979), Dott (1974, 1979, 1985), Greene (1982), Marvin (1973), and Mayo (1985) about the discovery of the North American craton, and the importance and use of the

idea of geosynclines, contractionism, oceanic permanency, and accretion of continents in attempts to explain the origin of the Appalachians.

3 Long was interviewed by Karen Brewster in April 2001. A transcription of the interview can be obtained from the Byrd Polar Research Center at Ohio State. The interview spans from Long's childhood in California to his current life in Alaska, covering his adventures as a mountain climber, his participation in IGY, his work at Ohio State, his teaching at Alaska Methodist University from 1965 until 1976, and his work as a consulting geologist after the financial collapse of Alaska Methodist University. His discovery of the Buckeye Tillite is only briefly described.

4 Long originally proposed the names "Buckeye Range" and "Buckeye Tillite" in honor of The Ohio State University's football team, the Buckeyes. The proposed name "Buckeye Range" was not deemed suitable; they settled on the Ohio Range. Although "Buckeye Range" did not get past the US Board on Geographic Names, Long also named the tillite the "Buckeye Tillite," which it remains.

5 By the late 1950s and early 1960s paleomagnetists had already sampled equivalent dolerite formations elsewhere in Antarctica. Initial results appeared in 1959 and 1960. See Volume II, §5.9.

6 Doumani obtained an undergraduate degree from Terra Sancta College in Jerusalem in 1947. He obtained a B.A. in geology from the University of California in 1956 and an M.A. in paleontology the following year. Like Long, Doumani worked in Antarctica during IGY, and became a Research Associate at IPS.

7 Doumani also had a paper included in the proceedings of the Cape Town meeting, but did not attend the symposium. I do not know if he too applied to IPS to attend the symposium, but Mirsky presumably read Doumani's paper.

8 The Beacon Sandstone with sporadic plant fossils overlies basement with Paleozoic dykes and is intruded by Ferrar dolerite (Bull and Irving, 1960).

9 Hamilton clearly had little time. Had he spent longer in Canberra and visited the Geophysics Department of the Australian National University he likely would have encountered Colin Bull or Ted Irving working on the paleomagnetism of the Ferrar dolerites (II, §5.6).

10 I would be remiss to not mention Albert Wolfson's advocacy of continental drift based on his explanation of bird migration (Wolfson, 1948, 1955). Wolfson, a zoologist at Northwestern University was not a member of the North American community of Earth scientists. He became interested in drift in 1945 while researching bird migration. Wolfson wrote his own excellent retrospective of the development of his ideas and the resistance he encountered from G. G. Simpson and other paleontologists (Wolfson, 1985). His 1948 paper caused a stir. Seven letters appeared in the December 24 issue of *Science* for 1948. None supported his drift explanation although his paper was described by one author as "stimulating" and another said his hypothesis was "brilliant." The most significant letter, the only one from a paleontologist, was written by Dean Amadon, New York City's American Museum of Natural History. Amadon, citing primarily G. G. Simpson, dismissed Wolfson's account. As Wolfson himself reported in his retrospective paper, Amadon even quoted the following passage from Simpson, 1948.

The fact that almost all paleontologists say that paleontological data opposed the various theories of continental drift should obviate further discussion of this point and would do so were it not that the adherents of these theories all agree that paleontological data do support them. It must be almost unique in scientific history for a group of students admittedly without special competence in a given field thus to reject the all but unanimous verdict of those who do have such competence.

(Amadon, 1948: 706)

So much for Wolfson; he was not a paleontologist, just a zoologist. Amadon, however, did raise legitimate objections, challenging Wolfson's claim that migratory routes of some birds appeared to offer no survival value. Citing one of Wolfson's favorite examples, the arctic tern, he cited a 1946 study by B. Kullenberg who explained its migration across the Atlantic in terms of its partiality to always remaining in plankton-rich cold waters.

Wolfson later encountered Simpson directly. Wolfson's colleagues in geology at Northwestern arranged an informal debate between him and Simpson at a GSA meeting. Wolfson did not identify the meeting, but recounted what transpired.

> I was young, enthusiastic, and eager. I had read most of the papers and felt pretty confident about the arguments. I said yes . . . I was flattered and pleased that Simpson had agreed to discuss the issues. Of course, I went first. I got up, made my fifteen minute presentation. . . and sat down. They then called on Dr. Simpson. He got up, walked to the front of the group – the memory of it is still vivid – and with his extremely dignified appearance and gracious tone he said, "Dr. Wolfson has his opinions and I have mine." Then he sat down. So there wasn't much scientific debate that day . . . I was greatly disappointed.
>
> *(Wolfson, 1985: 185)*

Is Wolfson's strong advocacy of mobilism to explain bird migration a counter to regionalism and the strong opposition to mobilism? Not at all; indeed, Wilson's story fully supports it. Wilson was an outsider. He did not know Marshall Kay's work; he knew little geology. With no training in geology, he was not predisposed against continental drift. Drift provided him with what he believed was the only satisfactory explanation of lengthy bird migrations.

8

Reception and development of mobilism in Europe: 1920s through the 1950s

8.1 Introduction

Now I examine mobilism's reception among Europeans. Generally, in continental Europe neither mobilism nor fixism enjoyed a decided advantage in explaining regional problems, but there were regional variations. Workers in the western Alps, where there was abundant evidence of lateral movements, generally favored mobilism; workers in Germany or the eastern Alps, where vertical movements appeared to prevail, were fixists; and workers in Soviet Russia, where the "stable" Russian Platform dwarfs all other features and where vertical motions are the most apparent, were fixists. There was also a group of European mobilists who studied more than one disjunctive segment of Caledonides in Western Europe, North America, or East Greenland who argued that they were best explained mobilistically by joining them into a single orogenic belt closing the Atlantic. The Caledonide Orogeny culminated in the Late Silurian.

I then turn to the reception of mobilism in the United Kingdom. Many Britons undertook geological investigations of their worldwide empire where mobilism often explained better the local geology; it is no surprise therefore that a good number of Britons who came to prefer mobilism had worked extensively in what was then India (now India, Pakistan, and Bangladesh), South Africa, or the Falkland Islands. Also, mobilism was generally preferred by specialists to whom it offered better solutions; botanists and paleobotanists, especially those working on *Glossopteris* flora, were apt to support mobilism. There was, however, one prominent Briton, E. B. Bailey, who supported mobilism and yet spent almost his entire career working on regional problems of Great Britain. He specialized on the Scottish Caledonides, which extend across Scotland, Northern Ireland, northern Wales, and northwestern England (but which I refer to as the Scottish Caledonides as the key evidence is there). However, even he did not become a mobilist until after visiting their equivalents in North America.

Finally, I examine the views of Dutch geologists who attempted to explain the geology of the Dutch East Indies (now Indonesia). At first, much of it was best explained in terms of mobilism, and the generation of Dutch geologists who worked there at the time Wegener's theory was emerging supported continental drift. But the next generation rejected mobilism, which may seem contrary to what would be expected if regionalism dominates, but it was not; new data, as then interpreted, conflicted with the early mobilistic explanation for the East Indian archipelago, causing Dutch geologists to change their minds and become predominately fixist. This generational difference shows that neither teaching nor tradition explains why members of the next generation changed their minds. Their preference for fixism, like their predecessors' preference for mobilism, is explicable in terms of what was known about the region of study when each generation made their studies.

8.2 Continental Europe: preliminary comments

The regions in which the leaders did their training fieldwork seems in turn decisive in shaping their philosophies: Argand's overemphasis on the plasticity of rocks can be traced down to his training in the Pennine Zone of the Alps where ductile recumbent folds of immense dimensions are the dominating structure types; Stille's geosyncline-craton separation and his emphasis on short-duration orogenic phases separated in time by anorogenic periods stem directly from his mapping experience in north-central Germany; the views of Russian masters of the oscilla-tionist school have developed in their homeland, where indeed such movement appear to be predominant. So, while studying a given tectonic (or any, for that matter) theory it seems essential to know the scientific background of the people who developed it.

(Şengör, 1979/1982a)

In summary, the reaction in Continental Europe was extremely diversified and dominated by an association of strong post World War I politics, the language barrier, the stifling of academic authority, passions of individuals, and regionalism of geology.

(Carozzi, 1985: 122)

Şengör, Carozzi, and also Schwarzbach have discussed the reception of Wegener's theory, primarily restricting themselves to the years between the wars. Şengör (1979/1982a, b) and Schwarzbach (1980/1986) described how Wegener's views were strongly rejected or simply ignored by leading German tectonicists. Carozzi (1985) showed that differing attitudes toward mobilism and fixism among European Earth scientists are partially explained by regionalism. I think they are correct, and I rely very much on their work. Regionalism had its greatest effect among tectonicists. Those, such as Argand, Staub, Collet, and Salomon-Calvi, who worked on mountain belts with well-described and extensive overthrusts were mobilists. Those who began their careers working on orogenic regions where movements appeared to be mostly vertical instead of horizontal were generally fixists; included are almost all German and Russian tectonicists, and others who specialized on the Eastern Alps. Those geologists who either played an important role in discovering the Caledonides along Greenland's eastern coast, or worked in at least two regions where other Caledonian fragments are prominent (Norway, Spitsbergen, Scotland, and northeastern North America), tended to be mobilists. Just as du Toit worked in both South Africa and

South America, these geologists, notably E. Wegmann, H. Bütler, H. Aldinger, and O. Holtedahl, had firsthand experience of more than one now-separated Caledonides segment.

After briefly describing the reception of Wegener's theory in Germany, concentrating on the views of Hans Cloos and Hans Stille, and reviewing the attitudes of some who participated in a 1939 Frankfurt symposium where mobilism was debated, I shall consider the views of Alpine geologists and of the Dutch, many of whom worked in the Dutch East Indies. I shall utilize the important work of William Fraser Hume (1867–1949). Educated at the Royal College of Science, London, under John Wesley Judd (1840–1916), Hume spent most of his career in Egypt. He joined the Geological Survey of Egypt in 1898, two years after its creation, and became Director in 1911. He later served as Technical Counsellor of the Survey, was elected President of the Royal Geographical Society of Egypt, and was awarded the Lyell Medal of the Geological Society of London in 1919. One of Hume's books, *Terrestrial Theories: A Digest of Various Views as to the Origin and Development of the Earth and their Bearing on the Geology of Egypt*, which went to press in 1941 but was not published until 1948, contains textually close summaries of many papers by participants in the mobilist controversy. Hume was particularly interested in tectonics, and reported the views of many European tectonicists, even corresponded with them, and had them read over what he had written about their work.

8.3 Fixists from continental Europe: Stille and Cloos

Mobilism had a poor reception in Germany among tectonicists and a mixed reception among others (Carozzi, 1985). Schwarzbach (1980/1986: 109), who concentrated primarily on tectonicists, declared, "It was actually in Germany that Wegener found the least acceptance – 'a prophet without honor in his own country'." Both noted Max Semper's 1917 vitriolic attack, which matches anything that would later be said against drift at the 1926 AAPG symposium in the United States (§4.3).

It is certain that Wegener's theory was established with a superficial use of scientific methods, ignoring the various fields of geology . . When we find out that based upon this kind of performance Wegener has the audacity to state that "old theories as presented by E. Suess led to absurd consequences," then we can only try to keep our distance and beg him not to deal with geology any longer, but to look for other topics which until today have forgotten to write above their doors; Oh holy Saint Florian, protect this house but burn down the others! In regard to the often repeated excuse that the theory still needs to be changed and is incomplete, one should recall Goethe who spoke about amateurs who never finish because they never know how to start in the first place!

(Semper, 1917: 157–158; Carozzi's translation, 1985: 124)

Many followed prominent fixists such as Hans Stille (1876–1967) and Hans Cloos (1885–1951). Schwarzbach (1980/1986: 110) claimed that Stille found nothing appealing about Wegener, while Cloos thought that Wegener's views were interesting and inventive, but never accepted them; Schwarzbach considered Stille and Cloos "the two towering figures of German geology," and that it was because of them that Wegener's ideas were largely ignored not only in Germany but throughout much of Europe. Schwarzbach (1980/1986: 108) also noted that Hermann von Ihering, a landbridger, argued strongly against Wegener (§3.4). Carozzi (1985) mentioned several German geologists who were sympathetic toward Wegener, but did not mention von Ihering, Cloos, or Stille. In summarizing the reaction of Wegener's critics, he said:

critics of Wegener behaved, in general, as relatively narrow-minded specialists. They were structural geologists, paleogeographers, paleontologists, and paleoclimatologists, geophysicists being excluded. They rejected the theory on the basis of single arguments, or very local facts, without seeing the broad picture. Criticisms were usually well-documented by personal studies and travel experience, or by a thorough knowledge of the appropriate literature. They failed, nevertheless, to appreciate the worldwide scale of continental drift.

(Carozzi, 1985: 126)

Sounds like many geologists of the period. Truly knowledgeable globalists were rare.

Şengör (1979/1982a and 1982b) identified Stille as one of Europe's leading fixists.[1] Stille, whose views Şengör traced back to Élie de Beaumont and Dana, was a contractionist, believing that mountains form by crustal shortening. He divided Earth's crust into geosynclines, belts of high mobility, and cratons, regions of low mobility. Cratons are stable, and fixed relative to one another. He distinguished between high or continental cratons, and low cratons. Low cratons make up parts of the ocean floor, and are sunken high cratons (Şengör, 1979/82a: 41). Stille (1955: 183) claimed that the Pacific and part of the Atlantic Ocean existed before the end of the Precambrian, and that the Indian Ocean formed later "by subsidence of a large part of Gondwanaland." Mountains form in geosynclinal regions when symmetrically squeezed by surrounding cratons. The image often used was that the converging cratons acted like opposing jaws of a vice crushing the geosyncline between them; some lateral motion was allowed but very little. Stille distinguished two types of mountain ranges: Alpine-type formed from compressed geosynclines, which he later called orthogeosynclines, and German-type (originally called Saxonian mountain ranges), which were less compressed geosynclines, and which he called parageosynclines.

Stille worked on the Saxonian Mountains in 1913. Five years later, he drew the distinction between these two types of mountain ranges. Şengör explains:

Stille was a convinced contractionist and following the footsteps of Dana ascribed both the building of the geosyncline and its eventual demise during the orogenic phase(s) to lateral compression caused by contraction. Because the Mesozoic-Cainozoic deformation of the Central European ground was interpreted as having been a result of lateral compression, the enormous difference in style between what Stille came to call *Saxonien* (= Saxony) and the "normal" orogenic belts such as the Alps needed explanation. This is the subject of an extended work, *Über Hauptformen der Orogenese und ihre Verknüpfung* (On main forms of orogeny and their connections) (1918), where Stille emphasized that the nature of the ground was decisive in the generation of various kinds of tectonic styles. A fourfold classification of orogenic styles was proposed, from the "highest" forms to the "lowest": 1. *Deckengebirge* (= nappe mountains). 2. *Faltengebirge* (= fold mountains). 3. *Bruchfaltengebirge* (= fault-fold mountains). 4. *Blockgebirge* (= block mountains).

These different styles were argued to result not from any difference in the applied horizontal pressure, which must have been the same everywhere as it resulted from the contraction of the planet, but from the different nature of the ground affected by this pressure. Stille later called the classes 1 and 2 "Alpinotype mountains", and 3 and 4 "Germanotype mountains". With this step Stille could explain almost all tectonic phenomena on the Earth by means of the contraction theory.

(Şengör, 1979/1982a: 26–27)

Stille's ideas stem from his early fieldwork in mountains of central Germany, where vertical movements appeared to dominate. With this as his base he then explained Alpine-type mountains without requiring huge horizontal movements; it was a question of greater or lesser compressibility; he did not need mobilism to explain mountain-building, and he found his own contractionist model more satisfactory.

Stille envisioned repeating cycles of long anorogenic and short orogenic stages (Şengör, 1979/1982a). During anorogenesis, epeirogenic (vertical) movements dominated. Cratons were bent but not broken in a wave-like fashion producing downwarps (geosynclines) and upwarps (geanticlines). Stille called these epeirogenic movements "undulations." With the transport of sediment from geanticlines into geosynclines and continued secular contraction, the highly mobile orthogeosynclines thicken, melt at their base, and Alpine-type mountains arise. Ensuing epeirogenic movements create less compressible parageosynclines flanking already consolidated Alpine-type mountain belts. Parageosynclines fill with sediment, are compressed, and German-type mountains form on either side of the now eroded Alpine-type mountains. This cycle repeated itself adding parageosynclines to the growing cratons, after which only German-type mountains can form. Stille (1955) thought that when no more orthogeosynclines remained, Earth went through a major regeneration (*Algonkischer Umbruch*). He argued this had happened once before following the Algoman orogeny at the close of Huronian (Early Proterozoic) time, and speculated that we may now be "on the threshold of a third tectonic era which would include hundreds of millions of years to come" (Stille, 1955: 172). Stille extended his regionally developed theory around the world and used it to unravel Earth's tectonic history, turning to the Americas in 1936 and 1941, to the Atlantic in 1939, and to the Pacific in 1955 (Stille, 1936, 1939, 1941, and 1955).

Stille presented his fixist account of the formation of the Atlantic (summarized by Hume, 1948: 360–369) at the 1939 Frankfurt symposium. Drawing on the recent findings of the German oceanographic *Meteor* Expedition (1925–7), he found a host of transverse relationships and symmetries which he thought showed that the cratonic regions of the continents surrounding the Atlantic had not moved horizontally relative to each other since before the Atlantic, according to Wegener, had opened. He stressed the symmetry between the north and south parts of the Atlantic Basin (Figure 8.1). A line passing from Guinea to San Roque divides the Atlantic into its northern and southern basins. Both basins have westerly convex swells that face easterly convex orogenic belts (Antilles) flanked by ocean deeps. Stille described the current orogenic activity in each of these as Alpine-type, claiming that both orogens were being strongly compressed in a north–south direction. Lines SN and SS symmetrically divided the northern and southern Atlantic basins, and Stille emphasized the ridge segment and orogenic regions of each. Extending the line dividing the northern and southern basins, Stille not only found that it intersected the point where the trends of the Andes and western margin of South America bend, but also that it bisected the angle of their trend-change, further supporting the view that the cratons in question had not moved laterally relative to each other.

He claimed that both basins of the Atlantic were old; the northern already existed in the Cambrian Period and the southern since the Paleozoic Era. He also claimed that a broad trans-Atlantic landbridge separated the northern and southern Atlantic during the Triassic and Jurassic Periods. He claimed that the Mid-Atlantic Swell was extensional and probably volcanic in origin (Şengör, 1979/82a: 28).

By emphasizing the above transverse symmetries, he provided fixists with arguments to counter the congruencies between opposing continental margins that were appealed to by mobilists (RS1). He also provided paleontologists with a trans-Atlantic landbridge. Stille, the tectonicist, was not too concerned about when the landbridge existed, and deferred to paleontologists. He admitted that isostasy presented a difficulty to sinking continents or broad landbridges but noted that geophysics was still a young science. Hume explains:

Much more difficult is the problem of the sinking of former continents . . . Numerous possibilities for such changes suggest themselves, for instance they may result from mass-displacements in the regions of greater mobility in the deep-seated regions of the crust. Stille agrees . . . that geophysical science is still in its infancy and because it cannot give a final explanation of the sinking of such extended areas, does not mean the impossibility of such an occurrence.

(Hume, 1948: 360)

Mobilism, as Stille well knew, had its own unresolved mechanism difficulty.

Although Stille drew heavily on the findings from the *Meteor* expedition, he paid little attention to the fact that the congruencies between opposing continental margins are echoed in the trend of the Mid-Atlantic Ridge. Perhaps he thought there was no reason to call attention to other pro-mobilist arguments.

Hans Cloos, like Stille, developed a global theory that was based on his own fieldwork mainly in Germany and Africa, including South Africa, and on his overwhelming interest in structural geology and tectonics. To explain Cloos' rejection of

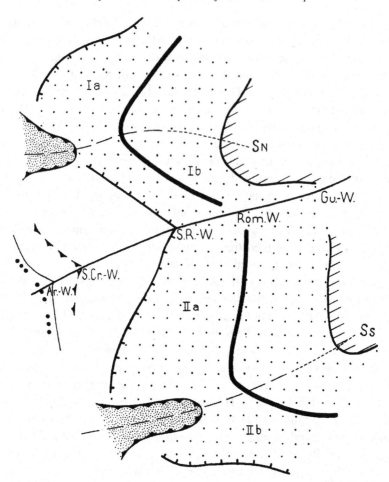

Figure 8.1 Stille's Figure 4 (1939: 321) of the Atlantic showing his accented symmetries. Ar.-W., Arica Angle; S.Cr.-W., Santa Cruz Angle; S.R.-W., San Roque Angle; Rom.-W., Romanche Angle; Gu.W., Guinea Angle. The Gu.-W.–S.R.-W. is the line of symmetry dividing the North Atlantic (I) from the South Atlantic (II). The bold line between the American and African Atlantic margins is the Mid-Atlantic Swell (Ridge). The two densely dotted areas projecting from the American margin into the Atlantic are the northern and southern Antillean orogenic regions. They are deep cratonic areas between the American high cratons and the Atlantic median swells (ridges). SN, North Atlantic axial line of symmetry between the areas Ia and Ib of the North Antillean–Mid-Atlantic Structure; SS, South Atlantic axial line of symmetry between the areas IIa and IIb of the South Antillean–South Atlantic Structure. Ar.-W. marks the angle where the Andes change their trend. S.Cr.-W. is the angle where the Pacific margin of South America changes its trend.

mobilism in terms of regionalism is a challenge because most who participated in the controversy and who worked in South Africa supported mobilism because it solved local problems there (§6.3–§6.12). Thus it is important to determine just how Cloos came to his ideas. I shall argue that regionalism goes a long way towards explaining his preference for fixism. Although Cloos offered a tentative solution to the origin of Permo-Carboniferous glaciation and biotic disjuncts, he concentrated almost entirely on regional tectonics.

Hans Cloos liked Wegener personally, later describing their first encounter, which likely occurred just before the outbreak of World War I.[2]

One day while I was teaching at Marburg a man came to me, whose fine features and penetrating, gray-blue eyes I was unable to forget. He had developed an extraordinary theory in regard to the structure of the earth. He asked me whether I, a geologist, was prepared to help him, a physicist, by contributing pertinent geological facts and concepts. I liked the man very much, even though I was skeptical of his ideas. Thus began a loose co-operation on a subject in which the Red Sea rapidly assumed a central position. This man was Alfred Wegener.

(Cloos, 1947/1953: 395–396)

Cloos was probably skeptical of Wegener's ideas from the very beginning. He believed that Earth's crust was divided into very ancient polygonal plates (fields or blocks) of considerable depth. These basement blocks, as he later called them, are separated by "geofractures" or "geosutures" (Cloos, 1948: 99). Blocks are rigid and move as units. Groups of blocks can also move together as a unit. When elevated they form continents; when depressed they become ocean basins. Block boundaries are ancient features, and assume different forms: mountains, geosynclines, faults, or grabens. Cloos developed his ideas from work in Germany, then in Norway and Sweden, Africa, and western North America, and later examined bathymetric maps of the Red Sea and Atlantic, which, he argued, confirmed them.

The author's own contributions to this problem date back to 1910, when he first investigated the influence of old Tertiary faults on younger Tertiary folds in the Swiss Jura mountains near Basel. Later on he dealt with the Rhine graben, the primary boundary of the Rhenish Schiefergebirge, the pre-Cambrian stages of important block boundaries in Norway and Sweden, with the old structures in Silesia and the Bohemian mass, the Harz mountains, northeast and South Africa, certain areas in North America, the Atlantic structure in the Azores, and with the Red Sea.

(Cloos, 1948: 99)

Cloos' theory precluded mobilism, which he first rejected in 1937. He also used his model to outline fixist solutions to the origin of Permo-Carboniferous glaciation and biotic disjuncts. Hume later explained:

The structure of the areas [polygonal basement blocks] leaves plenty of room for them to undergo horizontal changes and makes it easier to have a mechanical conception of such movements. Cloos mentions having himself described such horizontal movements, but adds that the Wegener theory of the free displacements of whole continents is not compatible with a complete areal structure durchgehender Felderbau.

Cloos in thought turns back rather to the pre-Wegener conception of important vertical movements . . . This offers a convenient explanation of certain geographical homologies and especially that of the Permian glaciation in the Southern Hemisphere. This principle regarding areas (Felderprinzip) offers other simplifications of problems. Thus oceanic regions with high boundaries need only to be slightly raised relatively in order to give rise to dams or bridges across which fauna and flora can wander.

(Hume, 1948: 280–281; my bracketed addition)

Cloos reconsidered Wegener's theory in 1939, when he introduced a fixist account of the origin of the Red Sea, which he had first seen in 1909 on his way to map the Erongo Mountains in the German colony of South-West Africa (now Namibia). Approximately thirty-five years later, Cloos recalled a conversation he and Wegener had had in Marburg, about the Red Sea.

His hypothesis became famous all over the world, for it placed an easily grasped and sensational idea on a semi-scientific basis. His theory loosened the continents from the terrestrial core, and changed them into icebergs of gneiss floating on a sea of basalt. He let them float and drift, tear apart from and bump into one another. Fissures, rifts, graben were left where they had pulled apart; where they collided, mountains folded. "Just look at Arabia!" Wegener cried heatedly, and let his pencil fly over the map. "Is that not a clear example? Does the peninsula not turn on Sinai to the northeast like a door on a hinge, pushing the Persian mountain chains in front of it, attaching them on the two hooks of Syria and Oman like drapes! In the rear, the Arabian table has been torn off Africa. It has moved away from the Abyssinian angle, opening a rift 200–250 miles wide, exactly the amount of narrowing suffered by the Persian mountain chains."

"But the triangle of Danakil in the southeast corner of the rift, how does it fit into your movement?

"It is lava." Wegener replied, "which welled up from the bottom of the graben."

"Very well, but how can lava float on lava?"

"The lava does not float; it is carried by lighter blocks of gneiss that had been split off before."

. . . The Red Sea was only part of the general subject; South Africa also figured largely in our talks, especially the mountains of the Cape Colony, which terminate abruptly in the Atlantic, and seem to continue unchanged 4000 miles west in South America.

The "drift theory" appeared in scientific journals . . . It even found its way into textbooks. The Red Sea moved ever more clearly and firmly into the center of the argument . . .where it properly belonged. Three experienced and successful structural geologists – a Swiss from Neuchâtel [Argand], another from Zürich [R. Staub], and an Austrian [F. E. Suess] – tried to apply Wegener's hypothesis to the Alps. The hinge of Suez and the revolving door of Arabia were received with enthusiasm, and widely used to explain the Persian folds and the Red Sea trench. Solid, heavy Africa was not very much inconvenienced. Then two insignificant events occurred, the first ten years, the second twenty years, later, which undermined these too loosely cemented pillars of the Wegener hypothesis.

(Cloos, 1947/1953: 396–397; my bracketed additions)

The first "insignificant event" was his explanation of the Rhine Graben, which separates the Black Forest in Germany and the Vosges Mountains in northern

France near its border with Germany. With the completion of a tunnel on the German side in 1928, Cloos had the opportunity to study the now exposed marginal fault. He discovered that the rift block was a downward narrowing wedge that had sunk relative to either side. He then began making small-scale models, and found one that worked. Instead of pulling either side apart as required by mobilism, he arched them upwards. He later recalled:

I produced the same subsidence with all its enchanting natural and quasi-natural refinements, without any pull, but simply by arching up the artificial crust; this automatically stretched the upper side. We now had a model of the sloping surfaces of the two opposite mountain ranges, which anyone would recognize who had ever traveled from the Black Forest toward Swabia, and from the Vosges toward the Paris basin. The two marginal blocks . . . had risen, and some sort of deep volcanic expansion had forced the whole area upward like a tremendous bubble.

(Cloos, 1947/1953: 312–313)[3]

He incorporated his solution into his overall theory. The rift valley was the common boundary of two basement blocks that made up the Rhine shield. The shield domed upward, fractured along a geo-fracture and the central rift-block sank.

Cloos, who had first visited the Red Sea in 1909, returned in 1936. Had it too been formed by uplifting of the Arabian and African shields, or had they pulled apart creating the Red Sea as Wegener had proposed? Cloos recalled:

After some years I took a new and, incidentally, last trip though the great graben. This time I came from the south, from the African highlands, and carried with me the impression of the mighty uplift of the African continent. To shorten the long sea voyage, I followed the course of the ship on the captain's new and detailed sea charts and became aware of the deep, narrow groove cut into the center of the graben floor, which seemed to exclude the idea of an open fissure of the full width of the Red Sea. I returned to the laboratory impressed by the large vertical and slight lateral displacements which seemed responsible for the great graben.

In a new test we pushed vertically instead of sideways. A dome took shape and behold! – if one moved along and slowly enough, so that the experimental material subsided from the crustal tensions under the sole influence of the gravitational pull, a wide, sharply outlined graben appeared in the crest of the dome. The subsiding block was wedge-shaped and narrowed downward. It followed beautiful, smooth fault planes, and was bound to the high ground on both sides by terraces, or steps, and the wedge sank! The result was not only like nature; it was beautiful as well.

(Cloos, 1947/1953: 398)

His fixist solution was published in 1939. He later wrote:

Would an extension of the crust in the dome by means of uplift suffice to open the rift? So the experiment seemed to indicate. Other smaller troughs in East Africa, on the Rhine, in Norway and Sweden – each of these representing a subsidence corresponding in depth to a definite amount of crustal swelling – seemed to confirm the theory. The piece of land called Danakil, which had been a stumbling block to Wegener's drift theory, now had its organic place in the dynamic process which had formed the dome.

(Cloos, 1947/1953: 398–399)

The Danakil triangle, northwest of Djbouti, juts outward into the Red Sea and destroys the congruency with Yemen, creating a difficulty for Wegener (RS2). His solution, he claimed, explained the Danakil triangle whereas Wegener's did not (RS3).

The second of these "insignificant events" was the publication of a new bathymetric map of the Gulf of Aden based on recent soundings. Cloos recalled:

Shortly before the outbreak of World War II, a second and modest factor came into play: a new sea chart of the Gulf of Aden appeared . . . If Wegener were right, the observer [Cloos] argued, in the Gulf of Aden we could expect to find tear fractures caused by the movement of Arabia north or northeast and away from Africa. This requirement would be fulfilled by fissures running more or less parallel to the edges of the Gulf, viz. running east or southeast. But if the Gulf of Aden is a graben, a subsidence in the center of a Nubian-Arabian dome or shield, the dome should be traversed in its long direction, from northwest to southeast, like the Red Sea, by steps, fissures, and smaller troughs. Alas – we are no wiser than before!

(Cloos, 1947/1953: 399–400; my bracketed addition)

As he later recalled, he (Cloos, 1942) did eventually find a solution in terms of his doming hypothesis.

Finally I visualized a solution, based on underground structures. Now the Gulf of Aden Graben could be plausibly accounted for by the elevation of a shield. I assembled the result and published it, with the notation that the explanation was by no means definitive or completely satisfactory.

(Cloos, 1947/1953: 400)

Cloos again argued that his overall solution was better than what mobilists offered. But the newly revealed structure of the Gulf of Aden was not exactly what he had expected. This he explained, not entirely to his satisfaction, by irregular stretching of the Nubian–Arabian shield as it domed upward. As for mobilism, Cloos concluded:

The structures, newly discovered through echo soundings can be shown to fit the drift hypothesis only through the most awkward wriggling. The properties of the Gulf of Aden contradict the drift hypothesis.

(Cloos, 1942: 363; rendering, Clancy Martin)

Despite his work in Africa, Cloos remained a regionalist, localizing his early structural studies there and in Germany; his lack of appreciation of the import of global stratigraphy, including Permo-Carboniferous glaciation, distribution of *Glossopteris* flora and other biotic disjuncts, colored his world view and led him to reject mobilism. His early fieldwork in Germany fixed his fixist ideas.

He went to Africa a fixist before Wegener had proposed his theory of continental drift. In South Africa he concentrated on structural problems, and offered a fixist explanation for the origin of the Cape Fold Belt. His interests were almost exclusively in structural geology, and when confronted with the Red Sea problem, sought inspiration from his home ground, Rhine Graben, and his own fixist explanation of it.

8.4 The 1939 pro-fixist Frankfurt symposium

While Cloos was its editor-in-chief, *Geologische Rundschau* published the proceeding of a symposium on the history of the Atlantic, held in Frankfurt am Main, January 7 and 8, 1939. Cloos decided to publish all thirty-eight papers of invitees even though some did not attend. Most participants were German, Swiss, or Dutch; du Toit from South Africa, Bucher and Willis from the United States, and Keidel from Argentina also contributed. Hume (1948) summarized most of the papers.[4]

Most participants were fixists. Of those who wrote primarily about tectonics (van Bemmelen, Nölke, Sonder, Stille, Staub, Leuchs, Lehmann, and van der Gracht), only van der Gracht favored mobilism. Van Bemmelen (1939) presented his oscillation (undulation) theory (§8.14). Nölke (1939), a sedimentologist from Bremen, claimed that continental drift was impossible, and favored Stille's contractionism. He argued that the geodetic measurements of Greenland's longitude were inconclusive, and, citing Jeffreys, he repeated yet again that there are no known forces able to move continents (Hume, 1948: 349–351). Nölke also claimed that isthmian connections adequately explain trans-Atlantic biotic disjuncts. Sonder (1939), a contractionist from Zurich, argued for the accretion of continents through build-up of successive peripheral orogenic belts, and explained biotic disjuncts by landbridges. Agreeing with those who argued that landbridges could not sink beneath oceans, Sonder appealed to the existing topography of the Atlantic (partly revealed by the *Meteor* Expedition), arguing that sea-level changes exposed migratory routes (Hume, 1948: 357–359).[5] Staub (1939) described similarities between the Sierra Madre Oriental Mountains in Mexico and the western Alps, but still he favored Cloos' fixist account of orogeny (Hume, 1948: 369–370). Leuchs (1939), a fixist, disagreed with Argand's mobilist account of the formation of the Alps and argued that Africa had not overridden part of Europe (Hume, 1948: 370–371). Lehmann (1939), from Essen, rejected Wegener's mobilism because of its mechanism difficulties.

At the Frankfurt symposium, Stille presented his fixist account of various symmetries associated with the Atlantic Basin as described above. Rittmann (1939a, b), a vulcanologist and fixist from Basel, thought that not enough was known about the early history of the Atlantic to determine its origin, but generally preferred Haarmann and van Bemmelen's oscillationist view; he also thought that progress had been impeded by over-specialization: some ignored findings from seismology and geophysics; others took little heed of geology (Hume, 1948: 335–337). Knetsch, from

Bonn, favored fixism, although as Hume (1948: 335) reported, he claimed "that there was no decisive evidence for or against the Wegener hypothesis." Knetsch (1939) admitted that there are many similarities between South America and South Africa, but argued that they could be explained by a direct landbridge, or migration via Antarctica; he also agreed with Cloos and Krenkel that there appeared to be strong similarities between the floor of the Atlantic and neighboring continents, and favored Haarmann and van Bemmelen's oscillation theory (Hume, 1948: 335). Von Bubnoff, from Greifswald, Germany, questioned some of the purported similarities between South America and South Africa, and favored Cloos' fixist views about the origin of the Atlantic (Hume, 1948: 337). Hennig (1939), from Tübingen, adamantly opposed mobilism, arguing that structural and stratigraphic disjuncts across the Atlantic had been exaggerated; admitting the reality of faunal disjuncts between Africa and South America, he favored Schuchert and Willis' isthmian links and believed that Antarctica might also have served as a migratory route (Hume, 1948: 386).

Kummerow (1939), a paleontologist from Rudersdorf, strongly opposed mobilism; specializing in ostracods, he argued that the facts of paleontology are inconsistent with the possibility of bringing together Europe and North America, and showed instead that an Atlantic Ocean already existed in Paleozoic times (Kummerow, 1939: 95). Behrmann, a geographer from Frankfurt, claimed that fixism offered a better explanation of past climates than mobilism; unperturbed by Permo-Carboniferous glaciation, he (1939: 120) agreed with A. Penck that it could be explained by a lowering of temperature over the whole Earth (Hume, 1948: 401).

Krenkel, a geologist from Leipzig, also defended fixism. Although his (1939) contribution to the proceedings was minimal, he (1925) had already argued against mobilism; du Toit (1937: 31) singled him out as one of the "principal contributors adverse to the hypothesis" of continental drift. Krenkel claimed that Africa and its neighboring regions have the same geological structure and that the continent and seafloor are of similar composition (Hume, 1948: 190; Umbgrove, 1947: 220; du Toit, 1937: 31). Even though Krenkel agreed that the continental margins of Western Africa and Brazil were somewhat congruent, he thought the purported geological similarities did not warrant juxtapositioning them (Hume, 1948: 187–188). He claimed that up to the beginning of the Carboniferous, Gondwana, a vast landmass, had existed, but then broke apart by crustal collapse, forming intervening ocean basins. Krenkel was unclear about how they collapsed, but he did think that the same tectonic forces shape both continents and oceans (Hume, 1948: 190). Despite opposing large-scale continental drift, Krenkel sanctioned small movements; Africa and Europe, bound together by a folded crustal zone (the Tethys geosyncline), have moved toward each other since the early Paleozoic forming mountains belts such as the Caledonides, Atlas, and Alps (Hume, 1948: 190–191). Unless he formulated his fixist views as a result of early work on the symmetry of mountain systems or he had concerns only for structure and tectonics, his fixist analysis of Africa's geological history does not support my regionalism account.

Contributing oceanographers also preferred fixism. W. Wüst (1890–1977), who had taken over leadership of the *Meteor* Expedition after Alfred Merz died, argued (1939) that its sonic survey of the Atlantic revealed a basin-swell structure in harmony with Cloos' division of the crust into polygonal areas (Hume, 1948: 403). On the shorter timescale, W. Schoot (1939) another *Meteor* Expedition participant, ruled out Köppen and Wegener's extravagant claim (erroneous, as Hospers (II, §2.9) was later to show) that the position of the poles and equator had been different so recently as the last glacial period; he claimed that the distribution and nature of recent deep-sediments in the equatorial region of the Atlantic Basin indicated that the climatic belts had remained unchanged since the initiation of the last glacial period (Hume, 1948: 405). C. Troll (1939), another oceanographer, also favored Cloos' view that continents and oceans other than the Pacific are similar. Hume neatly summarized Troll and Cloos' position.

In the opinion of the above writer [Troll], Cloos' conception of the polygonal area structure of High Africa and of the Atlantic Ocean has a special future. In Cloos' opinion, continent and ocean floor are not so fundamentally in contrast to one another, the unity of Gondwana is more sharply emphasized, old land connections or island bridges as called for by biogeographers and palaeontologists can be more easily accepted as unstable articulations between land blocks, and finally the accordant form between part of the former Gondwanaland becomes more acceptable.

(Hume, 1948: 406–407; my bracketed addition)

So Wüst and Schott both argued that the new finding about the topography of the ocean floor as revealed through echo soundings supported fixism. They were not alone: Cloos, Stille, Sonder, and von Bubnoff all believed that the new findings showed that there was no fundamental difference in structure between the Atlantic Basin and the continents. Looking ahead to the continued development of oceanography after World War II and its critical importance in furnishing strong support for mobilism, it is worth remembering that both fixists and mobilists claimed that the results of mapping of ocean floors up to the late 1930s supported their very different points of view.

The last fixist who contributed to the Frankfurt proceedings, important because of his later unbending resistance to mobilism and wide influence, was Walter Bucher: educated in Germany, he had emigrated to the United States. Like many above, Bucher (1939a, b, 1940) argued that the new observations of the ocean floor failed to support mobilism.

It looks as if large areas of the earth's surface had torn apart and separated. Wegener ... had the courage to say it does not only look that way, but this is precisely what happened. A single parent landmass tore, and its fragments drifted apart. He had to postulate that the rocks of the ocean floors were materially different from those of the continents in their behavior toward long-applied stresses. But the experience of the seafloor by means of echo sounding has revealed large areas where as much vertical relief exists as on land.

(Bucher, 1939b: 426–427)

Like Cloos, Krenkel, and others, Bucher (1939a, 1939b, and 1940) claimed that the new technique of echo sounding showed that the South Atlantic, like continents, is divided into basins and swells.

Even more significant is the fact that nonlinear elements of similar pattern dominate the present surface of the oceans. Here the revolutionary method of echo sounding is rapidly filling the enormous gaps in our knowledge. A study of the new map of the Southern Atlantic which has resulted from the cruise of the Meteor a decade ago is illuminating (Pratje, 1928). Great basins, such as the Brazilian and Argentinean basin on the west side, and the Cape Verde, Sierra Leone, Guinea, Angola, and Cape basins on the east side, are separated by swells which exhibit a distinct tendency to trend northeastward . . . The Mid-Atlantic ridge differs from the ordinary swell in consisting of a number of such swells between the basins on the west side of Africa, as Krenkel (1925), over a decade ago, and lately Hans Cloos (1937) have shown.

(Bucher, 1939b: 423)

When discussing the North Atlantic a year later, Bucher elaborated further.

But have we reason to think that the physical behavior of the sub-oceanic surface as a whole is essentially different anywhere from that of the continents? (Please note that this question does not refer to geologic history or to lithologic character, but solely to physical behavior.) The map of the North Atlantic shows that the rises which separate the individual basins consist of lines of swells comparable to a certain extent to that which in Eastern North America runs from southwestern Ontario through the Cincinnati and Nashville domes, separating the Appalachian from the East-central sedimentary basin. The whole pattern of the ocean floor seems, in fact, comparable to that of the basins and swells of the continental areas outside the great orogenic belts, although the scale is larger both horizontally and vertically on the oceanic surfaces. It seems that the same process that divides the continental platforms and probably also the ancient coigns (Cloos, 1937) into basins and swells produces similar changes on the ocean floor.

(Bucher, 1933; Wüst, 1939a)

The profiles constructed from echo-soundings obtained by the Meteor in the South Atlantic show also conclusively that in detail the surface of the basin floors does not maintain a uniform elevation for any distance and is not smooth as Wegener thought it to be. Like that of the continental margins, the behavior of the ocean floor is contrary to the expectations of Wegener.

(Bucher, 1940: 492–493; Wüst, 1939a is my Wüst, 1939; other references are the same)

Bucher applauded Cloos, arguing that he had presented the best explanation of the Mid-Atlantic Ridge.

Cloos has given a convincing structural analysis of this topographic pattern in his newest brilliant paper on *Hebung-Spaltung-Vulkanismus*. He interprets the whole as the result of an arching-up of the seafloor, possibly above a laccolithic intrusion. The lines of intensive volcanic activity indicate lines of fracturing which in an upwarp of this shape may be expected to parallel the main axis. The similarity between the actual condition and the pattern of fractures developed in one of Cloos' model experiments is quite convincing.

(Bucher, 1940: 495)

The US geologists Willis and Kay, as well as Bucher (now American), were much influenced by Cloos and Stille, and found that much of what they and other German structural geologists had to say about their own mountains applied to the Appalachians. I shall continue to monitor Bucher's views (§8.8, §9.6; III, §1.10, §6.3).

I turn now to mobilist symposiasts. Besides du Toit and van der Gracht, there were only five, Kurt Wegener, Alfred's brother, and four others who presented paleontological and biogeographical arguments. None appealed to the geology of Germany to support mobilism. Kurt Wegener, like his brother, argued that Danish geodesists had established the westward drift of Greenland relative to Europe (see §3.13). He (1939) also emphasized mobilism's ability to explain Carboniferous climates without drastic distortion of climatic belts (Hume, 1948: 338). H. Gerth (1939), a sedimentologist from Amsterdam, who already (1935) had written a book on South American geology, reviewed the geological history of the South Atlantic region, and was somewhat supportive of mobilism. He definitely thought it offered solutions to a number of problems such as the similarities among South African and South American Mesozoic reptiles and ammonites (Hume, 1948: 379–384). Wittmann, a biogeographer from Lorrach, Germany, thought that mobilism was the best explanation of the biotic disjuncts in South Africa, South America, and other Gondwana fragments (Hume, 1948: 392–396). Rutsch (1939), a paleontologist from Basel, who specialized in tropical American Tertiary mollusks, favored mobilism, noting that the free-swimming planktonic larvae of Tertiary mollusks found on both sides of the Atlantic could have crossed only a narrow Atlantic (Hume, 1948: 396–399). Kirsch, an orchid specialist, argued (1939) that the occurrence of certain species of orchids found in South Africa, Madagascar, the Seychelles Islands, and peninsular India require continental drift; they live symbiotically with certain fungi needed for seed germination, and he thought it highly unlikely that both the seed and fungi could survive trans-oceanic voyages (Hume, 1948: 344–345).

A few participants reserved their opinion. Becker, an astronomer from Glasgow University, suggested that the application of centripetal forces (flight from the poles) could, given certain assumptions, bring about polar wandering if they were strong enough, but did not say whether they were or not (Hume, 1948: 338–339). Von Huene (1939), a paleontologist/geologist from Tübingen, discussed reptilian distribution on both sides of the Atlantic from Lower Permian through Upper Cretaceous. He claimed that the distribution of Early Permian *Mesosaurus*, which appeared to be restricted to brackish-water and found only in southwestern Africa and eastern South America, required a land connection between them (Hume, 1948: 389). He further claimed that similarities between reptilian fauna on both sides of the Atlantic during the Upper Triassic did not require a direct connection between South America and Africa, but did suggest a connection through Antarctica. Von Huene also argued that South American reptilian fauna in the Upper Cretaceous Epoch were much more like those from North America than Africa, so that whatever connection there might have been between them had disappeared by the Upper Cretaceous (Hume, 1948: 390–391). H. Keidel (1877–1954), former Director of the Geological Survey of Argentina, wrote a lengthy paper on the geology of Argentina in which he noted that du Toit (1927, 1933) and Wegener (1922) had used du Toit's discovery of Permian glacial deposits and *Glossopteris* flora in Argentina in their defenses of mobilism.[6] Keidel (1939), who had helped du Toit (1927: 5) while he was in South America, was sympathetic to mobilism but had doubts. He discussed the many similarities between South America and South Africa, but he objected to du Toit's fit of the two continents (Hume, 1948: 410–411).

The oceanographer O. Pratje, who was on the *Meteor* Expedition, declared (1939) that all islands associated with the Mid-Atlantic Ridge, except for St. Paul's Island, were volcanic, and supposed "an intimate relationship between ridge structure and volcanic activity in the Atlantic Ocean region" (Hume, 1948: 405). He said nothing about mobilism.

The final contribution to the Frankfurt symposium was F. Bernauer's, "Island und die Frage der Kontinentalverschiebungen" (Iceland and the question of continental displacement). He proposed a relationship between volcanic activity, formation of the Mid-Atlantic Ridge, sub-surface currents, and small-scale continental drift. According to Bernauer (1939), underlying currents split the crust, creating a gap and exposing the magma beneath. Hume, who translated the paper, wrote:

The explanation can only be sought in a symmetrical stretching arising from the lower layer, which extension can only follow as a result of tearing, due to what is described as *"Reibungs-kuppelung"* – friction-coupling – of the adjoining hard covering layer. Such occurrences are to be observed where thick pasty slaggy masses, spread over still flowing lavas, are broken up into strips or even widely-separated islands as a result of the action of the subjacent currents ["Unterströmung"]. It is remarked as noteworthy that cracks and bendings of a fissure are always imitated by neighbouring fissures and only die away at a considerable distance.

(Hume, 1948: 378 translation of Bernauer, 1939: 357; my bracketed addition)

The exposed magma reacts with seawater to form palagonite.

The result of this flattened stretching is first of all a sinking of the upper surface in the stretched area. The sea can penetrate into the latter and comes into contact in the depths of the cracks with the ever-increasingly exposed magma. The result is rocks of palagonitic type, the numerous basalt cones, veins and small intrusions with glassy borders, small columnar jointing, water-containing tuffs, which point to a submarine pouring forth of a highly fluid magma.

(Hume, 1948: 378–379 translation of Bernauer, 1939: 357)

The newly formed palagonitic rocks rise isostatically.

Gradually the magma "becomes aware" of the diminution of the load over a wide area above it and begins by upward rising to reestablish equilibrium. Its compact mass smoothes out the defect of gravity, which is to be expected in the young volcanic zone because of the palagonite tuffs of light specific gravity . . . In consequence of the inertia (Trägheit) of the movement of ascension, there is a tendency to the formation of conspicuous forms along the fissure net of pillars and ridges, both on the small and on the large scale, the latter being in the Reykjanes peninsula.

(Hume, 1948: 379 translation of Bernauer, 1939: 357–358).

Bernauer then made the important suggestion that the Reykjanes Ridge and Mid-Atlantic "Swell" had a common origin.

The prolongation of the Icelandic volcanic zone is the equally volcanic Reykjanes ridge and the Atlantic Swell. It is easy to ascribe a similar structure to all these. The similarity of the trend to that of the two continental boundaries suggest a common cause, perhaps tearing by an undercurrent in the above-mentioned direction.

(Hume, 1948: 379 translation of Bernauer, 1939: 358)

The idea that the whole Atlantic "Swell" had formed the same way led to the question of continental drift. Bernauer, who assumed attachment of the ocean floor to the adjacent continents, thought that continents would drift apart, but not nearly to the extent envisioned by mobilists.

The continents would then have wandered from one another, but far from the extent assumed by Wegener; much more one would expect to find many small and large strips and masses of the former continuous floor of the Atlantic.

(Hume, 1948: 379 translation of Bernauer, 1939: 358)

He was not proposing seafloor spreading. The process he was invoking did not create the whole seafloor; it was only the Mid-Atlantic Ridge that was created. He rejected large-scale continental drift. If the continents, originally contiguous, had split apart along what became the Mid-Atlantic Ridge, and drifted to their current positions, then the whole seafloor should be made up of strips of palagonite rock, just like the Mid-Atlantic "Swell." Bernauer did not think it was. He claimed that the amount of drift was only the width of Iceland's young volcanic region, which he estimated to be from 100 to 200 km (Hume, 1948: 379; Bernauer, 1939: 357).

8.5 Some other fixist Europeans

At the Frankfurt meeting, most German Earth scientists rejected mobilism. Every structural geologist and tectonicist did so. Any support for mobilism that was expressed had nothing to do with German geology. Except for Bernauer, who claimed only very modest lateral motion, geologists who expressed an opinion about the structure of the Atlantic Basin sided with fixism, and agreed with Cloos, Haarmann, Stille, or one of the oscillationists. The day was still far off when oceanographic data would lead to the general acceptance of mobilism.

Fixism also held sway among tectonicists who worked on the Eastern Alps. An Austrian, Leopold Kober (1883–1970), rejected Wegener's theory. Şengör (1979/1982a) argued that Kober (and Stille) developed their views "largely based on their personal field experience in central Europe in the then poorly understood eastern Alps and Alpine chains of the eastern Mediterranean." Both Faill (1985) and Carozzi (1985) argued that Kober supported fixism because of his work on the Eastern Alps where overthrusts are not nearly as apparent as in the Western Alps. Şengör (1979/1982a) also noted similarities between Kober and Stille. Like Stille, Kober was a contractionist but eschewed the idea of overthrusts. He argued that mountains formed in regions of high mobility that he called "orogens." Orogens were surrounded by stable regions that Kober called "kratogens," which were roughly equivalent to Stille's cratons. Kober's orogens were symmetrical and gave rise to symmetrical mountains when subject to lateral compression from adjacent kratogens.

Despite Kober's influence, one of his younger colleagues, Franz Heritsch (1923) from the University of Graz, first rejected Wegener, but came to agree with him in 1927 after reviewing the development of Nappe Theory; he became a mobilist by way of Argand and Staub. Heritsch (1927) also appealed to O. Ampferer and R. Schwinner's idea that convection could cause drift (see note 21, §5.7 for Wegener's appeal to Ampferer and Schwinner).[7] In the 1929 English translation of his work, he wrote:

If we turn these views over in our minds, we observe that the simple-squeezing out of a root-syncline is no longer the "Nappist" conception, but that the movement of great blocks produced the nappe-mountains. In the conception of Argand and Staub it is the forward drive of the African shield which created the Alps. As a cause of these great movements we can draw upon Wegener's daring ideas, or we can bear in mind the "sub-flow" theory founded by Ampferer and developed by Schwinner. This, as Ampferer has shown, is indeed in no way irreconcilable with Wegener's hypothesis of Continental Drift.

(Heritsch, 1929: 194)

As already noted (§7.3), Kay and other Appalachian tectonicists had their counterparts in Germany who supported "vertical" instead of "horizontal" tectonics. Because of the influence of Stille and to a lesser extent of Kober on Kay, one might believe that his views were from them and not based on Appalachian studies. This, I think, would be a mistake. Kay's views were grounded in his own fieldwork. Of course, Stille and Kober influenced his interpretation, but even if he had simply adopted their ideas and not developed them himself, he thought they explained what he saw in the Appalachians. Kay's views were grounded in his own fieldwork.

Şengör (1979/1982a: 42) briefly describes another fixist view of mountain building, the oscillation theory. The German tectonicist Erich Haarmann (1882–1945) developed the theory and presented it in his 1930 *Die Oszillationstheorie*. Oscillationists rejected any large-scale horizontal motions and forces, even the sideways compression of geosynclines by cratons. Instead radial forces excited by vertical oscillations lead to mountain building. "According to this conception, the motive power for the formation of overthrusts nappes is the force of gravitation" (van Bemmelen, 1949: 287). Only Soviet Earth scientists fully embraced Haarmann's ideas.

During the 1931 Berlin meeting of the German Geological Society there appeared such strong opposition to his views that, except for van Bemmelen, who took Haarmann's theory and developed it further into his own undation theory, nobody in the Western world took them seriously. However, in Soviet Russia Haarmann's ideas, coupled with the concepts of Stille and

Kober, gave rise to the Russian school of primary vertical tectonics, especially through M. M. Tetyayev in the early 1940s and then through Vladimir Vladimirovich Beloussov. The Russian school was particularly impressed by the oscillatory movements of the crust, as expressed in their platform-dominated homeland and believed that such movements must be considered the fundamental mode of tectonism. They also divided the Earth's crust into stable regions (platforms) and mobile regions (geosynclines), but even in the case of geosynclinal deformation that leads to orogeny, all primary movements are assumed to be vertical.

(Şengör, 1979/1982a: 42)

The hostile attitude Soviet tectonicists adopted toward mobilism was driven by regionalism; they could explain the geology of the Russian Platform and vast tracts of Siberia in terms of oscillation theory without recourse to large lateral motions. Vladimir Vladimirovich Beloussov then took this regionally developed tectonics and, like Stille, applied it elsewhere. Soviet paleomagnetists, who began supporting mobilism in the late 1950s, had no effect on Beloussov's views. Others also adopted or proposed different versions of oscillation theory. Van Bemmelen (§8.14) developed his undulation theory in response to his fieldwork in Indonesia. Bucher (1933, 1939b) had his pulsation theory. Hume (1948: xlix) was sympathetic to oscillation theory.

Two other fixists, Pierre Termier of France and Hans Schardt of Switzerland, both Alpine geologists, deserve mention. They made important contributions to Nappe Theory but objected to mobilism (Greene, 1982: 207–220; Bailey, 1935; Trümpy, 1991). I shall consider their objections later (§8.9).

8.6 Mobilists from continental Europe

I first consider European tectonicists who were mobilists. All worked extensively in the Western Alps, where it was becoming evident that horizontal motions dominate the landscape. They include the Swiss Émile Argand, Léon Collet, Rudolf Staub, Elie Gagnebin, Albert Heim, and Arnold Heim, and the Germans Wilhelm Salomon-Calvi and Helmut de Terra. I then examine a second group of European geologists who worked in at least two sections of the Caledonides, which are in Norway, Spitsbergen, Greenland, Scotland (and adjacent areas), and northeastern North America, and which they recognized as tectonic disjuncts. They include Eugene Wegmann (Swiss), H. Bütler (Swiss), Herman Aldinger (German), and Olaf Holtedahl (Norwegian). Then I shall consider the work of Franz E. Suess (Austrian, and Eduard Suess' son) who became a mobilist as a result of his intensive fieldwork in metamorphic terraces of the Hercynides. He argued that they, like the Alps he had previously studied, were the result of pre-Alpine mobilism. Later he argued for mobilism because of strong similarities between Caledonide and Hercynide fragments on either side of the North Atlantic. The Hercynides resulted from orogeny that culminated in the Late Carboniferous, and their equivalent in North America is the Appalachian orogenic belt. Argand, Wegmann, and F. E. Suess especially made contributions to mobilism; although they did not appreciably affect the eventual outcome of the controversy, their work was prescient and, in large measure, correct. Argand provided mobilism with a sophisticated, general account of mountain building. Wegmann, Argand's successor at Neuchâtel, worked in the Western Alps and in areas dominated by Caledonide fragments. He documented the existence of east Greenland's Caledonides, and developed an innovative way to reconstruct Greenland's likely displacement relative to Scandinavia. Although Wegmann's most important work was in the Caledonides, his mobilistic view arose from early fieldwork under Argand's tutelage in the Western Alps; he is therefore a member of both groups. F. E. Suess applied the new techniques of metamorphic petrology to the deeply eroded Hercynides of the Bohemian Massif, recognized three tectonic zones, and showed that his zonal model applied also to the Alps and Caledonides. He went on to identify structural similarities among Caledonide fragments on both sides of the Atlantic.

8.7 Argand and his synthesis

Émile Argand (1879–1940) gave a mobilistic account of mountain-building, making it at least respectable to Alpine tectonicists. He also developed a mobilistic explanation for mountain systems that predicted the breakup of Wegener's Pangea, proposing, as already explained (§7.3), a proto-Atlantic. Argand's contribution was elegant and far-reaching.

Born in Geneva, Argand first honored his father's request to study architecture, but changed to medicine and philology in 1889.[8] While vacationing in the Alps, he met Maurice Lugeon (1870–1953), a leading Alpine geologist, who taught geology at the University of Lausanne, Switzerland. Argand showed Lugeon sketches he had made displaying the structure of the surrounding region. Lugeon reportedly afterwards said, "A man 25 years old [Argand] talked to me with the science of a master" (Carozzi, 1977: xiv; my bracketed addition). In return, Argand was impressed by Lugeon and tectonics, and began studying geology under him in 1905 at Lausanne. Within three years, he had mapped the Dent Blanche region, showing that the Dent Blanche nappe overrides the Monte Rosa nappe. He was awarded the Prix William Huber. In 1911, Argand, only thirty-two years old, became professor of geology, mineralogy, and paleontology at the University of Neuchâtel, Switzerland. He continued to work on the Western Alps. Gifted at three-dimensional analysis, he was able to recognize metamorphic structures that had disguised the stratigraphic layering. Adding time to his analysis, each map became a montage in his motion picture of Alpine tectonics. In 1915, Argand (1916) completed a general study of the Western Alps. Northward horizontal movements of a non-rigid crust explained the building of nappes and recumbent folds. Deeply buried nappes were pushed more forcefully than those at the surface. The newly formed Alps were pushed against the older Hercynides, which resisted their northward migration (Thalmann, 1943: 154–155).

As already noted (§3.11), Argand was a very early supporter of Wegener, present-ing his views at a November 1916 meeting of the Neuchâtel Society of Natural Sciences (Carozzi, 1977: xvii). He argued that Wegener's mobilism offered better solutions to the problems of biotic disjuncts and the Permo-Carboniferous glaciation than fixist theories.

[He] discussed contemporary views on the origin of continents and oceans. He discussed the theories of Wegener, who attempted to replace the hypothesis of sinking with dislocating. This new conception explains much better than the older one the similarity of fauna and flora among regions at present separated by vast oceans; it also is supported by the study of late Paleozoic glaciations.

(Anonymous, 1916: 115; Carozzi translation)

Unlike Cloos, who remained conceptually within the confines of his Rhine Graben and field of structural geology, Argand, perhaps taking his cue from Wegener, reached out across Earth and to fields beyond structural geology.

Argand presented an extensive defense of mobilism in his 1924 *La Tectonique de l'Asie* that has become a classic. This work, which Argand presented in his inaugural address at the 13th International Geological Congress at Brussels in August of 1922, was a unique synthesis of global tectonics in mobilist terms. It brought him worldwide recognition. His talk was remembered as being brilliant and the applause after his lecture exceeded half an hour (Thalmann, 1943: 159). Argand, using the analytical tools he had developed in explaining the Western Alps, presented an account of global tectonics that provided further support for Wegener.

Argand's outstanding publication, *La Tectonique de l'Asie* (1924), belongs today to the classical literature of geology. It can be regarded as a continuation of the epochal work of Ed. Suess, *The face of the earth*. Many if not most of Argand's results were attained independently of Wegener's hypothesis of continental drift and support marvelously the latter's grandiose ideas. For this work Argand was awarded in 1927 the "*Prix Marcel Benoist*" in Switzerland, and the Academy of Science in Paris offered him in the same year the "*Prix Curvier*" for the whole of his works. Many honors came to him, among them the election as a Foreign Correspondent of the Geological Society of London in 1923 (elected Foreign Fellow in 1938), and as Foreign Correspondent of our Society [GSA] at the New York Meeting in 1935.

(Thalmann, 1943: 159; my bracketed addition)

Argand published little in geology after *La Tectonique de l'Asie* (hereafter, *La Tectonique*). He still taught but became interested in philosophy, mathematics, phys-ics, and linguistics. Apparently he worked in private on finding a mechanism for continental drift, but left nothing to indicate what progress, if any, he made (Carozzi, 1977: xxiii).

Although Argand was viewed by some as a genius and widely recognized for his achievement, his work seems to have persuaded few, if any, who did not work on the Western Alps to convert from fixism to mobilism. Some accepted his early ideas about the formation of the Western Alps (Argand, 1916) but rejected the link that he made between them and global mobilism.

His views on deformation and orogeny were as follows: he (1924/1977: 30) claimed that sialic continents are highly plastic and therefore became folded through the action of horizontal (tangential) forces acting over long periods of time; even continental shields have some plasticity and still are susceptible to folding (Argand, 1924/1977: 105). All deformation is essentially horizontal. Vertically acting forces had little place in his world. Argand introduced the idea of deep-seated or basement folds (*plis de fond*):

Basement folding is the folding of the continental mass itself, from its plastic depths to its higher zones, where a lower plasticity predominates in varying degrees.

(Argand, 1924/1977: 102)

Complexes of basement folds form continental shields, and forming them requires much energy (Argand, 1924/1977: 105). They involve folding of old peneplaned mountains covered with metamorphosed sediments. When these continental shields are pushed horizontally, basement folds break, forming nappes. Argand developed his idea of deep-seated folds, in part, to explain the great size of the nappes of the Pennine Alps. Basement folds form when continents collide, Gondwana with Eurasia, for example, creating the Alps and Himalayas. Collision is an important way of building mountains. Basement folds are reworked during subsequent mountain-building episodes. They also form within the interiors of continents as they drift, their basal sialic layer interacting with underlying sima. Some continents are more plastic than others; those crisscrossed by basement folds such as continental shields are less plastic than those with fewer basement folds. Thus, the arrangement of basement folds influences the extent and direction of future deformation.

Argand contended that mountains also form through geosynclinal folding (*chaînes géosynclinals*). Geosynclinal depressions are caused by thinning of sialic crust resulting from horizontal traction that differentially stretches sialic masses of varying plasticity (1924/1977: 134). As geosynclines fill with sediments and magma rises from the underlying sima, they become compressed, folding occurs and mountains form. Compression occurs as approaching continents squeeze intervening "oceanic" geosynclines or as opposing continental blocks compress an intra-continental geosyncline. Much less energy is needed to fold geosynclinal sediments than to create deep-seated continental folds (Argand, 1924/1977: 32).

Using these concepts, he explained the orogenic history of Eurasia as the "Duel between Indo-Africa and Eurasia," the central factor in the formation of Alpine–Himalayan mountains. He then introduced continental drift in this interesting fashion:

To arrive at this conclusion, it was not necessary to use such ideas as have been put forward concerning the fixism of continents or the ability of continents to undergo considerable horizontal displacement. Therefore, the preceding remarks are independent of those theoretical viewpoints. None of them, strictly speaking, implies the necessity of a choice between these different points of view. But right now, we can clearly see them leading to a view that agrees well with the second hypothesis.

(Argand, 1924/1977: 117–118)

The evidence he had garnered for his theory of mountain building provided new and independent support for mobilism.

With mobilism introduced, Argand (§7.2), proposed a proto-Atlantic to explain the formation of the North American Caledonides in terms of his own theory of orogeny. His proto-Atlantic, which he referred to as "My *Proto-Atlantic*," was in effect an Early Paleozoic geosyncline (Argand, 1924/1977: 139). The collision of the Americas and Euro-Africa caused geosynclinal and basement folding, leading to formation of the Caledonides, fragments of which are now on both sides of the Atlantic. Argand also referred to his proto-Atlantic as a Caledonian geosyncline (Argand 1924/1977: 21, 139). Returning to Eurasia, he gave several examples to show that his own account of mountain formation along the Tethys required mobilism.

The penetration of Africa into the middle of Europe, of which I have spoken in the past under other circumstances, seems to me today – considering only concrete facts – to be unaccountable without continental drifts, particularly since the amplitude of the overthrusts, considering only what can be seen of it in the repeated soles of the Austro-Alpine nappes, is enormous even when we take into account some cause of error by excess that accompany this kind of estimation.

(Argand, 1924/1977: 142)

Wegener had proposed that Africa and Europe had moved toward each other; Argand agreed, adding that Africa had overthrust Europe closing the Tethys, which later reopened to form the modern Mediterranean as the two continents retreated away from each other. In working out the details, Argand (1924/1977: 145) also proposed, anticipating Carey (II, §6.7), that the Iberian and Italian peninsulas had rotated counter-clockwise relative to northern Europe. Wegener proposed that India moved northward, eventually pushing up against Asia; Argand proposed that India underthrust Asia. Wegener proposed that Arabia pushed against the Anatolian–Iranian region; Argand proposed both underthrusting and overthrusting of the colliding masses.

Argand then turned his attention to the Andes and other mountain ranges that Wegener envisioned to have formed on the leading edge of drifting continents. Argand introduced another type of folds, which he called marginal chains (*chaînes marginales*). These are new basement folds that form at leading edges of drifting continents through the reworking of previously formed basement folds as sial meets resistance from sima. He mentioned several examples: the Australian and New Zealand Alps (p. 149), the North American Cordillera (pp. 25–27), and the South American Andes, saying this about the last:

The basement folds of the South American Andes are due essentially to the resistance that the Pacific sima presented to the westward drift of Gondwana, and later on to that of South America, when it became separated from Gondwana by the opening of the Atlantic. One cannot say as yet . . . how many main generations of basement fold have contributed to the construction of the South American Andes during the Alpine cycle; the phases have, at least, been numerous. Taking into account the reused dead material [i.e., already formed, basement folds], one can distinguish Alpine basement folds consisting of Paleozoic folded material, which built the eastern chains of Argentina, Bolivia, and Peru, and Alpine [newly formed] basement folds that built, along a more western and longer zone extending from Venezuela to Cape Horn, the major part of the Andes proper: their dead material consists predominantly of Andean folds, that is, folds that were new at the end of the Jurassic or in the Early Cretaceous.

(Argand, 1924/1977: 149–150; my bracketed additions)

So Argand provided Wegener's mobilism with a general account of mountain build-ing. By combining them, he increased the problem-solving effectiveness of both his own theory of orogeny and Wegener's theory of continental drift (RS1). Comparing the relative merits of mobilism and fixism, and restricting himself to tectonics, he repeated the objection that contractionist theories of mountain building could not provide sufficient crustal shortening to account for observed folding, then went on to remark that this difficulty paled compared to the inability of contractionism to account for basement folds (RS2).

But, the difficulties of the contraction theory – assuming it is related to fixism – are greatly increased by the necessity of accounting for basement folds. Some have already considered contraction as inadequate to explain the total shortening due to ordinary folding. A lot could be said on this question and even more on the preliminary question that pertains to the means of estimating the shortening. But, I shall not insist. Basement folds display, as do new chains, large thrusts that do not facilitate either the estimation of the shortening or the task of fixist contractionism. But they imply, in particular, an energetic expense that surpasses by far anything one has wished to consider until today. To estimate the tonnage of the slightly ordered basement folds for the entire Earth to be about ten times that of the new chains is an assumption that remains certainly well below reality. Since the great old shields are essentially very broad basement folds, large basement *brachyanticlines*, one should say fifty or sixty times more. And after all, this is only the volumetric ration; in order to proceed to the energetic ration, one has to take into account the high degree of induration [i.e., hardening of soft sediments] and multiply everything by a coefficient that cannot be precisely given today but that certainly should have a high value. These are indeed numerous subjects for medita-tion, presented in a few words.

(Argand, 1924/1977: 126; my bracketed addition)

As Argand (1924/1977: 133) said, "The mobilistic theory accounts at least qualita-tively for basement folding. Sialic rafts are deformed and folded while drifting." Fixism does not (RS1, RS3).

He also argued that mobilism explained the relative amounts of resistance encoun-tered during formation of the Alps and Himalayas, the Cordillera of western North America, and the island arcs along Asia's Pacific coastline (RS1). Developing as a measure of resistance, the concept of sial-tonnage per unit distance travelled, he estimated that the East Asian island arcs met with less resistance as they formed than did the North America Cordillera, while the resistance along the Alpine–Himalayan belt varied, being especially great in the Himalayan segment, but everywhere exceeding that of the North American Cordillera. Mobilism, he argued, readily explained why resistance to the formation of the Alpine–Himalayan belt varied and was greater than that met by the formation of the North American Cordillera and East Asian island arcs. The first required initial movement, collision, and continued movement of large sialic masses, and resistance varied because continental blocks along the Alpine–Himalayan belt vary in size. The formation of the North American Cordillera and East Asian island arcs did not involve moving sial against sial, but only sial moving through sima, and therefore met with less resistance, sial being more resistant than sima.

The present-day mobilistic theory accounts, without difficulty, for the facts pertaining to the distribution of the tonnage and for their immediate interpretation. According to this theory, the relatively homogeneous and yielding medium that occupies the Pacific is the sima. Of course, the behavior of the sima with respect to the sial can present great differences according to the various manners in which the plasticity or viscosity of the media and the duration and intensity of the efforts intervene. But, for continental drift to occur the deformations in which the sima behaves as a yielding medium must predominate in the course of time over those in which the sima resists and compels the folding of the sial. From a predominantly statistical viewpoint that encompasses the entire globe and the whole duration of the deformation, it cannot be otherwise if there is a sima over which large rafts of sial are drifting.

(Argand, 1924/1977: 131)

Like Wegener, he explained island arcs as sialic fragments detached from trailing margins of drifting continents. Their formation required less energy than creation of the North American Cordillera.

The mobilistic theory explains easily . . . the energetic inferiority of East Asia versus North America is expressed. It admits bow stresses that compress and fold the sial against the sima, under certain circumstances. It also admits stern stresses that consist of a retraction of the sial from which results, for that material, the more or less complete interruption of folding with the predominance of traction effects: distensional fractures, buttonholelike tearings creating marginal seas, releases or cordilleras that from then on lag behind the continent in the form of somewhat severed festoons. The sima, on the other hand, is compelled to adapt itself to so many new conditions, and it rises under the marginal seas and in the spaces recently abandoned by the festoons, on the stern side . . . Since the mobilistic theory requires that bow stresses should have been predominant along the western margin of America, and the stern stresses should have lasted for an appreciable length of time in East Asia, the superiority of the former and the inferiority of the second with respect to folded tonnage are self-evident.

(Argand, 1924/1977: 132)

He was pleased with his arguments, but was wise enough not to claim them as proof of mobilism.

The elegance with which the theory of continental drift explains these fundamental facts, unknown at the time it was conceived, is certainly a great argument in its favor. None of these facts rigorously demonstrates the mobilistic theory or simply the hypothesis of the existence of the sima, but all of them are perfectly in agreement with both, to the extent of making them very likely to be true.

(Argand, 1924/1977: 132)

Argand (1924/1977: 162) acknowledged, "Almost nothing is known about the forces responsible for continental drift." He tantalizingly mentioned "passive transportation of the sial by currents of the sima," but did not elaborate, and gave no hint that he ever had in mind sub-crustal convection cells.

Argand's aim was not to provide a causal theory. Although he wanted to present a "dynamic tectonics," he used "dynamic" to distinguish his approach from "static tectonics," and not to imply that the forces were known. He did not pretend to "reduce tectonics to physics." He wanted to give as complete a descriptive history as he could of Earth's changing tectonic landscape.

One might say that there is a static tectonics and a dynamic tectonics. The first is the art of defining the present state of the structures; this merely requires correct observations, complemented – through means I shall not discuss here – by adequate comparisons and interpretations. This kind of tectonics, because of its static attitude toward the world, is not self-sufficient since the world does not stop. The dynamic tectonics would be, in its final expression, a completed tectonics, that is, a continuous history of the deformations of the planet in which all the evidences would relate to each other without gaps . . . Naturally, this is an effort to understand, although in an entirely different sense from that implied by the original force of the word "theory." My ambition, which is more realistic, is to reinterpret in a more precise fashion static tectonics and to reveal dynamic tectonics. I do not pretend to reduce tectonics to physics – this is a matter for the future.

(Argand, 1924/1977: 2)

He thought that mobilism had not been disproved and that the difficulties raised against it had been answered, yet he was not ready to accept it fully.

The nonexistence of a refutation should not, strictly speaking, be considered as a proof. Still strictly speaking, positive testimonies may have only the value of arguments. New facts may be discovered that have the strength of an unyielding obstacle. Other theories may be born, or reborn under rejuvenated forms.

(Argand, 1924/1977: 128)

He preferred mobilism because it delighted the senses more than fixism.

But presently there seems to be no fact sufficiently contradictory to prevent us from enjoying, on those rafts on which our fates float, the delight of the sensible transports to which A. Wegener invites us.

(Argand, 1924/1977: 128)

With the completion of *La Tectonique*, Argand left the development and defense of mobilism to others.

8.8 Reception of Argand's synthesis internationally

Argand's mobilism received a mixed reception among those influential mobilists who were not themselves focused on tectonics. Wegener and Holmes were impressed with his contributions. Du Toit was ambivalent. Wegener (1929/1966) welcomed Argand's support, and his explanation of mountain building. After admitting that his own explanation of the origin of the Alpine–Himalayan Mountains "might seem fantastic," he singled out Argand's *La Tectonique* as confirmation of it (Wegener, 1929/1966: 84). Redrawing Argand's Figure 13, which showed India underthrusting Asia, he agreed with Argand's analysis (Wegener, 1929/1966: 84–85). He (1929/1966: 85–86) adopted Argand's explanation of the formation of the Alps and Andes, and appreciated (1929/1966: 86–87) Argand's "tonnage" argument, quoting some of the above passages. Wegener did not, however, mention Argand's proto-Atlantic.

Holmes read Argand's *La Tectonique* before he first began supporting mobilism in print (§5.3), and (1926: 308) declared that Argand had liberated tectonics "from the deadening conception of fixity," and "revealed a world of eloquent testimony." A few years later, Holmes referred to Argand's work in his review of the AAPG symposium (§5.4.3). He (1928b: 431; 1929: 206–207) found "Argand's conception of varying plasticity" to be a "valuable corrective to the exactly fitting coast lines of Wegener's too dogmatic maps." In 1929, Holmes praised Argand's solution to the formation of the Himalayas, and reproduced the same figure (Figure 13) from *La Tectonique* that Wegener had redrawn.

Argand's bold and eloquent picture of the structure [of India underthrusting Asia], here reproduced from his *Tectonique de l'Asie*, shows graphically that the mechanism portrayed thickens the sial to the appropriate order required for isostatic equilibrium. Wild though this conception may seem, it is nevertheless in harmony with the extravagant geological history of Tibet and the Himalayas; it explains the mysterious disappearance of the Oman arc; and it recognizes the almost incredible northerly drift of India that is implied by the facts (a) that in late Carboniferous times India was glaciated from the south; whereas (b) from the Eocene onwards it became the active site of laterite formation.

(Holmes, 1929: 343; my bracketed addition)

After Holmes introduced his mantle convection theory, he (1931a) continued to praise Argand, and suggested that convection could explain movements of Africa relative to Europe and India relative to Asia.

The hypothesis of [mantle convection] has other significant implications. It indicates that superimposed on the general crustal drift there would be a relative approach between Africa and Europe . . . Although these currents would be directed towards the interior of Africa, and would thus place Africa under compression, the resultant would be a drive in a northerly direction. Here we have exactly the conditions postulated by Argand and other Alpine geologists as essential to account for the structure of the Alps.

(Holmes, 1931a: 585–586; my bracketed addition)

Again acknowledging Argand's solution to the Alps, Holmes noted:

For the Tethys-belt Argand's bold diagrams clearly illustrate the conception that ranges thrust over a foreland from a geosyncline imply an underthrust of the foreland in a direction having a component opposed to that of the surface overfolds and nappes. As in the case of compression of a plastic substance between the closing jaws of a vice, so here the energy is supplied by the movements of the forelands; and as the vice is actuated by an external source of energy, so here the forelands are carried forward by deep-seated rock-flowage maintained by currents in the substratum.

(Holmes, 1931a: 588)

Like Wegener, Holmes did not comment on Argand's proto-Atlantic.

Du Toit was not so impressed. He (1937: 21) referred to Argand's *La Tectonique* as monumental but added somewhat pejoratively that it had a "frankly tectonic outlook," adding that "he gives no clear-cut pictures of the land-masses and geosynclinal seas of the past as disclosed by the stratigraphical evidence." Du Toit was a stratigrapher, not a tectonicist; stratigraphy he thought was more important than structure and tectonics in understanding the geology of South Africa, and in reconstructing Gondwana. Du Toit thought that Argand had attributed too much plasticity to the continents: "The 'plasticity' of the continents postulated by Argand has rightly come in for strong criticism" (du Toit, 1937: 54). "Argand, outdoing Suess, has given a spectacular though questionable picture of a highly 'plastic' Asia pouring outwards" (du Toit, 1937: 313). Notwithstanding, he welcomed Argand's thrusting of India beneath Tibet. "All this supports Argand's view that India underlies much of Tibet, which means that the block so involved must be far larger than present-day India" (du Toit, 1937: 184). Extending his conditional support to the tectonicist, he declared:

Though apparently erring in ascribing an undue degree of "plasticity" to the continental blocks and assuming movements more precise and detailed than the data would seem to warrant, Argand's views have profoundly influenced geological thought and have greatly furthered the hypothesis of Continental Drift. Wisely Argand has made no attempt to explain the cause of such displacements, which he merely accepts as the unescapable deduction from the wealth of geological evidence available to the unfettered mind.

(du Toit, 1937: 21)[9]

Du Toit also said nothing about Argand's proto-Atlantic.

To gain further understanding of Argand's reception, I turn to Edward B. Bailey, who at the time had become sympathetic to mobilism, and Walter Bucher, who was implacably hostile. Bailey, who announced support for some aspects of mobilism in 1925, and became inclined toward it after seeing for himself both the American and Scottish Caledonides in 1927 (§5.3), probably was more familiar with and affected by contemporary studies of Alpine geology than any other British tectonicist. Bailey first voiced his opinion of Argand's work at a December 1925 meeting of the Royal Geographical Society of London. His presentation followed that of Léon Collet, who had become inclined toward mobilism through his work on the Western Alps. Collet argued in favor of mobilism and introduced Argand's *La Tectonique* to this audience. Bailey (1926a: 308) held Argand in high regard, praising him more effusively than he would later do. Indeed, Bailey, who mapped the Scottish Caledonides and had been guided through the Alps by Collet, perhaps captivated by Collet's enthusiasm for mobilism, supported "some sort of continental drift."

Then of course we have just heard from Professor Collet of Professor Argand's fascinating interpretation of the history of the Alps, starting from the first days long before the great period of Tertiary movement. Of course there is a good deal in this interpretation that is much more speculative than the mere tracing of the structure of the country; but a lot that Argand has told us about the development of the Alps will last for ever. This attempt to build up the history of the growth of the mountain chain is not new; it goes back for about one hundred years. We find it in the early writings of Studer on the Alps; it is a delightful idea, that of the mountain chain traveling forward, and the pebbles going forward from the chain as it grows, and then the chain riding over its own débris. One finds that in the early writings of Studer, and also in the story told by Argand; and already, difficult as the subject is, I feel convinced that there is a very great measure of truth in what Argand has given us. Professor Argand is, perhaps, at the moment the most brilliant of all the brilliant school of Alpine tectonists. Finally, if we take the current Swiss view of the structure of the Alps, and the similar view . . . for the Scandinavian chain, and again the similar view attained by Scottish geologists in their Highlands, we find that some sort of continental drift becomes a matter not of interpretation but of observation.

(Bailey, 1926a: 308–309)

However, Bailey's enthusiasm for Argand soon waned. Reviewing *La Tectonique*, he (1926b: 863) acknowledged Argand's contributions to unraveling the Pennine Alps, which "ensure a respectful hearing for" his new work, but ended by recalling Henslow's damning advice to Darwin.

One might continue this summary indefinitely, and discuss the making of the Mediterranean or the bow and stern phenomena of drifting continental rafts as exhibited along their oceanic margins. Space forbids. Let us merely recall the advice given to young Darwin preparing for his voyage on the *Beagle*. Lyell's "principles" had just appeared. "Get the book," said orthodox old Henslow, "study it but don't believe it." Argand has succeeded in correlating an enormous number of geological phenomena, and has thus given them a realizable unity. It may be wise, for the present, to consider his coordinating principles with a perfectly open mind; but this should not render us any the less grateful for the amount of the accomplishment of a singularly arduous and helpful piece of research.

(Bailey, 1926b: 864)

Bailey remained uneasy about Argand's synthesis even after he had argued that North America and Eurasia were once joined. Later (1935) he wrote a delightful history of Alpine tectonics, again praising Argand for his work on the Pennine nappes, but denying him full acceptance.

In 1916 Argand published *Sur l'arc des Alpes Occidentales*. I mention this paper because it is very largely concerned with the Pennides, though it also deals with the western Alps as a whole. The central idea is a presentation of the developmental history of the chain from Carboniferous times onwards. It is a motion-picture of Alpine tectonics. Personally I never feel safe of my foothold when I try to follow Argand into the dim recesses of the past; but this does not lessen my admiration for a guide who has opened up so many secure routes through the hitherto almost trackless Pennides.

(Bailey, 1935: 126)

Walter Bucher (1933) attacked Argand's mobilism in his *Deformation of the Earth's Crust*. As already noted (§3.8, §6.10, §7.3, §8.4), he was one of mobilism's most outspoken and unrelenting opponents. Because of his unusual background, Bucher played an important role in familiarizing North American Earth scientists with the views of continental European geologists, which W. H. Bradley (1969), his biographer, thought one of his greater contributions. Born in the United States, Bucher moved with his family to Germany when he was five. He was educated there, receiving his Ph.D. in 1911 from the University of Heidelberg, working under Salomon-Calvi, who, unlike his student, later became favorably inclined toward mobilism. Bucher returned to the United States, first to the University of Cincinnati and later to Columbia University. He became a prominent figure among North American Earth scientists. Elected to the NAS in 1938, he was twice President of the AGU (1948–53) and President of the GSA (1955).

Bucher praised Argand for his brilliance but criticized his alleged disregard of facts, a rhetorically effective way of saying not to take Argand seriously.

This is Argand's picture. It is a grandiose spectacle and a brilliant idea. The writer wishes that he could accept it . . . Many things would be so much easier. But before his eyes rises this new obstacle: the absence of any noticeable evidence of the unheard-of-distortion which many parts of the Mediterranean lands must have suffered if events had been as Argand pictures them.

(Bucher, 1933: 258)

Spain had rotated anti-clockwise and the Apennines had pivoted. Bucher was so sure that Italy had not pivoted in this way and that Argand had overestimated the plasticity of sial near the surface, that he did not take the time to read the newest studies.

If this example seems unconvincing [that Argand is wrong], look at the pivotal areas about which the Apennines are supposed to have swung through an angle of 90°. The writer has not available, at this writing, the publications necessary to check the literature on the Ligurian and Etruscan Apennines. But he knows that neither in Suess' work nor in such pertinent writings as he happens to know is there anything to suggest evidence of the sort of strain which rocks in the outermost few miles of the crust would have to undergo if they were twisted in a horizontal plane through an angle of 90°.

(Bucher, 1933: 259; my bracketed addition)

Bucher continued to use the same patronizing technique, first declaring Argand a genius, then dismissing his tectonic synthesis. First the praise:

The discussion [by Wegener] of orogenesis in the light of his theoretical views is most unsatis-factory. In 1922, E. Argand read an extensive address before the Thirteenth International Geological Congress at Brussels. In this vivid, though involved paper, he gives a bold synthesis of the concrete realities of Eurasian tectonics interpreted in terms of "mobilism" as he calls it, in contrast to the traditional "fixism." His interpretation rests on the two fundamental concepts of Wegener's reasoning. The first is that of continental "floes" of "salic" matter adrift in the "sima"; the second is the essential plastic behavior of all crustal materials under long-continued stress. The systematic application of this idea to continental structure is the specific contribution of Argand's genius.

(Bucher, 1933: 71; my bracketed addition)

Then, the dismissal: Bucher argued that Argand's explanation of intra-geosynclinal folds was inconsistent with a stable Pangea during Late Paleozoic. Given Argand's account, geosynclinal folds require initial thinning of the sialic crust to create the geosynclinal depression, which is later folded through compression. Argand claimed that crustal thinning is caused by a drifting apart of sialic masses and compression by their drifting together. Bucher claimed that this oscillatory motion would lead to an unstable Late Paleozoic Pangea, which, in turn, would destroy the key paleoclimatic and paleogeographical arguments in mobilism's favor.

Their [Argand's geosynclinal folds] explanation requires the introduction of an additional, auxiliary assumption. They are possible only if there is an oscillatory movement of parts of the continental floes. One part must be held by friction while the other drifts off, drawing out the crust thin along the line of parting. Later the straying portion must be held, while the lagging part catches up. How vital this auxiliary hypothesis is becomes evident when we apply it to the starting condition of Wegener's reasoning. The broad lines of geographical and climatological arguments which give weight to Wegener's interpretations in the minds of geologists, geographers, and climatologists, all converge in the fundamental concept of a parental land mass [Pangea] in which continents were originally united . . . Applying the mechanism provided by Argand's modification of Wegener's theory, we must assume the following events. The geosynclinal belts came into existence when parts of the parental continent began to pull away from the rest. The lines of yielding and pulling-out of the salic crust today are occupied by the belts of thin sediments folded in late Paleozoic time in the Appalachian region, the Hercynian system of Europe, and the belts in central Asia, eastern Australia, South Africa, western South America. Their arrangement would indicate a pulling away in all directions. But the fact that they have undergone intense folding indicates that the reverse movement has taken place repeatedly . . . All these reunions need not have all been simultaneous, but they must have occurred repeatedly and in all directions of the compass. More remarkable yet: when the parent continent ultimately split up it was not along these zones but along entirely new lines. This time, however, there was no significant thinning of the sialic mass nor vacillating movement that gave rise to folding, for there are nowhere late Paleozoic geosynclinal belts and folds along the new lines of parting, not on the east and west sides of Africa, not south of India, nor on the north and west sides of Australia . . . Beneath the folded sediments of late Paleozoic age lie peneplaned remnants of still older orogenies, of "Caledonian," Proterozoic, Archean age. This picture here drawn for the late Paleozoic

orogeny must not be thought of as exceptional, but must represent the typical process
of crustal deformation for the larger part of geological time.

(Bucher, 1933: 73–75; my bracketed additions)

Bucher claimed that Argand's theory of mountain building could not explain Hercynian–Appalachian orogenesis without a greatly divided Late Paleozoic Pangea that
would, he contended, undermine the paleoclimatic and paleogeographical arguments
in favor of mobilism. He made it sound as if he had struck at the very heart of the
Wegener–Argand tectonic synthesis. Not so. His criticisms were not consistent with
stratigraphic relationships as they were known at the time, and did not address the
central tectonic theme of that synthesis, namely the Mesozoic and later drift of
continents culminating in worldwide Tertiary orogenesis. However, in their enthusi-
asms, Wegener and Argand had overextended their arguments, which left them
vulnerable to Bucher's attack.

In the Wegener–Argand synthesis Pangea was a Permo-Carboniferous construct,
based importantly on the evidence of Permian and Late Carboniferous glaciation of
the present fragments of Gondwana. Wegener, tentatively and unwisely, extended
Pangea backwards into the Devonian (Köppen and Wegener's Figure 21, 1924: 143)
without good reasons for doing so. It was a step too far. Argand, on the basis of his
tectonic model, began reconstructing motions prior to and during the mid-
Carboniferous Hercynian orogeny, that is prior to the classic Permian and Late
Carboniferous Pangea. It too was a step too far because very little was then known
about the relative positions of continents prior to the mid-Carboniferous. Bucher
argued that pre- and syn-Hercynian motions would disrupt the climate of Pangea as
constructed by Wegener (Figures 35 and 36, Wegener, 1929: 137). But the two were
not compatible, the former were pre-Late Carboniferous and the latter, Late
Carboniferous and Permian, a critical distinction. Bucher obscured the issue by
referring to everything as Late Paleozoic that is Devonian through Permian. Regardless of the quality of Bucher's attack, I suspect it convinced many fixists who only
had a passing interest in Mediterranean geology that his book was not worth the
effort of close study.

Other fixists were less antagonistic to Argand. They took parts of Argand's
regional explanation for the Alps and Himalayas, but omitted his global mobilism.
T. C. Nicholas, who taught at Cambridge University, voiced precisely this reaction at
the December 1925 meeting of the Royal Geographical Society where Collet had
argued in favor of mobilism and introduced his British audience to Argand's *La
Tectonique.*

It is a great pleasure to me to be present to-night in order to express, even if in an inadequate
manner, the great debt which I owe to Professor Collet for the memorable experience of being
led by him through a part of the Alps during the summer before last. In the course of that
excursion, the pleasure of which was shared with me by Mr. Bailey and other British geologists,
we saw evidence for a good deal of what we have heard from the lecturer this evening,

particularly with regard to the structural phenomena of the Pennine Alps. And for my own part I have no hesitation in accepting the general truth of the views now current among Alpine geologists upon the formation of the pre-Alps, High Calcareous Alps, and Pennine Alps by recumbent folding. I think it is quite incontestable that the hinterland has moved for great distances towards the foreland, and that in doing so its long-continued pressure upon the intervening geosyncline has given rise to the immense recumbent fold which Professor Collet has demonstrated to us and of which he has shown us photographs. But though in my opinion the evidence for the movement of the hinterland towards the foreland is complete and overwhelming, I am not so sure that this evidence can be said to confirm the theory of Wegener, and I should prefer for the present to keep facts and theory separate.

(Nicholas, 1926: 309–310)

This guarded kind of response allowed fixists with a professional interest in the Alps to accept Argand's early regional account of them and still reject his later global mobilism.

Other fixists ignored Argand. At the 1926 AAPG symposium, van der Gracht (1928a: 69) was the only participant to refer to Argand's *La Tectonique*, calling it "remarkable." Argand is mentioned by only three symposiasts at the 1939 Frankfurt meeting; Bucher (1939a: 290) was dismissive, Kurt Leuchs (1939: 355–356) argued against Argand and others who claimed that Africa had overridden Europe to form the Alps; and G. Knetsch (1939: 254) merely noted that Argand and other Alpine tectonicists supported mobilism.

Retrospectively, the most important objection to Argand's theory of mountain building was that he attributed too much plasticity to the sialic continents. Few could believe the solid Earth was quite that mobile. Except for tectonicists of the Western Alps among whom Argand maintained great influence, his grand vision lay essentially dormant until revitalized by S. Warren Carey in the 1950s (II, §6.6–§6.12).

8.9 Reception of Argand's synthesis among tectonicists of Western Alps

I describe the reaction of nine prominent Alpine tectonicists, seven of whom reacted favorably. Staub (1890–1961), who was Swiss from Zurich, adopted Argand's Alpine analysis extending it to eastern Switzerland, and developed his own, somewhat idiosyncratic version of global mobilism. He (1924) wrote "Der Bau Alpen," adopting Argand's view that Africa's collision with Europe formed the Alps. Four years later, Staub presented his full version of mobilism, *Die Bewegungs-mechanismus der Erde* (*The Movement Mechanism of the Earth*). He (1928: 9–11) distinguished between marine or oceanic geosynclines (*ozeanischen geosynklinale*) and continental geosynclines (*geosynklinale kettenzone*). Closing the Tethys, an oceanic geosyncline, led to formation of the Alps, Apennines, and Sierras of southern Spain with Africa as the hinterland and Europe as the foreland. On either side of the Tethys there were continental geosynclines which were buckled and compressed forming the Pyrenees, Atlas, and Caucasus Mountains. Staub (1928: 163) reintroduced the idea that the

Pacific basin was formed by the separation of the Moon from Earth. He rejected westward drift of the Americas relative to Europe and Africa, but proposed a general westward drift of all continents. Laurasia and Gondwana moved to and from the poles; *polflucht* moved them together toward the equator; *poldrift* moved them apart toward the poles (Staub, 1928: 244). Earth's rotation caused *polflucht*; subcrustal streaming caused *poldrift*.

Collet (1879–1957), another Swiss tectonicist who worked on the Western Alps, adopted mobilism.[10] As already noted (§8.8), he summarized his support for mobilism and acknowledged Argand's and Staub's work during a talk in 1925 before the Geographical Society in London, which he repeated almost word for word a decade later in his *The Structure of the Alps* (1927/1935).

A southern continent (Africa or Gondwanaland) made of Sial, is separated from a northern continent (Europe or Eurasia), also made of Sial, by a geosyncline or the sea of Tethys of Suess. As a result of the northward drifting of the southern continent, the floor of Tethys thinly coated with Sial has been folded . . . northward, thus forming a mountain chain (the Alps) . . . The southern continent not only encountered the obstacles formed by the northern continent, but its frontal part has been thrust over it. By the employment of Wegener's ideas we could go a good deal further and – as shown by Argand and Staub – accept a northward drift of Europe, producing a distension to which the Mediterranean is due.

(Collet, 1935: 27; also see Collet, 1926: 301)

Collet noted that the discovery of radioactivity had raised a difficulty for contractionism and praised Joly's work. He was optimistic about the early acceptance of Wegener's theory.

As shown by Lord Rayleigh, many facts make it necessary to abandon Kelvin's classical theory of the cooling of the earth. The work done by Prof. Joly in this direction is of great value, and it will soon be impossible to resist Wegener's attractive idea.

(Collet, 1935: 27; also see Collet, 1926: 301)

Salomon-Calvi, another tectonicist and supporter of mobilism, was also familiar with the Western Alps. Born in Berlin in 1868, he obtained his doctorate in 1890 from Leipzig for his work on the Italian part of the Adamellogruppe, in northern Lombardy. He continued to work there until 1897 when he returned to Germany and took a position at the University of Heidelberg. Forced by the Nazis to resign from the University of Heidelberg, he settled in Turkey and took Turkish citizenship. He died in Ankara in 1941 (Şengör, June 15, 2004 email to author).

During his work in Italy he discovered what he first took to be a giant fault, but after reading Wegener, he considered it a zone of collision between drifting continents.[11] Hume, who met Salomon-Calvi, later explained.

When Professor Salomon-Calvi visited Egypt in the spring of 1938, I learnt that he was specially interested in the question of mountain formation as a result of Continental Collision. He subsequently sent me a series of his papers, one of which [Salomon-Calvi, 1937] contains an amplification of his earlier contributions to a study of the horizontal movements of continents. In 1891, he described a marked tectonic feature which he named the Tonale Line. In later studies, he traced this band westward under the plain of the Po to Savona [on the Gulf of Genoa] and obliquely across Corsica, while to the east it was followed as far as the Hungarian plain. At first he regarded it as a gigantic fault-line, but later after Wegener had published his famous theory, he concluded that this was a region lying between two continental areas which were being thrust forward toward one another, and the structure was termed by Salomon-Calvi a Synaphie [collision zone].

(Hume, 1948: 107; my bracketed additions)

Salomon-Calvi's adoption of mobilism is therefore traceable back to his early field-work in northern Italy. Like Argand and R. Staub, he envisioned a collision of Africa and Europe. He later argued that he had discovered a similar synaphie in Turkey (Şengör, 1979/1982a: 35).

The Swiss structural geologist Elie Gagnebin (1891–1949) also favored mobilism. He worked extensively on the Prealps, and taught at the University of Lausanne, where he succeeded Lugeon. Gagnebin (1922) introduced Wegener to French geologists (Carozzi, 1985: 126–127). He thought that Wegener tried to explain too much with his theory, but thought it was a promising working hypothesis. Rudolf Trümpy, once senior assistant to Gagnebin at Lausanne (1947–53), recalled that Gagnebin remained "an outspoken Wegenerian drifter," noting that he continued to defend it in his 1946 popular booklet, *Histoire de la Terre et des Êtres vivants*. Gagnebin had many interests outside geology, and was "a very likeable man, friend of Stravinsky, Cocteau and other members of the post-WWI cultural scene" (Trümpy, August 24, 2004 email to author).

The Swiss geologist Albert Heim (1849–1937) reportedly favored mobilism near the end of his illustrious career. Born just two years after Marcel Bertrand, he helped to document huge Alpine overthrusts, even though he first rejected Bertrand's analysis of the Glarus as a huge single overthrust. Heim studied under Arnold Escher, and in 1874 succeeded him as professor of geology at the Zurich Institute of Technology, which became Eidgenössische Technische Hochschule Zürich (hereafter, ETH Zurich). A year later he moved to the University of Zurich. He received honorary doctorates from Bern, Oxford, and Zurich, the Wollaston Medal from the Geological Society of London in 1904, and was elected a Foreign Member of the Royal Society (London) in 1896. He retired in 1911, but remained active for many years to come. He was President of the Swiss Geological Commission from 1894 until 1926.[12] Staub, who eventually occupied the chair once held by Heim, reported in his 1928 *Der Bewegungsmechanismus* that Heim, formerly a fixist, had changed his mind.

It is important to point out that recently also Albert Heim, one of the brilliant defenders of contraction theory, in his old age and with rare flexibility of mind, accepted Wegener's concept of the mobility of continents.

(Staub, 1928: 22; see Carozzi, 1985: 130)

Thus Heim, perhaps influenced by Staub, became a mobilist.[13] With the senior Heim's switch to mobilism, few Alpine geologists of the Western Alps remained fixists, illustrating the strong influence that this region of study had on their global view.

Albert Heim's son, Arnold Heim (1882–1965), whose father was disappointed that he was not selected to succeed him at ETH Zurich and the University of Zurich, where he lectured from 1908 to 1911 and 1924 to 1928, argued in favor of mobilism at the 16th International Geological Congress, held in 1933 in Washington, DC. Familiar with the Alps, he also knew the Himalayas and Andes firsthand. Although Arnold Heim (1936: 921) agreed with Argand and Staub's explanation of the Alpine diastrophism, it was not the only reason he preferred mobilism. He also appealed to the Permo-Carboniferous glaciation, and other paleoclimatic support offered by Köppen and Wegener.

Helmut (Hellmut) de Terra (1900–81) also argued in favor of mobilism in 1933 at the Washington congress. Born and educated in Germany, he obtained his Ph.D. at the University of Munich in 1924, where he studied geology and geography.[14] He participated on E. Trinkler's Central Asia Expedition (1927–8). He lectured at several universities in the United States, and emigrated there in 1930, becoming a naturalized US citizen in 1937. He obtained a position at the Peabody Museum of Yale University where he remained until 1939. While at Yale, he organized and directed two expeditions (1932–3, 1935) to the Himalayas. He also worked with Teilhard de Chardin during the Harvard–Carnegie Expedition to Burma (1937–8). He headed the Regional Geographic Section of the Geographic Board in Washington, DC during World War II. He later organized and directed archeological expeditions to Mexico, Spain, Italy, and the United States. De Terra had wide interests and wrote biographies of Teilhard de Chardin and Alexander Humboldt.

De Terra's mobilism arose through his experience of the Alps, the Himalayas, where he spent two-and-a-half years on two expeditions, and his reading of Eduard Suess and Argand.[15] He praised Suess and Argand for their speculations on the Himalayas but noted that their information was limited. After his time in the Himalayas, he thought he had information to warrant their comparison with the Alps.

Hitherto comparisons between the Alps and the Himalayas have suffered from our incomplete knowledge of the latter region. The geological Survey of India and some private explorations have advanced our geologic knowledge in the last decade so considerably that an attempt may now be made to summarize the stratigraphic evidence of Himalayan diastrophism. This evidence seems regionally most extensive and complete in the northwestern portion of the Himalayan belt, including the eastern Karakoram – a region with which I am personally acquainted.

(de Terra, 1936: 860)

After marshaling evidence in support of the many similarities between the Alps and Himalayas, he (1936: 869–870) concluded:

Whatever hypothesis may best prove fitted to explain orogenies, the similarity between Himalayan and Alpine folding lends support to the idea that the shifting of larger crustal masses in the sense that Wegener suggested, is a primary factor in mountain making.

The Swiss geologist Rudolf Trümpy was another Alpine mobilist. Born in 1921, he studied at ETH Zurich. His teachers included R. Staub, P. Niggli (1888–1953), and A. Jeannet (1883–1962). As already mentioned, Trümpy became senior assistant of Gagnebin and Lugeon at Lausanne University from 1947 to 1953. He succeeded Jeannet as professor of stratigraphy in 1953 at ETH Zurich, and later as professor of geology at ETH Zurich and Zurich University, where he remained until 1986. He served as president of the International Union of Geological Sciences from 1976 to 1980. Trümpy worked mainly in the Alps, studying relations between sedimentation and structure.

Trümpy, who had read Wegener's book as a schoolboy, proclaimed his support for mobilism during the winter 1941–2 when he was twenty-one. He gave a seminar in which he concluded that some sort of continental drift offered the only satisfactory explanation of mountain building (Trümpy, 2001: 480). Trümpy made fun of Haarmann's and van Bemmelen's views as well as of "Staub's Poldrift-Polflucht couple, which I called a perpetuum mobile." Staub was not displeased.

Staub apparently liked my talk and suggested that I submit a manuscript to the *Geologische Rundschau*. I declined, as this would have meant much more library work and taken time off a geology student's favourite occupation, hammering at mountains.

(Trümpy, August 24, 2004 letter to author)

Trümpy continued to support mobilism, arguing for it in his classes at ETH Zurich.

In the winter term 1953–1954, my very first at the ETH, I lectured on the Gondwana System, about which I knew next to nothing; this made me read a lot of literature and confirmed my views.

(Trümpy, August 24, 2004 letter to author)

Several years later, he had the opportunity to work in Greenland, which he thought of as part of North America.

I spent the summer of 1958 in East Greenland, the only stretch along the eastern coast of North America with Upper Permian to Lower Cretaceous marine sediments, partly in order to find out whether this really implied the existence of an old Atlantic, and I was relieved to find that this was not the case, the rocks having been laid down in a graben open to the north.

(Trümpy, August 24, 2004 letter to author)

He (1960) wrote a paper for the *Bulletin of the Geological Society of America* "to explain part of the Alpine story to non-Alpine readers" (Trümpy, August 24, 2004 letter to author). He questioned Argand's notion of embryotectonics, the idea that "geanticlines" or linear welts within geosynclines are embryonic anticlinal folds and represent an early stage of nappe development. Although silent on mobilism, he still thought it correct. "My critique of THIS aspect of Argand's theory did in no way imply a rejection of continental drift, quite on the contrary" (Trümpy, August 24, 2004 letter to author).

I suspect that Trümpy's attitude toward mobilism was shared by most Swiss Alpine geologists. They believed that mobilism best explained the formation of the Alps, but had nothing new to contribute to the larger debate, and therefore published few papers in which they actively supported mobilism. In describing the behavior and attitude of Alpine geologists toward mobilism, Trümpy declared:

Most Alpine geologists were happy to map "their" mountains, worrying (perhaps too) little about theories. Some of them accepted mobilism, without insisting; none of them (except, of course, in Germany and Austria) were openly against it.

(Trümpy, August 24, 2004 letter to author)

But, there may be more. Looking ahead, Alpine tectonicists played no role in marshaling support for mobilism during the 1960s. Indeed, their support of mobilism lessened. Again, I appeal to Trümpy. In a refreshingly honest comment Trümpy claimed:

On the other hand, the Alpine geologists must assume a fair part of the blame. [This also includes the author of this paper, who had started to write a book on the Alps and who lacked the energy and courage to complete it.] Since about 1935, they had disposed of the data and ideas advanced by fellow highlanders, such as Argand, Ampferer and Staub, and by lowlanders, such as Wegener, Holmes and du Toit. This would certainly not have allowed them to construe a global model, but at least a coherent interpretation of Alpine structure and evolution. Would such an attempt have been headed by the geophysicists who did establish the plate tectonics theory? We cannot tell.

(Trümpy, 2001: 481)

Trümpy (2001: 480) has shown that Alpine tectonicists started to become cautious during the 1940s and began "to downplay the importance of Alpine crustal shortening." Staub even "reverted to a neo-fixist stance" during the 1950s (Trümpy, 2001: 480). Alpine tectonicists lost their nerve, and were reluctant to follow "the consequences of their own observations" (Trümpy, 2001: 481).

Alpine tectonicists in the 1950s ignored the new paleomagnetic results favorable to mobilism. Agreeing with the suggestion that Alpine tectonicists cared about mobilism but not about its paleomagnetic support, Trümpy summarized his own and the general reaction of Alpine tectonicists.

We were badly prepared to accept or to criticize the early results of the early paleomagnetists; it is not for nothing that they were dubbed "paleomagnicians" until quite recently . . . Somewhat later, we admired Rutten's work on the reversals in Iceland; but the Alpine chapters in his book on Western Europe are quite awful. I regret that I cannot provide more specific information on this point.

(Trümpy, August 24, 2004 email to author)

Although the vast majority of fixists (and hence the vast majority of geologists) ignored or paid little attention to early paleomagnetic support for mobilism, as I shall show in Volume II, mobilists generally welcomed it. This lack of interest shown by

Alpine tectonicists in the early, pro-mobilist paleomagnetic results was unique among identifiable groups of mobilists.

I conclude this survey of mobilism's reception among those geologists who worked on the western Alps by discussing the views of Pierre Termier (1859–1930) and Hans Schardt (1858–1931). Both were contemporaries of Albert Heim and, like him, both played significant roles in documenting that "the Alps do not merely contain overthrusts, they *are* overthrusts" (Greene, 1982: 205). Unlike Heim, however, they rejected mobilism.

Pierre Termier, an eldest son, first studied mathematics, literature, and philosophy before switching to geology after mountaineering in the western Alps. After studying at the École des Mines in Paris, working as a mine inspector, and teaching at the École des Mines in Saint-Étienne, he was invited by Bertrand in the late 1880s to work with him on the crystalline massifs and metamorphic zone of Savoy. They became friends, and Bertrand became his mentor. Termier documented huge overthrusts in the southern Alps, and in 1903 showed that the eastern Alps are a series of overthrusts that override the western Alps (Greene, 1982: 213–220). He was rewarded by being appointed Director of the Service de la Carte Géologique de la France (1911), Inspector General of Mines (1914), a member of the Académie des Sciences (1909), and a commander of the Legion of Honor (1927). He was thrice President of the Société Minéralogique and the Société Géologique of France, and Vice-President of the Académie des Sciences.[16]

Termier rejected Wegener's mobilism in a talk he presented at the Institut Océanographique of Paris in February 1924 (Carozzi, 1985: 127). A written version, published the same year, was translated into English and published by the Smithsonian Institution (Termier, 1924, 1925). With more than a touch of condescension he criticized continental drift and those who supported it.

Such a theory has the possibility of being extremely convenient, an advantage which is not without grave danger, the danger of making superficial minds believe that enigmas are solved when they are simply displaced and replaced by those more general and much more irresolvable. Yes, it may seem very convenient to unite two continents or to separate them at will; to join them to explain the migration of the fauna and flora from one to the other, or the extension from one to another of some line of structure, for example, a chain of mountains; then separate them to explain on another occasion the dissimilarity of biological conditions that one observes there, or the difference which manifests itself in the geological history of two continents during an indefinite period of time. It is also very convenient to admit, with regard to the continents, that the terrestrial poles could shift. This, by a stroke of the pen, does away with all difficulty relating to the distribution of climate during the different geological periods.

(Termier, 1925: 224)

Wegener was a poet, his theory enticing, but there was nothing there.

The theory of Wegener is to me a beautiful dream, the dream of a great poet. One tries to embrace it, and finds that he has in his arms but a little vapor or smoke; it is at the same time both alluring and intangible.

(Termier, 1925: 236)

Wegener's theory had "undeniable attractions" and "deceptive power" but was "misleading" (Termier, 1925: 227).

Termier pointed to mobilism's lack of mechanism. The continents advance "under the power of one knows not what, an irresistible force" (Termier, 1925: 229). Unable to restrain his sarcasm, he suggested that geophysicists would find a solution.

To ameliorate Wegener's theory and render it lasting, to do away with the gross improbabilities of which I have spoke, we can trust to the geo-physicists. They will conceive by new hypotheses other details of the "machine," as Pascal remarks, details which perhaps will not be more correct than the first and which, in any case, will not be more possible of verification.

(Termier, 1925: 231)

Termier (1925: 233–234) also thought little of Joly's theory. Even if his account successfully explained longitudinal changes in the relative position of the continents, it failed to explain latitudinal changes, and therefore left mobilism without the means to close Tethys, and form the Alps and Himalayas. He also questioned Wegener's solution to the origin of mountain belts. He first argued, as others such as Jeffreys, Longwell, Willis, and Bowie were to do, that Wegener's mountain building and drift mechanisms were inconsistent (§4.3).

But who could believe, for example, in the formation of mountain chains by the reaction of the liquid *sima* on the advancing continent? If the *sima* is capable of opposing such a resistance to the movement of the floating mass, how is it that this mass is not held by the sima and how can it move? In the hypothesis of mobility, what becomes of the debris of the solid simique pellicle which forms the bottom of the ocean? Should it not accumulate in a thick fold of dark heavy rocks under the prow of the great ship? But nothing resembling it appears. How can the deep foldings of the solid continental *sal*, under the thrust of the liquid *sima*, transform itself in rising toward the surface and form these folds and beds that we see in our mountains and which suggest the idea of superficial wrinkles much more than that of a very deep seated disturbance propagated in a vertical direction.

(Termier, 1925: 230)

Despite his many objections, Termier announced that he would not completely reject Wegener's theory.

But in all reality we can not conclude, we can not say, that there is really nothing in Wegener's theory; neither can we affirm that it does not contain some truth. Our knowledge is very limited. It is always necessary to close a lecture on geology in humility. On the ship earth which bears us into immensity toward an end which God alone knows, we are steerage passengers. We are emigrants who know only their own misfortune.

(Termier, 1925: 236)

Not much of a concession, Termier seemed unwilling to entirely reject any geological theory.

As far as I know, Termier's attack is the harshest offered by an Alpine tectonicist, especially by one who played such an important role in documenting Alpine overthrusts. Du Toit later remarked that his attack

Reveals the gloomy spirit of its author, who, although recognising and indeed dilating on the merits of Continental Drift, finds himself overwhelmed by apparent contradictions, and, without attempting their solution, capitulates with the cry "The Theory of Wegener is to me a beautiful dream."

(du Toit, 1937: 31)

Du Toit is right. Termier had become pessimistic: "We are emigrants who know only their misfortune." His gloom may have been connected with the tragedies in his private life. Since his stunning demonstration in 1903 that overthrusts of the eastern Alps override the western Alps, his oldest son had died in 1906 in an accident, his wife had died of Parkinson's disease in 1916, and his son-in-law, the geologist Jean Boussac, died from injuries in World War I.

Termier did not mention Argand's mobilism. He gave his talk several months before the publication of *Tectonique*, but it is hard to believe that he did not know Argand's views. Even if he did not attend Argand's presentation at the 1922 International Geological Congress in Brussels, he surely would have been told about them.[17] Assuming this to be so, the following passage suggests that Termier's attack included Argand.

Supposing that one could accept Wegener's orogenic hypothesis for the Cordilleras of western America, or south Africa, or eastern Australia, it would be necessary to find another to explain the Alps, the Apennines, the Caucasus, the great chains of Central Asia. This is indeed what Wegener tries to do; but who is the tectonician who would consent to accepting two entirely different orogenesis, one for the Andean pile and those resembling it, another for the immense transverse chains which are the highest summits of our world and which have replaced the ancient Tethys?

(Termier, 1925: 230)

Who was that tectonician? Was Termier referring to Argand? Perhaps he was.

Hans Schardt, born in Basel, first studied pharmacy before switching to geology. His most important work, helping to unravel the Prealps, was completed while he was at the University of Neuchâtel. Schardt took over from Heim at the University of Zurich in 1911 and stayed for the rest of his career. He discussed mobilism the year of his retirement at a meeting on March 12, 1928, at which Staub, his successor at Zurich, spoke first in favor of mobilism. Schardt (1928) followed with his "Zur Kritik der Wegenerschen Theorie der Kontinenten-verschiebung" ("Critique of Wegener's theory of continental displacement") in which he declared that the fracturing and large displacement of crustal blocks demanded by Wegener's theory were impossible (Schardt, 1928: xiv). Earth's crust is too stiff to allow for movements of continents relative to one another (Schardt, 1928: xv). Schardt preferred contractionism.

Despite the fixism of Termier and Schardt, who came to evaluate mobilism in the twilight of their careers, I believe that the generally favorable reaction to mobilism by tectonicists who worked in the western Alps is evidence of the strong effect that region of study had. In the early 1920s Argand led the way to mobilism, and almost all workers in the western Alps followed.

8.10 The peri-Atlantic Caledonides: Wegmann

Of these geologists who worked on the Caledonides and in the present context of the mobilism debate, Eugene Wegmann's (1896–1996) work was the most important with Holtedahl's a close second. Wegmann (1935) documented the existence of a large Caledonide fragment along Greenland's eastern coast, thereby greatly strengthening Wegener's (1924: 57) claim that Greenland and Norway were once close together. (Wegener supposed that the Blooseville Coast of Greenland's eastern coastline lay next to Norway's coastline in the vicinity of Møre and Trøndelag.[18]) Although a student of Argand's and sympathetic to mobilism because of his own work in the western Alps, it was his comparative work in Norway and Greenland that led him to favor mobilism. He fitted together Greenland, northwestern Europe, Spitsbergen, and several other islands, and suggested ways to test his reconstruction. He stressed the importance of further work on the seafloor, and made one of the most interesting contributions to the 1939 Frankfurt symposium on mobilism, which I shall discuss at the close of this chapter.

Wegmann began studying with Argand at the University of Neuchâtel in 1915, and became his principal assistant while Argand worked on his *La Tectonique*. Wegmann probably was the student closest to Argand, and eventually succeeded him at Neuchâtel in 1940. Wegmann not only learned geology from Argand, but was infected by his mentor's zeal and energy in attacking problems (Schaer, 1995: 13). He received his Ph.D. in 1923. His thesis studies, *Zur Geologie der St. Bernharddecke im Val d'Hérens (Wallis)*, gave him detailed first-hand knowledge of the Western Alps (Wegmann, 1922).

In 1924 Wegmann began fieldwork in Norway, spending several summers working there under the helpful tutelage of J. J. Sederholm (1863–1934), the leading expert on Precambrian metamorphic granitic rocks of the Baltic Shield. They developed a close father–son relationship (Schaer, 1995: 14). Besides becoming an expert on granitic rocks, Wegmann familiarized himself with the way in which the Pre-Cambrian rocks of the Scandinavian Caledonides had been metamorphosed during the much later Caledonian orogeny.

Wegmann took part in the 1931–4 Danish Expedition under the leadership of Lauge Koch to Christian X Land, Greenland, well-prepared to understand what he saw. He dedicated his work in Greenland to Sederholm.

This work indicated the lines of investigation and the objects of research in other pre-Cambrian areas, too. I mention it here, because I have tried to apply the principles which my beloved teacher and fatherly friend, J. J. Sederholm, used at this time to reveal the great leading lines in the chaos of granites, gneisses, and other crystalline rocks which once consti-tuted the bed-rock of Fennoscandia. I wish, therefore, to dedicate this work to his memory as a token of gratitude for all that he has taught me, which I could not but think of every day during my work in the field.

(Wegmann, 1938: 7)

He confirmed the suggestion, partly attributable to Koch (1929), that the Caledonides occurred on Greenland's eastern coast. Wegmann (1935: 5) recalled, "In November, 1934, Dr. Lauge Koch asked me: 'Do you believe in a Caledonian orogeny in East Greenland; could you soon give me a short report on this problem?'" The answer came a year later when Wegmann (1935) presented a detailed analysis of part of East Greenland. He described numerous similarities between the sedimentary rocks, granites, and structures of Scandinavia and the Scottish Caledonides where near horizontal thrusts and large nappes, familiar in the Alps, had been recognized. He also noted similarities with Caledonide fragments of Scotland and northeastern North America. This led him (1935: 48–50) to argue that these and various other fragments from around the North Atlantic were once united, and had resulted from the closing of a huge geosyncline – Argand's proto-Atlantic with East Greenland added.

Accordingly a collector-trough, a kind of forerunner of the North-Atlantic and Scandic, must be assumed to have existed in pre-Cambrian times bordering on the old Greenlandic, Hebridian, Canadian mass on one hand, and Fennoscandia on the other hand. The Scandinavian part of the Caledonian geosyncline was considered already by Argand (1, p. 191–192) to be a forerunner of the Atlantic; at his time the Caledonian of Greenland was an unknown factor. It is only as a result of Koch's expeditions that the late Pre-Cambrian geosyncline was discovered.

> *(Wegman, 1935: 49; the Argand reference is to* La Tectonique; *the Scandic Sea is roughly equivalent to the Norwegian and Greenland Seas)*

The remnants are found in East Greenland, Spitsbergen, the coastal districts of Norway, parts of Scotland, Newfoundland, the Maritime provinces of Canada, and the New England states of the United States. Breakup of the landmass led to the formation of the present North Atlantic and Norwegian Sea. Wegmann felt that his extensive fieldwork in Eastern Greenland and Scandinavia gave him authority to propose:

During the last few years various interpretations as regards the interconnections of the North Atlantic mountain complexes have been published, unfortunately by scientists who know part of the literature, but nothing of the territory; so these suggestions of a geologist who has traveled and worked for several years in the territory of the Greenlandic and Scandinavian Caledonian may perhaps be of some interest.

> *(Wegmann, 1935: 510)*

Although the deformed and often metamorphosed rocks that Wegmann was dealing with were very different from the unaltered, little disturbed sequences of the Karroo and Santa Catarina Basins of South Africa and South America, there was nevertheless a strong parallel between his work and that of du Toit. I do not know if prior to going there Wegmann expected to find matching mountain belts in East Greenland, but when he did, he certainly knew what that meant.

He (1938: 134) continued to argue for the former assembly of now scattered Caledonide fragments, and later added a new twist (Wegmann, 1943, 1948). Not content with simply documenting their similarities, he attempted to track the locus of

Greenland's movement relative to Europe, noting that Wegener, du Toit, and
F. E. Suess (whose views I shall soon discuss) had offered slightly different reconstructions, and he set out a plan to test them (Wegmann, 1948: 18–19). He gave a new
one, arguing that Greenland moved westward relative to Europe along the De Geer's
line. De Geer's line extends

> northward from Vesteraalen in Norway, following the continental slope off Bear Island and
> Spitsbergen, it will cut across the Nansen threshold between Spitsbergen and Greenland, and
> subsequently follow the continental margin north and west of the Arctic archipelago as far as
> the mouth of Amundsen Sound.
>
> *(Wegmann, 1948: 21)*

Wegmann's solution is interesting: his perspicacious presentation of it, his understanding of the importance of cooperation among specialists from different fields,
and his realization of the need for more ocean floor exploration. He traced De Geer's
line on a globe and found that it was an arc of a small circle. De Geer (1858–1943), a
Swedish geologist, wrote primarily on the geology of Spitsbergen and post-glacial
Scandinavia. Wegmann wanted a way to map the simplest path of displacement of
Greenland relative to northwestern Europe as one with minimum distortion. Realizing that "a Mercator's projection whose equator is parallel with De Geer's line"
would provide this, in 1943 he asked Edmond Guyot, Director of the Astronomical
Observatory, Neuchâtel, to construct one (Wegmann, 1948: 22). Guyot (1943)
obliged. Greenland's past movement relative to Europe appeared as a straight line
parallel to the top and bottom of the map (Figure 8.2).

Looking twenty years ahead, this is what McKenzie and Parker (1967) did when
representing the movement of the Pacific Plate relative to a bordering plate containing North America and Kamchatka (IV, §7.13). However, this, I believe, is
where the similarity stops. Neither Wegmann (1943, 1948) nor Guyot (1943) introduced Euler's theorem, although Wegmann (1948: 17, 42) did describe the motions
as rotational. But he invoked other types of movements: parallel displacements
producing gaps between one or more fracture planes, rotational movements about a
stationary center either outside or inside the unit moved and a combination of these
two. Moreover, he (1948: 17, 43) characterized the movement of Greenland relative
to Europe as parallel, and therefore his use of the projection technique did not
imply use of Euler's theorem. Nonetheless, his and Guyot's realization that he was
dealing with movements on a spherical surface and their desire to find a clear
geometrical representation on a plane of such movements were significant steps
forward.

Following his reconstruction of Greenland's movement relative to Europe, Wegmann made several predictions about the rocks to be expected in places that were
once together. Again, it is not his particular predictions that are now important, but
his foresight in stressing the need of having different specialists work together in
checking. Only in this way could "the great problems" be solved.

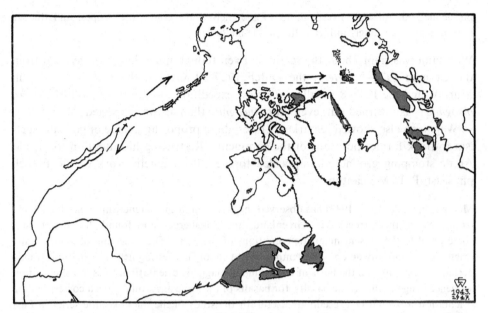

Figure 8.2 After Wegmann's Plate 2 (1948). Mercator projection whose equator is parallel to the De Geer's line – essentially an updated version of his original 1943 figure (Wegmann, 1943). De Geer's line, depicted as dashed line running from Norwegian coast to mouth of Amundsen Sound. Arrows along De Geer's line depict westward (eastward) drift of the American–Greenlandic block (European block) relative to eastward (westward) drift of European (American–Greenlandic) block, is a Mercator projection whose horizontal borders parallel Greenland's postulated movement along the De Geer Line. Proposed southward movement of North America relative to the Pacific, suggested by movement along San Andreas fault, is depicted by northeast–southwest trending arrows along western margin of North America. Caledonide fragments of Scandinavia, Spitsbergen, Greenland, Grant Land, the British Isles, and North America are shown as dark gray. Arrows associated with Scottish Caledonide fragment show their transverse slicing.

The great problems with modern geology should be solved by team work. This work should be carried out according to well prepared programmes which entrust to the individual specialists their part of the work in order that the object may be attained through a well organised collaboration.

(Wegmann, 1948: 41)

In discussing the "jig-saw puzzle-method" of putting all the continents back together again, Wegmann (1948: 42) accented the need to investigate the ocean floor, which he categorized as *terra incognita*.

The method most commonly employed is the jig-saw puzzle-method, by which arcs now separated are regarded as parts of an earlier unit and are put together like the pieces in a jig-saw puzzle. The gap between the pieces is the present sea bottom. The fact that the substratum of the sea bottom is almost a *terra incognita* plays the most important role in the puzzle-method. All new investigations of the shelves, the continental slopes, and the sea bottom must accordingly influence the puzzle-method.

Wegmann was right about the seafloor. Even though much had been learned from the German 1925–7 *Meteor*, the Dutch 1929–30 *Snellius*, the British 1933 John Murray, and the 1947–8 Swedish *Albatross* expeditions, ocean floors still were *terra incognita* – as it turned out, even more *incognita* than anyone expected.

Wegmann also showed remarkable foresight in proposing the use of paleomagnetism to track former positions of the continents. Restricting his attention to his old Arctic stomping grounds, he appealed to the paleomagnetic work of the French physicist, P. L. Mercanton.

Mercanton (1926, 1931, 1932) has observed the direction of the magnetism in basalts derived from Iceland, the Faroes, East Greenland, and Spitsbergen and found that it does not correspond to the present magnetic conditions of the earth. The collecting of magnetically orientated and locally determined samples might supply interesting new information. It may perhaps be possible, on the basis of the changing magnetic orientation, to establish a certain sequence in age of the Arctic basalts, for basalts of the same age would have a corresponding orientation. It would be exceedingly useful for the study of the thick basaltic series in East and West Greenland as well as within the whole Arctic region and would enable us to make a more detailed division of the diastrophism.

(Wegmann, 1948: 41; Mercanton references are the same as mine; for importance of Mercanton in 1950s development of paleomagnetism see II, §2.10, §4.11, §5.8)

Following Mercanton, Wegmann also restricted his appeal to basalts, and did not mention sampling any rocks. In 1948 very little was known about the magnetization of sediments; he just would not have known, and went with what he had. He also had no idea that within just a few years, paleomagnetism would be used to track the ancient latitudes of continents relative to each other, and that within a dozen years, paleomagnetists would show that continental drift had occurred (see Volume II).

The history of Wegmann's fieldwork explains his adoption of mobilism. Weaned on the Pennine Alps by Argand, taught by Sederholm to understand the intricacies of deformed and metamorphosed Precambrian rocks, he was well prepared to understand and compare the Caledonides of eastern Greenland and Scandinavia. Beginning as a regionalist, he then worked in other regions, becoming a globalist. Certainly his reconstruction of the Caledonide tectonic disjuncts distributed around the North Atlantic stands as a major achievement of the classical stage of the mobilist debate.

Wegmann's prescience about what was needed to resolve the question of mobilism was also remarkable. Seeing the need for work outside of his area of expertise, he, unlike other Alpine tectonicists, wanted to see investigators, following Mercanton, use paleomagnetism to test drift. He also had the foresight to see that knowledge of the seafloor might also settle the question. He had the insight to see the need to describe relative movements of continents on a sphere, which led him to propose using Mercator projections making the rotational movements appear as straight lines. Even if he did not explicitly introduce or even know about Euler's theorem, he described motions as rotational about an axis of rotation. Finally, he became something of a generalist in that he not only welcomed but encouraged future work in fields outside his own.

8.11 The peri-Atlantic Caledonides: mainly Holtedahl

Olaf Holtedahl (1885–1975) spent most of his long and fruitful career at the University of Oslo. He was a major figure among Norwegian geologists, one of that heroic generation of Norwegian scientist/explorers, whose investigations took them to high latitudes in both hemispheres. He worked in Spitsbergen in 1909 and 1910, spent summers from 1914 through 1917 studying Finmark, the northernmost district of Norway, and the summer 1918 investigating Bear Island. With permission from the Soviet government, he organized in 1921 the first geological expedition to Novaya Zemlya. He undertook his last high-latitude expedition six years later, this time to Antarctica.

Holtedahl received his Ph.D. in 1913 from the University of Oslo, and was hired as a docent in paleontology and historical geology. In 1920 he became professor in historical geology. He spent part of 1912 and 1913 studying paleogeography under Schuchert at Yale. Before returning to Norway, he attended the 12th International Geological Congress in Toronto in 1913 where he met E. B. Bailey (§8.13). They became lifelong friends. In 1928, Holtedahl examined the Scottish Caledonides under Bailey's guidance, and revisited them about ten years later just before Bailey and he (1938) co-authored a monograph on the Caledonides of northwestern Europe in which they argued for mobilism. He retired from the University of Oslo in 1955, but remained professionally active for another decade. In 1951 he was awarded the Wollaston Medal from the Geological Society of London, and elected a foreign member of the Royal Society of London in 1961 (Størmer, 1976).

Holtedahl's abiding interest was in the paleogeography of northwestern Europe. In his early papers (1913, 1920a, 1920b) he suggested that a former landmass, now sunk beneath the Norwegian and Greenland Seas, had once connected Scotland, Scandinavia, Spitsbergen, and Greenland; that Greenland and North America had been joined; and that a similar former landmass had existed off the east coast of North America. In this early work he (1920a) explicitly rejected Wegener's continental drift. I shall trace in some detail his conversion to mobilism.

Holtedahl (1913) speculated about the existence of this former landmass in his presentation at the 12th International Geological Congress in Toronto, and was in a good position to do so because of his knowledge and firsthand experience of several Caledonide segments. Drawing on his personal experience of Spitsbergen, he discussed similarities between the Silurian and Devonian Old Red Sandstone Series in Spitsbergen and Scotland. Expanding his comparison to include the Caledonides in Scandinavia, he argued that a former landmass had once connected them.

In Greenland, Scotland and Norway we know only of continental conditions during Devonic time ... The Devonic coarse conglomerate and sandstone beds of western Norway show directly

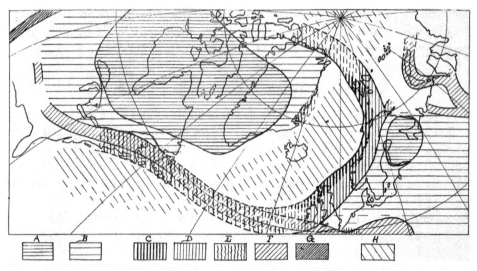

Figure 8.3 Holtedahl's Figure 12 (1920a: 18). A (continental shields) and B (sediment-covered rigid continental areas) are stable areas. C through G are post-Ordovician zones of folding: C, pre-Downtonian time [late Silurian]; D, pre-Devonian; E, Devonian; F, Late Paleozoic; G, post-Paleozoic. H, areas of postulated greatest vertical movement (Appalachia and former land in the Greenland Sea).

in their structure that they were deposited on a surface dipping to the southeast, an indication of high land in the opposite direction. Secondly, the same view comes naturally from consideration of the Siluric-Devonic geology of northwestern Europe. We know from investigations, both in Great Britain and in the Scandinavian countries, that in general in Ordovicic and Siluric time a land of great extent lay to the northwest, that is, on the Atlantic side. The influence of this continent is very strongly marked in late Siluric time. In Scotland and the north of England, we find immense masses of coarse clastic material coming from this land, and in Norway the Siluric limestones of the Kristiania area [southern Norway] are replaced to the northwest by clastic material that was derived from the west. We are able to see how the western upheaval gradually spread toward the southeast, accompanied by a deposition of red sandstones along the south-eastern border . . . If we now consider the intimate relation that no doubt exists between the Caledonian folding throughout Scotland, Scandinavia and Spitzbergen, and the distribution of the red sandstones, it seems very probable that also in the northern part of the folded area the high land was on the Atlantic side.

(Holtedahl, 1913: 710–711; my bracketed addition)

Seven years later, Holtedahl returned to the paleogeography of northwestern Europe expanding his analysis to North America. He again argued that a former land had existed in what is now the Norwegian Sea connecting Greenland, Spitsbergen, Scandinavia, and Great Britain, and compared it to Appalachia, the purported former landmass east of North America that Schuchert and others had already proposed (Holtedahl, 1920a: 19–22). He illustrated this in Figure 8.3.

The postulated land areas provided sediments that collected in the geosynclinal depression and were later deformed and metamorphosed during the Caledonian orogeny. Holtedahl realized that his account clashed with the American belief in the permanency of ocean basins, but thought postulation of a former landmass in the Norwegian Sea unavoidable.

It certainly will be hopeless for the advocates of permanency of the ocean basins to apply the theory to the Norwegian Sea. In places, e.g., at Spitzbergen, the shelf . . . is very narrow, and great depths are found not far to the west of the place where the huge continental sandstones of Devonian time – many thousand feet in thickness – are still to be seen. Land must have existed where there is now deep sea. A consequence of this view is that the greater density of the sub-oceanic crust areas must be regarded as due to secondary processes during and after the subsidence.

(Holtedahl, 1920a: 21)

There was, of course, a recently proposed alternative, but it was too daring and Holtedahl dismissed it in a footnote.

I shall not enter here into any discussion of the very daring – in my opinion far too daring – hypothesis of Wegener (*Die Entstehung der Kontinente und Ozeane*, Braunschweig, 1915) according to which the separation of America and Europe was caused by a horizontal movement of the upper or sialic zone of the crust, a movement that, so far as the northern part of the Atlantic area is concerned, is supposed to have taken place in Quaternary time. Far from being more easily explained in the light of this hypothesis (which postulates the existence of a coherent continental mass embracing both the new and the old world, with no Atlantic "fissure," in pre-Cenozoic time!), several of the facts mentioned in the preceding pages do not at all harmonize with it.

(Holtedahl, 1920a: 24)

He did not elaborate.

Turning to the Scandinavian Caledonides, he agreed with what was becoming generally accepted, that they were much like the Alps with huge overthrusts; he argued that there had been thrusting of sediments and igneous rock and "highly meta-morphic, often gneissose, unfossilized rocks above slightly-altered fossiliferous Cambro-Silurian sediments" from a region west of the Norwegian coastline (Holte-dahl, 1920b: 387). However, his recognition of the extensive overthrusting from what is now the Atlantic did not turn him into a mobilist. Instead he (1920b: 401) approvingly referred his readers to his previous paper (1920a) with its fixist paleogeography.

Holtedahl (1925) later presented a more detailed comparison of the geological similarities of Spitsbergen, Scandinavia, the British Isles, and eastern North America. He (1925: 9) documented the "quite astonishing number of similar features" in the Caledonides of Spitsbergen and Scotland.[19] Still using the fixist idiom, he (1925: 18) described Europe and North America as "tangentially stable areas" where there is "the tendency of the younger ranges to form at a greater distance from the old nuclei, the shields, than the older ones." The following year, he presented a paper at the 14th International Geological Congress in Madrid in *Pan-American Geologist*, which shows that he was beginning to think more about mobilism, but still retaining his fixist paleogeography; sinking former continental regions beneath oceans faced objections, but so too did mobilism.

Now, of course, we have the geophysical and other objections towards transforming a large continental area into a deep-sea basin; on the other hand, assuming a drifting one asks for facts of a structural nature indicating very recent tangential crust movements.

(Holtedahl, 1926: 271)

He constructed a geological map of the lands bordering the Norwegian Sea, noted their many similarities, but remarked on the poor fit of the Greenland and Norway margins.

I have prepared a map, plate XV, from which the distribution of various formations, according to our present knowledge, may be seen. A study of this map will show that if we make the two sides go together, conditions will not fit very well.

(Holtedahl, 1926: 272)

Holtedahl also noted that earlier work had suggested the occurrence of Caledonide-style tectonics along Greenland's eastern margin, but he wanted to wait for more information before speculating on drift.

A rather remarkable feature in the geology of the east Greenland coastal regions is the north and south belt of folded and, to some extent, metamorphosed Cambro-Siluric rocks (studied by Nathorst in 1899), lying below nearly horizontal Old Red sandstones, at about 72°–73° north latitude. It looks as if we have here a parallel to the Caledonian deformation so typical of the east side of the Norwegian Sea, and no doubt this is a feature that might be said to point towards a near relationship between the two sides of the ocean. We must hope that further investigations soon may enlighten us concerning the true character of that Greenland folded zone, and especially as to its continuation to the north and to the south.

(Holtedahl, 1926: 272)

Wegmann (1935) had yet to carry out his definitive work there (§8.10).

Holtedahl switched to mobilism in 1936, and two years later he and Bailey (1938) joined forces. Christoffer Oftedahl (1975) commented on Holtedahl's conversion.

After more years of study of the general geology of the Caledonian orogenic zone of southern Norway, Holtedahl presented a revolutionary new view at a meeting of Nordic scientists in Helsinki in 1936. Large-scale overthrusting had been postulated by the Swedish professor A. E. Törnebolm in 1896, then inspired by results from Scotland. But Törnebolm studied only the marginal parts of the Caledonian folded and metamorphosed rocks, and he relied especially on fieldwork in marginal parts in Sweden where the thrusting of metamorphic rocks over unmetamorphic Cambrian and Silurian rocks were most clearly observable. In southern Norway the central area of Jotunheimen with the highest peaks of Norway presents a dominating problem: Gabboric to intermediate igneous rocks of the mountains form an upper layer above rocks that must be Cambrian and Ordovician, but is the situation stratigraphic or tectonic? W. M. Goldschmidt, who described these igneous rocks in a classic paper in 1916, discussed the age problem and preferred an Ordovician age for the igneous activity, with some thrusting a little later. But Holtedahl discussed the possibility that the Jotun rocks represent thrust nappes from the Atlantic. He theorized that the downfolding of these rocks and their Cambrian-Ordovician base as well as the Precambrian were only a later phase of folding . . .

The sketchy views of 1936 were enlarged upon in a general treatise of the Caledonides, in which Holtedahl wrote the Scandinavian part and E. B. Bailey wrote the British part (1938). In those years Holtedahl was so modern that he gave a sketch map of Caledonian geography with Greenland in the pre-drift position, according to the theory of Wegener – a sketch that is inescapable in any present-day regional discussion of Caledonian geology. In the late 1930s this was quite daring, and I remember that it nearly automatically called for a smile among Norwegian geologists.

(Oftedahl, 1977: 2)

If I read this correctly, Oftedahl thought that Holtedahl accepted mobilism when he realized in the Scandinavian Caledonides the extent of metamorphism and over-thrusting from the Atlantic. This, I think, is not the whole story. Already in 1926, Holtedahl had recognized extensive metamorphism and overthrusting there yet he did not invoke mobilism. There may be another reason for him doing so. Surely he must have known or had some early inkling of Wegmann's work on East Greenland, first reported in 1935 (§8.10), which showed its intimate ties with Norway. A decade earlier he (1926) was eager to see what work in East Greenland would reveal, so he would have been alert to reports from there. Although when announcing his conver-sion he did not refer to Wegmann's work in Greenland, Holtedahl (1936: 144) indicated the presence of Caledonides along Greenland's eastern margin in his diagram showing Greenland's westward drift relative to Europe. Perhaps it was Wegmann's (1935) work that really tipped the balance for Holtedahl.

Regardless of precisely why he switched to mobilism and the idea of getting rid of the Greenland Sea in the Early Paleozoic, Holtedahl's work on different segments of the Caledonides was crucial. Furthermore, and very interestingly, Bailey and Holte-dahl (1938) explained the formation of the Caledonides of northwestern Europe by continental collision. They announced their adherence to mobilism at the beginning of their monograph. Their argument centered on the many similarities among Caledonide segments found now around the periphery of the North Atlantic.

The structural likeness between the south-eastern belt of the Caledonian zone in part of Scandinavia and its counterpart in the North-west Highlands of Scotland indicates the pres-ence of a north-west foreland, somewhat similar in character to that of South-east Scandi-navia. In present day geography the north-west foreland of Scotland appears to lie in a zone between the Caledonian belts of North-west Europe and East Greenland, a rather inappropri-ate position. There are, however, close resemblances between the geography of North Norway and Spitsbergen, on the one hand, and of Greenland, on the other, which seem to suggest a more logical constellation. We may perhaps, with Wegener, regard the Greenland mass as having been formerly connected with Europe. With this adjustment we obtain one zone of Caledonian orogeny in the North Atlantic area, with Central Greenland as the north-western foreland. The distribution of facies in the Caledonian geosynclinal sediments seems to point strongly toward such a conclusion (cf. p. II 72).

(Bailey and Holtedahl, 1938: 3; the page reference is to their
Figure 16 reproduced as my Figure 8.4)

Their reconstruction of the region prior to opening of the North Atlantic follows Wegener and is given in Figure 8.4. Greenland is shown as having undergone a clockwise rotation and westward movement relative to Europe. The Caledonide fragments in Scotland, Scandinavia, Spitsbergen, and Greenland were originally juxtaposed and their present positions are explained in terms of continental drift.

They explained the formation of the Scandinavian Caledonides by invoking an Early Paleozoic convergence of Greenland and Scandinavia. They envisioned a Lower Cambrian sea (geosyncline) between the Precambrian cratons of Greenland and Scandinavia, where, in Ordovician times, basic lavas arose, peridotite and gabbro intruded, and islands arose providing clastic sediments. They compared what he had in mind with "the present East Indian-Australian region" (Bailey and Holtedahl, 1938: II. 25). As the two landmasses converged the geosyncline was squeezed and extensively metamorphosed and overthrusting followed. Now prepared to believe some of what Argand had said, prophetically they suggested that in the British segment the extensively overthrusted Scottish Caledonides had formed by the collision of northwest Scotland (with its North American style Precambrian basement overlain by Early Cambrian shelf limestones of indisputable American affinity) with roughly southern Scotland, Wales, and England. They added that Europe has collided with other landmasses on three occasions since the Cambrian.

Argand . . . has pictured geosynclines as determined by stretching, by continental-drift apart, which attenuates the sial layer and eventually permits sima to reach the bottom of the sea at bathyal or abyssal depths. If such drift-separation continues, a new ocean bed may be developed, covered with products of submarine eruptions. If, as has thrice happened in the post-Cambrian history of Europe, a drift of separation gives place to a drift of approach, then a folded mountain chain comes into being.

(Bailey and Holtedahl, 1938: 49)

Presumably they had in mind successive collisions to form the Caledonide, Hercynide, and Alpine belts.

H. Bütler, a Swiss geologist, was encouraged by Wegmann and Lauge Koch to take part in the Danish 1931–4 Greenland Expedition. He spent approximately five months in 1934 investigating the Devonian stratigraphy and tectonics along a section of Greenland's eastern coastline (Bütler, 1935a). He (1935b: 31) suggested that if islands within and lands surrounding the Scandic (Greenland and Norwegian) Sea are brought together in accordance with the theory of continental pushing (*Kontinentalverschiebung*), then Greenland's Devonian Red Sandstones match up with those in Spitsbergen, Scotland, and Bear Island. He (1935a) claimed that he had identified an Argand-type basement fold beneath the Devonian Basin adjacent to

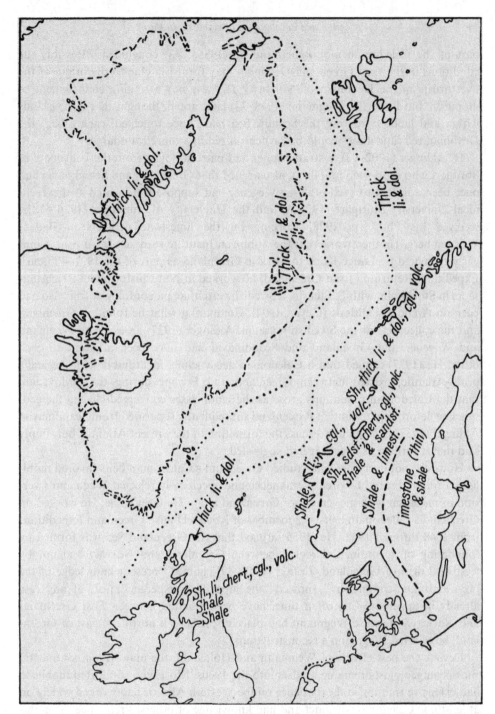

Figure 8.4 Bailey and Holtedahl's Figure 16 (1938: II. 72). Greenland is rotated and moved closer to northwestern Europe, and is superimposed over Iceland and Bear Island. Greenland also is shown in its current position. Matching Ordovician sediments are shown, but metamorphosed regions are not indicated. There are thick limestone-dolomites (thick li. & dol.), volcanics (volc.), limestone conglomerates (li. cgl.), shale (Sh. and Shale), sandstones (sdst. and sandst.), and chert.

part of the Caledonian mountains, and he (1935a: 33) concluded: "Possibly the conditions in the East Greenland Devonian may furnish a clue to the origin of the overthrusts in the Devonian of Norway." His was not a ringing endorsement of mobilism, but Bütler, as Wegmann (1948: 34) later noted, thought that if the islands within and lands bordering the Scandic Sea had "once touch[ed] each other" the Devonian red sandstones could be "synchronized in a more rational way."

H. Aldinger (1902–93), a stratigrapher and paleontologist from the University of Stuttgart who supported mobilism, also argued that Greenland and Scandinavia had once been adjacent to each other as Wegener had supposed. Educated at the Technical University Stuttgart (1921–6) and the University of Tübingen (1926–8), he received his Ph.D. in 1928, working on the limestone deposits in Baden-Württemberg. He then worked at the Tübingen Institute studying fossil fish. Aldinger was invited by Lauge Koch to work in Greenland as part of the 1931–4 Danish Expedition. Returning from Greenland, he worked at Naturhistorischen Reichsmuseum in Stockholm with E. Stensiö, a Swedish vertebrate paleontologist who also had been on Arctic expeditions (Geyer, 1994). Comparing what he found in Greenland and the collections at the Swedish museum, Aldinger (1937) agreed with Wegmann and Wegener that Greenland and Scandinavia had once been adjacent to each other. He (1937) argued that a Caledonian geosyncline, identifiable with Argand's proto-Atlantic, existed between Greenland and Europe during the Ordovician, that the Caledonian mountains arose as the landmasses converged closing the geosyncline deforming its content. Greenland subsequently separated from Scandinavia. Aldinger (1937: 127) did not discuss the formation of the present Atlantic, but simply said that it formed as Wegener had suggested.

However, not everyone who studied Greenland's Caledonian belts favored mobilism. Erdhart Fränkl favored permanence and specifically rejected Wegmann's version of mobilism with movement of Greenland along De Geer's line. He worked on Greenland's Caledonian belts as a member of Koch's Danish Greenland Expeditions from 1948 through 1953. He (1956) argued that the Greenland Sea was formed by foundering of a continental region between Greenland and Norway and not by westward drift of Greenland. Fränkl (1956: 39) claimed "present knowledge of the pre-Cambrian stratigraphy" showed "the area of Spitsbergen must, at any rate already in upper pre-Cambrian time, have been situated between East Greenland and Norway." Because Wegmann had placed Spitsbergen north not east of Greenland, he rejected Wegmann's reconstruction.

Despite the best efforts of Wegmann and Holtedahl, the majority of continental European geologists remained fixists into the 1960s. Most who supported mobilism had extensive training and experience on the Western Alps or had worked widely on at least one Caledonide disjunct and had knowledge of others. Preference for mobilism among European geologists with interests in tectonics was strongest in those with one or other of these experiences. Of course, there were tectonicists who worked on the Western Alps or on Caledonide disjuncts but who remained fixists.

8.12 Hercynides/Variscides and Caledonides: F. E. Suess

I now turn to Franz Eduard Suess (1867–1941), son of Eduard Suess, and a leading European mobilist during the 1930s. Although he worked on the Alps, and later examined various disjuncts of the Caledonides, his interest in mobilism had another source, the Variscan (Hercynian)[20] Late Paleozoic orogenic belt of Central Europe, where he carried out extensive fieldwork and studies of metamorphism. Agreeing with Argand's and Staub's mobilistic interpretation of the Alps, he claimed (1926) that the Variscides had a similar origin. Three years later (1929), he concurred with Argand's mobilistic interpretation of the tectonics of Asia. Later still (1931), he argued that the Caledonides were comparable to the Variscides and the Alps; all three were highly asymmetric and had resulted from continental collisions. Familiar with the literature on the Scandinavian Caledonides, he suggested that they were once part of the same chain as the Variscides. Five years later, he (1936a, 1936b) compared the Caledonide and Hercynide of Europe with their extensions in North America, and, following Bailey and van der Gracht, he argued that the similarities between the new and old world segments provided support for continental drift. Suess (1938b) then extended his analysis of the Caledonides to East Greenland, and, following Holtedahl and Wegmann, used it to argue for continental drift. He offered a reconstruction of Spitsbergen, Greenland, Scandinavia, and the British Isles before the opening of the Atlantic Ocean and Norwegian Sea. Finally, he (1937, 1938a, 1939) argued for mobilism in his three-part, monumental *Bausteine zu einem System der Tektogenese* in which he drew together his previous work on tectonics and metamorphism.

Suess entered the University of Vienna in 1886. He worked primarily with Viktor Uhlig, receiving his Ph.D. in 1891 for work on the basin clays of Austria and Bavaria. He became an assistant at the German Technical University in Prague, helping Uhlig and C. Diener with their work on the Himalayas and Urals. He then became interested in petrology, studying with the petrologist Friedrich Becke (1855–1931) in Prague, becoming adept at microscopy, especially of the metamorphic rocks of orogenic belts. During the summers of 1892 and 1893, he helped F. Frech survey the region west of the Brenner Pass in the Central Alps, applying his new petrological skills. He then began to work at the Imperial Geological Institute in Vienna, where he hoped to continue his Alpine studies, but was ordered by its Director to map basement folds in the Bohemian Massif, then part of the Austro-Hungarian Empire. Suess transformed this undesirable assignment – it took him away from his beloved Alps – into a golden opportunity, spending much of the next thirty years unraveling the tectonic and metamorphic history of the European Variscides. Suess secured an academic post in 1908 at the University of Vienna, where he became Uhlig's successor in 1911, and remained there for the rest of his career.[21]

Suess eventually investigated many outcrops of the Variscides, from the Brünner intrusives of the Mähren, in the present Czech Republic, through the Bohemian

Massif to the Massif Central of France. He was particularly interested in their crystalline basement complexes and tectonically related metamorphism (syntectonic metamorphism). He (1923) compared what he found with the Alps, and developed a general model in which he recognized three zones within an orogen: an outer zone, non-metamorphosed and characterized by folds; a middle, syntectonic metamorphic and folded zone; and an inner, post-tectonic crystalline zone characterized by granitic intrusions (Suess, 1926). Granite intrusions are surrounded by an aureole of thermally metamorphosed rocks. The relative positions of the zones denote the direction of thrusting. The crystalline zone is always on the side from which pressure comes, the zone of non-metamorphosed folds is farthest away, and the metamorphosed zone lies between.

Suess compared the Variscides with the Alps. He selected the Moldanubian Block (the southwest portion of the Bohemian Massif) as his Variscian example in which little of the middle zone remained, and compared it with the East Alpine nappes and crystalline basement of the Dinarides to the southeast. He compared the nappes of the Erzgebirge (in the Saxothuringian Block within the Bohemian Massif east of the Moldanubian Block), and their intense folding, metamorphism, and overthrusting, with the Pennide zone of the Central Alps. The outermost zone of folded non-metamorphic rock is represented in the Variscides by the Devonian folded sedimentary rocks of the Rhine Slate Mountains (Rhinish Schiefergebirge) in Luxembourg, Belgium, and Germany, extending as far eastward as the Harz Mountains and the Thüringer Forest. He compared this zone with the Helvetic nappes in Switzerland (Suess, 1926, 1931; the latter is a good account in English of his model).[22]

From his documentation of the strong similarities between the Alps and Variscides, he argued (1926: 252) that the latter stood as high as the Alps and, just as Argand and Staub had invoked Wegener's theory to explain the origin of the Alps, he invoked mobilism to explain the Variscides. Agreeing with Argand's replacement in the Alpine instance of contractionist tectonics by continental collision, he (1926: 252–254) used that to explain the huge overthrusts of the much older Variscides (Hercynides).

Stepping outside continental Europe, he toured the Scottish Caledonides under Bailey's guidance, and argued (1931) that his three-zoned, orogen model applied there: the middle zone, the syntectonically highly metamorphosed and folded zone, corresponded to the Dalradian, the metamorphosed and post-tectonically crystalline Moine schists to the innermost zone, and the unmetamorphosed folded Devonian sedimentary rocks atop the Dalradian to his outermost zone. With its three-zoned Alpine structure, he argued (1931), following Bailey (1926), that the Scottish Caledonides are best explained in terms of mobilism, and argued that they are a continuation of the much larger Scandinavian belt.

Four years later, Suess attended the 1935 16th International Geological Congress in Washington DC, and used the opportunity to do fieldwork in New York State and Maryland in order to confirm what he had read about the area and test his zonal model. He (1936a) also presented a paper. Translated into English, the paper (1936b) was titled "Tectonic Affinities between European and North American Mountain Systems." He began by arguing that the structural similarities and crisscrossing of the Hercynides and Caledonides on both sides of the Atlantic provided the strongest support for a former close connection between North America and Europe.

The angular crossing of Caledonids and Hercynids, on both sides of the Atlantic ocean, furnishes, in conjunction with a variety of stratigraphic relationships, the most significant indication of the former connection of North America with Europe, in the sense of Wegener.

(Suess, 1936b: 82)

Using his knowledge of the European Hercynides and Caledonides, his zonal model, and a review of the literature, he sought to demonstrate their structural similarities and provide support for their mobilistic origins. He had no wish to appear arrogant, although he intended to persuade his largely North American audience that they had incorrectly analyzed their Appalachian orogenic belt.

An exchange of views from different quarters of the world is the chief objective of our gatherings; and when here I am able to weigh the matters of the other side of the Atlantic's tried experience with those obtained by me now from the literature of the described regions; so, if you please, what here is stated is not presented so much as categorical assertion as mere inquiry.

(Suess, 1936b: 82)

Leaving mere inquiry aside, he declared that the Appalachians south of New York State correspond to the European Hercynides.

In the European Hercynids three belts are joined in a single dynamic unit: (1) the generating Moldanubian belt in the south; (2) the sparsely pressed belts, with metamorphosed thrust-sheet structure (Erzgebirge, Spessart); and (3) the belt of nonmetamorphosed folds (Rhinish Schiefergebirge, Harz). In the American Hercynids, the so-called geosyncline corresponds to the third belt of Europe. The crystalline basement rocks, in the southeast, adjoin the geosyncline as the second belt. The crystalline facies and the wide-spread thrust-sheet and over-thrust structures in these regions lead to the assumption that a generating belt, corresponding to the Moldanubian, was adjoined . . . As in the European Hercynids, so, also in the American representatives, the granites are indigenous to the crystalline belt, and are pushed outward from it.

(Suess, 1936b: 81)

Continuing the comparison, he invoked mobilism and claimed that the Hercynian Appalachians were formed in the same manner as the Alps and the European Caledonides.

On the basis of their general structure, the Hercynian Appalachians are to be regarded as continental border-mountains, comparable to the Andean chains of both Americas. The associated continent, however, was not Laurentia, but an ancient land-mass which lay in the lap of the present Atlantic ocean. Through pressure the chain became welded to Laurentia, as the Alps were welded to the fragments of the Hercynids, and the European Caledonids to the Fennoscandian shield.

(Suess, 1936b: 81–82)

So Suess did more than argue that North America had once been united to Europe because of the similarities between the Hercynian belts of each continent; he claimed that the Appalachians were welded onto the advancing edge of North America through collision with some other landmass just as the Alps were welded onto remnant Hercynides in Europe. The European Caledonides formed in the same way. Suess, like Argand, was willing to suggest that the Early Paleozoic mountains of Europe and North America formed through collision of the two continents; he did not mention Argand's proto-Atlantic. Perhaps he did not think the continents had been very far apart before they began approaching each other.

Suess then turned more specifically to the North American Caledonides, connecting them to the Scottish Caledonides. The innermost zone of the Scottish Caledonides corresponds to the overthrust of the Logan line, and "extends from the St. Lawrence River, across Lake Champlain, nearly to New York City" (Suess, 1936b: 83). Identifying his two other zones proved more difficult. Nevertheless, Suess concluded, again appealing to Bailey (1926), in this way:

The essential facts which support, with greatest weight, the connection of Europe and North America, in the sense of Wegener, and as is already emphasized by Bailey, are the striking angular crossing, on both sides of the Atlantic, of the Caledonids and Hercynids; further, the sharp separation of the European Acadian faunal province from the Canadian shield; and the reappearance of the Laurentian facies in the Scottish Moine over-thrust, which thereby may be added to the Logan line. They permit of scarcely any other explanation for the relations in general.

(Suess, 1936b: 96)

He continued to connect Caledonide segments, adding eastern Greenland in a short paper (1938b) entitled "Der Bau der Kaledoniden und Wegener's Hypothese" (The building of the Caledonian and Wegener's hypothesis). Relying on Holtedahl's work in Spitsbergen and Norway, and Wegmann's in Greenland and Norway, he argued in favor of Wegener's idea of the North Atlantic being caused by later continental drift, although he proposed slightly different positions for Greenland, Spitsbergen, Scandinavia, and the British Isles (Figure 8.5).

Suess (1938b: 335) again used his zonal model to demonstrate the strong similarities among different Caledonide segments.

He also argued for mobilism in the third volume of his three-volume monumental *Bausteine zu einem System der Tektogenese (Building blocks for a system of tectogenesis)*. The third volume, *Part III. Der Bau der Kaledoniden und die Schollendrift im Nordatlantik* (The building of the Caledonian and block-drift in the North Atlantic), was itself divided into three parts; the first, "Die Kaledoniden in Schottland und Vergleiche" (The Caledonian in Scotland and comparisons), was published in 1939, and the others, "Der Bau der Kaledoniden und die

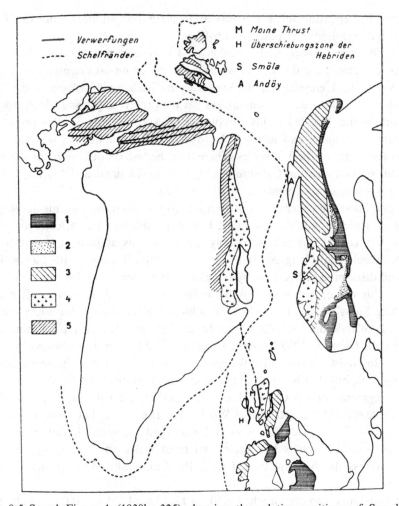

Figure 8.5 Suess' Figure 1 (1938b: 325) showing the relative positions of Scandinavia, Greenland, Spitsbergen, the British Isles, and Ellesmere Land in the sense of Wegener's hypothesis. 1 corresponds to foreland early Paleozoic sediments that were not moved but covered; 2 corresponds to a Sparagmite-Quartzite cover within the Scandinavian Caledonian; 3 corresponds to Suess' middle zone of syntectonic metamorphism; 4 corresponds to Suess' innermost part of an orogen with post-tectonic crystallization and katametamorphism; 5 corresponds to Precambrian and early Paleozoic folded and non-folded coastal chains in the Hekla-Hoek Formation in Spitsbergen.

Schollendrift im Nordatlantik" (The building of the Caledonian and block-drift in the North Atlantic) and "Die Kaladoniden in Goenland" (The Caledonian in Greenland), were posthumously published by the Geological Society of Austria in 1949 (Suess, 1949). Suess compared Caledonian disjuncts, drawing on his own work as well as that of Wegmann, Holtedahl, and Bailey. Referring to Wegener, Taylor, Argand, Staub, du Toit, and van der Gracht, he continued to defend mobilism by repeating his arguments and marshaling support from other fields such as paleontology and paleoclimatology.

Suess' progressive support for mobilism, I believe, supports regionalism. Suess' petrologic–tectonic zonal model of orogens was based on his extensive investigation of the Variscides/Hercynides, and armed with his firsthand knowledge of the Alps he was able to see the structural similarities between the younger and older orogenies. His work in the Alps led to his acceptance of Argand's and Staub's pro-mobilist explanation for them, and hence to his own belief that mobilism was required to explain the origin of the Variscides. Thereafter, he tested both his zonal theory and mobilism by investigating Caledonide disjuncts in Scotland and North American, and the Appalachian segments in North America.

Despite Suess' considerable contributions to mobilism through his development of Argand's idea of collision orogeny, I do not think he much affected the overall debate. Du Toit (1937: 161) favorably mentioned his address at the 16th International Geological Congress, but only to note that it agreed with Bailey. I have found no discussion of his work by Holmes or Wegener,[23] and Bucher (1933) failed to index his name, but praised him for his zonal model. However, after briefly describing Suess' model, he (1933: 278) added in a footnote, "But Suess speaks in terms of 'Wandernde Schollen'." Only Bucher and Knetsch cited him in the publications from the 1939 Frankfurt meeting, and he is not mentioned by anyone at the 1926 AAPG symposium. Wegmann, Bailey, and Holtedahl appreciated his contributions, but that is expected because they all argued that the dispersal of the various segments of the Caledonides made little sense without mobilism. Moreover, his death and the outcome of World War II ensured that his work would be read by few Earth scientists who did not read German. In addition, North American Earth scientists had their own fixist views about orogeny, and when looking across the Atlantic turned to Stille, Cloos, and Haarmann for support of them.

The Dutchman van der Gracht (§3.10, §3.12, §4.4, §4.10) was influenced by Suess. Like Suess and Bailey, he (1931) decided to examine North American orogenies from a European standpoint. Concentrating on the Late Paleozoic deformation of the Wichita and Quachita (Ouachita) Mountains in Arkansas, Oklahoma, and Texas, he wrote a 162-page monograph based on "about twelve years of work of the writer" as a petroleum geologist "in the southern Midcontinent of the United States" (van der Gracht, 1931: 3). He wrote it after he returned to Europe, where he visited the Alps and Variscides, and reviewed recent studies of them. Drawing heavily on Suess and Staub, and to a lesser extent on Argand, he (1931: 150) argued that there were "striking analogies between the Variscan orogeny of Europe and that of the American chains . . ." To explain both old and new world orogenies he invoked collisions between respective parts of Gondwana and Europe and North America. Noting the greater degree of deformation in Europe than in North America, he sought a mobilistic explanation for it.

In Europe, the Alpine compression, evidently, was far greater than in America. The great northern arc of the Alps and Carpathians encroaches far upon the Variscan structure. In America the Alpide chains remain altogether south of the Gulf of Mexico, in Chiapas, Guatemala and Honduras, and spread in great virgations through the West Indies . . . Surely, the hinterland of the Quachitas was not disturbed by any serious orogeny, at the time of the orogenic paroxysm of the European Alps, and, in comparison, deformation is insignificant. Yet, we note the same two phases which are so prominent in the Alps. It is only to be expected that the orogenic pressure in both the late Paleozoic and the Tertiary cycles affected the south-central States of North America less than southern Europe. We miss the tremendous vise jaw action between Gondwana and Eurasia in this extreme western portion of "Laurasia." Although the relations are far from cleared, it is evident that on the western side of both the American continents, entirely different conditions arise, which will surely have prevented the South America mass to play as great a part as the central bulk of Gondwana. Nevertheless, the general picture remains closely related.

(van der Gracht, 1931: 150–154)

Suess' model of repeated deformation of orogenic regions helped him see how North American orogenies fitted into a mobilist perspective.

This brief account of the impact of mobilism on tectonicists from continental Europe strongly suggests that their regions of study played a very significant role in the choices they made. Of course, there were many geologists who worked on or were familiar with more than one Caledonian disjunct or Variscan/Hercynian/Appalachian disjuncts, but who remained fixists. Notwithstanding, the pre-drift assembly of these tectonic disjuncts presently distributed around the North Atlantic stands as one of the major achievements of the classical stage of the mobilism debate. Nobody, however, claimed that they had presented a difficulty-free solution. There was still room for reasonable disagreement: there was still room for someone raised on the Western Alps to be a fixist, to accept the existence of huge overthrusts, yet deny global mobilism (Termier, 1924); it still remained reasonable for someone familiar with more than one of these tectonic disjuncts to prefer fixism (Fränkl, 1956).

8.13 Mixed reception in Britain and Ireland

The response to mobilism in Britain and Ireland was similar to that in continental Europe; support was split, with fixism remaining dominant (Marvin, 1985). If regionalism helped to shape preferences, there should be no strong preference for either mobilism or fixism in those regions where neither enjoyed an advantage; Earth scientists should defend their preference on grounds unrelated to the geology of their region, and with perhaps one exception, this was so in Britain and Ireland. As I have already described, Good (§3.5), Hutchinson (§3.5), Seward (§3.5), Walton (§3.5), Westoll (§3.5), Holland (§3.9), G. C. Simpson (§3.12), Thomas (§3.12), Joly (§4.9), Bull (§5.4), Holmes (§5.4), Rastall (§5.13), P. Evans (§6.13), Fox (§6.13), West (§6.13), and Fermor (§6.14) based their preference for mobilism on the geology of elsewhere, not Britain and Ireland. E. B. Bailey is the exception. His support for mobilism, to a considerable extent, grew out of his work on the Scottish Caledonides (§8.8, §8.12), which, as just seen, if viewed as a fragment of a once continuous orogenic belt extending from Spitsbergen and Greenland through Scandinavia, Scotland, Ireland, and eastern North America, provided mobilists with strong support. If regionalism is at play then support for mobilism based on the regional geology of Great Britain and Ireland should therefore come from those who devoted much time to the study of the Caledonides.

Other British Earth scientists did not prefer mobilism because of its ability to explain the regional geology of the British Isles, and I shall shortly mention several of them, but before doing so I need to stress the unique position occupied by Jeffreys and his response to mobilism. Throughout his long career he was adamant in his opposition, and his influence was profound and long-lasting. I have already described his early opposition to continental drift, and his response to Joly and Holmes (§4.11, §5.5). British workers perforce operated against a background of Jeffreys' relentless attack on continental drift.

Here I am concerned with the reception of mobilism among a number of British geologists beyond those just listed, beginning with W. B. Wright. He was impressed with mobilism's ability to explain geological and paleontological similarities on opposite sides of the Atlantic. Educated at Trinity College, Dublin, he began working for the Geological Survey of Great Britain and Ireland in 1901. During his long career, he worked under Lamplugh in Ireland, and Bailey in Scotland. His most important work, *The Quaternary Ice Age*, was first published in 1914. In 1921 he organized a branch office of the Geological Survey of Great Britain in Manchester, and worked on the Lancashire coalfields, subdividing them in terms of the freshwater mussels. He (1923a: 30) reported why he preferred mobilism during the 1922 meeting in Hull of the Geological Session of the BAAS on continental drift:

Mr. W. B. Wright pointed out that a critical comparison of the geological formations on the two sides of the North Atlantic shows on the whole a very remarkable correspondence, both stratigraphical and palaeontological, from the Archaean to the Cretaceous, and in particular brings to light certain facts even more strikingly indicative of a former approchement between the two continents than any pointed out by Wegener.

Several months later, Wright (1923b) defended Wegener against what he took to be a frivolous attack on Wegener's hypsometric curve, which statistically displayed Earth's preference for two levels, the average level of continents and ocean basins. Wright also participated in another discussion on mobilism thirteen years later. Asked to open the discussion, he mentioned Joly's and Argand's contributions to mobilism (Anonymous, 1935: vi). The reporter of the meeting mentions nothing about whether or not Wright still preferred mobilism. Wright's mention of Argand is interesting, given his association with Bailey. Perhaps Wright's work in Scotland with Bailey, who no doubt discussed Caledonide disjuncts with him, led Wright to support mobilism. Wright said that "listening to Bailey was like hearing the wheels grinding inexorably along" (Stubblefield, 1965).

John W. Evans (1857–1930) studied geology at the Royal College of Science, London, and in 1912 became lecturer in petrology at Imperial College where he remained until he retired in 1927. Elected FRS in 1919, he was president of the Geological Society of London from 1924 to 1926. He also spent much time working in South America and India before encountering Wegener (W. W. Watts, 1931) (§6.13). His principal reason for accepting mobilism was the similarities among various Gondwana fragments.

The evidence, based on similarity of lithological characters and fossil contents of the rocks, that South America east of the Andes and the Falkland Islands were once in much closer proximity to Africa, is to my mind conclusive, and scarcely less is that of a former association of a great part of India and of Australia with Africa.

(Evans, 1923: 393)

Evans also wrote the introduction to the English translation of the third edition of *The Origin* and in it he emphasized Wegener's explanation of the distribution of the Permo-Carboniferous glaciation.

Previous authors had suggested that the glaciation in South America, India, and Australia at the end of the Carboniferous or beginning of the Permian was due to the neighbourhood of the antarctic pole, but the difficulty had to be faced that there was no possible position of the pole that was not distant at least 70° from one of the glaciated areas. This difficulty ceases to exist if Professor Wegener is right in supposing that all of them were at that time in close proximity to one another, instead of being separated by thousands of miles of sea as at present.

(Evans, 1924: x)

Herbert Baker preferred mobilism because of its success in explaining the geological similarities between South Africa and the Falkland Islands, where he had served as Government Geologist.

Since my return from the Falkland Islands a few months ago I have followed with great interest the course of the discussion in the columns of NATURE which has ensued upon the publication by Prof. Wegener of his revolutionary views on the flotation and drifting of continental masses. During my recent geological survey of the Falkland Islands, I was very greatly impressed by the extraordinary similarity of the geology of the Islands to that of Cape Colony. The geological succession comprises rocks ranging in age from Archaean to Permo-Carboniferous, although rocks of Cambrian, Ordovician, and Silurian age appear to be absent. The oldest rocks closely resemble some of the Archaean rocks of Cape Colony; and from the Devonian to the Permo-Carboniferous the lithological and palaeontological succession is practically identical in the two areas. The post-Triassic dolerite dykes of the Falklands are also very like the intrusions of the same age in Cape Colony. The east and west folding so evident in the southern part of Cape Colony makes the most conspicuous feature in the Falkland Islands. The only notable point of difference in the two areas is that whereas in Cape Colony the lowest division of the Cape System (Devonian), namely the Table Mountain Series, is much folded, the corresponding rocks in the Falkland Islands have escaped such disturbance and lie almost horizontal, or with only a gentle dip, over an area of many square miles. The equivalents of the middle and upper members of the Cape System (Bokkeveld Series and Witteberg Series) are, however, intensely folded in the Falkland Islands. From the orthodox point of view one has to believe in the persistence, in minute detail, of a stratigraphical sequence representing the passage of a great period of geological time, across the 5000 miles of ocean which separate Cape Colony from the Falkland Islands, and, in face of the array of facts marshaled into such an orderly effective host by Wegener and again by Du Toit, this becomes, on a sudden, an unexpected strain upon one's faith.

(Baker, 1923: 80)

T. Wemyss Fulton (1923) invoked continental drift to explain the prolonged migration of European freshwater eels. There are American and European fresh-water species. Both spawn in the North Atlantic in the Sargasso Sea much closer to America than to Europe. The American species mature for approximately a year before entering fresh water, the European species take three years to do so, which is atypically long for freshwater eels. Fulton (1923) explained.

It is scarcely possible to understand this unique phase in the life cycle of the European eel on the hypothesis that the geographical conditions were formerly the same as now exist. But if Wegener's theory be accepted, the explanation is simple. As the coasts slowly receded from one another the larval life on what became the European species was more and more prolonged by natural selection in correspondence with the greater distance to be traversed.

(Fulton, 1923: 360)

Twenty years later, G. D. Hale Carpenter at the 1943 joint meeting of the Linnean and Zoological Societies of London also "drew attention to the breeding place of the European and American freshwater eels as furnishing one of the strongest biological arguments for Wegener's theory" (Anonymous, 1943: 124).

R. B. Seymour Sewell (1880–1964) developed an interesting pro-drift argument based on his study of Copepoda, a subclass of aquatic crustaceans, which typically inhabit brackish or fresh water. Sewell spent most of his career in India. Trained as a zoologist at Cambridge, he joined the Indian Medical Service in 1908, and in 1910 became Surgeon-Naturalist to the Marine Survey of India. After military service in World War I, he returned to India as Superintendent of the Zoological Survey and became its head in 1925. He retired in 1933, and was appointed leader of the John Murray Expedition to the Indian Ocean. He and colleague D. H. Wiseman reported some of its results at the 1937 symposium on continental drift at the 24th Indian Science Congress, but was silent about mobilism (§6.13). However, four years later he argued on the basis of his copepod studies for continental drift. According to his biographer, C. F. A. Pantin (1965: 150), his preference was partly inspired by the work of the expedition. Sewell first presented his ideas at the 1943 joint meeting of the Linnean and Zoological Societies of London and further developed them in his 1956 Presidential Address before the Linnean Society. He offered an ingenious explanation of why there are today more species of copepods living in brackish or fresh waters of the estuaries of rivers bordering the Indian and Atlantic Oceans than bordering the Pacific. On drift theory the river estuaries of the Indian and Atlantic Oceans would become progressively more distant from one another, ideal for speciation.

In the late Carboniferous era the originally single continent of "Pangea" began, according to the drift theory, to fragment by the formation of a series of rifts; at their inception these must have resembled the Rift Valley of Africa, and have been filled with fresh water. At the commencement of the Tertiary epoch narrow gulfs had formed between America and Africa and between Africa and India, and from the southern ends of these gulfs a connection narrow sea ran eastward to the Pacific, being bounded on the southern side by the combined mass of Antarctica and Australia. In these two narrow gulfs the conditions probably very much resembled those of the Baltic Sea of today for owing to the large number of great rivers that flowed into them, the surface water must have been brackish . . . but, as the oceans gradually widened, brackish-water areas slowly became cut off from each other by true marine regions, and thus, possibly as a result of isolation, new species were evolved, so that we now find distinct species inhabiting the mouths of nearly all the great rivers. So recently did this occur that in many instances the species is confined to a single river, though some have been able to spread for a short distance along the coast, while others have penetrated up the river into fresh water. We thus get distinct sets of species inhabiting the Ganges and Brahmaputra estuaries in the Bay of Bengal and Quilimana River in East Africa, all in the Indian Ocean, and others in the Congo and Niger Rivers of West Africa, and the Amazon, Mississippi and St. Lawrence Rivers of America in the Atlantic Ocean.

(Anonymous, 1943: 120, 122)

By contrast, according to drift theory, no such process was at work around the much larger Pacific Ocean, which consequently now has fewer species.

It would thus appear that the drift theory can afford a valid explanation of the distribution of these warm-water epiplanktonic Copepoda, whereas the older theory of the permanence of the older basins fails to do so.

(Anonymous, 1943: 122)

Three other zoologists opined about mobilism at this same 1943 joint meeting of the Linnean and Zoological Societies of London. One was A. Tindell Hopwood (1897–1969), a vertebrate paleontologist at the British Museum, who, in 1933, coined the name "Proconsul" for the African Miocene ape fossil he found in Olduvai Gorge. In general, he preferred Matthew's fixist explanation (§3.6) for the distribution of Tertiary and recent mammals. However, he believed that continental drift had some virtues.

Continental drift does afford the most satisfactory explanation of the present curious distribution of the late Carboniferous and early Permian glacial deposits, and also provides a mechanism for the post-Middle Pleistocene earth movements, which not only elevated still further the great mountain ranges of the western Americas, and of the Alpine-Himalayan girdle, but, in addition, powerfully affected various other features such as the great Rift Valley system of Africa.

(Anonymous, 1943: 123)

Hopwood's concluding comment suggests that he was somewhat inclined toward mobilism and he did not think it was in conflict with Matthew's explanation of the arrival of mammals in the Southern Hemisphere.

The explanation of this apparent conflict between the geological and zoological evidence is that the break-up of the two primitive continents of Laurasia and Gondwanaland took place in pre-Tertiary times. Insufficient is known of the Mesozoic mammals to make their evidence concerning that earlier period of any value.

(Anonymous, 1943: 123)

A. C. Hardy (1896–1985), a plankton expert who had been the zoologist on the *Discovery* expedition to the Antarctic (1925–7), held the Regius Chair in Natural History at Aberdeen University, and later became Sir Alastair Hardy, Professor of Zoology at Oxford. He sounded much like Hopwood.

Whilst I can see so much in the distribution of the lower animals to support the hypothesis of continental drift, I always feel that the mammalian fauna of Australia is a stumbling block.

(Anonymous, 1943: 123)

Hardy then launched into an excellent critique of mobilism's solution to the origin of marsupials, foreshadowing some and repeating earlier difficulties raised by Australian biogeographers (Pearson, 1940; Abbie, 1941; Troughton, 1959; Ride, 1962) (§9.4), and of course by G. G. Simpson (1943) (§3.6, §3.7). He argued that phylogenetically South American and Australian marsupials were not as close as formerly thought, and that a land connection between Australia and South America would not explain the presence of monotremes in Australia and their absence in South America; had there been such a connection, placental mammals most surely would have accompanied marsupials to Australia (Anonymous, 1943: 123). Urging that these objections be discussed, he did not reject mobilism outright. It is hard to determine his overall attitude.

John Brooke Scrivenor (1876–1950), who spent most of his career in the Federated Malay States, was pro-drift. He graduated in geology from Oxford, where he studied under J. J. Sollas and Herbert H. Thomas. He was appointed Government Geologist of the Federated Malay States in 1903, and later Director. His original contract was for three years, but he remained there until retirement in 1931. During his last few years as Director, he wrote *The Geology of Malayan Ore-deposits* (1928) and *The Geology of Malaya* (1931) (Willbourn, 1950).

In 1941, a decade after retiring, Scrivenor wrote a review of geological research in the Malay Peninsula and what he referred to as the Indo-Australian Archipelago, which extended eastward from the Dutch East Indies (now Indonesia) to the Bismarck Archipelago. He argued that Wegener's theory offered the best solution to certain peculiarities in the shape and position of various islands in the eastern part of the Archipelago. When asked to speak at the 1942 meeting of the Linnean Society of London on the biogeographic division of the Indo-Australian Archipelago, he again supported mobilism. At the meeting, Airy Shaw, himself a mobilist, asked Scrivenor to describe the general attitude of geologists toward Wegener's theory. Scrivenor is reported to have said:

He thought it was felt that the theory explained so much that there was good reason for believing in it; but, at the same time, it must be confessed that it rested on nothing more substantial than the possibility of land-areas fitting together if moved towards one another. No proof of the theory has yet been produced and the speaker recalled a criticism of many years ago, when it was pointed out that although we had considerable knowledge of the land surfaces our knowledge of the ocean beds was very slight. The extension of echo-sounding would remedy this deficiency.

(Anonymous, 1942)

Indeed, it would be studies of the ocean floors during the 1950s and 1960s that would do more than anything else to establish mobilism, but ironically it was work during the late 1930s on the seafloor around the Dutch East Indies by Kuenen, Umbgrove, and Vening Meinesz which led them to argue against it and carry Dutch opinion with them (§8.14). Scrivenor (1941: 144) noted Kuenen's objection to mobilism but still thought it the best explanation for the eastern part of the Archipelago.

E. Zeuner (1905–63) was born in Berlin, and studied geology, paleontology and mineralogy at the Universities of Berlin, Tübingen, and Breslau. Obtaining his Ph.D. at the University of Breslau in 1927, he remained there as Privatdozent until 1931 when he became lecturer and senior assistant at the University of Freiburg-im-Brisgau. Realizing that his life would become intolerable under Hitler, Zeuner left Germany with his family in 1934 and settled in England, becoming a research associate in paleontology at the British Museum. He also became an honorary lecturer in geochronology at the London Institute of Archeology (1935), and eventually became the first professor of environmental archeology at the University of London after the Institute became part of the University. Zeuner was known primarily as a Pleistocene geologist and paleontologist, with a strong interest in geochronology. He had a lifelong interest in entomology and mammalian paleontology. His three books, *The Pleistocene Period* (1945), *Dating the Past* (1946, 1950, 1952, 1958), and *A History of the Domestication of Animals* (1963), reflect his wide interests (Hollingworth, 1965).

Zeuner embraced mobilism because of his entomological work, in particular on butterflies, and he (1942a) detailed their distribution in what he referred to as the Indo-Australian Archipelago (Indonesia, New Guinea, and the Melanesian Islands). One group, the *Ornithoptera* group, was prevalent in the Moluccas and Solomon Islands in the Pliocene, but had only recently populated New Guinea. Another group, the *Troides* group, which had arrived only recently in New Guinea where there was only a single species, populated the northern Moluccas after the *Ornithoptera*, and failed to reach the Solomon Islands. He argued that mobilism best explained both their distributions.

According to this theory [continental drift], the Australian block has been drifting northwards during and since the Tertiary. The theory was applied to the Australasian Archipelago for the first time by Wegener . . . then by Smit Sibinga . . . and du Toit. The conceptions of these workers differ in details but they agree in the following important points . . . Whether one admits the major movements of Australia from the far south or not, a withdrawal of New Guinea out of the island chains towards the south . . . would suffice to straighten out the disturbed island chains and bring the northern Moluccas near to the Solomons . . . The advance of New Guinea to its present position . . . would have been . . . late enough for the *Ornithoptera*-group to have reached the Solomons before the island chain Moluccas-Solomons was broken. Furthermore, the northern Moluccas would have occupied a position farther east until the end of the Pliocene, and this would solve the second problem, namely, the late arrival of *Troides* in the northern Moluccas and their early occupation by *Ornithoptera* instead.

(Zeuner, 1942a: 173; my bracketed addition)

Without mobilism, even the absence until very recently of both groups from New Guinea, let alone their curious distribution elsewhere, could not be explained.

In 1942 Zeuner (1942b) presented a paper at a meeting of the Linnean Society of London on the biogeographic division of the Indo-Australian Archipelago. Expanding his concerns beyond butterflies in the closing discussion, he argued:

As regards the Indo-Australian Archipelago, there is good geological evidence suggesting that Australia approached Asia in the course of the upper Tertiary and the Pleistocene. Faunal and floral distribution in the Archipelago, on the other hand, suggest that the intermingling of Asiatic and Australian elements is a comparatively recent phenomenon. Geological and biological evidence, therefore, corroborate the fact of horizontal displacement, but this has nothing to do with the geophysical premises of Wegener's theory.

(Anonymous, 1942: 165)

Once again, drift's problematic mechanism was in the background.

Zeuner participated in the 1943 joint meeting of the Linnean and Zoological Societies of London where his analysis of the geographical distribution of certain mammals and insects of the Indo-Australian Archipelago led him to conclude:

at the end of the Tertiary, New Guinea lay at least 750 miles south-east of its present position. No other geological or climatological theory would account for the facts of distribution equally well.

(Anonymous, 1943: 123)

Zeuner consistently sought drift solutions. In all four editions (1946–58) of *Dating the Past*, he estimated annual rates of drift between Newfoundland and Ireland, Buenos Aires and Cape Town, peninsular India and South Africa, and Tasmania and Wilkes Land. After citing Wegener, Taylor, du Toit, and Holmes, to whom he attributed "an up-to-date treatment . . . in his *Principles of Physical Geology*," Zeuner declared:

The chief problem of the theory of continental drift is not the principle involved (horizontal movements of some kind or other being too obvious to be denied as such) but rather their intensity and rate. The figures which Wegener gives . . . for the average rate of horizontal movement are based (a) on the geological phase during which, according to him, the separation of two blocks began, and (b) on estimates for the duration of geological periods which were considered too small, judged by the recent results of the radioactivity method.

(Zeuner, 1958: 355)

Zeuner (1958: 355) then did the obvious: he recalculated drift-rates by using Holmes' radiometrically determined ages; they were much more reasonable than Wegener's. The rate for the separation of Newfoundland and Ireland (0.2 to 0.15 meters per year) he thought was still too high, but the other rates ranged from 0.07 to 0.09 meters per year, which he argued were much more reasonable and in keeping with horizontal displacements along the San Andreas fault in California and the Alpine fault in New Zealand.[24]

The British botanist H. K. Airy was especially interested in problems posed by New Guinea's flora. Relying on the data collected by Dutch and US workers, and a pro-drift analysis by the Dutch botanist H. J. Lam (founding editor of *Blumea*, a new and influential botanical journal), Shaw argued that drift offered the best explanation for New Guinea's flora. Lam (1934) had observed (i) that New Guinea had no extant flora that had descended from plants that had immigrated to the island a long time ago, (ii) that its flora was almost entirely Asian, and (iii) that its flora was younger than Borneo's Asiatic flora. Lam argued that if the Australian–New Guinea continental block had originally been at much higher latitude and far south of the Malay Peninsula and Dutch East Indies, it would have had an Australian not an ancient Asian flora. With progressive movement to its present position, Asiatic flora would have spread to New Guinea, replacing its Australian flora which would have had difficulty adapting to the new equatorial forest environment. Thus Wegener's drift explained (i) and (ii), and because Borneo had not, he thought, changed its position relative to Asia, it also explained (iii). Lam appealed to the work of Smit Sibinga, who, like other early Dutch geologists, had employed mobilism to explain the geology of the Dutch East Indies (§8.14).

Shaw (1942) endorsed Lam's 1934 analysis. He favorably discussed Köppen and Wegener's ideas on paleoclimates and noted G. C. Simpson's (not G. G. Simpson, who was a fixist) support of mobilism (§3.13). He admitted that he might be too enamored with mobilism, and confessed to a family connection as a reason for favoring Wegener's theory.

It may be that I am unduly under the spell of Wegener; if so, I hope that I may be pardoned, since I cannot entirely suppress a certain clannish sense of satisfaction that the first germs of the theory of isostatic compensation, or flotation of the earth's crust, upon which Wegener's theory so largely depends for its feasibility, were originally suggested, in 1885, by my great-grandfather (Airy, 1885); cf. Hinks (1931). I firmly believe, however, that Wegener's hypothesis, or extensions or modifications of it, will ultimately give us the clue to much that seems at present inexplicable in the distribution of plants and animals.

(Shaw, 1942: 153)

The next British mobilist I discuss is Arthur Wade (1878–1951), a petroleum geologist, who approved (§6.13) of Gutenberg's version of mobilism (III, §2.3).[25] Wade, a Yorkshireman, graduated B.Sc. from the Royal College of Science, London (which became in 1907 Imperial College, University of London), winning its Murchison Medal and Prize (1904), and was awarded a D.Sc. from London University in 1911. Wade was a fellow of the Geological Society of London, Geological Society of America, and the Institute of Petroleum, London. From 1909 until 1913 he taught petroleum geology at Imperial College, and began consulting on petroleum exploration in the Middle East. When oil seeps were discovered in 1913 in Western Papua, Wade was commissioned by the British Government to advise the Australian Government if it was worth drilling. Appointed director of oilfields to the Australian Government, he remained working in New Guinea until 1919. He continued working as a petroleum geologist for various governments and old companions throughout the world, eventually settling in Australia (Montgomery and Raggatt, 1951).

Wade first argued in mobilism's favor at the 1933 World Petroleum Congress, where he presented the paper that caught the attention of P. Evans, another petroleum geologist and mobilist (§6.13). Assuming that most petroleum comes from strata rich in marine phytoplankton, which thrive more in tropical than in cold waters, he (1934) argued that the Late Paleozoic oil fields in the United States could not be explained unless North America had then been in the tropics and partially covered with shallow tropical seas. Because mobilists placed the United States in the tropics during the Late Paleozoic, he argued in its favor.

He viewed his idea as a working hypothesis.

It is in this spirit that the author puts forward this paper. He has attempted to apply Gutenberg's theory to the special subject of the distribution of oilfield areas on the earth's surface and is himself astonished with its success as a working hypothesis for this purpose.

(Wade, 1934: 76–77)

He (1934, 1935) was attracted to Gutenberg's idea that most mountain belts formed as Laurasia and Gondwana were pushed together, and thought the giant overthrusts he had observed in New Guinea had been formed as the northward drifting New Guinea–Australian block met outliers of the southward moving Asian block. "The author's personal observations in New Guinea, Timor and the East Indies tend to confirm Gutenberg's views so far as this region is concerned" (Wade, 1934: 76).

In 1941 Wade turned his attention to Antarctica. After describing what was then known about Antarctica's geology, he began arguing for mobilism by drawing similarities between Antarctica and other Gondwana fragments, partly based on his own travels in the southern continents. He studied *Glossopteris* beds in Western Australia and Madagascar.

These beds [in Antarctica] contain a Glossopteris flora, a flora which is characteristic of strata of this age wherever they occur in the Southern Hemisphere – Australia, India, Madagascar, South Africa, and South America. The sequence is remarkably similar to that found in parts of

Australia and South Africa. The succession from Lower Cambrian upwards is closely paralleled in the Kimberley District of Western Australia (Wade, 1924 and 1938).

(Wade, 1941: 27; my bracketed addition)

Again drawing on his own work in Australia and Madagascar, he discussed several other similarities.

The leucite bearing lavas of Mount Gauss, a volcano in the coastal area of Kaiser Wilhelm II. Land, are of similar [Tertiary] age. Leucite bearing lavas are of uncommon occurrence on the earth's surface. They occur associated with the extinct volcanoes of the West Kimberley District of Western Australia, which are certainly not older than Mesozoic and may be of Tertiary age (Wade and Prider, 1939). In almost every feature of rock occurrence, succession, and even to the ancient and complete peneplanation of the Precambrian (basement) rocks, Western Australia closely resembles Antarctica. To a very marked degree the same may be said of Madagascar, on the opposite side of the Indian Ocean (Wade, 1931).

(Wade, 1941: 28; my bracketed addition)

Having expanded his horizons to include Antarctica and similarities among Gondwana's fragments, he became more of a generalist, and became firmer in his support of mobilism, as his last paper on the question makes clear.

The weight of evidence is in favour of the movement of the continental masses. The conditions existing in the make-up of the earth's crust are considered to be such that movement of the light sial masses is possible, and even probable. In land masses of the Southern Hemisphere similarities exist in types of sedimentation, sequence of deposits, structural features, distribution of plant and animal life as indicated by fossils, and glaciation originating in a common centre of ice-cap which must have been located around the South Pole. The reasonable explanation of such similarities is that all these land masses in the Southern Hemisphere were in close connexion not far from the polar area in Upper Carboniferous and Permian times. The group formed Gondwanaland, and the present Antarctic continent was part of it, though not, at that time, in its present position. This old continental area started to break up in the Mesozoic period, the various parts drifting off to form Antarctica, Australia, India, Africa, and South America.

(Wade, 1941: 34)

Wade's attraction to and defense of mobilism had nothing at all to do with the geology of England, his home country. He first supported mobilism because of his work in New Guinea and interest in the latitudinal distribution of oil fields. His support was fueled by his work in Australia and Madagascar, which gave him firsthand knowledge of similarities between them, and what he had read about Antarctica.

I now turn to William Joscelyn Arkell (1904–58). Arkell favored mobilism in the late 1940s, but when at the Sedgwick Museum in Cambridge changed his mind; this was during the first half of the 1950s just as the first paleomagnetic evidence favoring mobilism began to be obtained in the nearby Department of Geodesy and Geophysics (II, §3.5–§3.10). Arkell entered Oxford University in 1922, earning first-class honors in geology in 1925. Never in robust health, he pursued a distinguished research career, and taught at Oxford until the mid-1940s. He was elected to the Royal Society (London) in 1947. Offered a university lectureship in the Department of Geology at Cambridge, he declined because of poor health. He then accepted a senior research fellowship at Trinity College, Cambridge, and was given an office at the Sedgwick Museum. He died in 1958 (Cox, 1958).

Arkell's first major work, *Jurassic System in Great Britain*, published in 1933 when he was still in his twenties, secured him an international reputation as an authority of Jurassic stratigraphy and paleontology. In the years to come, he studied Jurassic rocks worldwide in the field and in museum collections, and by the late 1940s he became the foremost living authority on Middle and Upper Jurassic ammonites. His *Jurassic Geology of the World* appeared in 1956. Arkell received the Mary Clark Thompson Gold Medal in 1944 from the NAS, the Lyell Medal from the Geological Society of London in 1949, and the von Buch Medal from the German Geological Society in 1953. He was a specialist par excellence.

Arkell (1949) argued in favor of mobilism because he thought it offered a better explanation of the distribution of Jurassic ammonites than fixism; he (1949: 413) claimed that their distribution indicated a more compact arrangement of continents as required by mobilism. He (1949: 413–415) thought that plotting occurrences of ammonites in accordance with du Toit's Gondwana during the Early Jurassic and Early Cretaceous made their distribution more compact, more understandable.

Arkell reconsidered mobilism in his 1956 *Jurassic Geology of the* World. Acknowledging his earlier support, he now claimed that differences between ammonite fauna on opposites of the Atlantic ruled against a former closing of the North Atlantic.

> I have pointed out previously (Arkell, 1949, p. 415) that a more compact arrangement of the continents, as envisaged by advocates of continental drift, would have the world-wide dispersal of so many Jurassic ammonites easier to understand, but any such general considerations are outweighed by these particular discrepancies between contemporaneous faunas on opposite sides of the Atlantic. The evidence here is positive and it weighs heavily against the drifting apart of the New and Old Worlds since the Jurassic.
>
> *(Arkell, 1956: 604)*

Not only were contemporaneous ammonite faunas on opposite sides of the Atlantic decidedly different, but there were strong similarities among faunas from the Mediterranean region and Nepal, Indonesia, and New Guinea.

> It happens that in Portugal and in Cuba there are two of the richest faunas of late-Upper Oxfordian ammonites known anywhere. Both comprise enough links with faunas in England and elsewhere to make their dating and contemporaneity virtually certain. Yet these two faunas are decidedly different: there is no identity at specific level, and Cuba has two subgenera of *Perisphinctes* which cannot be matched in Europe, and an endemic genus, *Vinalesphinctes*, which may possibly be a local, parallel, development to *Ringsteadia* and *Pictonia*, but is quite different from anything in Portugal or elsewhere. Again, in *Kimeridgian* of Mexico, there are abundant links which make dating virtually certain, but the swarm of *Idoceras* all have a

peculiar local stamp, and the endemic general *Mazapilites, Epicephalites* and *Subneumayria* are altogether original . . . Considering how uniform these faunas are throughout the Mediterranean region, and how Tithonian faunas in the eastern Tethys extend from Spiti and Nepal to Indonesia and New Guinea, and taking into account the evidence of the Oxfordian faunas also, it is impossible to believe that the Gulf of Mexico region was substantially closer to the Mediterranean region in the Jurassic than it is now.

(Arkell, 1956: 603)

He really was quite firm about it: Cuba and Portugal (and points east) had such very different faunas and it was hard to reconcile them with their close juxtaposition as required on drift reconstructions. I do not know how this difficulty that Arkell raised against continental drift has been resolved, but it clearly must have been. Perhaps there was a slender land barrier between them at the time before the new and old world had begun to part and before a substantial ocean had developed between them.

As I shall later describe, Arkell participated in a symposium during spring 1955 at Cambridge on polar wandering and paleomagnetism (II, §3.10). Although there was at the time support for polar wandering, Arkell remained silent on the subject. Arkell was a careful, dedicated, and scholarly man who struggled hard with the question of what bearing the distribution of his beloved ammonites had on the drift debate; clearly there was no unique answer forthcoming from these free-swimming, wide-ranging pelagic mollusks.

I now remark generally on the reception of mobilism among geologists in Britain and close with an extended account of the role of E. B. Bailey. Wade, like other British supporters of mobilism, did not become supportive of mobilism because of British geology. Both Arkell's initial preference for mobilism and his later preference for fixism were globally based. Likewise, those who were opposed did not do so because of its inability to explain the geology of the British Isles. Jeffreys raised mechanism difficulties against mobilism, and explained the origin of mountains by contractionism (§4.3, §4.11, §5.5). Gregory argued that mobilism could not explain similarities among circum-Pacific past and present-day biota, and believed in the existence of former landbridges (§3.4). Lake (1922b, 1923) argued that mobilism could not explain the former presence of *Glossopteris* in Russia and other places in the Northern Hemisphere (§3.4). Lake (1922a, 1922b, 1923) and Jeffreys (II, §6.11; III, 2.10; IV, 3.4) believed that Wegener had greatly over-valued the congruency of continental margins on opposite sides of the Atlantic. Lake (1922b, 1923) also thought that Wegener had overestimated geological similarities on opposite sides of the Atlantic. Grenville A. J. Cole (1922) wrote that Wegener's theory "must not be taken in the spirit of a jest," and then did just that.

There is a concluding figure in many Bantu dances – it survives even in folk-dances at Skansen – where two partners turn back to back, bump, and part again. The possibility of this figure on a continental scale is thrilling and attractive. If Africa once parted from America, she may woo her mate again as years pass by. The hand of the philosopher may be laid on the great land blocks, and the occurrence of Glossopteris in India or of *Geomalacus maculosus* in Kerry may be explained by a single process of "Vershiebung." If the fitting is not sufficiently accurate, some placidity is granted to the sial blocks, and "Umwalzung" is also possible. Wegener's conception, however, must not be taken in the spirit of a jest.

(Cole, 1922: 799)

He added (1922: 801), "Nothing daunts so bold a champion. The hand of the master presses on the sial blocks or on the polar axis, and all goes well with the hypothesis." Cole (1922) also referred readers to Lake and Jeffreys' attacks on Wegener. E. R. Roe-Thompson (1922) agreed with Lake: Wegener had overestimated the goodness of fit of continents, and had distorted continents to match similar geological structures found in Africa and South America, and also in Europe and North America; his sins seemed endless, he could do nothing right. A. Morley Davis (1923) objected to Fulton's mobilist explanation of the extended larval stage of the European freshwater eel; he argued that sinking landbridges could equally well explain the European eel's lengthy larval stage and long migration. Thus, those in Britain who opposed mobilism, like those who upheld it, did not do so on the basis of the geology of their home country.

If neither fixism nor mobilism had an advantage in solving regional problems, then perhaps the effect of specialization should play a prominent role in theory preference. It did, at least for those who preferred mobilism; many preferred it because it solved problems in their specialization. The British paleoclimatologist G. C. Simpson (§3.13) became a mobilist because it explained Permo-Carboniferous glaciation; paleobotanists Seward, Walton, Hamshaw Thomas, Good, and Hutchinson likewise supported mobilism because it explained the distribution of *Glossopteris* and other Southern Hemisphere floras; Shaw was strongly inclined toward mobilism because it solved New Guinea's phytogeographical problems; Westoll (§3.8) adopted mobilism because it explained the distribution of the freshwater fish and the primitive amphibians he studied. C. E. P. Brooks (§3.12) was an exception. A meteorologist, he rejected mobilism, believing rather in changing climatic zones, and developed his own fixist solution to the origin of the Permo-Carboniferous glaciation. Wade and P. Evans were attracted to mobilism partly because it offered an explanation for the distribution of the world's oil fields.

E. B. Bailey is of special interest because, unlike most other Earth scientists who spent their professional career working almost entirely on British geology, he favored mobilism. Unlike other British mobilists, his opinion arose from his work on the Scottish Caledonides; perhaps the most important regional problem that mobilism neatly solved.[26] In an earlier section I have partially described his collaboration with Collet (§8.9) and Holtedahl (§8.11).

Sir Edward Battersby Bailey (1881–1965) was one of those people who are larger than life. The third of six sons, he described himself as a sickly child, and recalled that he was bullied at school when very young.[27] In an attempt to overcome this disadvantage, he improved his toughness by playing team sports, boxing, and cross-country running. In 1898 he won an open scholarship to Clare College,

Cambridge University, which he entered the following year. Again wanting to toughen himself up, he would sleep with his bedroom window wide open, played rugby and lacrosse, rowed, and continued cross-country running and boxing, even winning the Freshman's Heavyweight Medal for boxing.[28] Still managing to find time to study, he obtained a first in Part I of the Natural Science Tripos. At the start of his third year, he decided to concentrate on geology with physics in Part II of the Tripos, and achieved a first class. His favorite bedtime reading was Suess' *Antlitz der Erde*, which he read in French! As Bailey later remarked, "I well remember my first introduction to his work in de Margerie's famous edition, *La face de la Terre*. Many a time I listened to the morning song of the birds before reluctantly closing its inexhaustible pages" (Bailey, 1935: 9–10). Following graduation, Bailey was offered a position with the Geological Survey of Great Britain. There were two vacancies; one in England, the other in Scotland. J. B. Scrivenor was given first choice, and to Bailey's delight, left him the Scottish post. Scotland had its Caledonides, and Bailey brought to them the spirit although not yet the experience of an Alpine tectonicist. Bailey worked in various capacities at the Survey from 1902 until 1929, becoming a recognized expert on the Scottish Caledonides. He obtained the chair of geology at the University of Glasgow in 1929, and remained there until 1937 when he was appointed Director of the Geological Survey. He retired in 1945, and was knighted that year.

Bailey was one of the most respected British Earth scientists to support mobilism before World War II. He was elected (1930) FRS, and received (1943) the society's Royal Medal. He also was elected a foreign member of the Norwegian Academy of Science and Letters (1938), honorary fellow of the National Institute of Science of India (1941), foreign associate of the NAS (1944), associate of the Royal Academy of Belgium (1946), and honorary member of the Swiss Academy of Sciences (1948). He was awarded the Murchison (1935) and Wollaston (1948) medals from the Geological Society of London, and received honorary degrees from Harvard, Birmingham, Glasgow, Belfast, Cambridge, and Edinburgh. In 1928, he was President of BAAS.

As already noted (§8.8), Bailey (1926a) first accepted mobilism in December 1925 after hearing Collet argue in favor of Argand's account of the origin of the Alps and Himalayas. He claimed that if Argand and Collet's view of the Alps is accepted and applicable to the Scottish and Scandinavian Caledonides then drift is inevitable. What led Bailey to continental drift, while other workers on the Scottish Caledonides remained fixists even though they agreed with him that they were, like the Alps, extensively overthrust? And, what led him eventually to present his own argument in favor of mobilism based on the crisscrossing and continuation into North America of the Caledonides and Hercynides of Europe? The answers surely must be Bailey's fieldwork outside Great Britain, in the Alps, North America, and Scandinavia.

Bailey toured the Alps twice under Collet's guidance, the Prealps in summer 1909 and the Pennides in 1924. These trips probably strengthened Bailey's belief that the

Scottish Caledonides were Alpine-type mountains, and helped him understand and appreciate Argand's arguments. After Collet had presented his pro-drift interpretation of the Alps before the Geographical Society of London in December 1925, Bailey spoke about the importance to him of seeing the Pennine Alps.

I was one of a party, and there are others here, who were led by Professor Collet through the much more difficult region of the Pennine range, which had been interpreted by Lugeon and Argand. And here we were on very different ground indeed. It is not, as a matter of fact, difficult to see the great folds. One sees enormous recumbent folds in one's walks about, and one can follow them out with the map; but the difficulty in this case is that the rocks have been so tremendously altered by movement and heat that they are, for the most part, in the condition of crystalline schists. There may be some British geologists who would be shocked at the very idea of calling rocks Mesozoic that have garnets in them – that have lost their fossils and have become crystalline schists – but I may say that all Swiss geologists are satisfied that the *schists lustrés*, even when they have become thoroughly metamorphic rocks, are of Mesozoic age. And when one looks at those great folds in the Pennine Alps with their Palaeozoic cores and their Mesozoic envelope – one realizes that here are folds made in the Tertiary movement that has developed the Alps.

(Bailey, 1926a: 308)

Seeing may not be believing, but it certainly helps. I suspect that most geologists who had not seen the western Alps thought that mobilism was not really needed to explain them. Bailey saw them, and became as close to being an Alpine geologist as he could be without spending his career atop them. The young Bailey kept Suess' *La Face de la Terre* by his bedside; the older Bailey wrote *Tectonic Essays Mainly Alpine*. Bailey also saw Caledonide overthrusts at the time of the 11th International Congress in Stockholm in 1910. Joining an excursion led by A. G. Högbom, who had helped A. E. Törnebolm map the huge overthrusts of the Scandinavian Caledonides, he went to Jämtland and saw them. Bailey, who pioneered the wearing of shorts in the field, was singled out for special mention in the excursion's report.

After a light lunch served on the top of the mountain, a small but eager party set off to visit the famous and classical section of Törnebolm, descending on the west side of the mountain. We continued down the north side of the mountain, till we gained the undisputedly Silurian stratas at the lower waterfall. Our wild Scotchman, barelegged and bareheaded now set off at a gentle trot over the broad and swampy lowland moors to reach Mullfjället, the substratum, over which the whole overthrust must have passed. He was back by midnight after what you might call a stiff day.

(Quensel, 1912: 1232)

Bailey's apprenticeship in the Scottish Caledonides and his extended fieldtrips in the Alps and Scandinavian Caledonides, and the friendships that grew from them, led him to mobilism. As he said:

Finally, if we take the current Swiss view of the structure of the Alps, and the similar view, worked out by that giant of the north, Törnebolm, for the Scandinavian chain, and again the similar view attained by Scottish geologists in their Highlands, we can find that some sort of continental drift becomes a matter not of interpretation but of observation.

(Bailey, 1926a: 309)

Nature does not lie. Go see for yourself, he was saying, and mobilism will become a matter of observation. Rhetoric perhaps, it was his way of stressing the importance of firsthand experience of field exposures.

When Bailey first announced for mobilism (1926a), he left (§8.8) unspecified whose version of mobilism he had in mind; either he had changed his mind about Argand by June 1926 (1926b) when he suggested that *La Tectonique de l'Asie* should be studied but not believed, or he did not have Argand's synthesis particularly in mind in the first place. Bailey took just one year (while in North America) to become a fully fledged mobilist, developing his own argument in its favor, and claiming that North America and Europe had once been connected. Again, his firsthand fieldwork was crucial.

Bailey and Collet were invited by R. M. Field to join faculty and students from Princeton University on their 1927 summer jaunt by train across Canada with a few stops in the United States. This allowed Bailey to study the Appalachian (Hercynian) and Taconic (Caledonian) orogenic belts, and enabled him to enhance the case for mobilism. Wegener had argued that the extension of the two mountain belts from Europe to North America testified to their former proximity (§3.2). Intimately familiar with the segments of both belts in Great Britain and Ireland, Bailey now observed that the two belts meet, crisscross and begin to separate in Wales, and then again meet and crisscross in New York State and separate in Pennsylvania. In his first presentation of this trans-Atlantic correspondence in *Nature*, he (1927: 647) claimed that Wegener might be a "true prophet." In his more extensive presentation marking his acceptance of "some type of" continental drift, he concluded:

The study that we have made of mountain chains with their folds and their thrusts, which individually may be of the order of 100 miles, involves a recognition of some type of continental drift. Of late years Wegener has developed this idea on a particularly grand scale. He has accounted for many recognized correspondences in the geology of the two sides of the Atlantic by supposing that the ocean has flowed in between the Old World and the New, as the two continental masses, with geological slowness, drifted asunder. One cannot help feeling that Wegner may perhaps be telling us the truth. The available evidence is crude and ambiguous; but it is certainly startling to be confronted on the coasts of Britain and America with what read like complementary renderings of a single theme: the crossing of Caledonian Mountains by Hercynian.

(Bailey, 1929: 76)[29]

Covering much of the same ground as Holtedahl (1925), as Bailey (1929: 66) himself later realized, and previewing some of what F. E. Suess (1936a, b) would later say, Bailey emphasized the numerous similarities among the Scandinavian, Scottish, and North American Caledonide disjunctive segments.

The pivot of our comparison is furnished by the mountain chain of Scandinavia. This chain is markedly symmetrical. On one side, in Sweden, it is carried forward along the Törnebolm thrust-zone on to undisturbed early Palaeozoic rocks of Baltica. On the other side, as exposed in the north-west Highlands, it has traveled along the Moine thrust-zone on to undisturbed early Palaeozoic rocks recognized by [Eduard] Suess as part of Laurentia. Across the Atlantic, the Scotto-Scandinavian mountains reappear in Newfoundland and Nova Scotia, and the Moine thrust-zone is represented by the well-known dislocation-belt of St. Lawrence and Lake Champlain.

(Bailey, 1927: 674; my bracketed addition)

He (1927: 674) visited the thrust-zone of St. Lawrence with Collet and Field, and they remarked on its similarities to the Moine thrust-zone (Bailey, Collet, and Field, 1928). Turning to the crossing of Hercynides over the Caledonides, he drew comparisons between North America and Europe. They seemed especially evident to him as a European geologist, and led him to think that Wegener may be correct.

Many thoughts spring to the mind of the European geologist who finds himself standing in Pennsylvania on Hercynian mountains *outside* the line of the type Caledonian chain.

(1) The westward convergence of the two Palaeozoic chains – so far apart in Poland and Lapland, already in contact in South Wales and Ireland – has led to their actual crossing in the United States.
(2) Not only have the mountains crossed, but also the stratigraphy. In Pennsylvania there is an immense concordant succession from Cambrian to Carboniferous. In the anticlines we find our Durness Limestone (Beekmantown) as if we still stood in the north-west Highlands of Scotland. In the synclines we discover Upper Carboniferous Coal Measures (Pennsylvanian) derived from the waste of a growing Hercynian chain, and our thoughts are transferred at once to South Wales, the Ruhr, and Poland.
(3) In much of the Canadian part of the Appalachian System, a limestone facies within the Lower Carboniferous serves as a punctuation mark between Caledonian and Hercynian movements, just as it does in the British Isles and in Belgium.
(4) It is as if the Atlantic did not exist or, in other words, as if Wegener, after all, were a true prophet.

(Bailey, 1927: 643–644)

Because of what he observed, Bailey, like Holtedahl and later Wegmann and Suess, became a du Toit of the Northern Hemisphere, actually detailing the geological similarities among areas presently separated by oceans, and then arguing for their former contiguity and subsequent disruption; he and they recognized the widely dispersed Caledonide segments as tectonic disjuncts whose place in the tectonic scheme of continental drift was analogous to the more familiar biotic disjuncts. Bailey and Holtedahl (1938), as already explained (§8.10), consolidated these arguments in their co-authored monograph on the Caledonides of northwestern Europe.

   Bailey returned to mobilism a year later as co-author of an elementary geology textbook (Bailey and Weir, 1939). The book arose from a first-year course that was offered at Glasgow University during the years 1930 to 1937. Bailey wrote all the chapters except those on paleontology. Despite the book's general title, *Introduction to Geology*, the two authors concentrated on the geology of the British Isles. Nevertheless, they occasionally did mention the geology of elsewhere and discussed mobilism. They introduced continental drift as a matter of current dispute among geologists, and Bailey himself, despite his endorsement nineteen years earlier of "some type" of continent drift, was unwilling to accept it.

One of the main questions at present discussed by geologists is: Do continents drift? To this no definite answer can be given, but it will be possible in the sequel to indicate the sort of evidence that suggests the possibility that, for instance, South Africa and South America once lay side by side (Fig. 307). Perhaps the matter may be settled by measurement, for a succession of longitude observations taken at a particular site in Greenland makes it appear possible Greenland is nowadays drifting at an appreciable rate towards America.

(Bailey and Weir, 1937: 7; Fig. 307)

They illustrated the breakup of Pangea, and Bailey wrote the following passage, probably his strongest endorsement of mobilism.

Let us now turn our attention to a hypothetical supercontinent that has been called Gondwanaland, after the Gondwana district of central India. Gondwana includes India, most of Africa, South America, Australia and Antarctica. It is treated as a unit because its constituent parts show *many extraordinary similarities* of geological history. Moreover, from Upper Carboniferous to Triassic times inclusive, the life-history of Gondwanaland was very different from that of the rest of the world in Asia, Europe and North America. Both these features are well displayed in Upper Carboniferous times, when extensive land-glaciation occurred throughout Gondwanaland. It is *astounding* to learn of ice-sheets in India, South Africa and Brazil contemporaneous with forests of tropical type in Britain. It *almost unanswerably suggests* that the position of the earth's crust, with reference to the poles and equator, has changed during geological time. As, however, the glaciated localities extend somewhat beyond the limits of a single hemisphere, it would appear that the earth's crust has not migrated as a whole, but that in late Carboniferous times Gondwanaland was a fairly compact unity clustered about the South Pole, and that its constituent parts have since drifted asunder into their present positions (Fig. 307). Gondwanaland had a very remarkable flora after its glaciation, that is, for the most part, in Permian times. The flora is as widespread as Gondwanaland itself, and yet is quite distinct from the contemporaneous floras of the Northern Hemisphere. Its commonest genus is *Glossopteris*.

(Bailey and Weir, 1939: 441; emphasis added)

Bailey's language is particularly strong for a textbook of that date, and indicates that he thought the Southern Hemisphere offered the stronger support for mobilism despite his own exclusively northern work; he offered the matching Permo-Carboniferous glaciation of Gondwana as stronger support for mobilism than the matching of the Caledonide orogenic belts of Laurasia. He ended the discussion of Gondwana by acknowledging the opposition.

All geologists are agreed as to the reality of the resemblances which unite Gondwanaland, but many ridicule the idea that its constituent parts have drifted apart. They postulate former land-bridges, as they are called, across the intervening oceans, and suppose that these bridges have foundered.

(Bailey and Weir, 1939: 441–442)

Bailey did not think too highly of landbridges.

Bailey, as expected, discussed orogenesis. Dismissing contractionism by noting, "All one can say is that at present there is no evidence that folded mountains are the product of a cooling earth" (Bailey and Weir, 1939: 124), he introduced mobilism as an alternative.

The folded mountains with their immense thrusts are clear evidence that different portions of the earth's crust sometimes move towards one another for considerable distances. Reference has already been made to the hypothesis of drifting continents. It is thought by many that folded mountain chains are a symptom of drift; that the Alps, for instance, were raised during a collision of the European and African blocks.

(Bailey and Weir, 1939: 124)

Bailey went on to explain that mobilists do not have an adequate mechanism, and added with a wonderful double-entendre that mantle convection as a possible cause is beyond the ken of ordinary geologists.

Those who adopt the drifting of continents as an explanation of folded mountains have still to find a satisfactory motive force. Convection currents in the layer below the continents have been postulated; but the subject is too deep for ordinary geologists.

(Bailey and Weir, 1939: 124)

Bailey admitted that "drifting of continents is a very hypothetical idea," but also noted that it was partly based on isostasy, which is itself well-established (Bailey and Weir, 1939: 124).

Bailey did not discuss the matching crisscrossing of the Caledonides and Hercynides on both sides of the Atlantic. He summarized them as follows:

Let us return to the resemblances which exist between the Cambrian, and probable Lower Ordovician, of the Scottish North-West Highlands and the contemporaneous rocks of parts of Newfoundland and the Appalachian Mountains. We find:

(1) The faunas are the same,
(2) The rock types, quartzite and limestone, are the same though they do not agree in their detailed arrangement.
(3) The structural position is the same. Both lie at the north-west edge of the Caledonian Mountain Chain, folded and thrust in Early or Mid-Palaeozoic times.

These resemblances certainly support the view that America and Europe have drifted apart; but they certainly do not prove any such proposition.

(Bailey and Weir, 1939: 395)

Bailey recognized the need to explain Britain's once warmer climate.

We have in these few instances [extensive Carboniferous coal measures and modest coral reefs] noted that Britain formerly enjoyed much warmer climates than at the present day. This supports the view that the continents are drifting and that Britain long ago was much farther south than it is to-day.

(Bailey and Weir, 1939: 19; my bracketed addition)

He also noted Greenland's rich Cretaceous flora was a "strong argument for drifting continents" (Bailey and Weir, 1939: 464).

Unlike other British mobilists, Bailey's support for drift developed from his work in the Scottish Caledonides; he and others reconstructed the Caledonide fragments as one continuous Early Paleozoic orogenic belt extending from Spitsbergen through Scandinavia and Greenland to Scotland, Ireland, northern England and Wales, and beyond to eastern North America. He knew from firsthand experience that although much older they shared many features in common with the Alps. Bailey's mobilism grew out of what might be described as his imaginative outward-looking regionalism. Then he turned his attention to Gondwana with its Permo-Carboniferous glaciation and ubiquitous *Glossopteris*, and culminating mafic magmatism, claiming that it offered even stronger evidence for mobilism. Unlike most geologists who worked exclusively or almost exclusively in the Northern Hemisphere and did not fully appreciate the importance of evidence from the Southern Hemisphere, Bailey did, even though he had come to favor mobilism before he wrote about Gondwana. Broadening his vision, he became both a globalist, appreciating the geology of both hemispheres, and a generalist, appreciating findings outside his own field of tectonics.

Thus, regionalism goes a long way to explain why many British geologists endorsed mobilism: eleven of them spent substantial portions of their careers in India, southern Africa, Falkland Islands, or Malay Peninsula. Arthur Wade, concerned with geological similarities among Gondwana fragments, worked in Madagascar, Australia, New Guinea, and Africa. John W. Evans worked in India and South America. Holland, Fox, and Fermor became directors of the Indian Geological Survey, and Sewell, West, and P. Evans worked in India. Hutchinson made several trips to the Southern Hemisphere to collect plants. Baker was Government Geologist for the Falkland Islands. Scrivenor was Director of the Geological Survey of the Federated Malay States. Although I have not attempted to determine the percentage of British Earth scientists who worked in India and who expressly favored mobilism, I have not found any who spent considerable time there after Wegener's exposé of continental drift, who argued against it. British Earth scientists, except for Bailey and Wright, who, I suspect, learned about the Scottish Caledonides from Bailey, found little evidence in their homeland itself that provided mobilism or fixism with a distinct advantage. In addition, Good, Hutchinson, Seward, Shaw, G. C. Simpson, Thomas, Walton, and Westoll preferred mobilism because of its success in solving problems in their specialty. Given the effects of regionalism and specialization, it is no surprise that the overall British response to mobilism was mixed, and that most British mobilists worked in specialties in which, or regions in which, it enjoyed the advantage in solving important problems.

8.14 The Dutch East Indies: the changing attitude of the Dutch

Because of the widespread colonies of the Netherlands and the prominence of the Royal Dutch Shell Oil Company, Dutch geologists and geophysicists like their British counterparts worked in many regions of the world during the first half of the twentieth century. They devoted considerable attention to the Dutch East Indies (now Indonesia). The first generation of them to study the archipelago overwhelmingly supported mobilism. Rachel Laudan (1980) and Martin Schwarzbach (1980/1986) correctly argue that this was so because it helped explain its geology. However, geodetic and certain oceanographic observations made during the late 1920s and early 1930s created difficulties for continental drift and led the next generation to prefer fixism. So from that time to just prior to the discovery of plate tectonics the Dutch were predominately fixists. These swings of opinion are what would be expected if their opinions tracked the evolving geological knowledge of the archipelago.

The first-generation supporters of mobilism included G. A. F. Molengraaff (1916), Hendrik A. Brouwer (1917, 1919a, b), Johannes Wanner (1921), I. N. Wing Easton (1921a, b), Gerard L. Smit Sibinga (1927, 1933), and Bernd George Escher (1933). All worked in the oil-rich Dutch East Indies. Molengraaff also worked in the Transvaal in South Africa (§4.12). The botanist H. J. Lam gave an early account of the evolution of New Guinea's flora on the basis of Smit Sibinga's mobilistic interpretation of the archipelago. Van der Gracht (1928a), Wegener (1929/1966), and du Toit (1937) noted the link between the region studied by these workers and their enthusiasm for mobilism; the geology they saw led them to mobilism. Van der Gracht, who, in 1913–14, had served as advisor for the Netherlands Colonial Government Expedition in the Dutch East Indies, also found the evidence for mobilism impressive.[30] Without knowing its cause, he saw, in the physiographic and geological evidence, New Guinea driving northward.

The extremely active manner in which New Guinea has quite recently pressed, and possibly is still now boring, to the north into the great Soenda festoons, is so obvious that *the fact* cannot be denied by any one who is really familiar with these regions, regardless of its causal explanation. This is the reason why the Dutch geologists (Molengraaff, Brouwer, Wing Easton) who worked in the East Indies, are invariably favorably inclined to Wegener's hypothesis. I have also visited this area: the evidence is indeed striking. Without knowing why, we *see* that New Guinea drifts violently to the north.

(van der Gracht, 1928a: 57)

Wegener (1929/1966) devoted five pages (89–94) to summarizing the work of Dutch geologists who were mobilists, at least in part, because of their work in the Dutch East Indies.

Since just here in the Sunda Islands the consequences of drift theory seem so fantastic at first sight, it is certainly worth noting that the Dutch geologists who are working in the Sunda archipelago were among the first to take their stand on drift theory; the very first of these was Molengraaff, who came out for the theory as early as 1916; later also van Vuuren [1920], Wing Easton [1921a, b], Escher [1922] and recently Smit Sibinga [1927] in particular, who has given a complete presentation of the geological development of the Sunda Archipelago from the standpoint of drift theory, and in doing so has also solved the old problem of the region of the peculiar shapes of Celebes and Halmahera.

(Wegener, 1929/1966: 90; my bracketed additions; Wegener's references coincide with mine)

Du Toit (1937) described the evidence favoring mobilism from the Archipelago, noting that most Dutch workers there favored mobilism.

Carozzi summarized Dutch efforts and their use of continental drift to solve geological problems of the East Indies.

An almost instant and highly enthusiastic reaction to Wegener's theory came from Dutch geologists working in the East Indies, now Indonesia. They could hardly refrain from publishing a great number of papers which considered Wegener's ideas, in particular the westward drift of the block New Guinea-Australia against Southeast Asia, the ideal explanation by the geomorphology, the structure, and the volcanism of the Indonesian archipelago, not to mention the present bathymetry of its oceanic portions. These constructive contributions began with Gerard A. F. Molengraaff . . . In relation to his particular interest in coral reefs and isostasy, Molengraaff accepted "the possibility of horizontal movements of continental blocks such as Wegener assumes in his bold hypothesis about the origin of continents" . . . [Molengraaff's contribution to the AAPG symposium is discussed in §4.12]. Hendrik A. Brouwer followed with several fundamental papers . . . published in 1917 and 1919. He analyzed the crustal movements in the region of the East Indies archipelago. He interpreted the shape of Timor as the result of the collision between New Guinea and Southeast Asia and made a pertinent comparison with Argand's paper of 1916 on the arc of the Western Alps in which Wegener's mobilistic approach was just emerging. In another contribution . . . published in 1919, Brouwer explained the absence of active volcanism in an area of the East Indies where tectonic compression, as a result of drifting, would be at a maximum. Subsequently, Johannes Wanner (1921), I. N. Wing Easton (1921), and Gerard L. Smit Sibinga (1927), among others, used Wegener's theory as an efficient working tool in their investigation of the Dutch Indies. Smit Sibinga gave a complete presentation of the geological history of the Sunda archipelago as a function of the drift theory . . ..

(Carozzi, 1985: 131; my bracketed addition. Also see Wegener (1929/1966: 91–93)
for a fairly complete summary of Smit Sibinga (1927))

B. G. Escher (1885–1967), professor of geology at the University of Leyden (an older half-brother of the artist M. C. Escher), was perhaps the last of this early mobilist generation (Laudan, 1980). Unlike Molengraaff, Brouwer, and Smit Sibinga, he (in 1931) appealed to Holmes' mantle convection as the cause of continental drift. Escher became interested in Vening Meinesz's proposed downbuckled crust to explain the narrow belts of negative gravity anomalies he had observed over trenches south of Java and Sumatra (§5.6). Vening Meinesz suggested that abnormal accumulation of light crustal material (sial) within the underlying denser simatic layer could account for them and, further, that the downbuckled crust was caused by tangential forces. Escher proposed that mantle convection caused this down-buckling; the crust was being dragged down into the denser simatic layer by converging descending convection. Escher later learned that Holmes had independently made such a suggestion at the 1931 meeting where he, Jeffreys, and others had discussed Holmes' convection theory (§5.5).

On the first of September 1931 I attributed the formation of this lighter root [of sialic crust] to opposing currents in the substratum and four weeks later in a discussion of the earth's crust in the geographical section of the British Association, on September the 28th 1931, Holmes quite independently said [the same thing].

(Escher, 1933: 682; my bracketed additions)

Escher (1933: 679) recalled that he had first appealed to convection in 1931 as an explanation for the many volcanoes in what was then called the East Indies and Vening Meinesz's belt of negative anomalies. The East Indian Volcanological Survey had been undertaken to map comprehensively the many volcanoes in the East Indies, and its results were published in 1929 and 1930. Escher (1933: 679) conjectured that because the "parallelism of the rows of volcanoes and the belt of negative anomalies is so striking . . . some causal nexus . . . must be assumed," and that in September 1931 he had

pointed to such a common cause in the currents of the substratum, that lies below the earth's crust, thus making use of Ampferer and Schwinner, after already having given as my opinion in 1922 that currents of magma are the common cause of mountain forming, earthquakes and volcanism.

(Escher, 1933: 679)

Escher first learned of Holmes' work in 1931, and read his 1931a, 1931b, and 1933 papers. He preferred Holmes' convection to that of Ampferer and Schwinner, and liked its appeal to radiogenic heat as the cause. Escher (1933: 680) also adopted Holmes' model of the crust and substratum, his appeal to the peridotite-eclogite phase change, and his idea of convection as the cause of both continental drift and orogenesis. Relating volcanoes to the negative gravity anomalies, Escher, following Holmes, argued that the plutonic chambers supply magma for volcanoes through tensional cracks in the crust that is stretched due to the opposing movement of vortices.

. . . vortices that gradually evolve in the main current of the substratum, as Holmes believes, offer an acceptable explanation for the occurrence of stretching in the earth's crust and this would also render the formation of plutonic chambers more plausible, that turn to volcanic chambers during the process of crystallization.

(Escher, 1933: 683)

Escher illustrated his view with the diagram shown in Figure 8.6.

Again echoing Holmes, Escher (1933: 680) hypothesized:

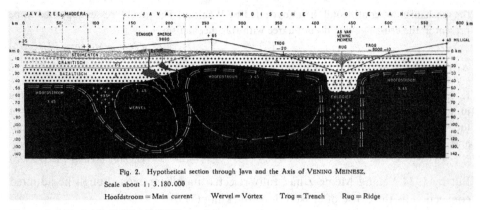

Fig. 2. Hypothetical section through Java and the Axis of VENING MEINESZ,

Scale about 1: 3.180.000

Hoofdstroom = Main current   Wervel = Vortex   Trog = Trench   Rug = Ridge

Figure 8.6 Escher's (1933: 681) Figure 2.

When continental and oceanic currents meet in the substratum, the earth's crust, that is carried along, is piled up. This piling up causes a thickening of the edge of the continent, by which the supply of heat is gradually increased. This in turn results in the forming of a vortex that exerts a horizontal tension on the crust, and at the same time at the side of it a horizontal compression which condenses basalt to the high-pressure facies eclogite. This eclogite being denser than peridotite sinks downwards.

Holmes had given Escher all he needed. The opposing action of the main and vortex currents explained the formation of volcanoes, and the combined action of downward convection currents and production of dense eclogite assisted the downbuckling of the sialic crust causing negative gravity anomalies.

This mobilist interpretation of the evolution of the Dutch East Indies was rejected essentially unanimously by the next generation of Dutch Earth scientists, especially these four major figures, Felix Andries Vening Meinesz, Philip Henry Kuenen, Johannes Herman Frederik Umbgrove and Reinout Willem van Bemmelen.

As already discussed (§5.6), Vening Meinesz (1930) appealed to downbuckling of crust to explain the origin of the low, narrow belt of negative gravity anomalies along the Java Trench that he had discovered. He agreed with his predecessors that the Dutch East Indies were the first stage of mountain formation.

This hypothesis is in harmony with the supposition of Molengraaff, Brouwer and numerous other geologists that the [East] Indian Archipelago represents the first stage of a mountain folding process.
(Vening Meinesz, 1930: 571; my bracketed addition)

But he also surmised that the downbuckle requires a simatic layer of sufficient strength to resist the moving sialic block and cause it to fold upon itself. This led him to reject Wegener's mechanism.

If the hypothesis given in this paper about big crustal foldings is right, the ocean floors appear to have enough strength for folding processes to take place. This appears to be contradictory to the mechanism of continental drift advocated by Wegener. The great stresses supposed to be present in the crust likewise do not agree with this mechanism.

(Vening Meinesz, 1931: 576)

At first, however, he did not dismiss mobilism entirely.

It is, however, another question whether the folding hypothesis is also contradictory to the principal subject of Wegener's theory, the continental drift itself. The horizontal movement of the Asiatic continent, which this hypothesis likewise assumes to occur, renders it not improbable that both principles may be reconciled.[31,32]

(Vening Meinesz, 1931: 576)

But by 1934 Vening Meinesz had fully rejected mobilism even though he adopted convection as the cause of crustal downbuckling (§5.6).

In the same year Ph. H. Kuenen also rejected mobilism. Kuenen was appointed geologist of the 1929–30 *Snellius* Expedition on the recommendation of Escher, his teacher at the University of Leiden.[33] Bathymetric charts based on 30000 new wire and echo soundings, a ten-fold increase, were published by van Riel in 1933, and Kuenen was the first geologist to take full advantage of them. He gave two preliminary analyses the following year, dismissing mobilism. However, he delayed detailed discussion of mobilism until 1935 when he prepared a comprehensive presentation of the bathymetric results. Here is what he wrote in 1934.

In my opinion the new data, therefore, point to the following relations. The Australian and Asiatic continents touch along a mobile belt, but they belong to one continuous mass in the same manner as Asia and southern British India. The deep basins and troughs are depressed areas of the continent situated in the mobile belt and are not the remains of a former oceanic connection between the India and Pacific oceans. It would lead us too far to enter into the question of the continental drift of Australia. In the first Part of Volume V of the Snellius Reports this subject is considered and the conclusion arrived at there from quite independent data, is also that Australia always touched the East Indian archipelago in approximately its present relative position.

(Kuenen, 1934: 193)

This new bathymetry had led Kuenen to reject the mobilist solutions, but it was not what he had set out to do. Of the four geologists he thanked for "kindly" giving him "helpful criticism" (Escher, Molengraaff, Brouwer, and Louis Martin Robert Rutten), only Rutten was then a fixist. He was a longtime fixist and, as will become clear, remained one for years to come (IV, §3.5).

Kuenen rightly devoted most of his attention to the account of the Dutch East Indies by G. L. Smit Sibinga (1933), his being the most complete and current. Smit Sibinga (1933) maintained, like Wegener, that the East Indian Archipelago had formed as a group of island arcs during the Mesozoic peeling off from Asia as it drifted westward. There they remained undisturbed until the end of the Quaternary when the northward-drifting New Guinea–Australian block plowed into them and bent them into their current form.

In his attempt to reconstruct the original position and shape of the highly deformed island arcs, Smit Sibinga had to assume various submarine connections and ruptures among them. He also tried to locate the seam along which the New Guinea–Australian and Asian blocks had actually collided. Kuenen (1935: 98–108) showed that where Smit Sibinga proposed breaks in an original island arc or group of them, the new sounding showed continuity, and where Smit Sibinga proposed transverse faults marking the breakup of originally attached island arcs, there were connecting ridges or separating troughs but no faults (RS2). As Kuenen put it:

The foregoing shows that most of the submarine connections and ruptures are definitely in contrast to the structural connections and ruptures assumed by Smit Sibinga and that a number of elevations exist for which there is no place allotted in the theory.

(Kuenen, 1935: 100–101)

Kuenen also showed that none of Smit Sibinga's purported seam lines between the Asiatic and New Guinea–Australian blocks made sense in light of the new bathymetry.

Reviewing the facts ascertained concerning the position of the joining line between Australia and Asia we have found [that] . . . A clear morphological dividing line does not exist.

(Kuenen, 1935: 105; my bracketed addition)

Kuenen also raised several theoretical difficulties with Wegener's and Smit Sibinga's interpretations. Although most did not depend on the new sounding, one of them did (RS2).

If the sial strips can float about *ad libitum* in the sima this must be highly fluid. The varied and steep submarine topography of most of the sima pools contradicts this conception of the nature of the sima. The deep between the Banda Arcs is 2500 meters greater than the normal uncovered sima. If the sima was so highly fluid it could not possibly fail to rise up between the floating islands.

(Kuenen, 1935: 103)

And it had apparently not done so.

Because of these new bathymetric data, Kuenen broke with the Dutch mobilists, precisely what is to be expected if regionalism was at work. Kuenen was not an anti-mobilist before 1935; he was led to become one because of new regional data.

Notwithstanding, Kuenen recognized mobilism's advantages, and tentatively proposed a northward drift of the New Guinea–Australian block that had not much affected the morphology of the islands.

It must be admitted that serious as the objections to the drifting of Australia may be, there are other strong arguments in its favour. The two principal ones are, first, the Australian faunistic relations to British India, and to the Antarctic with South America, secondly, the possibility of explaining the Permian ice age. Must these advantages of drift theory be discarded on account of the objections raised above, or can a compromise be made? . . . a possible way out of the difficulty will be put forward, as follows: Australia and New Guinea with Misool [a small island approximately 30 km east of the eastern tip of New Guinea] always formed part of the same continent. This continent always had roughly the same position in relation to the East Indies. All the movement made by Australia was obtained by tectonic compression in the East Indies in the same manner as all the movement made by British India was obtained by tectonic compression in the central mountains of Asia.

(Kuenen, 1935: 109; my bracketed addition)

Kuenen (1935: 109) realized that he was stretching a point.

I am far from claiming that this history of the East Indies and Australia is entirely satisfactory. The most doubtful points seem to be whether the structure of the East Indies is of such a nature that, flattened out in its original shape, it could move Australia sufficiently and in the correct sense to bring it beside the former position of Africa and India, and whether a satisfactory explanation of the gravity anomalies could be given. The only reason for proposing this bold and poorly founded hypothesis, is that it combines the most attractive elements in the drift theory as far as Australia is concerned, without showing the – to my mind insurmountable – obstacles of the original suggestions of Wegener for these parts.

(Kuenen, 1935: 109)

While admitting the possibility of continental drift, Kuenen preferred a fixist interpretation of the region. In the final chapter of his monograph, he returned to the question of Australian drift and concluded:

We still have to consider the part that was played by the Australian continental block. In order to explain the compression in the mobile belt it must be assumed, that it has decreased considerably in breadth . . . The two continents in this way approached each other. As, however, parts of the Australian mass belong to the same Mesozoic sedimentation basins as the Molukken Arcs, this continent did not drift a very great distance before it finally established a contact, but always lay in its present relative position.

(Kuenen, 1935: 116)

Any drifting of Australia relative to the East Indies was minor; their separation had changed little. Seeking a compromise and unwilling to abandon former ideas, he included them all.

Kuenen also suggested that the East Indies could be viewed as a developing isthmian link (§3.11) between Australia and Asia.

Ch. Schuchert and Bailey Willis have recently proposed a new theory to explain faunistic links between widely separated areas without assuming continental drift or the submergence of vast continental areas. They suggest that the missing links were formed by narrow geanticlinal ridges, for which they proposed the term isthmian links. Panama forms the type example . . . Although the principle of isthmian links can hardly be claimed to be new, it is certainly of great value that it has again been brought forward and shown to be an alternative to Wegener's theory. In the opinion of the present author it is especially important, when applying the principle to the East Indies, to weigh the possibility of former submergence of the present interrupted links to explain the isolation of Australia which is far more striking than its roads of communication. We are dealing with a link that has recently been established and may in future develop into a complete isthmian link.

(Kuenen, 1935: 109–110)

Kuenen returned to these proposals in his 1950 book *Marine Geology*. Although he noted that Indonesia could be incorporated into du Toit's 1937 reconstruction of Gondwana, he still viewed drift as highly speculative.[34]

Although the application of continental drift to the Indonesian regions, as suggested by Wegener and others, appears to lead to contradictions, an attempt can be made to apply the general principle in a modified form. The present writer proposed (1935) a tentative alternative to Wegener's reconstruction for southeastern Asia that now appears to fit Du Toit's maps of the Southern Hemisphere fairly well . . . Australia is supposed to have remained in contact with Asia via the Moluccas from the start, but to have swung around when it came apart from Gondwana and Antarctica. During this drift Indonesia must have been bent and crumpled. The deep-sea depressions would have been formed by folding and subsidence in and around the isthmus connecting Australia to Asia.

(Kuenen, 1950: 182)

Kuenen had come to believe his earlier hypothesis "farfetched."

. . . many geologists and geophysicists are strongly opposed to the theory of drift (see especially Umbgrove, 1947) and that the present writer is also skeptical of its merits. The survey just given shows that Indonesia, far from giving eloquent testimony in favor of Wegener's views, as some authors have held, presents many features strongly opposed to the theory of drift. Only by assuming the admittedly farfetched reconstruction outlined above can a flat contradiction be avoided.

(Kuenen, 1950: 182; the reference to Umbgrove (1947) is to the second edition of his The Pulse of the Earth)

He also thought that the more general evidence cited in favor of mobilism had not stood up to careful scrutiny; there were no difficulty-free solutions.

If the conclusions of adherents to drift in some form or other such as Du Toit, Wegmann, Gutenberg, and Kirsch, are confronted with the opinions, for instance of Bucher, Umbgrove, Stille, and Cloos, it becomes obvious that neither of the two camps can claim a decisive victory. But the evidence favorable to drift often proves illusive or at least open to serious doubt, on closer inspection. For the time being most geologists appear to have lost faith in continental drift as a sound working hypothesis.

(Kuenen, 1950: 129)

Although Kuenen (1950: 127) noted that it "is not the place here to treat the almost endless series of arguments that have been brought forward in the controversy around continental drift," he continued to raise difficulties that "stand in close relation to" marine geology. Appealing to seismic and gravimetric data, Kuenen argued that the seafloor was not sufficiently plastic to yield to moving continents as Wegener wanted (RS2).

Turning to another aspect, it can be pointed out that the gravity investigation of oceanic volcanic islands shows how these heavy masses are not compensated by light roots. The crust appears to bend elastically beneath the superimposed weight but does not react plastically (Vening Meinesz, 1944). The velocity of propagation of seismic waves across the Pacific floor further shows the great rigidity down to several dozens of kilometers in the suboceanic crust.

Both observations strongly militate against Wegener's conception of the plastic nature of the sima along the Pacific floor.

(Kuenen, 1950: 127; Vening Meinesz reference is the same as mine)

To this restatement of the mechanism difficulty, he added another based on the apparent abundance of sediments on the seafloor (RS2).

The theory of continental drift can be applied only if the postulate is made of a plastic seafloor that yields to the pressure of a sliding continent. Although geophysical evidence is strongly opposed to this contention, the fact remains that the solid crust of the ocean floor cannot be sampled. Hence, appeal can always be made to our lack of knowledge, when postulating properties for the suboceanic crust. Now that a thick cover of pelagic deposits has been deduced, however, a new difficulty to continental drift becomes apparent. These deposits have a much lower specific gravity than the underlying sima and are certainly not in a suitable physical condition to spread out laterally when the advancing American continents swept westwards. These advancing continental icebergs should act as a skimmer moving across the Pacific pool of sima and crowd together the superficial layer of sediment against their blunt bows. The western ranges of America are not composed of deep-sea sediments and cannot represent this Pacific cover. The westward drift of South America is supposed to have reached 3000 km, and the deep-sea deposits skimmed off the sima should amount to 9000 km^3 per km length of coast. Complete isostatic adjustment of this mass would still leave a volume of 1000 to 2000 km^3 above the ocean floor. If this mass of scum were to form a plateau at sea level its breadth would reach 300 km. Far from encountering anything of this sort, a deep-sea trough is found along the Pacific coast of South America.

(Kuenen, 1950: 399)

The idea of ocean-wide subduction on a massive scale, which eventually removed this difficulty, was still a decade and a half away. Kuenen (1950: 127) also reiterated Bucher's by now most familiar difficulty that Wegener's mechanisms for drift and mountain building were inconsistent.

Bucher (1933, p. 72) further points to an illogical postulate of Wegener: "The less plastic mass of the continent is pictured as thrown into folds by the 'resistance' of the more plastic substratum [plasticity postulated by Wegener, Ph. H. K.]. We shall be willing to entertain such seemingly illogical ideas only if the theory of drift as a whole proves to be a real key to the understanding of continental structure as a whole".

(Kuenen's bracketed addition; Ph. H. K. refers to Phillip Kuenen;
Bucher reference coincides with mine)

Although Kuenen was not willing to reject mobilism completely, he surely did not view it as a sound working hypothesis.

J. H. F. Umbgrove, another Dutch geologist who discussed the evolution of the Dutch East Indies, was an adamant fixist. Kuenen (1950: 129) was certainly right to single him out in the two passages quoted above and group him with Bucher (United States), Stille (Germany), and Cloos (Germany). Umbgrove (1951) spoke against mobilism at the 1950 symposium on continental drift held in Birmingham,

UK (§5.9), and attacked mobilism in various publications including the second edition of *The Pulse of the Earth* (1947).

In 1934, Umbgrove (1934b: 173) first voiced objection to mobilism's early interpretation of the Dutch East Indies, basing his attack on Kuenen's (1933) preliminary analysis of the bathymetric data from the *Snellius* Expedition. Four years later, he added further criticism. He argued (1938: 25) that sedimentary and bathymetric studies, primarily from the *Snellius* Expedition, indicated a partially submerged geosyncline that had continuously filled with sediments since the Triassic, or even Permian; this began in eastern Celebes, passed through the Banggai Archipelago, Ceram, Kei, Tajando, Tanimber, and Letti, and ended in Timor. Because sediments were so extensive, Umbgrove argued that they could not have come solely from within the Dutch East Indies; some must have been derived from northern Australia and New Guinea, indicating their proximity at least since the Permian. This, he concluded (1938: 25), "is incompatible with the line of thought about the origin of the East Indian Archipelago given originally by Wegener and worked out by other authors."

Umbgrove reconsidered the origin and development of the East Indies in a series of lectures he gave at the University of Cambridge in May 1946, published in 1949. Hardly deigning to argue against mobilistic interpretations, he repeated what he had said in 1938, this time downgrading its importance by placing it in parentheses.

(It may be noted in parenthesis that these and other deductions concerning the later history of the East Indies are incompatible with the line of thought about the origin of the East Indian archipelago given by Wegener and worked out by other authors.)

(Umbgrove, 1949: 36)

Returning to the subject the next year, Umbgrove began his discussion of various hypotheses about the East Indies by noting:

It is not my intention to mention all the theories bearing on the origin of island-arcs, which in the course of time have been presented by various authors ... There would, for instance be no sense in dwelling any longer on hypotheses involving large horizontal movements of island-festoons in the way suggested by Wegener, Du Toit, Wing Easton, Smit Sibinga a.o.

(Umbgrove, 1947: 147–148; a.o. is short for "other authors.")

To make sure that readers clearly understood his disdain of mobilism, Umbgrove reminded them of its mistaken understanding of Earth's crust.

Horizontal movements and crustal shortening of geosynclinal belts are phenomena associated with epochs of compression. The movements are, however, not of the sort suggested by drift-hypotheses which envisage sialic blocks floating in a simatic syrup. The earth is surrounded by a world-embracing rigid crust and the tangential pressures in the crust have to be described and figured in a quite different manner ... Accordingly these older theories may be left out of consideration for the present.

(Umbgrove, 1947: 148)

To explain biotic disjuncts, Umbgrove adopted isthmian connections, even going so far as to reproduce Willis' illustration of his trans-Atlantic example. Identifying himself as a permanentist, he related isthmian connections with the Mid-Atlantic Rise and Carlsberg Ridge, and also appealed to the view of Nölke (§8.4) and Cloos (§8.3).

Among the features which are liable to change in one or perhaps several respects are the details of the submarine relief. For example, here too we need only think of . . . the basins of the Atlantic and Indian Oceans . . . and the "rejuvenation" of the relief, which is so strikingly illustrated by the Mid-Atlantic Rise and the Carlsberg System. These movements remind one at once of the "isthmian links" of Schuchert and Willis (fig. 149) [showing Willis' trans-Atlantic isthmian link]. Reflections of a highly interesting though speculative nature were made by Willis and Nölke on the emersion and submersion of narrow trans-oceanic isthmi. Their origin and submersion will probably remain a mystery for some time to come, but the inference that such features had in fact been formed and then vanished would not seem to be an improbable one. Cloos, too, when dealing with the analogy of the characteristics of the continental and oceanic sectors (Atlantic and Indian) drew attention to the fact that the narrow zones surrounding the continental basins can look back upon a particularly agitated history, with important vertical upward and down-ward oscillations. I consequently incline towards the view that no essential objection can prevent us from accepting the idea of isthmian links as postulated by Schuchert and Willis. Of course, it cannot be denied that the existence of land-connections during certain given periods cannot be proved geologically. Each so-called reconstruction of a trans-oceanic land-bridge must neces-sarily retain a hypothetical character. Yet such land-bridges need not be discarded a priori as mere products of the imagination.

(Umbgrove, 1947: 238–239; my bracketed addition)

Umbgrove was also impressed with the Brooks–Schuchert–Willis solution to Permo-Carboniferous glaciation (§3.12). After raising the standard difficulties against mobilism, Umbgrove appealed to their paleogeographic reconstruction of the end of the Paleozoic, noting that it coincided with Schuchert and Willis' solution to biotic disjuncts, which he already had endorsed.

Schuchert and Willis later illustrated the supposed distribution of land and sea, as well as that of cold and warm ocean currents, and tried to explain the Upper Paleozoic glaciation without the occurrence of alterations in the present position of continental blocks. These authors, like Brooks, are of the opinion that a solution must take into account the existence of former land connections between Africa and Australia, and between Africa and South America. Many investigators have based similar conclusions on other grounds (i.e. on various geological arguments, and the distribution of land plants, animals and marine faunas). The preceding chapter explained that though each reconstruction of transoceanic land-bridges retains a purely hypothetical character, the idea that land-bridges existed in former times should not be wholly discarded. A "not impossible" reconstruction will be found in fig. 174, [a fixist reconstruction of geographic conditions during the Upper Paleozoic ice ages].

(Umbgrove, 1947: 281; my bracketed addition)

Umbgrove's reconstruction closely resembled Brooks' (see Fig. 3.5). In the late 1940s fixists across the world were embracing one another.

Van Bemmelen, a former mobilist and student of Brouwer and Molengraaff, later recalled his early enthusiasm for mobilism.

As a young student the present author listened to a lecture by Alfred Wegener at The Hague in the early twenties. This event, the books of Köppen and Wegener, Argand (1924), Staub (1924), and the influence of his teachers H. A. Brouwer and G. A. F. Molengraaff made him an enthusiastic supporter of the mobilistic concept of geodynamics. In his doctoral thesis on the Betic cordilleras of southern Spain (1927) he suggested that the front of the African shield had been thrust over the southern rim of the European continent, piling up the (just discovered) large nappes of the Alpujarrides and the Sierra Nevada.

(van Bemmelen, 1972: 16)

Van Bemmelen changed to fixism after working for several years in Indonesia when he came to believe that Haarmann's fixist theory of orogenesis explained what he saw. He began working for the Geological Survey of the Netherlands Indies in 1927. Recalling what happened, he said:

It was especially Haarmann's book (1930) which influenced the author's tectonic approach during his fieldwork in Indonesia. Haarmann explained the geodynamic phenomena in a fixistic way by his *bicausality* concept: The primary tectogenesis creates relief energy by differential vertical movements in situ, while the secondary tectogenesis tends to remove the potential gravitation energy by means of glide tectonics. Java and Sumatra provided fine case histories in favor of this concept (e.g., Van Bemmelen, 1934).

(van Bemmelen, 1972: 16)

He soon developed his own version of oscillationist (his undation or wave) theory, partly in response to objections that were raised to "Oszillationstheorie" at a meeting of the German Geological Society and partly in response to what he had seen in the Dutch East Indies.

The tectonic evidence found on the island arcs of Indonesia could apparently be explained by means of this *bicausality concept*. Since, however, several aspects of Haarmann's "Oszillationstheorie" had been severely attacked during a special meeting of the German Geological Society at Berlin in 1931, the present author was induced to introduce two major alterations, one on the energy problem and the other on the lateral shifting of the vertical movements during the orogenesis of the Alpine type.

(van Bemmelen, 1972: 16)

The second alteration helps explain why he developed his own fixist theory in light of his work in Indonesia.

Another major difference between the undation theory and the oscillation theory is based on the evidence of the orogenic evolution of the Sunda mountain system in Indonesia. That island-arc system developed in the course of time by a progressively sideward shifting of the differential vertical movements, eventually forming a series of parallel tectonic belts, consisting of a foredeep, a non-volcanic outer arc with radially outward directed overthrust nappes, an interdeep, a volcanic inner arc, and a backdeep. These arcuate orogenic belts were found to be arranged around a focal area of geodynamic disturbance. It is clear that instead of oscillations in situ we have in the Alpine type of orogenesis to deal with crustal waves (the Latin for wave is "*undus*") spreading from distinct regional centres.

(van Bemmelen, 1972: 17)

Van Bemmelen further developed his regional account of Indonesian tectonics in a book-length manuscript, which he began in 1937, but events during World War II led him to write a second book.

The author had been working in Bandung from 1937 to 1941 on a book on the geology of the Indian Archipelago. The first manuscript was almost completed when the war with Japan broke out. This MS with all the original drawings, photos and literature references had been committed to the charge of an Indonesian functionary of the Geological Survey, in order to save it from the Japanese when the author became a prisoner of [war] in 1942. This functionary refused to return the MS after the liberation of Java in 1945, and took it with him to the Republic of Djokja [what Indonesia was called between its declaration of independence two days after the Japanese surrendered and Holland's recognition of the republic as a sovereign state at the end of 1949]. Thereafter, the Director of the Bureau of mines, requested the author to write a new book on this subject. This work was started in Holland in the second half of 1946.

(van Bemmelen, 1949: ix; my bracketed addition)

He finished the book in May 1949, three years after he started. Armed with his undation theory and an abundance of new information, he concluded after 732 pages:

Starting from a number of continental nuclei the sialic crust has grown in thickness and extent during the geological evolution of this region. This process of growth of the sialic crust was accompanied by orogenic revolutions, which have welded new rigid belts to the continental shields. The Pacific Mountain System had its main development in the Lower Mesozoic and consolidated the continental area of SE Asia and the older circum-Australian girdle. In the intervening region between Asia and Australia the process of mountain building continued and is still in full swing. It is to be expected that in the geological future this region will also consolidate, forming a continental bridge between Asia and Australia.

(van Bemmelen, 1949: 732; his emphasis)

Van Bemmelen claimed that his undation theory explained well the region's abundant igneous activity, and its relation to orogeny.

Stress has been laid on the geochemical (viz. igneous) processes accompanying the orogenic development. It is impossible to tell the geological history of the Indian Archipelago without taking due account of the physico-chemical processes active below and inside the crust during its long history. This view point has been worked out in the author's "Undation theory" . . .

(van Bemmelen, 1949: 723)

It is no surprise that he should think that an understanding of igneous activity was crucial to understanding the evolution of the Archipelago; he was confident that his undation theory provided it.

Volcanism is a major phenomenon accompanying the structural evolution of the earth's crust. Its external products can be found in all geological epochs. Taking the conception of volcanism in its wider sense as the entire complex of internal and external processes of igneous activity, it is probably the most important factor in crustal evolution. The Indian Archipelago represents a striking instance of this thesis.

(van Bemmelen, 1949: 224)

He then raised several specific difficulties with mobilism (RS2).

According to Wegener's hypothesis the Sahul Shelf forms part of the Australian continental block, torn from the Gondwana Land, which subsequently drifted through the "sima-sea" towards the equator. In the course of this process it caused folding of the crustal layers in front of it, giving rise to orogenesis in the East Indies. However, it is remarkable that this drift has not pushed up a marginal mountain range along the front of the Australian block proper, opposite to Timor, as was suggested by Wegener for the Cordilleras de los Andes along the margin of westward drifting South America.

(van Bemmelen, 1949: 721)

He declared that a Jurassic geosyncline ran from Australia to Timor where it joined another geosyncline that continued to East Celebes, and that a late Mesozoic land-bridge had connected Southeast Asia with northwest Australia. He cited influential Dutch workers who had now come to believe that Australia and the East India Archipelago remained fixed relative to one another.

Although Smit Sibinga (1933) has attempted to apply Wegener's fundamental conception to the Indian Archipelago, the majority of the geologists familiar with the geology of this part of the earth's surface, reject the hypothesis of a drifting Australian continent (Umbgrove, 1934b, p. 172; P. Kuenen, 1935a, p. 98).

(van Bemmelen, 1949: 721; his Umbgrove, 1934b is my Umbgrove, 1934a;
his Kuenen, 1935a is my Kuenen, 1935; his Smit Sibinga (1933) is my Smit Sibinga (1933))

So far as the Dutch geologists were concerned, mobilism's possible application to the East Indian Archipelago after a brave start had by 1950 become a dead issue. The extensive bathymetric results of the *Snellius* Expedition and Vening Meinesz's interpretation of his discovery of the belts of negative gravity anomalies over the Java Trench led this second generation of Dutch workers to break ranks with their predecessors. Van Bemmelen's 1949 updated account kept them corralled in the fixist camp.

Dutch biogeographers too were affected. In 1934 the botanist H. J. Lam wholeheartedly supported mobilism, giving a mobilist solution of the biogeographical problems posed by New Guinea's flora (§8.13), even going to the extent of thanking the mobilist Smit Sibinga for looking approvingly over his 1934 paper (Lam, 1934: 118, 142). Four years later, Lam (1938: 147) had changed his mind, thanking the fixist Umbgrove for reviewing the geology of the Malaysian region with him. Lam (1938: 148) also referred to both Umbgrove's and Kuenen's work, and quoted Kuenen regarding a landmass that extended from Asia to Australia reaching its greatest extent at the beginning of the Tertiary.

The Dutch botanist C. G. G. J. van Steenis, a fixist (§3.4), later recalled Lam's renunciation of mobilism and used the opportunity to raise objections to the mobilist solution to the biogeographical problems posed by the flora of the East Indies.

Accepting geological arguments as decisive, Lam, in 1934 and 1935, has advanced an explanation on the basis of Wegener's theory, and made the bold speculation that Australia and New Guinea with their originally poor Subantarctic flora, had drifted together from the Subantarctic regions towards the NW and so had come, in the Upper Tertiary, into contact with the Malaysian tropical plant world. The Malaysian vegetation, then, overwhelmed this original Subantarctic flora and its remnants could only survive on the high mountains of New Guinea. This certainly contradicts the opinion of Hooker, Bentham, and Diels, who in their analyses have shown that the Australian flora is mainly an (early) derivative of Asiatic stock. Moreover, the flora of South Australia should show, according to Lam's hypothesis, a far more pronounced Subantarctic character than is actually the case. Further arguments against it are the large number of Asiatic genera in New Guinea which are not further distributed, and the occurrence of ancient forms of nearly all Asiatic families in New Guinea. Both arguments are against a very young contact and development; indeed, they point in the reverse direction. Dr. Lam himself admitted to me that he has abandoned his theory, and to a large extent, the view that plant geography must be tied to some geological theory.

(van Steenis, 1950: lxxiv–lxxv)

Looking ahead, it is interesting to see what another biogeographer said after the establishment of plate tectonics about Lam's initial support and later renunciation of mobilism, and van Steenis' approval of his renunciation.

It was H. J. Lam in the mid-thirties who first espoused Wegenerian views and propounded collision of Australia/New Guinea with western Malesia [the biogeographical province stretching from Sumatra and the Malay peninsula to the Bismarck Archipelago] in the late Tertiary (Lam, 1934). This proposition was so far ahead of accepted botanical dogma, reaching back to such nineteenth-century giants as Bentham, Diels, and Hooker, that he soon abandoned the idea (1938), later to be supported in his recantation by van Steenis (1950) . . . But as modern plate tectonic theory has developed in the last decade, the idea of a Laurasian/Gondwanic collision has reappeared in papers by . . . and most recently by van Steenis himself (van Steenis 1979).

(Whitmore, 1981: 70; my bracketed addition)

The application of drift solutions to biogeographical problems related to the flora of the East Indies was strongly influenced by new knowledge, not, however, always in the direction of what, in retrospect, is seen as progress.

8.15 Regionalists and globalists

The field geologist can survey an area of about 100 km (10^7cm) in diameter. The geodynamic processes occurring within the limits of this area are normally described by means of static (fixistic) models. Displacements of the entire field of study of the field geologist with respect to other areas of the earth's surface generally are of no real or direct importance for his restricted purposes of investigation. But phenomena which extend across this frame of direct observation are more easily described by dynamic or mobilistic models of evolution.

(van Bemmelen, 1972: 126)

With this comment, van Bemmelen nicely tied regional tectonics with fixism, and showed the limits of structural geology as usually practiced. Both were unconcerned with intercontinental connections, each region was treated in splendid isolation.

The influence of psychology on the preferred model of description is revealed by the following general tendencies in the geonomic concepts of our time. Contemporary Russian authors, such as Beloussov (1966, 1967), Subbotin *et al.* (1965), born and educated amidst the almost "endless" plains of the Eurasian continental shield – prefer fixistic concepts. Also, J. T. Wilson was for a long time opposed to the idea of drift, because his own early field-work in the Canadian Shield did not suggest anything other than long stability and continuity. This author [Wilson], however, is now converted to drift.

(van Bemmelen, 1972: 127)

The controversies between the fixistic and mobilistic schools appear to be deeply rooted. Wegmann (1956), for instance, lamented: "It is not only a single controversy, but we have to deal with entire knots of antinomies." Nevertheless, both groups have some perfectly sound scientific arguments for their opinions . . .

(van Bemmelen, 1972: 127)

From evidence given in this and the two previous chapters, it is clear that during the mobilism debate regionalism strongly affected theory preference, and I end this chapter with some remarks on its general influence. Regionalism had a positive effect on the reception of mobilism in South Africa, India, South America, the western Alps, and initially the East Indies; a negative effect in North America, Australia (see Chapter 9) and later in the East Indies. In those regions where neither mobilism nor fixism had an advantage in solving regional problems, theory preference was based on other factors, including participants' specialization, the other differentiating factor (§1.15).

Most geologists were regionalists, their preference shaped by their region of study. Globalist transcended region. Globalists were not only concerned about the geology of more than one region and their interrelations, they viewed Earth as a whole, including the geology of oceanic regions.

Globalists were few. Holmes was a globalist. So was Daly. Neither invested more effort in explaining the features of one region than another. Du Toit became a substantial globalist. Unlike Daly and Holmes, however, he began as a regionalist, favoring mobilism from the beginning because it made sense of the geology of southern Africa. Du Toit, the stratigrapher, developed his own version of continental drift, and appealed in its defense to specialties outside his own and to regions beyond southern Africa and Gondwana. Bailey, the tectonicist, did not develop his own version of mobilism; he eventually appealed to work outside his specialty and Scotland, his region of study. Joly was a globalist, and as I shall show later, so was Carey (II, §5.3). Argand also became a globalist; starting in the western Alps, his interests eventually embraced mountain belts on many continents. Wegmann, Holtedahl, and F. E. Suess also started out as regionalists; they were never fully fledged globalists, but became super-regionalists, experiencing for themselves the similarities among far-flung Caledonide segments and comparing the extensive over-thrusting that they exhibited with that of the western Alps. Among mobilists, Wegener was the most remarkable generalist. He was an outsider, an atmospheric not a solid Earth scientist, above all, a reader; his attitude to Earth study from the outset was that of a generalist. He was not wedded, as so many were, to his own region or his own data because he himself collected none that was directly relevant to settling the drift question. He had no regional turf to protect, no axe to grind. He did not judge a geological theory by its ability to solve problems specific to his region or specialization. Holmes, Daly, Joly, and Carey too, like Wegener, concentrated on the relative overall problem-solving effectiveness of mobilism and fixism; they kept their sights on the big picture; they saw, appreciated, accepted, and defended the global advantages of mobilism over fixism. Some fixists were globalists. Stille is a good example, extending his view of orogenesis to most continents and attempting to explain trans-Atlantic similarities without them. Jeffreys was also a globalist. Even though his knowledge of geology was very limited, he did not invest more effort in explaining the features of one region than another.

North America regionalists scrutinized the Appalachians and were not anxious to export their ideas as to their origin to other orogenic belts. Regionalists in South Africa realized that mobilism offered the best solution to their local problems and, given the opportunity, began working elsewhere in Gondwana to check the alleged similarities. Because mobilism was intercontinental, geologists who began regionally but had larger ambitions welcomed the opportunity to work in other regions. Du Toit, King, and Martin all made intercontinental comparative strati-graphic studies, and Plumstead welcomed the opportunity to work on the Permian flora of Antarctica. The more they learned about other continents the more they became convinced of mobilism's correctness. As King remarked about the different parts of Gondwana, "See one place, see one continent and you have seen the lot." But, of course, geologists had to see for themselves, and few who did not live or work on a former fragment of Gondwana immersed themselves in its geology and understood it properly; ignorance or partial knowledge of Gondwana's geology was common in most northern institutions. Consequently, even if geologists who worked exclusively or primarily on Northern Hemispheric problems were some-what familiar with the advantage mobilism enjoyed over fixism in explaining the geology of the Gondwana continents, they rarely appreciated just how substantial these advantages were. Neither William Long's lack of knowledge about Southern Hemispheric geology nor his initial negative attitude toward continental drift made him unique. What made Long special was his walk upon the boulder beds in Antarctica, his chance encounter with Adie which led to his realization that he had walked upon a Paleozoic tillite, and his willingness to shift from fixism to mobilism. Long became an expert on the geology of Antarctica and, helped by Adie, began to view Antarctica as part of Gondwana. At first a stranger in a strange land, he became an expert in Late Paleozoic glaciation of Antarctica, and, once he learned about the geology of other southern continents, he became a mobilist. Hamilton's journey from passive to active mobilism also shows the influence of regionalism. He read du Toit, and was impressed. He thought it fruitless to write in favor of mobilism as he had nothing new to offer, and found no new evidence of it while working in North America. He remained a closet mobilist; even with all his skills he was working in the wrong region to find new evidence that would at the time be persuasive. Antarctica provided him, as he had hoped and expected, with that striking new evidence. Returning home as an active mobilist, he saw North American geology with different eyes; he invoked drift to explain the Gulf of California, and later much of the Cenozoic tectonics of the western United States. En route to becoming a fully fledged globalist, he applied mobilism to the Urals and tectonics of Indonesia.

Notes

1 Şengör's (1979/1982a and 1982b) main concerns are to document Suess' influence among pre-1950 tectonicians, and to discuss philosophical and substantive differences among what he calls the Wegener–Argand school and the Kober–Stille school, a division that he thinks is central to understanding classical theories of orogenesis, his other major concern. They, he claims, "more or less correspond" to Argand's mobilists and fixists but he avoids labeling them in this way because he believes "the difference between the Wegener–Argand and Kober–Stille schools to have been profound philosophical differences that inhibited dialogue between their proponents rather than just differences about whether or not they accepted continental drift" (Şengör, 1979/1982a: 383). Because my interests are restricted to whether or not they accepted continental drift, I shall, when discussing those within the Kober–Stille school, consider only those aspects that directly relate to their rejection of mobilism and refer to them as fixists. Argand defined mobilism and fixism in terms of large-scale drifting, and linked mobilism with Wegener.

We have theories by the dozen, but their very number decreases the chance of agreement among them. Today, it seems that the dispute is focused on the theories that imply the fixism of continents and the hypothesis of large-scale drifting as visualized and powerfully presented by A. Wegener.

(Argand, 1924/1977: 124)

Van Bemmelen also used Argand's distinction between mobilists and fixists in a slightly different way from those who believe in continental drift and those who do not. Van Bemmelen (§8.14), who for a good part of his career defended his own "Undation theory" which allowed for only vertical movements and excluded large horizontal movement, included among mobilists those who appealed to horizontal movements even if they did not support or even argued against continental drift. Thus, van Bemmelen's mobilists also included Vening Meinesz and others who supported his downbuckling crust such as Hess, Kuenen, and Umbgrove, even though they did not believe in continental drift. Here is how he put it in 1949:

For instance, it might be accepted that the bulk of Earth's sialic crust was present since pre-archaeic times, only changing in distribution and structure during the geological evolution. Or we might follow the conception of the growth of the sialic crust in thickness and extent during this geological phase of evolution of our planet. If the first conception is followed, the orogenic processes can be considered as the result of lateral compressive forces in the crust, the narrowing of mobile geosynclinal areas between pre-existing continental blocks. An extreme solution in this line of thought is the hypothesis of continental drift, advanced by Taylor, Wegener, and others. Smit Sibinga (1933) attempted to work out this conception for the Indian Archipelago. The formation of mountain roots (the presence of which is attested by gravity measurements) might be explained by buckling downward of the crust beneath geosynclinal belts. This has been suggested by Vening Meinesz whose "buckling hypothesis" has been accepted by many authors (Escher, Hess, Kuenen, Umbgrove, etc.). The other line of thought, that of the growth of the continental (or sialic) crust in the course of the geological evolution, has been suggested by Bertrand (1887) and found also many followers (e.g., Stille, 1924, 1944; Kober, 1928; Kraus, 1928; Born, 1932). This conception professes the sialic crust to have grown in the course of geological evolution, starting from a number of archaeic nuclei. These primeval centres of the crust did not move along Earth's surface during the later evolution, but had a fixed position. The present author has called these two major conceptions respectively the "mobilistic" and the "fixistic" school (1931d, 1933i).

(van Bemmelen, 1949: 723)

Van Bemmelen may have wanted to redefine Argand's mobilism/fixism distinction because Argand had given "fixism" a particularly negative connotation.

Fixism is not a theory but a negative element common to several theories. All things being well considered, it is essentially the absence of position versus a problem that is precisely that of mobilism, and it can be defined only with respect to the latter. Strictly speaking, it can be neither demonstrated nor refuted; such is the fate of any idea that relies on the absence of testimonies.

(Argand, 1924/1977: 125)

2　Cloos did not arrive and begin teaching at Marburg until winter 1914–15 (Schwarzbach, 1980/1986: 22). I suspect Cloos and Wegener first met before the outbreak of World War I; they probably first met in late 1913 or early 1914.

3　For other accounts of Cloos's account of the Rhine graben see de Sitter (1964: 130–132) and Holmes (1965: 1046–1051).

4　This summary of the 1939 meeting is drawn from Hume's account and from Deborah Dysart-Gale's and Clancy Martin's translations of some of the papers, including several that Hume did not summarize. I am grateful to them for their help.

5　Sonder, who argued against mobilism at the 1939 Frankfurt meeting (§8.4), was one of the minority of Swiss geologists who favored fixism. R. Trümpy (August 24, 2004 letter to author) informed me: "Sonder was a rather isolated figure in Switzerland. Nobody took him seriously, except Augusto Gansser, who briefly became a fixist before accepting with great success plate tectonics."

6　H. (Hans) Keidel went by J. (Juan) Keidel when his papers were published in Spanish.

7　R. Trümpy (August 24, 2004 letter to author) noted: "Around 1940, Schwinner, from Graz, wrote a foul pamphlet denouncing the nappe structure of Alps as a French and Jewish invention." Trümpy also stressed Ampferer's theoretical contribution to mobilism, and offered an explanation as to why he was ignored.

It is not admissible to neglect Otto Ampferer, the most original and far-sighted one of the Austrian geologists, even if his works were written in bad German and nobody took the trouble to translate them into some sort of English . . . Ampferer and Holmes were the real precursors of plate tectonics . . .

I should add that Carozzi (1985: 130) introduced me to Heritsch's mobilism.

8　This summary of Argand's life is drawn from Thalmann (1943) and Carozzi (1977).

9　Argand's appeal to plasticity of continental blocks and du Toit's rejection of the idea show the influence of regionalism on both of them. Argand appealed to plasticity because of what he observed in the Western Alps; du Toit found no such evidence of plasticity from his work in South Africa.

10　Bailey (1958) provides an obituary of Collet.

11 Rudolf Trümpy (August 24, 2004 letter to author) informed me that Salomon-Calvi's first interpretation was the correct one.

12 See Bailey (1938).

13 Şengör talked to R. Trümpy, who succeeded Staub at Zurich University and Zurich Institute of Technology, about Staub's announcement that Heim had become a mobilist. They believe that Heim changed his mind because of Staub's influence. Heim also respected Argand.

I once discussed Staub's statement with Trümpy. Heim must have switched over under Argand's influence. He admired Argand enormously and confessed in his *Geologiie der Schweiz* that his IQ did not allow him to follow him completely.

(Şengör, August 6, 2004 email to author)

14 These biographical comments on de Terra are drawn from obituaries by Stewart (1982) and Savage (1981), and I would like to thank Barbara L. Narendra, archivist at the Peabody Museum of Natural History, for finding them.

15 Including de Terra in this discussion of those who supported mobilism because of their work in the western Alps is, I admit, questionable. However, although de Terra did not argue for mobilism directly from his work in the western Alps, he appealed to Argand, and argued for mobilism based on similarities between the Alps and Himalayas.

16 See Haller (1976), Greene (1982), and Bailey (1935) for more on Termier's life and work.

17 Rudolf Trümpy (August 24, 2004 letter to author) confirmed that Termier "certainly knew of Argand's 1922 presentation," and added that Argand's presentation "was widely discussed and less widely understood."

18 Wegener did not provide any details and his maps are too imprecise to tell exactly how close and just where Greenland and Norway almost touch. I have followed Wegmann's (1948: 18) reading of Wegener.

19 The British geologist Brian Harland, who became a strong supporter of mobilism during the 1950s primarily because of his work in Spitsbergen and also welcomed and helped obtain paleomagnetic support for mobilism (III, §1.11), later described Holtedahl's detailed comparison of Spitsbergen and Scotland in the following terms:

Holtedahl long ago made a comparison between Spitsbergen and Scotland (1925) in which the points of analogy seemed superficial in many respects, but the comparison was in a sense prophetic, even to the tectonic similarity of the Devonian graben in Spitsbergen with the Scottish Lowlands.

(Harland, 1960)

20 Hercynian and Variscan are the main Late Paleozoic orogenies of Europe. The Appalachian is their North American equivalent. They are generally considered to have culminated in the Late Carboniferous and, some have argued, extended into the Early Permian. They and their corresponding mountain belts (Hercynides and Variscides) were named in different places but are now essentially synonymous, usage being equally divided. I shall use them interchangeably, using the term employed by the author of the topic under discussion.

21 This synopsis of Suess' career is compiled from Kölbl (1949), Leuchs (1947), and Waldmann (1953).

22 K. Hsü (1995: 162–163) discusses F. E. Suess' model. Putting Suess' model in "plate-tectonic" terms, he explains that the innermost (non-folded, post-tectonically metamorphosed) zone is equivalent to the overriding plate; Hsu functionally defines it as the bulldozer. Suess' middle (folded and metamorphosed) zone is equivalent to the subducted plate; Hsu functionally defines it as the plate overridden by the bulldozer. Suess' outermost (folded but non-metamorphosed) zone is equivalent to the escaped part of the subducting plate; it is functionally defined as that which escapes (subduction) and has been pushed to the foreland by the bulldozer.

23 I may be unfair to Wegener. A. Hallam (September 9, 2004 email to author) noted
that "Wegener had died several years before F. E. Suess' most important work
was published, and he was not preoccupied with examining geological literature in
the years leading to his death."

24 Because Wegener delayed the opening of the North Atlantic well past the time when it is
now known to have opened, his estimates for the rate of separation of respective
landmasses remained too high even after recalibrating on modern timescales.

25 Le Grand (1988: 85–86) includes a discussion of Wade's mobilism in his section on the
reception of mobilism in Australia. I think Wade is better categorized as an English
petroleum geologist who came to support mobilism because of his work in New Guinea
and interest in the distribution of oil fields, and later found additional support for mobilism
from his work in other areas of Gondwana, and application of mobilism to Antarctica.
He just happened to live his later life in Australia. We agree about what attracted Wade to
mobilism.

26 The controversy over whether the Scottish Caledonides are nappes is interesting (Bailey,
1935; Greene, 1982; and Oldroyd, 1990).

27 This summary of Bailey's life is taken from C. J. Stubblefield's biographical memoir for the
Royal Society (London) (Stubblefield, 1965).

28 Rudolf Trümpy and Anthony Hallam both remarked about Bailey's toughness in the
field. Trümpy (August 24, 2004 letter to author) recalled that, when he took part in a
field trip in the Scottish Highlands, Bailey wore sandals and would "swim in the loch
every September morning, clad in the scantiest of all possible garbs." Hallam noted
(September 9, 2004 email to author), "Bailey was always incredibly energetic and Spartan
in his fieldwork, and often refused a sandwich lunch as a distraction."

29 Bailey presented his argument in favor of mobilism in several publications. After first
supporting it in his 1927 note to *Nature*, he gave his Presidential address on September 10,
1928, before Section C of the BAAS, and the talk was published the next year (Bailey,
1929). An anonymous and brief report of his forthcoming talk appeared in *Nature* three
days before he delivered it (Anonymous, 1928b). Bailey wrote a shortened version for
Nature which appeared on September 24, 1928 (Bailey, 1928a). Adapting to a different
audience, another version appeared in the *Scottish Geographical Magazine* (Bailey, 1928b).

30 For information on van Waterschoot van der Gracht's life, see Gulley (1944).

31 Vening Meinesz also objected to Wegener in 1934. Using a similar argument when
considering the huge negative gravity anomalies in the Dutch East Indies, he (1934c:
120–121) wrote:

Admitting that crust in the belt of negative anomalies has been subject to a lateral
shortening through compression . . . we may draw two further conclusions from the way
the belt is situated . . . In the second place, it is worth while to remark that the anomalies are
just as strong in the part of the belt that borders on the Indian ocean than in the part
bordering on the Australian continent. This appears to indicate that the resistance offered
by the ocean-floor of the Indian Ocean is not inferior to that offered by the Australian
continent . . . It seems to be in decided contradiction to Wegener's theory of continental
drift; we could not well imagine such a migration through the ocean-floor when this part
of the crust can so strongly resist pressure. Of course the actual facts only concern the floor
of the Indian Ocean south of Sumatra and Java and the floor of the Pacific east of the
Talmud Islands and so this conclusion ought not, without further proof, to be extrapolated
to all ocean-floors elsewhere.

With his discovery of huge negative gravity anomalies in the West Indies, Vening Meinesz
extended the argument to the Atlantic Ocean.

A second important result of the gravity investigations in the Atlantic is the irregularity
of the gravity field which is at once evident when we examine the profiles of Plate I.
This seems likewise to disagree with Wegener's theory. If the irregular topography of the

sea-bottom in the Atlantic, which has been found as well by the soundings of these expeditions as by those of previous ships and which is in good harmony with the similar results of the "Meteor" Expedition in the southern Atlantic, if this irregular topography was caused, as Wegener supposes, by sialic remains of the disruptive zone which are floating on the substratum, the gravity anomalies, after isostatic reduction, ought not to show great deviations. The gravity field, however, is irregular and this points in another direction than Wegener's supposition. It proves the presence of numerous mass-disturbances and so we are led to suspect tectonic action in this area. This surmise is corroborated by the presence of earthquake-epicentra. Both arguments point to a crust which is capable to transmit stress.

(Vening Meinesz, 1934d: 196)

32 Du Toit (1937) briefly discussed Vening Meinesz's discovery of the strong belt of negative gravity anomalies in the Dutch East Indies. He (1937: 188) suggested that the belt represented part of the boundary between the blocks of Laurasia and Gondwana.

33 Kuenen was very thankful for Escher's help in securing his appointment as geologist to the *Snellius* Expedition, which launched his highly successful career. Kuenen (1950) dedicated his *Marine Geology* to Escher.

To B. G. Escher, Professor of Geology, University of Leyden, inspiring teacher and generous friend, who first directed the writer's interest towards the problems of marine geology by insuring his appointment as geologist to the *Snellius* Expedition.

34 Du Toit (1937: 188) mentioned Kuenen's 1935 analysis of the bathymetric data from the *Snellius* East Indies Expedition, but said almost nothing about it because he had not actually seen it. What he knew about Kuenen's position came from his reading of Schuchert's six-page review, and Schucherts did not mention Kuenen's highly speculative drift alternative. Schuchert (1936: 295–296) quoted only Kuenen's summary statement which did not include his drift alternative.

Kuenen postulates further "that the Asiatic and the Australian continents at one time formed a more or less continuous mass" (to which the writer would add, surely not since the Paleozoic). Also, that the mobile belt has decreased considerably in breadth and has been pressed together. "The two continents in this way approached each other." Kuenen, however, thinks that the Australian mass "did not drift a very great distance" and that its movement affected only in a minor degree the resulting tectonic phenomena.

9

Fixism's popularity in Australia:
1920s to middle 1960s

"Carey, Carey, Carey, there is more to Australia than Carey."
Curt Teichert, comment to author

9.1 Introduction

The general theme of this chapter is that regionalist attitudes, as Frankel (1984a) and Le Grand (1988) have argued, were largely responsible for mobilism's mostly negative reception in Australia. Mobilists offered solutions to several problems in Australian geology, for instance, the origins of its Permo-Carboniferous glaciation and of its biotic disjuncts, especially its marsupials, but fixists did not find them persuasive. The fit between Australia and Antarctica was not as convincing as that between Africa and South America; the Great Australian Bight could be nestled up against Antarctica's Wilkes Land, but not unambiguously. Fixists thought they could account for the Permo-Carboniferous glaciation in Australia because the glaciation appeared to come from the south, and could have originated at higher latitudes even if Australia had not drifted. Also, there were uncertainties about the timing of the glaciation, and whether it was synchronous with that elsewhere in Gondwana. As for the origin of Australian marsupials, their resemblance to South American marsupials was questionable, and fixists could, with some confidence, appeal to Matthew and G. G. Simpson's alternative solution that has already been discussed (§3.8). Geosynclinal theories fitted comfortably with Australian geology. Australians were quite at home accreting Australia around ancient cratons, reconstructing its geological history as a stable continental shield to which marginal geosynclinal belts were added. A huge Precambrian shield forms the western half of the continent, and within it are Archean and Early Proterozoic nuclei. To the east, the shield is bordered by a series of Paleozoic orogenic belts, known collectively as the Tasman Geosyncline, which gets younger eastward. Australia's geological structure, like North America's, fitted well with fixist geosynclinal theory.

As a consequence, from the 1920s through the mid-1960s most leading Australian Earth scientists were fixists. Most never argued against mobilism in print; they seemed confident that fixism was true, and saw no need to defend it or often even to mention it.

496

Except for those who debated the origin of marsupials, most Australians' comments on mobilism were restricted to the classroom and informal discussion. E. S. Hills thought mobilism a topic of tearoom conversation, not formal debate (Le Grand, 1988: 83).

A few Australians, however, did favor mobilism and in the 1950s, Warren S. Carey became one of its most avid supporters, and readers may be surprised that he does not feature more prominently in this chapter, but I deal with his work later in detail (II, §6.5–§6.15). Edgeworth David and Leo A. Cotton, both at the University of Sydney, thought mobilism worthy of investigation, but they were not deeply involved in the mobilism debate and other than inspiring Carey as a student at the University, their pro-mobilist views had no general influence. Several biologists who worked on Australian biogeographical problems also supported mobilism, Launcelot Harrison and George Edward Nicholls, for example, argued for it during the 1920s and 1930s.

9.2 Geologists working on Australia's geology favorable to mobilism

Sir Edgeworth David (1858–1934) was the dominant figure in Australian geology and the most prominent to favor mobilism during the first third of the last century (Mawson, 1932–1935; Skeats, 1933; Andrews, 1942; Branagan, 1981; Voisey, 1991). Born near Cardiff, Wales, he studied classics at New College, Oxford, earning a first in 1880. Deciding to become a geologist, he attended the Royal School of Mines. David sailed to Australia in 1882 for a job at the Geological Survey of New South Wales. He was elected to the Royal Society of London in 1900 for his work on recent and ancient glaciation and the successful boring (1897) and interpretation of Funafuti's coral atoll. The year before his election he was appointed Chair of Geology at the University of Sydney where he built a strong department. He taught many leading geologists of the next generation, including E. C. Andrews, W. H. Benson, W. R. Browne, Leo A. Cotton, D. Mawson, T. G. Taylor, C. E. Tilley, A. B. Walkom, and W. G. Woolnough. At the age of fifty, David accompanied Shackleton on his *Nimrod* Expedition to Antarctica (1908–9), during which he led a party to the South Magnetic Pole. After retiring from the University of Sydney in 1924, he began work on a comprehensive book on Australian geology. Unable to complete it, he asked W. R. Browne to do so, and *The Geology of the Commonwealth of Australia* was published sixteen years later in 1950 in three volumes.[1]

David received numerous honors: the Bigsby and Wollaston medals of the Geological Society of London (1899, 1915), the Mueller Medal of the Australian and New Zealand Association for the Advancement of Science (hereafter, ANZAAS) (1909), the Conrad Malte-Brun Prize of the Geographical Society of France (1915), the Clarke Medal of the Royal Society of New South Wales (1917), and the Patron's Medal of the Royal Geographical Society (1926). He was awarded Sc.D. from Oxford University in 1911, and honorary Doctorates of Science from St. Andrews

(1926), Wales (1926), and Cambridge universities (1926). He was President of the ANZAAS in 1904 and 1913, and was unanimously elected the first president of the Australian National Research Council (1919).

David declared support of mobilism late in his life, too late to have any lasting influence. He (1928) gave a public lecture at the age of seventy in Sydney entitled "Drifting Continents: The Wegener Hypothesis" on June 12, 1928. He was almost certainly influenced by du Toit, with whom he had become friends in 1914 when they examined together Late Paleozoic glacial strata in Australia. David visited South Africa in 1926 en route to England. He discussed continental drift with du Toit, who gave him several chapters of his forthcoming book, *A Geological Comparison of South America with South Africa*.

Before leaving Johannesburg for the Cape David and Du Toit discussed the concept of continental drift, Du Toit having recently visited South America. David noted that Du Toit "has an important new book on the geology of South Africa now going through the press". David was especially grateful to Du Toit for giving him "chapters of the book" but Du Toit felt it was a reasonable return for the boomerang David had given him on his 1914 visit to Australia!

> *(Branagan, 1981: 39–40; Branagan identified the book as du Toit's* Geological Comparison, *and his quoted remarks by David are from letters published in the* Sydney Morning Herald *in March 1926)*

He also examined the Witwatersrand goldfields and compared them with similarly mineralized deposits in Australia (Branagan, 1981: 39).

Although David did not prepare a paper for publication based on his 1928 address, there is a report of it, which may have been David's notes or a preport of what he said. Assuming that it accurately reflects David's views, he certainly thought mobilism was worth seriously entertaining. He (1928: 60) began:

The whole theory appears fantastic at first sight, but a contemplation of the facts of science, as voiced by Sir Edgeworth David and shown in his slides, includes the hypotheses within the realm of probability.

And he (1928: 62) concluded: "Wegener's hypothesis is welcomed by botanists and zoologists. It explains also innumerable knotty problems of the geologist and others."

Sandwiched between these remarks, David marshaled evidence in favor of Wegener's theory. He appealed to geological similarities and accompanying congruencies between continental margins of South America and Western Africa.

Another great argument for the former contact of the continents is the direction of the trend lines of the folding of the crust. There is absolutely remarkable continuance of these if South America is fitted into Western Africa. Such a piecing together restores the lost "Atlantis." The vanished continent has not sunk beneath the sea. It has drifted apart and the newly filled Atlantic occupies the rift.

> *(David, 1928: 62)*

Turning to the other side of Africa and to his adopted country, he tied its drift with recent mountain building in New Guinea.

On the other margin of Africa, Wegener claims that Australia has broken away, and is to this day pushing up against New Guinea and crumpling rocks there as recent as Middle Miocene; the folds now are 15000 feet high. Nowhere else have we such young rocks exhibiting such great movement.

(David, 1928: 62)

David cited the strong similarity, known to gold miners, between the Nullagine in Western Australia and the Rand, South Africa.

Again, a study of rock structures supports the Drift Theory. The Nullagine conglomerates of W.A., with alluvial gold and diamonds are duplicated exactly in the "blanket" of the Rand, South Africa.

(David, 1928: 62)

Returning to South Africa and South America, David referred to the similarities du Toit had described between South America and South Africa that could be applied to Australia (§6.6). He singled out du Toit's appeal to the disjunctive distribution of *Mesosaurus*, but asked why it had not been found in Australia.

Du Toit is absolutely convinced of the former union of South America and South Africa, witnessing as evidence the occurrence of Mesosaurus in the beds of Upper Dwyka formation and identically the same fossil reptile in South Brazil and Argentine. This is a pro-Wegener argument of much weight; it is somewhat discounted by the fact that no Mesosauri have yet been found in Australia.

(David, 1928: 62)

David contrasted the abundance of radiolarian ooze in the Pacific with its scarcity in the Atlantic as further support of mobilism.

Again, the general prevalence and thickness of the deposits of Radiolarian ooze in the Pacific points to a permanence of that ocean basin as contrasted with the Atlantic where this is relatively scanty.

(David, 1928: 62)

David, who had helped discover and interpret Permo-Carboniferous glaciation in Australia and had seen India's Late Paleozoic glacial beds, appealed to mobilism.

But Wegener is again supported by glacial evidence of the Permo-Carboniferous Ice Age. The trend of the drift of the Polar Ice would be simply radial from Antarctica if Wegener's Pan Gea of that epoch could be accepted.

(David, 1928: 62)

However, he found the distribution of an extinct Permian shark troublesome.

An anti-Wegener argument is found in the occurrence of Helicoprion. This is a spiral saw-like arrangement from the jaw of an extinct shark, which must have been of a large size and a powerful swimmer. Helicoprion has been found north of Perth (W.A.), in Kashmir, in N.E. Russia, and in Japan. This presumes a large ocean for the shark where Wegener claims solid land.

(David, 1928: 62)

David, like Wegener, mistakenly thought that geodetic measurements of Greenland would soon determine if drift had occurred. He was aware of Joly's theory, and he thought Wegener's estimate of Greenland's westward movement excessive.

There is much controversy as to whether this very gradual movement [of the sialic continents] is westward or eastward, Wegener supporting the westward drift argument, and being opposed by Joly and his followers. Wegener claims that Greenland at the present day is moving westward at the rate of 32 meters per annum! If this is a fact, modern accuracy of measurement of longitude aided by wireless should soon establish the truth.

(David, 1928: 61; my bracketed addition)

David appreciated mobilism's success in solving problems both within and beyond Australia: the formation of mountains in New Guinea, the occurrence in Australia of the Gondwana-wide Permo-Carboniferous glaciation, and the similarities between the Nullagine of Western Australia and the Rand of South Africa. But he also was bothered by certain difficulties, the lack of *Mesosaurus* in Australia, and the presence of *Helicoprion*.

What he did not discuss is also of interest. He quite sensibly said nothing about the Jeffreys mechanism difficulty. Nor did he apply mobilism to any problems connected with Australian tectonics; nothing, for example, about the geological structure of Australia or the formation of its orogenic belts. The latter made sense because Australia's orogenic belts formed before Wegener's drift, and have been eroded to little more than high hills.

Leo Arthur Cotton (1883–1963) was another Australian geologist who thought mobilism worth investigating. It is hard to gauge his influence but it cannot have been great. Born in Nymagee, New South Wales, he was educated at the University of Sydney, obtaining a B.A. first-class honors in mathematics (1906), a B.Sc. first in geology with emphasis in physics (1908), an M.A. (1916) and a D.Sc. (1920) (Browne, 1963). Cotton, like David, was a member of the *Shackleton* Expedition to Antarctica (1908–9). He was appointed lecturer in geology at Sydney under David (1911), and promoted to assistant professor (1920). He succeeded David as Chair of Geology at Sydney (1924) and held that position until his retirement in 1948. Cotton was president of Section C of the ANZAAS (1928), and Clarke Lecturer (1946) and President of the Royal Society of New South Wales (1928).

With his strong background in maths and physics, Cotton was attracted to problems such as isostasy, diastrophism, polar wander, and the strength of Earth's crust (Browne, 1963: 82). As far as I know, he did not publish anything on the geological history of Australia, and his brief comments on continental drift were unrelated to Australian geology.

In 1923 Cotton considered polar wandering.

If the theory of polar wanderings be merely accepted as a working hypothesis no harm can be done. On the contrary, it may stimulate investigation into many fields of research so that, even if the practical applications of the theory be ultimately shown to be quantitatively small, such investigations may bear fruit in other ways.

(Cotton, 1923: 497)

He was closely familiar with the work of geophysicists and geophysically minded geologists of the day. As Le Grand noted (1988: 85), Cotton was aware that a number of German workers had seriously considered polar wandering, and he argued that Earth was sufficiently yielding to long-term forces to make it probable.

Mathematical theory, geodetic evidence, the testimony of seismology, and geological observations will converge towards the conclusion that the earth as a whole behaves like a highly viscous solid and is capable of closely approximating to the form appropriate to hydrostatic equilibrium for forces of secular duration. This view is not, however, opposed to the conception of the earth as a highly rigid body for stresses of short period such as those involved in the propagation of seismic waves and of tidal deformations. When these postulates are substituted in Darwin's analysis, instead of the postulate of a perfectly rigid earth, the mathematical investigation actually supports the view that polar wanderings are not only possible but probable.

(Cotton, 1923: 496)

He then went on to suggest that polar wander could explain climatic changes, which

would naturally attend any considerable displacement of the poles, and such changes might find, in part at least, their explanation in this hypothesis. The glacial and desert climates of the geological past, by virtue of their abundant record should prove an interesting field of investigation from this point of view.

(Cotton, 1923: 496)

Cotton (1923: 469–470) added mysteriously that polar wandering would produce huge tidal variations with a fourteen-month periodicity, which would help explain "the great prevalence of shallow water sedimentation in the stratigraphic record." It also explained the "linear disposition of folded mountains and volcanic belts," and "many difficult problems in distribution, migration and extinction of faunas and floras." He said nothing about continental drift. Perhaps he did not yet know about it.

Cotton (1929) first mentioned mobilism in his 1928 Presidential Address to Section C of the ANZAAS. But he said so little that it is hard to tell how seriously he had thought about it. Cotton introduced geologists to new theories of mountain building, especially those that utilized geophysics, because some "have appeared in various journals, many of which are not ordinarily read by geologists" (1929: 173). He examined theories of Taylor, Wegener, Daly, Jeffreys, and Joly, and thought Joly's view the best, describing it as "brilliant" and "as the most comprehensive attempt yet made to elucidate that grand problem – the cause of diastrophism – which lies at the root of all interpretations of geological history" (1929: 202, 210). He was familiar with Jeffreys' (§4.11) and Holmes' (§5.4) objections to Joly's theory, but nonetheless quoted a 1926 work of Holmes, saying that Joly's work was "a very great achievement" (Cotton, 1929: 212).

Cotton (1929: 198) grouped theories in terms of forces proposed. He rejected Taylor's capture of the Moon (§2.10). He argued that Daly failed to explain why Earth had once bulged outward around the equator and at the poles and inward along mid-latitudes. Like Holmes (§4.8), Cotton (1929: 197) claimed that Daly left unexplained the domes and furrows, which were "based rather upon the palaeogeographic evidence than upon any mechanical principle."

Although Cotton also rejected Wegener's appeal to extraterrestrial forces, he explicitly pointed out that this failure was not a reason to reject continental drift. Why? Joly's theory could supply the needed mechanism.

The gravitational interaction between the earth and sun which gives rise to the precession of the equinoxes has been invoked by Wegener as one cause partly responsible for the drifting of his sial continental masses through the oceans of sima. The force actually due to this influence, like that which he also postulates for the drift from the poles, has been demonstrated to be an almost infinitesimally small fraction of the value of gravity, and hence totally inadequate for the purpose. It must not, however, be concluded that because Wegener has failed to find an adequate cause for his system of continental displacements, that such movements are necessarily invalidated. The theory recently put forward by Joly, if correct, would lend material support as a basis for the mechanism of the Wegener theory.

(Cotton, 1929: 197)

So far as I can determine, Cotton never found an appropriate occasion to consider mobilism comprehensively – it flitted in and out of his discussions but was never center stage. Cotton (1945) mentioned continental drift as one of four possible explanations of strandline movements of vertical range greater than ± 300 feet. The others were changes in the configuration of ocean basins, volume changes of Earth's crust due to expansion and contraction, and outpourings of lava on ocean floors. This does not sound like a man who was strongly inclined toward mobilism. If he did think it worthy of serious investigation, he apparently did not consider himself the one to do it. He clearly thought highly of Holmes' work as evidenced by his many references to it, and he quoted him on the merits of Joly's theory. He certainly had ample opportunity to discuss Holmes' mobilism, and I find it puzzling that he did not. Cotton met Holmes in 1920 at the first Pan-Pacific Science Congress, and asked him to serve as an outside examiner for Carey's D.Sc. thesis (II, §6.5). Holmes, like Joly, embraced radioactivity, but Cotton should, I think, have found Holmes' mechanism preferable to Joly's. Cotton might have been captivated by continental drift for a while, but he wrote no more on it, his own research being unrelated to it. He did however encourage Carey (II, §6.11).

9.3 Geologists against mobilism

I turn now to fixists, beginning with E. C. Andrews (1870–1948), who was perhaps the most outspoken Australian anti-mobilist. He received his undergraduate degree (1894) from the University of Sydney where he came under the spell of David, but too early to be influenced by David's late interest in mobilism. David recommended him to Louis Agassiz at Harvard for the task of collecting coral-reef material from Fiji. Returning from Fiji, Andrews was hired by the Geological Survey of New South Wales (1899), where he remained for twenty-three years, eventually becoming Government Geologist. He retired in 1931 (Browne and Walkom, 1952). Andrews was President of the ANZAAS (1930), and President of the Linnean Society of New South Wales (1938). He received the David Syme Prize and Medal from the University of Melbourne in 1915, the Clarke Memorial Medal from the Royal Society of New South Wales (1928), and the Lyell Medal from the Geological Society of London (1931) (Cotton, 1948).

Andrews argued against mobilism in the late thirties and forties. In his 1938 Presidential Address before the Linnean Society of New South Wales, he began with a summary of Australia's structural history.

The present marked stability of Australia proper has been secured by the knitting together of several ancient nuclei into one massive unity during the Palaeozoic, whereas the outlying insular arcs have persistently retained their mobile and unstable structure. The ancient nuclei of the mainland are composed fundamentally of early pre-Cambrian structures ... In addition to these nuclei a long borderland appears to have extended, with a possible break in the Newcastle-Sydney district, from the north of Queensland to a point well south of Tasmania. The nuclei and the eastern borderland (or borderlands) were separated from each other by long geosynclines, the eastern troughs receiving sediments from the eastern borderland . . . The growth of the continent in stability was a slow and gradual process; it proceeded first as from west to east and thence north-easterly during the Caledonian movement, thus knitting together the masses ... By the powerful folding movements which closed the upper Ordovician and the upper Silurian the continental mass lying to the west of the meridian of Melbourne had been rendered stable; Tasmania had been attached firmly also to the growing continent at the close of the Silurian . . .

(Andrews, 1938: xi–xii)

He ridiculed Wegener by calling him a metaphysician and his theory an ancient epic.

Philosophical studies, however, even in geology, suffer amazingly from the ever present but alluring metaphysical element. Students, nurtured carefully in the principles of orthodox geology, indulge in speculations which appear not to be sanctioned altogether by the testimony of the rocks. Such flights of fancy, with increasing departure from the Earth, diverge more and more from actual facts of earthly processes. The Wegener Hypothesis, a high-level flight in matters geological, and one of great stimulation in many ways, nevertheless smacks suspiciously of the waxen wings of Icarus and courts a similar fate when leaving the earth too far below. However, by means of its exquisitely poetical presentation, it has made a powerful appeal to the love of the marvelous in man, scientist and non-scientist alike. It captivates the imagination with its semblance to an ancient epic, a veritable Odyssey, in which the continents and the Polar sites, like Homeric or Virgilian heroes, are driven by warring Fates from their grand old home of Pangea only to wander helplessly, like so much plankton, through the vast waste of Oceanus. Wegener, by this imaginative excursion, was led to infer the impermanence of position of the major structures of the Earth, such as continents and oceans.

(Andrews, 1938: v)

Andrews was slightly kinder to Argand and Joly, but he thought both went well beyond the evidence.

Argand, whose field studies in the western Alps amount to genius, and to whom students of geological structure have been placed under an everlasting debt of gratitude, arrives at the conclusion that the external Prealps of the Bernese Oberland represent a small portion of Africa resting upon Europe. The student naturally asks whether all the available evidence had been exhausted before arriving at this conclusion, and whether the confrontation of this idea with the known history of Europe during the pre-Cambrian, Caledonian, and Hercynian activities might not have led to a more simple explanation of the Prealpine occurrences . . . When Joly explains the great topographical revolutions as depending, in great measure, on the influence of accumulated radioactive energy in the subsurface regions, the reader naturally asks whether this explanation is the simplest available, in view of the paucity of direct evidence in support of his views.

(Andrews, 1938: v)

Andrews (1938: xxiii–xxv) criticized Wegener's explanation of the origin of mountains (formed by uplift of geosynclinal areas according to Andrews) and offered a fixist interpretation of the distribution of *Glossopteris*. *Glossopteris*, he thought, originated from a cosmopolitan Carboniferous flora that flourished in mild climates, and this ancestral flora evolved into *Glossopteris*, probably in Antarctica, as it underwent severe cooling. *Glossopteris* adapted for cold migrated northward to South America, Australia, Africa, and India. He did not say how, as terrestrial plants, they migrated across wide oceans.

Having criticized Argand's synthesis, he offered his own, describing Europe's structural history along the lines he had proposed for Australia, welding of ancient nuclei and accretion of marginal geosynclines.

The pre-Palaeozoic and later tectonic history of Europe . . . throws light . . . onto the problem. The story commences with the formation of a great borderland on the north and of a vast nucleus – Baltica, with the Russian Platform foundation and its great western extension into Britain – around which lay a long geosyncline on the north, north-west, and south-west, while the extensive borderland lay beyond it towards the Polar regions, and to the west of the British Isles. The complex Caledonian mountain system grew out of this mobile area and heavy thrustings were delivered both southwards and northwards. This activity resulted in the welding together of the ancient nucleus and borderland, a fragment of which appears to outcrop in the north-west of the British Isles. Although this grand effort at stabilization had occurred on the northern margin of the European nucleus, nevertheless mobile geanticlinal tracts still surrounded it on the east . . . and on the south . . . Then followed the close folding of the sediments of these geosynclines . . . welding together of the ancient blocks of western Siberia and of the Russian Platform . . . Later again, the Alpine foldings marked the addition of still another zone of stability to southern Europe. In this movement Spain and France were welded, or knit, together; the Atlas Mountains also marked a stable addition to Africa, with overfolding and thrusting southward.

(Andrews, 1938: xxx)

Regionalist that he was, Andrews simply applied his account of Australia's geology to that of Europe.

W. H. Bryan (1891–1966) was another staunch fixist. The first honors graduate (1914) in geology and mineralogy at the University of Queensland, he did a year of post-graduate work at the University of Cambridge, and returned to the University of Queensland where he became head of the Geology Department and remained until retirement in 1959. Bryan was President, Royal Society of Queensland (1924), and President, Geological Section of the ANZAAS (1946) (Hill, 1966).

Bryan discussed past and present structural relationships of the Australian continent to the Pacific Ocean in his 1944 Clarke Memorial Lecture to the Royal Society of New South Wales. Although he did not mention mobilism explicitly, his views precluded it. In step with other Australian structural geologists, Bryan believed that the Australian continent had grown mostly eastward from the Precambrian shield. He argued that Australia was originally contiguous with Asia, and extended eastward to the Marshall Line, which he thought marked the southwestern structural boundary of the Pacific basin. He maintained that the continent had completed its growth by Paleozoic times with the addition of the Tasman geosyncline and the large borderland, Tasmantis, east of the geosyncline. So the continent remained until early Tertiary times, when much of Tasmantis foundered, forming seas between Australia and Asia, and other unfoundered parts such as New Zealand.

E. Sherbon Hills (1906–86), probably Australia's most widely known and influential geologist during the 1940s and 1950s (Hill, 1987: 316), was strongly opposed to mobilism. He worked in several specialties including vertebrate paleontology with an emphasis on fossil fishes, stratigraphy, petrology, physiography, and structural geology. Hills was born near Melbourne and obtained his B.Sc. (1928), M.Sc. (1929), and D.Sc. (1938) from the University of Melbourne. After obtaining his M.Sc. he studied at Imperial College of Science and Technology, London, where he was awarded a Ph.D. in 1931. Returning to the University of Melbourne, he was lecturer in geology (1932–8), acting senior lecturer (1938–40), senior lecturer (1940–2), associate professor of geology (1942–3), and professor of geology and mineralogy (1944–71). Hills also became Dean of Science (1947–8), chairman of the Professorial Board and Pro-Vice-Chancellor (1959–62), and Deputy Vice-Chancellor (1962–71) of the University of Melbourne. He retired in 1971.

Recognized beyond Australia, Hills was elected to the Royal Society of London (1954). He received the Geological Society of London's Wollaston Fund Award (1942) and its Bigsby Medal (1951), and the Lomonosov Medal from the University of Moscow (1968). In his own country he was awarded the David Syme Prize from the University of Melbourne (1939), and the W. R. Browne Medal from the Geological Society of Australia (1979). He was Chairman of the Interim Council and then Foundation President of the Geological Society of Australia (1952–5), a Foundation Fellow of the Australian Academy of Science (1954), President of Section P (Geography) of the ANZAAS (1947), and President of the Royal Society of Victoria (1955–66).

Echoing what by now had become the generally accepted Australian view, Hills described its structural history in terms of the knitting together of several ancient nuclei to form the Australian shield. His endorsement gave Australian fixism its final form that remained unchallenged until the rise of paleomagnetism in the 1950s. He did this in his Clarke Memorial Lecture to the Royal Society of New South Wales in 1945, a year after Bryan's lecture.

In recognizing the existence of stable nuclei and relatively mobile zones in the Australian Shield, I have reached, though by a different route, similar conclusions to Bryan, Cotton and Andrews. Andrews regards the old nuclei as having been welded together by the compression of Paleozoic geosynclines; I would suggest, however, that the welding took place much earlier, although leaving zones of weakness between the nuclei and around the margins of the continent, and that the structure as a whole is perhaps but little more stable now than it was when it was formed.

(Hills, 1945: 90)

Once Australia's shield formed, the continent increased its size with the welding together of its shield and the accretion of the eastern Tasman geosyncline. Possible drift of Australia had no place in Hills' account.

Although Hills rarely discussed mobilism publicly, in 1955 he explicitly rejected King's view that Australia had formed from the breakup of Gondwana. According to Hills, King's mobilism was "contradicted by local geological evidence." Australia had formed by successive continental accretion.

King called this Mesozoic surface layer "Gondwana surface"; thereby assuming paleographical relationships between the continents which are not relevant (meaningless) in respect to the actual morphological problems of Australia. Furthermore, King's assertion that a unified [old layer] was divided by a single tectonic event, the beginning of the break up of Gondwanaland, is contradicted by local geological evidence. This evidence would dictate that all local rising and sinking that became [sediment basins] through filling in all parts of Australia would have the same geological history and the same structure. This is not the case, and we must consider the morphological perspective of Penck, which derives more from the facts than from generally accepted concepts.

(Hills, 1955: 201; translation from the German by Deborah Dysart-Gale; my bracketed additions)

Three years later Hills had another opportunity to criticize mobilism. Asked to contribute a paper to a memorial volume honoring D. M. S. Watson, who directed his work on fossil fishes while working in London on his Ph.D., he presented a review of Australian fossil vertebrates. Hills (1958: 86) began his essay by noting that Australian fossil vertebrates "have an important bearing on major problems of geology, notably on palaeogeographic reconstructions and in turn on tectonic theories, especially that of continental drift." He argued that Australia had not been part of Gondwana. He inferred from Australian Paleozoic land vertebrates:

There is . . . very little indication of land links between that continent [Australia] and Africa or South America. Rather, throughout this long period, are the continental faunal similarities with Northern Hemisphere, such as might be established even today by the obliteration of Wallace's Line, and some readjustment of climatic zones. Again the deficiencies in the Australian fauna, notably in the reptilia and amphibia of the Carboniferous and lower Permian, point to the existence of important barriers to migration from Africa.

(Hills, 1958: 93; my bracketed addition)

Marine Paleozoic faunas also indicated that Australia had not been part of Gondwana during the Paleozoic.

The new evidence for the occurrence of marine conditions as far south as latitude 25° S. in the Westralian geosyncline during Devonian and Carboniferous . . . and of Ordovician seas to latitude 19° S . . . is in line with evidence from the vertebrate faunas of the continued existence of the Australian continent as an entity, rather than as part of a larger Gondwana landmass, throughout Palaeozoic time.

(Hills, 1958: 93)

He argued that Australia's Mesozoic vertebrates gave little indication

of direct connexions with South Africa as part of a Gondwana continent, and . . . that they offer no support to the notion of faunistic distinction between the elements of a presumed Gondwanaland and the rest of the world.

(Hills, 1958: 97–98)

Although Hills knew about the very new support for mobilism arising from paleomagnetism, he said nothing about it.[2]

Hills did not acknowledge the success of mobilism in print, I understand, until 1970 when he added a section on seafloor spreading and plate tectonics to the second edition of his *Outlines of Structural Geology*, and announced in its Preface, written in 1970:

It was not long since, in an earlier work, it was apposite to refer to gravitational tectonics and flowage as a burgeoning branch of structural geology; but it must now be said that rigid-plate tectonics and ocean-floor spreading are even more remarkable concepts, which must have fundamental effects on regional structural geology and on our understanding of global phenomena.

(Hills, 1972: vii)

Indeed, Dorothy Hill (§9.4) was told that Hills never entirely accepted plate tectonics.

Hills never came to full acceptance of the plate tectonics concept, as he could not envisage the nature of the energy that would satisfy the requirements to move such huge masses as the continental slabs (Cozens, 1986). Cozens found this attitude puzzling, believing that plate tectonics offered the mechanism for the formation of many of the regional structures Hills described.

(Hill, 1987: 303)

I close this discussion of geologists who spent most of their careers studying Australian geology by returning to David through William Rowan Browne (1884–1975). Unlike his mentor, Browne argued against mobilism. Born in Ireland, the sixth of eight children, he entered Trinity College Dublin in 1903 where he planned to read classics (Vallance, 1978). But he got tuberculosis. At the time, British physicians often prescribed long sea voyages for those suffering from tuberculosis. Browne set sail for Australia alone in 1904. Two years later he was well enough to attend the University of Sydney, where he encountered Edgeworth David. Browne proceeded to win the University Prize for Geology and the department's prize for fieldwork. He obtained his B.Sc. in 1910 with a first in mathematics and geology. Except for two one-year stays at the University of Adelaide, one in its observatory (1910) and another in its geology department (1912), he remained a member of the Department of Geology at the University of Sydney, where David had in 1911 secured him a position.[3] Browne received his D.Sc. from the University of Sydney in 1922 for his petrological study of the Willyama complex, Broken Hill, New South Wales. He retired in 1949, but remained active, undertaking annual fieldtrips to Mt. Kosciusko, New South Wales, to study its Pleistocene glaciation.

Browne was primarily a petrologist, but his research encompassed stratigraphy, structural geology, glaciology, and economic geology. He was much honored for his work, receiving the W. B. Clarke Medal of the Royal Society of New South Wales (1942), The Royal Society of New South Wales Medal (1956), and the Mueller Medal, ANZAAS (1960). He was elected a foundation fellow of the Australian Academy of Science (1954). A founding member of the Geographical Society of New South Wales, he twice served as its president (1929–30, and 1948–9). He also was a founding member of the Geological Society of Australia, and its president from 1955 to 1956. Soon after his death the Geological Society of Australia established the W. R. Browne Medal for distinguished service to Australian geology.

His research accomplishments were substantial, but they were published in Australian journals and not well known outside. He is probably best remembered for completing David's *The Geology of the Commonwealth of Australia*. David asked Browne in March 1934 to finish the work. Loyal to his former teacher, he spent the next fifteen years doing so, adding all new work. Much of these three volumes must surely have been written by Browne. The work first appeared in July 1950, a year after Browne retired. Refusing to take co-authorship, he acknowledged himself as an editor, admitting to having "much supplemented" David's work. He returned to the UK "to see the book through the press," almost half a century after leaving as a very sick young man. As he explained "an early fright teaches one to take proper care."

He lived nine decades, did not drink, did not smoke, and never lost his Ulster brogue (Irving, August, 2010 personal communication to author).

Mobilism is mentioned only twice in this large work, first in its summary of Australia's Permian System. Although Australia and Gondwana's other fragments had been connected because of their similar Gondwana flora, marine faunas, and widespread glaciation, it rejected mobilism. The books argued that the absence of Permian terrestrial animals in Australia, so prevalent in South Africa and South America, created a difficulty for mobilism. Following Teichert, also a staunch fixist (§9.4), it proposed that shallow seas had separated Australia from the other continents, which would allow for migration of *Glossopteris* flora and marine fauna but prevent dispersal of land animals.

Whether or not one accepts the concept propounded by Wegener and supported by du Toit and others that India, Australia, Antarctica, Africa and South America were formerly joined together in one compact land-mass, there can be no question that some connexion existed in Permian time between these now widely separated lands. Such connexion may be inferred from the existence in them of the Gondwana flora, from the striking similarities of their marine faunas, and from the vestiges of a great glaciation that are common to them. The absence from Australia of all traces of the Permian land-animals that flourished so abundantly in South Africa and South America may, as Teichert has suggested, be indicative of a shallow separating sea extensive enough to act as a barrier to their passage but not sufficient to prevent the migration of marine invertebrates or the spread of the *Glossopteris* flora through the medium of winged seeds or spores.

(David, 1950: 398)

The second mention of mobilism is in the review of Australia's tectonic history. The account is fixist. Marsupials migrated from Asia, and Asia and Australia had not changed their positions relative to one another.

While in the strict geographical sense Australia is an isolated land-mass, inspection of a bathymetrical map will show that it stands upon the same submerged shelf or platform with New Guinea and Tasmania, and that the combined continental mass is separate and distinct from the continent of Asia. About the mutual relations of these two continents during geological time much difference of opinion exists, but there are certain grounds on which former land-connexions between them have been claimed, such as the presence in the Upper Permian and Triassic terrestrial bed of Australia of fossilized remains for reptiles which must have come from the Northern Hemisphere. There is reason to believe, as shown below, that for at least part of Palaeozoic time geosynclinal communications existed between Australia and New Guinea on the one hand and Asia on the other, and it is probable that land-bridges were formed when the sea was temporarily driven out by orogeny or emergence. It is usually considered that final severance from Asia took place not long after the close of the Mesozoic Era. The marsupials must have entered Australia overland from the north-west long before their phenomenal development in the Pleistocene, and their survival was due to the absence of the carnivorous placental mammals which in other continents dominated the land-fauna from Eocene time onwards. The testimony of the land-plants is in harmony with the view that overland migration into Australia was impossible after the Cretaceous Period.

On the other hand, exponents of the hypothesis of Continental Drift consider that the present contiguity of south-east Asia and Australia has only been attained since Meso-zoic time.

(David, 1950: 686)

The book rejected the mobilist idea that Australia had collided with the Dutch East Indies; even if mobilism explained many difficult geological problems, it did not explain the geological relations among Australia, New Guinea, the East Indies, and New Zealand.

On the hypothesis that Australia drifted northward in Pleistocene time and collided with the Dutch East Indies both Wegener and du Toit have attempted to explain the distribution of land and sea and the present relations of Australia to New Guinea, the East Indies and New Zealand. While one may subscribe to the general conception of large-scale horizontal move-ments of the earth's crust, and be attracted by the fascinating hypothesis of continental drift, which cuts the Gordian knot of so many difficult geological problems, it must be said that, as they stand, both of the explanations referred to, which incidentally differ widely from one another in detail, fail to interpret convincingly the known geological facts. On the other hand, the distribution and affinities of the marine fossil faunas seem to be adequately and simply explained by supposing the countries in question to have stood in substantially their present relations to each other since at least Middle Palaeozoic times.

(David, 1950: 688–689)

The sentiments expressed in David (1950) and almost certainly written by Browne were that although mobilism may seem attractive, it failed to explain much of the geology of Australia and surrounding regions, and what mobilism could explain, fixism could also. This critical, even cold attitude to continental drift contrasts with the open and warm way that it was initially welcomed by David himself in the 1920s (§9.2). These passages from the *Geology of the Commonwealth of Australia* (David, 1950) were surely written by Browne, who was reflecting not only the mood in Australia during the 1940s but also the increasing intolerance generally of mobilism during the 1930s and 1940s and before the rise of paleomagnetism. Carey often remarked that Browne's David was a betrayal of David in regard to continental drift (Irving, August 2010 personal communication to author). Carey had a point. Browne should have openly explained why his statements under the nominal authorship of David differed from the true David.

Browne was uneasy with the new paleomagnetic support for mobilism. During the middle 1950s, Browne greatly helped Irving and his student Ron Green find suitable rocks for paleomagnetic investigation giving them while in his mid-seventies con-ducted tours of key outcrops (II, §5.3). Irving found his help vital to the success of their work. Nevertheless, Browne was not convinced by their results. Indeed, in 1955 he refereed and rejected Irving's first pro-mobilist paper when submitted to the *Journal of the Geological Society of Australia*, and later published as Irving (1956) (Irving, 2004 email to author).

9.4 Paleontologists working in Australia reject mobilism

Leading paleontologists rejected mobilism, especially those whose interests and research extended beyond paleontology to other fields. Curt Teichert, A. A. Öpik, and Martin F. Glaessner were prominent members of this group. All were trained in Europe and emigrated to Australia where they spent much of their careers. This European triumvirate did much to advance paleontology in Australia. Teichert arrived in 1937, Öpik in 1948, and Glaessner in 1950. I shall first consider their views, and then those of Dorothy Hill, the most prominent Australian-born paleontologist during the middle decades of the twentieth century.

Teichert (1905–96) was born in Königsberg, East Prussia. Ironically, Teichert decided to study geology partly because he read Wegener's *Origin* (Crick and Stanley, 1997). He accepted a teaching position at the University of Freiburg in 1927, where he remained for one year. He obtained his Ph.D. (1928) from the University of Königsberg. Awarded a Rockefeller Foundation award (1930), he undertook paleontological studies in Washington, DC, New York City, and Albany, New York. Soon recognized as an expert on cephalopods, he joined a Danish expedition to Greenland in 1931–2, led by Lauge Koch (Wegmann went on the expedition (§8.9)). Teichert spent the next year in Germany working on his Greenland fossils. With the rise of Nazism, Teichert and his Jewish wife left Germany, and spent the next four years in Copenhagen. With the help of the Rockefeller Foundation, which aided displaced German scientists with transportation and two years support at any academic institution that would provide a position, Teichert and his wife emigrated to Perth. There he joined the Geology Department of the University of Western Australia (1937), becoming one of only six full-time professional paleontologists in Australia, and the only one in Western Australia (Crick and Stanley, 1997: 751). Teichert remained at the University of Western Australia for seven years, writing about eighty papers on fossils ranging in age from Ordovician to Eocene. He discussed issues in biostratigraphy, paleoclimatology, neotectonics, and Pleistocene reefs (Crick and Stanley, 1997: 751). Employed by the Geological Survey of Victoria and the University of Melbourne, he continued working in Australia until 1952, when he moved to the United States to work for the USGS. In 1964 he was offered a distinguished professorship at the University of Kansas, where he remained until 1975 when obliged to retire at age seventy. Becoming an adjunct professor at the University of Rochester, he managed to write fifty papers from 1977 to 1987. Teichert received the David Syme Prize from the University of Melbourne (1949) for his work in Australia, the Raymond C. Moore Medal from the Society of Economic Paleontologists and Mineralogists (1982) for excellence in paleontology, and its Paleontological Society Medal (1984).

From the 1940s through the 1970s he appealed to Australia's geology to argue against mobilism. There was, he said (1941), no need to invoke mobilism to account for the upper Paleozoic glaciation. He claimed:

The Permian glaciation of Australia was a single major event with the strongest refrigeration during the Artinskian [latter part of Early Permian]. Rich marine faunas arrived in Australia after the climax of the glaciation had been passed. The upper Paleozoic strata of Western Australia was deposited in a geosynclinal trough which was marginal to the pre-Cambrian shield and continuous with the Timor geosyncline of the East Indies. The relative geographical position of Australia and the East Indies in late Paleozoic time must have been very much the same as at present.

(Teichert, 1941: 371; my bracketed addition)

As for *Glossopteris* flora, the discovery by Virkki (1937, 1939) of winged "pollen-grains" associated with *Glossopteris* remains made it possible that long-distance wind transportation carried them between the Gondwana continents. This was the explanation that Dunbar offered at the 1949 New York meeting (§3.8). Given Caster's comment that Teichert had helped Dunbar prepare his published response to Caster (§3.8), Teichert possibly told Dunbar about Virkki's work.

In order to explain the existence of the *Glossopteris* flora in Australia it is . . . not necessary to assume the existence of any direct land connection between Western Australia and any land at the west. The whole problem of the distribution of this flora has been somewhat simplified since Virkki's discovery of winged "pollen-grains" of *Glossopteris* in the basal tillite series of India and Australia (1937, 1939).

(Teichert, 1941: 411; references to Virkki agree with mine)

Teichert admitted that it would not be easy for even "winged" spores to travel between Australia and India, and acknowledged that fragments of Gondwana must have been closer to Australia than now, but he had no patience with mobilism.

Attempts at a reconstruction of Gondwana in which Western Australia is placed alongside Peninsular India can be dismissed . . . No evidence is found in Western Australia in Tertiary time to support . . . Du Toit's supposed "northerly creep" of Australia in Tertiary time.

(Teichert, 1941: 412)

It is no wonder that Dunbar welcomed Teichert's views at the 1949 New York Meeting – three of Dunbar's thirteen references were to him. I suspect that Marshall Kay also appreciated Teichert's contribution, because Teichert's personal copy of the proceedings of the meeting was given to him by Marshall Kay, who wrote on the title page "Curt Teichert with the regards of Marshall Kay."[4]

Teichert continued to reject mobilism in 1952 in his presentation before the 19th International Geological Congress. Drawing on his many years of work on the Northwest Basin of Western Australia, he continued to argue that the Indian Ocean had existed at least since the Precambrian, and that Australia had not been directly connected to the west Gondwana (Teichert, 1952: 134).

In 1959, Teichert again argued that Australia had been bordered to the west by an ocean since the early Paleozoic, and had been an isolated continent since at least the Permian.

Along the western margin of the Australian continent there exist four major sedimentary basins, filled with predominantly marine rocks from Cambrian to Tertiary age, and up to 40 000 feet thick. Seaward these basins continue into depressions recognizable in the continental shelf and even the continental slope. Their very presence, the nature of their sediments and the composition and relationships of their fossil faunas indicate the existence of an open ocean to the west of Australia since early Paleozoic time. Composition of the Australian fossil land vertebrate faunas suggests isolation of the Australian continent since at least Permian time.

(Teichert, 1959: 562)

Teichert disdained mobilism.

Gondwanaland is a geological problem which stands and falls with the geological and geophysical evidence. It may not be impossible, perhaps by still another rearrangement of the southern continents on the basis of continental drift, to make allowance for some of the facts here presented. The writer admits that he is not interested in such exercises. In evaluating a scientific hypothesis the first fundamental question is whether it is necessary at all, not whether it is possible to find facts which may be fitted into it. As regards Australia, it seems that its paleogeography, its structure and the history of its life can be satisfactorily explained without the Gondwanaland hypothesis and on the assumption that Australia was always a separate continent.

(Teichert, 1959: 586)

With his attitude, it is no surprise that he did not even mention Carey's work or the growing paleomagnetic support for mobilism.

Teichert's anti-mobilist attitude continued well into the 1970s, even after the success of plate tectonics, and he probably never accepted mobilism. In 1971, he teamed up with A. A. Meyerhoff, a US petroleum geologist who made a name for himself opposing mobilism during the 1960s and 1970s. They argued that if the supercontinents of Gondwana and Laurasia had existed, neither extensive glaciation nor coal formation would have occurred because both require extensive moisture, which would not have been available within the interior of either supercontinent.

Therefore, until advocates of the new global tectonics find an alternate explanation for tillite and coal distribution, the spreading sea-floor, mobile-plate, and polar-wandering hypotheses will have to be regarded as interesting speculations which are supported by only a fraction of the known geological, paleontological, and paleoclimatological data.

(Meyerhoff and Teichert, 1971: 285)

Part of this criticism was not new. A. P. Coleman had presented the same difficulty about glaciation in the 1920s and 1930s (§3.12), but du Toit had suggested inland seas as a source for the needed moisture. By then paleomagnetism and oceanography had provided ample support for mobilism. None of this deterred Meyerhoff and Teichert.

Teichert (1974) again appealed to his work on marine sedimentation in Western Australia which he thought disproved any connection with a landmass to the west. Quoting from his 1959 paper discussed above, he recounted:

In 1959 I came to the conclusion that the extensive distribution of late Paleozoic marine rocks in Western Australia disproved ". . . any ideas of a former westward extension of the Australian continent across the present Indian Ocean . . . If there was a restricted Gondwanaland further west, Australia never formed part of it."

(Teichert, 1959)

Making sure that nobody misunderstood the importance of the evidence against mobilism, Teichert (1974: 387) quoted from an article he had read in *The New Yorker Magazine*.

Some of the facts presented in this paper may be "consistent with" hypotheses of continental drift and "Gondwanaland," but very many obviously are not, and it will not be possible to ignore them in perpetuity in discussions about the new global tectonics. As has been aptly stated, ". . . it is always possible to choose among competing theories if you are willing to disregard half the evidence."

(Donald U. Wise, as quoted in The New Yorker, *April 4, 1970)*[5]

Even though by 1970 mobilism was well entrenched, Teichert remained unable to reevaluate his work in Western Australia nor to develop an alternative account consistent with it.

Armin Alexsander Öpik (1898–1983) was born in Lontova, Estonia (Shergold, 1985). He studied geology and mineralogy at the Estonian State University at Tartu. He obtained the degree of Magister Mineralogiae (1926), became reader in geology and mineralogy (1926–30), earned his Ph.D. (1928), and became professor of geology and palaeontology and Director of the Geological Institute and Museum (1929–44). With the Russian occupation of Estonia in 1944, Öpik left and lived in displaced persons' camps in Germany until 1948. Through contact with Teichert, who had known Öpik since 1926, he secured a position with the (Australian) Bureau of Mineral Resources (1948), where he continued to work for many years after compulsory retirement at age sixty-five.

Öpik was a world expert on trilobites. By 1982, he had described over half of all known Australian trilobites. Because of his work on trilobites and Cambrian strata, he received the Sixth Charles Doolittle Walcott Medal of the NAS. He was elected inaugural President of the Commonwealth and Territories Division of the Geological Society of Australia (1952), and became a Fellow of the Australian Academy of Science (1962). He was the Clarke Memorial Lecturer of the Royal Society of New South Wales (1965).

Öpik spoke out against mobilism at the 1956 Hobart symposium on continental drift (II, §6.16). His talk was not included in the published proceedings. However, his attitude and objections to mobilism are preserved in the summary of the symposium that he and M. A. Condon prepared for the Record of the Bureau of Mineral Resources, Australia. The summary contains Öpik's own account of his talk as well as his comments on some of the others.

Öpik was not adamantly opposed to mobilism. Indeed, he once wrote somewhat favorably of it.

The distribution of Cambrian, Ordovician, and Silurian faunas in America and Northern Europe does not contradict the drift-theory of A. Wegner and may, together with other evidence, testify to the correction of this theory.

(Öpik, 1939: 66; this translation from the German is from Wegmann (1948: 10))

Although he thought that neither fixism nor mobilism had been established, he had by the mid-1950s come to favor fixism just as paleomagnetic evidence to the contrary had begun to emerge. Referring to himself in the third person, he reported that he had maintained:

Wegener originally started from two facts: 1) the geographical homology of the coasts of Africa and South America, and 2) the occurrence of floras and faunas of temperate or tropical climate in the present Arctic and Antarctic latitudes. Beginning with these facts, the hypothesis of Continental Drift was, and still is, applied to explain palaeoclimatological problems. At the same time it serves to explain phenomena of inorganic geology and distribution of fossil and living organisms on the earth. On the contrary, Öpik regards such a monistic explanation of geological history, based on a single cause, as an assumption not warranted by the present state of knowledge. Mobilism (drift) and fixism (no-drift) when accepted axiomatically can be "proved" by the same set of geological facts, or selection of facts.

(Condon and Öpik, 1956: 11)

Öpik claimed that convection currents would not cause continents to drift but only to uplift and subside. He dismissed the use of paleoclimates and biogeography because the prime cause of their variability was "cosmic."

Palaeoclimatology should be excluded from consideration, because the dominant cause of climactic changes is the variability of solar radiation. Terrestrial causes are masked by the larger cosmic causes. Biological evidence, fossil and recent, is ambiguous and cannot support any hypothesis.

(Condon and Öpik, 1956: 12)

He elaborated on why appealing to biogeography in support of mobilism was fruitless (see §3.9 and §9.5 for similar assessments). After summarizing the talk, "Insects and Continental Drift," by J. W. Evans in which he argued in favor of a land connection between Australia and Antarctica (§9.4), Öpik opined:

It is perhaps of significance that biological material and interpretations of geographic distribution are not compatible with palaeontological materials and interpretations. Biology has to consider hundreds of thousands of species; interpreters work with higher categories of the system (as families, orders) which have been considered abstractions since Linneaus. The study of the distribution of such "abstractions" cannot give reliable results about actual migration and spreading . . . (A.A.O.)

(Condon and Öpik, 1956: 8)

Drawing on his worldwide knowledge of Cambrian stratigraphy through his work on trilobites, he saw no reason to support any version of mobilism that had already been proposed to explain Australia's "Palaeozoic history." Moreover, those that had, created difficulties.

The present position of Australia presents no difficulty in explaining its Palaeozoic history. In the models its positions in the Arabian Sea or on the edge of Gondwana make difficulties and even impossibilities, as for example the distribution of Acado-Baltic and Pacific Cambrian faunas . . . It may be necessary to investigate the possibility of Australia being composed of two parts (eastern and western originally geographically remote in Palaeozoic times and united by drift along the 138th meridian and forming part of Gondwana. Postulated impossibilities in the models beget more impossibilities . . .

(Condon and Öpik, 1956: 12)

The proposed fit between Australia and Antarctica raised other problems.

Another great difficulty in the models of Wegener 1924, du Toit, 1937, King, 1950 and Carey, is the fit of Australia and Antarctica in which the Adelaidian Geosyncline, which should be open is fitted against the Antarctic shield. The westward bend of the geosyncline at Kangaroo Island indicates a gap between Australia and Antarctica in the Cambrian, and evidently in later times. The Cambrian fauna of South Australia is unique and cannot be connected with the Andine Cambrian.

(Condon and Öpik, 1956: 12)

Öpik ended his talk by repeating the now standard objection that glaciation would not occur within the interior of so large a supercontinent.

A choice of models is already possible (several models by Wegener, by du Toit, by L. King, by S. W. Carey). Öpik prefers the model with disconnected Laurasia and Gondwana and a sea between them; geological events are easier to understand from an assumption of smaller rather than large land masses. A Gondwana, for example, six to ten times as large as Australia would never have had a central ice cap, because of atmospheric dynamics and absence of supply of winter precipitation. A single ice divide as shown in King's model is even more improbable.

(Condon and Öpik, 1956: 12)

Thus Öpik not only claimed that mobilism was not needed to explain Australia's Paleozoic history but also argued that applying it had strange consequences.

As far as I have been able to tell, Öpik did not address the question of mobilism in any of his many subsequent publications. This is not surprising because almost all his research was directed toward trilobites and other Cambrian marine fauna whose distribution had little direct bearing on Wegener's theory, which concerned only Late Paleozoic and later events.

Martin Fritz Glaessner (1906–89) was born in northwestern Bohemia (now part of the Czech Republic).[6] Developing an early interest in natural history, at sixteen he became a research associate of the Museum of Natural History in Vienna. Glaessner entered the University of Vienna in 1925, where he obtained two doctorates, one in law (1926), at the insistence of his parents, and a second in geology and paleontology (1931). After spending 1930–1 at the British Museum in London, he moved to Moscow where he set up a micropaleontological laboratory as a senior research officer of the USSR's Academy of Sciences. Glaessner became a worldwide expert on foraminifera and other marine microfossils. Forced to choose between Soviet citizenship or leaving the Soviet Union, he returned to Vienna at the end of 1937. Glaessner, whose father was Jewish, was arrested in 1938. Fortunately, G. M. Lees, Chief Geologist of the Anglo-Iranian Oil Company, offered Glaessner a job in New Guinea. Glaessner remained there for the next decade, working on regional geological and paleontological problems. His *Principles of Micropalaeontology* was published in 1945. Glaessner accepted a position at the University of Adelaide, where he remained until retirement in 1971. At Adelaide, Glaessner's research interests shifted to the peculiar Late Precambrian fossils in the Ediacaran Hills, near Adelaide. He spent much of his remaining career doing fundamental work on the Ediacaran fauna, now recognized worldwide. Glaessner (1962) argued that most Ediacaran organisms were ancestors of still-living animal phyla, in contrast to the view of A. Seilacher (1989) and others who claimed that most Ediacaran organisms were a failed evolutionary experiment.[7]

Like Öpik and Teichert, Glaessner became a leading figure among geologists in his adopted country. He was elected to the Australian Academy of Science (1957), was a member of its Council (1960–2), and Chair of the National Committee of Geological Sciences (1962–77). He was the Clarke Memorial Lecturer, Royal Society of New South Wales (1953); winner of the Walter Burfitt Prize, Royal Society of New South Wales (1962); the Verco Medal, Royal Society of South Australia (1970); the ANZAAS Medal (1980). Recognized outside of Australia and New Zealand, he received the Charles Doolittle Walcott Award and Medal, NAS (1982); the Lyell Medal, Geological Society of London; the Suess Medal, Geological Society of Austria (1986).

Adamantly opposed to mobilism until nearly the end of his life, Glaessner wrote little on the subject, and then only disparagingly. He wrote a paper entitled "Geotectonic position of New Guinea," which was partly based on his extensive fieldwork on the island, and which proffered a fixist account of the geological history of Australia and New Guinea similar to that of Hills (Glaessner, 1950: 867–868). He dismissed mobilism in a footnote. Referring to Carey's mobilist interpretation of New Guinea's geology, he made it quite clear where he stood.[8]

Carey (unpublished thesis, 1938) recognized the presence of sigmoidal arcs in this region and the occurrence of outward migration with inward overfolding of mobile belts. His constructions and interpretations, however, are based on the hypothesis of large-scale continental migration and differ fundamentally from those presented here.

(Glaessner, 1950: 878)

A decade later, he again dismissed mobilism at a 1959 symposium in Melbourne marking the centenary of Darwin's *Origin of Species*. Writing on the geological history of Australian fauna, he gave a fixist account, advocating migrating fauna not drifting continents. He favored an Asiatic origin for Australian marsupials (Glaessner, 1962: 248). He dismissed mobilism.

The hypothesis of continental drift which is frequently applied to the history of the Australian fauna will not be considered here. The "assemblies" of continents drawn by various adherents of this hypothesis (Wegener, du Toit, King, Carey) and even successive versions produced by individual authors, differ so widely that they are mutually exclusive. Complete sequences of assumed positions of continents in time which could be checked against known facts of the geological history of any particular area have not been produced. The outlines of unchanging areas "drifted" over the surface of the globe are in general the present outlines of continents, with the addition of shelves in the oceanographic sense of the term. At most periods during its geological history the Australian Continent would have had entirely different outlines. These lines and their fittings are therefore arbitrary. The basic hypothesis may be adjustable to fit certain facts of historical biogeography but it does not contain basic and independent factual geological evidence which would explain them.

(Glaessner, 1962: 244)

These remarks at the 1959 symposium show just how strongly he opposed mobilism. Wegener, du Toit, King, and Carey may have disagreed about precisely when and where drift had occurred; however, they agreed about the major movements, and it was unfair to imply otherwise. Moreover, mobilists developed different versions, in part, because they checked their views against an expanding body of facts concerning the geological history of particular areas. After all, if disagreement among proponents was reason for rejection, Glaessner should have rejected fixism, because fixists disagreed about many things; Australian fixists even disagreed about when and where Australia grew.

McGowran, once a student of Glaessner, referring to the above passage, discussed Glaessner's strong resistance to mobilism. McGowran also mentioned Glaessner's dismissal of mobilism's paleomagnetic support, and compared Glaessner with the fixists G. G. Simpson, Teichert, and Maurice Ewing – McGowran knew and was influenced by Teichert and Simpson.

Glaessner was an outright sceptic. He had no need for that hypothesis. What he said about the geological advocates of mobilism as in the Leeper volume [Glaessner (1962)] – that they had mutually conflicting patterns to be ignored until they came up with a cogent story – he said more forcefully about the palaeomagnetists in the late 1950s at a couple of seminars that I attended, as well as in conversation. He was not impressed by Irving or Runcorn *et al.* until their polar wandering patterns became cogent – not least consistent. Glaessner himself had a curious (to me at the time and still!) similarity to two others in my own pantheon, Simpson and Teichert – immensely knowledgeable about the hard data of palaeobiology, aware of the big picture, yet cautious to the point of being reactionary about these big ideas.

(McGowran, January 25, 2003 email to author; my bracketed addition)

Indeed, Glaessner and Teichert were more rigidly opposed to mobilism than either Simpson or Ewing. As will be later shown (III, §1.12), Simpson took notice of paleomagnetism's support of mobilism, and both he and Ewing eventually accepted mobilism (IV, §1.14). Moreover, Creer, Irving, Runcorn, and other paleomagnetists changed the apparent polar wander curves they developed for different landmasses as new data appeared, which is exactly what they should have done. By 1959 they had developed a cogent story, but Glaessner did not recognize, took no notice, or simply did not accept their story.

He certainly took notice of some new work. Referring to Hills' 1958 paper on Australian vertebrate fauna (see above), Glaessner agreed with his fellow fixist.

The progress of geological exploration not only in Australia but also in Timor, New Guinea, New Caledonia and New Zealand has made it possible to establish the fact of Australia's isolation at least since Permian time, with a short-lived continental link at the end of the Cretaceous. This explains not only the presence of marsupials but also the extraordinary scarcity of late Palaeozoic and Mesozoic land tetrapods which contrast strongly with other southern continents, and on which Hills (1958) has recently commented.

(Glaessner, 1962: 248)

He also reported on new findings from Antarctica, arguing that they were an obstacle for mobilism.

Nothing has been said about possible connections with Antarctica, not because the question is considered unimportant but because very little factual geological evidence bearing on it is available. Apart from possible structural connections between New Zealand and the folded mountain zone of East Antarctica, the similarity between the rocks of the West Australian Shield and the old rocks of West Antarctica lying to the south of it has been considered as indicating a former unity of the two continents. The first radioactive datings of Antarctic rocks published in Russia tend to raise serious doubts about this unity. The old rocks, though similar, seem to differ very significantly in age. Furthermore, recent geotectonic compilations place a mobile zone, probably of Palaeozoic age between the two continents. The floors of the Indian Ocean and of the Tasman Sea are now being studied with modern methods of marine geology and geophysics and important data on possible former land areas in these regions are to be expected.

(Glaessner, 1962: 248)

He was keeping up with the new work on Australia brought about by surveys carried out by IGY. But he apparently ignored or was unaware of Hamilton's (1960) establishing a connection between the newly discovered Transantarctic Mountains and the Adelaide geosyncline (§7.7).

Glaessner clearly thought that new findings supported fixism and was confident they would continue to do so.

While the factual evidence is being assembled gradually and laboriously, but with increasing speed and reliability, the historical biogeography of this region should be studied mainly from the viewpoint of island links and oceanic dispersal rather than with the aid of *ad hoc* assumptions of vertical and horizontal movements of continents.

(Glaessner, 1962: 248)

Apparently, paleomagnetism and the new work in Antarctica were not to be relied on.

Glaessner eventually accepted mobilism, but his acceptance was late. McGowran recounted Glaessner's (and his own) conversion.

As to Glaessner's "conversion": I have to confess that this is entirely "Personal Communication" – I had hundreds of conversations with him, from brash student in 1955 to driving him home after a Peter Vail lecture shortly before his death in 1989. In the 1960s he paid close attention to Bill Menard's Marine geology of the Pacific and to the ocean-floor revolution generally (with Heezen and Ewing on opposite sides). He was unimpressed when I reported Tuzo Wilson's paper at the Hobart ANZAAS in 1965 – the first report on transform faults, I think. But he was paying attention when the new Adelaide Professor, Royce Rutland arrived in 1966 and brought modern tectonics including a pretty open mind on drift. By 1968 – I was lucky enough to spend a semester in Princeton itself – Hess, Morgan, Vine, McKenzie, Fischer and all at that white-hot time – and I told him all about it as a somewhat tardy but enthusiastic convert myself – by 1968 Martin was totally convinced that there was a mechanism. After all, he had been discussing and writing tectonics since the late 1920s and knew enough about the Alpine belts to see how they could fit the new model.

(McGowran, January 23, 2003 email to author)

McGowran's remark also suggests that Glaessner erroneously thought that plate tectonics offered a mechanism for continental drift.

This triumvirate of European geologists, who made major contributions to Australian and world paleontology, became conservative mandarins within the community that adopted them. They all argued that Australia's geology and its history ruled against mobilism.

Dorothy Hill, a specialist on fossil corals, was Australia's most respected Australian-born paleontologist during the 1950s and 1960s. Hill (1907–97) graduated in geology from the University of Queensland in 1928 and obtained her Ph.D. at the University of Cambridge (1932). In 1932 she returned to the University of Queensland, where she remained until retirement in 1972. She received the Lyell Medal from the Geological Society of London (1964), The Clarke Medal from the Royal Society of New South Wales (1966), served as President, Geological Society of Australia (1973–5), President of the Australian Academy of Science (1970), and was the first Australian woman elected to the Royal Society of London (1965) (Campbell and Jell, 1999).

Although Hill applied her work on corals only occasionally to mobilism, it is clear that she favored fixism. She wrote three papers discussing mobilism. Two (1957 and 1959b) were on coral distributions during the upper Paleozoic and Silurian. In her other paper (1959a, but completed in March 1957) she constructed world maps showing the amount of the inundation of continents by the sea during the Sakmarian (Early Permian), basing her conclusions on paleontological and stratigraphic data.

There are two ways to explain the presence of corals in latitudes that now have climates too cold for them to flourish: either move the landmass closer to the equator or broaden the warm and temperate climatic zones. The former may happen because of polar wandering or continental drift, and the second by global warming or increase in the poleward circulation of warm ocean currents. In her 1957 Presidential Address to the geology section of the ANZAAS on the distribution of Devonian coral reefs she dismissed the first.

If we argue, as indeed seems reasonable, that such associations [of calcareous sedimentation with vertically extended masses known as bioherms and with laterally extensive masses known as biostromes] indicated then, as now, shallow waters and tropical or subtropical seas, we can consider that in the Middle Devonian warm climates extended further polewards than now, i.e. that the earth's surface was warmer. In early Upper Devonian times, their extension towards the pole was ever greater - 70°N in both Canada and Nova Zembla. Since no upper Middle or early Middle Devonian marine sediments are known in the southern continents south of 32°S, we cannot hold the absence of colonial corals and biostromes and bioherms there to support any hypothesis of polar wandering; and it seems to me from the radial arrangement of the northern limits of the corals around the north pole, that we have no occasion to consider that either the poles or the continents were differently arranged then.

(Hill, 1957: 57; my bracketed addition)

In her 1959 paper on Silurian corals, she again preferred warm Northern Hemisphere, high latitude Silurian oceans to explain abundant reefs as far north as 70°, and she did not invoke polar wander or continental drift.

After a review of the Silurian coral faunas of the different continents . . . it is concluded that . . . development of rich coral-stromatoporoid-algal reefs in northern latitudes between 50° and 70° N. suggests to the author that at least the northern Silurian oceans were warmer than those of today.

(Hill, 1959b: 151)

Referring to work of Ma, a Chinese expert on corals, who argued for continental drift, she explicitly declared her preference for warming oceans over mobilism.

Ma (1956) to explain this distribution invoked sudden total displacements of the solid earth shell and accompanying drift of continents. But it seems to me that a rise in sea temperatures is a more likely explanation.

(Hill, 1959b: 164)

Expanding her scope beyond corals to a wide variety of stratigraphical and paleontological evidence, Hill (1959a) constructed maps of the continents showing the extent of their inundation by sea in the Sakmarian (Early Permian). Because during this Early Permian period there was glaciation and *Glossopteris* flora in the Southern Hemisphere and peninsular India, the question of mobilism naturally arose. She offered a fixist interpretation, but she thought it was not without its difficulties. She (1959a: 622) again argued that Northern Hemisphere Sakmarian seas were very much enlarged and its lands correspondingly reduced; Southern Hemisphere Sakmarian lands were slightly reduced. Evaluating evidence for glaciation, she argued that alleged glaciation in North America, the Squantum strata and conglomerates in Kansas, was probably not glacial, that the glaciation in Southern Rhodesia (now Zimbabwe) and the Belgian Congo (now the Democratic Republic of the Congo and the Republic of the Congo) was in upland valleys and not continental, and that the Sakmarian glaciation in India was not continental but alpine (1959a: 604, 611, and 600). With this reduction of continental low-latitude glaciation in regions near the equator and in the Northern Hemisphere, she (1959a: 622) concluded that continental glaciation was restricted to the Southern Hemisphere in a belt between 40° S and 20° S. In this way she dismissed the paleoclimatic argument for continental drift. Hill (1959a: 622) concluded, "The roughly concentric development of the glacials about the present South Pole suggests that they form a climatic zone about a Sakmarian Pole approximately in today's position." According to Hill, fixism could explain the unusual climate during the Sakmarian. However, Hill (1959a: 622) admitted, "the present evidence gives no clear-cut support for either fixity or mobility of poles or continents." Nonetheless, as I shall later show, her commitment to fixism was so strong that she was unwilling to accept mobilism even as late as 1970 (III, §1.18).

9.5 Biologists working in Australia disagree about mobilism

Mobilism enjoyed support among a few researchers interested in the geography of Australian biota; their training was primarily or exclusively in biology rather than geology. Wegener offered biogeographers interested in the origin of Australian fauna an explanation of its tri-partite division (§3.3). The oldest group showed similarities with the fauna of India, Madagascar, and Africa; similarities that arose Wegener thought when they were united in Gondwana. The second group contained marsupials, and other animals similar to those found in South America; it evolved at the time after India, Madagascar, and Africa had separated from Antarctica leaving South America and Australia still attached to Antarctica. The most recent group, made up of mammals such as humans, bats, rats, and the dingo, came from Southeast Asia during the Quaternary once Australia and New Guinea had arrived at their present positions. At first, some Australian biogeographers favored mobilism to fixism, but by the 1940s, fixism became more popular, and remained so until after the rise of paleomagnetism in the late 1950s. Some biologists in the 1950s favored a southern entry of Australian marsupials, but did not explicitly opt for mobilism.

I begin with Launcelot Harrison (1880–1928). He was the first and most vehement mobilist among all Australian biogeographers before the advent of plate tectonics, and he wrote engagingly about his work. An eldest child, he was born in Wellington, New South Wales. An excellent student, he could not afford to continue his education. After working in insurance for ten years, he entered the University of Sydney, encountered David, and won prizes in zoology, botany, and paleontology. Earning a B.Sc. with distinction in 1913, he won a scholarship to continue studies at Emmanuel College, Cambridge University, where he graduated in 1916 with a B.A. (Research). Sent by the War Office to Mesopotamia as advisory entomologist to the British forces during World War I, he devised procedures for preventing the spread of insect-carried diseases to the troops. Unfortunately, Harrison contracted typhus and a severe form of malaria, which eventually led to his death at age forty-seven. Harrison was appointed lecturer and demonstrator in zoology at the University of Sydney (1918), professor of zoology (1920), and succeeded to the Challis chair in zoology (1922). He was President of the Royal

Zoological Society of New South Wales (1923–4) and President of the Linnean Society of New South Wales in 1927. Nominated to continue as president of the society, he died before its next annual meeting (Anonymous, 1928a).

Harrison was a biologist, not a geologist. Primarily an entomologist, he had broad interests in zoology, evolution, and biogeography. He championed the idea that parasites could help to clarify phylogenetic relationships among hosts. He also was an expert in Australian frogs, and had a great working knowledge of New South Wales' biota. It was his interest in biogeography that led him to Wegener.

Harrison's contributions to biogeography and continental drift are described in three presidential addresses to scientific societies in Australia. In the first of these, he addressed the origin of Australian fauna in his Presidential Address before the Royal Zoological Society of New South Wales (1924). He disagreed with Matthew's 1915 view (§3.6) that Australia's fauna originated in the Northern Hemisphere, and argued that Matthew's failure to address Gondwana and its flora was a serious omission.

Matthew dismisses Gondwana Land as being outside the scope of his discussion, but to leave this question of Antarctic connections without discussing the Mesozoic floral distribution is to offer Hamlet without the prince.

(Harrison, 1924: 252)

Both animals and plants had to be considered; biogeography was about biota, not about fauna or flora in isolation, because the distribution of one affected the other.

Harrison argued that marsupials had reached Australia from Antarctica across a landbridge. Admitting that the similarities between South American and Australian marsupials had been exaggerated, he still thought they were sufficient to conclude that Australia's marsupials had come from there. Acknowledging that no marsupial fossils had been found in Antarctica, he countered by noting that none were known from Asia. Turning to Gondwana flora, he argued that Antarctica possessed a rainforest flora through the Jurassic into the Tertiary, and that Gondwana's other fragments shared many floras from the Late Paleozoic through the Tertiary. If there was a rich flora there likely also was a rich fauna. He proposed:

If Antarctica possessed a rain forest flora through the length of time indicated, from the Jurassic into the Tertiary, it seems equally certain that it must have possessed a fauna. Some happy chance may in the future throw some palaeontological light upon the nature of this fauna, but it seems most likely that the mantle of ice and snow which covers the continent will keep the matter a mystery for us. But when we know quite definitely that certain kinds of plants occurred there, which are still found in South America, in New Zealand, and in Australia, and when we see that many kinds of animals are common to these three regions, and afford no evidence whatsoever of northern origin, despite the fact that we have not found them in Antarctica, it seems to me that we are well justified in assuming their existence there in times past.

(Harrison, 1924: 253)

He also thought that the nature of many Australian invertebrates supported land-bridges between Australia and South America *via* Antarctica.

I have not time to traverse the large body of evidence relating to the invertebrates, but in almost every group there are members which can most plausibly be accounted for by a southern origin. Benham (1909) has put forward a particularly good case for the Oligochaetes [earthworms], which also holds to some extent for groups like the land planarians and land nemerteans [ribbon worms], all creatures for which it is difficult to imagine a drift transportation. Australian systematic zoology teems with forms exhibiting South American affinities. Curiously enough, many of these forms find their nearest relatives in Chili [Chile], but this is probably due to the similar nature of the physical environment, and does not necessarily argue for a direct trans-Pacific connection.

(Harrison, 1924: 253; my bracketed additions)

Placing the origin of Australian marsupials within the framework of Australian biogeography, he attempted to answer a difficulty facing his solution – the absence of marsupials in New Zealand and Africa. He suggested that Antarctica's

connection with New Zealand post-dated that with Australia, and may have occurred after our supposed Antarctic marsupials had been killed off by increasing cold. But the data available at present are not sufficient to allow of putting a period to any of these hypothetical land connections. All we can say is that Africa and Australia, common possessors of a Proteacean flora [a Gondwana flora that originated in the Cretaceous], would seem at some time to have been connected, probably through the south; that there is some evidence for a separate connection with Madagascar; and also that Australia, New Zealand, and South America would appear to have had connections with Antarctica, though not contemporaneously.

(Harrison, 1924: 253; my bracketed addition)

Appealing to his previous work on parasite–host relations, he buttressed his argument about the closeness between Australian and South American marsupials by observing that trematode (flatworm) and Mallophagan (lice) parasites from Australian and South American marsupials were very closely related. He furthered the need for a connection between Australia and South America through Antarctica by appealing to the distribution of crayfish and frogs, two groups that he had been studying.

He was, however, bothered by the inconsistency between his landbridge solution and isostasy. Harrison (1924: 249) responded by suggesting that "no great degree of elevation of the Antarctic Hemisphere would bring Australia, New Zealand, South America, and Africa into connection with an extended Antarctica." This man was ready for mobilism.

Harrison announced his conversion to mobilism in his second (1926) presidential address, this time to the Sydney University Science Society. This was two years *before* David spoke in its favor. Harrison recounted his conversion and his initial reaction to Wegener.

Here, then is a veritable impasse. The zoologist still cries for his land bridges, and the geologist will not let him have them. "How to do to avoid?" – as a distinguished Cambridge zoologist of Polish origin once asked me. Up till quite recently I refused to let my land bridges go, and argued that no great degree of interference with the main ocean basins was involved in joining up Antarctica with South America, New Zealand, Australia, and Africa.

(Harrison, 1926: 12)

The impasse he referred to was the inconsistency between sinking landbridges and isostasy. Wegener offered a way out, but Harrison thought mobilism was fantasy until he read Skerl's translation of Wegener's *The Origin* (§3.1, note 1).

Then I read Wegener, at first in a brief notice in *Nature*, at which I scoffed light-heartedly. The ideas of the world as a jigsaw puzzle, and of the solemn continents sliding around the globe upon their bottoms seemed utterly fantastic. In 1924, however, an English translation of the third German edition of his book appeared under the title, "The Origin of Continents and Oceans." Before this I have remained to pray. It seems to offer a working hypothesis which will lift the biologist out of most of his troubles.

(Harrison, 1926: 13)[9]

Applying mobilism to Australasian zoogeographical problems, he noted it faced difficulties.

I have begun to test its application to Australasian problems, and it seems, so far, to open up a bright vista of hope. Nevertheless, a complete application bristles with difficulties, which only time and patience can elucidate or prove insuperable. I make no claim that Wegener is correct. I merely suggest that his hypothesis taken broadly seems to offer the basis for a working hypothesis of great value for zoogeographical problems.

(Harrison, 1926: 13)

But he was already predisposed to mobilism; being opposed to Matthew's views, which were dead set against his own, he probably felt that Wegener was too good to be true. Even though it still "bristled with difficulties," Wegener's theory of continental drift was not inconsistent with isostasy.

Knowing something of Antarctica's ancient flora, he knew that it must have supported a fauna that cohabitated successfully with it for it was absurd to propose there to be plants but no animals, and this would likely have included marsupials, frogs, southern crayfishes, and "a host of other creatures . . . common to Australia and South America" (Harrison, 1926: 12). After summarizing Wegener's history of Gondwana, he asked (1926: 14) his audience the same question he must have been asking himself: "What utterly ridiculous nonsense you say. But is it?" He noted that the "distinguished geologist, Du Toit, accepts it for Africa," that "Brouwer and Molengraaff . . . who know more about the East Indian Archipelago than anyone else . . . gravely accept it." He himself (1926: 14) even took the fact that "large numbers of scientific men have clamoured against it" as "a hopeful sign of its ultimate verity."

Harrison listed problems mobilism solved.

Let us examine for a moment the general solutions that it affords. It gives us a change of relations of continental masses which does not offend against isostasy. It give us a southern land mass from which the glaciers which scratched the surfaces of Africa, India, Australia, and South America in Permo-Carboniferous time may well have been derived, and it explains their contemporaneous glaciation, which is difficult to account for if they were always as wide apart as at present. It gives us a comprehensible Gondwana Land. It gives us a cold temperate

Antarctica with relations to the other southern continental masses. It gives us the Lemuria and Atlantis that so many palaeogeographers have demanded. It explains the long divorce between southeastern Asia and Australasia, which is so obvious to Australian students, and was baffling in the light of present distribution of land and water.

(Harrison, 1926: 15)

He was particularly impressed with mobilism's ability to explain the many differences between the Southeast Asian (Oriental) and Australasian fauna, which could not, he thought, be explained if the latter had always been close to Southeast Asia. He described differing placements of the hypothesized line of demarcation between the two faunal regions, and he noted that they had had very different floras.

The East Indian Archipelago is classic ground, and is well known biologically. Following upon the investigations of Wallace, Huxley gave the name Wallace's Line to a line passing between Bali and Borneo to the west, and Lombok and Celebes to the east, and then bending eastwards to pass south of the Philippines, which was held to mark the divide between the Oriental and the Australasian faunal regions. Later Pelseneer gave the name Weber's Line to a similar line passing between Timor and Australia, the Kei Islands and the Aru Islands, Celebes and Buru, the Philippines and Halmahera. Merrill has recently shown on botanical grounds that Wallace's Line should pass between Palawan and the main Philippines. The belt between these lines marks a transition zone.

(Harrison, 1926: 16)

After identifying the half-dozen known Asiatic and Australasian animals that had crossed from one region to the other before the Pleistocene and noting the recent invasion of Oriental fauna that had "poured" into Australia from Asia, Harrison argued that large geographical changes were implied.

This can mean one thing only: that there had been no landway open since the beginning of the Tertiary. And yet, in recent time, a flood of Oriental flora and fauna has poured into Australia from the north, composed of forms which did not require absolute land connections. Why only recently? Why not during the whole of the Mesozoic and Tertiary if the way were open? We are almost forced to the conclusion that the relations of land and sea in this area have not always been as at present.

(Harrison, 1926: 16)

For Harrison landbridges were no longer enough. Wegener offered the solution; Australia and New Guinea were widely separated from Asia until recent times. Flora and fauna of the two regions remained distinct until Australia and New Guinea had almost reached their current positions relative to Asia.

Harrison hoped that modifications to Wegener's theory might remove some of its difficulties, and suggested that Joly's mobilism might help.

It is possible, however, that modifications of Wegener's views may meet with a more general approval. Thus, Joly's suggestion that the sima becomes fluid only at periodic intervals owing to the accumulation of heat due to radio-activity may offer the basis of a more acceptable hypothesis, and the tidal force on this temporarily fluid layer might be sufficient to account for the pulling away of continental masses in an easterly direction.

(Harrison, 1926: 15)

But Harrison (1926: 15) recognized that he "was not competent to discuss the geology and geophysics of the hypothesis."

It was "as a biologist, and something of a zoogeographer," that he found that Wegener's mobilism offered "a hope of solution of the problems of distribution in the Australasian Region, and it was from that point of view that" he proposed "to examine it" (Harrison, 1926: 15). Harrison also sought additional problems that Wegener's theory explained, problems that it initially had not been designed to solve (RS1), because

Against all this [Wegener's success] we must keep in mind that the hypothesis has been built up very largely from the body of facts which it now purports to explain, and therefore might be expected to explain them.

(Harrison, 1926: 15; my bracketed addition)

Harrison devised the following test:

Wegener shows us a way [to further test his mobilism]. Australia, with New Guinea, has swung northwards in the Quaternary, but prior to that was widely separated from South-eastern Asia. If this happens as Wegener suggests, and New Guinea really pushed the Molluccas and New Britain aside, then we should be able to settle the matter quickly, for some forms should show a distribution on either side of New Guinea, but not appear in New Guinea itself.

(Harrison, 1926: 15; my bracketed addition)

With regard to its principle, this test foreshadowed Sahni's 1936 application of mobilism to explain the juxtaposition of the *Glossopteris* flora of India from the Sino-Sumatran *Gigantopteris* flora (§3.5). India, like New Guinea and Australia, drifted northward toward Asia, bringing its own endemic biota. Harrison found several examples.

Amblycephalid snakes, which are found throughout Southeastern Asia, Malaya, the Archipelago to Celebes, and reappear in the Solomon Islands . . . are absent from New Guinea. The Ranid genus *Cornufer* has a somewhat similar distribution, and extends through the Solomons to Fiji. The sub-genus *Discodeles* of *Rana* widely distributed in India, curiously enough misses Malaya and the Archipelago as well as New Guinea, but has three species in the Solomons and one in the Bismarks. It is further remarkable that the Solomon Islands should have two endemic genera of Ranid frogs, which show no relationship with the recent Ranas of New Guinea.

(Harrison, 1926: 17)

Harrison (1926: 17) closed his talk by announcing that although he had not completed his investigations, he had already shown "the advantage to Australian zoogeography which Wegener's hypothesis affords," and had emphasized that Wegener offered biogeographers a working hypothesis regardless of what fixist geologists thought. "Whether he prove right or wrong, whether his hypothesis can be licked into a better shape by geologists or not, is immaterial. As a working hypothesis for bio-geography, it has great value."

Elected President of Section D of the ANZAAS for 1926, he, in what became his third presidential address, described his further application of Wegener's hypothesis to an understanding of the composition and origin of Australian fauna. This was published posthumously (Harrison, 1928). It ran to seventy-four pages.

After summarizing mobilism's ability to explain the distribution of *Glossopteris* flora, the Permian glaciation of South America, South Africa, India, and Australia, and the origin and composition of Australia's fauna, Harrison (1928: 335) repeated his cautionary reminder that "it must be remembered that the facts which it [mobilism] purports to explain are identical with the data from which it has been constructed." He wanted new biogeographic data against which mobilism could be tested. Recalling that his attraction to Wegener's hypothesis arose from its ability to explain both the similarities between Australian and South American fauna and the differences between Australian and Southeast Asian fauna, he proclaimed that he had now found more biogeographical applications of Wegener's hypothesis – ones that had not been used in its initial construction (RS1). Again he emphasized that he was concerned only with mobilism's biogeographical success.

Since I have held a faith in Antarctic derivation, and have not been able to see any evidence for the very common assumption that Australia has been connected with Asia in the past, I originally had the intention of examining Wegener's hypothesis on these two points. I have been carried away beyond my original intentions; and have been led to put forward some slight extensions of Wegener's views, imperfect, doubtless in their present form, but such as may, under criticism, develop into a useful working hypothesis. That is all I wish at present to suggest. I am not concerned with geological, geophysical, or geodetic proofs of the truth or otherwise of the drift hypothesis.

(Harrison, 1928: 335)

His "slight extension" of Wegener was that he thought that Australia had undergone not one but two separate invasions of biota from the North. Harrison (1928: 346) called the earlier Pliocene invasion, the pan-tropical. It consisted of Polynesian and Lemurian elements, which arrived in Australia when New Guinea impinged "upon an ancient arc extending from Africa and Madagascar to Polynesia, but separated from the southeastern corner of Asia." The second and truly Indo-Malayan invasion occurred during the Pleistocene when New Guinea continued moving northward, and crumpled the ancient arc up against Asia. He constructed a map (Figure 9.1) to show his proposed migration routes.

Celebes was the key. He (1928: 354) claimed that Celebes had once been "the pivoting point of a series of four rays, namely, a Philippine arc, a Micronesian arc, a Melanesian arc, and an arc which provided a portion only of the Lemurian complex."

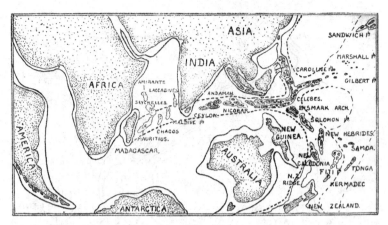

Figure 9.1 Harrison's Figure 20 (1928: 391). Reconstruction of supposed Migration Routes. Various segments of the ancient arcs met at Celebes. Hypothesized migratory routes linked Lemuria with Polynesia, Micronesia, and Asia through the Philippines. The migration routes provided Celebes with a mixed biota containing Lemurian and Pacific elements, which migrated to Australia through New Guinea during the Pliocene.

Let us suppose that the intrusion of New Guinea did take place. Then I would suppose that Celebes, too, was included in the displacing process, was forced westwards and slightly northwards, and had its northern peninsula bent into its present form. The two southern peninsulas represent the fractured end of the double arc. The eastern peninsula would have joined the Buru, Cream, Timor arc, and the western main chain of the Lesser Sundas. Celebes itself indicates clearly enough its origin from two components of a double arc by the north to south lowlands separating its east and west central highlands. As to the fate of these arcs when continued west and south, no data sufficient for discussion are available.

(Harrison, 1928: 354)

He also presented a more detailed map (Figure 9.2) showing the geography of Celebes, and surrounding islands before New Guinea impinged on them.

He thought that his reconstruction might explain the closeness of the Southeast Asian fauna of Bali and the Australasian fauna of Lombok – the two islands where Wallace had first observed great differences between the two faunal realms.

That the present relation of Lombok to Bali has always been the same is frankly incredible. Perhaps they lay further south, with a general south-west north-east trend, such as Timor has to-day. Their lost south-western ends would have linked up with some part of the Lemurian arc (Fig 7).

(Harrison, 1928: 354)

Harrison's application of Wegener's mobilism as a hypothesis for explaining the composition and origin of Australia fauna was the most forceful defense and interesting extension of Wegener's mobilism penned by any Australian scientist until

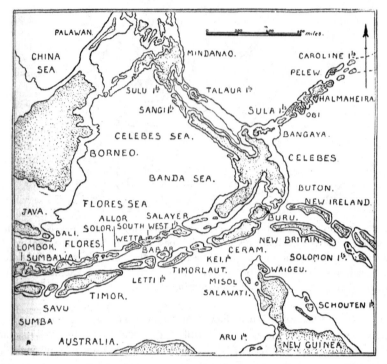

Figure 9.2 Harrison's Figure 7 (1928: 355). Relations of Celebes and Borneo before the upward thrust of New Guinea.

Carey. He based his assessment entirely on its ability to explain the biogeography of Australia's fauna, and did not allow his enthusiasm for mobilism to be dampened by the predominantly fixist opinions of Australian geologists. It is a tragedy that he died in his prime. With his drive, imagination, reasonableness, and wit, who knows what effect he might have had on the tide of fixism that arose in Australia and worldwide in the 1930s through the 1950s.

George Edward Nicholls (1878–1953), the next zoologist to support mobilism, was born in England and educated at King's College, London. He taught zoology at the University of Allahabad, India, before obtaining a Beit Memorial Fellowship at King's College, which enabled him to return to London in 1914, where he was soon hired as reader in zoology at King's College for Women (Waring, 1953). Appointed chair of biology at the University of Western Australia in 1921, he remained there until retirement in 1947. Elected a Fellow of the Linnean Society of London in 1908, he served on its Council from 1916 to 1919. He also was awarded his D.Sc. from King's College in 1924.

Nicholls recognized as an opportunity that Western Australia's fauna was so little known. He concentrated on the systematics of freshwater and terrestrial crustacea, and became the leading expert on Phreatoicidea, a suborder of small freshwater crustaceans (Nicholls, 1943, 1944). It was this interest in the origin of Western Australia's fauna that prompted his interest in continental drift.

He (1933) argued in favor of mobilism in his 1932 Presidential Address to the zoological section of ANZAAS. He gave a lengthy description of the marine and terrestrial fauna of Western Australia, and identified numerous mutual similarities with faunas of South America, South Africa, the East Indies, and islands in the Indian Ocean. He thought these similarities were sufficient to require former land connections. He opted for drift, not landbridges.

An escape from this necessity of numerous hypothetical land bridges is offered in the hypothesis of drifting continents suggested by Taylor and independently by Wegener. If it be accepted that such a crustal movement is possible (and many geologists are prepared to accept it), if further, the drift may be at different rates in different regions with fragmenting of the hinder margin of the continental masses leaving behind island festoons, many of the problems of biogeography meet with the simplest of solutions. For in this hypothesis the continents are equally accepted as permanent land platforms but the essential difference is that Wegener believed that these continents once formed part of a practically continuous land mass – subject, of course to partial and temporary overflow by shallow seas – which he calls Pangea.

(Nicholls, 1933: 136)

Nicholls (1933: 137) closed his talk by repeating Harrison's caution that Wegener's theory should explain new facts, facts that it was not designed originally to explain. He acknowledged that drift faced difficulties. One was that it provided no early connection between the Australia-New Guinea and the Indo-Malayan region, which was needed to explain affinities between ancient flora found in northeastern Australia and Indo-Malaya (RS2).

A difficulty that presents itself is one concerning the time at which was effected the entrance into the ancient Australasian land mass, from Indo-Malaya, of the floral element which characterises the northern Euronotian. [The Euronotian was that part of the ancient Australian flora found in northeastern Australia that had many affinities with similarly aged ancient flora of Indo-Malaya.] According to Wegener, the rift which widened into the Indian Ocean first appeared north of Australia, in late Palaeozoic or early Mesozoic. By the Cretaceous this is represented as having become of very considerable width, and the Western Pacific Islands, if these were broken fragments lagging behind the receding Australian continent, would also have become separated by a wide interval of sea.

(Nicholls, 1933: 137; my bracketed addition)

Thus Nicholls and Harrison, both biologists, were unwilling to accept mobilism fully, but they thought it an excellent working hypothesis that explained the origin and composition of Australia's flora and fauna. Like many others who worked on biogeographical problems (§3.9), regardless of whether they were mobilists or fixists, they also believed that biogeographers often had to work with unreliable and incomplete data. Harrison thought these difficulties were particularly great for those who did not have easy access to major collections.

Another snare in the path of students who are far away from the center of things is comprised in the older and unrevised palaeontological records. No albatross is found in the North Atlantic to-day. It might be suggested that, if the albatrosses are a southern group, and the Atlantic is a young ocean, these birds have not succeeded in passing the tropics. But a fossil albatross is recorded from England. In this case Mr. Tom Iredale has examined the fossil, and is able to tell me that it is not an albatross at all, but a large petrel, and further that the so-called trogon fossils of Europe are not trogons. Ettingshausen's determinations of Australian Tertiary plants offer another example. These are now generally discredited, but a good deal of useless phytogeographical argument has been based upon them in the past. A large number of such records stands in need of revision.

(Harrison, 1928: 345)

According to Nicholls,

There are numerous pitfalls in the path of the zoogeographer. For all but the specialist, in any given group, the actual identity or distinctness of named forms is frequently in doubt; in different orders, genera may come to have widely different values . . . These difficulties are accentuated when the consideration is one of distribution in the past in regions where palaeontological records are incomplete.

(Nicholls, 1933: 131)

Charles Anderson (1876–1944), youngest of nine children, was born in the Orkney Islands, Scotland. He earned an M.A. and B.Sc. from Edinburgh University, where he was a medalist in many sciences including chemistry, crystallography, geology, mineralogy, physics, and zoology. In 1901 he was hired as a mineralogist at the Australian Museum in Sydney. Appointed Director of the Museum (1921), such was his versatility that he abandoned mineralogy for vertebrate palaeontology. Anderson was President of the Royal Society of New South Wales (1924), Linnean Society of New South Wales (1932), Anthropological Society of New South Wales (1930, 1931), and the Geographical Society of New South Wales (1941, 1942). He became an expert in Australian marsupials and, in 1937, discovered the first marsupial fossils from New Guinea.

Anderson (1940) reexamined the possibility that marsupials reached Australia from the north through Asia or from the south through Antarctica. He acknowledged that there was not enough information to decide which view is correct; both possibilities faced a missing data difficulty.

It is impossible to discuss the pros and cons of this question at great length, nor am I competent for such a task, which has led much better men than I to clasp their aching heads and exclaim, "A plague on both your houses." It is evident, however, that at the present time we have not the facts which are necessary to enable a sound conclusion to be reached, nor indeed, is our knowledge sufficient to allow us to interpret correctly such facts as we do possess, though recent investigations not infrequently clear up doubtful points. Thus it is now generally accepted that the resemblances of the Borhyaenids [extinct South American marsupial] to the Thylacine [Tasmania], and of the Caenotestes [living family of South

American marsupials] to the diprotodonts [sub-order of living Australian marsupials] are due to convergence, not to close affinity. On the other hand, there are numerous examples of resemblances between the flora and fauna of Australia and South America which cannot be lightly dismissed. Among these may be instanced the pleurodiran turtles, the tree frogs, the freshwater Crustacea, certain insects, earthworms; and other invertebrate groups.

(Anderson, 1940: 104; my bracketed additions)

Like Harrison and Nicholls, Anderson (1940: 105) noted that proponents on both sides used unreliable or conflicting data and "have sometimes been tempted to over-emphasize certain arguments, and have failed to maintain the judicial attitude."

He (1940: 106) turned to mobilism, and conceded, "Wegener's hypothesis of drifting continents has been hailed by many as the solution of many puzzling problems in zoogeography" and it "gets round the troublesome theory of isostasy," meaning sunken landbridges. Although he praised Harrison's use of drift, he found his predecessor's opinion, that it explained everything for which it was designed, both incorrect and naive.

In light of the Wegener hypothesis, the late Launcelot Harrison made a comprehensive and able study of possible land connections to the north and south of Australia and their bearing on the origin and composition of Australian plants and animals, although he naively remarked that "the hypothesis has been built up very largely from the body of facts which it now purports to explain, and therefore might be expected to explain them."

(Anderson, 1940: 106)

He also took more seriously than either Harrison or Nicholls the geological and geophysical objections that had been raised against Wegener's mobilism.

It may be that the Wegener hypothesis is the answer to the zoogeographer's prayer, but it has not yet attained to the dignity of a theory, and many geologists and geophysicists are not prepared to accept it.

(Anderson, 1940: 106)

Because Anderson was as much geologist as biologist, it is not surprising that he was more sensitive than Harrison or Nicholls to the mounting geological objections to mobilism. Although Anderson mentioned no geologist by name, he might have been influenced by Andrews' 1938 highly critical pro-fixist review (§9.3).

Joseph Pearson, a zoologist from the University of Tasmania, agreed with Anderson; the lack of knowledge about the early history of marsupials left marsupial systematics up in the air.

The truth is that we do not appear to have a sufficient knowledge of the early history of the marsupials upon which to base a satisfactory classification, and there are too many gaps in our knowledge of marsupial palaeontology to invoke its aid in solving the many problems which have arisen. We do not know how the Australian polyprotodonts are related to the American Didelphyidae or how the diprotodonts have arisen from the polyprotodonts. All that we are entitled to say is that the prototypal marsupial was a diadactylous polyprotodont which probably has its closest living counterpart in the American Didelphyidae.

(Pearson, 1940: 45)

For Pearson that was that and he remained undecided about whether Australian marsupials came from Asia or South America, and he did not venture a guess about drift.

A. A. Abbie (1905–76), another zoologist to discuss the origin of Australian marsupials, was born in England. He earned a scholarship to the University of London in 1922, but joined his family when they emigrated to Australia. After earning his B.Sc. in 1929 from the University of Sydney, he returned to England and earned a Ph.D. (1934) in anatomy at University College, London. In 1935 he became senior lecturer in anatomy at the University of Sydney, where he was awarded an M.D. in 1936. After serving in World War II, he took up the Elder chair of anatomy and histology at the University of Adelaide, where he spent the remainder of his career. In 1950 he began studying the anatomical features of the Australian Aborigines, and continued to do so until his retirement in 1970.

Abbie became interested in the origin, radiation, and taxonomy of Australian marsupials while teaching anatomy at the University of Sydney. He definitely preferred an Asian origin.

Australian marsupials probably own an American ancestry, but there is some doubt as to the path followed in the transfer from one continent to the other. Most authors favor an approach from the north over the land chain which once connected with Asia; but there is no wanting advocates for a route across the great Antarctic continent which formerly united the southern ends of America, Australia, and Africa. Some animals, for example wingless birds and certain invertebrates, appear to have used a southern land bridge, and it is conceivable that marsupials also passed this way; however, many objections may be urged against this view. Wingless birds are found in South America and Australia, but they occur also in New Zealand and South Africa – both lacking either marsupials or their remains. The connexion with New Zealand and South Africa may have been lost in the interval between birds and marsupials, but why did it persist between South America and Australia? These objections seem to offer insuperable difficulties to acceptance of the southern route hypothesis. No such objections have been raised against the northern route theory, and until they are raised the northern route must be accepted as that followed by marsupials in their passage to Australia.

(Abbie, 1941: 80)

Like Pearson, Abbie did not bother to discuss mobilism.

Three symposia bearing on continental drift took place in Australia during the 1950s. The first and most important was the 1956 Hobart symposium, which, except for Evans' contribution, will be discussed later (II, §6.16). The second symposium, sponsored in 1959 by the Commonwealth Scientific and Industrial Research Organization (hereafter, CSIRO), concerned biogeography and ecology in Australia. Allen Keast, Curator of Birds and Reptiles; E. Le G. Troughton, Past Curator of Mammals; J. W. Evans, Director; D. F. McMichael and Tom Iredale, all from the

Australian Museum in Sydney; and E. F. Riek and S. J. Paramonov, both from CSIRO, discussed biogeographical problems that had a bearing on whether continental drift had occurred or at least whether land connections were needed between Australia and other Gondwana fragments. The third symposium, sponsored in 1959 by the Royal Society of Victoria to mark the hundredth anniversary of both the society and publication of Darwin's *Origin of Species*, contained presentations by Glaessner (§9.4), B. E. Balme, and W. D. L. Ride that pertained to mobilism.

Troughton argued for fixism, and saw no evidence for a connection between Australia and Antarctica. Concentrating on the origin and radiation of Australia's marsupials, he agreed with Abbie's views, repeating his objections against a southern origin.

Recognizing the American ancestry of Australian marsupials, Abbie (1941) points out that most authors favor their northern approach along the land chain once connected with Asia, though some postulate a southern route across a great Antarctic continent, formerly uniting the southern ends of America, Australia, and Africa. Wingless birds, Abbie says, inhabit South America and Australia, but also occur in New Zealand and South Africa which lack marsupials or their remains. If the connexion with New Zealand and South Africa became lost in the interim (as between birds and marsupials) why did it persist between South America and Australia? And if it existed in the time of birds why did not some primitive premammals of South Africa reach Australia? Such objections seem to offer insuperable obstacles in accepting the southern route hypothesis. According to Abbie, no such objections have been raised against the northern approach, and until they are the northern route must be accepted as that followed in the passage of marsupials to Australasia.

(Troughton, 1959: 69–70)

Troughton (1959: 71) declared, "not biogeographical theory of drifting or 'colliding' land-masses" could explain "the rich radiation and inter-distribution of marsupials."

E. F. Riek wrote an essay on Australian freshwater Crustacea, concentrating on crayfish.

The assumed common origin and close affinity of all the crayfish of the southern continents to one another with their parasitic histriobdellid worms and ostracods and commensal temnocephalids is seen, on close examination, to be poorly based. Further, their present distribution could have been attained without the necessity of land connections between the southern continents.

(Riek, 1959: 257)

Harrison had argued in favor of a common origin of southern hemispheric crayfish, but Riek saw no reason to cite the deceased zoologist's work.

J. W. Evans (1906–90) was born in India, studied geology, botany, and zoology at Cambridge University (1923–6), took a position as an entomologist at CSIRO (1928), and became Director of the Australian Museum, Sydney (1955–66). Evans (1958) argued in favor of a former land connection between Australia and Antarctica at the Hobart 1956 symposium on mobilism (II, §6.16).

I have been asked by Professor Carey to present in an objective manner such evidence relating to the distribution of insects as has a bearing on Continental Drift, and, at the same time, to bring forward any other information of a biological nature which may be appropriate.

(Evans, 1958: 135)

Evans obliged with bonuses. He discussed the distribution of insects but also considered, often drawing on Harrison's work, *Nothofagus* (southern beeches); conifers common to Tasmania, New Zealand, and Chile; crayfish of a type found only in Tasmania, Australia, South America, and Madagascar; large flightless birds such as rheas and ostriches that shared the same species of parasites; and polychaete worms, crustacea, mollusks, and echinoderms that live exclusively in the temperate or cold regions of the Southern Hemisphere. Evans did not fulfill the other part of Carey's request; he said nothing about mobilism *per se*, and left unspecified whether Gondwana broke apart because of continental drift or partial foundering.

The available evidence would seem to suggest that up to some time during the middle or late Mesozoic period all these several land areas will either have been in direct contact with each other, or have been part of a large continental mass of which Antarctica formed part.

(Evans, 1958: 158)

In his next paper, he stated that as a biogeographer he was not going to discuss how Gondwana broke apart; he wanted only to talk about matters he knew about.

The evidence provided by the distribution of insects is thus overwhelmingly in favour of the existence in mid or late Mesozoic times of a large land area of which all the southern continents formed a part. The problem of how this supposed land mass became separated into its present components is not a biological one and hence needs no discussion here.

(Evans, 1959: 161)

Evans discussed whether marsupials entered Australia from Asia or Antarctica. He linked the entry of marsupials with the distribution of insects, arguing that their distribution eliminated an Asian and suggested an Antarctica connection.

The evidence furnished by insect distribution supports an early connection between Australia and India and it is probable that this was in existence during the Jurassic and may have extended into the Lower Cretaceous. It supports also a land connection with the Indo-Malayan region at the close of the Tertiary. It lends no support to a land connection with Asia during the Upper Cretaceous [when marsupials presumably entered Australia] . . . On the other hand, there is a great deal of evidence based on insect distribution which suggests that South America and Australia may have been in land continuity, by way of Antarctica, during the very period when marsupials are believed to have entered Australia. This being so, surely it is more logical to suppose that marsupials may have entered Australia from the south and not from the north.

(Evans, 1959: 162; my bracketed addition)

He considered the difficulty with a southern migratory route: why, if marsupials entered Australia from Antarctica, are there no marsupials in New Zealand and Africa? Repeating Harrison's 1924 rebuttal, Evans (1959: 162) declared, correctly as later determined, that Africa had severed its connection with Antarctica before South America and Australia. Turning to New Zealand, he suggested:

So far as marsupial absence from New Zealand is concerned, nothing can be deduced from this; many groups of insects, of otherwise almost universal occurrence, also lack representation in New Zealand; but it is the presence in and not absence from land areas of organisms which calls for reasoned explanations.

(Evans, 1959: 162)

The New Zealand question was not so easy, and was not answered until almost twenty years later when the dating of seafloor spreading in the Tasman Sea determined that New Zealand rifted from eastern Australia in the Late Cretaceous.

S. J. Paramonov, another entomologist who worked for CSIRO, specialized in Australian flies. Like Evans, he argued for a southern Mesozoic land connection but gave no indication whether he preferred landbridges or mobilism.

As will be seen later, there is such a close relationship between the representatives of certain old groups in the [Diptera] faunas of Australia and South America, that the author is convinced that an ancient geological connection must have existed. This connection was probably via the Antarctic continent which in the past had quite a different climate.

(Paramonov, 1959: 169; my bracketed addition)

McMichael and Iredale found that two groups of Australian freshwater mollusks and one of land mollusks showed marked affinities with those in South America and Africa and were unlike any mollusks in the Northern Hemisphere; they argued (1959: 243) that Antarctica had linked Australia with South America and Africa. Again, they said nothing about whether they preferred drift or landbridges.

B. E. Balme, a palynologist from the Department of Geology at the University of Western Australia, raised a difficulty against the mobilist interpretation of the disjunct *Glossopteris* flora. He worked on spores associated with *Glossopteris* flora. Although Balme (1962: 262) acknowledged that Seward's mobilist interpretation of *Glossopteris* flora was reasonable, and was aware of Plumstead's work on *Glossopteris*, his discovery of "suites of Permian aged spores and pollen grains" in North America and Europe, which he thought were "essentially similar" to those associated with *Glossopteris* flora, led him to propose that *Glossopteris* flora had not been restricted to India and continents in the Southern Hemisphere.

Almost everyone who worked on Australian biogeographical problems ignored the new paleomagnetic support for mobilism. Even Evans said nothing, and he knew about it because he and Irving were friends, and he had heard Irving's talk on the paleomagnetic support for mobilism at the Hobart 1956 symposium (II, §5.3, §6.5, §6.6).

I have found only three Australian biogeographers who even mentioned paleomagnetism up through 1960, even though much new data had appeared from Australia and elsewhere over the previous four years. Keast discussed possible land connections between Gondwana fragments, summarizing current opinion, but keeping silent about his own. He was perhaps the first Australian biologist to mention paleomagnetism.

Postulated land connections (Mesozoic or pre-Mesozoic) between Australia and the other southern continents, suggested by various biologists to account for present-day resemblances in the floras and faunas of these areas remain, like the Wegener hypothesis itself, without present supporting geological evidence. Many workers (reviews in this book) have stressed that the Australian flora and fauna are, basically, a composite of several elements. The persistence of many plant and animal groups in the southern continents only is generally accepted today as indicating nothing more than that they are relics of formerly cosmopolitan groups. A number of workers in the Southern Hemisphere are, however, strong advocates of a common southern origin for them. Interest in Antarctic radiation is now being revived by geomagnetic findings that the position of the South Pole has changed at various times during the Earth's history.

(Keast, 1959: 28)

Keast said nothing more about paleomagnetism. He provided no references, mentioned no particular findings, and kept his own opinion about paleomagnetism to himself. But unlike the other Australian biologists, at least he alluded to the new evidence.

Keast returned to this topic in 1963. Having moved to the Department of Biology at Queen's University, Kingston, Ontario, he organized and gave the opening talk at a symposium on mammalian evolution on the southern continents at the XVI International Congress of Zoology. After providing a short summary of those in favor of mobilism, fixism, or undefined land connections between Antarctica and other Gondwana fragments, Keast again mentioned mobilism's support from paleomagnetism, this time referring to it by name. He also discussed recent work by the New Zealand biogeographer, Lucy Cranwell (III, §1.12), who supported mobilism, based on her microfossil studies of *Nothofagus* and Proteaceous pollens in Antarctica.

Paleomagnetism, with its findings that the magnetic, and hence true poles have shifted position during the Earth's history, and newer discoveries of fossil pollens in Antarctica, raise intriguing prospects. *Nothofagus* and Proteaceous pollens are abundant in Upper Cretaceous deposits in Graham Land (Cranwell, 1961), "as rich as anything in New Zealand," while undated material occurs at McMurdo Sound (pers. comm.). These, plus the "amazing assemblage of early conifers and members of the Proteaceae known from Kerguelen," believed to be "older than mid-Tertiary" (pers. comm.) suggest warmer conditions or movement of these lands southward.

(Keast, 1963: 39; Keast's "southward" should be "northward";
Keast's Cranwell, 1961 is my Cranwell, 1963)

He did not speak about the paleomagnetic evidence for relative movement among Gondwana's fragments. His concluding remarks, which followed the above passage, show that he was not committed to fixism or mobilism.

Mammalogists retain an open mind with respect to drift. Fundamental considerations, as noted previously, still apply: historically the mammal faunas of the 3 continents [Africa, Australia, and South America] have little, or nothing (Africa), in common; distributional facts are explained by northern colonization; and mammals essentially postdate the likely time of drifting. The Wegener theory has little relevance to the present Symposium.

(Keast, 1963: 39; my bracketed addition)

Keast sat on the fence and avoided the issue. Continental drift may have occurred. But, even if it did, it need not be discussed because the past distribution of mammals is immaterial to it. Keast still preferred a northern entry of marsupials to Australia, but continental drift might still have occurred.

Returning five years later to the origin of marsupials, he still preferred migration from Asia. Keast (1968b: 374) certainly thought that the "consensus" in 1968 was still "that the marsupials colonized Australia from Asia, making their way down the island chains." He (1968b: 374) noted that in conflict with the paleomagnetic support for a southern origin "are the findings of structural geology suggesting that Australia has been in its present position relative to Asia since at least the late Mesozoic (Audley-Charles, 1966)." In another paper, Keast (1968a) seemed undecided about mobilism. After summarizing its support, he again referred to Audley-Charles' views and added Teichert's for good measure.

In conflict with evidence that the continents have changed position are certain data that they have been stable relative to each other since at least the mid-Mesozoic. Of greatest significance here is what can be inferred about the relationship of Australia to Asia from the study of structural geology, since some of the more extensive suggestions about drifting have been made about this continent. Most interesting here is the conclusion by Audley-Charles (1966), after many detailed stratigraphic studies on the island of Timor, that the spatial relationships of Australia, Timor, and the rest of the Indonesian archipelago have not greatly altered since the Triassic. He thus supports the views of Teichert (1941). Admittedly postulated reconstructions of this sort always have an inferential or circumstantial element to them but it is significant that, apparently, as strong a case can be made that Australia did not drift during the Tertiary as that it did. Certainly stratigraphic data would appear to be as persuasive as paleomagnetic data. Audley-Charles stresses that his information strongly contradicts any contentions of drift.

(Keast, 1968a: 250)

I find Keast's view remarkable in light of the maturity of the paleomagnetic case for continental drift, especially the movement of Australia, and surely he must have known about the 1966 confirmation of the Vine–Matthews hypothesis. As an expert on mammals and birds and well acquainted with Australian geology, the combination of his specialization and regional studies seems to have made it difficult for him to accept mobilism. He relied on structural geologists engaged in regional studies, not listening to what paleomagnetists and the marine geologists were saying.

Nancy T. Burbidge, a botanist with the Division of Plant Industry, CSIRO, examined the geography of Australia's flora. She was slightly inclined toward fixism, was willing to entertain mobilism, but was unwilling to commit to either.

The author's views may, therefore, be defined as an unwillingness to accept any hypothesis without reservation. Continental drift does not solve all the difficulties, especially within the available portion of the time scale. This does not imply that the importance of the relationships with the flora of other southern lands is denied. On the contrary the strong generic affinities with the flora of South America and, by contrast, the lack of generic affinities with Africa and India (except indirectly) represent a feature that the reconstructions of Pangea (Carey 1958) fail to explain. Similarly the "tropical" elements of long standing in the Australian flora suggest that the region must have been in contact or communication with areas other than that of the Antarctic prior to the early Tertiary. These facts must be explained without forgetting that certain genera in the Australian flora are apparently related to those of certain temperate areas in the north hemisphere. Attempts to correlate these facts with the movements proposed by those favouring continental drift result in complexities both geographical and climatological in nature. The plants are present but the fact that the paths by which they arrived remain obscure seems an unsatisfactory basis for accepting the drift hypothesis, especially when affinities with South Africa are so remote.

*(Burbidge, 1960: 156; the Carey reference is to his reconstruction
presented at the Hobart 1956 symposium (II, §6.7))*

She also thought the new paleomagnetic findings in favor of mobilism were important.

At present workers seem to be divided into those who fully accept the Wegenerian hypothesis (usually with some of the modifications proposed by later authors) and those who believe in the essential stability of continental masses. These two points of view must, owing to recent developments, be seen in the light of the data resulting from research in palaeomagnetism. The information concerning polar migrations, with or without associated continental movement, has obvious implications for every plant geographer.

(Burbidge, 1960: 156)

So she recognized the importance of paleomagnetism, but was unwilling to take mobilism as a given and then attempt to remove the difficulties facing mobilism's biogeographical solutions. Perhaps the new findings made her doubt fixist biogeography, but they were not enough for her to reinvestigate Australian biogeography from a mobilist standpoint. Nevertheless, she went further than most.

W. D. L. Ride, a marsupial expert at the Western Australian Museum, presented a most interesting paper at the 1959 meeting of the Royal Society of Victoria; he thought the new work in paleomagnetism made mobilism reasonably certain, but still preferred G. G. Simpson's northern sweepstakes route from Asia (§3.6). Unlike most Australian workers, Ride not only read the paleomagnetic results, he was most impressed with them.

Drifting continents have also been evoked to explain distributions. Such a southern bridge may have existed and in recent years it has become reasonably certain that continental drift has occurred. This has resulted from the work of geophysicists. Studies of palaeomagnetism (Jaeger, 1956; Du Bois, Irving, Opdyke, Runcorn and Banks, 1957; Chang and Nairn, 1959) and even knowledge of wind directions in past ages (Opdyke and Runcorn, 1959) have combined to give us a picture of continents which moved relative to each other.

(Ride, 1962: 291)

Ride correctly realized that paleomagnetists had not yet determined in 1959 if Australia had remained joined to Antarctica until after the Late Mesozoic or mid-Tertiary when marsupials had migrated to Australia. "Unfortunately, it is not yet clear as to whether continental drift occurred sufficiently late in time to allow it to affect the distribution of mammals" (Ride, 1962: 291). More specifically, it was not yet known whether Australia had separated from Antarctica after marsupials had reached Antarctica. There was not enough paleomagnetic or biogeographical information; in this youthful subject there was too much missing data. "In this case, our facts have failed us and the possibilities appear to be these" (Ride, 1962: 291). He then identified three possible solutions, and favored the first because it was the simplest.

Either marsupials alone entered by a sweepstakes route from the north, *or* there was an intercontinental connection in the south (or north) and marsupials evolved before placentals, *or* there was an intercontinental connection and early placentals crossed it with marsupials but later became extinct in Australia . . . Applying this principle to the possibilities arising from the present state of our knowledge, I would favour the northern sweepstakes route (it requires a single hypothesis) over intercontinental connections (they require two).

(Ride, 1962: 292)

Ride also raised one of the by now standard difficulties against a southern (northern) intercontinental connection: if marsupials entered Australia through a direct connection with Antarctica (Asia) then placentals also should have done so by the same route. He suggested but did not endorse a radical way to avoid the difficulty: that marsupials evolved before placentals. Ride (1962: 292) predicted, "In the near future, the date of continental separation may be established and we will then have to reconsider the position."

The future arrived quickly, and Ride (1964: 98) announced at the beginning of his next paper, "I also consider the current status of the Continental Drift hypothesis and conclude that it provides an adequate explanation for anomalies in the composition of the Australian mammalian fauna." Ride first recalled his previous rejection of a southern route.

During the early part of this century many responsible zoologists believed that the long isolation of Australia, taken together with the fact that the greater component of the Australian fauna was marsupial and similar to that of South America, could best be explained by the theory of Continental Drift. Further palaeontologists working on South American marsupials argued for the existence of special relationships between separate parts of the South American Marsupial fauna and parts of the Australian fauna; in particular between *Thylacinus* and Borhyaenidae, and between Diprotodonta and Paucitubercula. I have reviewed this work, and that of others, and conclude that these relationships cannot be supported.

(Ride, 1964: 126)

Ride once again summarized the paleomagnetic support for mobilism that now included the missing data from Australia.

It is now known that polar wandering curves of the various continents do not agree and that the pole positions given by various continents for any particular age can only be made to agree by displacing the continents relative to each other, i.e., by "drifting" them. Until very recently, it was not known whether Continental Drift as postulated from these data occurred late enough in time for it to effect the composition of the Australian Mammal fauna. However, the work of Irving *et al.* (1963) has made it abundantly clear that Australia only achieved its present latitude during the late Tertiary and that since the Mesozoic it has been moving slowly northwards across what is now the Southern Ocean.

(Ride, 1964: 126)

Irving and company (1963) showed that Australia had remained at a high latitude (~70° S for southern Australia) until at least Early Cretaceous based on the very careful work of Robertson (1963) on the 90 million-year-old Mt. Dromedary Complex, which convinced Ride that Australia had not separated from Antarctica until it had been occupied by marsupials and from where they had migrated to Australia.

He accepted the hypothesis that Australian marsupials had entered via Antarctica. Indeed, he was one of the few who changed his mind about mobilism because of paleomagnetic support. He was one of the very few who had read the very new paleomagnetic evidence and fully understood its implications. It is to Ride's credit that he was willing to change his mind because of relevant work in a very new field. Recognizing that there was insufficient biogeographical data to determine whether marsupials had entered Australia from Antarctica or Asia, he turned to findings in a different field that had settled the issue. By acknowledging that workers in an entirely different field had solved his problem, Ride showed more humility and wisdom than most other fixists did. He showed no resentment that paleomagnetists had more decisive data than he had for determining whether or not Australian marsupials had originated in South America or Asia.

Ride said nothing about the difficulty he had raised two years earlier about why placentals also did not reach Australia at the same time. He did not know why placentals failed to accompany marsupials but he felt confident that marsupials had entered Australia from Antarctica. As far as he was concerned, the question, "From where did Australia's marsupials come?" had been answered, and there were now the new unsolved problems of "Why did placentals not accompany marsupials?" and "What types of marsupials first entered Australia?" He viewed paleomagnetism's establishment of continental drift as an opportunity for biogeographers to apply their data to new problems. Unfortunately, they did not have the needed paleontological data, for as Ride (1964: 126) remarked, "Fossil marsupials do not yet help us." So they could concentrate on finding them in Antarctica.

The "placental" problem is still not solved. The first fossil land mammal was not discovered in Antarctica until 1982. Woodburne and Zinsmeister (1982, 1984) found an extinct marsupial belonging to the Polydolopidae family, which had been known for many years to have been common in South America. They found the fossil in strata about 40 million years old just off the Antarctic Peninsula on Seymour Island. Their discovery showed that marsupials dispersed from South America to Antarctica. But it did not explain why placentals did not accompany marsupials to Australia nor what kind of marsupials first entered Australia. As Woodburne and Zinsmeister put it:

[The discovery of the fossil] raises the question as to which, if any, other family groups from South America colonized Antarctica, or vice versa. The Seymour Island polydolopids invite discussion of the nature of the biogeographic filter between the Antarctic peninsula and the east Antarctic continent during the early Cenozoic and perhaps make more perplexing the absence of any evidence that placental land mammals (known contemporaneously in South America) were present on either the Antarctic or Australian continents during that time.

(Woodburne and Zinsmeister, 1982: 285; my bracketed addition)

In an attempt to explain why placentals never reached Australia, Woodburne and Zinsmeister (1984: 943) proposed that the placentals living in South America when marsupials migrated to Antarctica were primarily large herbivorous ungulates, and they failed to do so because they were less well equipped to make the long journey than the insectivorous-omnivorous marsupials. But they recognized that their solution faced a severe difficulty, for not all South American placentals were of large size and some may have been opportunistic feeders. "Many problems remain, then, when trying to account for the past distribution of marsupial versus placental mammals in the southern continents" (1984: 943). By 1994, there were nine different hypotheses about why marsupials but not placentals reached Australia from South America by crossing Antarctica (Szalay, 1994: 425). But then things got even more complicated. Godthelp *et al.* (1992) claimed that they had discovered the fossil remains of a Late Paleocene non-volant placental mammal, *Tingamarra porterorum*, which they identified as condylarth-like, which apparently did migrate along with marsupials to Australia in the Early Tertiary but did not flourish (Godthelp *et al.*, 1992). If their identification, which was based on a single tooth, is correct, then at least one placental mammal migrated to Australia via Antarctica. However, Woodburne and Case (1996) have disputed their interpretation, and claim that the *Tingamarra porterorum* is a marsupial. In addition, Early Cretaceous ancestral marsupial (Luo *et al.*, 2003) and placental mammalian fossils have been found in China (Ji *et al.*, 2002). It now appears that marsupials and placentals diverged in Asia no later than in the Early Cretaceous, that marsupials eventually migrated through North America to South America, and then to Australia via the Antarctic (Luo *et al.*, 2003). Imagine if the Early Cretaceous marsupial fossils in China had been discovered before the mobilism debate was resolved. Fixists would have quite reasonably made much of the discovery. See Kemp (2005) for further discussion of the evolution of marsupials and their colonization of Australia.

9.6 Regionalism in Australia

There is no question that mobilism's reception in Australia was largely negative. Harrison (1926, 1928), David (1928), Cotton (1929), and Nicholls (1933) welcomed mobilism at first, but then by the early 1940s it seems that all influential geologists and paleontologists had rejected mobilism. Biogeographers too began to part company with mobilism, and some supported unspecified land connections between Australia and Antarctica. Anderson (1940) and Pearson (1940) were neutral, and Burbidge (1960) was neutral, leaning slightly toward fixism. Abbie (1941), Riek (1959), Troughton (1959), and Balme (1962) were fixists. Evans (1958, 1959), McMichael and Iredale (1959), and Paramonov (1959) favored land connections between Australia and Antarctica, but left unspecified the type of land connection, and fell short of favoring mobilism. By the late 1930s leading geologists and paleontologists were decidedly negative. Andrews (1938), Bryan (1944), Hills (1945, 1955), and Browne, in his supplement to David (1950) were strongly opposed. Although David (1928) and Cotton (1929) had earlier been supportive of mobilism, even they were unwilling to accept it, considering it only worthy of investigation. All leading paleontologists, Teichert (1941, 1952, 1959, 1971 (with Meyerhoff), 1974), Glaessner (1950, 1962), Öpik (1956), and Hill (1957, 1959a, 1959b) opposed mobilism; Teichert and Glaessner adamantly opposed. By the 1950s the majority of the leaders of Australian Earth science had made it quite clear that they strongly opposed mobilism. Only a small minority showed some sympathy and they thought it only worth seriously considering. S. W. Carey, whom I consider in some detail in the second volume, was the only major Australian geologist to write in favor of mobilism during the 1950s. He must have felt an outsider with such a strong negative reception of his work among the elite of Australia's Earth science. Fortunately, John C. Jaeger was newly elected to this elite, and he, as head of a new Geophysics Department, was looking for innovative projects in the new university (Australian National University). This very negative attitude to drift did not prevent him from appointing Irving in 1954 to carry out the Australian part of the global paleomagnetic test of continental drift (II, §3.13). As I shall show, Jaeger was immune to geological regionalism, having recently entered Earth science from applied mathematics; he began writing in favor of mobilism in 1956 (II, §6.17).

Biogeographers were more supportive of mobilism than geologists and paleontologists. Of the fifteen I have considered, three (20%) explicitly supported mobilism (Harrison, Nicholls, and Ride), five (33%) were fixists (Abbie, Balme, Keast, Riek, and Troughton), three (20%) were neutral (Anderson, Burbidge, and Pearson), and four (27%) favored unspecified land connections between Australia and Antarctica (Evans, McMichael and Iredale, and Paramonov).

Do the constraints imposed by having a regional outlook explain mobilism's Australian reception? I think they do. The most vehement fixists were primarily interested in Australia's geological and tectonic history. Paleontologists either contributed to the fixist interpretation of Australia's geological history, or were sensitive to it. Everyone who opposed mobilism argued that it was just not needed to explain either Australia's geology or biota, and most argued that it created unnecessary difficulties. In contrast, David and Cotton, the two influential geologists who had some sympathy with mobilism, did not emphasize its ability to solve Australian problems, but saw drift as concerned with broader issues. David emphasized mobilism's success in explaining intercontinental similarities, regardless of whether they pertained to Australia. I agree that he discussed mobilism's success in explaining some aspects of Australia's geology, but when doing so he dealt with problems that also involved some other region; he also raised difficulties that mobilism faced in explaining some aspects of Australia's biota. Cotton's interest and mild support of mobilism were unrelated to Australia's geology and biogeography. Australian geology had a major influence among those (the great majority) who rejected mobilism, and played a minor or no role at all among the few who were sympathetic to it.

Mobilism received a mixed reception among Australian biogeographers, which is understandable given the diverse origins of its remarkable biota. There were biogeographical difficulties with both fixism and mobilism, and some acknowledged that there was insufficient data to settle issues firmly. Indeed, Ride accepted mobilism, and the invasion of marsupials from the south, because of its paleomagnetic support, even though he did not know why placentals apparently failed to accompany marsupials as they migrated out of South America across Antarctica into Australia. Moreover, almost all these biogeographers had backgrounds in biology, and were little concerned with Australia's geology. However, several biogeographers who called for landbridges were concerned, but left the questions of land connections and mobilism unspecified.

Most Earth scientists who worked in Australia ignored the new paleomagnetic results or thought little of them, and it is tempting to think that this was an attitude symptomatic of regionalism. It was not. As I shall later show, almost every fixist acted as if paleomagnetism were worthless or simply did not exist. Ride was a rarity among Australian scientists, but fixists who, like him, became mobilists because of its paleomagnetic support, were rarities in any region where fixism reigned.

9.7 Regionalism, rationality, and wisdom: an interim summary

This concludes my account of the classical period of the mobilism debate, during which no difficulty-free solutions emerged. In this final section I want to comment on regionalism, on the varying status of the difficulties that mobilism encountered, and on the scourges of tunnel vision and group-think. During this period, the majority of scientists preferred fixed to moving continents because they saw this as a better way to explain what they knew best, the geology of their study region, and because they imagined that in the absence of an acceptable mechanism for drift they could not do otherwise. More confident in their own work, and much more familiar with the geology of their own or neighboring regions than of others, participants became unduly influenced by those theories that they thought best explained the geology of their own region, country, or continent; this is tunnel vision. Tunnel vision of individuals, group-think of the community, both feed on each other.

Field geology is highly interpretative, and interpretations are often presented as facts. Practitioners were often unwittingly influenced by the theories they supported; they were often skeptical of field reports of others from other regions that were not in agreement with their views; they raised difficulties with solutions that were not in agreement with their own (RS2). In contrast, if they read reports containing solutions to problems in regions they themselves had not studied first-hand, and if these solutions agreed with their own, they adopted them and grafted them onto their own for their own region, which were thereby reinforced (RS1).

During the classical period, neither mobilism nor fixism was worthy of full acceptance. Participants could, with reason, prefer either. But was it reasonable of them to assume that their own regional analyses provided *the* critical evidence on global theories? I do not think so; the confidence with which some regionalists rejected mobilism was not reasonable and certainly it was unwise. I believe that many who did so, did so too readily, dismissing and sometimes not altogether understanding mobilism's ability to solve problems in regions with which they themselves were not familiar; the same applies to mobilism's ability to solve problems in specialist studies – biogeographical and paleoclimatological studies, for example. It was natural that workers attached great importance to their own work on which they had expended great effort and for which they anticipated recognition, but it was dangerously vain of them to assume that their work was paramount. It was this pervasive vanity that gave the discourse its fixist bias.

This does not mean that workers should not have raised difficulties with opposing theories (RS2), worked to remove difficulties from their own theories (RS1), and argued strenuously in favor of them (RS3). They were correct to do so, but it would have been a wiser course of action, which they could have taken as early as the 1930s, to learn more about the geology of regions in which mobilism more readily provided explanations, and to have been more skeptical of their own rejection of it. This might have led a fixist, such as Kay, to reevaluate the geological and faunal similarities on either side of the North Atlantic, and ask how much of his theory of mountain building could be retained, and grafted onto mobilism without inconsistency. Here I mention Kay (§7.3; III, §1.14) and G. G. Simpson (§3.6, §3.7; III, §1.12) because even though they did not start to rethink their position until the mid-1960s, they were less rigid than many other old-time fixists who remained professionally active during the 1950s and 1960s. In principle, Argand's 1924 concept of a Proto-Atlantic could have provided Kay with the opportunity to graft onto mobilism his ideas about the evolution of mountains. Indeed, once seafloor spreading had been confirmed and plate tectonics introduced, he (1969: 970) hinted at such a possibility by suggesting that the Appalachian–Caledonian mountain belt had formed in an ancient eugeosynclinal belt which "may have been oceanic, though not of the breadth or depth of the present Atlantic." One might ask, why did he not do so earlier? If, before the 1950s, Kay and other fixists had been open to the possibility that they just might be wrong and mobilism right, and had thought more deeply about whether their observations and at least some of their ideas could be recast within a mobilist framework, they might have been more open to the next major development in the drift–mobilist controversy, namely the rise of paleomagnetism. Alas, as I shall describe in the next volume, they were not: most ignored or simply dismissed the paleomagnetic evidence, and the few such as Kay who eventually did take heed, did not do so until the mid-1960s, a decade after paleomagnetists had begun presenting strong arguments in mobilism's favor; had they done so earlier, they might have dissuaded G. G. Simpson from declaring so emphatically that mammalian distributions were relevant to the mobilism debate. Looking ahead to the mid-1960s (III, §1.12), Simpson (1965) at least acknowledged that the new paleomagnetic findings cast serious doubt on fixism and that pre-Cenozoic drift was a legitimate possibility, but not Cenozoic drift for which at the time there was strong paleomagnetic evidence. These reactions by Kay and Simpson, even though less rigid than almost all other fixists, still show that they were unable to admit that they were not in as good a position as other workers in other fields to determine if continents had drifted; they held their own experience paramount. Neither Kay nor Simpson had the humility and wisdom of the biogeographers Ride and Darlington. Ride, in Australia following carefully the new paleomagnetism, switched to mobilism in 1959 (§9.5). The American Darlington, as I shall later describe (III, §1.12), became a mobilist in 1963 after spending almost two years acquainting himself firsthand with evidence from the Southern Hemisphere, and also like Ride learning about paleomagnetism. He then openly acknowledged that paleomagnetism had shown that drift had probably occurred. They showed an admirable clarity of thought and fair-mindedness.

Regionalism hindered mobilism and helped fixism. Because much support for mobilism during the classical stage of the debate was gained through identifying and further substantiating intercontinental disjuncts, regionalists never could meaningfully debate their purported similarities. Because of the differing skills of geologists, because of substantive interpreting sometimes pre-built into their field-work, and because of ambiguities in their complicated descriptive vocabulary, Earth scientists and, in particular, regionalists tended to be skeptical or dismissive of field reports from elsewhere. This was particularly true of reports by workers who were personally unknown to them or even more so by unknowns trained by unknowns in distant lands. Because most fixists had little opportunity or no interest in comparative intercontinental fieldwork, it was usually done by mobilists or by workers who were soon to become mobilists, and their reports were criticized by fixists who argued that their mobilistic biases caused them to describe intercontinental similarities where none existed or to exaggerate those that did. At bottom, regionalists were simply not much interested in global questions; they neglected to properly evaluate intercontinental similarities when determining whether mobilism or fixism was the better global theory. It was a truly vicious circle of distrust, which, by 1950, remained unbroken.

I want now to comment on how the perceived importance of difficulties evolved over time. It is intriguing. Some difficulties were deemed make or break difficulties; without solutions, no further progress on large issues was possible. Some were deemed less critical, others were reclassified as unsolved problems requiring no immediate answer; they could be set aside for the time being. Downgrading or reclassification of a difficulty confronting a solution to an unsolved problem arising from a solution occurs, I believe, only when the solution is imbedded within a theory that has another solution that is difficulty-free or is deemed difficulty-free by those who downgrade the difficulty to an unsolved problem.

I have just described how the absence of placental mammals in Australia became a real stumbling block to deriving Australian marsupials from South America via Antarctica, and hence to the mobilist explanation of their disjunctive distribution. Once Ride embraced the new paleomagnetic results, and accepted that they provided a difficulty-free mobilist solution, in particular that Australia had remained attached to Antarctica late enough for marsupials to have migrated there from South America, he inferred that in all probability they had done so. But Ride also realized that he still could not explain why placental mammals did not accompany marsupials on their journey across Antarctica. Because he had changed his mind and accepted the Antarctic route, the question "Why did placentals not accompany them?" became, for Ride, a new unsolved problem, a problem for the future, not an obstacle to the idea that marsupials colonized Australia from South America via Antarctica. By fiat, Ride reclassified the former difficulty as an unsolved problem. Ride's decision was wise, for as already noted, even the question of whether placentals accompanied marsupials is still not sorted out.

As this volume has shown, the most widely cited difficulty during the classical period of the mobilism debate was that there was no acceptable mechanism for continental drift. Jeffreys raised this difficulty in 1922 and then repeatedly over the next fifty years (Jeffreys, 1979), and it was widely cited, regardless of whether those doing so had special knowledge of global mechanics. It essentially remained the leitmotif of fixists up to the 1966 confirmation of the Vine–Matthews hypothesis, the 1966 confirmation of Wilson's idea of ridge-ridge transform faults, and was only laid to rest (except by Jeffreys and a few others) by the 1967/8 advent of plate tectonics. Mobilists such as Daly (§4.6) and Holmes (§5.4) attempted to provide mobilism with an adequate mechanism, but they were even unable to convince other mobilists that they had succeeded (§4.8, §5.7, §5.9). Other mobilists, such as du Toit, van der Gracht, Wegener, King, and Rastall, attempted to reclassify the mechanism difficulty as an unsolved problem (§5.13), but fixists ignored their arguments or were unconvinced by them. Fixists could do so because mobilists had not demonstrated by physical measurement that continents had actually drifted; they had yet to produce a difficulty-free solution. Without one, fixists could with reason reject their plea to reclassify the mechanism difficulty as an unsolved second-order problem about the cause of continental drift (§5.13). Nonetheless, fixists who simply rejected mobilism because it was deemed impossible by themselves or others, were being disingenuous. After all, they themselves had no difficulty-free solution of their own.

In contrast, there were those like J. W. Evans (§9.5) who considered it their task to carefully document essentially irrefutable links of a wide variety of biota between Australia and South America, and were content to do so. Evans resolutely refused to become involved in the geophysical aspects of the landbridge–continental drift debate because as an entomologist he did not feel competent to do so. But he was more than willing to entertain continental drift. He, and others like him, did not at the time have wide influence, but theirs were authentic voices that were not always heeded.

Ride had recognized by 1963 that paleomagnetism provided mobilism with a difficulty-free solution to the problem of marsupial disjuncts (§9.5). However, he was, as I shall show in Volume II, one of only a few who realized that paleomagnetism had done so, acceptance lagging behind well-documented discovery. Indeed, the failure of most Earth scientists to recognize that it had settled the issue of whether or not continental drift had occurred is one of the main topics of Volume II. With this failure of recognition, most Earth scientists continued to believe that continental drift either could not happen or was highly unlikely. In the end, plate tectonics, the geometrification of geology, provided irrefutable justification for mobilism without discussing forces, without the need or intent to provide a mechanism which for so long had been considered a necessity. How this huge change was brought about is one of the main topics of Volume IV.

Participants in the mobilist debate, especially regionalists, often suffered individually from tunnel vision and collectively from group-think. This point was humorously made by Eugene Wegmann (who will reappear in II, §1.16, §6.5; IV, §1.5) in his contribution to the 1939 Frankfurt symposium on the Atlantic basin (§8.4). Wegmann drew two cartoons. Cloos, as editor of *Geologische Rundschau*, included them at the end of the proceedings. One of them, reproduced in Figure 9.3, provides a particularly appropriate way to end this account of the classical stage of the mobilist controversy.

Individual scientists worldwide have the tunnel vision common to their region, which on a collective basis is the group-think of their regional community. The worldwide chaotic directions of the telescopes of individual scientists have a regional order, and individual scientists as members of their regional community have their telescopes pointing in the same direction.

Tunnel vision is insidious because the afflicted are unaware of it. Specialists and regionalists alike found it hard to appreciate work in other fields and regions that conflicted with their viewpoint because those in their respective communities think in

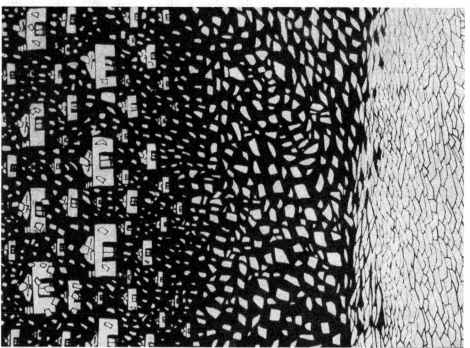

Figure 9.3 *Wegmann* (1939). The caption of the cartoon on the left is *Die wissenschaftliche Literatur* (The scientific literature); the one on the right is *Die wissenchaftlichen Gesichtspunkte* (The scientific points of view).

the same way, producing reports that agree with each other and pile atop each other as illustrated in the other cartoon in Figure 9.3.

Walter Bucher (who also will reappear in II, §2.14, §3.7, §4.2, §4.13; III, §1.4, §1.10, §2.19, §6.3; IV, §7.2), also appreciated Wegmann's cartoon, although he proposed a more generous view of scientists.

With wise humor the brilliant edition concludes the second *Atlantic Heft* of the *Geologische Rundschau* with a cartoon drawn by the facile pen of the Swiss geologist, C. E. Wegman. It shows the men of science, each peering intently into one of innumerable narrow tubes, each clinging desperately to his "point of view" (Wegman, 1939). A more generous mood might liken the narrow lines of vision of the individuals to the prisms in an arthropod's eye, each of which catches only a minute fraction of the reality without, with "science" playing the role of the retina which records the partial images, and of the brain that integrates them into a comprehensive picture. While most of our life must be spent peering through one prism, the real satisfaction comes when we make use of our privilege to step back and see what is visible of the whole.

(Bucher, 1940; Bucher dropped one of the n's in Wegmann)

Bucher thought he had escaped the risk of tunnel vision, having stepped back to see the "greater reality," the "whole" picture.

Such an attempt was made in the preceding pages, which try to sketch the still shifting outlines of the emerging pattern of a fascinating part of the great reality that we are studying, in which we find the ultimate justification of our individual labors.

(Bucher, 1940: 508–509)

Just what had Bucher done? He had presented a fixist account of the formation of the Atlantic basin in which he argued that oceanographic exploration had revealed an Atlantic basin whose structure was much like that of surrounding continents. This was what he had expected, and it fit with his fixism (§8.4). Bucher began his paper this way:

During the last quarter century, the Atlantic Ocean, more than any other region on earth, has given direction to new and fruitful lines of thought in geology. While it was destined by location to play this role among oceans, the impetus that started the new lines of investigation came to a large extent from the bold reasoning of Alfred Wegener, who challenged most of the concepts on which attempts to understand the dynamics of crustal deformation had so far been based. This critical evaluation of Wegener's ideas made it imperative that more accurate information on the topographic configuration of the seafloor be secured. The menace of the U-boat in the [First] World War led to the development of sonic sounding methods simultaneously in the United States, in England, and in Germany. The new tool applied to the numerous shipping lanes soon made the floor of the North Atlantic the best-known submarine topography. It consists of a double series of irregular basins separated by swells. Results of the *Meteor* expedition have shown that the same pattern dominates the floor of the South Atlantic in a better-defined manner than had been known before. The nature of these basins and swells represents a first great problem.

(Bucher, 1940: 489; my bracketed addition)

The *Meteor* had taken sonic readings on thirteen crossings of the South Atlantic. Although enough had been learned about the Mid-Atlantic Ridge in the South Atlantic to make discussion worthwhile, Bucher chose to restrict his vision of the "great reality," of the "whole," to the North Atlantic.

> The parallelism that exists between the opposite shores of the Atlantic and the remarkable central line of swells, the Mid-Atlantic ridge, constitutes a closely related problem. Since it is most tangibly developed in the South Atlantic and requires for its discussion a comparative study of other ocean basins, it will not be touched upon in this paper.
>
> *(Bucher, 1940: 490)*

Bucher did not think that discussion of the Mid-Atlantic Ridge and other features of the North Atlantic required comparisons with other ocean basins; it alone was sufficient. By setting aside further discussion of the South Atlantic, he avoided the very strong support for mobilism that the entire Atlantic basin and surrounding continents had to offer. How did he begin his discussion of the North Atlantic? Not by discussing the many trans-Atlantic similarities but by arguing (1940: 490) that geodetic measurement "destroys the evidence for a 'drift' of Greenland as a whole." He had allowed himself to be ambushed by tunnel vision. Avoiding such an ambush is particularly difficult when everyone around you agrees with your point of view. Students who witnessed Bucher's debate with King in 1957 at Columbia University believed that King had clearly won the debate (§6.10). They recognized Bucher's tunnel vision, and perhaps the group-think of their fixist professors. Did Bucher realize it too? Did King awaken him from his dogmatic slumber? I suspect not.

Notes

1 Branagan (1981) provides an excellent account of David's failure to complete *The Geology of the Commonwealth of Australia*.

2 Hills communicated R. Green and E. Irving's paper "The Palaeomagnetism of the Cainozoic basalts from Australia," which provided evidence of drift between Europe and Australia to the Royal Society of Victoria on July 11, 1957 (Green and Irving, 1958).

3 Browne used to say that he made his greatest discovery in Adelaide, namely, a student named C. E. Tilley. Tilley followed Browne back to Sydney, where he studied with Browne and David. Tilley eventually became Head of the Department of Petrology at the University of Cambridge, where he befriended Keith Runcorn (II, §3.11).

4 I met Teichert when he was at Kansas University in the early 1980s. I knew little about his own participation in the controversy, except for the fact that he had been an adamant fixist, and had spent part of his career in Australia. I asked him about Carey, and his enthusiasm for mobilism. He photocopied the title page to his personal copy of the 1949 meeting, told me to get a copy of the proceeding and read it.

5 Wise was not speaking about mobilism when he made the comment that Teichert quoted. Wise was quoted by Henry S. F. Cooper, Jr. (1970), who wrote an article, "Letter from the Space Center," for *The New Yorker Magazine*. Cooper covered NASA's 1970 Apollo 11 Lunar Science Conference that was held in Houston. Wise was a geologist who taught at the University of Massachusetts. He was a panelist at a meeting on the origin of the Moon, and he made the comment in reference to a talk by the Australian geochemist Ringwood. Wise was displeased with Ringwood who argued his theory.

6 Biographical information on Glaessner is drawn from Brian McGowran's (1994) sensitive biographical memoir. McGowran studied under Glaessner during the 1950s at the University of Adelaide, where he is a member of the Department of Geology and Geophysics. Following in his teacher's footsteps, McGowran has had a very successful career working in micropaleontology, biostratigraphy, paleooceanography, and evolution. McGowran has helped me understand Glaessner's anti-mobilism.

7 I think Glaessner is now thought to have been correct; Ediacaran fossils are generally interpreted as being related to living Cnidaria (Knoll, 2003: 164–173).

8 McGowran informed me that Glaessner thought highly of Carey's fieldwork in New Guinea; his disagreement with Carey was one of interpretation.

9 Harrison could not have read anything by Wegener in *Nature* in 1922 because nothing by Wegener was published in the journal. Wegener (1922) wrote a paper, "The origin of continents and oceans" which appeared in *Discovery*. Lake (1922a), Roe-Thompson (1922) and Cole (1922) wrote highly negative assessments in *Nature* of Wegener's mobilism (§8.13). Two anonymous neutral reviews (1922a, 1922b) also appeared in *Nature*, and the latter referenced Wegener's paper in *Discovery*. Perhaps Harrison read Wegener's paper in *Discovery*, or simply read about Wegener in *Nature*. Reading Lake, Roe-Thompson, and Cole's attacks certainly would have given Harrison the idea that Wegener's mobilism was fantastical.

References

Abbie, A. A. 1941. Marsupials and the evolution of mammals. *Aust. J. Sci.*, **4**: 77–92.

Adie, R. J. 1952. The position of the Falkland Islands in a reconstruction of Gondwanaland. *Geol. Mag.*, **89**: 401–410.

Agharkar, S. P. 1937. Wegener's theory of continental drift with reference to India and adjacent countries (General discussion). *Proceedings 24th Indian Science Congress*, Hyderabad, 502–520.

Airy, G. B. 1885. On the computation of the effect of attraction of mountain-masses, as disturbing the apparent astronomical latitude of stations in geodetic survey. *Phil. Trans. Roy. Soc. London*, **145**: 101–104.

Aldinger, H. 1937. Das ältere Mesozoikum ostgonlands. *Geologische Rundschau*, **28**: 124–128.

Aldrich, M. L. 1976. Frank Bursley Taylor 1860–1938. In Gillispie, C. C., ed., *Dictionary of Scientific Biography, XIII*. Charles Scribner's Sons, New York, 269–272.

Aldrich, M. L. 1979. American state geological surveys, 1820–1845. In Schneer, C., ed., *Two Hundred Years of Geology in America*. University Press of New England, Hanover, NH, 133–143.

Allègre, C. 1988. *The Behavior of the Earth*. Translated by Kurmes van Dam, D. Harvard University Press, Cambridge, MA.

Allwardt, A. O. 1990. *The Roles of Arthur Holmes and Harry Hess in the Development of Modern Global Tectonics*. University Microfilms, Ann Arbor, MI.

Almeida, F. F. M. 1953. Deformacòes causadas pelos gelos na Série de Tubarào em São Paulo. *DNPM, Div. Geol. Min.*, No. 64: 1–8.

Almeida, F. F. M. 1954. Botucatú, um deserto triássico da America do Sul. *Notas prelim. estudos, Rio de Janeiro, Div. Geol. Min.*, No. 86.

Amadon, D. 1948. Continental drift and bird migration. *Science*, **108**: 705–707.

Ampferer, O. 1906. Über das Bewegungsbild von Faltengebirgen. *Jahrb. Geol. Reichsanstalt*, **56**: 539–622.

Ampferer, O. 1925. Über kontinentverschiebungen. *Die Naturwissenschaften*, **13**: 669–675.

Anderson, C. 1940. Origin and migration of Australian marsupials. *Aust. J. Sci.*, **2**: 103–106.

Andrews, E. C. 1938. Some major problems in structural geology. *Proc. Linn. Soc. NSW*, **63**: v–xi.

Andrews, E. C. 1942. The heroic period of geological work in Australia. *J. Proc. Roy. Soc. NSW*, **76**: 96–128.

Andrews, H. N. 1947. *Ancient Plants and the World They Lived In.* Comstock Publishing Co., Ithaca, NY.

Anonymous. 1916. Séance du vendredi 3 November 1916. *Soc. Neuchatel Sci. Nat. Bull.*, **42**: 115.

Anonymous. 1922a. Wegener's displacement theory. *Nature*, **109**: 202–203.

Anonymous. 1922b. The flotation of continents. *Nature*, **109**: 757.

Anonymous. 1923. The distribution of life in the Southern Hemisphere and its bearing on Wegener's hypothesis. *Nature*, **111**: 131.

Anonymous. 1928a. Launcelot Harrison. *Aust. J. Zool.*, **5**: 132–137.

Anonymous. 1928b. The Palaeozoic mountain systems of Europe and America. *Nature (Supplement)*, **122**: 365–366.

Anonymous. 1935. Discussion upon the hypothesis of continental drift. *Proc. Geol. Soc. London*, **91**: vi–xi.

Anonymous. 1942. Discussion at the 16 April 1942 general meeting of the Linnean Society. *Proc. Linn. Soc. London*, **154**: 163–165.

Anonymous. 1943. Discussion at the 8 April 1943 general meeting of the Linnean Society and Zoological Society of London. *Proc. Linn. Soc. London*, **155**: 119–125.

Argand, É. 1916. Sur l'arc des Alpes occidentales. *Ecologae Geol. Helv.*, **14**: 145–191.

Argand, É. 1924/1977. La tectonique de l'Asie. *Proceedings of the XIIIth International Geological Congress*, **1** (part 5): 171–372. English translation by Carozzi, A. V. *The Tectonics of Asia.* Hafner Press, New York.

Arkell, W. J. 1949. Jurassic ammonites in 1949. *Science Progress*, **37**: 401–417.

Arkell, W. J. 1956. *Jurassic Geology of the World.* Oliver and Boyd, Edinburgh.

Arldt, T. 1901. *Die entwickling der kontinente und ihrer lebewelt.* Englemann, Leipzig.

Arnold, C. 1947. *An Introduction to Paleobotany.* McGraw Hill, New York.

Audley-Charles, M. G. 1966. Mesozoic paleogeography of Australasia. *Palaeogeogr. Palaeoclimatol. Palaeoecol.*, **2**: 1–25.

Axelrod, D. I. 1952. Variables affecting the probabilities of dispersal in geologic time. In *The Problem of Land Connections Across the South Atlantic, with Special Reference to the Mesozoic. Bull. Am. Mus. Nat. Hist.*, **99**: 177–188.

Bailey, E. B. 1926a. Comments at 14 December 1925 meeting of the Royal Geographical Society, London on the Alps and Wegener's theory. *Geogr. J.*, **67**: 307–309.

Bailey, E. B. 1926b. Structural features of the Earth. *Nature*, **117**: 863–864.

Bailey, E. B. 1927. Across Canada with Princeton. *Nature*, **120**: 673–675.

Bailey, E. B. 1928a. The Palaeozoic mountain systems of Europe and America. *Nature*, **122**: 811–814.

Bailey, E. B. 1928b. The ancient mountain-systems of Europe and America. *Scot. Geogr. Mag.*, **44**: 321–324.

Bailey, E. B. 1929. The Palaeozoic mountains systems of Europe and America. *Rep. Brit. Assoc.* 1928: 57–76.

Bailey, E. B. 1935. *Tectonic Essays Mainly Alpine.* Clarendon Press, Oxford.

Bailey, E. B. 1938. Professor Albert Heim. *Obit. Not. Fell. Roy. Soc.*, **II**: 471–474.

Bailey, E. B. 1939. Dr. W.B. Wright. *Nature*, **144**: 775–776.

Bailey, E. B. 1958. Prof. Léon W. Collet. *Nature*, **181**: 17.

Bailey, E. B. and Holtedahl, O. 1938. Northwestern Europe, caledonide. In *Regionale Geologie der Erde*, **2**, pt. 2: 1–76.

Bailey, E. B. and Weir, J. 1939. *Introduction to Geology.* MacMillan and Co., Ltd., London.

Bailey, E. B., Collet, L. W., and Field, R. M. 1928. Paleozoic submarine landslips near Quebec City. *J. Geol.*, **36**: 577–614.

Baker, H. A. 1923. Letter about Wegener's theory of drifting continental masses. *Nature*, **111**: 80–81.

Baker, H. B. 1912–1914. The origin of continental forms. *Ann. Rep. Michigan Acad. Sci.*, 1912: 116–141; 1913: 26–32, 107–113; 1914: 99–103.

Baker, H. B. 1932. *The Atlantic Rift and Its Meaning*. Privately published, Detroit.

Balme, B. E. 1962. Some palynological evidence bearing on the development of the *Glossopteris-Flora*. In Leeper, G. W., ed., *The Evolution of Living Organisms*. Melbourne University Press, Melbourne, 269–280.

Barrell, J. 1914. The strength of the earth's crust. *J. Geol.*, **22**: 425–433, 441–468, 655–683.

Behrmann, W. 1939. Einwande der geographie gegen die Wegeneresche theorie der kontinetalverschiebungen. *Geologische Rundschau*, **30**: 111–120.

Beloussov, V. V. 1966. Modern concepts of the structure and developments of the earth's crust and the upper mantle of continents. *Q. J. Geol. Soc. London*, **69**: 293–314.

Beloussov, V. V. 1967. Against continental drift. *Sci. J.*, **3**, 56–61.

Benioff, H. 1959. Circum-Pacific orogeny. *Ottawa Dominion Obs. Pub.*, **20**: 395–402.

Benndorf, H. 1931. Alfred Wegener. *Gerlands Beitr. Geophys.*, **31**: 337–377.

Bernauer, von F. 1939. Island und die Frage der Kontinentalverschiebungen. *Geologische Rundschau*, **30**: 357–358.

Berry, E. W. 1922. Outlines of South American geology. *Pan-Am. Geol.*, **38**: 187–216.

Berry, E. W. 1928. Comments on the Wegener Hypothesis. In van Waterschoot van der Gracht, W., ed., *Theory of Continental Drift: A Symposium on the Origin and Movement of Land Masses both Inter-continental and Intra-continental, as Proposed by Alfred Wegener*. American Association of Petroleum Geologists, Tulsa, OK, 194–196.

Bigarella, J. J. and Salamuni, R. 1961. Early Mesozoic wind patterns as suggested by dune bedding in botucatú sandstone of Brazil and Uruguay. *Geol. Soc. Am. Bull.*, **72**: 1089.

Billings, M. P. 1942. *Structural Geology*. Prentice-Hall, Inc., New York.

Billings, M. P. 1954. *Structural Geology* (second printing). Prentice-Hall, Inc., Englewood Cliffs, NJ.

Billings, M. P. 1959. Memorial to Reginald Aldworth Daly. *Proc. Vol. Geol. Soc. Am.* 1958, 114–121.

Birch, F. 1951. Remarks on the structure of the mantle, and its bearing upon the possibility of convection currents. *Trans. Am. Geophys. Union*, **32**: 533–534.

Birch, F. 1960. Reginald Aldworth Daly. *Biogr. Mem. Natl. Acad. Sci.*, **34**: 31–64.

Black, G. W. Jr. 1979. Frank Bursley Taylor – forgotten pioneer of continental drift. *J. Geol. Ed.*, **27**: 67–70.

Blackett, P. M. S., Clegg, J. A., and Stubbs, P. H. S. 1960. An analysis of rock magnetic data. *Proc. Roy. Soc. London, A*, **256**: 291–322.

Bowie, W. 1925a. Isostasy at Madrid meeting of International Geophysical Union. *Pan-Am. Geol.*, **43**: 337–344.

Bowie, W. 1925b. Scientists to test "drift" of continents. *New York Times*, September 6, p. 5.

Bowie, W. 1928. Comments on the Wegener hypothesis. In van Waterschoot van der Gracht, W., ed., *Theory of Continental Drift: A Symposium on the Origin and Movement of Land Masses both Inter-continental and Intra-continental, as*

Proposed by Alfred Wegener. American Association of Petroleum Geologists, Tulsa, OK, 178–186.

Bowie, W. 1935. The origin of continents and oceans. *Scientific Monthly*, **41**: 444–449.

Bowie, W. 1936. The place of geodesy in geophysical research. *Trans. Am. Geophys. Union*, **17**: 15–20.

Box, J. F. 1978. *Fisher, the Life of a Scientist*. John Wiley & Sons, New York.

Bradley, W. H. 1969. Walter Hermann Bucher. *Biogr. Mem. Natl. Acad. Sci.*, **40**: 19–34.

Branagan, D. 1981. Putting geology on the map: Edgeworth David and the geology of the Commonwealth of Australia. *Hist. Records Aust. Sci.*, **5**: 31–58.

Brewster, K. 2001. *Interview of Dr. William Long*. Byrd Polar Research Center at Ohio State, Columbus, OH.

Bridgman, P. W. 1922. *Dimensional Analysis*. Yale University Press, New Haven, CT.

Brock, B. B. 1957. World patterns and lineaments. *Trans. Proc. Geol. Soc. S. Afr.*, **60**: 127–175.

Brock, B. B. 1972. *A Global Approach to Geology: The Background of a Mineral Exploration Strategy Based on Significant Form in the Patterning of the Earth's Crust*. A. A. Balkema, Cape Town.

Brooks, C. E. P. 1926. *Climate Through the Ages*: Ernest Benn Limited, London.

Brooks, C. E. P. 1930. Discussion on geological climates. *Proc. Roy. Soc. London, B*, **106**: 312–314.

Brooks, C. E. P. 1949. *Climate Through the Ages* (2nd edition). Ernest Benn Limited, London.

Brouwer, H. A. 1917. Uber gebirgsbildung und vulkanismus in den Molukken. *Geologische Rundschau*, **8**: 197–209.

Brouwer, H. A. 1919a. On the non-existence of active volcanoes between Pantar and Dammer (East Indian Archipelago), in connection with the tectonic movements in this region. *Proc. Kon. Akad. Wetensch. Amsterdam*, **21**: 795–802.

Brouwer, H. A. 1919b. On the crustal movements in the region of the curving rows of islands in the eastern part of the East-Indian Archipelago. *Proc. Kon. Akad. Wetensch. Amsterdam*, **22**: 772–782.

Brouwer, H. A. 1921. The horizontal movement of geanticlines and the fractures near their surface. *J. Geol.*, **29**: 561–577.

Browne, W. R. 1963. Professor L. A. Cotton. *Aust. J. Sci.*, **26**: 81–82.

Browne, W. R. and Walkom, A. B. 1952. Ernest Clayton Andrews. *Proc. Linn. Soc. NSW*, **77**: 98–103.

Brush, S. G. 1986. Early history of selenogony. In Hartmann, W. K., Phillips, R. J., and Taylor, G. J., eds., *Origin of the Moon*. Lunar and Planetary Institute, Houston, TX, 9.

Brush, S. G. 1996a. *Fruitful Encounters*. Cambridge University Press, Cambridge.

Brush, S. G. 1996b. *Transmuted Past*. Cambridge University Press, Cambridge.

Bryan, W. H. 1944. The relationship of the Australian continent to the Pacific Ocean: now and in the past. *J. Proc. Roy. Soc. NSW*, **78**: 42–62.

Bubnoff, S. von. 1939. Zur frage transozeanischer kontinentalverbindungen. *Geologische Rundschau*, **30**: 303.

Bucher, W. H. 1933. *The Deformation of the Earth's Crust*. Princeton University Press, Princeton, NJ.

Bucher, W. H. 1939a. Versuch einer analyse der großen bewegungen der Erdkruste. *Geologische Rundschau*, **30**: 285–296.

Bucher, W. H. 1939b. Deformation of the Earth's crust. *Geol. Soc. Am. Bull.*, **50**: 412–423.

Bucher, W. H. 1940. Submarine valleys and related geologic problems of the North Atlantic. *Geol. Soc. Am. Bull.*, **51**: 489–511.

Bucher, W. H. 1952. Continental drift versus land bridges. In Mayr, E., ed., *The Problem of Land Connections Across the South Atlantic, with Special Reference to the Mesozoic. Bull. Am. Mus. Nat. Hist.*, **99**: 93–104.

Bull, A. J. 1921. A hypothesis of mountain building. *Geol. Mag.*, **58**: 365–367.

Bull, A. J. 1927. Some aspects of the mountain building problem. *Proc. Geol. Assoc.*, **38**: 145–156.

Bull, C. and Irving, E. 1960. The Palaeomagnetism of some hypabyssal intrusive rocks from South Victoria Land, Antarctica. *Geophys. J.*, **3**: 211–224.

Bullard, E. C. 1951. Remarks on deformation of the Earth's crust. *Trans. Am. Geophys. Union*, **32**: 520.

Bullard, E. C. 1960. Response by E. C. Bullard. *Geol. Soc. Am. Proc.* 1959: 92.

Bullard, E. C. 1975a. The emergence of plate tectonics: a personal view. *Rev. Earth Planet. Sci.*, **3**: 1–30.

Bullard, E. C. 1975b. The effect of World War II on the development of knowledge in the physical sciences. *Proc. Roy. Soc. London A*, **343**: 519–536.

Burbidge, N. T. 1960. The phytogeography of the Australian region. *Aust. J. Bot.*, **8**: 75–211.

Bütler, H. 1935a. Some new investigations of the Devonian stratigraphy and tectonics of East Greenland. *Meddel. Grønland,* **103**, Nr. 2: 1–35.

Bütler, H. 1935b. Die mächtigkeit der kaledonischen molasse in Ostgrönland. *Mitt. Naturforsch. Ges. Schaffhausen*, **12**: 17–33.

Camp, W. H. 1952. Phytophyletic patterns on lands bordering the South Atlantic basin. In Mayr, E., ed., *The Problem of Land Connections Across the South Atlantic, with Special Reference to the Mesozoic. Bull. Am. Mus. Nat. Hist.*, **99**: 205–212.

Campbell, C. D. and Runcorn, S. K. 1956. The magnetization of the Columbia River basalts in Washington and Northern Oregon. *J. Geophys. Res.*, **61**: 449–458.

Campbell, D. H. 1944. Living fossils. *Science*, **100**: 179–181.

Campbell, K. S. W. and Jell, J. S. 1999. Dorothy Hill, A.C., C.B.E. *Biogr. Mem. Fell. Roy. Soc.*, **45**: 197–217.

Carey, S. W. 1955. The orocline concept in geotectonics. *Roy. Soc. Tasmania,* **89**: 255–288.

Carey, S. W. 1958. A tectonic approach to continental drift. In Carey, S. W., Convener, *Continental Drift: A Symposium.* University of Tasmania, Hobart, 177–355.

Carozzi, A. V. 1977. Editor's introduction. In *La tectonique de l'Asie. Proceedings of the XIIIth International Geological Congress*, **1** (part 5): 171–372. English translation by Carozzi, A. V. *The Tectonics of Asia.* Hafner Press, New York, xiii–xxvi.

Carozzi, A. V. 1985. The reaction in continental Europe to Wegener's theory of continental drift. *Earth Sci. Hist.*, **4**: 122–137.

Caster, K. E. 1952. Stratigraphic and paleontological data relevant to the problem of Afro-American ligation during the Paleozoic and Mesozoic. In Mayr, E., ed., *The Problem of Land Connections Across the South Atlantic, with Special Reference to the Mesozoic. Bull. Am. Mus. Nat. Hist.*, **99**: 105–152.

Caster, K E. 1977. Response by Kenneth E. Caster upon receiving the Paleontological Society Medal, November 9, 1976. *J. Paleontol.*, **51**: 648–651.

Caster, K E. 1981. My involvement with continental drift. Unpublished 12-page paper written in response to questions asked by Frankel, H.

Caster, K. E. and Mendes, J. C. 1948. Du Toit's geological comparison of South America with South Africa after twenty years (Abstract). *Geol. Soc. Am. Bull.*, **58**: 1173.

Chamberlin, R. T. 1928. Some of the objections to Wegener's theory. In van Waterschoot van der Gracht, W., ed., *Theory of Continental Drift: A Symposium on the Origin and Movement of Land Masses both Inter-continental and Intra-continental, as Proposed by Alfred Wegener.* American Association of Petroleum Geologists, Tulsa, OK, 83–87.

Chamberlin, R. T. 1938. Review of du Toit's *Our Wandering Continents. J. Geol.*, **46**: 791–792.

Chamberlin, T. C. 1890. The method of multiple working hypotheses. *Science*, **15**: 92–96.

Chamberlin, T. C. 1897. The method of multiple working hypotheses. *J. Geol.*, **5**: 837–848.

Chaney, R. W. 1940. Tertiary forests and continental drift. *Geol. Soc. Am. Bull.*, **51**: 469–488.

Cloos, H. 1937. Zur grosstektonik hochafrikas und seiner umgebung, eine fragestellung. *Geologische Rundschau*, **28**: 333–348.

Cloos, H. 1939. Hebung-spaltung-vulkanismus. *Geologische Rundschau*, **30**: 405–409.

Cloos, H. 1942. Tektonische bemerkungen uber der boden des Golfs von Aden. *Geologische Rundschau*, **33**: 354–363.

Cloos, H. 1947/1953. *Gesprach mit der Erde.* 1947. *Conversation with the Earth.* 1953. Translated from the German by Garside, E. B.; slightly abridged by Closs, E. and Dietz, C., eds. Knopf, New York.

Cloos, H. 1948. The ancient European basement blocks: preliminary note. *Trans. Am. Geophys. Union*, **29**: 99–103.

Colbert, E. H. 1952. The Mesozoic tetrapods of South America. In Mayr, E., ed., *The Problem of Land Connections Across the South Atlantic, with Special Reference to the Mesozoic. Bull. Am. Mus. Nat. Hist.*, **99**: 237–249.

Cole, G. A. J. 1922. Wegener's drifting continents. *Nature*, **110**: 798–801.

Coleman, A. P. 1916. Dry land in geology. *Geol. Soc. Am. Bull.*, **27**: 175–193.

Coleman, A. P. 1923. The Wegener hypotheses: discussion. *Nature*, **111**: 30–31.

Coleman, A. P. 1924. Ice ages and drift of the continents. *Am. J. Sci. 5th ser.*, **7**: 398–404.

Coleman, A. P. 1925. Permo-Carboniferous glaciation and the Wegener hypothesis. *Nature*, **115**: 603.

Coleman, A. P. 1926. *Ice Ages, Recent and Ancient.* MacMillan, New York.

Coleman, A. P. 1933. Ice ages and the drift of continents. *J. Geol.*, **41**: 409–417.

Coleman, A. P. 1938. Antarctica and glacial ages. *Nature*, **142**: 998–999.

Collet, L. W. 1926. The Alps and Wegener's theory. *Geogr. J.*, **67**: 301–312.

Collet, L. W. 1927. *The Structure of the Alps.* Edward Arnold, London. 1935. 2nd edition, Butler & Tanner Ltd., London.

Compton, R. H. 1923. Botanical aspects of Wegener's hypothesis. *Nature*, **111**: 533.

Condon, M. A. and Öpik, A. A. 1956. *Summary of Continental Drift Symposium, Hobart*, March, 1956. Unpublished manuscript from Öpik's papers, Basser Library, Australian Academy of Sciences.

Cooper, H. S. F. Jr. 1970. Letter from the space center. *The New Yorker*, April 4: 80–112.

Cotton, L. A. 1923. Some fundamental problems of diastrophism and their geological corollaries with special reference to polar wanderings. *Am. J. Sci.*, **6**: 453–501.

Cotton, L. A. 1929. Presidential address: Causes of diastrophism and their status in current geological thought. *Report of the Nineteenth Meeting of ANZAAS* (*Hobart, 1928*), 171–210.

Cotton, L. A. 1945. The pulse of the Pacific. *J. Proc. Roy. Soc. NSW*, **79**: 41–76.

Cotton, L. A. 1948. Ernest Clayton Andrews. *Aust. J. Sci.*, **11**: 48–49.

Cox, L. R. 1958. William Joscelyn Arkell. *Biogr. Mem. Roy. Soc.*, **4**: 1–14.

Cozens, B. 1986. Letter to D. Hill, August 25.

Cranwell, L. M. 1963. Nothofagus: living and fossil. In Gressitt, J. Linsley, ed., *Pacific Basin Biogeography*. Bishop Museum Press, Honolulu, 387–400.

Crick, R. E. and Stanley, G. D., Jr. 1997. Curt Teichert. *J. Paleontol.*, **71**: 750–752.

Crookshank, H. and Auden, J. B. 1956. Lewis Leigh Fermor. *Biogr. Mem. Fell. Roy. Soc.*, **2**: 101–116.

Crowell, J. C. 1952. Probable large lateral displacement on San Gabriel fault, southern California. *Am. Assoc. Petrol. Geol. Bull.*, **36**: 2026–2035.

Crowell, J. C. 1960. The San Andreas fault in southern California. *21st International Geological Congress Report, Part* **18**, 45–52.

Daly, R. A. 1923a. A critical review of the Taylor-Wegener hypothesis. *J. Washington Acad. Sci.*, **13**: 447–448.

Daly, R. A. 1923b. The Earth's crust and its stability. *Am. J. Sci.*, **5**: 349–371.

Daly, R. A. 1923c. Decrease of the Earth's rotational velocity. *Am. J. Sci.*, **5**: 372–377.

Daly, R. A. 1925. Relation of mountain-building to igneous action. *Proc. Am. Phil. Soc.*, **64**: 283–307.

Daly, R. A. 1926. *Our Mobile Earth*. Charles Scribner's Sons, New York.

Daly, R. A. 1929. *Our Mobile Earth* (2nd printing). Charles Scribner's Sons, New York.

Daly, R. A. 1933. *Igneous Rocks and the Depths of the Earth*. McGraw-Hill, New York.

Daly, R. A. 1938. *Architecture of the Earth*. Appleton-Century, New York.

Daly, R. A. 1945. A review of Holmes on physical geology. *Am. J. Sci.*, **243**: 572–575.

Daly, R. A. 1951. Relevant facts and inferences from field geology. In Gutenburg, B., ed., *Internal Constitution of the Earth* (2nd edition). Dover Publications, New York, 23–49.

Darlington, P. J., Jr. 1948. The geographical distribution of cold-blooded vertebrates. *Q. Rev. Biol.*, **23**: 1–26, 105–123.

Darlington, P. J., Jr. 1949. Beetles and continents. *Q. Rev. Biol.*, **24**: 342–345.

Darlington, P. J., Jr. 1957. *Zoogeography*. John Wiley & Sons, New York.

Darrah, W. C. 1939. *Textbook of Paleobotany*. Appleton-Century, New York.

David, E. 1928. Drifting continents: The Wegener hypothesis. *Aust. Geogr.*, **1**: 60–61.

David, T. W. E. (edited and much supplemented by Browne, W. R.) 1950. *The Geology of the Commonwealth of Australia*. Edward Arnold & Co., London.

Davies, A. M. 1923. The life-cycle of the eel in relation to Wegener's hypothesis. *Nature*, **111**: 496.

Davies, A. M. 1931. Comments. *Geogr. J.*, **78**: 537.

Day, A. L. 1910. Mineral relations from laboratory viewpoint. *Geol. Soc. Am. Bull.*, **21**: 141–178.

de Beaufort, L. F. 1951. *Zoogeography of the Land and Inland Waters*. Sidgwick and Jackson, London.

Demhardt, I. J. 2006. Alfred Wegener's hypothesis on continental drift and its discussion in Petermanns Geographische Mitteilungen (1912–1942). *Polarforschung*, **75**: 29–35.

de Sitter, L. U. 1964. *Structural Geology* (2nd edition). McGraw-Hill, New York.

de Terra, H. 1936. Himalayan and Alpine orogenies. *International Geological Congress, XVI, Reports*, **II**: 859–871.

Dey, A. K. 1937. Wegener's theory of continental drift with reference to India and adjacent countries (General discussion). *Proceedings 24th Indian Science Congress, Hyderabad*, 502–520.

Diels, L. 1936. The genetic phytogeography of the southwestern Pacific area, with particular reference to Australia. In Goodspeed, T. H., ed., *Essays in Geobotany in Honor of William Albert Setchell*. University of California Press, Berkeley, 189–194.

Diener, C. 1915. Die groszformen der erdoberflächen. *Mitt. k. k. Geol. Ges. Wien*, **58**: 329–349.

Dixey, F. 1939. Some observations on the physiographical development of central and southern Africa. *Trans. Proc. Geol. Soc. S. Afr.*, **41**: 113–171.

Dixon, H. H. 1934. John Joly. *Obit. Not. Fell. Roy. Soc.*, **1**: 259–314.

Doel, R. E. 1999. Hubbert, Marion King. In *The Handbook of Texas*. Online: www.tshaonline.org/handbook/online/articles/fhu85.

Doel, R. E., Levin, T. J., and Marker, M. K. 2006. Extending modern cartography to the ocean depths: military patronage, Cold War priorities, and the Heezen–Tharp mapping project, 1952–1959. *J. Hist. Geogr.*, **32**: 605–626.

Dott, R. H., Jr. 1974. The geosynclinal concept. In Dott, R. H., Jr., and Shaver, R. H., eds., *Modern and Ancient Geosynclinal Sedimentation: Proceedings of a Symposium Dedicated to Marshall Kay*. SEPM Special Publication, **19**: 1–13.

Dott, R. H., Jr. 1977. Memorial to Marshall Kay. *Mem. Geol. Soc. Am.*, **III**.

Dott, R. H., Jr. 1979. The geosyncline – first major geological concept made in America. In Schneer, C., ed., *Two Hundred Years of Geology in America*. University Press of New England, Hanover, NH, 238–267.

Dott, R. H., Jr. 1985. James Hall's discovery of the craton. In Drake, E. T. and Jordan, W. M., eds., *Geologists and Ideas: A History of North America Geology, GSA Centennial Special Volume 1*. Geological Society of America, Boulder, CO, 157–167.

Doumani, G. A. 1964. Volcanoes of the Executive Committee Range, Byrd Land. In *Antarctic Geology, SCAR Proceedings 1963*. North-Holland Publishing Co., Amsterdam, 666–675.

Doumani, G. A. and Long, W. E. 1962. The ancient life of the Antarctic. *Sci. Am.*, **207**: 163–184.

Dunbar, C. O. 1952. Discussion. In Mayr, E., ed., *The Problem of Land Connections Across the South Atlantic, with Special Reference to the Mesozoic. Bull. Am. Mus. Nat. Hist.*, **99**: 155–158.

Dunham, K. C. 1966. Arthur Holmes. *Biogr. Mem. Fell. Roy. Soc.*, **12**: 291–310.

Dunham, K. C. 1983. Sidney Henry Haughton. *Biogr. Mem. Fell. Roy. Soc.*, **29**: 245–267.

Dunkle, D. H. 1952. Discussion of the evidence of fresh-water fishes. In Mayr, E., ed., *The Problem of Land Connections Across the South Atlantic with Special Reference to the Mesozoic. Bull. Am. Mus. Nat. Hist.*, **99**: 235–236.

Du Rietz, G. E. 1940. Problems of bipolar plant distribution. *Acta Phytogeographica Suecica*, **14**: 215–282.

Du Rietz, G. E. 1960. Discussion: evidence for continental drift and polar wandering. *Proc. Roy. Soc. London*, **152B**: 668–669.

du Toit, A. L. 1912. *Physical Geography for South African Schools*. (2nd edition). Cambridge University Press, Cambridge.

du Toit, A. L. 1921a. Land connections between the other continents and South Africa in the past. *S. Afr. J. Sci.*, **18**: 120–140.

du Toit, A. L. 1921b. The carboniferous glaciation of South Africa. *Trans. Geol. Soc. S. Afr.*, **24**: 188–227.

du Toit, A. L. 1922. The evolution of the South African coast line. *S. Afr. Geogr J.*, **6**: 32–41.

du Toit, A. L. 1924. The contribution of South Africa to the principles of geology. *S. Afr. J. Sci.*, **21**: 52–78.

du Toit, A. L. 1927a. The fossil flora of the upper Karroo beds. *Ann. S. Afr. Mus.*, **22**: 289–420.

du Toit, A. L. 1927b. *A Geological Comparison of South America with South Africa.* Carnegie Institution, Washington.

du Toit, A. L. 1937. *Our Wandering Continents: An Hypothesis of Continental Drifting.* Oliver and Boyd, Edinburgh.

du Toit, A. L. 1939a. The origin of the Atlantic-Arctic-Ocean. *Geologische Rundschau*, **30**: 138–147.

du Toit, A. L. 1939b. Antarctica and glacial ages. *Nature*, **143**: 242–243.

du Toit, A. L. 1940. Observations on the evolution of the Pacific Ocean. In *Proceedings Sixth Pacific Science Congress*, Volume I: 175–183.

du Toit, A. L. 1944. Tertiary mammals and continental drift. *Am. J. Sci.*, **242**: 145–163.

du Toit, A. L. 1945. Further remarks on continental drift. *Am. J. Sci.*, **243**: 404–408.

du Toit, A. L. 1952. *Comparacao Geologica Entre a America do Sul e a Africa, do Sol.* Servico Grafico do Institutio Brasileiro de Geografia e Estatistica, Rio de Janeiro.

Dutton, C. E. 1889. On some of the greater problems of physical geology. *Bull. Phil. Soc. Wash.*, **11**: 51–64.

Edwards, W. N. 1955. The geographical distribution of past floras. *Rep. Br. Assoc. Adv. Sci.*, **46**: 165–176.

Elsasser, W. H. 1971. Sea-floor spreading as thermal convection. *J. Geophys. Res.*, **76**: 1101–1112.

Escher, B. G. 1922. *Over oorzaak en verband der inwendige geologische krachten.* Leyden.

Escher, B. G. 1933. On the relation between the volcanic activity in the Netherlands, East Indies and the belt of negative gravity anomalies discovered by Vening Meinesz. *Proc. Kon. Ned. Akad. Wetensch.*, **36**: 677–685.

Evans, J. W. 1923. The Wegener hypothesis of continental drift. *Nature*, **111**: 393–394.

Evans, J. W. 1924. Introduction. In Wegener, A., *The Origin of Continents and Oceans.* Translated from the 3rd edition by Skerl, S. G. A. Methuen and Company, London, v–xii.

Evans, J. W. 1958. Insect distribution and continental drift. In Carey, S. W., Convener, *Continental Drift: A Symposium.* University of Tasmania, Hobart, 134–161.

Evans, J. W. 1959. The zoogeography of some Australian insects. In Christian, C. S., Crocker, R. L. and Keast. A., eds., *Biogeography and Ecology in Australia.* Dr. W. Junk, The Hague, 150–163.

Evans, P. 1937. Wegener's theory of continental drift with reference to India and adjacent countries (General discussion). *Proceedings 24th Indian Science Congress*, Hyderabad: 502–520.

Ewing, M. 1952. The Atlantic Ocean. In Mayr, E., ed., *The Problem of Land Connections Across the South Atlantic, with Special Reference to the Mesozoic. Bull. Am. Mus. Nat. Hist.*, **99**: 87–92.

Faill, R. T. 1985. Evolving tectonic concepts of the central and southern Appalachians. In Drake, E. T. and Jordan, W. M., eds., *Geologists and Ideas: A History of North American Geology*, GSA Centennial Special Volume I. Geological Society of America, Boulder, CO, 19–43.

Fermor, L. 1944. Gondwanaland: a former southern continent. *Proc. Bristol Nat. Soc.*, **9**: 483–493.

Fermor, L. 1949a. Thomas Henry Holland. *Obit. Not. Fell. Roy. Soc.*, **6**: 83–114.

Fermor, L. 1949b. Garnets and moving continents. *Proc. Rhodesia Sci. Assoc.*, **42**: 11–17.

Fermor, L. 1951. The mineral deposits of Gondwanaland. *Trans. Inst. Min. Metall.*, **60**: 421–465.

Feynman, R. 1999. *The Pleasure of Finding Things Out*. Perseus Publishing, Cambridge, MA.

Fisher, O. 1881. *The Physics of the Earth's Crust*. Macmillan, London.

Fisher, O. 1883. Physics of the Earth's crust. *Nature*, **27**: 76–77.

Fowler, C. M. R. 2000. *The Solid Earth*. Cambridge University Press, Cambridge.

Fox, C. S. 1935. *A Comprehensive Treatise on Engineering Geology*. D. van Nostrand, New York.

Fox, C. S. 1937. Wegener's theory of continental drift with reference to India and adjacent countries (General discussion). *Proceedings 24th Indian Science Congress, Hyderabad*: 502–520.

Frankel, H. 1976. Alfred Wegener and the specialists. *Centaurus*, **20**: 305–324.

Frankel, H. 1978a. Arthur Holmes and continental drift. *Brit. J. Hist. Sci.*, **11**: 130–149.

Frankel, H. 1978b. The Non-Kuhnian nature of the recent revolution in the Earth Sciences. In Hacking, I. and Asquith, P. D., eds., *Proceedings of the Philosophy of Science Association 1978 Biennial Meeting, Vol. 2*. Philosophy of Science Association, East Lancing, MI, 197–214.

Frankel, H. 1979a. The reception and acceptance of continental drift theory as a rational episode in the history of science. In Mauskopf, S. H., ed., *The Reception of Unconventional Science: AAAS Selected Symposium*. Westview Press, Boulder, CO, 51–89.

Frankel, H. 1979b. The career of continental drift theory: an application of Imre Lakatos' analysis of scientific growth to the rise of drift theory. *Stud. Hist. Phil. Sci.*, **12**: 130–149.

Frankel, H. 1980a. Problem-solving, research traditions, and the development of scientific fields. In Asquith, P. D., ed., *Proceedings of the Philosophy of Science Association 1980 Biennial Meeting, Vol. I*. Philosophy of Science Association, East Lancing, MI, 26–40.

Frankel, H. 1980b. Hess's development of his seafloor spreading hypothesis. In Nickles, T., ed., *Scientific Discovery: Case Histories*. D. Reidel Publishing, Dordrecht, 345–366.

Frankel, H. 1981. The paleobiogeographical debate over the problem of disjunctively distributed life forms. *Stud. Hist. Phil. Sci.*, **12**: 211–259.

Frankel, H. 1982. The development, reception, and acceptance of the Vine-Matthews–Morley hypothesis. *Hist. Stud. Phys. Sci.*, **13**: 1–39.

Frankel, H. 1984a. Biogeography, before and after the rise of sea floor spreading. *Stud. Hist. Phil. Sci.*, **15**: 141–168.

Frankel, H. 1984b. The Permo-Carboniferous ice cap and continental drift. *Compte rendu de Neuvième Congrès International de Stratigraphie et de Géologie du Carbonifère*, **1**: 113–120.

Frankel, H. 1998. Continental drift and the plate tectonics revolution. In Good, G. A., ed., *Sciences of the Earth: An Encyclopedia of Events, People and Phenomena, Garland Encyclopedia in the History of Sciences III.* Routledge, New York, 118–136.

Fränkl, E. 1956. Some general remarks on the Caledonian mountain chain of East Greenland. *Meddel. Grønland*, **103**, Nr. 11: 1–43.

Friedman, R. M. 1982. Constituting the polar front, 1919–1920. *Isis*, **73**: 343–362.

Fulton, T. W. 1923. The life-cycle of the eel in relation to Wegener's hypothesis. *Nature*, **111**: 359–360.

Gagnebin, E. 1922. La dérive des continents selon la théorie d'Alfred Wegener. *Rev. Gen. Sci. Pures Appl.*, Paris, **36**: 139–142.

Gagnebin, E. 1946. *Histoire de la terre et des Êtres vivants*. Guilde du Livre, Lausanne, Switzerland.

Georgi, J. 1935. *MID-ICE: The Story of the Wegener Expedition to Greenland.* English translation by Lyon, F. H. E. P. Dutton and Co., New York.

Georgi, J. 1962. Memories of Alfred Wegener. In Runcorn, S. K., ed., *Continental Drift*. Academic Press, London, 309–324.

Gerth, H. 1935. *Geologie Südamerikas*. Berlin, Germany.

Gerth, H. 1939. Stratigraphische und faunistische Grundlagen zur geologischen geschichte des Sudatlantischen Raumes. *Geologische Rundschau*, **30**: 64–79.

Gevers, T. W. 1944. Acceptance speech for the Draper Memorial Medal. *Trans. Proc. Geol. Soc. S. Afr.*, **47**: liii–liv.

Gevers, T. W. 1949. The life and work of Dr. Alex L. du Toit: Alex L. du Toit Memorial Lecture. *Trans. Proc. Geol. Soc. S. Afr.* (Annexure to) **52**: 1–109.

Geyer, Otto F. 1994. Prof. Dr. Hermann Aldinger (1.2.1902–20.12.1993). *Dtsch. Geol. Ges. Nachr.*, **51**: 12–16.

Gilbert, G. K. 1896. The origin of hypotheses, illustrated by the discussion of a topographic problem. *Science, n.s.* **3**: 1–12.

Glaessner, M. 1945. *Principles of Micropalaeontology*. Melbourne University Press, Melbourne, in association with Oxford Univeristy Press.

Glaessner, M. 1950. Geotectonic position of New Guinea. *Bull. Am. Assoc. Petrol. Geol.*, **34**: 856–881.

Glaessner, M. 1962. Isolation and communication in the geological history of the Australian fauna. In Leeper, G. W., ed., *The Evolution of Living Organisms.* Melbourne University Press, Melbourne, 242–249.

Glaessner, M. 1984. *The Dawn of Animal Life: A Biohistorical Study.* Cambridge University Press, Cambridge.

Godthelp, H., Archer, M., Cifelli, R., Hand, S. J., and Gilkeson, C. F. 1992. Earliest known Australian Tertiary mammal fauna. *Nature*, **356**: 514–516.

Gold, E. 1965. Sir George Simpson, K.C.B., F.R.S. *Nature*, **205**: 1156.

Good, R. 1947. *The Geography of the Flowering Plants* (1st edition). Longmans, Green & Co., London.

Good, R. 1950. Present position of the theory of continental drift. *Nature*, **166**: 585–586.

Gordon, R. G. and Stein, S. 1992. Global tectonics and space geodesy (large scale measurements). *Science*, **256**: 333–342.

Graham, K. W. T. and Hales, A. L. 1957. Palaeomagnetic measurements on Karroo dolerites. *Adv. Phys.*, **6**: 149–161.

Green, R. and Irving, E. 1958. The palaeomagnetism of the Cainozoic basalts from Australia. *Proc. Roy. Soc. Victoria*, **70**: 1–17.

Greene, M. T. 1982. *Geology in the Nineteenth Century: Changing Views of a Changing World.* Cornell University Press, Ithaca, NY.

Greene, M. T. 1984. Alfred Wegener. *Soc. Res.*, **51**: 739–761.

Greene, M. T. 1998. Alfred Wegener and the origin of lunar craters. *Earth Sci. Hist.*, **17**: 111–138.

Greene, M. T. 1999. Archival versus canned history. *Earth Sci. Hist.*, **18**: 336–343.

Gregory, J. W. 1905. The face of the Earth. *Nature*, **72**: 193–194.

Gregory, J. W. 1928. Wegener's hypothesis. In van Waterschoot van der Gracht, W., *Theory of Continental Drift: A Symposium on the Origin and Movement of Land Masses both Inter-continental and Intra-continental, as Proposed by Alfred Wegener.* American Association of Petroleum Geologists, Tulsa, OK, 93–96.

Gregory, J. W. 1930. Discussion on geological climates. *Proc. Roy. Soc. London, B*, **106**: 309–312.

Griggs, D. 1939. A theory of mountain-building. *Am. J. Sci.*, **237**: 611–650.

Griggs, D. 1951. Summary of the convection-current hypothesis of mountain building. *Trans. Am. Geophys. Union*, **32**: 527–528.

Griggs, D. 1974. Response by David Griggs upon receiving the Arthur L. Day Medal. *Geol. Soc. Am. Bull.*, **85**: 1342–1343.

Guimarães, D. 1964. *Geologia do Brasil.* Ministério das Minas e Energias, Rio de Janeiro.

Gulley, M. G. 1944. Willem A. J. M. van Waterschoot van der Gracht (1873–1940). *Bull. Am. Assoc. Petrol. Geol.*, **28**: 1066–1070.

Gutenberg, B. 1951. Summary: colloquium on plastic flow and deformation within the Earth. *Trans. Am. Geophys. Union*, **32**: 539–543.

Gutenberg, B. and Richter, C. F. 1954. *Seismicity of the Earth.* Princeton University Press, Princeton, NJ.

Guyot, E. 1943. L'emploi d'une projection cylindrique oblique. *Soc. Neuchatel Sci. Nat. Bull.*, **68**: 105–111.

Hales, A. L. 1936. Convection in the Earth. *Mon. Not. Roy. Astron. Soc.*, **3**: 372–379.

Hales, A. L. 1986. Geophysics on three continents. *Am. Rev. Earth Planet. Sci.*, **14**: 1–20.

Hallam, A. 1973. *A Revolution in the Earth Sciences: From Continental Drift to Plate Tectonics.* Clarendon Press, Oxford.

Hallam, A. 1983. *Great Geological Controversies.* Oxford University Press, Oxford.

Haller, J. 1976. Termier, Pierre. In Gillispie, C. C., ed., *Dictionary of Scientific Biography*, XIII. Charles Scribner's Sons, New York, 283–286.

Hamblin, J. D. 2005. *Oceanographers and the Cold War: Disciples of Marine Science.* University of Washington Press, Seattle.

Hamilton, W. 1960. *New Interpretation of Antarctic Tectonics.* USGS Professional Paper 400-B: 379–380.

Hamilton, W. 1961. Origin of the Gulf of California. *Am. Bull. Geol. Soc.*, **72**: 1307–1318.

Hamilton, W. 1963a. Polar wandering and continental drift: an evaluation of recent evidence. In Munyan, A. C., ed., *Polar Wandering and Continental Drift.* Society of Economic Paleontologists and Mineralogists, Special Publication No. 11: 74–93.

Hamilton, W. 1963b. Tectonics of Antarctica. *Am. Assoc. Petrol. Geol., Mem.*, **2**: 4–11.

Hamilton, W. 1964a. Tectonic map of Antarctica: a progress report. In *Antarctic Geology, SCAR Proceedings 1963.* North-Holland Publishing Co., Amsterdam, 676–680.

Hamilton, W. 1964b. Discussion of paper by D. I. Axelrod, fossil floras suggest stable, not drifting, continents. *J. Geophys. Res.*, **69**: 1666–1671.

Hamilton, W. 1965. *Diabase Sheets of the Taylor Glacier Region, Victoria Land, Antarctica*. USGS Professional Paper 465-B: B1–B71.

Hamilton, W. 1966. Formation of the Scotia and Caribbean Arcs: *Can. Geol. Survey Paper*, 66–15: 178–187.

Hamilton, W. 1967. Tectonics of Antarctica. *Tectonophysics*, **4**: 555–568.

Hamilton, W. 1968. Cenozoic climatic change and its cause. *Am. Meteorol. Soc. Meteorol. Monogr.*, **8**: 128–133.

Hamilton, W. 1969. Mesozoic California and the underflow of Pacific mantle. *Geol. Soc. Am. Bull.*, **80**: 2409–2430.

Hamilton, W. 1970. The Uralides and the motion of the Russian and Siberian Platforms. *Geol. Soc. Am. Bull.*, **81**: 2553–2576.

Hamilton, W. 1988. Plate tectonics and island arcs. *Geol. Soc. Am. Bull.*, **100**: 1503–1527.

Hamilton, W. 1998. Archean magmatism and tectonics were not products of plate tectonics. *Precambrian Res.*, **91**: 143–179.

Hamilton, W. 2002. The closed upper-mantle circulation of plate tectonics. In *Plate Boundary Zones*, AGU Geodynamics Series, **30**: 359–410.

Hamilton, W. and Hayes, P. T. 1959. United States Geological Survey field work in South Victoria Land, 1958–1959. *Polar Record*, **9**: 575.

Hamilton, W. and Myers, W. B. 1966. Cenozoic tectonics of the western United States. *Rev. Geophys.*, **4**: 509–549.

Harland, W. B. 1960. An outline structural history of Spitsbergen. In Raasch, G. O., ed., *Geology of the Arctic*. University of Toronto Press, Toronto, 68–132.

Harrison, L. 1924. The migration route of the Australian marsupial fauna. *Aust. Zool.*, **3**: 247–263.

Harrison, L. 1926. The Wegener hypothesis from a biological standpoint. *Sydney Univ. Sci. J.*, **10**: 10–19.

Harrison, L. 1928. The composition and origins of the Australian fauna, with special reference to the Wegener Hypothesis. *Rep. Aust. Assoc. Adv. Sci.*, **18**: 332–396.

Haughton, S. H. 1953. Gondwanaland and the distribution of early reptiles: Alex. L. du Toit Memorial Lectures No. 3. *Trans. Proc. Geol. Soc. S. Afr.* (Annexure to), **56**: 1–30.

Haughton, S. H. 1949. Alexander Logie du Toit. *Obit. Not. Fell. Roy. Soc.*, **VI**: 385–395.

Haughton, S. H. 1962. Fifty years of geology in parts of Africa. *Proc. Geol. Soc. S. Afr.*, **65**: vii–xxii.

Hebda, R. J. and Irving, E. 2004. On the origin and distribution of Magnolias: Tectonics, DNA, and climate change. In *Timescales of the Paleomagnetic Field*, AGU Geophysical Monograph Series, **145**: 43–57.

Heim, A. 1936. Energy sources of the earth's crustal movements. *International Geological Congress, XVI, Reports*, **II**: 909–924.

Hendrix P. F. (ed). 1995. *Earthworm Ecology and Biogeography in North America*. Lewis Publishers, Boca Raton, FL.

Hennig, E. 1939. Daten aus dem Werdegange des Sudatlantischen Raumes. *Geologische Rundschau*, **30**: 80–85.

Heritsch, F. 1923. *Die Grundlagen der Alpinen tektonik*. Gebrüder Borntraeger, Berlin.

Heritsch, F. 1927. *Die deckentheorie in den Alpen*. Fortschritte der geologie und palaeontologie, 6. Gebrüder Borntraeger, Berlin. English translation (1929) by Boswell, P. G. H., *The Nappe Theory in the Alps (1905–1928)*. Methuen, London.

Hess, H. H. 1939. Island arcs, gravity anomalies and serpentinite intrusions, a contribution to the ophiolite problem. *Reports 17th International Geological Congress (Moscow)*, **2**: 262–283.

Hess, H. H. 1951. Comment on mountain building. *Trans. Am. Geophys. Union*, **32**: 528–531.

Hill, D. 1957. The sequence and distribution of Upper Palaeozoic coral faunas. *Aust. J. Sci.*, **19**: P42–P61.

Hill, D. 1959a. Sakmarian geography. *Geologische Rundschau*, **47**: 590–629.

Hill, D. 1959b. Distribution and sequence of Silurian coral faunas. *J. Proc. Roy. Soc. NSW*, **92**: 151–173.

Hill, D. 1966. Walter Heywood Bryan. *J. Geol. Soc. Aust.*, **13**: 613–618.

Hill, D. 1970. Clarke Memorial Lecture for 1971: The bearing of some Upper Palaeozoic reefs and coral faunas on the hypothesis of continental drift. *J. Proc. Roy. Soc. NSW*, **103**: 92–102.

Hill, D. 1987. Edwin Sherbon Hills. *Biogr. Mem. Fell. Roy. Soc.*, **33**: 291–323.

Hill, M. L. and Dibblee, T. W., Jr. 1953. San Andreas, Garlock and Big Pine faults, California. *Geol. Soc. Am. Bull.*, **64**: 443–458.

Hills, E. S. 1945. Some aspects of the tectonics of Australia. *J. Proc. Roy. Soc. NSW*, **79**: 67–91.

Hills, E. S. 1955. Die landoberfläche Australiens. Translated by Loewe, F. *Die Erde, Z. Ges. Erdk.* Berlin, 3–4: 195–205.

Hills, E. S. 1958. A brief review of Australian fossil vertebrates. In Westoll, T. S., ed., *Studies on Fossil Vertebrates Presented to David Meredith Seares Watson*. University of London Athlone Press, London.

Hinks, A. R. 1931. Comments. *Geogr. J.*, **78**: 433–440; 536.

Hinton, H. E. 1951. The Wegener-Du Toit theory of continental displacement and the distribution of animals. *Adv. Sci.*, **8**: 74–78.

Hoeg, E. 1937. Plant fossils and paleogeographical problems. *Compt. Rend. 2 Cong. Avanc. Etud. Stratigr. Carbonifere*. Van Aelst, Maastricht.

Holland, T. H. 1941. The evolution of continents: a possible reconciliation of conflicting evidence. *Proc. R. Soc. Edinburgh, Section B*, **61**: 146–166.

Holland, T. H. 1944. The theory of continental drift. *Proc. Linnaean Soc.*, **155**: 112–125.

Hollingworth, S. E. 1944. Review of principles of geology. *Geogr. J.*, **104**: 119–122.

Hollingworth, S. E. 1965. Frederick Everard Zeuner. *Proc. Geol. Soc. London*, **1618**: 122–123.

Holmes, A. 1913. *The Age of the Earth*. Harper, London.

Holmes, A. 1915a. Radioactivity and the Earth's thermal history: Part I, The concentration of the radioactive elements in the Earth's crust. *Geol. Mag.*, **2**: 60–71; 1915b. Part II, Radioactivity and the Earth as a cooling body. *Geol. Mag.*, **2**: 102–112; 1916. Part III, Radioactivity and isostasy. *Geol. Mag.*, **4**: 264–274; 1925a. Part IV, A criticism of Parts I, II, and III. *Geol. Mag.*, **62**: 504–515; and 1925b. Part V, The control of geological history by radioactivity. *Geol. Mag.*, **62**: 529–544.

Holmes, A. 1920. *The Nomenclature of Petrology*. Murby, London.

Holmes, A. 1925c. Radioactivity and Earth history. *Geogr. J.*, **65**: 528–532.

Holmes, A. 1926. Contributions to the theory of magmatic cycles. *Geol. Mag.*, **63**: 306–329.

Holmes, A. 1927. Some problems of physical geology and the Earth's thermal history. *Geol. Mag.*, **64**: 263–278.

Holmes, A. 1928a. Radioactivity and continental drift. *Geol. Mag.*, **65**: 236–238.

Holmes, A. 1928b. Radioactivity and continental drift. *Nature*, **122**: 431–433.

Holmes, A. 1929. A review of the continental drift hypothesis. *Mining Mag.*, **40**: 205–209, 286–288, 340–347.

Holmes, A. 1931a. Radioactivity and earth movements. *Trans. Geol. Soc. Glasgow (for 1928–29)*, **18**: 559–606.

Holmes, A. 1931b. Problems of the earth's crust. *Geogr. J.*, **78**: 445–451, 541–542.

Holmes, A. 1933. The thermal history of the earth. *J. Wash. Acad. Sci.*, **23**: 169–195.

Holmes, A. 1945. *Principles of Physical Geology*. Thomas Nelson, London (American Printing, Ronald Press, New York).

Holmes, A. 1953. The South Atlantic: land bridges or continental drift? *Nature*, **171**: 669–671.

Holmes, A. 1965. *Principles of Physical Geology*. The Ronald Press Company, New York.

Holtedahl, O. 1913. On the old red sandstone series of Northwestern Spitzbergen. *XIIth International Geological Congress, Toronto, Compte-Review*, 707–712.

Holtedahl, O. 1920a. Paleogeography and diastrophism in the Atlantic-Arctic region during Paleozoic time. *Am. J. Sci.*, **49**: 1–25.

Holtedahl, O. 1920b. The Scandinavian 'Mountain Problem'. *Geol. Soc. London, Q. J.*, **76**: 387–402.

Holtedahl, O. 1925. Some points of structural resemblance between Spitsbergen and Great Britain, and between Europe and North America. *Avh. Nor. Vidensk.-Akad. Oslo, Mat.-Naturvidensk. Kl.*, No. 4.

Holtedahl, O. 1926. Tectonics of Arctic regions. *Pan-Am. Geol.*, **46**: 257–272.

Holtedahl, O. 1936. Trekk av det skandinaviske fjellkjedestrøks historie. *Nordiska (19 Skandinaviska) Naturforskaremöten I Helsingfors*, 129–145.

Hora, S. L. 1937. Wegener's theory of continental drift with reference to India and adjacent countries (General discussion). *Proceedings 24th Indian Science Congress*, Hyderabad, 502–520.

Hospers, J. 1953. Reversals of the main geomagnetic field, part I. *Proc. Roy. Neth. Acad. Sci., Amsterdam, Series B*, **56**: 467–476.

Hospers, J. 1954. Rock magnetism and polar wandering. *Nature*, **173**: 1183–1184.

Hospers, J. 1955. Rock magnetism and polar wandering. *J. Geol.*, **63**: 59–74.

Hsü, K. J. 1995. *The Geology of Switzerland*. Princeton University Press, Princeton, NJ.

Hubbert, M. K. 1937. Theory of scale models as applied to the study of geologic structures. *Am. Bull. Geol. Soc.*, **48**: 1949–1520.

Huene, F. von. 1939. Fossile landfaumen und das Atlantis-problem. *Geologische Rundschau*, **30**: 86–89.

Hume, W. F. 1948. *Terrestrial theories: A Digest of Various Views as to the Origin and Development of the Earth and Their Bearing on the Geology of Egypt*. Government Press, Cairo.

Hutchinson, J. 1946. *A Botanist in Southern Africa*. P. R. Hawthorn, Ltd., London.

Irving, E. 1956. Palaeomagnetic and palaeoclimatological aspects of polar wandering. *Geofis. Pura Appl.*, **33**: 23–48.

Irving, E. 2008. Why Earth became so hot 50 million years ago and why it then cooled. *Proc. Natl. Acad. Sci. USA*, **105**: 16061–16062.

Irving, E., Roberston, W. A., and Stott, P. M. 1963. The significance of the paleomagnetic results from Mesozoic rocks of eastern Australia. *J. Geophys. Res.*, **68**: 2313–2317.

Isacks, B. L., Oliver, J., and Sykes, L. 1968. Seismology and the new global tectonics. *J. Geophy. Res.*, **73**: 5855–5899.

Jeffreys, H. 1924. *The Earth: Its Origin, History and Physical Constitution* (1st edition). Cambridge University Press, Cambridge.

Jeffreys, H. 1926a. On Professor Joly and the Earth's thermal history. *Phil. Mag.*, **1**: 923–931.

Jeffreys, H. 1926b. The Earth's thermal history, and some related problems. *Geol. Mag.*, **63**: 516–525.

Jeffreys, H. 1927. The Earth's thermal history. *Geol. Mag.*, **64**: 444–446.

Jeffreys, H. 1928. On Professor Joly and the Earth's thermal history. *Phil. Mag.*, **5**: 208–214.

Jeffreys, H. 1929. *The Earth: Its Origin, History and Physical Constitution* (2nd edition). Cambridge University Press, Cambridge.

Jeffreys, H. 1931. Problems of the Earth's crust. *Geogr. J.*, **78**: 451–453.

Jeffreys, H. 1935. *Earthquakes and Mountains* (1st edition). Methuen & Co. Ltd., London.

Jeffreys, H. 1979. *The Earth: Its Origin, History and Physical Constitution* (6th edition). Cambridge University Press, Cambridge.

Jensen, P. F. 1923. Ekspeditionen til Vestgrönland Sommeren 1922. *Meddel. Grönland*, **63**: 205–283, Copenhagen.

Ji, Q., Luo, Z.-X., Yuan, C.-X., Wible, J. R., Xhang, J.-P., and Georgi, J. A. 2002. The earliest known eutherian mammal. *Nature*, **416**: 816–822.

Joleaud, L. 1924. L'histoire biogeographique de l'Amerique et la theorie de Wegener. *J. Soc. Am. Paris*, **16**: 325–360.

Joly, J. 1923a. Continental flotation and drift. *Nature*, **11**: 80–81.

Joly, J. 1923b. The surface movements of the Earth's crust. *Nature*, **111**: 603–606.

Joly, J. 1923c. The movements of the Earth's surface crust. *Phil. Mag.*, **45**: 1167–1188.

Joly, J. 1925. *The Surface-History of the Earth* (1st edition). Oxford University Press, London.

Joly, J. 1926. The surface history of the Earth. *Phil. Mag.*, **1**: 932–939.

Joly, J. 1927. Dr. Jeffreys and the Earth's thermal history. *Phil. Mag.*, **4**: 338–348.

Joly, J. 1928a. Continental drift. In van Waterschoot van der Gracht, W., ed., *Theory of Continental Drift: A Symposium on the Origin and Movement of Land Masses Both Inter-continental and Intra-continental, as Proposed by Alfred Wegener.* American Association of Petroleum Geologists, Tulsa, OK, 88–89.

Joly, J. 1928b. The Earth's thermal history. *Phil. Mag.*, **5**: 215–221.

Joly, J. 1930. *The Surface-History of the Earth* (2nd edition). Oxford University Press, London.

Jongmans, W. J. 1937. The flora of the Upper Carboniferous of Djambi (Sumatra, Netherl., India) and its possible bearing on the paleogeography of the Carboniferous. *Compte Rendu, 2e Congress de Stratigraphie Carbonifère, Heerlen, 1935*, **1**, 345–362.

Joyce, J. R. F. 1951. The relation of the Scotia Arc to Pangea. *Adv. Sci.*, **8**: 82–88.

Just, T. 1952. Fossil floras of the southern hemisphere and their phytogeographical significance. In Mayr, E., ed., *The Problem of Land Connections Across the South Atlantic, With Special Reference to the Mesozoic. Bull. Am. Mus. Nat. Hist.*, **99**: 189–204.

Kay, M. 1941. Classification of the Artinskian series in Russia. *Bull. Am. Assoc. Petrol. Geol.*, **29**: 1396–1404.

Kay, M. 1942. Development of the Allegheny syclinorium and adjoining regions. *Bull. Am. Geol. Soc.*, **53**: 1601–1658.

Kay, M. 1944. Geosynclines in continental development. *Science*, **99**: 1944.

Kay, M. 1947. Geosynclinal nomenclature and the craton. *Bull. Am. Assoc. Petrol. Geol.*, **31**: 1289–1293.

Kay, M. 1951. *North American Geosynclines*. Geological Society of America Memoir, **48**.

Kay, M. 1952a. Stratigraphic evidence bearing on the hypothesis of continental drift. In Mayr, E., ed., *The Problem of Land Connections Across the South Atlantic, With Special Reference to the Mesozoic. Bull. Am. Mus. Nat. Hist.*, **99**: 159–162.

Kay, M. 1952b. Modern and ancient island arcs. *Palaeobotanist*, **1**: 281–283.

Kay, M. 1969. Continental drift in North America. In Kay, M., ed., *North Atlantic: Geology and Continental Drift*, AAPG Memoir, **12**: 965–973.

Kay, M. 1974. Reflections. In Dott, R. H., Jr. and Shaver, R. H., eds., *Modern and Ancient Geosynclinal Sedimentation: Proceedings of a Symposium Dedicated to Marshall Kay*, SEPM Special Publication, **19**: 377–380.

Kay, M. 1976. Stille, Wilhelm Hans. In Gillispie, C. C., ed., *Dictionary of Scientific. Biography, XIII*. Charles Scribner's Sons, New York, *63–65*.

Kay, M. and Colbert, E. 1965. *Stratigraphy and Life History*. John Wiley & Sons, New York.

Kearey, P., Klepeis, K. A., and Vine, F. J. 2009. *Global Tectonics* (3rd edition). Wiley-Blackwell, Oxford.

Keast, A. 1959. The Australian environment. In Christian, C. S., Crocker, R. L., and Keast, A., eds., *Biogeography and Ecology in Australia*. Dr. W. Junk, The Hague, 16–35.

Keast, A. 1963. Mammalian evolution on the southern continents. *Proceedings 16th International Congress on Zoology*, **4**: 37–39.

Keast, A. 1968a. Australian mammals: zoogeography and evolution. *Q. Rev. Biol.*, **43**: 373–408.

Keast, A. 1968b. Introduction: the southern continents as backgrounds for mammalian evolution. *Q. Rev. Biol.*, **43**: 225–233.

Keidel, H. 1939. Über die 'Gondwaniden' Argentiniens. *Geologische Rundschau*, **30**: 148–249.

Kemp, T. S. 2005. *The Origin and Evolution of Mammals*. Oxford University Press, New York.

King, L. 1950. Speculations upon the outline and disruption of Gondwanaland. *Geol. Mag.*, **87**: 353–359.

King, L. 1953. Necessity for continental drift. *Bull. Am. Assoc. Petrol. Geol.*, **37**: 2163–2177.

King, L. 1958a. A new reconstruction of Laurasia. In Carey, S. W., Convener, *Continental Drift: A Symposium*. University of Tasmania, Hobart, 13–23.

King, L. 1958b. The origin and significance of the great sub-oceanic ridges. In Carey, S. W., Convener, *Continental Drift: A Symposium*. University of Tasmania, Hobart, 62–102.

King, L. 1958c. Basic palaeogeography of Gondwanaland during the Late Palaeozoic and Mesozoic eras. *Geol. Soc. London, Q. J.*, **14**: 47–70.

King, L. 1958d. Reply to comments. *Geol. Soc. London, Q. J.*, **14**: 75–77.

King, L. 1961. The palaeoclimatology of Gondwanaland during the Palaeozoic and Mesozoic eras. In Nairn, A. E. M., ed., *Descriptive Paleoclimatology*. Interscience Publishers Inc., New York, 307–331.

King, L. 1964. Summary of symposium. In *Antarctic Geology, SCAR Proceedings 1963*. North-Holland Publishing Co., Amsterdam, 727–732.

King, L. 1965. Geological relationships between South Africa and Antarctica. Alex L. du Toit Memorial Lecture No. 9. *Geol. Soc. S. Afr.* (Annexure to), 67: 1–31.

King, W. B. R. 1945. Review of principles of physical geology. *Geol. Mag.*, 82: 46–47.

Kirsch, G. 1928. *Geologie und Radioaktivität*. Springer, Vienna and Berlin.

Kirsch, G. 1939. Diskussionsbemerkungen. *Geologische Rundschau*, 30: 310–311.

Knetsch, G. 1939. Atlantis (Zur geologie des sudatlantischen ozeans). *Geologische Rundschau*, 30: 250–283.

Knoll, A. H. 2003. *Life on a Young Planet*. Princeton University Press, Princeton, NJ.

Kober, L. 1921. *Der bau der Erde*. Gebrüder Borntraeger, Berlin.

Koch, J. P. 1917. Danmark-Ekspeditionen til Grönlands Nordöstkyst 1908/08 under Ledelsen af L. Mylius-Erichsen, 6. *Medd. Grønland*, 46.

Koch, L. 1929. The geology of East Greenland. *Medd. Grønland*, 73, Nr. 2: 1–204.

Kölbl, L. 1949. Franz Eduard Sueß. *Mitt. Geol. Ges Wien*, 36–38: 267–284.

Köppen, W. 1921. Polwanderungen, verschiebungen der kontinente und klimugeschichte. *Petermanns Geogr. Mitt.*, 67: 145–149, 191–194.

Köppen, W. and Wegener., A. 1924. *Die Klimate der Geologischen Vorzeit*. Gebrüder Borntraeger, Berlin.

Krenkel, E. 1925. *Geologie Afrikas.*, Volume 1. Gebrüder Borntraeger, Berlin.

Krenkel, E. 1939. Oberkretazische alkalimagmen am Atlantischen saume Afrikas. *Geologische Rundschau*, 30: 61–63.

Krige, L. J. 1926. On mountain building and continental sliding. *S. Afr. J. Sci.*, 23: 206–215.

Krige, L. J. 1930. Magmatic cycles, continental drift and ice ages. *Proc. Geol. Soc. S. Afr.*, 32: 21–40.

Krishnan, M. S. 1943. *Geology of India and Burma*. Madras Law Journal Office, Madras; 2nd edition, 1949; 3rd edition, 1958. Higginbothams (Private) Ltd, Madras; 4th edition, 1960. The Associated Printers (Madras) Private Ltd, Madras.

Krishnan, M. S. 1959. History of the Indian Ocean. In Sears, M., ed., *International Oceanographic Congress*. American Association for the Advancement of Science, Washington, DC, 34–35.

Kuenen, Ph. H. 1933. De Beweging van Australië ten opzichte van Nederlandsch Indië. *Vakbl. Biol.*, 14.

Kuenen, Ph. H. 1934. Relations between submarine topography and gravity field. In *Gravity Expeditons at Sea 1923–1932*, Volume II. Netherlands Geodetic Commission, Delft, 182–194.

Kuenen, Ph. H. 1935. Geological interpretation of the bathymetrical results. *The Snellius Expedition, V, pt. 1*. Kemink en Zoon N. V., Utrecht.

Kuenen, Ph. H. 1950. *Marine Geology*. John Wiley & Sons, New York.

Kullenberg, B. 1946. Über verbreitung und wanderungen von vier *Sterna*-arten. *Arkiv Zool.*, 38: 1–80.

Kummerow, E. H. E. 1939. Paläontologie und drifthypothese. *Geologische Rundschau*, 30: 95–99.

Lake, P. 1922a. Wegener's displacement theory. *Nature*, 110: 77.

Lake, P. 1922b. Wegener's displacement theory. *Geol. Mag.*, 59: 338–346.

Lake, P. 1923. Wegener's hypothesis of continental drift. *Nature*, 111: 226–228.

Lam, H. J. 1934. Materials towards a study of the flora of the island of New Guinea. *Blumea*, **1**: 115–159.

Lam, H. J. 1938. On the phylogeny of the Malaysian burseraceae-canarieae in general and of haplolobus in particular. *Blumea*, **3**: 126–158.

Lambert, W. D. 1921. Some mechanical curiosities connected with the Earth's field of force. *Am. J. Sci.*, **2**: 129–158.

Lambert, W. D. 1923. The mechanics of the Taylor-Wegener hypothesis of continental drift. *J. Wash. Acad. Sci.*, **13**: 448–450.

Laporte, L. F. 1985. Wrong for the right reasons: G. G. Simpson and continental drift. In Drake, E. T. and Jordan, W. M., eds., *Geologists and Ideas: A History of North American Geology, GSA Centennial Special Volume I*. Geological Society America, Boulder, CO, 273–284.

Laudan, R. 1980. Oceanography and geophysical theory in the first half of the twentieth century: The Dutch school. In Sears, M. and Merryman, D., eds., *Oceanography: The Past*. Springer-Verlag, New York, 656–666.

Laudan, R. 1985. Frank Bursley Taylor's theory of continental drift. *Earth Sci. Hist.*, **4**: 118–121.

Lawson, A. C. 1932. Insular arcs, foredeeps and geosynclinal seas of the Asiatic coast. *Bull. Geol. Soc. Am.*, **43**: 353–381.

Le Grand, H. E. 1986. Specialities, problems and localism. *Earth Sci. Hist.*, **5**: 84–95.

Le Grand, H. E. 1988. *Drifting Continents and Shifting Theories*. Cambridge University Press, Cambridge.

Lehmann, K. 1939. Pingen-forschung und Wegenersche theorie. *Geologische Rundschau*, **30**: 352.

Leme, A. B. P. 1929. État des connaissances géologiques sur le Brésil. *Bull. Soc. Géol. France*, **29**: 35–87.

Leuchs, K. 1939. Tektonik und lithogenese. *Geologische Rundschau*, **30**: 353–356.

Leuchs, K. 1947. Franz Eduard Sueß. *Akademie der Wissenschaften in Wien, Almanach für das 1945*: 319–323.

Leverett, F. 1939. Memorial to Frank Bursley Taylor. *Proc. Geol. Soc. Am. 1938, May issue*: 191–200.

Leverett, F and Taylor, F. 1915. *The Pleistocene of Indiana and Michigan and the History of the Great Lakes*. USGS Monograph 53.

Lewis, C. 2000. *The Dating Game*. Cambridge University Press, Cambridge.

Lindroth, C. H. 1957. *The Faunal Connections Between Europe and North America*. John Wiley & Sons, New York.

Long, W. E. 1962a. Permo-Carboniferous glaciation in Antarctica. *Geol. Soc. Am. (Abstr.) Special Paper*, **68**: 314.

Long, W. E. 1962b. Sedimentary rocks of the Buckeye Range, Horlick Mountains, Antarctica. *Science*, **136**: 319–321.

Long, W. E. 1964. The stratigraphy of the Horlick Mountains. In *Antarctic Geology, SCAR Proceedings 1963*. North-Holland Publishing Co., Amsterdam, 352–363.

Longwell, C. R. 1927. Review of *Our Mobile Earth*. *Am. J. Sci.*, **13**: 524–525.

Longwell, C. R. 1928. Some physical tests of the displacement hypothesis. In van Waterschoot van der Gracht, W., ed., *Theory of Continental Drift: A Symposium on the Origin and Movement of Land Masses both Inter-continental and Intra-continental, as Proposed by Alfred Wegener*. American Association of Petroleum Geologists, Tulsa, OK, 145–157.

Longwell, C. R. 1938. Continents adrift: review of Alex du Toit's *Our Wandering Continents. Geogr. Rev.*, **28**: 704–705.

Longwell, C. R. 1944a. Some thoughts on the evidence for continental drift. *Am. J. Sci.*, **242**: 218–231.

Longwell, C. R. 1944b. Further discussion of continental drift. *Am. J. Sci.*, **242**: 514–515.

Longwell, C. R. 1944c. Determinations of geographic coordinates in Greenland. *Science*, **100**: 403–404.

Luo, Z.-X., Ji, Q., Wible, J. R., and Yuan, C.-X. 2003. An Early Cretaceous tribosphenic mammal and metatheira evolution. *Science*, **302**: 1934–1940.

Ma, T. Y. H. 1956. A reinvestigation of climate and the relative positions of continents during the Devonian. *Research on the Past Climate and Continental Drift*, **9**.

Maack, R. 1934. Die Godwanaschichten in Suedbrasilien und ihre Beziehungen zur Kaokoformation Suedwestafrikas. *Z. Ges. Kdrdlk. Erdkunde Berlin*, **5/6**: 194–222.

Maack, R. 1953. O desenvolvimento das camadas Gondwanicas do sul do Brasil e suas relacoes com as formacoes Karru da Africa do Sol. *Arq. Biol. Tecnol.*, **VII**: 201–253.

MacBride, E. W. 1938. Antarctica and glacial ages. *Nature*, **142**: 97–99.

MacCarthy, G. R. 1926. Radioactivity and the floor of the oceans. *Geol. Mag.*, **63**: 301–305.

McGowran, B. 1994. Martin Fritz Glaessner 1906–1989. *Hist. Records Aust. Sci.*, **10**: 61–81.

McKenzie, D. P. 1969. Speculations on the consequences and causes of plate motions. *Geophys. J. R. Astron. Soc.*, **18**: 1–32.

McKenzie, D. and Parker, R. L. 1967. The North Pacific: an example of tectonics on a sphere. *Nature*, **216**: 1276–1280.

McMichael, D. F. and Iredale, T. 1959. The land and freshwater mollusca of Australia. In Christian, C. S., Crocker, R. L., and Keast, A., eds., *Biogeography and Ecology in Australia*. Dr. W. Junk, The Hague, 224–245.

Maguire, J. M. 1990. Obituary: Dr Edna P. Plumstead FRSSAf. *Trans. Roy. Soc. S. Afr.*, **47**, Part 3: 355–357.

Martin, H. 1961. The hypothesis of continental drift in the light of recent advances of geological knowledge in Brazil and in South West Africa: Alex. L. du Toit memorial lectures No. 7. *Trans. Proc. Geol. Soc. S. Afr.* (Annexure to), **64**: 1–47.

Marvin, U. B. 1973. *Continental Drift: The Evolution of a Concept*. Smithsonian Institution Press, Washington, DC.

Marvin, U. B. 1985. The British reception of Alfred Wegener's continental drift hypothesis. *Earth Sci. Hist.*, **4**: 138–159.

Marvin, U. B. 2001. Review of the rejection of continental drift. *Metascience*, **10**: 208–217.

Matthew, W. D. 1915. Climate and evolution. *Ann. N.Y. Acad. Sci.*, **24**.

Maud, R. R. 1998. Obituary: Lester Charles King FRSSAf. (1907–1998). *Trans. Roy. Soc. S. Afr.*, **74**: 209–210.

Mawson, D. 1932–1935. Sir Tannatt William Edgeworth David 1885–1934. *Obit. Not. Fell. Roy. Soc. (London) 1932–35*, **1**: 493–501.

Mayo, D. E. 1985. Mountain-building: The nineteenth-century origins of isostasy and the geosyncline. In Drake, E. T. and Jordan, W. M., eds., *Geologists and Ideas: A History of North American Geology, GSA Centennial Special Volume I*. Geological Society of America, Boulder, CO, 1–18.

Mayr, E. 1944. The birds of Timor and Sumba. *Bull. Am. Mus. Nat. Hist.*, **83**: 127–194.

Mayr, E. 1952. Conclusion. In Mayr, E., ed., *The Problem of Land Connections Across the South Atlantic with Special Reference to the Mesozoic, Bull. Am. Mus. Nat. Hist.*, **99**: 85, 255–258.

Mayr, E. 1953. Fragments of a Papuan ornithogeography. *Proceedings 7th Pacific Science Congress of the Pacific Science Association*. Whitcombe and Tombs Ltd, Auckland, 11–15.

Melton, F. A. 1925. Review of the origin of continents and oceans. *Science*, **62**: 14–15.

Menard, H. W. 1986. *The Ocean of Truth: A Personal History of Global Tectonics*. Princeton University Press, Princeton, NJ.

Mercanton, P. L. 1926. Inversion de l'inclinaison magnétique terrestre aux âges geologiques. *Soc. Suisse Geophys. Met. Astr. Fribourg*, **1926**: 345–349.

Mercanton, P. L. 1931. Inversion de l'inclinaison magnétique terrestre aux âges géologiques: Nouvelles observations I. *C. R. Acad. Sci. Paris*, **182**: 859–860.

Mercanton, P. L. 1932. Inversion de l'inclinaison magnétique terrestre aux âges géologiques: Nouvelles observations II. *C. R. Acad. Sci. Paris*, **194**: 1371–1372.

Meyerhoff, A. A. and Teichert, C. 1971. Continental drift, III: Late Paleozoic glacial centers, and Devonian-Eocene coal distributions. *J. Geol.*, **79**: 285–321.

Mirsky, A. 1964a. Reconsideration of the "Beacon" as a stratigraphic name in Antarctica. In *Antarctic Geology, SCAR Proceedings 1963*. North-Holland Publishing Co., Amsterdam, 364–378.

Mirsky, A. 1964b. Discussion. In *Antarctic Geology, SCAR Proceedings 1963*. North-Holland Publishing Co., Amsterdam, 733–735.

Miser, H. D. 1921. Llanoria, the Paleozoic land area in Louisiana and Eastern Texas. *Am. J. Sci. (5th series)*, **2**: 61–89.

Mitchell, P. C. 1930. Discussion on geological climates. *Proc. Roy. Soc. London, B*, **106**: 307–309.

Molengraaff, G. A. F. 1916. The coral reef problem and isostasy. *Proc. Kon. Akad. Wet. Amsterdam*, **19**: 610–627.

Molengraaff, G. A. F. 1928. Wegener's continental drift. In van Waterschoot van der Gracht, W., ed., *Theory of Continental Drift: A Symposium on the Origin and Movement of Land Masses both Inter-continental and Intra-continental, as Proposed by Alfred Wegener*. American Association of Petroleum Geologists, Tulsa, OK, 90–92.

Molnar, P. 1988. Continental tectonics in the aftermath of plate tectonics. *Nature*, **355**: 131–137.

Molnar, P. and Tapponnier, P. 1975. Cenozoic tectonics of Asia: Effects of a continental collision. *Science*, **189**: 419–426.

Montgomery, J. N. and Raggatt, H. G. 1951. Arthur Wade (1878–1951). *Bull. Am. Assoc. Petrol. Geol.*, **35**: 2643–2645.

Myers, G. S. 1938. Fresh-water fishes and West Indian zoogeography. *Annu. Rep. Smithsonian Inst.* 1937: 339–364.

Myers, G. S. 1953. Paleogeographical significance of fresh-water fish distribution in the Pacific. *Proceedings 7th Pacific Science Congress of the Pacific Science Association*, Whitcombe and Tombs Ltd., Auckland, 38–48.

Nevin, C. M. 1949. *Principles of Structural Geology* (4th edition). John Wiley & Sons, New York.

Newman, R. P. 1995. American intransigence: The rejection of continental drift in the great debates of the 1920s. *Earth Sci. Hist.*, **14**: 62–83.

Nicholas, T. C. 1926. Comments. *Geogr. J.*, **67**: 309–310.

Nicholls, G. H. 1933. The composition and biogeographical relations of the fauna of Western Australia. *Report of the Twenty-first Meeting of ANZAAS (Sydney)*, 93–138.

Nicholls, G. H. 1943. The phreatoicoidea. Part I. The amphisopidae. *Pap. Proc. Roy. Soc. Tasmania*, 1942: 1–145.

Nicholls, G. H. 1944. The phreatoicoidea. Part II. The phreatoicoidae. *Pap. Proc. Roy. Soc. Tasmania*, 1943: 1–156.

Nölke, F. 1939. Zur tektonik des Atlantischen beckens. *Geologische Rundschau*, **39**: 21–27.

Norlund, N. E. 1937. Astronomical longitude and azimuth determinations (George Darwin Lecture, 1937). *Mon. Not. Roy. Astron. Soc.*, **97**: 489–506.

Oftedahl, C. 1977. Memorial to Olaf Holtedahl, 1885–1975. *Geol. Soc. Am. Memorials*, **VII**.

Oldroyd, D. 1990. *The Highlands Controversy*. University of Chicago Press, Chicago, IL.

Oliveira, A. I., de and Leonardos, O. H. 1940. *Geologia do Brasil*. Rio de Janeiro. 2nd edition, 1943.

Oliver, J. and Isacks, B. 1967. Deep earthquake zones, anomalous structures in the upper mantle, and the lithosphere. *J. Geophys. Res.*, **72**: 4259–4275.

Öpik, A. A. 1939. Paläontologie, Arctisforschung und kontinentalverschiebung. *Mitt. Naturf. Ges. Schaffhausen*, **16**: 47–69.

Oreskes, N. 1988. The rejection of continental drift. *Hist. Stud. Phys. Sci.*, **18**: 311–348.

Oreskes, N. 1999. *The Rejection of Continental Drift*. Oxford University Press, New York.

Oreskes, N. 2001. Author's response to Marvin. *Metascience*, **10**: 217–222.

Pantin, C. F. A. 1965. Robert Beresford Seymour Sewell. 1880–1964. *Biogr. Mem. Fell. Roy. Soc.*, **11**: 146–155.

Paramonov, S. J. 1959. Zoogeographical aspects of the Australian dipterofauna. In Christian, C. S., Crocker, R. L., and Keast, A., eds., *Biogeography and Ecology in Australia*. Dr. W. Junk, The Hague, 164–191.

Pearson, J. 1940. The relationships of the marsupials. *Aust. J. Sci.*, **3**: 43–47.

Pekeris, C. L. 1926a/1935. Thermal convection in the interior of the Earth. *Phil. Mag.*, **1**: 923–931. *Mon. Not. Roy. Astron. Soc.*, **3**: 343–367.

Plumstead, E. 1952. Description of two new genera and six new species of fructifications borne on Glossopteris leaves. *Trans. Geol. Soc. S. Afr.*, **55**: 281–328.

Plumstead, E. 1956a. Bisexual fructifications borne on *Glossopteris* leaves from South Africa. *Palaeontographica, Abt. B*, **108**: 1–25.

Plumstead, E. 1956b. On ottokaria, the fructification of gangamopteris. *Trans. Geol. Soc. S. Afr.*, **59**: 211–236.

Plumstead, E. 1958a. Further fructifications of the glossopteridae and a provisional classification based on them. *Trans. Geol. Soc. S. Afr.*, **61**: 51–79.

Plumstead, E. 1958b. The habit of growth of glossopteridae. *Trans. Geol. Soc. S. Afr.*, **61**: 81–96.

Plumstead, E. 1961. Ancient plants and drifting continents. *S. Afr. J. Sci.*, **57**: 173–181.

Plumstead, E. 1962. *Geology. 2. Fossil Floras of Antarctica*. Trans-Antarctic Expedition, 1955–58, Scientific Report, 9.

Plumstead, E. 1982. April 13, 1982 letter to Henry Frankel. Unpublished.

Poole, J. H. J. 1931. Comments. *Geogr. J.*, **78**: 442–444.

Pratje, O. 1928. Beitrag zur bodengestaltung des sudatlantischen ozeans. *Centralbl. Mineral. B*: 129–152.

Pratje, O. 1939. Aussprachebemerkung zu den ozeanographischen vorträgen. *Geologische Rundschau*, **30**: 383.

Pretorius, D. A. 1975. Memorial to Byron Birtton Brock. *Geol. Soc. Am. Mem.*, **IV**.

Quensel, P. 1912. Tectonic features and eruptives of northern Sweden. *Congrès Géologique International Compte Rendu de XIme Session, Stockholm, 1910*, **2**: 1227–1232.

Radhakrishna, B. P. 1996. Lakshmeshwar Rama Rao: A centenary tribute. In Sahni, A., ed., *Cretaceous Stratigraphy and Palaeoenvironments: L. Rama Rao Volume*, Memoir 37, Geological Society of India, Bangalore, v–x.

Radhakrishna, B. P. 1999. A tribute to William Dixon West. In Subbarao, K. V., ed., *Decan Volcanic Province I, West Volume*. Geological Society of India, Bangalore, xii–xx.

Rainger, R. 2000. Science at the crossroads: The Navy, Bikini Atoll, and American oceanography in the 1940s. *Hist. Stud. Phys. Sci.*, **30**: 349–371.

Rao, L. Rama. 1937. Wegener's theory of continental drift with reference to India and adjacent countries (General discussion). *Proceedings 24th Indian Science Congress*, Hyderabad, 502–520.

Rastall, R. H. 1929. On continental drift and cognate subjects. *Geol. Mag.*, **66**: 447–456.

Read, H. H. 1944. Geology without tears, a review of principles of physical geology. *Nature*, **154**: 720–721.

Read, H. H. 1949. *Geology: An Introduction to Earth-History*. Oxford University Press, London.

Regan, C. T. 1930. Discussion on geological climates. *Proc. Roy. Soc. London, B*, **106**: 314–316.

Reid, H. F. 1922. Drift of the Earth's crust and displacement of the pole. *Geogr. Rev.*, **12**: 672–674.

Reynolds, D. 1968. Memorial of Arthur Holmes. *Am. Mineral.*, **53**: 560–566.

Rice, W. R. 1938. Review of du Toit's *Our Wandering Continents*. *Am. J. Sci.*, **35**: 391–393.

Ride, W. D. L. 1962. On the evolution of Australian marsupials. In Leeper, G. W., ed., *The Evolution of Living Organisms*. Melbourne University Press, Melbourne, 281–309.

Ride, W. D. L. 1964. A review of Australian fossil marsupials. *J. Roy. Soc. West. Aust.*, **47**: 97–131.

Riek, E. F. 1959. The Australian freshwater crustacea. In Christian, C. S., Crocker, R. L., and Keast, A., eds., *Biogeography and Ecology in Australia*. Dr. W. Junk, The Hague, 246–273.

Rittmann, A. 1939a. Bemerkung zur "Atlantis-Tagung" in Frankfurt im Januar 1939. *Geologische Rundschau*, **30**: 284.

Rittmann, A. 1939b. Über die herkunft der vulkanischen energie und die entstehung des sials. *Geologische Rundschau*, **30**: 52–60.

Robertson, W. A. 1963. Paleomagnetism of some Mesozoic intrusives and tuffs from Eastern Australia. *J. Geophys. Res.*, **68**: 2299–2312.

Roe-Thompson, E. R. 1922. Wegener's displacement theory. *Nature*, **110**: 214.

Romer, A. S. 1945. The later carboniferous vertebrate fauna of Kaunova (Bohemia) compared with that of the Texas redbeds. *Am. J. Sci.*, **243**: 417–442.

Romer, A. S. 1952. Discussion. In Mayr, E., ed., *The Problem of Land Connections Across the South Atlantic with Special Reference to the Mesozoic. Bull. Am. Mus. Nat. Hist.*, **99**: 250–254.

Rutsch, R. 1939. Entwicklung tropisch-amerikanischer tertiarfauen und kontinentalverschiebungs-hypothese. *Geologische Rundschau,* **30**: 362–372.

Sahni, B. 1926. The southern fossil floras: A study in plant geography of the past. Presidential Address, *Proceedings 13th Indian Science Congress, Bombay,* 229–254.

Sahni, B. 1935a. The Glossopteris flora in India. *Sixth International Botanical Congress, Amsterdam,* **II**: 245–247.

Sahni, B. 1935b. Permo-carboniferous life-provinces, with special reference to India. *Curr. Sci.,* **IV**: 385–390.

Sahni, B. 1936. Wegener's theory of continental drift in the light of palaeobotanical evidence. *Indian Bot. Soc. J.,* **15**: 319–332.

Sahni, B. 1937. Wegener's theory of continental drift with reference to India and adjacent countries (General discussion). *Proceedings 24th Indian Science Congress,* Hyderabad: 502–520.

Sahni, B. 1938. Recent advances in Indian paleobotany. *Lucknow Univ. Studies,* no. 2: 1–98.

Sahni, M. R. 1952. Birbal Sahni: a biographical sketch of his personal life. *Palaeobotanist,* **1**: 1–8.

Salomon-Calvi, W. 1937. Die fortsetzung der tonalelinie in kleinasien. *Anz. Akad. Wiss.,* **74**: 117–119.

Sankaran, A. V. 1998. M. S. Krishnan: geologist *par excellence. Curr. Sci.,* **75**: 1084–1085.

Savage, D. 1981. Helmut de Terra, 1900–1981. *Soc. Vertebrate Paleontol., News Bull.,* no. 123: 54.

Schaeffer, B. 1952. The evidence of the fresh-water fishes. In Mayr, E., ed., *The Problem of Land Connections Across the South Atlantic with Special Reference to the Mesozoic. Bull. Am. Mus. Nat. Hist.,* **99**: 227–234.

Schaer, J. P. 1995. Eugene Wegmann (1896–1982): Vie et oeuvre d'un géologue européen. *Mem. Soc. Geol. France,* **168**: 13–23.

Schardt, H. 1928. Zur kritik der Wegenerschen theorie der kontinentenverschiebung. *Vierteljahrsschr. Naturforsch. Ges. Zurich,* **73**: 14–15.

Schlee, S. 1973. *The Edge of an Unfamiliar World: A History of Oceanography.* E. P. Dutton, New York.

Schoot, W. 1939. Paläogeographie und Tiefseeseimente des Atlantischen Ozeans. *Geologische Rundschau,* **30**: 382.

Schuchert, C. 1926a. The paleogeography of Permian time in relation to the geography of earlier and later periods. *Proceedings Second Pacific Science Congress of the Pacific Science Association* Volume II: 1079–1091.

Schuchert, C. 1926b. Review of *Our Mobile Earth. Science,* **64**: 624.

Schuchert, C. 1928a. The hypothesis of continental displacement. In van Waterschoot van der Gracht, W., ed., *Theory of Continental Drift: A Symposium on the Origin and Movement of Land Masses both Intercontinental and Intracontinental as Proposed by Alfred Wegener.* American Association of Petroleum Geologists, Tulsa, OK, 104–144.

Schuchert, C. 1928b. Sites and nature of the North American geosynclines. *Geol. Soc. Am. Bull.,* **34**: 165.

Schuchert, C. 1932. Gondwana land bridges. *Geol. Soc. Am. Bull.,* **43**: 875–916.

Schuchert, C. 1936. Geologic interpretation of the bathymetry of the East Indian Archipelago. *Am. J. Sci.,* **32**: 292–297.

Schwarzbach, M. 1980/1986. *Alfred Wegener, the Father of Continental Drift.* Translated by Love, C. Science Tech, Inc., Madison, WI.

Schweber, S. S. 1988. The mutual embrace of science and the military: ONR and the growth of physics in the United States after World War II. In Mendelsohn, E., Smith, M. R., and Weingart, P., eds., *Science, Technology and the Military.* Kluwer, Dordrecht, 3–45.

Schwinner, R. 1919. Vulkanismus und gebirgsbildung: ein versuch. *Z. Vulkanol.*, **5**: 175–230.

Scrivenor, J. G. 1928. *The Geology of Malayan Ore-deposits.* Macmillan, London.

Scrivenor, J. G. 1931. *The Geology of Malaya.* Macmillan, London.

Scrivenor, J. G. 1941. Geological research in the Malay Peninsula and Archipelago. *Geol. Mag.*, **78**: 125–150.

Scrivenor, J. G. 1942. Geological and climatic factors affecting the distribution of life in the archipelago. *Proc. Linn. Soc. London*, **154**: 120–126.

Seilacher, A. 1989. Vendozoa: organismic construction in the Proterozoic biosphere. *Lethaia*, **22**: 229–239.

Semper, M. 1917. Was ist eine arbeitshypothese? *Zentralbl. Mineral. Geol. Palaeontol.*: 146–163.

Şengör, A. M. C. 1975. The origin of lunar craters. *The Moon*, **14**: 211–236. Translation of Wegener's 1921 *Die entstehung der mondkrater.* Vieweg & Sohn, Braunschweig.

Şengör, A. M. C. 1979/1982a. Classical theories of orogenesis. In Miyashiro, A., Aki, K., and Şengör, A. M. C, eds., *Orogeny.* John Wiley & Sons, New York, 1–48. English edition translated from 1979 Japanese edition. Iwanami Shoten, Tokyo.

Şengör, A. M. C. 1982b. Eduard Suess' relations to the pre-1950 schools of thought in global tectonics. *Geologische Rundschau*, **71**: 381–420.

Şengör, A. M. C. 1983. Gondwana and Gondwanaland: a discussion. *Geologische Rundschau*, **72**: 397–399.

Şengör, A. M. C. 1985. Professor Isan Ketin: an appreciation. In Şengör, A. M. C., ed., with assistance from Yilmaz, Y., Okay, A. I., and Gorur, N., *Tectonic Evolution of the Tethyan Region.* Kluwer Academic, Dordrecht, xxxi–xxxvi.

Şengör, A. M. C. 1991. Difference between Gondwana and Gondwana-Land. *Geology*, **19**: 387–388.

Şengör, A. M. C. 2003. *The Large-wavelength Deformations of the Lithosphere: Materials for a History of the Evolution of Thought from the Earliest Times to Plate Tectonics.* GSA Memoir 196. Geological Society of America, Boulder, CO.

Şengör, A. M. C. and McKenzie, D. 1998. Memorial to İhsan Ketin, 1914–1995. *Geol. Soc. Am. Memorials for 1997*, **28**: 31–35.

Seward, A. 1924. The later records of plant life. *J. Geol. Soc. London*, **80**: xlxii–xcvii.

Seward, A. 1929. Botanical records of the rocks: with special reference to the early *Glossopteris* flora. *Brit. Assoc. Adv. Sci. Rep.*, **97**: 199–216.

Seward, A. 1930. Discussion on geological climates. *Proc. Roy. Soc. London, B*, **106**: 303–307.

Seward, A. 1931. *Plant Life Through the Ages.* Cambridge University Press, Cambridge.

Seward, A. 1933. *Plant Life Through the Ages* (2nd edition). Cambridge University Press, Cambridge.

Seward, A. 1938. Review of *Our Wandering Continents: An Hypothesis of Continental Drifting* by Alex L. du Toit. *Geol. Mag.*, **75**: 319–323.

Seward, A. and Conway, V. 1934. A phytogeographical problem: fossil plants from the Kerguelen Archipelago. *Ann. Bot.*, **48**: 736–737.

Shaw, H. K. A. 1942. The biogeographic division of the Indo-Australian Archipelago: 5. Some general considerations from the botanical standpoint. *Proc. Linn. Soc. London*, **154**: 148–154.

Shergold, J. H. 1985. Armin Alexsander Öpik (1998–1983). *BMR J. Aust. Geol. Geophys.*, **9**: 69–81.

Simpson, G. C. 1923. The Wegener hypothesis: discussion. *Nature*, **111**: 30–31.

Simpson, G. C. 1929. Past climates. *Manchester Mem. Lit. Phil. Soc.*, **74**: 1–34.

Simpson, G. C. 1930. Discussion on geological climates. *Proc. Roy. Soc. London, B*, **106**: 299–302.

Simpson, G. G. 1940a. Antarctica as a faunal migration route. *Proceedings Sixth Pacific Science Congress of the Pacific Science Association*, 755–768.

Simpson, G. G. 1940b. Mammals and land bridges. *J. Wash. Acad. Sci.*, **30**: 137–163.

Simpson, G. G. 1943. Mammals and the nature of continents. *Am. J. Sci.*, **241**: 1–31.

Simpson, G. G. 1947. Holarctic mammalian faunas and continental relationships during the Cenozoic. *Geol. Soc. Am. Bull.*, **58**: 613–637.

Simpson, G. G. 1952. Probabilities of dispersal in geologic time. In *The Problem of Land Connections Across the South Atlantic with Special Reference to the Mesozoic. Bull. Am. Mus. Nat. Hist.*, **99**: 163–176.

Simpson, G. G. 1965. *The Geography of Evolution*. Chilton Books, Philadelphia, PA.

Singewald, J. T. Jr. 1928. Discussion of the Wegener theory. In van Waterschoot van der Gracht, W., ed., *Theory of Continental Drift: A Symposium on the Origin and Movement of Land Masses both Inter-continental and Intra-continental, as Proposed by Alfred Wegener*. American Association of Petroleum Geologists, Tulsa, OK, 189–193.

Sitholey, R. V. 1950. Professor Birbal Sahni, F.R.S. *J. Indian Bot. Soc.*, **XXIX**: 1–5.

Skeats, E. W. 1933. *Some Founders of Australian Geology*. Australian Research Council, Sydney.

Skottsberg, C. 1940. Nagra drag av den Antarktiska kontinentens biologiska historia. *Norske vidensk. selek. forhandl.*, **12**: 45–55.

Skottsberg, C. 1960. Remarks on the plant geography of the southern cold temperate zone. *Proc. Roy. Soc. London, B*, **152**: 447–457.

Smit Sibinga, G. L. 1927. Wegener's theorie en het outstaan van den Oostelijken O. I. Archipel. *Tijdschr. K. Ned. Aardr. Genoots., Ser. 2*, **44**: 581–598.

Smit Sibinga, G. L. 1933. The Malay double (triple) orogen. *Proc. Kon. Akad. Wet.*, **36**: 202–210, 323–330, 447–453.

Smuts, J. C. 1925. Presidential address delivered July 6, 1925. *S. Afr. J. Sci.*, **22**: 1–19.

Snider, A. 1858. *La Création et Ses Mysteres Dévoilés*. Franck et Dentu, Paris.

Sonder, R. A. 1939. Zur tektonik des Atlantischen Ozeans. *Geologische Rundschau*, **30**: 28–51.

Staub, R. 1924. Der Bau der Alpen. *Beitr. Geol. Karte Schweiz, Neue Folge*, **52**: 1–272.

Staub, R. 1928. *Der bewegungsmechanismus der Erde*. Gebrüder Borntraeger, Berlin.

Staub, W. 1939. Ost-Mexiko, das Nordwest-Ende der Mediterranen orogenen zone. *Geologische Rundschau*, **30**: 346–351.

Steenis, C. G. G. J. van. 1950. The delimitation of Malaysia and its main plant geograpical divisions. *Flora Malesiana, Ser. 1*, **1**: lxx–lxxv.

Steenis, C. G. G. J. van. 1962. The land-bridge theory in botany. *Blumea*, **11**: 235–372.

Steenis, C. G. G. J. van. 1963. Transpacific floristic affinities. In Gressitt, J. L., ed., *Pacific Basin Biogeography*. Bishop Museum Press, Honolulu, 219–231.

Steenis, C. G. G. J. van. 1971. Nothofagus, key genus of plant geography, in time and space, living and fossil, ecology and phylogeny. *Blumea*, **29**: 65–98.

Steenis, C. G. G. J. van. 1979. Plant-geography of east Malesia. *Bot. J. Linn. Soc.*, **79**: 97–178.

Stewart, T. D. 1982. Helmut de Terra, 1900–1981. *Am. Antiquity*, **47**: 793–794.

Stille, H. 1918. Über Hauptformen der Orogenese und ihre Verknüpfung. *Nachr. K. Ges. Wiss. Gottingen, Math-Phys. Kl.*: 1–32.

Stille, H. 1936. Present tectonic state of the Earth. *Bull. Am. Assoc. Petrol. Geol.*, **20**: 847–880.

Stille, H. 1939. Kordillerisch-Atlantische wechselbeziehungen. *Geologische Rundschau*, **30**: 129–139.

Stille, H. 1941. *Einführung in den bau Amerikas*. Gebrüder Borntraeger, Berlin.

Stille, H. 1955. Recent deformations of the Earth's crust in the light of those of earlier epochs. *Geol. Soc. Am. Special Paper*, **62**: 171–191.

Stommel, H. M. 1994. Columbus O'Donnell Iselin. *Biogr. Mem. Nat. Acad. Sci.*, **65**: 165–186.

Størmer, L. 1976. Olaf Holtedahl. *Biogr. Mem. Fell. Roy. Soc.*, **22**: 193–205.

Stubblefied, C. J. 1965. Edward Battersby Bailey. *Biogr. Mem. Fell. Roy. Soc.*, **11**: 1–21.

Stubblefied, C. J. 1970. Darashaw Nosherwan Wadia. *Biogr. Mem. Fell. Roy. Soc.*, **16**: 543–562.

Subbotin, S. I., Naumchik, G. L., and Rakhimova, I. S. 1965. Structure of the earth's crust and mantle processes in the upper mantle: Influence of upper mantle processes on the structure of the earth's crust. *Tectonophysics*, **2**: 111–209.

Suess, E. 1875. *Die entstehung der Alpen*. Wilhelm Braumuller, Vienna.

Suess, E. 1904–1909. *The Face of the Earth*. Translated by Sollas, H. B. C. (5 volumes). Vol. 1, 1904; Vol. 2, 1905; Vol. 3, 1906; Vol. 4, 1909; Vol. 5, 1920. Clarendon Press, Oxford. First published in German as *Das Antlitz der Erde*. Vol. Ia, 1883; Vol. Ib, 1885; Vol. II, 1888; Vol. 3/1, 1901; Vol. 3/2, 1904. Tempsky, Prague.

Suess, F. E. 1923. Zum Vergleiche zwischen alpinem und variszischem Bau. *Geologische Rundschau*, **14**.

Suess, F. E. 1926. *Intrusionstektonik und wandertektonik im variszischen grundgebirge*. Gebrüder Borntraeger, Berlin.

Suess, F. E. 1929. The European Altaids and their correlation to the Asiatic structure. In Gregory, J. W., ed., *The Structure of Asia*. Methuen, London, 35–57.

Suess, F. E. 1931. A suggested interpretation of the Scottish Caledonide structure. *Geol. Mag.*, **68**: 71–81.

Suess, F. E. 1932. Crystalline schists of the moldanubian type. *Geol. Mag.*, **69**: 431–432.

Suess, F. E. 1936a. Europäische und nordamerikanische gebrigszuammenhänge. *International Geological. Congress XVI, Reports*, **II**: 815–828.

Suess, F. E. 1936b. Tectonic affinities between European and North American mountain systems. *Pan-Am. Geol.*, **65**: 81–96.

Suess, F. E. 1937. Bausteine zu einem system der tektogenese. I. Periplutonische und enorogene Regionalmetamorphose in ihrer tektogenetischen Bedeutung. *Fortschr. Geol. Palaeontol.*, **13**: 1–86.

Suess, F. E. 1938a. Bausteine zu einem system der tektogenese. II. Zum bewegungsbilde des alteren Mitteleuropa; hypokinematische regionalmetamorphose. *Fortschr. Geol. Palaeontol.*, **13**: 87–238.

Suess, F. E. 1938b. Der bau der Kaldoniden und Wegener's hypothese. *Zentralbl. Mineral. Geol. Palaeontol.*, *Abt. B*: 321–327.

Suess, F. E. 1939. Bausteine zu einem system der tektogenese. III. Die bau der Kaledoniden und die schollendrift im Nordalantik; A, Die Kaledoniden in Schottland und Vergleiche. *Fortschr. Geol. Palaeontol.*, **13**: 239–376.

Suess, F. E. 1949. Bausteine zu einem system der tektogenese. III. Die bau der Kaledoniden und die schollendrift im Nordalantik; B, Die Kaledoniden in Skandinavien; C, Die Kaledoniden in Groenland, *Mitt. Geol. Ges. Wien*, **36–38**: 29–230.

Szalay, F. S. 1994. *Evolutionary History of the Marsupials and an Analysis of Osteological Characters*. Cambridge University Press, Cambridge.

Tapponnier, P., Peltzer, G., Armijo, R., Le Dain, A.-Y., and Cobbold, P. 1982. Propagating extrusion tectonics in Asia: new insights from simple experiments with plasticine. *Geology*, **10**: 611–616.

Taylor, F. B. 1898. *An Endogenous Planetary System*. Archer Publishing Co., Fort Wayne, IN.

Taylor, F. B. 1910. Bearing of the Tertiary mountain belt on the origin of the Earth's plan. *Geol. Soc. Am. Bull.*, **21**: 179–226.

Taylor, F. B. 1925. Movement of continental masses under action of tidal forces. *Pan-Am. Geol.*, **43**: 15–50.

Taylor, F. B. 1926. Greater Asia and isostasy. *Am. J. Sci.*, **12**: 47–67.

Taylor, F. B. 1931. Letter to Professor Robert E. Martin.

Teichert, C. 1941. Upper Paleozoic of Western Australia: correlation and paleogeography. *Bull. Am. Assoc. Petrol. Geol.*, **25**: 371–415.

Teichert, C. 1952. Carboniferous, Permina, and Jurassic in the Northwest Basin, Western Australia. In Teichert, K., ed., *XIX Congres Geologique International, Symposium Sur les series de Gondwana*. Algiers, 115–135.

Teichert, C. 1959. Australia and Gondwanaland. *Geologische Rundschau*, **47**: 526–590.

Teichert, C. 1974. Marine sedimentary environments and their faunas in Gondwana area. In Kahl, C. F., ed., *Plate Tectonics: Assessments and Reassessments*, AAPG Memoir 23: 361–398. American Association of Petroleum Geologists, Tulsa, OK.

Termier, P. 1924. La derive des continents. *Rev. Sci., Paris*, **69**: 257–267.

Termier, P. 1925. The drifting of the continents. *Smithson. Inst. Ann. Rep.* 1924, 219–236.

Thalmann, H. E. 1943. Memorial to Émile Argand. *Proc. Vol. Geolog. Soc. Am.* 1942: 153–165.

Thomas, H. H. 1925. The Caytoniales, a new group of angiospermous plants from the Jurassic rocks of Yorkshire. *Phil. Trans. Roy. Soc. London B*, **213**: 299–363.

Thomas, H. H. 1930. Discussion on geological climates. *Proc. Roy. Soc. London, B*, **106**: 316–317.

Thomas, H. H. 1941. Albert Charles Seward. 1863–1941. *Obit. Not. Fell. Roy. Soc.*, **3**: 867–880.

Thomas, H. H. 1952. A Glossopteris with shorled leaves. *Palaeobotanist*, **1**: 435–438.

Thomas, H. H. 1958. Lidgetonia, a new type of fertile Glossopteris. *Bull. Brit. Mus. (Nat. Hist.) Geol.*, **3**: 179–189.

Thompson, M. D. and Bowring, S. A. 2000. Age of the Squantum "Tillite," Boston Basin, Massachusetts: U-Pb zircon constraints on terminal Neoproterozoic glaciation. *Am. J. Sci.*, **300**: 630–655.

Totten, S. 1981. Frank B. Taylor, plate tectonics and continental drift. *J. Geol. Ed.*, **29**: 212–220.

Troll, C. 1939. Bemerkungen zum Atlantischen problem, geäussert im Anschluss an die drei ozeanographischen Beiträge. *Geologische Rundschau*, **30**: 384–386.

Troughton, E. LeG. 1959. The marsupial fauna: its origin and radiation. In Christian, C. S., Crocker, R. L., and Keast, A. eds., *Biogeography and Ecology in Australia*, Dr. W. Junk, The Hague, 69–88.

Trümpy, R. 1991. The Glarus Nappes: a controversy of a century ago. In McKenzie, J., Muller, D., and Weissert, H., eds., *Controversies in Modern Geology*. Academic Press Ltd., New York, 385–404.

Trümpy, R. 2001. Why plate tectonics was not invented in the Alps. *Geologische Rundschau*, **90**: 477–483.

Twidale, C. R. 1992. King of the plains: Lester King's contributions to geomorphology. *Geomorphology*, **5**: 491–509.

Umbgrove, J. H. F. 1934a. The relation between geology and gravity field in the East Indian Archipelago. In *Gravity Expeditions at Sea 1923–1932*, Volume II. Netherlands Geodetic Commission, Delft, 140–162.

Umbgrove, J. H. F. 1934b. A short survey of theories on the origin of the East Indian Archipelago. In *Gravity Expeditions at Sea 1923–1932*, Volume II. Netherlands Geodetic Commission, Delft, 163–182.

Umbgrove, J. H. F. 1938. Geological history of the East Indies. *Bull. Am. Assoc. Petrol. Geol.*, **22**: 1–70.

Umbgrove, J. H. F. 1942. *The Pulse of the Earth* (1st edition). Nijhoff, The Hague.

Umbgrove, J. H. F. 1947. *The Pulse of the Earth* (2nd edition). Nijhoff, The Hague.

Umbgrove, J. H. F. 1949. *Structural History of the East Indies*. Cambridge University Press, Cambridge.

Umbgrove, J. H. F. 1951. The case for the crust-substratum theory. *Adv. Sci.*, **8**: 67–71.

Vallance, T. G. 1978. William Rowan Browne 1884–1975. *Records Aust. Acad. Sci.*, **4**: 65–81.

van Bemmelen, R. W. 1934. Ein beispiel fur sekundartektogenese auf Java. *Geologische Rundschau*, **25**: 175–194.

van Bemmelen, R. W. 1939. Das permanenzproblem nach der undationstheorie. *Geologische Rundschau*, **30**: 10–20.

van Bemmelen, R. W. 1949. *The Geology of Indonesia, Volume 1A: General Geology of Indonesia and Adjacent Archipelagoes*. (Special edition of the Bureau of Mines in Indonesia) Government Printing Office, The Hague.

van Bemmelen, R. W. 1972. *Geodynamic Models*. Elsevier, Amsterdam.

van der Gracht, W. A. J. M. 1928a. Introduction: the problem of continental drift. In van Waterschoot van der Gracht, W., ed., *Theory of Continental Drift: A Symposium on the Origin and Movement of Land Masses both Inter-continental and Intra-continental, as Proposed by Alfred Wegener*. American Association of Petroleum Geologists, Tulsa, OK, 1–75.

van der Gracht, W. A. J. M. 1928b. Remarks regarding the papers offered by the other contributors to the symposium. In van Waterschoot van der Gracht, W., ed., *Theory of Continental Drift: A Symposium on the Origin and Movement of Land*

Masses both Inter-continental and Intra-continental, as Proposed by Alfred Wegener. American Association of Petroleum Geologists, Tulsa, OK, 197–222.

van der Gracht, W. A. J. M. 1931. The permo-carboniferous orogeny in the south-central United States. *Verh. K. Akad. WetAmsterdam, Afd. Natuurkd,* **27**: 1–170.

van der Gracht, W. A. J. M. 1935. Discussion. *Bull. Am. Assoc. Petrol. Geol.,* **19**: 1816–1818.

van Riel, P. M. 1934. The bottom configuration in relation to the flow of the bottom water. *The Snellius-Expedition,* Volume II, Part 2, Chapter II. Kemink en Zoon N.V., Utrecht.

van Vuuren, L. 1920. *Het Gouvernement Celebes Proeve eener Monographie, 1.*

Vening Meinesz, F. A. 1930. Maritime gravity survey in the Netherlands East Indies, tentative interpretation of the provisional results. *Proc. Kon. Ned. Akad. Wetensch.,* **33**: 566–577.

Vening Meinesz, F. A. 1934a. Gravity and the hypothesis of convection-currents in the Earth. *Proc. Kon. Ned. Akad. Wetensch.,* **37**: 37–45.

Vening Meinesz, F. A. 1934b. Interpretation of gravity results: Theoretical considerations. In *Gravity Expeditions at Sea 1923–1932,* Volume II. Netherlands Geodetic Commission, Delft, 7–64.

Vening Meinesz, F. A. 1934c. Interpretation of gravity anomalies in the Netherlands East Indies. In *Gravity Expeditions at Sea 1923–1932,* Volume II. Netherlands Geodetic Commission, Delft, 116–139.

Vening Meinesz, F. A. 1934d. Interpretation of gravity anomalies in other areas. In *Gravity Expeditions at Sea 1923–1932,* Volume II. Netherlands Geodetic Commission, Delft, 195–208.

Vening Meinesz, F. A. 1941. Gravity over the Hawaiian Archipelago and over the Madeira area: Conclusions about the Earth's Crust. *Proc. Kon. Ned. Akad. Wetensch.,* **44**: 2–12.

Vening Meinesz, F. A. 1944. De Verdeling van continenten en oceanen over het Aardoppervlak. *Versl. Kon. Akad. V. Wetensch. Amsterdam, Afd. Nat.,* **53**: 151–157.

Vening Meinesz, F. A. 1951. Convection currents in the mantle. *Trans. Am. Geophys. Union,* **32**: 531–533.

Vening Meinesz, F. A. with the collaboration of Umbgrove, J. H. F. and Kuenen, Ph. H. 1934. *Gravity Expeditions at Sea 1923–1932, Volume* II. Netherlands Geodetic Commission, Delft.

Virkki, C. 1937. On the occurrence of winged pollen grains in the Permo-Carboniferous rocks of India and Australia. *Proc. Indian Acad. Sci. B,* **6**: 428–431.

Virkki, C. 1939. On the occurrence of similar spores in a lower Gondwana glacial tillite from Australia and in lower Gondwana shales in India. *Proc. Indian Acad. Sci. B,* **9**: 7–12.

Voisey, A. H. 1991. *Sixty Years on the Rocks: The Memoirs of Professor Alan H. Voisey.* The Earth Sciences History Group, Geological Society of Australia Inc., Sydney.

Von Engeln, O. D. and Castor, K. E. 1952. *Geology.* McGraw-Hill, New York.

Von Ihering, H. 1931. Land bridges across the Atlantic and Pacific Oceans during the Kainozoic Era. *Q. J. Geol. Soc. London,* **87**: 376–391.

Wade, A. 1924. Petroleum prospects, Kimberley District of Western Australia and Northern Territory. *Parl. Pap. Melbourne,* 1–10.

Wade, A. 1931. The geology of part of Western Madagascar. *J. Inst. Petrol. Technol.,* **17**: 357–361.

Wade, A. 1934. The distribution of oilfields from the view-point of the theory of continental spreading. *Proceedings World Petroleum Conference*, **1**: 73–77.

Wade, A. 1935. New theory of continental spreading. *Bull. Am. Assoc. Petrol. Geol.*, **19**: 1806–1816.

Wade, A. 1938. The geological succession in the West Kimberley District of Western Australia. *Rep. Aust. N. Z. Assoc. Sci.*, **23**: 93–96.

Wade, A. 1941. The geology of the Antarctic Continent and its relationship to neighbouring land areas. *Proc. Roy. Soc. Queensland*, **52**: 24–35.

Wade, A. and Prider, R. T. 1939. The leucite bearing rocks of the West Kimberley Area, Western Australia. *Rep. Aust. N. Z. Assoc. Sci.*, **24**: 99.

Wadia, D. N. 1919. *Geology of India for Students*. Macmillan, London. 2nd edition, 1939; 3rd edition, 1953; 4th edition, 1961.

Wadia, D. N. 1936. The trend-line of the Himalayas: its north-west and south-east limits. *Himal. J.*, **8**: 63–69.

Wadia, D. N. 1965. Address by Dr. D. N. Wadia, President of the XXII session, at the first meeting of the General Assembly. *Indian Minerals*, **XIX**: 35–36.

Waldmann, L. 1953. Das lebeswerk von Franz Eduard Sueß. *Geol. Bund. (Vienna) Jahr*, **96**: 193–216.

Walton, J. 1953. *An Introduction to the Study of Fossil Plants* (2nd edition). Adam & Charles Black, London.

Wanenacker, J. M., Twenhofel, W. H., and Raasch, G. O. 1934. The Paleozoic strata of the Baraboo area, Wis. *Am. J. Sci.*, **28**: 1–30.

Wanner, J. 1921. Zur tektonik der Molukken. *Geologische Rundschau*, **12**: 155–165.

Waring, H. 1953. George Edward Nicholls. *Aust. J. Sci.*, **16**: 56–57.

Watts, W. W. 1931. Obituary, John William Evans: 1867–1930. *Proc. Roy. Soc. London, B*, **107**: xxvii–xxx.

Wegener, A. 1912a. Die entstehung der kontinente. *Geologische Rundschau*, **3**: 276–292.

Wegener, A. 1912b. Die entstehung der kontinente. *Petermanns Geogr. Mitt.*, **58**: 185–195, 253–256, 305–309. Jacoby, E. E. 2001. Translation of Wegener's 1912 Die enstehung der kontinente. *J. Geodyn.*, 32: 29–63.

Wegener, A. 1915. *Die Entstehung der Kontinente und Ozeane*. Friedrich Vieweg & Sohn, Braunschweig, 1st edition, 1915; 2nd edition, 1920, 3rd edition, 1922; 4th edition, 1924; 4th revised edition, 1929; 5th edition, revised by Wegener, K., 1936.

Wegener, A. 1921. *Die Entstehung der Mondkrater*. Rriedr. Vieweg & Sohn, Braunschweig.

Wegener, A. 1922. The origin of continents and oceans. *Discovery*, **3**: 114–118.

Wegener, A. 1924. *The Origin of Continents and Oceans*. Translated from the 3rd edition by Skerl, S. G. A., Methuen and Company, London. (1929/1966) Translated from the 4th revised German edition by Biram, J., Dover Publications, New York.

Wegener, A. 1927. Die geophysikalischen grundlagen der theorie der kontinenten Verschiebung. *Scientia*, **41**: 103–116.

Wegener, A. 1928. Two notes concerning my theory of continental drift. In van Waterschoot van der Gracht, W., ed., *Theory of Continental Drift: A Symposium on the Origin and Movement of Land Masses both Inter-continental and Intra-continental, as Proposed by Alfred Wegener*. American Association of Petroleum Geologists, Tulsa, OK, 97–103.

Wegener, E. (with assistance of Loewe, Fritz), eds., 1939. *Greenland Journey: The Story of Wegener's German Expedition to Greenland in 1930–31 as Told by*

Members of the Expedition and the Leader's Diary. English translation from the 7th German edition by Deans, W. M., Blackie & Sons Limited, London.

Wegener, E. 1960. *Alfred Wegener: Tagebucher, Briefe, Erinnerungen*. F. A. Brockhaus, Wiesbaden.

Wegener, K. 1939. Die geophysikalischen grundlegen der verschiebungstheorie. *Geologische Rundschau*, **30**: 3–5.

Wegmann, C. E. 1922. Zur geologie der St. Bernharddecke im Val d'Herens (Wallis). Ph.D. thesis. *Soc. Neuchatel Sci. Nat. Bull.*, **47**.

Wegmann, C. E. 1935. Preliminary report on the Caledonian orogeny in Christian X's Land, *Medd. Grønland*, **103**, Nr. 3: 1–59.

Wegmann, C. E. 1938. On the structural divisions of Southern Greenland. *Medd. Grønland*, **113**, Nr. 23: 1–148.

Wegmann, C. E. 1939. Zwei bilder fur das arbeitszimmer eines geologen. *Geologische Rundschau*, **30**: 389.

Wegmann, C. E. 1943. Sur un controle géologique de la dérive des continents. *Soc. Neuchatel Sci. Nat. Bull.*, **68**: 97–104.

Wegmann, C. E. 1948. Geological tests of the hypothesis of continental drift in the Arctic regions. *Medd. Grønland*, **144**, Nr. 7: 1–48.

Werenskiold, W. 1931. Comments. *Geogr. J.*, **78**: 537–538.

West, W. D. 1937. Wegener's theory of continental drift with reference to India and adjacent countries (General discussion). *Proceedings 24th Indian Science Congress*, Hyderabad: 502–520.

West, W. D. 1965. D. N. Wadia: An appreciation. In Jhingran, A. G., ed., *Dr. D. N. Wadia Commemorative Volume*. Mining, Geological and Metallurgical Institute of India, Calcutta, 1–9.

Westoll, T. S. 1944. The Haplolepidae: A new family of Late Carboniferous bony fishes. *Bull. Am. Mus. Nat. Hist.*, **83**: 5–121.

White, D. 1928. Continental drift. In van Waterschoot van der Gracht, W., ed., *Theory of Continental Drift: A Symposium on the Origin and Movement of Land Masses both Inter-continental and Intra-continental, as Proposed by Alfred Wegener*. American Association of Petroleum Geologists, Tulsa, OK, 187–193.

Whitmore, T. C. 1981. Wallace's line and some other plants. In Whitmore, T. C., ed., *Wallace's Line and Plate Tectonics*. Oxford University Press, Oxford, 70–80.

Willbourn, E. S. 1950. Mr. J. B. Scrivenor, I.S.O. *Nature*, **165**: 791–792.

Willis, B. 1893. The mechanics of Appalachian structure. *Report of United States Geological Survey*, **13**, Part 2: 211–281.

Willis, B. 1907. A theory of continental structure applied to North America. *Geol. Soc. Am. Bull.*, **18**: 394–395.

Willis, B. 1928. Continental drift. In van Waterschoot van der Gracht, W., ed., *Theory of Continental Drift: A Symposium on the Origin and Movement of Land Masses both Inter-continental and Intra-continental, as Proposed by Alfred Wegener*. American Association of Petroleum Geologists, Tulsa, OK, 76–82.

Willis, B. 1932. Isthmian links. *Geol. Soc. Am. Bull.*, **43**: 917–952.

Willis, B. 1944. Continental drift, ein marchen. *Am. J. Sci.*, **242**: 509–513.

Wilson, J. T. 1949. The origin of continents and Precambrian history. *Trans. Roy. Soc. Can. (Ser. III)*, **XLIII**: 157–184.

Wilson, J. T. 1952a. Some considerations regarding geochronology with special reference to Precambrian time. *Trans. Am. Geophys. Union*, **33**: 195–203.

Wilson, J. T. 1952b. Orogenesis as the fundamental geological process. *Trans. Am. Geophys. Union*, **33**: 444–449.

Wilson, J. T. 1959. Geophysics and continental growth. *Am. Sci.*, **47**: 1–24.

Windhausen, A. 1931. *Geología Argentina* (Segunda Parte). Buenos Aires.

Wing Easton, N. 1921a. On some extensions of Wegener's hypothesis and their bearing upon the meaning of the terms geosyncline and isostasy. *Verh. Geol. Mijnbouwkd. Genoot. Ned. Kolonien Geol. Ser.*, **5**: 113–133.

Wing Easton, N. 1921b. Het ontstaan van den maleischen Archipel, bezien in het licht van Wegener's hypothesen. *Tijdschr. K. Ned. Aardrijkskd. Genoot.*, **38**: 484–512.

Wiseman, J. D. and Seymour Sewell, R. B. 1937. Wegener's theory of continental drift with reference to India and adjacent countries (General discussion). *Proceedings 24th Indian Science Congress, Hyderabad*: 502–520.

Wolfson, A. 1948. Bird migration and the concept of continental drift. *Science*, **108**: 23–30.

Wolfson, A. 1955. Origin of North American bird fauna: critique and reinterpretation from the standpoint of continental drift. *Am. Midland Nat.*, **53**: 353–380.

Wolfson, A. 1985. Bird migration and the concept of continental drift. *Earth Sci. Hist.*, **4**: 182–186.

Woodburne, M. O. and Case, J. A. 1996. Dispersal, vicariance, and the Late Cretaceous to early Tertiary land mammal biogeography from South America to Australia. *J. Mamm. Evol.*, **3**(2): 121–161.

Woodburne, M. O. and Zinsmeister, W. J. 1982. Fossil land mammal from Antarctica. *Science*, **218**: 284–286.

Woodburne, M. O. and Zinsmeister, W. J. 1984. The first land mammal from Antarctica and its biogeographic implications. *J. Palaeontol.*, **58**: 913–948.

Wooldridge, S. W. 1951. The bearing of Late-Tertiary history on vertical movements of the continents. *Adv. Sci.*, **8**: 80–82.

Wooldridge, S. W. and Morgan, R. S. 1937/1959. *An Outline of Geomorphology: The Physical Basis of Geography* (1st edition)/(2nd edition). Longmans, Green and Co Ltd., London.

Wright, W. B. 1914. *The Quaternary Ice Age*. Macmillan, London.

Wright, W. B. 1923a. The Wegener hypothesis. *Nature*, **111**: 30–31.

Wright, W. B. 1923b. The Wegener frequency curve. *Geolog. Mag.*, **60**: 239–240.

Wulff, E. V. 1950. *An Introduction to Historical Plant Geography*. Translated by Brissenden, E. Chronica Botanica Co., Waltham, MA.

Wüst, G. 1939. Die grossgliederung des Atlantischen tiefeseebodems. *Geologische Rundschau*, **30**: 132–137.

Zeuner, F. E. 1942a. Studies in the systematics of Troides Hubner (Lepidoptera Papilisonidae) and its allies; distribution and phylogeny in relation to the geological history of the Australasian Archipelago. *Trans. Zool. Soc. London*, **25**: 107–181.

Zeuner, F. E. 1942b. The divisions as indicated by the distribution of insects in relation to geology. *Proc. Linn. Soc. London*, **154**: 157–163.

Zeuner, F. E. 1945. *The Pleistocene Period, Its Climate*. Printed by the Ray Society, sold by B. Quaritch, Ltd., London.

Zeuner, F. E. 1946. *Dating the Geological Past: An Introduction to Geochronology*. Methuen, London. 2nd edition, 1950; 3rd edition, 1952; 4th edition, 1958.

Zeuner, F. E. 1963. *A History of Domesticated Animals*. Hutchinson, London.

Index

Printed in the United States
By Bookmasters